Higher-Order Techniques in Computational Electromagnetics

Mario Boella Series on Electromagnetism in Information and Communication

Piergiorgio L.E. Uslenghi, PhD – Series Editor

The Mario Boella series offers textbooks and monographs in all areas of radio science, with a special emphasis on the applications of electromagnetism to information and communication technologies. The series is scientifically and financially sponsored by the Istituto Superiore Mario Boella affiliated with the Politecnico di Torino, Italy, and is scientifically cosponsored by the International Union of Radio Science (URSI). It is named to honor the memory of Professor Mario Boella of the Politecnico di Torino, who was a pioneer in the development of electronics and telecommunications in Italy for half a century, and a vice president of URSI from 1966 to 1969.

Published Titles in the Series

Fundamentals of Wave Phenomena, 2nd Edition
by Akira Hirose and Karl Lonngren (2010)

Scattering of Waves by Wedges and Cones with Impedance Boundary Conditions
by Mikhail Lyalinov and Ning Yan Zhu (2012)

Complex Space Source Theory of Spatially Localized Electromagnetic Waves
by S.R. Seshadri (2013)

The Wiener-Hopf Method in Electromagnetics
by Vito Daniele and Rodolfo Zich (2014)

Higher-order Techniques in Computational Electromagnetics
by Roberto Graglia and Andrew Peterson (2015)

Forthcoming Titles

Slotted Waveguide Array Antennas
by Sembiam Rengarajan and Lars Josefsson (2016)

Higher-Order Techniques in Computational Electromagnetics

ISMB Series

Roberto D. Graglia
Politecnico di Torino

Andrew F. Peterson
Georgia Institute of Technology

SciTech Publishing
an imprint of the IET

theiet.org

SciTech
PUBLISHING
an imprint of the IET

Published by SciTech Publishing, an imprint of the IET
www.scitechpub.com
www.theiet.org

First published 2015

ISBN 978-1-61353-016-0 (hardback)
ISBN 978-1-61353-037-5 (PDF)

Typeset in India by MPS Limited

To our wives,
Cinzia and Debra

To our children,
Matteo, Giulia, Kitt, Kinsey, and Katieanne

To our Editor Dudley Kay without whose
never-failing sympathy and encouragement
this book would have never been finished.

Contents

Preface

The use of computational tools in all engineering disciplines is ubiquitous. However, in high-frequency electromagnetics, including antennas, microwave devices, and radar scattering applications, most of the techniques in widespread use today are best described as "low-order" methods. While robust, existing approaches tend to employ piecewise-constant or piecewise-linear functions to represent the fields or currents used as the primary unknown quantities. The principal limitation of low-order techniques is that the error in the computational result is reduced only gradually as additional computational effort is brought to bear.

Research carried out during the past two decades suggests that improvements in accuracy, computational cost, and reliability can be achieved through the use of "high-order" techniques. The purpose of the present text is to present high-order basis functions, explain their use, and illustrate their performance. The specific basis functions under consideration were developed by the authors and include scalar and vector functions for use with equations such as the vector Helmholtz equation and the electric field integral equation. Until now, the details of these basis functions were only available in the form of journal papers. It is the authors' hope that the present text will stimulate their broader acceptance and more widespread use throughout the computational electromagnetics community.

While the majority of the text is focused on piecewise-polynomial functions for representing fields and currents on or near smooth structures, it is also necessary to consider special singular basis functions for treating fields and currents near geometrical edges and corners. For those features, singular basis functions can enhance the accuracy and efficiency far more effectively than higher-order polynomial basis functions. The development of singular expansion functions is far less mature, in general, than the development of polynomial expansion functions. We include one chapter to introduce singular basis functions to the readers.

Foreword

The Mario Boella series contains textbooks and research monographs in all areas of Radio Science, with a special emphasis on the applications of electromagnetism to information and communications technologies. The series is scientifically and financially sponsored by the Istituto Superiore Mario Boella affiliated with the Politecnico di Torino, Torino, Italy, and is scientifically co-sponsored by the International Union of Radio Science (URSI). It is named to honor the memory of Professor Mario Boella of the Politecnico di Torino, who was a pioneer in the development of electronics and telecommunications in Italy for half a century and was a vice president of URSI from 1966 to 1969.

This advanced research monograph is devoted to high-order basis functions in computational electromagnetics. It is co-authored by two international experts in the field, Professor Roberto D. Graglia of the Politecnico di Torino, Torino, Italy and Professor Andrew F. Peterson of the Georgia Institute of Technology, Atlanta, GA, United States of America. These two scientists have published extensively on this topic for the past two decades. The present monograph contains not only a compendium of their previous results, but also a substantial amount of novel research in this very important area of numerical applications of electromagnetism. It will constitute an indispensable reference work for all future research on computational techniques.

Piergiorgio L. E. Uslenghi
ISMB Series Editor
Chicago, June 2015

CHAPTER 1

Interpolation, Approximation, and Error in One Dimension

In science and engineering, one is often required to provide the value $\tilde{f}_i$ reached by the quantity $f(s)$ at s_i out of the knowledge of n samples $f_k = f(s_k)$ obtained for $k = 1, 2, \ldots, n$, with $s_k \neq s_i$ for all k. This is an *interpolation* problem if s_i is located within the same range defined by the n isolated points s_k and an *extrapolation* problem whenever s_i is outside of this range. The interpolation problem is a particular case of the more general *curve fitting* (or *approximation*) problem where the function $\tilde{f}(s)$ re-constructed out of the discrete set of known data points goes exactly through these data points.

The task of interpolating a function is closely related to the problem of representing an unknown function $f(s)$ within a numerical solution procedure, which is the subject of subsequent chapters.

1.1 Linear Interpolation and Triangular Basis Functions

For example, a very simple rule is the *linear interpolation* rule where one joins with straight lines the values $f_1, f_2, \ldots, f_n$ obtained for the successive values of $s = s_1, s_2, \ldots, s_n$ and then uses these $(n-1)$ lines to get the value of $\tilde{f}(s)$ at any intermediate point

$$\tilde{f}(s) = \frac{s_{m+1} - s}{s_{m+1} - s_m} f_m + \frac{s - s_m}{s_{m+1} - s_m} f_{m+1} \quad \text{for } s_m \leq s \leq s_{m+1} \tag{1.1}$$

The previous formula can easily be written in terms of a series involving the n coefficients f_m (for $m = 1, \ldots, n$) each one multiplied by a *basis function* B_m that is independent of the value of its associated coefficient

$$\tilde{f}(s) = \sum_{m=1}^{n} f_m\, B_m(s) \quad \text{for } s_1 \leq s \leq s_n \tag{1.2}$$

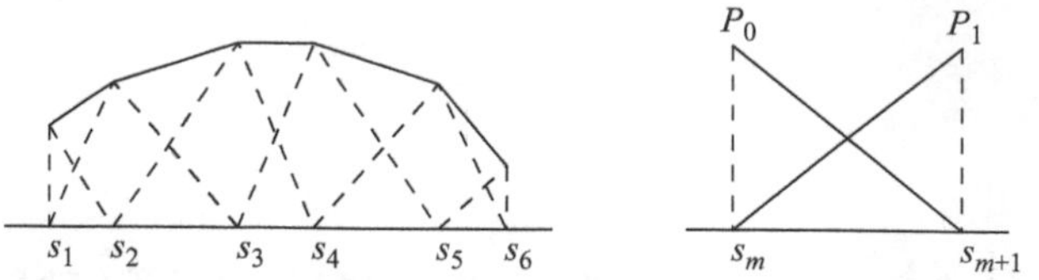

Figure 1.1 Typical linear interpolation summation (at left). The two linear basis functions that span the interval $\{s_m \le s \le s_{m+1}\}$ are the Lagrange polynomials $P_0(1,\, s-s_m)$, $P_1(1,\, s-s_m)$ shown at right.

with

$$B_m(s) = \begin{cases} \dfrac{s-s_{m-1}}{s_m - s_{m-1}} & \text{for } s_{m-1} \le s \le s_m \\ \dfrac{s_{m+1}-s}{s_{m+1}-s_m} & \text{for } s_m \le s \le s_{m+1} \\ 0 \ \ \text{otherwise} & \end{cases} \tag{1.3}$$

and where for $m=1$ (or $m=n$) one has to discard the first (or the second) expression on the right-hand side of (1.3). The functions (1.3) form the so-called *triangular basis function* family because each $B_m(s)$ has the shape of either a half triangle or full triangle along the s variable, with a maximum value (equal to unity) at $s=s_m$. Figure 1.1 (at left) pictorially represents a typical linear interpolation summation (1.2) obtained by using the triangular basis functions (1.3); the series terms $f_m\, B_m(s)$ are shown by dashed-lines in Figure 1.1 (at left). This figure also shows that on each interval $\{s_m \le s \le s_{m+1}\}$ $f(s)$ is linearly interpolated by

$$\tilde{f}(s) = f_m\, P_0(1,\, s-s_m) + f_{m+1}\, P_1(1,\, s-s_m) \tag{1.4}$$

where

$$\begin{aligned} P_0(1,\, z) &= 1 - \frac{z}{s_{m+1}-s_m} \\ P_1(1,\, z) &= \frac{z}{s_{m+1}-s_m} \end{aligned} \tag{1.5}$$

are the two first-order Lagrange polynomials shown in Figure 1.1 (at right), with $P_0(1,\, z) + P_1(1,\, z) = 1$. Notice that the first variable of $P_i(p,\, z)$ is the order p of the Lagrange polynomial that attains a value of unity at $z/(s_{m+1}-s_m) = i/p$.

In case of equally spaced samples, one has

$$\ell = s_m - s_{m-1} = s_{m+1} - s_m \tag{1.6}$$

for all m; this simplifies (1.2) into

$$\tilde{f}(s) = \sum_{m=1}^{n} f_m\, B(s-s_m,\, \ell) \quad \text{for } s_1 \le s \le s_n \tag{1.7}$$

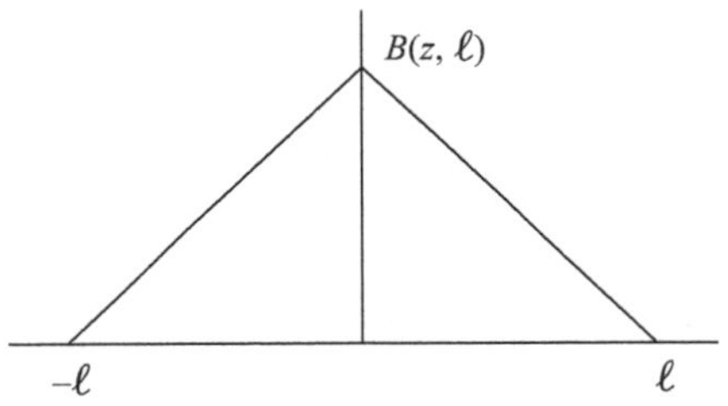

Figure 1.2 The basis function $B(z, \ell)$ of (1.8).

with

$$B(z,\ell)=\begin{cases}1-\dfrac{|z|}{\ell} & \text{for } |z|\leq\ell \\ 0 & \text{otherwise}\end{cases} \tag{1.8}$$

and where, once again, it is understood that, while using (1.7), one has to take $B(z,\ell)=0$ for $(z<0, m=1)$, and for $(z>0, m=n)$. Figure 1.2 shows the (simplified) triangular basis function (1.8).

Although it is not always granted that the *best* interpolation will be obtained for equally spaced subintervals, it is certainly true that the general expression (1.3) for the triangular basis function is less appealing than (1.8), which is valid in case of equally spaced samples and, in numerical implementations, one would certainly prefer to use the latter rather than (1.3).

In order to use the same simple expression for all the basis function defined on intervals of different lengths we could first define the Lagrangian basis functions (1.5) on a unitary length interval, which we call the *parent* interval, and then map this interval and its two Lagrange basis functions onto the true used interval, which we call the *child* interval. Let us assume that the parent interval is along the ξ parent-axis and that its end-points are $\xi=0$ and $\xi=1$. On the parent interval, the two linear basis functions (1.5) read

$$\begin{aligned}P_0(1,\xi)&=1-\xi\\ P_1(1,\xi)&=\xi\end{aligned} \tag{1.9}$$

with

$$\begin{bmatrix}P_0(1,0)\\ P_0(1,1)\end{bmatrix}=\begin{bmatrix}1\\ 0\end{bmatrix};\quad \begin{bmatrix}P_1(1,0)\\ P_1(1,1)\end{bmatrix}=\begin{bmatrix}0\\ 1\end{bmatrix} \tag{1.10}$$

which readily proves that the linear transformation

$$s(\xi)=P_0(1,\xi)\,s_m+P_1(1,\xi)\,s_{m+1} \tag{1.11}$$

is the one that maps the end-points (0, 1) of the parent interval into the end-points

$$s(0)=s_m;\quad s(1)=s_{m+1} \tag{1.12}$$

of the child interval. The Jacobian of the transformation (1.11)

$$\mathcal{J} = \frac{\mathrm{d}s}{\mathrm{d}\xi} = s_{m+1} - s_m \tag{1.13}$$

coincides with the length of the child interval while the ξ parent point associated with the child point s is obtained by the inverse transformation

$$\xi = \frac{s - s_m}{\mathcal{J}} \tag{1.14}$$

A similar *mapping* procedure can be applied with minor modifications not only to interpolate quantities along straight s-lines defined by *linear* transformations of the form of (1.11), as we just did, but also quantities defined along curved lines of curvilinear abscissa s, for which the Jacobian of the transformation is not a constant; these lines can for example be defined in a Cartesian-frame (x, y, z) by the functions $x = X(s)$, $y = Y(s)$, and $z = Z(s)$. In this connection, since it could seem that we have simplified the expression of the basis functions (as in (1.9)) at the expense of a more complex procedure to find the parent variable ξ for any given value of s, we point out that interpolation is often used to compute a fixed number of new samples $\tilde{f}$ inside each child interval. This computation does not require us to use the inverse mapping formula (such as (1.14)) if the parent interval is subdivided into equal-length subintervals, but just the use of the direct mapping formula (such as (1.11)) to find the child s values.

1.2 Interpolation and Basis Functions of Higher Polynomial Order

1.2.1 Lagrange Interpolation

The importance of the mapping procedure discussed at the end of Section 1.1 is due to the fact that it can be easily generalized to define interpolation procedures of (local) higher polynomial order. For example, let us assume that the parent interval $\{0 \le \xi \le 1\}$ is subdivided into p equal-length subintervals defined by the end-points

$$\xi_k = \frac{k}{p}, \quad \text{for } k = 0, 1, \ldots, p \tag{1.15}$$

As shown in Figure 1.3, when we map the parent into the child interval according to the mapping formulas (see (1.9) and (1.11))

$$s(\xi) = (1 - \xi)\, s_a + \xi\, s_b \tag{1.16}$$

the subinterval end-points set becomes

$$s_k = s_a + \frac{k}{p}\,(s_b - s_a), \quad \text{for } k = 0, 1, \ldots, p \tag{1.17}$$

The length of each parent domain subinterval is $1/p$ while, as shown in Figure 1.3, we continue to use the symbol ℓ to denote the length $(s_b - s_a)/p$ of each child *subinterval* by assuming they are of equal length (as it happens for a rectilinear child interval).

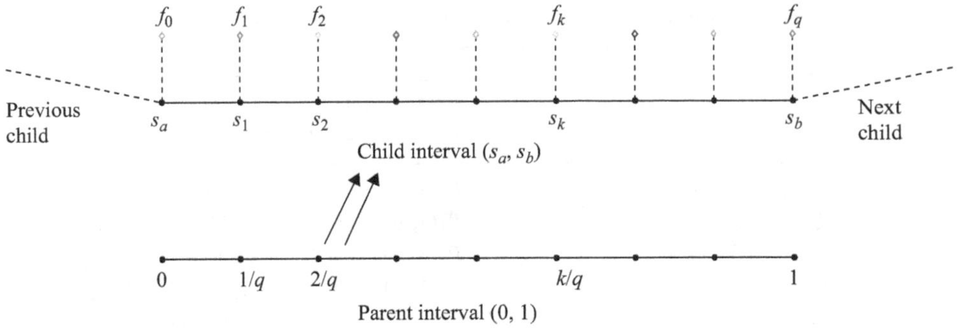

Figure 1.3 Parent to child mapping. If the parent interval is subdivided into p equal-length subintervals the child interval is also subdivided into p subintervals.

Now let us assume that the $(p+1)$ end-points (1.17) are used as interpolation points. In this case, the Lagrange polynomials

$$P_i(p, \xi) = \prod_{\substack{j=0 \\ j \neq i}}^{p} \frac{p\xi - j}{i - j} \tag{1.18}$$

obtained for $i = 0, 1, \ldots, p$ directly define, in the parent domain, the $(p+1)$ basis functions of order p needed to interpolate a function at these points. In fact, $P_i(p, \xi)$ is unity at $\xi = i/p$ while it is equal to zero on all the remaining end-points (1.15) obtained for $k \neq i$.

By using the interpolatory basis functions (1.18), one can compute the value

$$\tilde{f}(s) = \sum_{k=0}^{p} f_k\, P_k(p, \xi) \tag{1.19}$$

of a quantity $f(s)$ at any point $s(\xi)$ of the child interval out of the knowledge of the $(p+1)$ samples $f_k = f(s_k)$, with s_k given in (1.17).

The interpolatory polynomial families (1.18) are reported in Table 1.1 up to the fourth p-order. Recall that $P_0(p, \xi)$ indicates the polynomial that reaches a unitary value at the first end-point $\xi = 0$ of the parent interval [0, 1], while $P_p(p, \xi)$ is the polynomial that reaches a unitary value at the last end-point $\xi = 1$ of the parent interval.

Notice also that all the polynomials (1.18) span the whole parent interval $\{0 \leq \xi \leq 1\}$ and are therefore different from zero on all its p subintervals. In (1.19), the polynomials $P_0(p, \xi)$ and $P_p(p, \xi)$ are associated with the first (f_0) and the last (f_p) child-interval sample, respectively. Each of these two samples is eventually associated with another interpolatory polynomial defined on the previous (for f_0) or on the next (for f_p) child interval (see Figure 1.3); that is, the domain of the basis functions associated with the first and last samples could be larger than that defined by the p subintervals considered here. Conversely, the domain of the basis functions associated with the samples f_k for $k = 1, 2, \ldots, (p-1)$ always coincides with the p subintervals defined here, meaning that those basis functions are exactly equal to zero in the region external to the whole child interval considered here.

Table 1.1 Lagrange interpolatory polynomial families up to the fourth order, for $0 \leq \xi \leq 1$.

$P_0(1, \xi) = 1 - \xi$	$P_0(2, \xi) = (1 - 2\xi)(1 - \xi)$
$P_1(1, \xi) = \xi$	$P_1(2, \xi) = 4\xi(1 - \xi)$
	$P_2(2, \xi) = -\xi(1 - 2\xi)$
$P_0(3, \xi) = (1 - 3\xi)(2 - 3\xi)(1 - \xi)/2$	$P_0(4, \xi) = (1 - 4\xi)(1 - 2\xi)(3 - 4\xi)(1 - \xi)/3$
$P_1(3, \xi) = 9\xi(2 - 3\xi)(1 - \xi)/2$	$P_1(4, \xi) = 16\xi(1 - 2\xi)(3 - 4\xi)(1 - \xi)/3$
$P_2(3, \xi) = -9\xi(1 - 3\xi)(1 - \xi)/2$	$P_2(4, \xi) = -4\xi(1 - 4\xi)(3 - 4\xi)(1 - \xi)$
$P_3(3, \xi) = \xi(1 - 3\xi)(2 - 3\xi)/2$	$P_3(4, \xi) = 16\xi(1 - 4\xi)(1 - 2\xi)(1 - \xi)/3$
	$P_4(4, \xi) = -\xi(1 - 4\xi)(1 - 2\xi)(3 - 4\xi)/3$

At this point, we ought to make the following remarks:

- First of all one can map the parent into the child interval to make the latter coincide with the entire interpolation domain. In this manner, the procedure just described is equivalent to interpolation of order p on p equally spaced samples. However, high-degree polynomials are generally unsuitable for interpolation with equidistant nodes over the entire interpolation interval. In fact, if the length of the interpolation interval exceeds a certain value, it could happen that in the attempt to improve the approximation by increasing the number $(p + 1)$ of the equidistant interpolation points together with the order p of the used interpolant, the maximum difference between the given function and the approximant increases (it could even become infinite as $p \to \infty$). This phenomenon, known as *Runge's phenomenon*, is simply avoided by using piecewise polynomials, that is, different polynomials (also called spline curves) in adjacent child intervals. Obviously, another possible remedy to the Runge problem (whenever it occurs) is to avoid using equidistant points, although this is not a general remedy since the best choice of where to locate the interpolation points depends on the behavior of the functions one needs to interpolate.
- The technique described above is readily modified to deal with non-equally spaced samples. In this case one does not really need to map a parent into any child interval since it is possible to replace the Lagrange polynomials $P_k(p, \xi)$ that appear in (1.19) with the following ones, directly defined in terms of the s variable

$$P_i(p, s) = \prod_{\substack{j=0 \\ j \neq i}}^{p} \frac{s - s_j}{s_i - s_j} \tag{1.20}$$

- The mapping technique described in this subsection is particularly convenient in applications where the samples f_k are not immediately available, but unknowns to be measured or computed by some numerical solution process, such as those described in the rest of this book.

1.2.2 Hermite Interpolation

Unfortunately, the derivatives of the interpolant $\tilde{f}(s)$ obtained with linear or higher order Lagrange interpolations do not agree with the derivatives of the original function $f(s)$ except in rare circumstances. In addition, note that the first derivative of $\tilde{f}(s)$ obtained using those interpolation schemes is generally discontinuous at the ends of each child interval. Since it is typically observed that the error of $\tilde{f}(s)$ becomes unacceptably large on some regions whenever the slope of the polynomial interpolant $\tilde{f}(s)$ differs markedly from that of $f(s)$ at some of the interpolation points, its seems a natural remedy, in those cases, to force the interpolant to match both f and its first derivative at the interpolation points. Approximants that match a given function and its first nth derivatives at the interpolation points are obtained by using Hermite interpolatory polynomials.[1]

In the following, we discuss in detail the general case of interpolatory schemes that match the given function and its first nth derivatives at all the $(p+1)$ interpolation points obtained by mapping into the child interval the parent interval subdivided into p equal-length subintervals. This is in line with our primary aim to derive basis functions for numerical solution algorithms where the function samples are the problem *unknowns*; at the same time, it does not prohibit one to deal with cases of arbitrarily distributed samples by setting $p=1$. To further elaborate on Hermite interpolation of functions with an arbitrary distribution of *known* samples, the interested reader is referred to numerical analysis books particularly devoted to this subject [3–6].

In order to add the condition that the first derivative of $\tilde{f}(s)$ is equal to $f'(s)$ at the extreme points of the child interval, as well as at the other internal interpolation points, we must replace the previously defined Lagrange basis functions with others of higher order, since each interpolation point is now associated with two degrees of freedom (DoFs), one for the value $\tilde{f}$ of the interpolant and the other for the value $\tilde{f}'$ of the derivative of the interpolant at the interpolation point.

For example, we can substitute the triangular basis functions B_m discussed in Section 1.1 with piecewise cubic basis functions $\mathcal{H}_m^{(1)}$ and $\mathcal{H}_m^{(2)}$, and replace the linear interpolation (1.2) with

$$\tilde{f}(s)=\sum_{m=1}^{n}\left[f_m\,\mathcal{H}_m^{(1)}(s)+f'_m\,\mathcal{H}_m^{(2)}(s)\right]\quad \text{for } s_1\le s\le s_n \tag{1.21}$$

where $\mathcal{H}_m^{(1)}(s)$, $\mathcal{H}_m^{(2)}(s)$ and their first derivative are zero at all the interpolation points with the exception of $s=s_m$, where one sets $\mathcal{H}_m^{(1)}=1$ and $\mathrm{d}\mathcal{H}_m^{(2)}/\mathrm{d}s=1$.

Once again, the expression of $\mathcal{H}_m^{(1)}(s)$ and $\mathcal{H}_m^{(2)}(s)$ is more conveniently obtained by working in the parent interval defined by the two end-points $\xi=0$ and $\xi=1$. On *each parent interval*, instead of the two linear Lagrange basis functions previously reported in (1.9), we now interpolate with the following four Hermite polynomials

$$h_0^{(1)}(1,\,\xi)=(1+2\,\xi)\,(1-\xi)^2 \tag{1.22}$$

$$h_1^{(1)}(1,\,\xi)=\xi^2\,(3-2\,\xi) \tag{1.23}$$

---

[1]For example, Hermite interpolatory polynomials and cubic splines have been used to study the surface currents induced by obliquely incident waves on an impedance body of revolution (BOR) [1]. For such a rotationally symmetric structure, currents are represented by a Fourier series of uncoupled terms in the azimuthal angle, and unknowns are limited to those required to provide a one-dimensional representation along the generating arc. The BOR approach saves a tremendous amount of computation time and memory [2], and the use of higher order models and Hermite basis functions enhances the resulting accuracy.

$$h_0^{(2)}(1, \xi) = \xi\,(1-\xi)^2 \tag{1.24}$$

$$h_1^{(2)}(1, \xi) = -\xi^2\,(1-\xi) \tag{1.25}$$

with

$$\begin{cases} h_i^{(1)}(1, j) = \delta_{ij} \\ \left.\dfrac{\mathrm{d}h_i^{(1)}(1, \xi)}{\mathrm{d}\xi}\right|_{\xi=j} = 0, \quad \text{for } i,j = 0, 1 \end{cases} \tag{1.26}$$

and

$$\begin{cases} h_i^{(2)}(1, j) = 0 \\ \left.\dfrac{\mathrm{d}h_i^{(2)}(1, \xi)}{\mathrm{d}\xi}\right|_{\xi=j} = \delta_{ij}, \quad \text{for } i,j = 0, 1 \end{cases} \tag{1.27}$$

where

$$\delta_{ij} = \begin{cases} 1 & \text{for } i = j \\ 0 & \text{for } i \neq j \end{cases} \tag{1.28}$$

is the Kronecker delta. Notice here that we do not use the *standard* symbol normally used for the Hermite polynomials, but rather we consider them as polynomials of two variables where, as already done for the Lagrange polynomial families, the first variable is the number p of equal-length intervals in which the parent interval has been subdivided (p is unity for the polynomials reported above).

From (1.26), it is rather clear that the basis functions $h_0^{(1)}$ and $h_1^{(1)}$ interpolate the function $f(s)$ at $s(\xi = 0)$ and $s(\xi = 1)$, respectively, while (1.27) show that the functions $h_i^{(2)}$ are able to interpolate the first derivative of f with respect to the child variable s when properly re-scaled by a constant.

The basis functions $h_i^{(2)}$ must in fact be re-scaled by the Jacobian

$$\mathcal{J} = \frac{\mathrm{d}s}{\mathrm{d}\xi} \tag{1.29}$$

of the transformation from parent to child interval evaluated at the interpolation point, since in the child space, one has

$$f'(s) = \frac{\mathrm{d}f}{\mathrm{d}s} = \frac{\mathrm{d}f}{\mathrm{d}\xi}\,\mathcal{J}^{-1} \tag{1.30}$$

In fact, the first derivative of the re-scaled functions

$$H_i^{(2)} = \mathcal{J}|_{\xi=i} \times h_i^{(2)}(1, \xi) \tag{1.31}$$

made with respect to the child variable s are

$$\frac{\mathrm{d}H_i^{(2)}}{\mathrm{d}s}\bigg|_{s_j} = \frac{\mathrm{d}h_i^{(2)}}{\mathrm{d}\xi}\bigg|_{\xi=j} = \delta_{ij}, \quad \text{for } i,j=0,1 \tag{1.32}$$

Recall that in case of linear mapping (1.11) the Jacobian is constant on the child interval (see (1.13)).

It should now be clear that if we increase the number of the interpolation points by subdividing the parent interval into p equal-length subintervals we need to define:

- $(p+1)$ polynomial basis functions $h_i^{(1)}(p,\ \xi)$ (obtained for $i=0,1,\ldots,p$), with

$$\begin{cases} h_i^{(1)}(p, j/p) = \delta_{ij} \\ \dfrac{\mathrm{d}h_i^{(1)}(p,\ \xi)}{\mathrm{d}\xi}\bigg|_{\xi=\frac{j}{p}} = 0, \quad \text{for } j=0,1,\ldots,p \end{cases} \tag{1.33}$$

- $(p+1)$ polynomial basis functions $h_i^{(2)}(p,\ \xi)$ (obtained for $i=0,1,\ldots,p$), with

$$\begin{cases} h_i^{(2)}(p, j/p) = 0 \\ \dfrac{\mathrm{d}h_i^{(2)}(p,\ \xi)}{\mathrm{d}\xi}\bigg|_{\xi=\frac{j}{p}} = \delta_{ij}, \quad \text{for } j=0,1,\ldots,p \end{cases} \tag{1.34}$$

The general expression of these polynomials, whose degree is never higher than $(2p+1)$, is

$$h_i^{(1)}(p,\xi) = \left[1 - 2(p\xi - i)\sum_{\substack{j=0\\j\neq i}}^{p}(i-j)^{-1}\right] P_i^2(p,\ \xi) \tag{1.35}$$

$$h_i^{(2)}(p,\xi) = \frac{(p\xi - i)}{p} P_i^2(p,\ \xi), \quad \text{for } i=0,1,\ldots,p \tag{1.36}$$

where

$$P_i^2(p,\ \xi) = \prod_{\substack{j=0\\j\neq i}}^{p}\left(\frac{p\xi - j}{i-j}\right)^2 \tag{1.37}$$

is the square of the Lagrange polynomial (1.18), of order $2p$. Notice also that (1.33) immediately yields

$$\sum_{i=0}^{p} h_i^{(1)}(p,\xi) = 1 \tag{1.38}$$

on the following grounds: (1) the subset of the functions $h_i^{(1)}(p,\ \xi)$ is complete to zeroth order; (2) at each interpolation point only one function of this subset does not vanish and is equal to unity; and (3) all the derivatives of a constant (such as the right-hand side of (1.38)) are zero

Table 1.2 Hermite interpolatory polynomial families that match a function and its first derivative at the end-points of p equal-length subintervals (up to $p=3$), for $0 \le \xi \le 1$.

$h_0^{(1)}(1,\xi)=(1+2\xi)(1-\xi)^2$	$h_0^{(2)}(1,\xi)=\xi(1-\xi)^2$
$h_1^{(1)}(1,\xi)=\xi^2(3-2\xi)$	$h_1^{(2)}(1,\xi)=-\xi^2(1-\xi)$
$h_0^{(1)}(2,\xi)=(6\xi+1)(1-2\xi)^2(1-\xi)^2$	$h_0^{(2)}(2,\xi)=\xi(1-2\xi)^2(1-\xi)^2$
$h_1^{(1)}(2,\xi)=16\xi^2(1-\xi)^2$	$h_1^{(2)}(2,\xi)=-8\xi^2(1-2\xi)(1-\xi)^2$
$h_2^{(1)}(2,\xi)=\xi^2(1-2\xi)^2(7-6\xi)$	$h_2^{(2)}(2,\xi)=-\xi^2(1-2\xi)^2(1-\xi)$
$h_0^{(1)}(3,\xi)=\frac{1}{4}(1+11\xi)(1-3\xi)^2(2-3\xi)^2(1-\xi)^2$	$h_0^{(2)}(3,\xi)=\frac{1}{4}\xi(1-3\xi)^2(2-3\xi)^2(1-\xi)^2$
$h_1^{(1)}(3,\xi)=\frac{243}{4}\xi^3(2-3\xi)^2(1-\xi)^2$	$h_1^{(2)}(3,\xi)=-\frac{27}{4}\xi^2(1-3\xi)(2-3\xi)^2(1-\xi)^2$
$h_2^{(1)}(3,\xi)=\frac{243}{4}\xi^2(1-3\xi)^2(1-\xi)^3$	$h_2^{(2)}(3,\xi)=-\frac{27}{4}\xi^2(1-3\xi)^2(2-3\xi)(1-\xi)^2$
$h_3^{(1)}(3,\xi)=\frac{1}{4}\xi^2(1-3\xi)^2(2-3\xi)^2(12-11\xi)$	$h_3^{(2)}(3,\xi)=-\frac{1}{4}\xi^2(1-3\xi)^2(2-3\xi)^2(1-\xi)$

while, at the same time, the second entry of (1.33) enforces a zero first derivative for (1.38) at each one of the $(p+1)$ interpolation points of the child interval.

In this case, the re-scaled basis functions that interpolate the derivative of f with respect to the child variable s are

$$H_i^{(2)}(p,\xi)=h_i^{(2)}(p,\xi)\,\mathcal{J}_i(p) \tag{1.39}$$

where

$$\mathcal{J}_i(p)=\left.\frac{\mathrm{d}s}{\mathrm{d}\xi}\right|_{\xi=\frac{i}{p}} \tag{1.40}$$

indicates the value of the Jacobian $\mathcal{J}$ at $\xi=i/p$. Table 1.2 reports the expression of the polynomials (1.35, 1.35) up to a number p of subintervals equal to 3.

The preliminary results presented so far clarify that the Lagrange interpolation discussed in Section 1.2.1 is the simplest form of Hermite interpolation, obtained without imposing any condition on the derivative of the interpolants.

The expression of the Hermite polynomials required to build interpolants that could match a function and its first nth derivatives at all the interpolation points of the *child interval* is obtained by generalizing the previous results of this subsection. These polynomials, whose degree is never higher than $n+p(n+1)$, are

$$\begin{cases} h_i^{(1)}(p,\xi) = P_i^{n+1}(p,\xi)[1+Q_i^{(1)}(p,\xi)] \\ h_i^{(2)}(p,\xi) = P_i^{n+1}(p,\xi)\,Q_i^{(2)}(p,\xi) \\ \cdots \quad \cdots \\ \cdots \quad \cdots \\ h_i^{(n+1)}(p,\xi) = P_i^{n+1}(p,\xi)\,Q_i^{(n+1)}(p,\xi) \end{cases} \tag{1.41}$$

with

$$\begin{cases} Q_i^{(r)}(p,\xi) = \sum_{k=1}^{n} a_k^{(r)} (p\xi - i)^k \\ \dfrac{d^k}{d\xi^k} Q_i^{(r)}(p,\xi) = p^k \sum_{m=0}^{n-k} a_{k+m}^{(r)} \dfrac{(k+m)!}{m!} (p\xi - i)^m \end{cases} \tag{1.42}$$

$$\begin{cases} Q_i^{(r)}\Big|_{\xi=i/p} = 0 \\ \dfrac{d^k}{d\xi^k} Q_i^{(r)}\Big|_{\xi=i/p} = p^k a_k^{(r)} k! \end{cases} \tag{1.43}$$

for $i = \{0, 1, \ldots, p\}$, $r = \{1, 2, \ldots, n+1\}$, $k = \{1, 2, \ldots, n\}$, and with

$$P_i^{n+1}(p, \xi) = \prod_{\substack{j=0 \\ j\neq i}}^{p} \left(\frac{p\xi - j}{i - j}\right)^{n+1} \tag{1.44}$$

The n coefficients $a_k^{(r)}$ of the nth-degree polynomial $Q_i^{(r)}$ are determined by imposing the appropriate values (zero or one) for the successive derivatives of the polynomials (1.41) at $\xi = i/p$. Notice in fact that the particular expression (1.41) already guarantees that *these interpolatory polynomials vanish together with their first n derivatives* at all the interpolation points $\xi = k/p$ for all $k \neq i$; for $\xi = i/p$ the polynomials vanish as well, with the exception of $h_i^{(1)}$ for which one gets $h_i^{(1)} = 1$ at $\xi = i/p$. Thus, to find the coefficients $a_k^{(r)}$ one has to differentiate (1.41) with respect to ξ, then set $\xi = i/p$ and solve a linear system of equation whose unknowns are the $a_k^{(r)}$ coefficients. The values of the successive derivatives of (1.41) are obtained by application of the chain-rule (Leibniz formula)

$$D^k[g_a(\xi)\, g_b(\xi)] = \sum_{m=0}^{k} \binom{k}{m} D^m[g_a(\xi)]\, D^{k-m}[g_b(\xi)] \tag{1.45}$$

where the operator $D^k[\ \ ]$ is defined by

$$D^k[g(\xi)] = \begin{cases} g(\xi) & \text{for } k = 0 \\ \dfrac{d^k}{d\xi^k} g(\xi) & \text{for } k \text{ positive integer} \end{cases} \tag{1.46}$$

Equation (1.45) together with (1.41, 1.43) immediately yields the following linear system of n^2 equations, obtained for $k, r = 1, 2, \ldots, n$

$$\frac{\mathrm{d}^k}{\mathrm{d}\xi^k}\, h_i^{(1)}(p,\xi)\Big|_{\xi=\frac{i}{p}} = k!\,p^k \sum_{m=0}^{k-1} a_{k-m}^{(1)} \frac{D^m\left[P_i^{n+1}(p,\,\xi)\right]_{\xi=\frac{i}{p}}}{m!\,p^m}$$
$$+D^k\left[P_i^{n+1}(p,\,\xi)\right]=0 \tag{1.47}$$

$$\frac{\mathrm{d}^k}{\mathrm{d}\xi^k}\, h_i^{(r+1)}(p,\xi)\Big|_{\xi=\frac{i}{p}} = k!\,p^k \sum_{m=0}^{k-1} a_{k-m}^{(r+1)} \frac{D^m\left[P_i^{n+1}(p,\,\xi)\right]_{\xi=\frac{i}{p}}}{m!\,p^m} = \delta_{rk} \tag{1.48}$$

where δ_{rk} is the Kronecker delta (1.28), and where one uses

$$\begin{cases} \dfrac{D^1\left[P_i^{n+1}(p,\,\xi)\right]}{p} = P_i^{n+1}(p,\,\xi)\,\eta_i(1,\,p,\,\xi) \\[2ex] \dfrac{D^2\left[P_i^{n+1}(p,\,\xi)\right]}{p^2} = P_i^{n+1}(p,\,\xi)\left[\eta_i^2(1,\,p,\,\xi)-\eta_i(2,\,p,\,\xi)\right] \\[2ex] \dfrac{D^3\left[P_i^{n+1}(p,\,\xi)\right]}{p^3} = P_i^{n+1}(p,\,\xi)\Big[\eta_i^3(1,\,p,\,\xi) \\ \qquad -3\,\eta_i(1,\,p,\,\xi)\,\eta_i(2,\,p,\,\xi)+2\,\eta_i(3,\,p,\,\xi)\Big] \\ ::::::::::::::::::::::::::::::::::::::: \\ \dfrac{D^n\left[P_i^{n+1}(p,\,\xi)\right]}{p^n} = P_i^{n+1}(p,\,\xi)\;[\cdots\cdots\cdots] \end{cases} \tag{1.49}$$

with

$$\eta_i(k,\,p,\,\xi) = (n+1)\sum_{\substack{j=0\\ j\neq i}}^{p} (p\,\xi - j)^{-k} \tag{1.50}$$

$$\eta_i(k,\,p,\,\xi)|_{\xi=\frac{i}{p}} = (n+1)\sum_{\substack{j=0\\ j\neq i}}^{p} (i-j)^{-k} \tag{1.51}$$

Notice here that we wrote (1.49) to bring out the same common factor $P_i^{n+1}(p,\,\xi)$ that is unity at $\xi = i/p$.

For the more general case represented by (1.41), it is also readily proven that (1.38) holds true, while the re-scaled basis functions that interpolate the successive derivatives of f with respect to the child variable s are

$$H_i^{(m+1)}(p,\xi) = h_i^{(m+1)}(p,\xi)\,\mathcal{J}_i^{\,m}(p), \quad \text{for } m = 1, 2, \ldots, n \tag{1.52}$$

where

$$\mathcal{J}_i(p) = \left.\frac{\mathrm{d}s}{\mathrm{d}\xi}\right|_{\xi=\frac{i}{p}} \tag{1.53}$$

indicates the value of the Jacobian $\mathcal{J}$ at $\xi = i/p$.

In order to help the reader follow the construction procedure just discussed by working out a new example on his own, we report below the results obtained for the Hermite polynomials that match a given function and its first and second derivatives at the $(p+1)$ end-points of the p equal-length subintervals of the child interval. Those Hermite polynomials have degree no higher than $(3p+2)$ and are given by the expressions

$$h_i^{(1)}(p,\xi) = \left\{1 - 3(p\xi - i)\left[\alpha_i - \beta_i \frac{(p\xi - i)}{2}\right]\right\} P_i^3(p,\xi) \tag{1.54}$$

$$h_i^{(2)}(p,\xi) = \frac{(p\xi - i)}{p}\left[1 - 3\,\alpha_i\,(p\xi - i)\right] P_i^3(p,\xi) \tag{1.55}$$

$$h_i^{(3)}(p,\xi) = \frac{1}{2}\left(\frac{p\xi - i}{p}\right)^2 P_i^3(p,\xi) \tag{1.56}$$

with

$$\alpha_i = \sum_{\substack{j=0\\ j\neq i}}^{p} (i-j)^{-1} \tag{1.57}$$

$$\beta_i = 3\,\alpha_i^2 + \sum_{\substack{j=0\\ j\neq i}}^{p} (i-j)^{-2} \tag{1.58}$$

and where

$$P_i^3(p,\xi) = \prod_{\substack{j=0\\ j\neq i}}^{p} \left(\frac{p\xi - j}{i-j}\right)^3 \tag{1.59}$$

is the cube of the Lagrange polynomial (1.18), of order $3\,p$. Table 1.2.2 reports the expression of these polynomials up to a number p of subintervals equal to 2.

Table 1.3 Hermite interpolatory polynomial families that match a function and its first and second derivatives at the end-points of p equal-length subintervals (up to $p = 2$), for $0 \le \xi \le 1$.

$h_0^{(1)}(1,\xi) = (1+3\xi+6\xi^2)\,(1-\xi)^3$	$h_0^{(2)}(1,\xi) = \xi\,(1+3\xi)\,(1-\xi)^3$	$h_0^{(3)}(1,\xi) = \frac{1}{2}\,\xi^2\,(1-\xi)^3$
$h_1^{(1)}(1,\xi) = \xi^3\,(10-15\xi+6\xi^2)$	$h_1^{(2)}(1,\xi) = -\xi^3\,(4-3\xi)\,(1-\xi)$	$h_1^{(3)}(1,\xi) = \frac{1}{2}\,\xi^3\,(1-\xi)^2$
$h_0^{(1)}(2,\xi) = (1+9\xi+48\xi^2)\,(1-2\xi)^3\,(1-\xi)^3$	$h_0^{(2)}(2,\xi) = \xi\,(1+9\xi)\,(1-2\xi)^3\,(1-\xi)^3$	$h_0^{(3)}(2,\xi) = \frac{1}{2}\,\xi^2\,(1-2\xi)^3\,(1-\xi)^3$
$h_1^{(1)}(2,\xi) = 256\,\xi^3\,(1-3\xi+3\xi^2)\,(1-\xi)^3$	$h_1^{(2)}(2,\xi) = -32\,\xi^3\,(1-2\xi)\,(1-\xi)^3$	$h_1^{(3)}(2,\xi) = 8\,\xi^3\,(1-2\xi)^2\,(1-\xi)^3$
$h_2^{(1)}(2,\xi) = -\xi^3\,(1-2\xi)^3\,(58-105\xi+48\xi^2)$	$h_2^{(2)}(2,\xi) = \xi^3\,(1-2\xi)^3\,(10-9\xi)\,(1-\xi)$	$h_2^{(3)}(2,\xi) = -\frac{1}{2}\,\xi^3\,(1-2\xi)^3\,(1-\xi)^2$

1.3 Error in the Representation of Functions

1.3.1 Interpolation Error

Linear interpolation on equally spaced intervals of very small length ℓ is used in many applications, for example to plot functions on definite intervals with high fidelity. There are however several other important numerical applications where small ℓ values are unfeasible. For example, these occur whenever the samples ought to be obtained by first solving numerically a system of equations whose order increases for increasing number of samples. In these cases, to reduce the system order, one would like to use fewer samples separated by larger intervals. To do that, one has to predict the interpolation error in order to use the "largest" possible intervals.

Since bounded functions on definite intervals can be represented in terms of trigonometric series we may draw several important conclusions simply by considering the sinusoidal function $f(x) = \sin(x)$ and the behavior of the interpolated functions $\tilde{f}(x)$ over the first full-wavelength interval $\lambda_0 = 2\pi$. Notice that for the sake of simplicity we consider in this subsection functions that vary along a *rectilinear* x-axis while always locating the first sample at $x=0$ for the sake of comparison. Figure 1.4 shows the linear, quadratic, cubic, and quartic interpolation of sin(x) obtained by using *Lagrange* basis functions on equally spaced intervals of length $\ell = \lambda_0/10$, $\lambda_0/7$, $\lambda_0/6$, and $\lambda_0/6$, respectively. In particular, the interpolations of polynomial order higher than first were performed according to the technique described in Section 1.2.1 thereby using *child intervals* of length equal to 2ℓ $(=2\lambda_0/7)$, 3ℓ $(=\lambda_0/2)$, and 4ℓ $(=2\lambda_0/3)$ for quadratic, cubic, and quartic interpolation, respectively.

The right-hand column of Figure 1.4 shows that the error $\upsilon(x) = \tilde{f}(x) - \sin(x)$ contains several spurious high-frequency components of the form $\upsilon_\alpha(x) = \Upsilon_\alpha \sin(\alpha x)$ (with $\alpha > 1$). The effect of these spurious components will be discussed in Section 1.3.2. These spurious components are obviously present also in the interpolated functions shown on the left-hand side of Figure 1.4, since they are absent in the given function sin(x). However, for the interpolation orders considered in Figure 1.4, one always has $|\upsilon(x)| < 0.05$ on the entire period λ_0, which signifies that if one chooses 0.05 as the maximum error level one can easily reduce the number of samples by enlarging the interpolation step and increasing the interpolation order at the same time.

In case of a complex function $g(x)$, one may have to choose the interpolation step to guarantee a certain error level for the magnitude $|\tilde{g}(x)|$ of the function obtained from the interpolated approximants of the real and the imaginary parts of g. In this case, provided the error level is relatively low, it is in general sufficient to choose the interpolation step that guarantees the required error level on both the interpolated real and imaginary parts of g. This can be appreciated by considering for example Figure 1.5 where we report the reconstructed magnitude of the complex function $g(x) = \exp(jx)$ interpolated with basis functions of increasing order (from linear up to the fifth order) over one full-wavelength interval $\lambda_0 = 2\pi$. In particular, notice that the magnitude of $\tilde{g}$ reconstructed after *linear* interpolation always looks like a sequence of U-shaped curves; this implies that linear interpolation of a complex function between samples does not control the error in the magnitude of that function to the same extent that linear interpolation of a real-valued function does.

However, it may be misleading to consider the interpolation error on a point by point basis because the errors have higher magnitude in some isolated regions while remaining rather small on the rest of the interpolation interval. A better parameter to assess the quality of the interpolated functions versus the interpolation order and sampling length is given by the integral

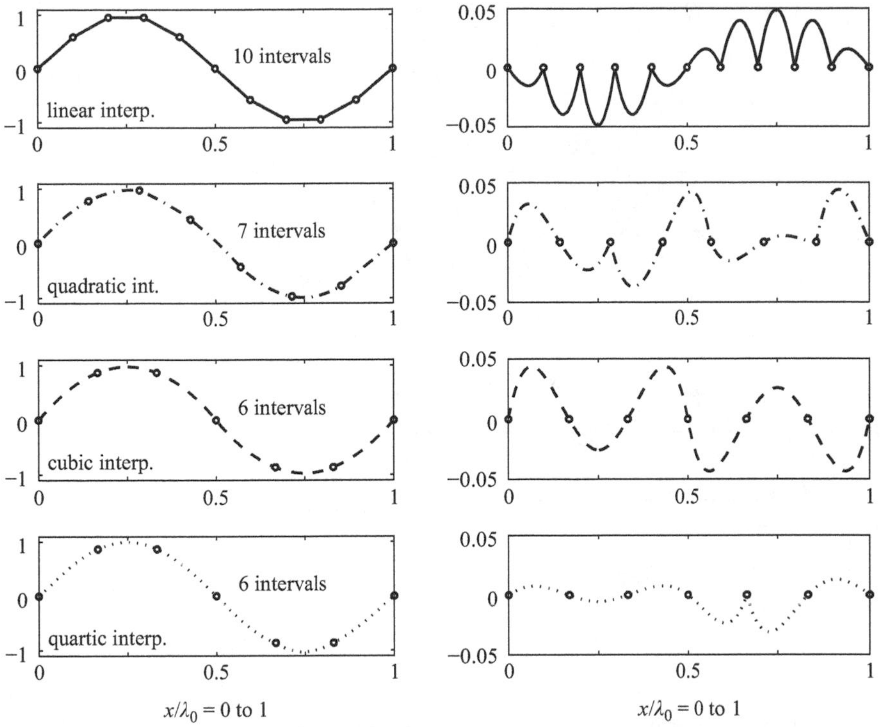

Figure 1.4 At left, one cycle of a sinusoidal function is interpolated by linear up to quartic interpolation. The errors with respect to the exact sinusoidal function are shown at right.

of the error magnitude over the whole interpolation interval. For example, we may use as a global parameter the following L_2 relative error

$$\varepsilon = \sqrt{\frac{\displaystyle\int_{\lambda_0} \left[\tilde{f}(x) - f(x)\right]^2 \mathrm{d}x}{\displaystyle\int_{\lambda_0} f^2(x)\mathrm{d}x}} \tag{1.60}$$

where λ_0 is the entire interpolation interval. For small ℓ values of the interpolation step (including the limiting case of $\ell \to 0$), the error has the following asymptotic behavior

$$\varepsilon \simeq \alpha_p \left(\frac{\ell}{\lambda_0}\right)^{(p+1)} \tag{1.61}$$

where the multiplicative constant α_p depends on the order p of the interpolation and, obviously, on the function $f(x)$ that one interpolates. Recall that p is the number of subintervals of each child interval as well as the polynomial order of the Lagrange basis functions in use.

For the trigonometric functions sin(x) and cos(x), the interpolation error (1.60) is equal to unity for $\tilde{f}(x) = 0$ (a case one may easily consider to be among the worst ones) while its

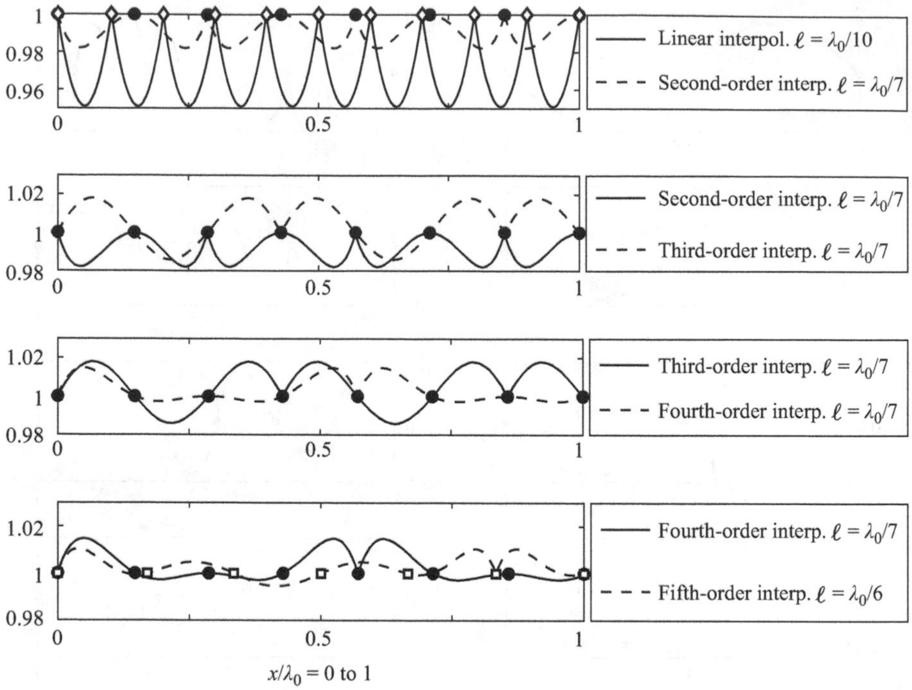

Figure 1.5 The real and imaginary parts of the complex function $g(x) = \exp(jx)$ interpolated over one full-wavelength interval $\lambda_0 = 2\pi$ with different basis functions, from linear up to the fifth order, and with interpolation step $\ell = \lambda_0/n$ reported in the figure legend. The interpolation steps were chosen to provide decreasing maximum error for increasing order. The figures, reported with the same scale, show the magnitude of the interpolated functions $\tilde{g}(x)$ which ideally should be equal to unity. For the first two interpolation orders, the magnitude is always less or equal to one while for higher orders it oscillates around unity.

asymptotic behavior (1.61) for the first six interpolation orders is reported in Table 1.4 and shown in Figure 1.6.

The error (1.60) is predicted quite accurately by the asymptotic expression (1.61) for interpolation steps smaller than a certain threshold value, which depends on the order p of the interpolation as well as on the function one interpolates. As a matter of fact, for the trigonometric functions $\sin(x)$ and $\cos(x)$, one finds that for interpolation steps

$$\ell < \frac{\lambda_0}{2(p+1)} \tag{1.62}$$

the error (1.60) behaves as in Figure 1.6 and is given by the asymptotic expression (1.61). The threshold (1.62) valid for trigonometric functions corresponds to a minimum threshold $n_{\min}$ (for the number n of equal-length subintervals distributed over a λ_0-period) given by

$$n > n_{\min} = 2(p+1) \tag{1.63}$$

Table 1.4 Asymptotic behavior of the relative interpolation error.

The relative error (1.60) for the pth-order interpolation of the functions $\sin(x)$ and $\cos(x)$ over an interval of length $\lambda_0 = 2\pi$ is

$$\varepsilon \simeq \alpha_p (\ell/\lambda_0)^{(p+1)}$$

The coefficients α_p for the first six interpolation orders are reported below followed by the numbers n_p of equally spaced samples required to obtain $\varepsilon = 10^{-3}$ and $\varepsilon = 10^{-6}$, with corresponding interpolation step $\ell = \lambda_0/(n_p - 1)$

$\alpha_1 = 3.6$	$\alpha_2 = 11.3$	$\alpha_3 = 40$	$\alpha_4 = 157$	$\alpha_5 = 641$	$\alpha_6 = 2{,}990$
$n_1 = 60$	$n_2 = 24$	$n_3 = 15$	$n_4 = 12$	$n_5 = 11$	$n_6 = 10$
$n_1 = 1{,}899$	$n_2 = 226$	$n_3 = 81$	$n_4 = 45$	$n_5 = 31$	$n_6 = 24$

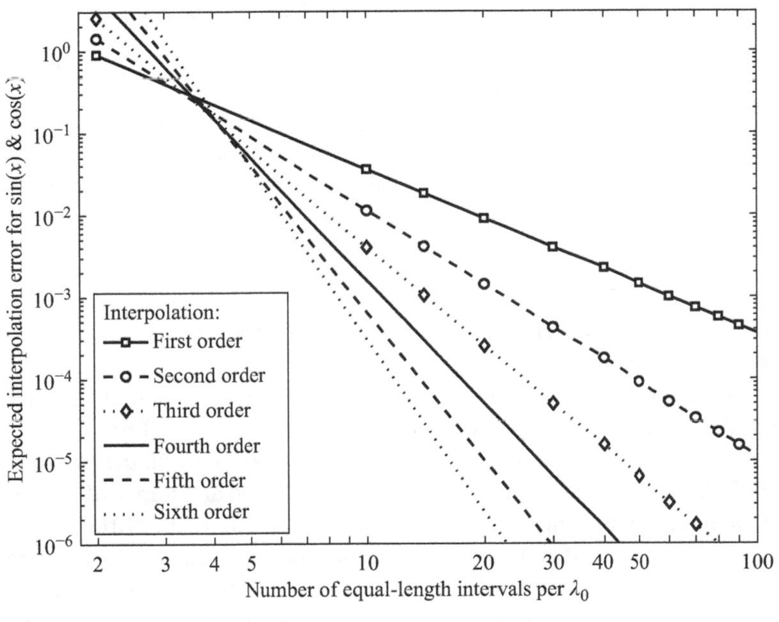

Figure 1.6 Asymptotic behavior of the relative error (1.60) for the trigonometric functions $\sin(x)$ and $\cos(x)$ computed with $\lambda_0 = 2\pi$. The figure shows the results for the first six interpolation orders.

In this connection, (1.62) guarantees that the errors ε also follow their asymptotic behavior (1.61) when very large child intervals of length

$$\ell_{\text{child}} = p\,\ell = \frac{p}{p+1}\,\frac{\lambda_0}{2} \tag{1.64}$$

are employed, up to a maximum child-interval length always in the range

$$\lambda_0/4 \le \max(\ell_{\text{child}}) < \lambda_0/2 \tag{1.65}$$

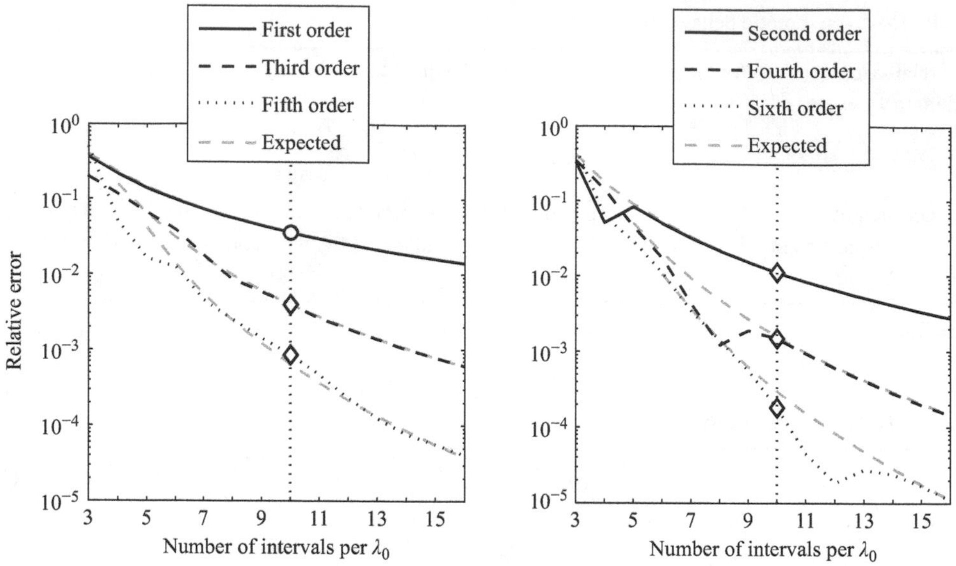

Figure 1.7 The trigonometric function sin(x) interpolated by linear up to sixth-order interpolation over the interval $\lambda_0 = 2\pi$. The figures show the error (1.60) with respect to the exact sinusoidal function together with the expected behavior of the error as reported in Table 1.4 (shown by gray dashed-lines).

for all possible p values. For example, with a second-order interpolation (i.e., for $p = 2$), the asymptotic behavior is matched for all $\ell_{\text{child}} \leq \lambda_0/3$.

The error behavior for interpolation steps larger than these threshold values is shown in Figures 1.7 and 1.8 for the sin(x) and cos(x) functions, respectively. From these figures, one sees that errors of the order of the 4% are obtained for $\ell = \lambda_0/10$ with the first-order (i.e., the linear) interpolation, $\ell = \lambda_0/7$ with interpolation of the second and third orders, $\ell = \lambda_0/6$ with fourth order of interpolation, and $\ell = \lambda_0/5$ with interpolation of the fifth and sixth orders. Interpolation steps of $\ell = \lambda_0/4$ or less are usually not recommended because for $\ell = \lambda_0/4$ the interpolation error (1.60) is about 10% even when using a sixth-order interpolation.

In summary, the results and the figures of this subsection clearly show that whenever an error of a few percent is acceptable, higher polynomial interpolation orders permit us to reduce the total number of samples by at least 30% with respect to the number of samples required by linear interpolation. This reduction factor exceeds 40% when the maximum acceptable error is smaller than 0.1%.

1.3.2 Spectral Integrity and Other Frequency Domain Considerations

An alternative perspective is offered by an analysis of the interpolation error in the Fourier transform domain. Consider a linear interpolation performed by using the triangular basis functions described in Section 1.1.

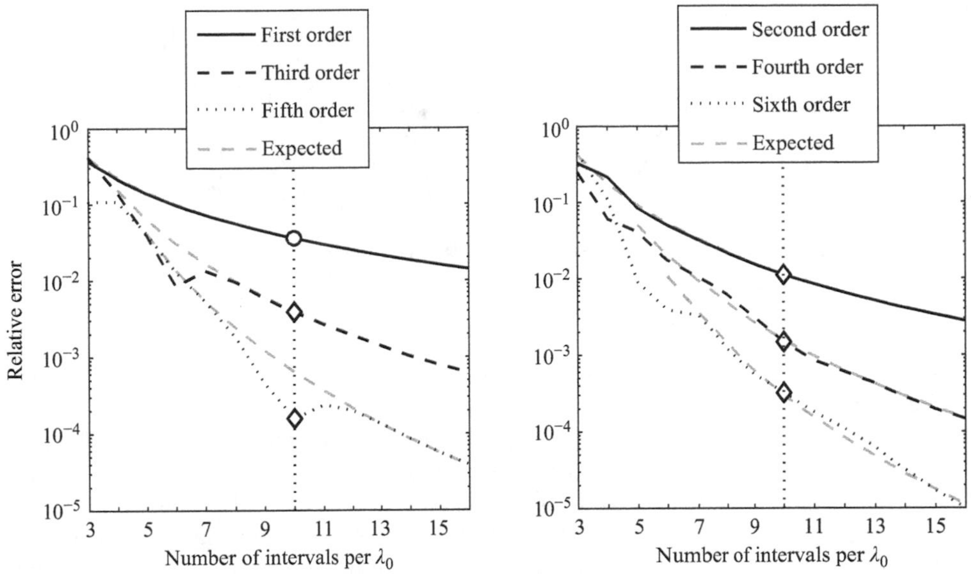

Figure 1.8 The trigonometric function $\cos(x)$ interpolated by linear up to sixth-order interpolation over the interval $\lambda_0 = 2\pi$. The figures show the error (1.60) with respect to the exact sinusoidal function together with the expected behavior of the error as reported in Table 1.4 (shown by gray dashed-lines).

The x-interval on which each $B_m(x)$ is different from zero is called the *support* of $B_m(x)$. Since $B_m(x)$ has a limited support, its Fourier transform has an unbounded support. For example, the Fourier transform of the simplified triangular basis function (1.8) is

$$\mathsf{B}(k_x) = \mathcal{F}[B(x,\ell)] = \int_{-\infty}^{+\infty} B(x,\ell)\exp(-jk_x x)\,\mathrm{d}x = \ell\,\mathrm{sinc}^2\left(\frac{k_x \ell}{2}\right) \tag{1.66}$$

with

$$\mathrm{sinc}(z) = \sin(z)/z \tag{1.67}$$

and where, from now on, the symbol $\mathcal{F}[g]$ indicates the Fourier integral of g which, in turn, is a function $\mathsf{g}(k_x)$ of the frequency k_x.

Equation (1.66) implies that the Fourier transform of the linear interpolation (1.7) on equal-length intervals is

$$\begin{aligned}\tilde{\mathsf{f}}(k_x) = \mathcal{F}[\tilde{f}(x)] &= \sum_{m=1}^{n} f_m\,\mathcal{F}[B(x - x_m, \ell)] \\ &= \ell\,\mathrm{sinc}^2(k_x\ell/2)\sum_{m=2}^{n-1} f_m \exp(-jk_x x_m)\end{aligned} \tag{1.68}$$

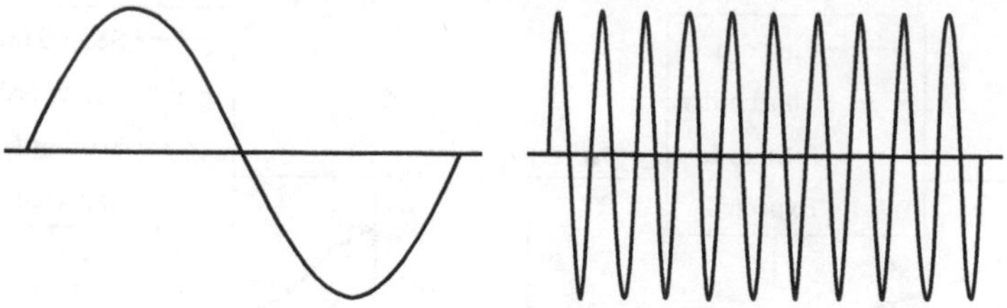

Figure 1.9 The pulse-modulated functions of (1.69) for $q=10$, $n=10$ (at left) and $q=10$, $n=1$ (at right).

whenever the first (f_1) and the last (f_n) samples of f are zero. The Fourier integrals (1.66, 1.68) are zero at $k_x=2\pi/\ell$, which is the angular frequency that corresponds to a wavelength $\lambda_x=2\pi/k_x$ exactly equal to the sampling length ℓ. Thus, a linear interpolation on equally spaced intervals is clearly unable to reconstruct periodic components of the form $F\sin[2\pi(x-x_1)/\ell]$. In fact, the sampling length ℓ must always be chosen according to the value of the *highest* frequency component one wants to *catch* with the interpolation procedure. We will shortly see that one has to set $\ell<\lambda/2$ regardless of the order of the interpolation used; λ being the wavelength of the highest frequency component one wants to catch.

Since interpolation is typically performed on definite domains, with no loss of generality we may draw the most important conclusions by simply considering the case of pulse-modulated functions of the form

$$g_{nq}(x)=\frac{2}{q\,\lambda_0}\sin\left(\frac{k_0\,x}{n}\right)[u(x)-u(x-q\lambda_0)] \tag{1.69}$$

$$\mathcal{F}[g_{nq}(x)]=\mathsf{g}_{nq}(k_x)=-2j\,\frac{\exp(-j\pi\,q\,k^-)}{n\,k^+}[\mathrm{sinc}(\pi\,q\,k^-)-\mathrm{sinc}(2\pi\,q/n)\exp(-j\pi\,q\,k^+)] \tag{1.70}$$

with

$$\lambda_0=2\pi/k_0 \tag{1.71}$$

$$k^{\pm}=\frac{k_x}{k_0}\pm\frac{1}{n} \tag{1.72}$$

The function $g_{nq}(x)$ given in (1.69) is written in terms of the unitary step function

$$u(z)=\begin{cases}0 & \text{for } z<0\\ 1 & \text{for } z>0\end{cases} \tag{1.73}$$

If n, q, and q/n are integers greater than unity, $g_{nq}(x)$ consists only of the first q/n complete cycles of the sinusoidal function $\sin(k_0\,x/n)$ (see Figure 1.9). In this case, the Fourier spectrum of $g_{nq}(x)$ simplifies into

$$|\mathsf{g}_{nq}(k_x)|=2\left|\frac{\mathrm{sinc}[q\pi(k_x/k_0-1/n)]}{n\,k_x/k_0+1}\right| \tag{1.74}$$

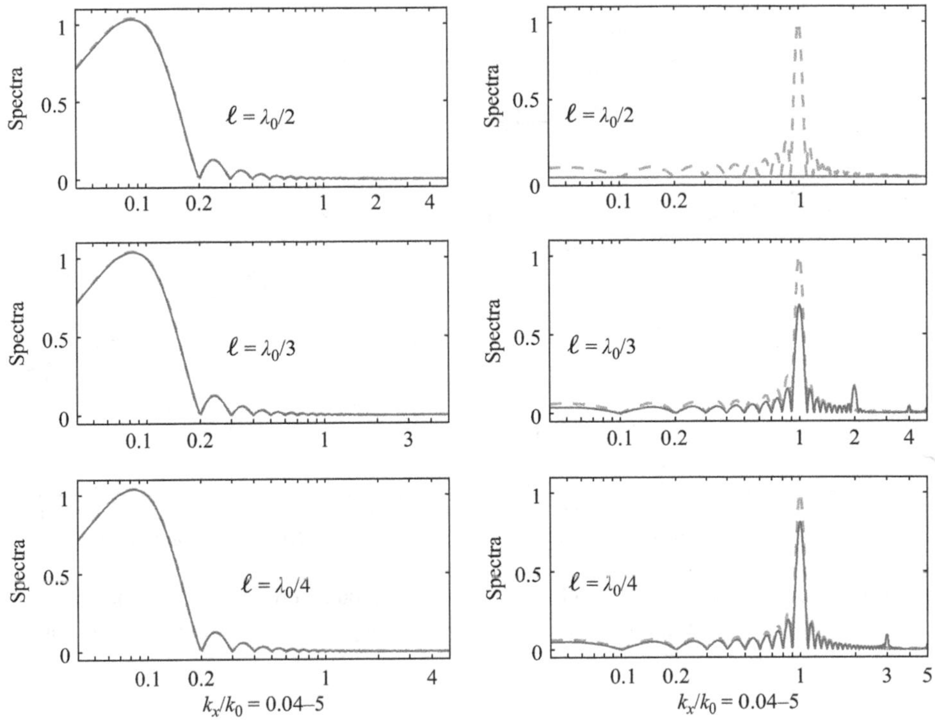

Figure 1.10 Spectra of the pulse-modulated function of the form (1.69) obtained for $q = 10$. The figures at left show the lower frequency carrier-case results obtained for $n = 10$ (one sinusoidal cycle in the pulse). At right we report the higher frequency carrier-case results obtained for $n = 1$ (10 sinusoidal cycles in the pulse). The exact spectrum is reported by gray dashed-lines; the spectra obtained by linear interpolation are reported by black solid-lines.

and reaches its maximum value for $k_x \lesssim k_0/n$ while it remains significantly different from zero on the interval $\{0 \leq k_x/k_0 \leq 1/n + \alpha(q)\}$ with a decreasing upper bound for increasing q values.

Figure 1.10 reports for positive values of the normalized frequency k_x/k_0 the spectrum (1.74) obtained for $q = 10$ in the two cases of $n = 10$ (lower frequency carrier case, shown at left) and $n = 1$ (higher frequency carrier case, shown at right); recall that $|\mathrm{g}_{nq}(-k_x)| = |\mathrm{g}_{nq}(k_x)|$. Exact spectra are reported in Figure 1.10 by gray dashed-lines while the black solid-lines show the spectra $|\widetilde{\mathrm{g}_{nq}}|$ of the linearly interpolated functions $\widetilde{g_{nq}}(x)$ for three different sampling lengths $\ell = \lambda_0/2$, $\lambda_0/3$, and $\lambda_0/4$. The black-line spectrum reported on the top-right part of the figure is relative to the linearly interpolated pulse-function modulated by the highest carrier-frequency obtained by using equally spaced intervals of length $\ell = \lambda_0/2$. This spectrum is zero for all k_x and is indeed reported to show that linear interpolation on equally spaced intervals is able to *catch* the periodic components of the sampled function up to the frequency $k_0 = 2\pi/\lambda_0$ only provided one uses a sampling length $\ell < \lambda_0/2$. It is worth noticing that the spectral error of the linearly interpolated pulse-function modulated by the lowest carrier-frequency is always

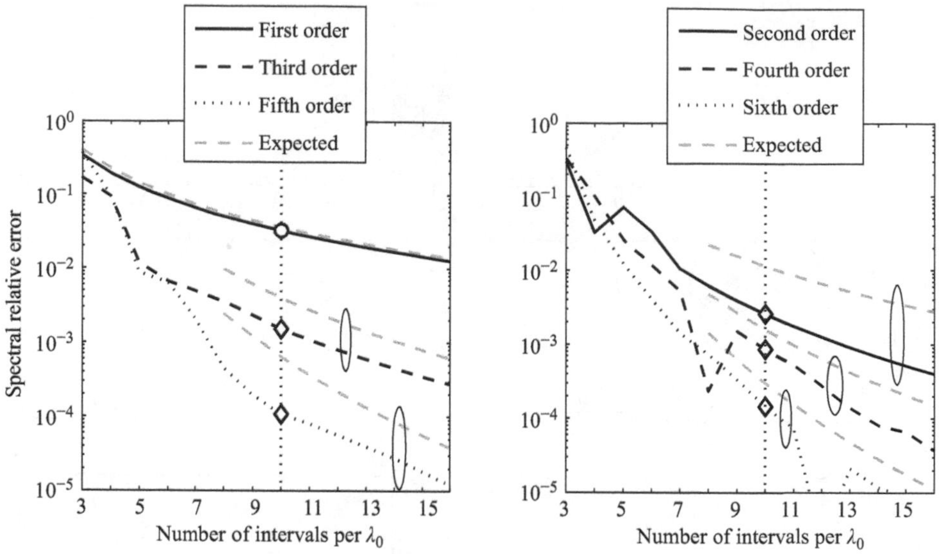

Figure 1.11 Spectral relative error (1.75) for a pulse-modulated function of the form (1.69) obtained with $q = 10$, $n = 1$ (higher frequency carrier case). The error is computed for $k = 2\,k_0$.

less than that relative to the highest frequency carrier. Notice also that the spectra of the interpolated functions shown in Figure 1.10 contain *incorrect* spikes of amplitude and height in inverse proportion to the frequency at which they are centered; the first of these spikes is always centered at $k_x = k_0(\lambda_0/\ell - 1)$.

To quantitatively discuss the spectral performances of higher order interpolations, we may compute the following spectral relative error

$$e_{L2}(k) = \sqrt{\frac{\displaystyle\int_0^k \left|\widetilde{\mathsf{g}}_{nq}(k_x) - \mathsf{g}_{nq}(k_x)\right|^2 \mathrm{d}k_x}{\displaystyle\int_0^k \left|\mathsf{g}_{nq}(k_x)\right|^2 \mathrm{d}k_x}} \tag{1.75}$$

by using interpolation of different orders on equally spaced subintervals for the pulse-modulated function (1.69) in case of $n = 1$, $q = 10$ (higher frequency carrier case). Figure 1.11 shows the behavior of the spectral error (1.75) computed for $k = 2\,k_0$ together with the *expected* error already shown in Figures 1.6–1.8. It is of importance to observe that by using linear interpolation the spectral error is almost the same as (although a little bit smaller than) the relative error (1.60) while, for higher order of interpolations, the spectral error is in general always smaller than the relative error (1.60).

Similarly, the behavior of the spectral error (1.75) computed for $k = 10\,k_0$ is shown in Figure 1.12 together with the *expected* error already shown in Figures 1.6–1.8. Notice that in this case the asymptotic behaviors of the spectral error (as well as the corresponding thresholds) are the same ones already discussed in Section 1.3.1.

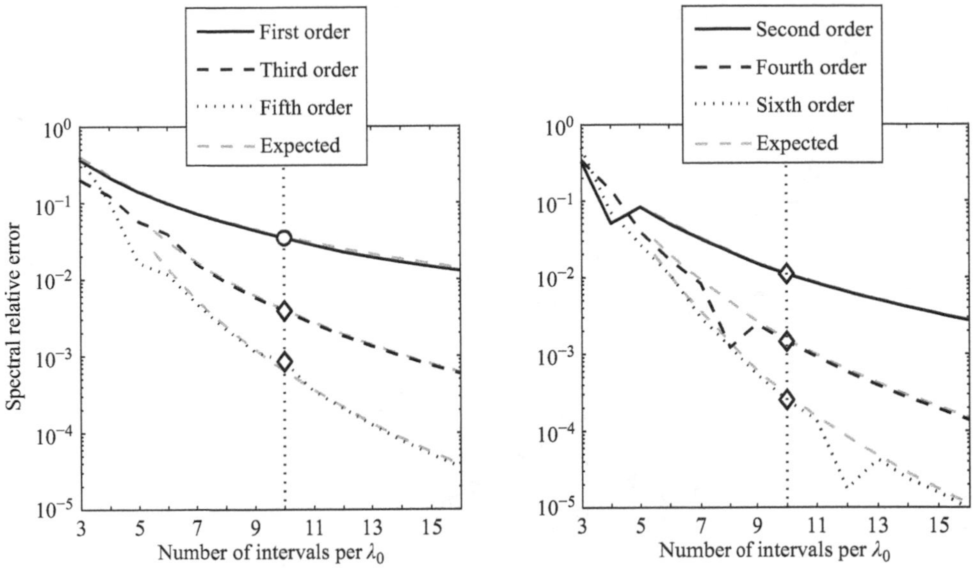

Figure 1.12 Spectral relative error (1.75) for a pulse-modulated function of the form (1.69) obtained with $q = 10$, $n = 1$ (higher frequency carrier case). The error is computed for $k = 10\,k_0$.

Since 99.88% of the energy of the modulated signal is contained in the frequency range $k_x < 2\,k_0$ (while more than 99.99% of the energy is contained in the frequency range $k_x < 10\,k_0$) one may easily conclude that the interpolation errors spoil the highest frequency part of the Fourier spectrum more than the lower frequency part and, even more important, that in general the interpolation errors are not enhanced by the Fourier integrals.

In summary, once again, the results and the figures of this subsection clearly show that whenever an error of a few percent is acceptable, higher polynomial interpolation orders easily permit us to reduce the total number of samples by at least 30% compared to the number of samples required by linear interpolation. This reduction factor becomes larger than 40% if the maximum acceptable error is less than 0.1%.

1.4 Approximation of Functions with Border Singularities

There are applications where one has to represent (on a given region and in terms of known basis functions) a quantity that is singular (that is, with a value approaching infinity) on some border $\boldsymbol{r}_s$ of the region. In electromagnetic applications, for example, it may happen that one has to deal with singularities of the electromagnetic field, or of the current and charge densities. The singular behavior of these quantities, when present, is *mitigated* by the fact that the energy associated with the electromagnetic fields in bounded regions is always bounded. Thus, in two- and three-dimensional electromagnetic problems, one may typically encounter logarithmic singularities of the form $\ln(\boldsymbol{r} - \boldsymbol{r}_s)$, or singularities of the form $(\boldsymbol{r} - \boldsymbol{r}_s)^{\nu-1}$, where ν is a frequency-independent singularity coefficient known *a priori* (typically, with $1/2 \leq \nu < 1$).

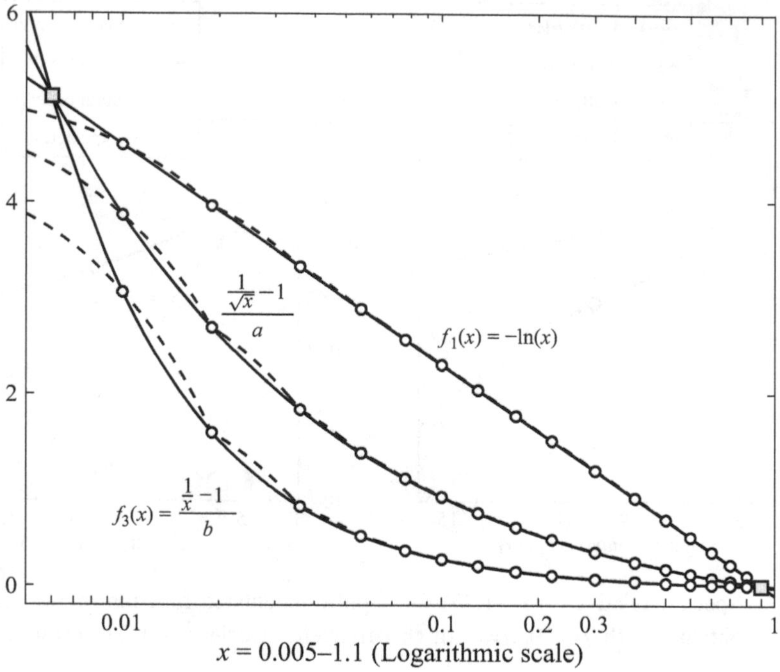

Figure 1.13 Linear approximation of the functions $f_1(x) = -\ln(x)$, $f_2(x) = (1/\sqrt{x} - 1)/a$, and $f_3(x) = (1/x - 1)/b$ on the interval $[0.005 \leq x \leq 1.1]$. The approximate functions (dashed-line) are compared with the exact ones (solid-lines).

Unfortunately, polynomial basis functions are not well suited for modeling singular behavior, despite the fact that it is common practice to decrease the size of the approximation subintervals in the neighborhood of $\boldsymbol{r}_s$ in an attempt to better model that behavior.

The results that follow, although relative to one-dimensional scalar quantities, permit one to assess the quality of *polynomial* approximations of different orders on functions of different (and increasing) singularity orders. For example, Figures 1.13 and 1.14 show on a semi-logarithmic scale the results of a linear (first-order) and of a quartic (fourth-order) approximation of the singular functions

$$
\begin{aligned}
f_1(x) &= -\ln(x) \\
f_2(x) &= \frac{1/\sqrt{x} - 1}{a} \\
f_3(x) &= \frac{1/x - 1}{b}
\end{aligned}
\tag{1.76}
$$

on the interval $[0.005 \leq x \leq 1.1]$.

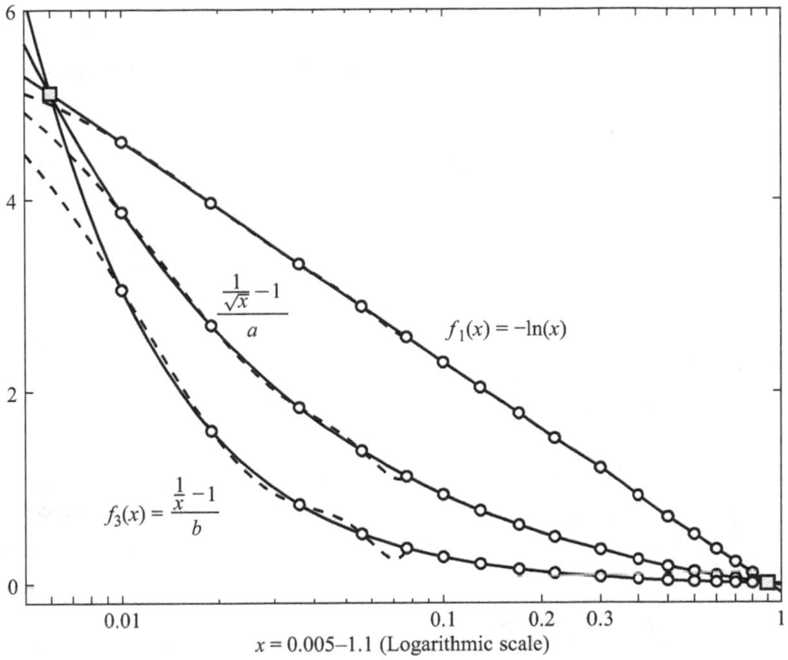

Figure 1.14 Fourth-order approximation of the functions $f_1(x) = -\ln(x)$, $f_2(x) = (1/\sqrt{x} - 1)/a$, and $f_3(x) = (1/x - 1)/b$ on the interval $[0.005 \leq x \leq 1.1]$. The approximate functions (dashed-line) are compared with the exact ones (solid-lines).

In the following, we consider the order of the approximation to be the order of the polynomial *basis functions used to obtain the approximated function. This definition is extended to include the cases when singular expansion functions are added to the polynomial expansion function set.*

To provide a fair comparison of the results over the chosen approximation interval, the functions f_2 and f_3 are scaled by properly choosing the constant coefficients a and b to get $f_1 = f_2 = f_3$ at $x = 0.006$, while $f_1 = f_2 = f_3$ remain equal to zero at $x = 1$. The square markers of Figures 1.13 and 1.14 indicate the two points at which one has $f_1 = f_2 = f_3$. The circular markers denote the unequally spaced interpolation points $\boldsymbol{x}_{\text{int}}$ that gather together in the vicinity of the singular point $x = 0$, with

$$\boldsymbol{x}_{\text{int}} = [0.01,\ 0.019,\ 0.036,\ 0.056,\ 0.077,\ 0.1,\ 0.13,\ 0.17,\ 0.22,\ 0.3,\ \ldots]$$

and $\Delta x = x_{\text{int}}[k] - x_{\text{int}}[k-1] = 0.1$ for $k \geq 11$. The approximate values of Figures 1.13 and 1.14 are in fact obtained with polynomial interpolation for $x \geq 0.01$ whereas polynomial *extrapolation* is used on the first subinterval for $x < 0.01$, where the same polynomial approximation obtained by interpolating on the adjacent subinterval to the right of $x = 0.01$ is retained.

These figures show that the approximation error increases with the order of the singularity and with proximity to the singular point. In fact, the approximation errors turn out to be higher on the first sixth subintervals (for $x < 0.1$) while they remain rather small on the remaining (eleven) subintervals, for $0.1 < x \leq 1.1$. If one changes the order of the approximation from

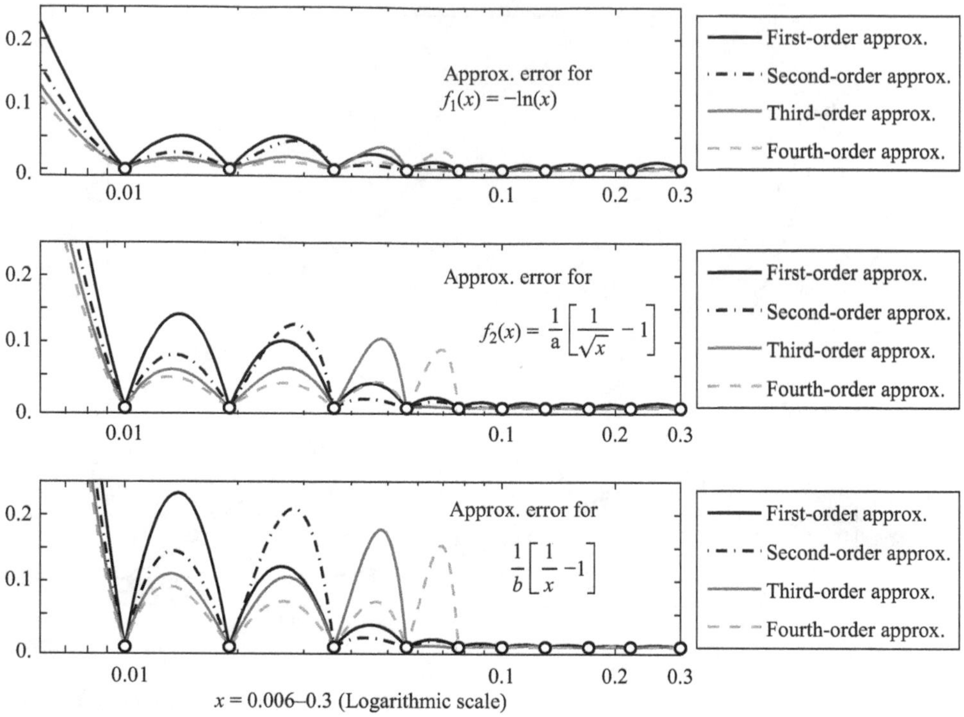

Figure 1.15 Absolute errors obtained by approximating the functions $f_1(x) = -\ln(x)$, $f_2(x) = (1/\sqrt{x} - 1)/a$, and $f_3(x) = (1/x - 1)/b$ (from top to bottom) with the lowest approximation orders $n = 1, 2, 3$, and 4. For $x < 0.01$ extrapolation has been used thereby retaining the same polynomial approximation obtained by interpolating on the adjacent subintervals to the right of $x = 0.01$.

1 to 4 in the attempt to diminish the error, the error on the fourth interpolation subinterval increases. This behavior is clarified in Figure 1.15, where we report the errors obtained by using approximations of order n = 1, 2, 3, and 4. From this figure, one can appreciate that, typically, the error of the nth-order approximation is maximum on the nth *interpolation* subinterval, provided the interpolation points are properly spaced.

In other words, it doesn't pay to increase the order of a polynomial approximation in the neighborhood of a singular point, but it is rather more convenient to use a first-order approximation on a set of samples that gets more dense in the vicinity of this singular point (obviously, one cannot interpolate the latter). Furthermore, polynomial approximations fail in the subinterval that contains a singular point; on this subinterval, extrapolation may be used although the error always tends to infinity at the singular point.

1.4.1 Singular Expansion Functions

However, thankfully, to drastically reduce the error on intervals bounded by a singular point there is a simple technique that is applicable provided one knows the singularity coefficient ν

or, more in general, whenever the asymptotic behavior of the to-be-approximated quantity at the singular point is known in advance. In fact, for example, by setting

$$\xi = \frac{x}{\ell} \tag{1.77}$$

the following identities are readily proved

$$\begin{bmatrix} g_1(x) \\ g_2(x) \end{bmatrix} = \begin{bmatrix} \ln(x) \\ x^{-\beta} \end{bmatrix} = \begin{bmatrix} \alpha_{s1}\ \ln(\xi) \\ \alpha_{s2}\ (\xi^{-\beta} - \xi) \end{bmatrix} + \begin{bmatrix} \alpha_{01} \\ \alpha_{02} \end{bmatrix} (1-\xi) + \begin{bmatrix} g_1(\ell) \\ g_2(\ell) \end{bmatrix} \xi \tag{1.78}$$

with

$$\begin{bmatrix} \alpha_{s1} \\ \alpha_{s2} \end{bmatrix} = \begin{bmatrix} 1 \\ \ell^{-\beta} \end{bmatrix}; \quad \begin{bmatrix} \alpha_{01} \\ \alpha_{02} \end{bmatrix} = \begin{bmatrix} \ln(\ell) \\ 0 \end{bmatrix} \tag{1.79}$$

The last two terms on the right-hand side of (1.78) take the form

$$\alpha_0\,(1-\xi) + g(\ell)\,\xi$$

and can be regarded as the linear interpolation on the interval $\{0 \leq x \leq \ell\}$ of a g-quantity equal to α_0 at $\xi = x = 0$. It is therefore rather clear that the singular behavior of the functions $g_1(x)$, $g_2(x)$ is exactly modeled in (1.78) by the two new singular expansion functions $\ln(\xi)$ and $(\xi^{-\beta} - \xi)$, respectively (with $\beta > 0$ understood).

In this connection, notice that the singular logarithmic function is modeled by a value of $\alpha_{01} = \ln(\ell) \neq 0$. That is to say that, in general, to better model a singular behavior one really needs to *add* new DoFs to those normally used to model bounded functions in terms of interpolatory polynomials. In our previous example, we have added one new DoF represented by the value of the coefficient α_s appearing in (1.78).

The latter clarification is of importance because the fact that

$$\lim_{\beta \to 0} \left(\xi^{-\beta} - \xi\right) = 1 - \xi \tag{1.80}$$

may lead one to think that it is possible to substitute the expansion function $(1-\xi)$ with the singular function $(\xi^{-\beta} - \xi)$ when dealing with non-logarithmic singularities. Unfortunately, this does not work whenever the quantity one has to approximate has a behavior more complex than that of the function $g_2(x)$ in (1.78), as we will see shortly in the following paragraphs.

Finally, we observe that the singular function $(\xi^{-\beta} - \xi)$ has been introduced here just to discuss the possible substitution of the regular function $(1-\xi)$ with the singular one. In applications, this singular function can be replaced with no damage by the singular function $(\xi^{-\beta} - 1)$, which is equal to zero in the limit for $\beta = 0$.

1.4.2 Singular Functions that Comply with Exact Approximations in Terms of Singular Plus Polynomial Basis Functions

To immediately show the improvement of the approximations obtained by introducing and using singular expansion functions, we can just consider the function $f_3(x) = (1/x - 1)/b$

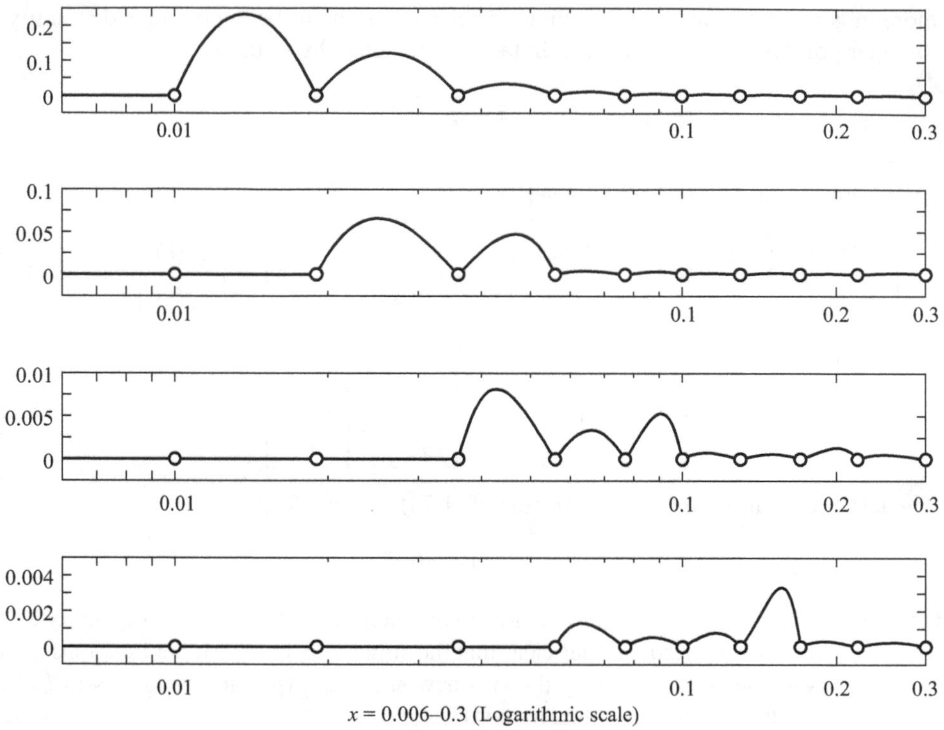

Figure 1.16 Absolute errors for the nth-order approximation of the singular function $f_3(x) = (1/x - 1)/b$, for $n = 1, 2, 3$, and 4 (from top to bottom). The approximations are obtained by using piecewise Lagrangian interpolatory basis functions of the nth order together with the singular basis function $(x/\ell_n)^{-1} - x/\ell_n$, which straddles the first n subintervals $(0 \le x \le \ell_n)$ and vanishes at $x = \ell_n$, ℓ_n being the total length of the first n subintervals. Notice in the figure the use of different vertical scales.

on the same interval considered before to draw Figures 1.13–1.15. (Recall that the absolute error relative to the approximations before obtained for f_3 is higher than those obtained for f_1 and f_2.)

Figure 1.16 shows the errors for the approximations of order $n = 1, 2, 3$, and 4 obtained by adding to the expansion function set the singular function $(x/\ell_n)^{-1} - x/\ell_n$ which straddles the first n subintervals $(0 \le x \le \ell_n)$ and vanishes at $x = \ell_n$. In this case, the approximation error is exactly equal to zero for all $x \le \ell_n$, and it became square integrable in spite of the fact that f_3 is not square integrable.

Most of all, the results of Figure 1.16 show that there is really no need to reduce the subintervals' length in the neighborhood of the singular point whenever the appropriate singular expansion function is added to the expansion set. If this is done, larger subintervals (i.e., defined by a smaller number of bounding points) can be used and it become really advantageous to increase the expansion function set with higher order polynomial functions. For example, one can equally subdivide the parent interval $\{0 \le \xi \le 1\}$ into equal-length subintervals and then map the parent into the child interval $\{0 \le x \le \ell\}$.

To deal with functions that are singular at the two extreme points $x = 0$ and $x = \ell$ of the interval $\{0 \le x \le \ell\}$ we can easily extend the results (1.77, 1.78, 1.79) by introducing two dependent parent variables

$$\xi_1 = \frac{x}{\ell}, \quad \xi_2 = 1 - \frac{x}{\ell} \tag{1.81}$$

with dependency relation

$$\xi_1 + \xi_2 = 1 \tag{1.82}$$

together with the following singular expansion functions

$$S_{i,\ln}(\xi_i) = \ln \xi_i \tag{1.83}$$

$$S_{i,\nu-1}(\xi_i) = \xi_i^{\nu-1} - \xi_i \tag{1.84}$$

defined on the parent interval $\{0 \le \xi_i \le 1\}$, with $\xi_i = \xi_1$ or $\xi_i = \xi_2$. On the interval $\{0 \le x \le \ell\}$, any singular function

$$g(x) = a \ln(x) + b \ln(\ell - x) + \sum_\nu c_\nu \, x^{\nu-1} + \sum_\nu d_\nu (\ell - x)^{\nu-1} + Q(x) \tag{1.85}$$

obtained by linearly combining a regular bounded function $Q(x)$ with singular functions defined by all possible ν values less than unity is then approximated by

$$\begin{aligned} \tilde{g}(x) = {} & a \ln(\xi_1) + b \ln(\xi_2) + \sum_\nu \ell^{\nu-1} c_\nu \, (\xi_1^{\nu-1} - \xi_1) \\ & + \sum_\nu \ell^{\nu-1} d_\nu \, (\xi_2^{\nu-1} - \xi_2) + \sum_{k=0}^{p} \alpha_k \, P_k(p, \xi_1) \end{aligned} \tag{1.86}$$

for $p \ge 1$ and with

$$\alpha_k = (a + b) \ln(\ell) + \frac{k}{p} \sum_\nu \ell^{\nu-1} c_\nu + \left(1 - \frac{k}{p}\right) \sum_\nu \ell^{\nu-1} d_\nu + Q(k\ell/p) \tag{1.87}$$

where $P_k(p, \xi_1)$ indicates the interpolatory pth-order Lagrange polynomial given in (1.18). The α_k coefficients (1.87) are directly derived by exploiting the interpolatory property of the Lagrange polynomials $P_k(p, \xi)$. Notice that the representation given in (1.86) is *exact* for any $p \ge \max(1, q)$ if the bounded function $Q(x)$ appearing in (1.85) is an arbitrary qth-order polynomial; as a matter of fact, (1.78) and (1.79) represent a very simple particular case of the more general result just reported above.

As said before, in applications, one is free to replace the singular basis functions $S_{i,\nu-1}(\xi_i) = (\xi_i^{\nu-1} - \xi_i)$ with $(\xi_i^{\nu-1} - 1)$ to then obtain a new representation equivalent to (1.86); the derivation of this representation is left as an exercise to the reader.

1.4.3 Singular Functions that Do Not Permit Exact Approximations in Terms of Singular Plus Polynomial Basis Functions

So far we have discussed singular functions that can be *exactly modeled* by using singular and polynomial basis functions altogether. We now turn our attention to more complex singular functions where this is not possible. We start by considering a *first-order* approximation of the functions

$$\begin{aligned} h_1(x) &= (\sin x - \ln x)/2 \\ h_2(x) &= \cos x + 1/\sqrt{x} \\ h_3(x) &= \cos x + 1/x \end{aligned} \tag{1.88}$$

on the interval $\{0 \le x \le \ell\}$. In terms of the normalized parent variable $\xi = x/\ell$ and of the singular basis functions

$$S_{\ln}(x) = \ln \xi \tag{1.89}$$

$$S_{\nu-1}(\xi) = \xi^{\nu-1} - \xi \tag{1.90}$$

the approximations are given by

$$\begin{aligned} \widetilde{h}_1(x) &= \alpha_0\, P_0(1,\xi) + \alpha_1\, P_1(1,\xi) + \alpha_s\, S_{\ln}(\xi) \\ \widetilde{h}_2(x) &= \beta_0\, P_0(1,\xi) + \beta_1\, P_1(1,\xi) + \beta_s\, S_{-\frac{1}{2}}(\xi) \\ \widetilde{h}_3(x) &= \gamma_0\, P_0(1,\xi) + \gamma_1\, P_1(1,\xi) + \gamma_s\, S_{-1}(\xi) \end{aligned} \tag{1.91}$$

with expansion coefficients

$$\begin{bmatrix} \alpha_0 & \alpha_1 & \alpha_s \\ \beta_0 & \beta_1 & \beta_s \\ \gamma_0 & \gamma_1 & \gamma_s \end{bmatrix} = \begin{bmatrix} -(\ln \ell)/2 & (\sin \ell - \ln \ell)/2 & -1/2 \\ 1 & \cos \ell + 1/\sqrt{\ell} & 1/\sqrt{\ell} \\ 1 & \cos \ell + 1/\ell & 1/\ell \end{bmatrix} \tag{1.92}$$

obtained after imposing a zero approximation error at $x = 0$ and at $x = \ell$.

Once again, notice that β_0 (or γ_0) is different from zero, which prevents the replacement of a regular expansion function $P_0(1,\xi) = 1 - \xi$ with the singular basis function $S_{-1/2}(\xi)$ (or $S_{-1}(\xi)$) in order to work out the approximation problem with only two DoFs per function.

Furthermore, since one condition by which the expansion coefficients are obtained is to impose zero approximation error at the singular point, one may think of this as an interpolation process, although mathematical limits must be used to equate the approximated function to the given one at the singular point.

The results obtained by first-order approximation of the functions h_1, h_2, h_3 on the interval $\{0 \le x \le 2\pi\}$ subdivided into 10 equal-length subintervals are shown in Figure 1.17. For these approximations, the expansion coefficients associated with the first subinterval $\{0 \le x \le \pi/5\}$

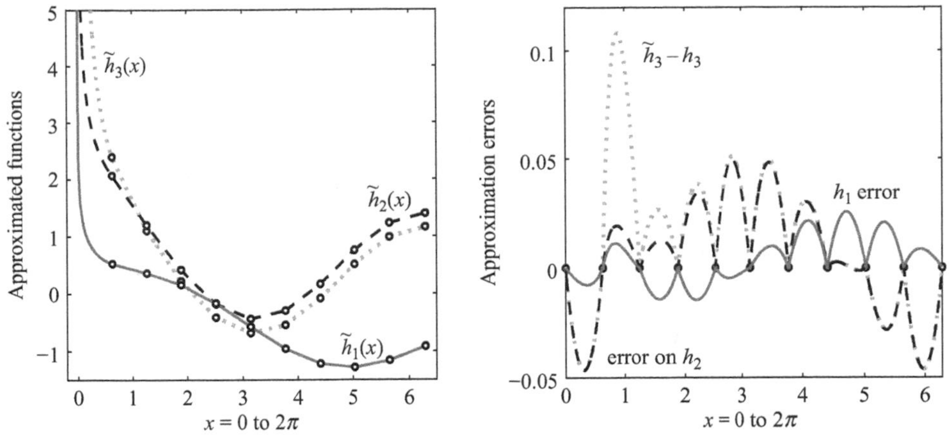

Figure 1.17 First-order piecewise approximation of the function $h_1(x) = (\sin x - \ln x)/2$ (solid-line), $h_2(x) = \cos x + 1/\sqrt{x}$ (dashed-line), and $h_3(x) = \cos x + 1/x$ (dotted-line) on the interval $[0, 2\pi]$ subdivided into 10 equal-length subintervals. The figure at left shows the approximated functions. The approximation errors are reported at right.

are obtained by setting $\ell = \pi/5$ into (1.92). In particular, from this figure one may notice that the errors are square-integrable and vanish at the singular point $x = 0$, even when the approximated function is not square-integrable (like for h_2 and h_3).

These approximation errors can be easily reduced by increasing the order of the approximation and by enlarging at the same time the length of the used subintervals, just as it happens when interpolating bounded (non-singular) functions on equal-length subintervals. For the sake of brevity, we discuss only the errors obtained with high-order approximation of the function $h_3(x) = \cos x + 1/x$, since we know from Figure 1.17 that the maximum approximation error is expected for h_3.

Figure 1.18 shows the absolute errors obtained by approximating $h_3(x)$ on the interval $[0, 2\pi]$ subdivided into equal-length subintervals. The number of used subintervals is 10, 7, 6, and 6 for first-, second-, third-, and fourth-order approximations, respectively. For the nth-order approximation, the singular basis function $S_{-1}(\xi) = (\xi/\ell)^{-1} - \xi/\ell$ straddles the first n subintervals, each of length ℓ_i for $i = 1, 2, \ldots, n$, and with $\ell = \ell_1 + \ell_2 + \cdots + \ell_n$. Notice in particular how third- and fourth-order approximations are able to drastically reduce the approximation error and, at the same time, the number of subintervals (this latter by more than the 30%).

A final but not less important issue that remains to be addressed is related to the method one uses to evaluate the expansion coefficient associated with each singular basis function. The choice of the singular functions to be included in the expansion function set is usually known in advance, at least in electromagnetic applications, and is either of the logarithmic kind or of the form of $(\boldsymbol{r} - \boldsymbol{r}_s)^{\nu-1}$. Conversely, in any numerical solution procedure, the expansion coefficient of a singular basis function cannot be evaluated by "taking a limit" as we did in the previous subsections because the singular behavior may not always be excited, depending on several different parameters of the problem considered (e.g., polarization and/or symmetry properties of the incident field or of the driving source, material parameters, etc.). In these applications, the expansion coefficient of a singular basis function is evaluated by minimizing some integral

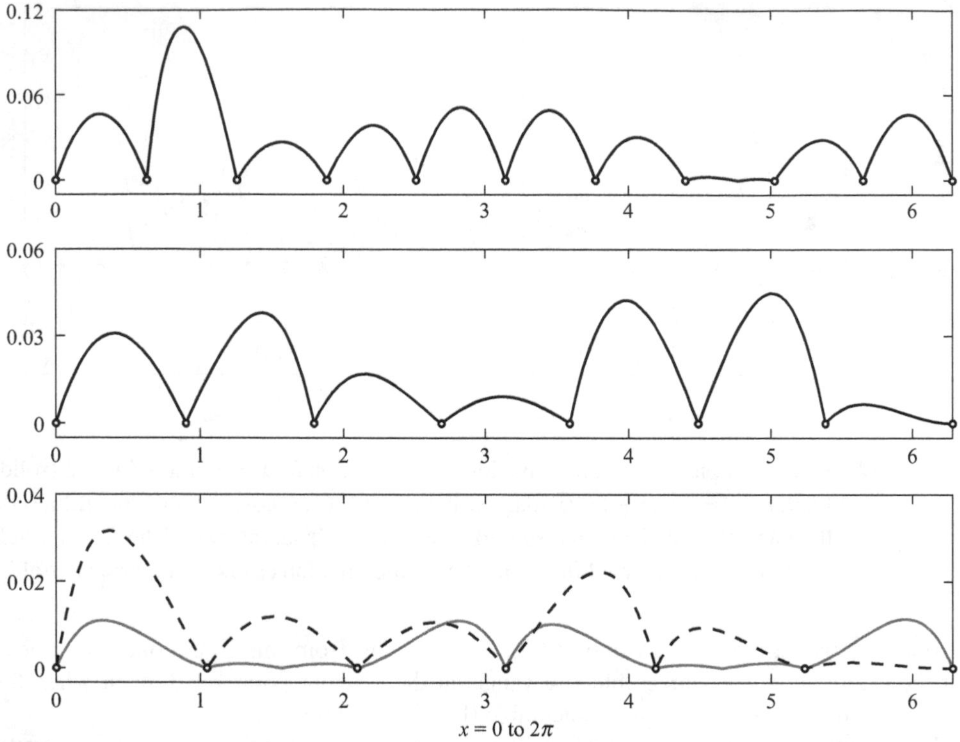

Figure 1.18 Absolute errors obtained by approximating the singular function $h_3(x) = \cos x + 1/x$ on the interval $[0, 2\pi]$ subdivided into equal-length subintervals. The figure reports the error obtained with the first-order approximation on 10 subintervals (at top), the error obtained with the second-order approximation on 7 subintervals (in the middle), and the error obtained with the third-order (solid-line) and fourth-order (dashed-line) approximations on 6 subintervals (at bottom). Notice in the figure the use of different vertical scales.

expression or by solving a system of linear equations, or both. Thus, the coefficients associated with singular basis functions are always evaluated numerically and always in error with respect to the *exact* ones. The primary effect of the error on the coefficient of the singular basis function is to yield an approximation error that does not vanish but actually goes to infinity at the singular point. However, the error remains square-integrable if the function to-be-approximated is square-integrable and, as said, this is always the case in electromagnetic applications.

To illustrate the effect of the error of the expansion coefficient of a singular basis function we continue to consider our worst-case test-function $h_3(x) = \cos x + 1/x$ (that is not square integrable) because it is the one with the highest degree of singularity so far considered. In this case, the exact value of the coefficient γ_s associated with the singular basis function $S_{-1}(\xi) = (\xi/\ell)^{-1} - \xi/\ell$ is always equal to $1/\ell$ for any nth-order approximation (see for example (1.92)), with $\ell = \ell_1 + \ell_2 + \cdots + \ell_n$ and where ℓ_i is the length of the ith subinterval. (Recall that for the nth-order approximation the singular basis function $S_{-1}(\xi)$ straddles the first n subintervals.)

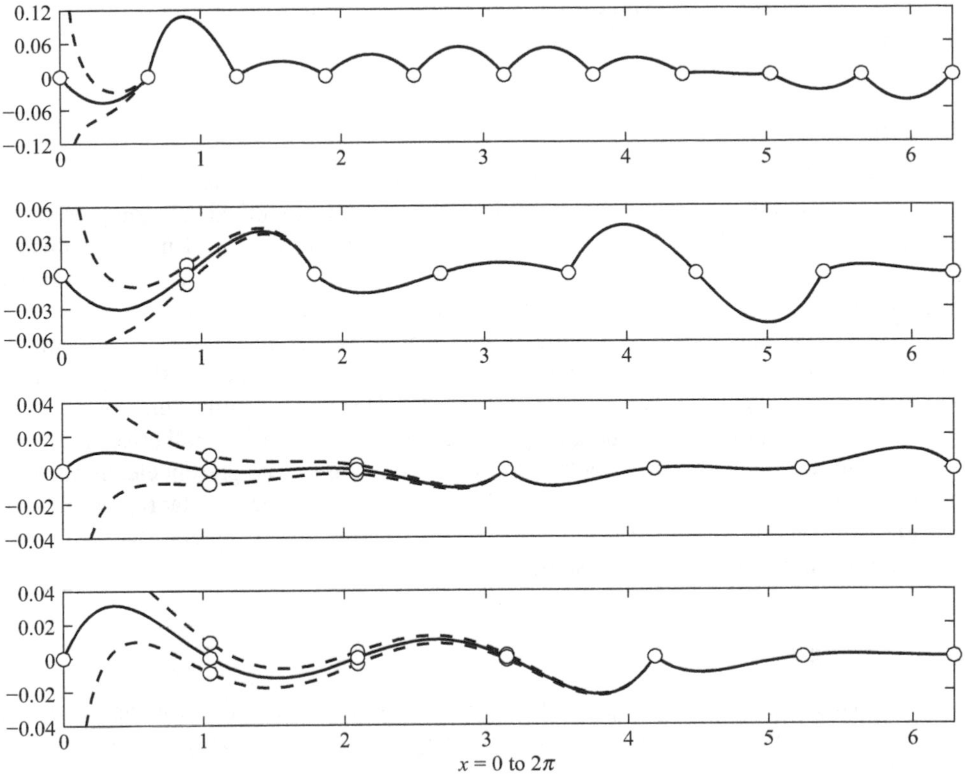

Figure 1.19 Errors obtained by approximating the singular function $h_3(x) = \cos x + 1/x$ on the interval $[0, 2\pi]$ subdivided into equal-length subintervals. The figure reports from top to bottom the error obtained with the first-order approximation on 10 subintervals, the error obtained with the second-order approximation on 7 subintervals, the error obtained with the third-order and fourth-order approximation on 6 subintervals. The errors obtained by using the exact value of the expansion coefficient for the singular basis function are reported with solid-lines. Dashed-lines show the errors obtained by varying this expansion coefficient by $\pm 1\%$. Notice in the figure the use of different vertical scales.

The effect of the numerical error of γ_s can now be simulated by modifying its exact value by $\pm 1\%$; the numerical *corruption* of this datum is quite conservative although, in numerical applications, the evaluation of the coefficients of the singular basis functions usually requires the numerical evaluation of singular integrals with a degree of precision to the state of the art. Figure 1.19 shows the approximation error $\tilde{h}_3(x) - h_3(x)$ obtained by corrupting only the value of γ_s. If γ_s is wrong, the approximation error is not bounded anymore and, in the limit for $x = 0$, it tends to $-\infty$ or $+\infty$ for $\gamma_s = 1/\ell - 1\%$ or $\gamma_s = 1/\ell + 1\%$, respectively. Notice however how for higher order approximations the error on γ_s does not deteriorate too much the results on the subintervals to the right of the first one (obviously the error on γ_s has no effect on the subintervals that do not belong to the domain of the singular basis function). Above all, notice how an error on γ_s does not prevent the use of high-order approximation on equal-length subintervals, which still remains the most convenient approach to reduce the number of data required to represent a given function, whether singular or not.

1.5 Summary

This chapter addressed one-dimensional polynomial interpolation, which is closely related to the use of polynomials to represent unknown quantities within numerical solution procedures. The convenience of defining interpolation functions on parent cells and mapping them to child cells was illustrated, and the error associated with the interpolation was investigated. The error analysis demonstrates that higher order polynomial interpolation requires fewer samples than linear interpolation when maximum errors are to be limited to a few percent. The representation of singular functions was also investigated. In some situations, singular behavior can be accurately modeled by the combination of polynomials and singular functions; in other cases, it may not be possible to obtain a representation with arbitrarily small error.

The remaining chapters of this book address the issue of how to represent electromagnetic currents and fields with subsectional polynomials. Chapter 2 considers scalar representations. Since electromagnetic fields are vector quantities, the following chapters focus on vector representations in two or three dimensions.

References

[1] R. D. Graglia, P. L. E. Uslenghi, R. Vitiello, and U. D'Elia, "Electromagnetic scattering for oblique incidence on impedance bodies of revolution," *IEEE Trans. Antennas Propag.*, vol. 43, no. 1, pp. 11–26, Jan. 1995.

[2] D. Z. Ding, Z. J. Li, and R. S. Chen, "Fast analysis of electromagnetic scattering from body of revolution using high-order basis functions," *IET Microwaves Antennas Propag.*, vol. 6 , no. 14, pp. 1542–1547, 2012.

[3] R. L. Burden, J. D. Faires, and A. M. Burden, *Numerical Analysis,* 10th ed., Boston, MA: Cengage Learning, 2015.

[4] R. H. Bartels, J. C. Beatty, and B. A. Barsky, *An Introduction to Splines for Use in Computer Graphics and Geometric Modeling,* San Francisco, CA: Morgan Kaufmann, 1995.

[5] F. B. Hildebrand, *Introduction to Numerical Analysis,* New York, NY: McGraw-Hill, 1956.

[6] G. Szegö, *Orthogonal Polynomials,* 4th ed., Providence, RI: American Mathematical Society, 1975.

CHAPTER 2

Scalar Interpolation in Two and Three Dimensions

Chapter 1 examined the representation of functions of one variable in terms of polynomials. This chapter extends the approach to scalar functions of two or three variables. In one dimension, a domain is easily subdivided into intervals; in multiple dimensions, intervals are replaced by adjacent, non-overlapping subdomains of simple shapes such as triangles, quadrilaterals, and tetrahedrons. These subdomains are often called *elements* or *cells*. In the three-dimensional case, a cell is defined by its bounding *faces*, *edges*, and corner *nodes*. Corner nodes are the end-points of the bounding edges. Apart from the corner nodes, other nodes lying on the cell borders and in the interior of the cell may be necessary to define curved elements. (In some situations involving curved cells, the role of nodes may be replaced by *control points* that may actually be located outside the cell.)

The collection of all the cells is called a grid or a *mesh*. The long-standing term "mesh" dates back to the time when most of the problems dealt with were two-dimensional in nature; a two-dimensional domain split into elements resembles a wire mesh, hence the name. The principal canonical cell shapes are the triangle and quadrilateral in 2D, and the tetrahedron, hexahedron or brick, and triangular prism in 3D.

To represent scalar functions on the global 2D or 3D domain, polynomial basis functions are defined on the cells in a manner similar to that used in Chapter 1. As in Chapter 1, the representations under consideration all impose at least first order or C_0 continuity across cell boundaries. C_0 continuity is continuity of the function but not necessarily of its derivatives. As a consequence, the cells in the mesh must be aligned to permit a continuous representation, and the basis functions defined on these cells must be properly constructed so to ensure the continuity of the approximated function across the cell interfaces. These issues will be explored in the following sections.

2.1 Two- and Three-Dimensional Meshes and Canonical Cells

The process of obtaining an appropriate mesh is termed *mesh generation* [1–3]. Nowadays several mesh generator codes based on fully automatic mesh generation procedures are available on the open market. Many of these codes also feature very powerful post-processing tools for graphic representation of the numerical results obtained by the *solver* codes. Access to a good software package for mesh generation and expertise in using this software are vital to the success of a modeling effort.

2.1.1 Conforming Meshes and Fundamentals of Geometrical Database Structure

A mesh is defined by the nodes, edges, and faces of the cells and *at a minimum* by the following four geometrical sets[1]:

1. The *node*-set formed by all the *different* nodes of the mesh. The nodes are distinguished by a *global* number, and the coordinates of each node are stored in appropriate coordinate arrays, ordered according to the numbered list of the nodes.
2. The *edge*-set formed by the numbered list of all the *different* edges of the mesh. Integer arrays are then used to store the order numbers of the nodes lying on each edge of the mesh.
3. The *face*-set formed by the numbered list of all the *different* faces of the mesh. Integer arrays are used to store the order numbers of the edges bounding each *face*.
4. The *volumetric* set formed by the numbered list of all the volumetric cells of the mesh. The order numbers of the faces bounding each *cell* are stored in appropriate integer arrays.

A mesh is said to be *non-conforming* whenever one of the following events happens

- a node lying in the *interior* of an edge (or face) of the mesh is a *corner* node of a different edge (or face) of the mesh (see Figure 2.1a and b);
- there is an edge lying in the *interior* of a face that does not bound that face (see Figure 2.1c).

If none of the previous circumstances is verified the mesh is, by exclusion, *conforming*. Although the previous definitions seem to be rather complex, they actually summarize in words a very simple concept that is better illustrated by the examples shown in Figures 2.1 and 2.2.

Use of conforming meshes and of appropriate cell basis functions facilitates the continuity of the approximated function across common interfaces in adjacent subdomains in a rather simple manner. Conversely, the continuity of the approximated function is normally lost when non-conforming meshes are used (there are applications where continuity is of less importance) since the basis function sets used on adjacent cells sharing a non-conforming boundary can seldom be equal along that boundary. In these cases, if required, the continuity of the approximated function could be imposed in a weak sense; for example by equating on the common non-conforming boundary the average integral values of the approximated function computed by using its expression on the left- and right-hand sides of the common boundary. Clearly, before doing this, a certain number of DoFs must still be available, and this could possibly involve the use of appropriate basis functions on each cell lying along the non-conforming

[1]Note that this implies that each different node, edge, face, and volumetric cell in a mesh is distinguished by a *global* number, while each node, edge, or face of any given cell is associated with a *local* number. The relationship between local and global numbers is usually stored in appropriate integer *pointer* arrays, so the geometrical information for each cell can be obtained from a unique database relative to the global mesh only. For the sake of brevity, in this book we omit any discussion on the various techniques used to optimize the structure of the geometrical database.

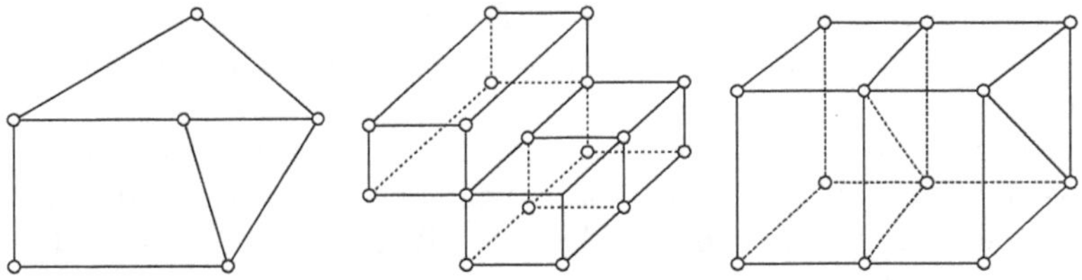

Figure 2.1 Examples of non-conforming meshes: (a) One corner node lies in the middle of an edge of the top triangular cell. (b) A corner node in common to the two bricks at right lies in the middle of an edge of the brick at left. (c) An edge in common to the two prisms at right "cuts" a quadrilateral face of the brick at left.

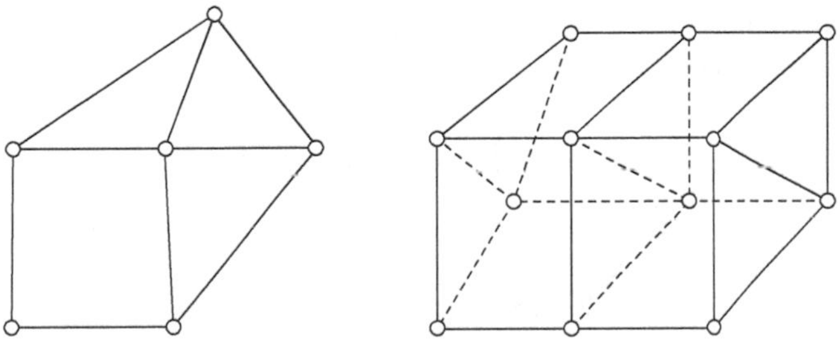

Figure 2.2 Examples of two-dimensional (a) and three-dimensional (b) conforming meshes.

boundary. Since our representations are C_0 continuous, numerical models and approximation procedures based on the use of non-conforming meshes are not considered in this book.

In addition to the geometrical sets and arrays already discussed on page 36, the mesh database may include additional (integer) arrays that are used to verify the mesh conformity and find the boundaries of the meshed domain, and that could also be used to check if the meshed domain does not contain any crack or cleft. For example, one sometimes needs additional "connectivity" matrices directly linking faces to adjacent volumetric cells, edges to adjacent cells, edges to adjacent faces, nodes to adjacent cells, faces, or edges, etc.

In this chapter, we define scalar basis functions on a variety of canonical *parent* or *reference* cell shapes, under the assumption that those basis functions will be subsequently mapped to a *child* cell in the actual (x, y, z) coordinates of interest. The parent cells are typically of unit dimension. We employ a set of appropriate normalized parent variables $\boldsymbol{\xi} = (\xi_1, \xi_2, \ldots, \xi_n)$ that depends on the cell shape. Each cell of a two- or three-dimensional *meshed* domain is thus defined by a different mapping

$$\boldsymbol{r} = \boldsymbol{r}(\boldsymbol{\xi}) \tag{2.1}$$

from a parent $\boldsymbol{\xi}$-space to the *child* space $\boldsymbol{r}$. The functions of the parent variables that define the geometry of a cell in the child domain are usually called *shape* functions. These shape functions are often the same basis functions used to approximate the function of interest within that cell. Depending on the specific mappings, the cells in the child domain may have boundaries that are linear or planar or boundaries that are curved. For most of the present chapter, we will focus on the basis function definition in the parent domain, and defer the details of the mappings until they are needed.

The interpolation/approximation procedure developed here is an extension of the same technique discussed in Chapter 1 for the high-order Lagrangian interpolation of a scalar function over a one-dimensional domain subdivided into small adjacent subdomains. The continuity of the approximate function across adjacent cells is readily enforced on a conforming mesh whereas the derivatives of the function are in general discontinuous on each cell's border, as in the one-dimensional case.

The interpolatory bases discussed in this chapter (and in the rest of the book) are only those with interpolation points defined on a regular interpolation grid on each cell. The reader should however keep in mind that the interpolation grids can actually be chosen *at pleasure*, provided one does not violate *a priori* the continuity of the approximate function across adjacent cells, whenever continuity is required. Interpolatory basis functions on a regular array of interpolation points are effectively obtained by using the so-called interpolatory polynomials of Silvester discussed in the next section.

2.2 Interpolatory Polynomials of Silvester

Interpolatory basis functions are usually defined in terms of Lagrange interpolation polynomials. For the canonical cells considered in this book, these polynomials are conveniently written using the interpolatory polynomials of Silvester [4] of one variable

$$R_i(p,\xi) = \begin{cases} \dfrac{1}{i!}\displaystyle\prod_{k=0}^{i-1}(p\xi - k) & \text{for } 1 \le i \le p, \\ 1 & \text{for } i = 0 \end{cases} \tag{2.2}$$

where, for each canonical element, the normalized coordinate variable ξ ranges over the interval [0,1]. These $(p+1)$ polynomials have the following properties ($0 \le i \le p$ is understood):

1. They are polynomials of degree i in ξ.
2. The integer parameter p (≥ 1) indicates the number of uniform subintervals into which the ξ-interval [0,1] is divided.
3. $R_i(p,\xi)$ is unity at $\xi = \frac{i}{p}$.
4. $R_i(p,\xi)$ has i zeros in the ξ-interval $\left[0, 1 - \frac{1}{p}\right]$ and no zero outside of this interval.
5. For $i \ge 1$ the zeros of $R_i(p,\xi)$ are at $\xi = 0, \frac{1}{p}, \frac{2}{p}, \ldots \frac{i-1}{p}$.

To construct the interpolatory *vector* functions that will be discussed later in Chapter 4, we also use the $(p+1)$ *modified* or *shifted* Silvester polynomials [5]

$$\begin{aligned}\hat{R}_i(p,\xi) &= R_{i-1}\left(p, \xi - \tfrac{1}{p}\right) \\ &= \begin{cases} \dfrac{1}{(i-1)!}\displaystyle\prod_{k=1}^{i-1}(p\xi - k) & \text{for } 2 \le i \le p+1, \\ 1 & \text{for } i = 1 \end{cases}\end{aligned} \tag{2.3}$$

that have the following properties ($1 \le i \le p+1$ is understood):

1. They are polynomials of degree $(i-1)$ in ξ.
2. The integer parameter $p(\ge 1)$ indicates the number of uniform subintervals into which the ξ-interval [0,1] is divided, as done with the (unshifted) Silvester polynomials (2.2).
3. $\hat{R}_i(p,\xi)$ is unity at $\xi = \frac{i}{p}$.
4. $\hat{R}_i(p,\xi)$ has $(i-1)$ zeros in the ξ-interval $\left[\frac{1}{p}, 1\right]$ and no zero outside of this interval.
5. For $i \ge 2$ the zeros of $\hat{R}_i(p,\xi)$ are at $\xi = \frac{1}{p}, \frac{2}{p}, \ldots \frac{i-1}{p}$.

By comparing (2.3) with (2.2) for the same number of zeros (i.e., for the same order), the array of interpolation points of (2.3) appears shifted because for the modified Silvester polynomials $\hat{R}_i(p,\xi)$ there is no zero at $\xi = 0$ (see Figure 2.3).

In numerical applications, these polynomials and their first derivatives with respect to ξ are conveniently computed by using very simple routines (as a matter of fact, just a couple of subroutines) which implement the following recurrence relations

$$\begin{cases} R_{i+1}(p,\xi) = \dfrac{p\xi - i}{i+1} R_i(p,\xi) \\ \hat{R}_{i+1}(p,\xi) = \dfrac{p\xi - i}{i} \hat{R}_i(p,\xi) \end{cases} \tag{2.4}$$

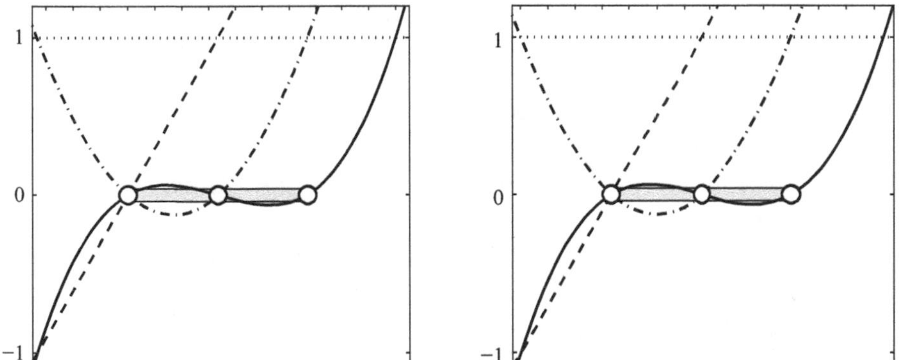

Figure 2.3 Silvester and *shifted* Silvester polynomials obtained with $p=3$. The Silvester polynomials shown at left are $R_0(3,\xi)$ (dotted-line), $R_1(3,\xi)$ (dashed-line), $R_2(3,\xi)$ (dash-dot line), and $R_3(3,\xi)$ (solid-line). The modified Silvester polynomials shown at right are $\hat{R}_1(3,\xi)$ (dotted-line), $\hat{R}_2(3,\xi)$ (dashed-line), $\hat{R}_3(3,\xi)$ (dash-dot line), and $\hat{R}_4(3,\xi)$ (solid-line). The gray boxes immediately show how, for the same number of zeros (i.e., for the same order), the array of interpolation points of the modified polynomials is right-shifted.

Table 2.1 The Silvester polynomial families $R_i(p,\xi)$ and $\hat{R}_i(p,\xi)$, for $p = 1, 2, 3$.

$R_0(1,\xi) = 1$	$\hat{R}_1(1,\xi) = 1$
$R_1(1,\xi) = \xi$	$\hat{R}_2(1,\xi) = \xi - 1$
$R_0(2,\xi) = 1$	$\hat{R}_1(2,\xi) = 1$
$R_1(2,\xi) = 2\xi$	$\hat{R}_2(2,\xi) = 2\xi - 1$
$R_2(2,\xi) = \xi(2\xi - 1)$	$\hat{R}_3(2,\xi) = (2\xi - 1)(\xi - 1)$
$R_0(3,\xi) = 1$	$\hat{R}_1(3,\xi) = 1$
$R_1(3,\xi) = 3\xi$	$\hat{R}_2(3,\xi) = 3\xi - 1$
$R_2(3,\xi) = 3\xi(3\xi - 1)/2$	$\hat{R}_3(3,\xi) = (3\xi - 1)(3\xi - 2)/2$
$R_3(3,\xi) = \xi(3\xi - 1)(3\xi - 2)/2$	$\hat{R}_4(3,\xi) = (3\xi - 1)(3\xi - 2)(\xi - 1)/2$

$$\begin{cases} \dfrac{\mathrm{d}R_{i+1}(p,\xi)}{\mathrm{d}\xi} = \dfrac{p}{i+1} R_i(p,\xi) + \dfrac{p\xi - i}{i+1} \dfrac{\mathrm{d}R_i(p,\xi)}{\mathrm{d}\xi} \\ \dfrac{\mathrm{d}\hat{R}_{i+1}(p,\xi)}{\mathrm{d}\xi} = \dfrac{p}{i} \hat{R}_i(p,\xi) + \dfrac{p\xi - i}{i} \dfrac{\mathrm{d}\hat{R}_i(p,\xi)}{\mathrm{d}\xi} \end{cases} \tag{2.5}$$

that begin with

$$R_0(p,\xi) = \hat{R}_1(p,\xi) = 1$$
$$\frac{\mathrm{d}R_0(p,\xi)}{\mathrm{d}\xi} = \frac{\mathrm{d}\hat{R}_1(p,\xi)}{\mathrm{d}\xi} = 0 \tag{2.6}$$

The first three families of the Silvester polynomials, for $p = 1$, 2, and 3, are reported in Table 2.1.

Lagrangian interpolation on the interval $\{0 \le \xi \le 1\}$ can effectively be performed by using Silvester polynomials after introducing two dependent parent variables

$$\xi_1 = \xi$$
$$\xi_2 = 1 - \xi \tag{2.7}$$

related by the dependency relation

$$\xi_1 + \xi_2 = 1 \tag{2.8}$$

It is in fact readily proven (see Figure 2.4) that the pth-order Lagrange interpolation polynomial $P_i(p,\xi)$ already discussed in Chapter 1 (see (1.18)) can now be written by using a two-indexing scheme as

$$P_{ij}(p,\xi_1,\xi_2) = R_i(p,\xi_1) R_j(p,\xi_2) \tag{2.9}$$

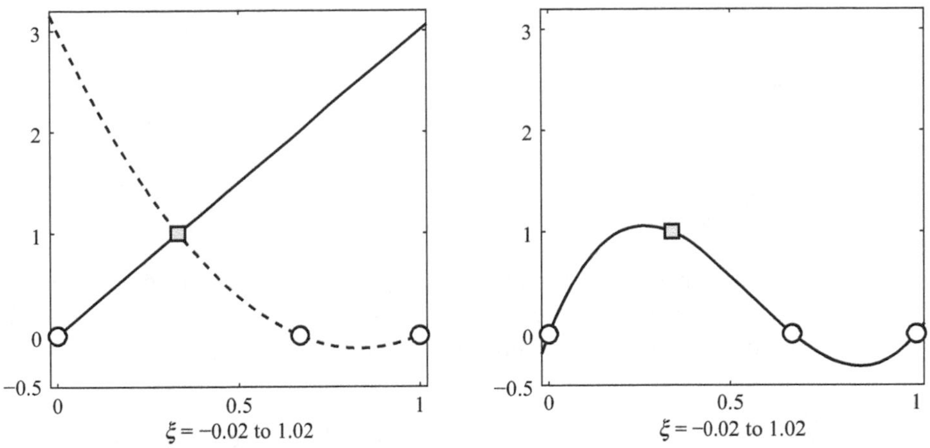

Figure 2.4 The behavior of $R_1(3,\xi)$ (solid-line) and $R_2(3,1-\xi)$ (dashed-line) is shown at left. The figure at right shows the behavior of the third-order polynomial $R_1(3,\xi)R_2(3,1-\xi)$ that is exactly equal to the Lagrange interpolatory polynomial $P_1(3,\xi)$.

with i and j related by the dependency relation

$$i+j=p \tag{2.10}$$

In the two-coordinate system (ξ_1,ξ_2) introduced with (2.7) and (2.8), the interpolation point $\xi=i/p$ where the Lagrange polynomial $P_i(p,\xi)$ attains a value of unity is now the point $\left(\frac{i}{p},\frac{j}{p}\right)$, with $i+j=p$.

2.3 Normalized Coordinates for the Canonical Cells

In this section, we summarize properties and notation for the normalized coordinates in terms of which both geometrical quantities and basis functions are defined on the canonical cells: the triangle, quadrilateral, tetrahedron, brick, and prism. In the following, to simplify the presentation, we mainly deal with rectilinear cells, i.e., with cells whose boundaries are formed by lines and planes. Geometrical quantities and bases defined for rectilinear cells extend quite naturally to curvilinear cells and bases, as we will see later on. As said at the beginning of this chapter (see (2.1)), it is convenient to view each cell as a mapping from some standard rectilinear *parent* cell of the same shape.

For all the canonical elements considered here, we find the following standardizations of the normalized coordinates convenient:

- Each edge of a two-dimensional cell or face of a three-dimensional cell defines a zero-coordinate surface for a normalized coordinate ξ_i where i indices edges or faces, respectively. The zero-coordinate edges for the triangular and quadrilateral cells are shown in Figure 2.5; the zero-coordinate surfaces for the tetrahedral, brick, and prism cells are shown in Figure 2.6.

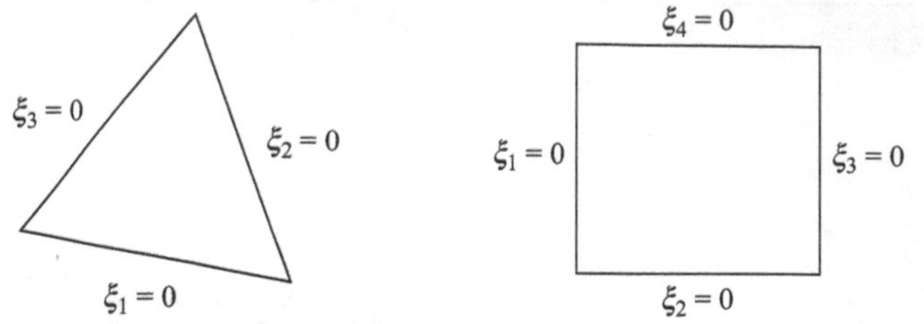

Figure 2.5 Zero-coordinate edges for triangular and quadrilateral cells. The triangle (at left) is described by the three normalized coordinates $\{\xi_1, \xi_2, \xi_3\}$; the quadrilateral (at right) is described by the four normalized coordinates $\{\xi_1, \xi_3; \xi_2, \xi_4\}$.

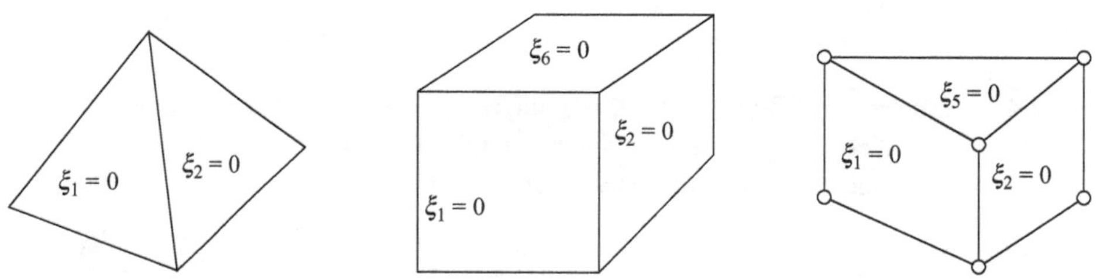

Figure 2.6 Zero-coordinate faces for the tetrahedral, the brick, and the prism cell. The tetrahedral cell at left is described by the four normalized coordinates $\{\xi_1, \xi_2, \xi_3, \xi_4\}$. The brick cell in the middle is described by the six normalized coordinates $\{\xi_1, \xi_4; \xi_2, \xi_5; \xi_3, \xi_6\}$, with the $\xi_{i+3} = 0$ face opposite to the $\xi_i = 0$ face. The prism cell at right is described by the five normalized coordinates $\{\xi_1, \xi_2, \xi_3; \xi_4, \xi_5\}$; its two opposite triangular faces are the $\xi_4 = 0$ and the $\xi_5 = 0$ face while the quadrilateral faces of the prism are the zero-coordinate surfaces $\xi_1 = 0$, $\xi_2 = 0$, and $\xi_3 = 0$.

- Each coordinate ξ_i varies linearly across a cell, attaining a value of unity at the face, edge, or vertex opposite its zero-coordinate surface. The coordinate gradient, $\nabla\xi_i$ (computed in the child space $\boldsymbol{r}$), is therefore perpendicular to the zero-coordinate surface and, on this surface, it points toward the interior of the cell.
- We index the coordinates so that a *right-handed* coordinate system is formed in the sense that $\hat{\boldsymbol{n}} \cdot (\nabla\xi_1 \times \nabla\xi_2)$ for two-dimensional elements and $\nabla\xi_3 \cdot (\nabla\xi_1 \times \nabla\xi_2)$ for three-dimensional elements is positive, where $\hat{\boldsymbol{n}}$ is a unit vector normal to a two-dimensional element. (An exception to this convention is the prism element discussed in Section 2.8, for which the usual dependent triangle area coordinates ξ_1, ξ_2, and ξ_3 are used to parameterize the prism's triangular cross section and ξ_4, with $\nabla\xi_4 \cdot (\nabla\xi_1 \times \nabla\xi_2)$ positive, is taken as the third independent coordinate.)
- The remaining coordinates are dependent coordinates, each of which may be associated with one or more independent coordinates to form a group of dependent coordinates.

In a list of the coordinates, or indices used to indicate these coordinates, it is convenient to highlight dependencies by writing groups of dependent coordinates together, separating each group by a semicolon. Whenever we must deal with independent coordinates, all but the last variable in a group will be taken as *"independent"* coordinates and the last variable will be taken as the *"dependent"* coordinate. Thus, in the coordinate listing $\{\xi_1, \xi_2, \xi_3, \xi_4\}$, we view ξ_1, ξ_2, and ξ_3 as the independent coordinates and ξ_4 as the dependent coordinate; in the listing $\{\xi_1, \xi_2, \xi_3; \xi_4, \xi_5\}$, the independent coordinates are ξ_1, ξ_2, and ξ_4, while ξ_3 and ξ_5 are the dependent coordinates. A similar grouping scheme is used in the following to indicate dependent indices corresponding to sampled values of the coordinates.

The Lagrange interpolation schemes and the scalar basis functions of each canonical cell are derived in the following sections. A detailed discussion is provided for the triangular cell to clarify the derivation procedure while, for the other cells, we present only the main results.

2.4 Triangular Cells

A domain that can be meshed with triangular cells is in general a non-flat surface of the three-dimensional (child) space $\boldsymbol{r}$. While our goal is to interpolate a given scalar function f and, possibly, compute an approximation of the gradient of f on this domain, it is first necessary to consider how well the domain can be approximated. If the domain is planar with a piecewise-linear boundary, rectilinear triangles can be used to obtain an exact polygonal representation of the domain. If the surface is not planar or does not have a polygonal contour, an approximation of the geometry can be obtained with rectilinear triangles of sufficiently small size. On the other hand, for curved surfaces, mappings that produce curvilinear cells can effectively reduce the geometrical error with fewer cells.

2.4.1 Cell Geometry Representation and Local Vector Bases

Each triangular cell of a *meshed* domain is defined by a different mapping $\boldsymbol{r} = \boldsymbol{r}(\boldsymbol{\xi})$ from the parent ξ-space (or reference cell) to the child space $\boldsymbol{r}$ (or actual (x, y, z) space). The three normalized parent coordinates $\boldsymbol{\xi} = (\xi_1, \xi_2, \xi_3)$ for triangles are related by

$$\xi_1 + \xi_2 + \xi_3 = 1 \tag{2.11}$$

These coordinates are often called *area coordinates* because of the geometrical interpretation given in Figure 2.7. They are also known as *simplex* coordinates, *triangle* coordinates, and *barycentric* coordinates.

The coordinates of a rectilinear triangle may be parameterized as

$$\boldsymbol{r} = \xi_1 \boldsymbol{r}_1 + \xi_2 \boldsymbol{r}_2 + \xi_3 \boldsymbol{r}_3 \tag{2.12}$$

where $\boldsymbol{r}_i$ is a vertex position vector for vertex i, and where ξ_1, ξ_2, and ξ_3 are the linear shape functions used to define a triangular cell of the child domain.

Similarly, a linear approximation $\widetilde{f}$ of a scalar function f on a triangular cell may be parameterized as

$$\widetilde{f} = \xi_1 f_1 + \xi_2 f_2 + \xi_3 f_3 \tag{2.13}$$

where f_i is the value of f at vertex i. The previous linear approximation for the function f is given in terms of the triangle parent coordinates and does not depend on the shape of the triangle in the child space; it is therefore valid also for curved triangles. Along the

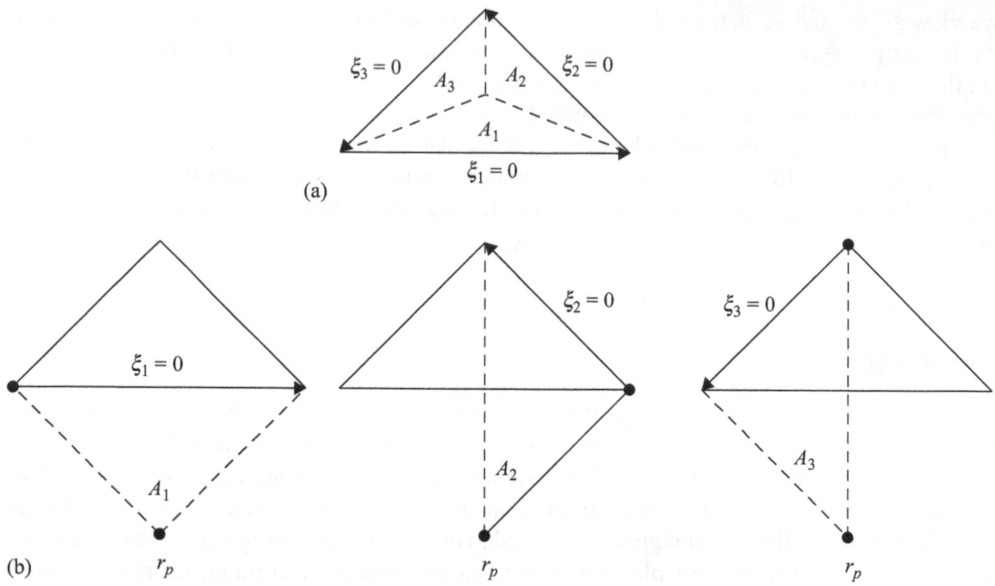

Figure 2.7 (a) The normalized coordinates ξ_1, ξ_2, and ξ_3 of a point $\boldsymbol{r}_p$ lying *within* a rectilinear triangle of area $A = A_1 + A_2 + A_3$ are: $\xi_1 = A_1/A$, $\xi_2 = A_2/A$, $\xi_3 = A_3/A$; with $\xi_1 + \xi_2 + \xi_3 = 1$ and where A_i is the area of the ith rectilinear subtriangle obtained by joining the point $\boldsymbol{r}_p$ with the two triangle vertices located on the $\xi_i = 0$ edge. The points lying on a line parallel to the $\xi_i = 0$ edge have the same ξ_i coordinate, and each point inside the triangular element is individuated by three non-negative parent coordinates. (b) If the point $\boldsymbol{r}_p$ lies *outside* of the triangle, one or two of its parent coordinates is negative. The sign associated with the ith parent coordinate of $\boldsymbol{r}_p$ and with the area A_i of the ith subtriangle is determined, for example, by counter-clockwise orienting the edges of the triangular cell (of area A) and by *walking* along the edges of the ith subtriangle in the counter-clockwise direction. The sub-area A_i is considered to be positive if the orientation of the $\xi_i = 0$ edge of the triangle is in the *path* direction; it is negative otherwise. In (b), the area (A_1) of the first subtriangle is negative while the areas of the second (A_2) and third (A_3) subtriangle are positive, with $(A_1 + A_2 + A_3)/A = 1$ and $\xi_1 + \xi_2 + \xi_3 = 1$.

$\xi_i = 0$ edge of the triangle, the dependency relation (2.11) simplifies (2.13) into the linear representation

$$\tilde{f} = f_{i+1} + \xi_{i-1}\,(f_{i-1} - f_{i+1}) = f_{i-1} + \xi_{i+1}\,(f_{i+1} - f_{i-1}) \tag{2.14}$$

to immediately match the linear approximation of f used in any cell attached to the triangle that shares this same edge, individuated by the global end-points $\boldsymbol{r}_{i-1}$ and $\boldsymbol{r}_{i+1}$. In the previous expression (2.14), the subscripts are counted modulo three, that is with $i + 1 = 1$ if $i = 3$, and $i - 1 = 3$ if $i = 1$.

To approximate the gradient of f over a meshed surface, we need to introduce a vector basis associated with each cell. The vector basis and the vector operators for curvilinear parametric coordinates are discussed in detail in Chapter 3 (Section 3.11). The mapping function

$$\boldsymbol{r} = \boldsymbol{r}(\xi_1, \xi_2, \xi_3) \tag{2.15}$$

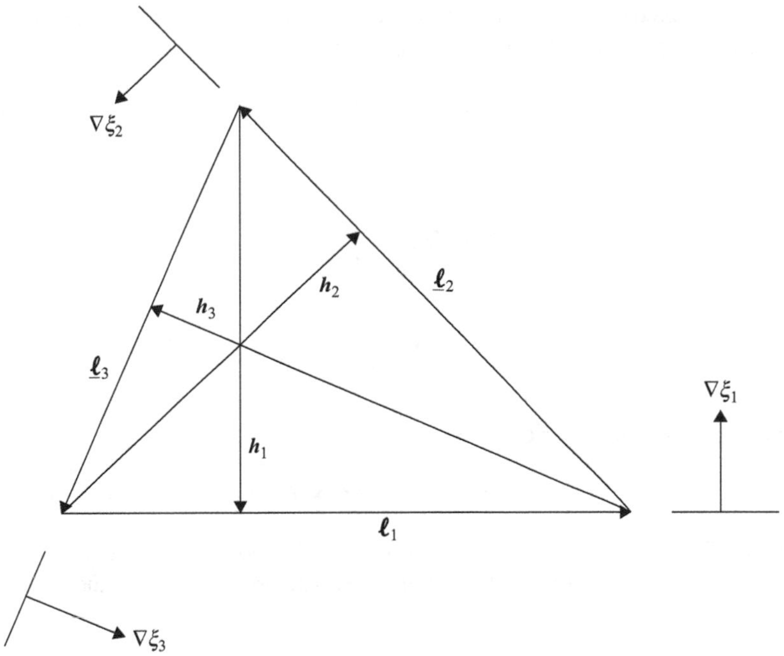

Figure 2.8 Edge, height, and gradient vectors for triangular elements.
© 1997 IEEE. Reprinted with permission from R. D. Graglia, D. R. Wilton, and A. F. Peterson, "Higher order interpolatory vector bases for computational electromagnetics," special issue on "Advanced Numerical Techniques in Electromagnetics," *IEEE Trans. Antennas Propag.*, vol. 45, no. 3, pp. 329–342, Mar. 1997.

permits one to derive from the *independent* coordinates ξ_1 and ξ_2 (being $\xi_3 = 1 - \xi_1 - \xi_2$) the following *unitary base vectors* for the triangular cell

$$\boldsymbol{\ell}^1 = \frac{\partial \boldsymbol{r}(\xi_1, \xi_2)}{\partial \xi_1} = \frac{\partial \boldsymbol{r}(\xi_1, \xi_2, \xi_3)}{\partial \xi_1} + \frac{\partial \boldsymbol{r}(\xi_1, \xi_2, \xi_3)}{\partial \xi_3}\frac{\partial \xi_3}{\partial \xi_1} = \left[\frac{\partial \boldsymbol{r}}{\partial \xi_1} - \frac{\partial \boldsymbol{r}}{\partial \xi_3}\right]_{\xi_2=\text{constant}} \tag{2.16}$$

$$\boldsymbol{\ell}^2 = \frac{\partial \boldsymbol{r}(\xi_1, \xi_2)}{\partial \xi_2} = \frac{\partial \boldsymbol{r}(\xi_1, \xi_2, \xi_3)}{\partial \xi_2} + \frac{\partial \boldsymbol{r}(\xi_1, \xi_2, \xi_3)}{\partial \xi_3}\frac{\partial \xi_3}{\partial \xi_2} = \left[\frac{\partial \boldsymbol{r}}{\partial \xi_2} - \frac{\partial \boldsymbol{r}}{\partial \xi_3}\right]_{\xi_1=\text{constant}} \tag{2.17}$$

from which the *edge vectors* (shown in Figure 2.8)

$$\begin{aligned} \boldsymbol{\ell}_1 &= -\boldsymbol{\ell}^2 \\ \boldsymbol{\ell}_2 &= \boldsymbol{\ell}^1 \\ \boldsymbol{\ell}_3 &= \boldsymbol{\ell}^2 - \boldsymbol{\ell}^1 \end{aligned} \tag{2.18}$$

are found,[2] with

$$\boldsymbol{\ell}_1 + \boldsymbol{\ell}_2 + \boldsymbol{\ell}_3 = 0 \tag{2.19}$$

[2]For rectilinear cells, each unitary base vector always coincides with at least one of the cell's edge vectors.

The gradients (evaluated in the child space) of the coordinate variables constitute the reciprocal base vectors and are determined from the edge vectors (see Figure 2.8 and Section 3.11) as

$$\nabla\xi_i = \frac{\hat{\boldsymbol{n}} \times \boldsymbol{\ell}_i}{\mathcal{J}} \tag{2.20}$$

with

$$\nabla\xi_1 + \nabla\xi_2 + \nabla\xi_3 = 0 \tag{2.21}$$

and where

$$\hat{\boldsymbol{n}} = \frac{\boldsymbol{\ell}^1 \times \boldsymbol{\ell}^2}{\mathcal{J}} = \frac{\boldsymbol{\ell}_i \times \boldsymbol{\ell}_{i+1}}{\mathcal{J}} \tag{2.22}$$

is the unit vector normal to the triangle while

$$\mathcal{J} = |\boldsymbol{\ell}^1 \times \boldsymbol{\ell}^2| = |\boldsymbol{\ell}_i \times \boldsymbol{\ell}_{i+1}| \tag{2.23}$$

is the Jacobian of the transformation from parent to child coordinates. The following biorthogonality relationship exists between the unitary base vectors ($\boldsymbol{\ell}^1$, $\boldsymbol{\ell}^2$) and the gradient vectors ($\nabla\xi_1$, $\nabla\xi_2$)

$$\boldsymbol{\ell}^j \cdot \nabla\xi_i = \delta_{ij} = \begin{cases} 1 & \text{for } i = j \\ 0 & \text{for } i \neq j \end{cases} \tag{2.24}$$

where δ_{ij} is the Kronecker delta. Since (2.18) and (2.24) yield

$$\begin{aligned} \boldsymbol{\ell}_i \cdot \nabla\xi_i &= 0 \\ \boldsymbol{\ell}_{i\pm1} \cdot \nabla\xi_i &= \pm 1 \end{aligned} \tag{2.25}$$

the edge and the gradient vectors are the convenient vector bases to express a vector

$$\boldsymbol{v} = E_{i-1}\,\boldsymbol{\ell}_{i-1} + E_{i+1}\,\boldsymbol{\ell}_{i+1} = G_{i-1}\,\nabla\xi_{i-1} + G_{i+1}\,\nabla\xi_{i+1} \tag{2.26}$$

and to compute its components

$$\begin{aligned} E_{i\mp1} &= \pm\boldsymbol{v} \cdot \nabla\xi_{i\pm1} \\ G_{i\pm1} &= \pm\boldsymbol{v} \cdot \boldsymbol{\ell}_{i\mp1} \end{aligned} \tag{2.27}$$

By recalling that the coordinate gradient $\nabla\xi_i$ is perpendicular to the ith edge and, on this edge, it points toward the interior of the triangular cell, we sometimes find it convenient to write

$$\nabla\xi_i = -\frac{\hat{\boldsymbol{h}}_i}{h_i} \tag{2.28}$$

where $1/h_i$ is the magnitude of the gradient, and $\hat{\boldsymbol{h}}_i$ is a unit vector opposite the direction of the gradient. On the ith edge $\hat{\boldsymbol{h}}_i$ is the unit outward normal to the element. The *height vector*

$$\boldsymbol{h}_i = h_i\,\hat{\boldsymbol{h}}_i = -h_i^2\,\nabla\xi_i \tag{2.29}$$

is also convenient since, for normalized coordinates and rectilinear cells, its magnitude measures the height of the cell perpendicular to the ith edge.

Figure 2.8 illustrates the various edge, gradient, and height vectors defined for a rectilinear triangular element. For curvilinear elements, all these geometrical quantities, including the Jacobian, vary with position.

The vector operators for curvilinear parametric coordinates are discussed in detail in Section 3.11 of Chapter 3. The gradient of a scalar function f on a given triangular cell is obtained as

$$\nabla f = \sum_{i=1}^{3} \frac{\partial f}{\partial \xi_i} \nabla \xi_i \tag{2.30}$$

The previous expression readily permits one to evaluate an approximation of the gradient of f out of the knowledge of the value of this function at three or more points. For example, for the linear interpolation (2.13), the gradient ∇f is approximated by the vector

$$\nabla \widetilde{f} = f_1 \nabla \xi_1 + f_2 \nabla \xi_2 + f_3 \nabla \xi_3 \approx \nabla f \tag{2.31}$$

which is constant if the triangular cell is rectilinear. This result shows that a linear approximation of a function f on a *conforming* triangular mesh that consists of rectilinear triangles yields a piecewise-constant approximation for the gradient of the function.

2.4.2 Lagrangian Basis Functions, Interpolation, and Gradient Approximation

We can now proceed to extend the previous discussion and consider high-order approximations of scalar functions on rectilinear or curvilinear triangles. This extension is particularly simple with a Lagrangian interpolation scheme, which is the only one considered in this chapter. The Lagrangian functions that we define in the rest of this chapter are also called "shape functions" because they are readily used to map a parent cell into child domain cells.

2.4.2.1 The Shape Polynomials $\alpha(p, \xi)$ to Interpolate All the Edges

By using interpolating polynomials of the Silvester form and a triple indexing scheme for the interpolation points, a pth-order Lagrangian interpolation of a function f on the triangle is given by[3]

$$\widetilde{f} = \sum_{i,j,k=0}^{p} f_{ijk}\, \alpha_{ijk}(p, \xi_1, \xi_2, \xi_3) \tag{2.32}$$

with

$$i + j + k = p \tag{2.33}$$

[3] $f, \widetilde{f}$, and the samples f_{ijk} are here considered as functions of the ξ-parent variables. The location of the sampling points $\boldsymbol{r}_{ijk} = \boldsymbol{r}(\boldsymbol{\xi}_{ijk})$ in the child space is obtained by the mapping (2.1), with $\boldsymbol{\xi}_{ijk} = \left(\frac{i}{p}, \frac{j}{p}, \frac{k}{p}\right)$.

and where

$$\alpha_{ijk}(p,\xi_1,\xi_2,\xi_3) = R_i(p,\xi_1)R_j(p,\xi_2)R_k(p,\xi_3) \tag{2.34}$$

is a pth-order Lagrange polynomial interpolating points within a triangle whose normalized coordinates are

$$(\xi_1,\xi_2,\xi_3) = \left(\frac{i}{p},\frac{j}{p},\frac{k}{p}\right) = \boldsymbol{\xi}_{ijk} \tag{2.35}$$

with $f_{ijk} = f(\boldsymbol{\xi}_{ijk})$. For the triangular cell, a pth-order complete set of Lagrangian scalar basis functions is thus formed by the $(p+1)(p+2)/2$ polynomials (2.34)[4] obtained for $i,j,k = 0,1,\ldots,p$, with $i+j+k=p$. The three indices that label α_{ijk} uniquely individuate the interpolation points of the triangular cell. In fact, the ath edge of the triangle is the coordinate-line $\xi_a = 0$ (for $a = 1,2,3$) while, in the parent domain, all the coordinate-lines $\xi_a = k/p$ are parallel to the $\xi_a = 0$ edge, with $\xi_a = p/p = 1$ on the vertex opposite to the $\xi_a = 0$ edge (see Figure 2.9). Notice also that (2.32) simplifies into (2.13) for $p = 1$.

It is now apparent that within our interpolation scheme the interpolation *nodes* are labeled by three ordered indices, with two of these indices equal to zero at the corner nodes of the triangular cell. If none of the indices are zero, the interpolation node is interior to the triangular element while, if only one of the indices is zero, the node is located on the cell boundary but not at a corner.

To guarantee the continuity of $\widetilde{f}$ across adjacent cells while using interpolatory polynomials, it is sufficient to associate a unique set of samples with the set of equally spaced interpolation points defined on each mesh-side which, eventually, coincides with the common edge of two or more adjacent cells. In this manner, along each mesh-side, $\widetilde{f}$ can also be written in terms of the *standard* Lagrange polynomials $P_n(p,\xi)$ discussed in the first chapter of this book. More specifically, on each of the three edges of the triangle one of the following results can be used

$$\begin{aligned}\xi_1 = 0 \text{ edge: } \widetilde{f} &= \sum_{n=0}^{p} f_{0,n,p-n}\, P_n(p,\xi_2) = \sum_{n=0}^{p} f_{0,p-n,n}\, P_n(p,\xi_3)\\ &\text{with } \xi_2 + \xi_3 = 1\end{aligned} \tag{2.36}$$

$$\begin{aligned}\xi_2 = 0 \text{ edge: } \widetilde{f} &= \sum_{n=0}^{p} f_{p-n,0,n}\, P_n(p,\xi_3) = \sum_{n=0}^{p} f_{n,0,p-n}\, P_n(p,\xi_1)\\ &\text{with } \xi_3 + \xi_1 = 1\end{aligned} \tag{2.37}$$

$$\begin{aligned}\xi_3 = 0 \text{ edge: } \widetilde{f} &= \sum_{n=0}^{p} f_{n,p-n,0}\, P_n(p,\xi_1) = \sum_{n=0}^{p} f_{p-n,n,0}\, P_n(p,\xi_2)\\ &\text{with } \xi_1 + \xi_2 = 1\end{aligned} \tag{2.38}$$

---

[4]Obviously, (2.32) and (2.34) imply $p \geq 1$.

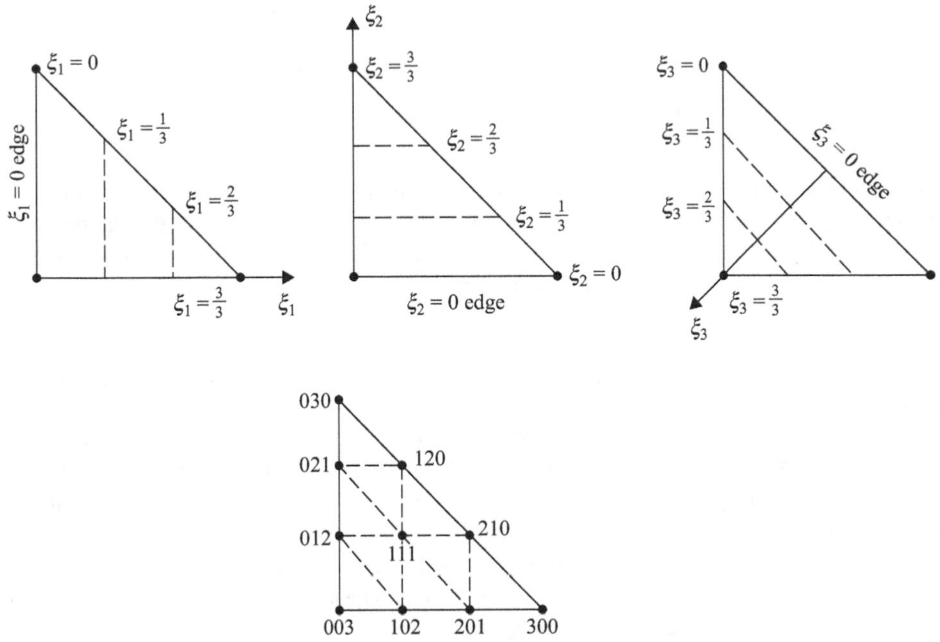

Figure 2.9 Interpolation nodes on triangular cells obtained for a Lagrange interpolation of order $p=3$. In the parent domain, the coordinate-lines $\xi_a = \frac{k}{p}$ (for $a = 1, 2, 3$) are parallel to the $\xi_a = 0$ edge, with $\xi_a = p/p = 1$ on the vertex opposite to the $\xi_a = 0$ edge (figures at top). The array of the interpolation points is shown at bottom. The three indices (ijk) that label the Lagrange interpolatory polynomial $\alpha_{ijk}(p, \xi_1, \xi_2, \xi_3)$ uniquely individuate the interpolation point $(\xi_1, \xi_2, \xi_3) = (\frac{i}{p}, \frac{j}{p}, \frac{k}{p})$.

The above-reported expressions are readily proven because, for $\xi_a + \xi_b = 1$, the *standard* Lagrange polynomial $P_n(p, \xi)$ reads as follows (see Section 2.2)

$$P_n(p, \xi_a) = P_{p-n}(p, \xi_b) = R_n(p, \xi_a) R_{p-n}(p, \xi_b) \tag{2.39}$$

with

$$R_m(p, 0) = \begin{cases} 1 & \text{for } m = 0 \\ 0 & \text{for } m \geq 1 \end{cases} \tag{2.40}$$

while (2.39) and (2.40) simplify the expression (2.34) of the interpolatory polynomial α_{ijk} (for $i+j+k=p$) on the triangle edges (ξ_1, ξ_2, or $\xi_3 = 0$) into

$$\begin{cases} \alpha_{ijk}(p, 0, \xi_2, \xi_3) = 0 & \text{for } i \geq 1 \\ \alpha_{ijk}(p, \xi_1, 0, \xi_3) = 0 & \text{for } j \geq 1 \\ \alpha_{ijk}(p, \xi_1, \xi_2, 0) = 0 & \text{for } k \geq 1 \end{cases} \tag{2.41}$$

$$\begin{cases} \alpha_{0jk}(p,0,\xi_2,\xi_3) = P_j(p,\xi_2) = P_{p-j}(p,\xi_3) \\ \alpha_{i0k}(p,\xi_1,0,\xi_3) = P_i(p,\xi_1) = P_{p-i}(p,\xi_3) \\ \alpha_{ij0}(p,\xi_1,\xi_2,0) = P_i(p,\xi_1) = P_{p-i}(p,\xi_2) \end{cases} \tag{2.42}$$

In this connection, it should also be clear that one can easily represent $\tilde{f}$ in terms of *standard* Lagrange polynomials $P_n(p, \xi)$ along any of the coordinate-lines of the triangle (ξ_i = constant, for $i = 1, 2$, or 3); the proof is left as an exercise to the reader. Table 2.2 reports the interpolating functions $\alpha_{ijk}(p, \boldsymbol{\xi})$ for triangular cells, up to $p = 3$.

Table 2.2 The first three orders of the shape polynomials interpolating all the edges of a triangular cell.

The $(p+1)(p+2)/2$ polynomials $\alpha_{ijk}(p,\boldsymbol{\xi}) = R_i(p,\xi_1)R_j(p,\xi_2)R_k(p,\xi_3)$ that interpolate a triangular cell are obtained for $i,j,k = 0,1,\ldots,p$, with $i+j+k = p$ and $p \geq 1$. Each of these polynomials is of order p. Because of the symmetry of the triangular cell it is sufficient to report, for each family, only a smaller set of polynomials that is then "completed" by other polynomials, obtained from those of the smaller set by cyclic permutations of the parent variables $\{\xi_1, \xi_2, \xi_3\}$. The smaller set of polynomials is more conveniently written in terms of the dummy triangular parent variables $\{\xi_a, \xi_b, \xi_c\}$. Each pth-order complete interpolatory set is then obtained by setting, in the functions $R_i(p,\xi_a)R_j(p,\xi_b)R_k(p,\xi_c)$

$\{\xi_a, \xi_b, \xi_c\} = \{\xi_1, \xi_2, \xi_3\}$ first set of permuted variables;
$\{\xi_a, \xi_b, \xi_c\} = \{\xi_2, \xi_3, \xi_1\}$ second set of permuted variables;
$\{\xi_a, \xi_b, \xi_c\} = \{\xi_3, \xi_1, \xi_2\}$ third set of permuted variables.

Thus, all the triangular functions are defined by using three different sets of permuted variables; notice how the number of the permuted sets equals the number of the cell corner nodes.

$p = 1$ (3 functions in total)
Dummy expressions (use set 1–3):
ξ_a (3 functions)
By permutations, the functions are:

$\alpha_{100} = \xi_1$
$\alpha_{010} = \xi_2$
$\alpha_{001} = \xi_3$

$p = 2$ (6 functions in total)
Dummy expressions (use set 1–3):
$\xi_a(2\xi_a - 1)$ (3 functions)
$2^2\xi_a\xi_b$ (3 functions)
By permutations, the functions are:

$\alpha_{200} = \xi_1(2\xi_1 - 1)$
$\alpha_{020} = \xi_2(2\xi_2 - 1)$
$\alpha_{002} = \xi_3(2\xi_3 - 1)$
$\alpha_{110} = 2^2\xi_1\xi_2$
$\alpha_{011} = 2^2\xi_2\xi_3$
$\alpha_{101} = 2^2\xi_3\xi_1$

$p = 3$ (10 functions in total)
Dummy expressions:
$\xi_a(3\xi_a - 1)(3\xi_a - 2)/2$ (3 functions; use set 1–3)
$3^2\xi_a\xi_b(3\xi_b - 1)/2$ (3 functions; use set 1–3)
$3^2\xi_b\xi_a(3\xi_a - 1)/2$ (3 functions; use set 1–3)
$3^4\xi_a\xi_b\xi_c$ (1 function; use set 1)
By permutations, the functions are:

$\alpha_{300} = \xi_1(3\xi_1 - 1)(3\xi_1 - 2)/2$
$\alpha_{030} = \xi_2(3\xi_2 - 1)(3\xi_2 - 2)/2$
$\alpha_{003} = \xi_3(3\xi_3 - 1)(3\xi_3 - 2)/2$
$\alpha_{120} = 3^2\xi_1\xi_2(3\xi_2 - 1)/2$
$\alpha_{012} = 3^2\xi_2\xi_3(3\xi_3 - 1)/2$
$\alpha_{201} = 3^2\xi_3\xi_1(3\xi_1 - 1)/2$
$\alpha_{210} = 3^2\xi_2\xi_1(3\xi_1 - 1)/2$
$\alpha_{021} = 3^2\xi_3\xi_2(3\xi_2 - 1)/2$
$\alpha_{102} = 3^2\xi_1\xi_3(3\xi_3 - 1)/2$
$\alpha_{111} = 3^4\xi_1\xi_2\xi_3$

2.4.2.2 Gradient Approximation

For high-order approximations, a straightforward evaluation of the gradient of (2.32) on the triangle yields

$$\nabla\tilde{f} = \sum_{i,j,k=0}^{p} f_{ijk}\,\nabla\alpha_{ijk}(p,\xi_1,\xi_2,\xi_3) \tag{2.43}$$

where

$$\begin{aligned}\nabla\alpha_{ijk}(p,\xi_1,\xi_2,\xi_3) &= R_i'(p,\xi_1)\,R_j(p,\xi_2)\,R_k(p,\xi_3)\,\nabla\xi_1\\ &\quad + R_i(p,\xi_1)\,R_j'(p,\xi_2)\,R_k(p,\xi_3)\,\nabla\xi_2\\ &\quad + R_i(p,\xi_1)\,R_j(p,\xi_2)\,R_k'(p,\xi_3)\,\nabla\xi_3\end{aligned} \tag{2.44}$$

with $i+j+k=p$, and where

$$R_\ell'(p,\xi) = \frac{\mathrm{d}R_\ell(p,\xi)}{\mathrm{d}\xi} \tag{2.45}$$

denotes the first derivative of the Silvester polynomial $R_\ell(p,\xi)$ with respect to ξ.

By using the functions $\alpha(p,\boldsymbol{\xi})$ given in (2.34), the polynomial order of the interpolant $\tilde{f}$ given in (2.32) is equal to p, while the polynomial order of the gradient of $\tilde{f}$ given in (2.43) is equal to $p-1$. Because of this, in spite of the fact that the polynomial order of $\alpha(p,\boldsymbol{\xi})$ is equal to p, the order of the approximations (2.32) and (2.43) is sometimes denoted by the half-integer orders $(p-0.5)$; for example, in the following, we could say "order 1.5" to denote the use of scalar functions $\alpha(p,\boldsymbol{\xi})$ of order $p=2$.

2.4.3 Interpolation Error

As observed in Chapter 1, bounded functions on definite intervals can be represented in terms of trigonometric series, and important conclusions can be drawn by considering sinusoidal functions of the form $f(x,y) = \sin(a_x x)\sin(b_y y)$, defined in the Cartesian (x,y)-frame, and the behavior of the associated interpolated functions $\tilde{f}$.

For simplicity, in this and in the next subsection, we only study how the function $f(x,y) = \sin(x)\sin(y)$ is approximated over the square domain Q_{λ_0} having edge-length λ_0 equal to 2π. Thus, in the (x,y)–frame, the corner nodes are located at $(0,0)$, $(2\pi,0)$, $(2\pi,2\pi)$, $(0,2\pi)$. The square is uniformly subdivided into $n\times n$ square patches of edge-length λ_0/n; in turns, each sub-square is further subdivided into four equal right-angled triangles. The mesh obtained in this manner contains $4n^2$ triangles and is considered of rather poor quality since all the triangles are right-angled. The edge-lengths of each triangular cell are:

$$\lambda_0/n = 2\pi/n,\quad \lambda_0/n = 2\pi/n,\quad \sqrt{2}\lambda_0/n = 2\sqrt{2}\pi/n \tag{2.46}$$

On each triangle, a grid of $(p+1)(p+2)/2$ interpolatory points defines the pth-order interpolatory polynomials of Section 2.4.2.1. The interpolatory grid contains three corner points plus

$(p-1)$ points along each edge of the triangle that may be in common with other adjacent triangular cells. Thus, at order p, the mesh defined above requires $2(np)^2 + 2np + 1$ interpolatory (or sampling) points.

To assess the quality of the interpolated functions versus the interpolation order for a specific mesh, we consider the integral of the error magnitude over the interpolation square by using the L_2 relative error

$$\varepsilon = \sqrt{\frac{\iint_{Q_{\lambda_0}} \left[\tilde{f}(x,y) - f(x,y)\right]^2 \mathrm{d}x\,\mathrm{d}y}{\iint_{Q_{\lambda_0}} f^2(x,y)\,\mathrm{d}x\,\mathrm{d}y}} \tag{2.47}$$

The *global* relative error ε has the asymptotic behavior

$$\varepsilon \simeq \alpha_p \left(\frac{\ell}{\lambda_0}\right)^{(p+1)} = \frac{\alpha_p}{n^{(p+1)}} \tag{2.48}$$

where the multiplicative constant α_p depends on the order p of the piecewise interpolatory function $\tilde{f}$, on the quality of the mesh and, obviously, on the function f that one interpolates, where $\ell = 2\pi/n$ is the smallest triangle edge-length in (2.46).

Figure 2.10 shows the interpolation errors (2.47) for the first six interpolation orders. Observe that one can obtain very good results by increasing the interpolation order, even with poor-quality meshes.

The results of Figure 2.10 show that, for a fixed number of sampling points, the global error can be systematically reduced by increasing the interpolation order. This is exemplified in Figure 2.11 by reporting the results obtained with 313 sampling points (for $[p, n] = [1, 12]$, [2, 6], [3, 4], [4, 3]), 1,201 sampling points (for $[p, n] = [1, 24]$, [2, 12], [3, 8], [4, 6]), and 2,665 sampling points (for $[p, n] = [1, 36]$, [2, 18], [3, 12], [4, 9]).

2.4.4 Spectral Integrity and Other Frequency Domain Considerations

The effects of the interpolation error can also be observed in the Fourier domain. (In this connection recall that, in antenna applications, the far-field radiation integral is very similar to the Fourier integral.)

The spectrum of the Fourier transform of $f(x, y) = \sin(x)\sin(y)$,

$$\begin{aligned}\left|\mathsf{f}(k_x, k_y)\right| &= \left| -\frac{4\exp\left[-j\pi(k_x + k_y)\right]\sin(k_x\pi)\sin(k_y\pi)}{\left(k_x{}^2 - 1\right)\left(k_y{}^2 - 1\right)} \right| \\ &= 4\pi^2 \left| \frac{\mathrm{sinc}[(k_x - 1)\pi]}{k_x + 1} \, \frac{\mathrm{sinc}[(k_y - 1)\pi]}{k_y + 1} \right| \end{aligned} \tag{2.49}$$

with

$$\mathrm{sinc}(z) = \sin(z)/z \tag{2.50}$$

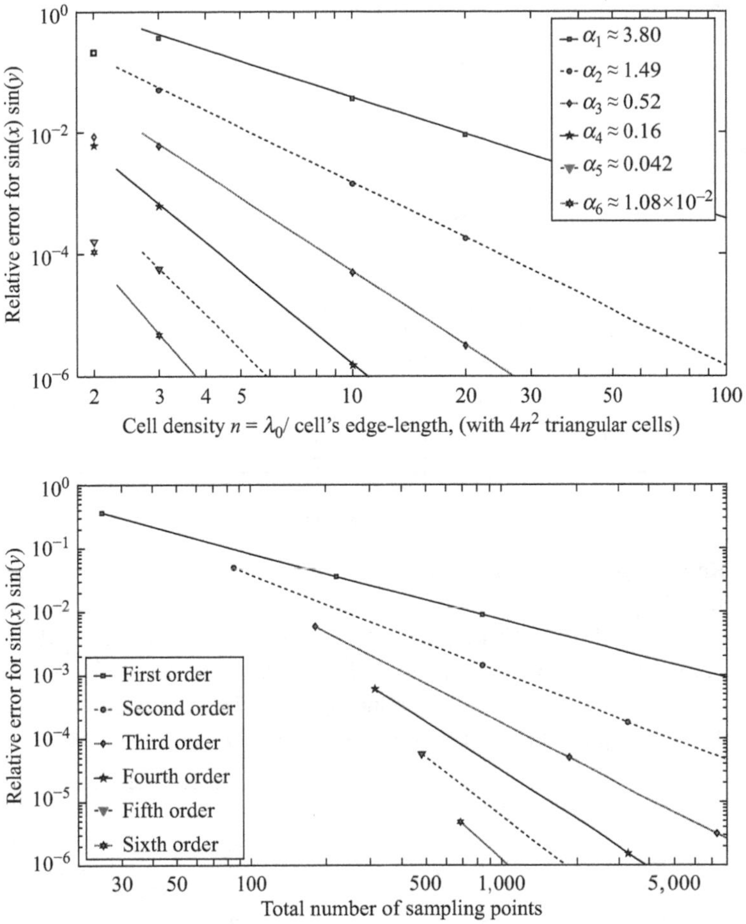

Figure 2.10 Relative approximation error (2.47) for the trigonometric function $\sin(x)\sin(y)$ numerically computed over the $\lambda_0 \times \lambda_0$ $(= 2\pi \times 2\pi)$ square Q_{λ_0}. The straight lines show the error asymptotic behaviors $\varepsilon \simeq \frac{\alpha_p}{n^{(p+1)}}$ for the first six interpolation orders, obtained with the α_p values reported in the legend (markers are at $n = 3$, $n = 5$, and $n = 20$). The figure at top shows the error versus the cell's density value n; the error versus the total number of sampling points is shown at bottom. For $n \geq 3$, the errors match almost perfectly the asymptotic behavior $\varepsilon \simeq \frac{\alpha_p}{n^{(p+1)}}$.

is symmetric with respect to the $k_x = k_y$ line and maximum at $k_x = k_y \approx 0.837472$, with $\left|\mathsf{f}(k_x, k_y)\right|_{\max} \approx 10.7113$. Thus, to quantitatively discuss the spectral effects of higher order interpolations of the function $\sin(x)\sin(y)$, we consider the spectral relative error

$$e_{L2}(k) = \sqrt{\frac{\int_0^k \left|\tilde{\mathsf{f}}(k_x, k_y) - \mathsf{f}(k_x, k_y)\right|^2_{k_y = k_x} \mathrm{d}k_x}{\int_0^k \left|\mathsf{f}(k_x, k_y)\right|^2_{k_y = k_x} \mathrm{d}k_x}} \tag{2.51}$$

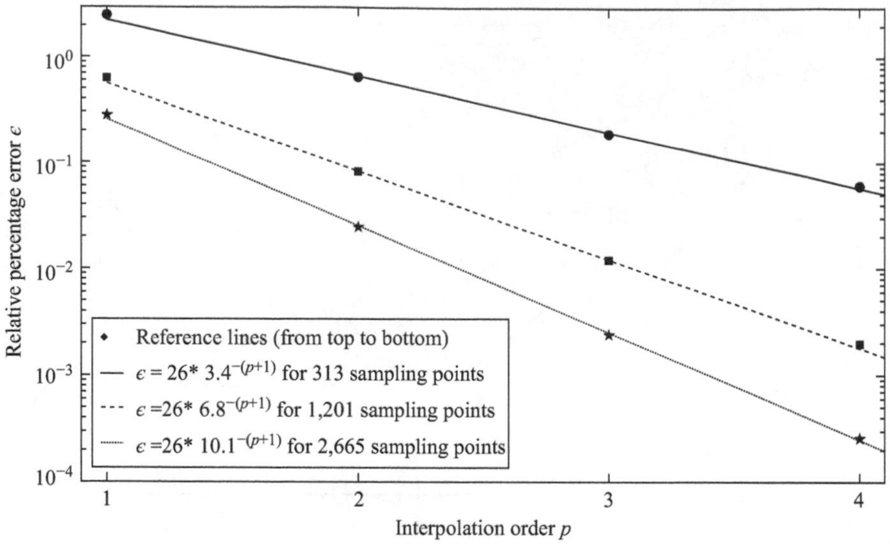

Figure 2.11 Relative percentage error ε versus the interpolation order p for a fixed number of sampling points. The circles mark the results obtained with 313 sampling points (for $[p, n] = [1, 12], [2, 6], [3, 4], [4, 3]$). The squares mark the results obtained with 1,201 sampling points (for $[p, n] = [1, 24], [2, 12], [3, 8], [4, 6]$), and the stars those obtained with 2,665 sampling points (for $[p, n] = [1, 36], [2, 18], [3, 12], [4, 9]$). The error diminishes at faster rates as the sampling rate is increased.

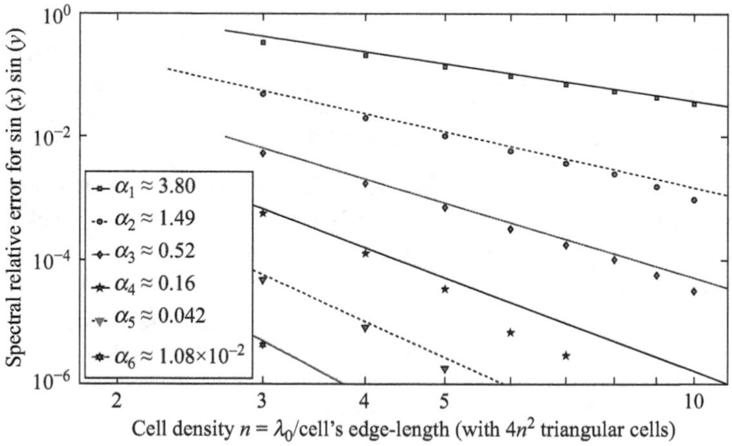

Figure 2.12 Spectral relative error for the function $\sin(x)\sin(y)$ interpolated over the $\lambda_0 \times \lambda_0$ $(= 2\pi \times 2\pi)$ square Q_{λ_0}. The error is computed with (2.51) for $k = 10$. Asymptotes are repeated from Figure 2.10 for comparison.

Figure 2.12 shows the behavior of the spectral error (2.51) computed for $k = 10$. Asymptotes of the expected error from Figure 2.10 are also repeated for comparison. In general, as shown by the results of Table 2.3, the spectral error at $k = 10$ is smaller than the expected relative error shown in Figure 2.10.

Table 2.3 Relative global error ε and spectral error e_{L2} for different meshes and orders.

Relative percentage error		Interpolation order	Number of sub-squares used	Number of sampling points	Percentage reduction of the # of sampling pts wrt the $p=1$ case
ε	e_{L2}	p	n		
2.488	2.261	1	12	313	–
2.122	1.930	1	13	365	–
2.153	1.978	2	4	145	60.3
0.576	0.525	1	25	1,301	–
0.650	0.584	2	6	313	75.9
0.586	0.523	3	3	181	86.1
0.400	0.365	1	30	1,861	–
0.411	0.368	2	7	421	77.4
0.189	0.172	3	4	313	83.2
0.100	0.091	1	60	7,321	–
0.142	0.095	2	10	841	88.5
0.078	0.071	3	5	481	93.4
0.061	0.055	1	77	12,013	–
0.061	0.057	4	3	313	97.4

Interpolation of the function $f(x,y) = \sin(x)\sin(y)$ defined on the $\lambda_0 \times \lambda_0$ $(=2\pi \times 2\pi)$ square Q_{λ_0}. The table reports the percentage reduction with respect to the linear interpolation ($p=1$) of the number of sampling points required by higher order interpolations to obtain equal or similar relative percentage errors ε and spectral errors $e_{L2}(k)$. The global errors ε are computed by using (2.47); the spectral errors $e_{L2}(k)$ are computed at $k=10$ by using (2.51). n is the number of sub-squares used to define the triangular meshes to get the results shown by each row. The triangular meshes are formed by $4n^2$ right-angled triangles. Notice in particular that by using 313 sampling points the global percentage error obtained with the linear (first-order) interpolation is 2.488, with second-order interpolation is 0.650%, with third-order interpolation is 0.189%, and at fourth order is 0.061%.

As noted in Chapter 1, the spectra of the interpolated functions contain incorrect spikes of amplitude and height in inverse proportion to the frequency at which they are centered. This phenomenon is difficult to observe in Figure 2.13; therefore it is highlighted by the results of Figure 2.14 that show the absolute error $|\mathrm{f}(k_x,k_y) - \hat{\mathrm{f}}(k_x,k_y)|$ obtained with different interpolation orders (and different meshes). Different vertical scales are used to build Figure 2.14. The incorrect spikes are the *signature* of the triangular-cell mesh and are due to the fact that the interpolated functions have discontinuous normal derivatives along the mesh edges.

In summary, for a two-dimensional interpolation, the results and the figures of this and of the previous subsection clearly show that whenever an error of a few percent is acceptable, higher polynomial interpolation orders easily permit one to reduce the total number of samples by at least 60% compared to the number of samples required by linear interpolation. This reduction factor becomes larger than 70% if the maximum acceptable error is less than 0.1%. Furthermore, the example discussed here clearly shows that poor-quality meshes can be used provided the interpolation order is sufficiently high. In other words, the negative effects of a poor-quality mesh are readily mitigated by increasing the interpolation order.

2.4.5 Curved Cells

A Lagrange parameterization of order $q(\geq 1)$ for a distorted or curved triangle can be expressed as

$$\boldsymbol{r} = \sum_{i,j,k=0}^{q} \boldsymbol{r}_{ijk}\, R_i(q,\xi_1)\, R_j(q,\xi_2)\, R_k(q,\xi_3), \quad i+j+k=q \tag{2.52}$$

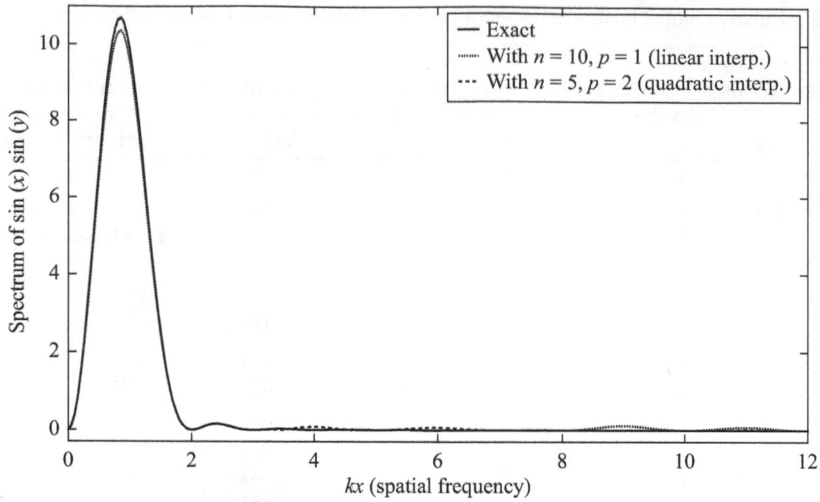

Figure 2.13 The exact spectrum $\left|\mathrm{f}(k_x, k_y)\right|$ of the function $f(x, y) = \sin(x)\sin(y)$ defined on $\lambda_0 \times \lambda_0$ ($= 2\pi \times 2\pi$) square Q_{λ_0} is compared, along the $k_x = k_y$ line, with the spectrum obtained by the first- (for $n = 10$) and second- (for $n = 5$) order interpolations discussed in subsection 2.4.3.

where a triple indexing scheme is again used to label the position vector[5] $\boldsymbol{r}_{ijk}$ interpolating the point with normalized coordinates $(\xi_1, \xi_2, \xi_3) = \left(\frac{i}{q}, \frac{j}{q}, \frac{k}{q}\right) = \boldsymbol{\xi}_{ijk}$.

Equation (2.52) maps the parent triangular simplex $T^2 \equiv \{0 \leq \xi_1, \xi_2, \xi_3 \leq 1; \xi_1 + \xi_2 + \xi_3 = 1\}$ into a distorted triangle of the child space $\boldsymbol{r}$, with a Jacobian matrix given by

$$\begin{bmatrix} \dfrac{\partial}{\partial \xi_1} \\ \dfrac{\partial}{\partial \xi_2} \end{bmatrix} = \begin{bmatrix} \dfrac{\partial x}{\partial \xi_1} & \dfrac{\partial y}{\partial \xi_1} & \dfrac{\partial z}{\partial \xi_1} \\ \dfrac{\partial x}{\partial \xi_2} & \dfrac{\partial y}{\partial \xi_2} & \dfrac{\partial z}{\partial \xi_2} \end{bmatrix} \begin{bmatrix} \dfrac{\partial}{\partial x} \\ \dfrac{\partial}{\partial y} \\ \dfrac{\partial}{\partial z} \end{bmatrix} \tag{2.53}$$

For the specific mapping in (2.52), the Jacobian matrix involves derivatives of the child variables with respect to the independent parent variables (ξ_1, ξ_2). The well-known condition for a "*one-to-one*" mapping is that the sign of the Jacobian determinant must remain unchanged at all the points of the domain mapped. In fact, one has always to avoid the so-called "*cell warping*" that occurs whenever the Jacobian determinant of the transformation from parent-to child space changes sign or vanishes; this typically could happen if the size of the cell is too large in a region where the curvature of the geometry is too strong. To choose the "appropriate" size of a curved (or distorted) cell, one has to consider the local curvature of the geometry

[5] The coordinates of the position vectors $\boldsymbol{r}_{ijk}$ appearing in (2.52) are often part of the mesh database prepared by the mesh generator although, in general, the whole geometry does not necessarily need to be "*piecewise*" defined by parameterizations such as (2.52). Mesh generators that use different ways to describe a complex geometry exist. For the sake of brevity, the various ways used to describe a geometry are not considered in this book.

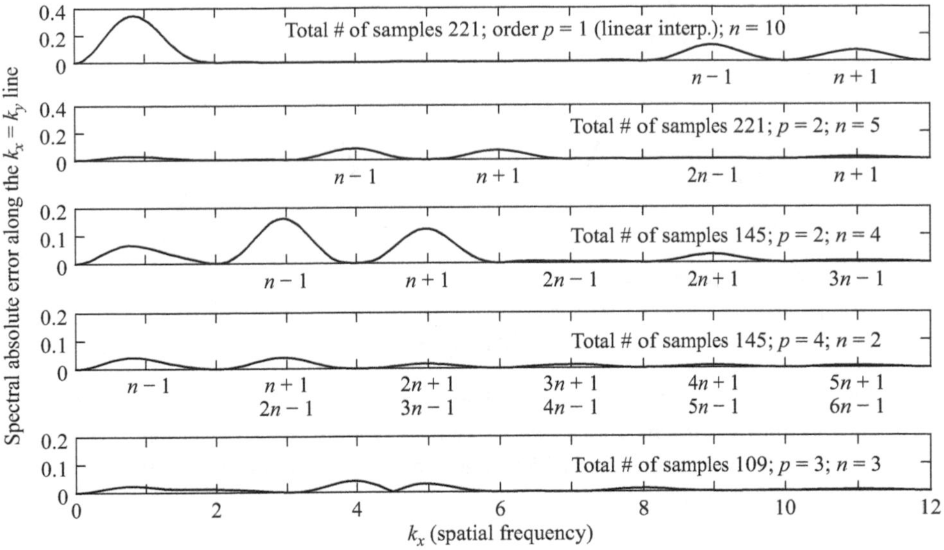

Figure 2.14 Absolute spectral error along the $k_x = k_y$ line. The *exact* spectrum is maximum at $k_x = k_y \approx 0.8$ (with $1 = 2\pi/\lambda_0$). The (exact) spectra of the interpolated functions contain incorrect spikes centered at $k_x = k_y = mn \pm 1$ (for $m \geq 1$ and integer), where n^2 is the number of square patches of edge-length λ_0/n used to define the triangular mesh in Section 2.4.3. Different vertical scales are used in the figure. The two sub-figures at top are obtained by interpolating sin(x) sin(y) with 221 samples. The third and fourth sub-figures are obtained by using 145 samples; the figure at bottom is obtained by third-order interpolation on 109 samples ($p = 3; n = 3$).

where the cell is located, as well as the order of the used parameterization of the cell. To further clarify what "cell warping" is, in Figure 2.15 we report two different *planar* triangles, both obtained by a parabolic mapping (that is obtained by setting $q = 2$ in (2.52)); the triangle at left is not warped while the triangle at right is warped.

Simple rules to avoid cell warping exist for first- and second-order distortion. For linear transformations (for example, for the triangles obtained by setting $q = 1$ in (2.52)), the necessary condition to avoid cell warping is that no internal angle of the mapped domain be greater than 180°. For quadratic (parabolic) transformations, it is necessary in addition to the previous requirement to ensure that the mid-side nodes are in the "middle third" of the distance between adjacent corners. (For the triangles obtained by setting $q = 2$ in (2.52), the mid-side nodes are $\boldsymbol{r}_{011}, \boldsymbol{r}_{101}$, and $\boldsymbol{r}_{110}$ while $\boldsymbol{r}_{200}, \boldsymbol{r}_{020}$, and $\boldsymbol{r}_{002}$ are the corner nodes). Conversely, numerical checks on the sign of the Jacobian determinant are necessary for cubic or higher order distortion to be sure that cell warping does not occur. A parabolic distortion of the cells is usually sufficient to describe curved, complex geometries whenever the size of the curved cells is not too large, although we ought to observe that the principal advantage of high-order basis functions is only realized when they are used with high-order representations of curved geometries to allow using cells of large size. When using cells of large size, the error due to a low-order description of a curved boundary should never be underestimated since it can easily be quite high.

For curvilinear (not-warped) elements, the edge, gradient, and height vectors retain their interpretations if we apply them to a tangent rectilinear element that can be constructed from the local edge and height vectors at each coordinate point within the element (cf. Figure 2.16).

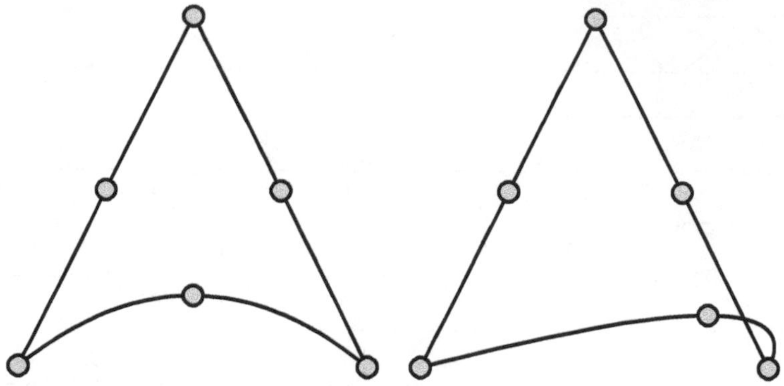

Figure 2.15 Curved, planar triangular cells. The triangle at left is not warped while the triangle at right is warped. Both triangles were obtained by using a parabolic (second-order) distortion with $q=2$ in (2.52); the circles mark the location of the six points $\boldsymbol{r}_{ijk}$ used in (2.52).

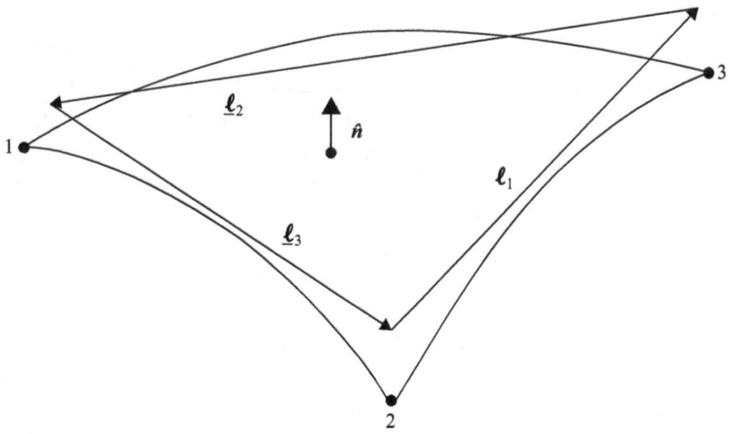

Figure 2.16 Triangle tangent to a curvilinear triangle at a point. The curvilinear and rectilinear tangent triangles have the same element coordinates, Jacobian, edge vectors, height vectors, and unit normal $\hat{\boldsymbol{n}}$ at the point of tangency.

© 1997 IEEE. Reprinted with permission from R. D. Graglia, D. R. Wilton, and A. F. Peterson, "Higher order interpolatory vector bases for computational electromagnetics," special issue on "Advanced Numerical Techniques in Electromagnetics," *IEEE Trans. Antennas Propag.*, vol. 45, no. 3, pp. 329–342, Mar. 1997.

2.5 Quadrilateral Cells

2.5.1 Cell Geometry Representation and Local Vector Bases

Figure 2.17 illustrates the various edge, height, and gradient vectors for a rectilinear quadrilateral element. The normalized coordinates for quadrilaterals are related by

$$\begin{aligned} \xi_1 + \xi_3 &= 1 \\ \xi_2 + \xi_4 &= 1 \end{aligned} \tag{2.54}$$

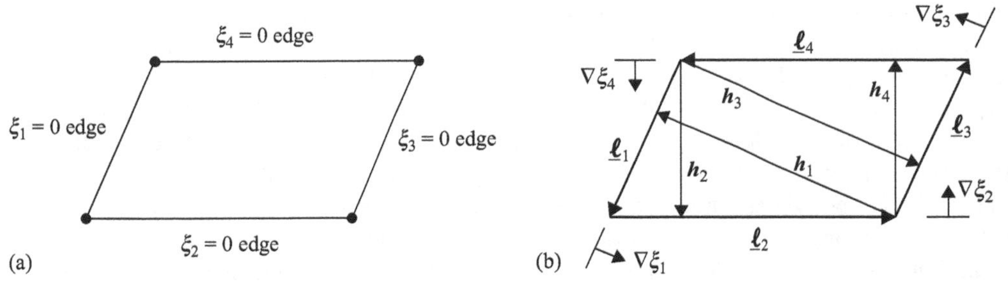

Figure 2.17 Quadrilateral elements: (a) zero-coordinate edges and (b) edge, height, and gradient vectors.

© 1997 IEEE. Reprinted with permission from R. D. Graglia, D. R. Wilton, and A. F. Peterson, "Higher order interpolatory vector bases for computational electromagnetics," special issue on "Advanced Numerical Techniques in Electromagnetics," *IEEE Trans. Antennas Propag.*, vol. 45, no. 3, pp. 329–342, Mar. 1997.

where we have taken ξ_1 and ξ_2 as independent coordinates. A Lagrange parameterization of order q for a curvilinear quadrilateral can be expressed in terms of the Silvester polynomials as

$$\boldsymbol{r} = \sum_{i,j,k,\ell=0}^{q} \boldsymbol{r}_{ik;j\ell}\, R_i(q,\xi_1)\, R_j(q,\xi_2)\, R_k(q,\xi_3)\, R_\ell(q,\xi_4) \tag{2.55}$$

with $i+k=j+\ell=q$ and where the quadruple index is used to label the position vector $\boldsymbol{r}_{ik;j\ell}$ interpolating the point with normalized coordinates

$$(\xi_1, \xi_3; \xi_2, \xi_4) = \left(\frac{i}{q}, \frac{k}{q}; \frac{j}{q}, \frac{\ell}{q}\right) = \boldsymbol{\xi}_{ik;j\ell} \tag{2.56}$$

Along the cell edges, (2.55) simplifies into the following qth-order polynomial vectors (written in terms of *standard* Lagrange polynomials)

$$\begin{cases} \boldsymbol{r}(\xi_1=0) = \sum_{n=0}^{q} \boldsymbol{r}_{0q;n(q-n)}\, P_n(q,\xi_2) = \sum_{n=0}^{q} \boldsymbol{r}_{0q;(q-n)n}\, P_n(q,\xi_4) \\ \boldsymbol{r}(\xi_3=0) = \sum_{n=0}^{q} \boldsymbol{r}_{q0;n(q-n)}\, P_n(q,\xi_2) = \sum_{n=0}^{q} \boldsymbol{r}_{q0;(q-n)n}\, P_n(q,\xi_4) \end{cases} \tag{2.57}$$

$$\begin{cases} \boldsymbol{r}(\xi_2=0) = \sum_{n}^{q} \boldsymbol{r}_{n(q-n);0q}\, P_n(q,\xi_1) = \sum_{n}^{q} \boldsymbol{r}_{(q-n)n;0q}\, P_n(q,\xi_3) \\ \boldsymbol{r}(\xi_4=0) = \sum_{n}^{q} \boldsymbol{r}_{n(q-n);q0}\, P_n(q,\xi_1) = \sum_{n}^{q} \boldsymbol{r}_{(q-n)n;q0}\, P_n(q,\xi_3) \end{cases} \tag{2.58}$$

which match the qth-order polynomial vector representation that eventually holds on the adjacent cells that share the same edge. However, the total degree of the polynomials appearing in the expression (2.55) is $2q$ since these polynomials are obtained by using a Cartesian product procedure that involves the multiplication of two polynomials of qth degree; namely $R_i(q,\xi_1)\,R_{q-i}(q,\xi_3) \times R_j(q,\xi_2)\,R_{q-j}(q,\xi_4) = P_i(q,\xi_1) \times P_j(q,\xi_2)$. It is therefore apparent that the Cartesian product procedure that we are going to use also for the three-dimensional cells is very effective but yields a number of interpolation points that is minimum only for simplices (that is, for triangular and tetrahedral cells). A parameterization of order q of a curvilinear quadrilateral can be done by using a number of interpolation points smaller than $(q+1)^2$, which is the number of points involved by (2.55). However, there is little or no advantage in using other parameterizations if these are not interpolatory and/or if they do not have simple, straightforward expressions. As a matter of fact, for a quadrilateral cell, one can obtain, with some effort, interpolatory parameterizations that involve the minimum number of points but, definitely, in this case these parameterizations do not have simple, straightforward expressions.

For the quadrilateral, the edge vectors derived from the *independent* coordinates ξ_1 and ξ_2 are (see Section 3.14)

$$\begin{aligned}
\boldsymbol{\ell}^1 &= \frac{\partial \boldsymbol{r}(\xi_1,\xi_2)}{\partial \xi_1} = \left[\frac{\partial \boldsymbol{r}(\xi_1,\xi_3;\xi_2,\xi_4)}{\partial \xi_1} - \frac{\partial \boldsymbol{r}(\xi_1,\xi_3;\xi_2,\xi_4)}{\partial \xi_3}\right]_{\xi_2,\xi_4 \text{ constant}} \\
\boldsymbol{\ell}^2 &= \frac{\partial \boldsymbol{r}(\xi_1,\xi_2)}{\partial \xi_2} = \left[\frac{\partial \boldsymbol{r}(\xi_1,\xi_3;\xi_2,\xi_4)}{\partial \xi_2} - \frac{\partial \boldsymbol{r}(\xi_1,\xi_3;\xi_2,\xi_4)}{\partial \xi_4}\right]_{\xi_1,\xi_3 \text{ constant}}
\end{aligned} \tag{2.59}$$

from which the edge vectors

$$\begin{aligned}
\boldsymbol{\ell}_3 &= -\boldsymbol{\ell}_1 = \boldsymbol{\ell}^2 \\
\boldsymbol{\ell}_2 &= -\boldsymbol{\ell}_4 = \boldsymbol{\ell}^1
\end{aligned} \tag{2.60}$$

of Figure 2.17 are found, with

$$\begin{aligned}
\boldsymbol{\ell}_1 + \boldsymbol{\ell}_3 &= 0 \\
\boldsymbol{\ell}_2 + \boldsymbol{\ell}_4 &= 0
\end{aligned} \tag{2.61}$$

In the special case where $\boldsymbol{\ell}^1$ and $\boldsymbol{\ell}^2$ are constant vectors, the quadrilateral is a planar parallelogram. The gradient vectors are determined in terms of the edge vectors as

$$\nabla \xi_i = \frac{\hat{\boldsymbol{n}} \times \boldsymbol{\ell}_i}{\mathcal{J}} \tag{2.62}$$

with

$$\begin{aligned}
\nabla \xi_1 + \nabla \xi_3 &= 0 \\
\nabla \xi_2 + \nabla \xi_4 &= 0
\end{aligned} \tag{2.63}$$

and where

$$\hat{\boldsymbol{n}} = \frac{\boldsymbol{\ell}^1 \times \boldsymbol{\ell}^2}{\mathcal{J}} = \frac{\boldsymbol{\ell}_1 \times \boldsymbol{\ell}_2}{\mathcal{J}} \tag{2.64}$$

is the unit vector normal to the quadrilateral, while

$$\mathcal{J} = |\boldsymbol{\ell}^1 \times \boldsymbol{\ell}^2| = |\boldsymbol{\ell}_1 \times \boldsymbol{\ell}_2| \tag{2.65}$$

is the Jacobian of the transformation from parent to child coordinates.

2.5.2 Lagrangian Basis Functions, Interpolation, and Gradient Approximation

2.5.2.1 *The Shape Polynomials $\alpha(p, \xi)$ to Interpolate All the Edges*

For the quadrilateral cell, a pth-order complete set of Lagrangian scalar basis functions is formed by the $(p+1)^2$ polynomials

$$\alpha_{ik;j\ell}(p, \xi_1, \xi_3; \xi_2, \xi_4) = R_i(p, \xi_1)\, R_k(p, \xi_3)\, R_j(p, \xi_2)\, R_\ell(p, \xi_4) \tag{2.66}$$

interpolating the point with normalized coordinates

$$(\xi_1, \xi_3; \xi_2, \xi_4) = \left(\frac{i}{p}, \frac{k}{p}; \frac{j}{p}, \frac{\ell}{p}\right) = \boldsymbol{\xi}_{ik;j\ell} \tag{2.67}$$

obtained for $i, j = 0, 1, \ldots, p$, with $i + k = j + \ell = p$. The interpolation nodes on quadrilateral cells obtained for a Lagrange interpolation of order $p = 2$ are shown in Figure 2.18. Table 2.4 reports the interpolating functions $\alpha_{ik;j\ell}(p, \boldsymbol{\xi})$ for quadrilateral cells, up to $p = 3$.

2.5.2.2 *Gradient Approximation*

For a pth-order Lagrangian interpolation of a function f on the quadrilateral one has

$$\tilde{f} = \sum_{i,j,k,\ell=0}^{p} f_{ik;j\ell}\, \alpha_{ik;j\ell}(p, \xi_1, \xi_3; \xi_2, \xi_4) \tag{2.68}$$

$$\nabla\tilde{f} = \sum_{i,j,k,\ell=0}^{p} f_{ik;j\ell}\, \nabla\alpha_{ik;j\ell}(p, \xi_1, \xi_3; \xi_2, \xi_4) \tag{2.69}$$

with $i + k = j + \ell = p$,

$$\begin{aligned} \nabla\alpha_{ik;j\ell}(p, \boldsymbol{\xi}) = {} & R_i'(p, \xi_1)\, R_j(p, \xi_2)\, R_k(p, \xi_3)\, R_\ell(p, \xi_4)\, \nabla\xi_1 \\ & + R_i(p, \xi_1)\, R_j(p, \xi_2)\, R_k'(p, \xi_3)\, R_\ell(p, \xi_4)\, \nabla\xi_3 \\ & + R_i(p, \xi_1)\, R_j'(p, \xi_2)\, R_k(p, \xi_3)\, R_\ell(p, \xi_4)\, \nabla\xi_2 \\ & + R_i(p, \xi_1)\, R_j(p, \xi_2)\, R_k(p, \xi_3)\, R_\ell'(p, \xi_4)\, \nabla\xi_4 \end{aligned} \tag{2.70}$$

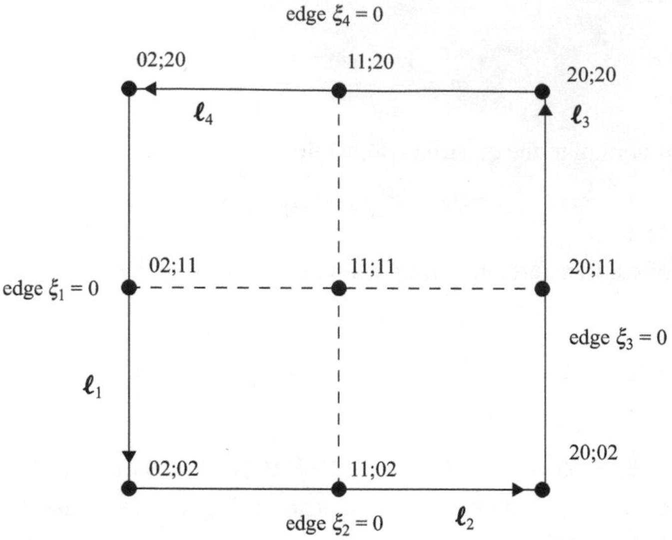

Figure 2.18 Interpolation nodes on quadrilateral cells obtained for a Lagrange interpolation of order $p = 2$.

and where the quadruple index is used to label the interpolation points with normalized coordinates $(\xi_1, \xi_3; \xi_2, \xi_4) = \left(\frac{i}{p}, \frac{k}{p}; \frac{j}{p}, \frac{\ell}{p}\right) = \boldsymbol{\xi}_{ik;j\ell}$.

2.6 Tetrahedral Cells

2.6.1 Cell Geometry Representation and Local Vector Bases

The local, normalized coordinates $\boldsymbol{\xi} = (\xi_1, \xi_2, \xi_3, \xi_4)$ for tetrahedrons (see Figure 2.19), also called *volume* coordinates,[6] *simplex* coordinates, etc., are related by

$$\xi_1 + \xi_2 + \xi_3 + \xi_4 = 1 \tag{2.71}$$

where we have taken ξ_1, ξ_2, and ξ_3 as independent coordinates.

A pth-order complete set of Lagrangian scalar basis functions for the tetrahedral cell is formed by the $(p+1)(p+2)(p+3)/6$ polynomials

$$\alpha_{ijk\ell}(p, \boldsymbol{\xi}) = R_i(p, \xi_1)\, R_j(p, \xi_2)\, R_k(p, \xi_3)\, R_\ell(p, \xi_4) \tag{2.72}$$

[6]The parent coordinates of a rectilinear tetrahedron are shown to be *volume* coordinates by following a reasoning similar to that used to prove that the parent coordinates of a rectilinear triangle are area coordinates. These details are omitted here.

Table 2.4 The first three orders of the shape polynomials interpolating all the edges of a quadrilateral cell.

The $(p+1)^2$ polynomials $\alpha_{ik;j\ell}(p,\boldsymbol{\xi}) = R_i(p,\xi_1)R_k(p,\xi_3)R_j(p,\xi_2)R_\ell(p,\xi_4)$ that interpolate a quadrilateral cell of order p are obtained for $i,j=0,1,\ldots,p$, with $i+k=j+\ell=p$ and $p\geq 1$. Each of these polynomials is of order $2p$. Because of the symmetry of the quadrilateral cell it is sufficient to report, for each family, only a smaller set of polynomials that is then "completed" by other polynomials, obtained from those of the smaller set by permutations of the parent variables $\{\xi_1,\xi_3\}$ and $\{\xi_2,\xi_4\}$. The smaller set of polynomials is more conveniently written in terms of the dummy quadrilateral parent variables $\{\xi_a,\xi_c;\xi_b,\xi_d\}$. Each pth-order complete interpolatory set is then obtained by setting, in the functions $R_i(p,\xi_a)R_k(p,\xi_c)R_j(p,\xi_b)R_\ell(p,\xi_d)$:

$\{\xi_a,\xi_c;\xi_b,\xi_d\} = \{\xi_1,\xi_3;\xi_2,\xi_4\}$ first set of permuted variables;
$\{\xi_a,\xi_c;\xi_b,\xi_d\} = \{\xi_1,\xi_3;\xi_4,\xi_2\}$ second set of permuted variables;
$\{\xi_a,\xi_c;\xi_b,\xi_d\} = \{\xi_3,\xi_1;\xi_2,\xi_4\}$ third set of permuted variables;
$\{\xi_a,\xi_c;\xi_b,\xi_d\} = \{\xi_3,\xi_1;\xi_4,\xi_2\}$ fourth set of permuted variables.

Thus, all the quadrilateral polynomials are defined by using four different sets of permuted variables; notice how the number of the permuted sets equals the number of the cell corner nodes.

$p=1$ (4 functions in total)

Dummy expressions (use set 1–4):

$\xi_a\,\xi_b$ (4 functions)

By permutations, the functions are:

$\alpha_{10;10} = \xi_1\xi_2,\quad \alpha_{10;01} = \xi_1\xi_4,$
$\alpha_{01;10} = \xi_3\xi_2,\quad \alpha_{01;01} = \xi_3\xi_4$

$p=2$ (9 functions in total)

Dummy expressions:

$\xi_a(2\xi_a-1)\,\xi_b(2\xi_b-1)$ (4 functions; use set 1 to 4)
$2^2\xi_a(2\xi_a-1)\,\xi_b\xi_d$ (2 functions; use set 1 & 4)
$2^2\xi_a\xi_c\,\xi_b(2\xi_b-1)$ (2 functions; use set 2 & 3)
$2^4\xi_a\xi_c\,\xi_b\xi_d$ (1 function; use set 1)

By permutations, the functions are:

$\alpha_{20;20} = \xi_1(2\xi_1-1)\xi_2(2\xi_2-1)$
$\alpha_{20;02} = \xi_1(2\xi_1-1)\xi_4(2\xi_4-1)$
$\alpha_{02;20} = \xi_3(2\xi_3-1)\xi_2(2\xi_2-1)$
$\alpha_{02;02} = \xi_3(2\xi_3-1)\xi_4(2\xi_4-1)$
$\alpha_{20;11} = 2^2\xi_1(2\xi_1-1)\xi_2\xi_4$
$\alpha_{02;11} = 2^2\xi_3(2\xi_3-1)\xi_2\xi_4$
$\alpha_{11;20} = 2^2\xi_2(2\xi_2-1)\xi_1\xi_3$
$\alpha_{11;02} = 2^2\xi_4(2\xi_4-1)\xi_1\xi_3$
$\alpha_{11;11} = 2^4\xi_1\xi_3\xi_2\xi_4$

$p=3$ (16 functions in total)

Dummy expressions (use set 1–4):

$\xi_a(3\xi_a-1)(3\xi_a-2)\,\xi_b(3\xi_b-1)(3\xi_b-2)/4$ (4 functions)
$3^2\xi_a(3\xi_a-1)(3\xi_a-2)\,\xi_b(3\xi_b-1)\xi_d/4$ (4 functions)
$3^2\xi_a(3\xi_a-1)\xi_c\,\xi_b(3\xi_b-1)(3\xi_b-2)/4$ (4 functions)
$3^4\xi_a(3\xi_a-1)\xi_c\,\xi_b(3\xi_b-1)\xi_d/4$ (4 functions)

By permutations, the functions are:

$\alpha_{30;30} = \xi_1(3\xi_1-1)(3\xi_1-2)\xi_2(3\xi_2-1)(3\xi_2-2)/4$
$\alpha_{30;03} = \xi_1(3\xi_1-1)(3\xi_1-2)\xi_4(3\xi_4-1)(3\xi_4-2)/4$
$\alpha_{03;30} = \xi_3(3\xi_3-1)(3\xi_3-2)\xi_2(3\xi_2-1)(3\xi_2-2)/4$
$\alpha_{03;03} = \xi_3(3\xi_3-1)(3\xi_3-2)\xi_4(3\xi_4-1)(3\xi_4-2)/4$
$\alpha_{30;21} = 3^2\xi_1(3\xi_1-1)(3\xi_1-2)\xi_2(3\xi_2-1)\xi_4/4$
$\alpha_{30;12} = 3^2\xi_1(3\xi_1-1)(3\xi_1-2)\xi_4(3\xi_4-1)\xi_2/4$
$\alpha_{03;21} = 3^2\xi_3(3\xi_3-1)(3\xi_3-2)\xi_2(3\xi_2-1)\xi_4/4$
$\alpha_{03;12} = 3^2\xi_3(3\xi_3-1)(3\xi_3-2)\xi_4(3\xi_4-1)\xi_2/4$
$\alpha_{21;30} = 3^2\xi_2(3\xi_2-1)(3\xi_2-2)\xi_1(3\xi_1-1)\xi_3/4$
$\alpha_{21;03} = 3^2\xi_4(3\xi_4-1)(3\xi_4-2)\xi_1(3\xi_1-1)\xi_3/4$
$\alpha_{12;30} = 3^2\xi_2(3\xi_2-1)(3\xi_2-2)\xi_3(3\xi_3-1)\xi_1/4$
$\alpha_{12;03} = 3^2\xi_4(3\xi_4-1)(3\xi_4-2)\xi_3(3\xi_3-1)\xi_1/4$
$\alpha_{21;21} = 3^4\xi_1(3\xi_1-1)\xi_3\,\xi_2(3\xi_2-1)\xi_4/4$
$\alpha_{21;12} = 3^4\xi_1(3\xi_1-1)\xi_3\,\xi_4(3\xi_4-1)\xi_2/4$
$\alpha_{12;21} = 3^4\xi_3(3\xi_3-1)\xi_1\,\xi_2(3\xi_2-1)\xi_4/4$
$\alpha_{12;12} = 3^4\xi_3(3\xi_3-1)\xi_1\,\xi_4(3\xi_4-1)\xi_2/4$

obtained for $i,j,k,\ell=0,1,\ldots,p$, with $i+j+k+\ell=p$ and $p\geq 1$, and where a quadruple indexing scheme is used since these polynomials interpolate the points with normalized coordinates

$$\boldsymbol{\xi}_{ijk\ell} = \left(\frac{i}{p},\frac{j}{p},\frac{k}{p},\frac{\ell}{p}\right) \tag{2.73}$$

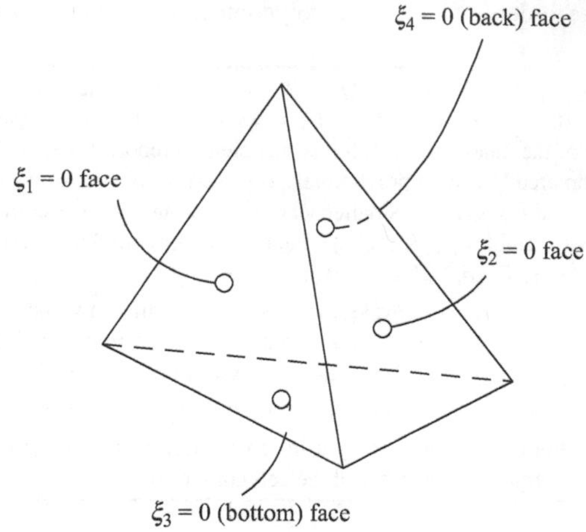

Figure 2.19 Zero-coordinate surfaces for the tetrahedral cell.

The Lagrangian basis functions $\alpha_{ijk\ell}$ can be used to map the parent tetrahedron into the child space. A Lagrange parameterization of order $q(\geq 1)$ for a curvilinear tetrahedron can for example be expressed as

$$\boldsymbol{r} = \sum_{i,j,k,\ell=0}^{q} \boldsymbol{r}_{ijk\ell}\, \alpha_{ijk\ell}(q, \xi_1, \xi_2, \xi_3, \xi_4), \quad i+j+k+\ell=q \tag{2.74}$$

where the quadruple index is used to label the position vector $\boldsymbol{r}_{ijk\ell}$ interpolating the point with normalized coordinates $\boldsymbol{\xi}_{ijk\ell} = \left(\frac{i}{q}, \frac{j}{q}, \frac{k}{q}, \frac{\ell}{q}\right)$. The coordinates of a *rectilinear* tetrahedron are obtained from the previous expression simply by setting $q = 1$, which yields

$$\begin{aligned} \boldsymbol{r} &= \xi_1 \boldsymbol{r}_{1000} + \xi_2 \boldsymbol{r}_{0100} + \xi_3 \boldsymbol{r}_{0010} + \xi_4 \boldsymbol{r}_{0001} \\ &= \xi_1 \boldsymbol{r}_1 + \xi_2 \boldsymbol{r}_2 + \xi_3 \boldsymbol{r}_3 + \xi_4 \boldsymbol{r}_4 \end{aligned} \tag{2.75}$$

where $\boldsymbol{r}_i$ is a vertex position vector for vertex i.

For a given geometrical parameterization of the tetrahedron, the *unitary base vectors*, derived from the *independent* coordinates ξ_1, ξ_2, and ξ_3, are (see Section 3.11)

$$\begin{cases} \boldsymbol{\ell}^1 = \dfrac{\partial \boldsymbol{r}(\xi_1, \xi_2, \xi_3)}{\partial \xi_1} = \left[\dfrac{\partial \boldsymbol{r}(\xi_1, \xi_2, \xi_3, \xi_4)}{\partial \xi_1} - \dfrac{\partial \boldsymbol{r}(\xi_1, \xi_2, \xi_3, \xi_4)}{\partial \xi_4}\right]_{\xi_2, \xi_3 \text{ constant}} \\ \boldsymbol{\ell}^2 = \dfrac{\partial \boldsymbol{r}(\xi_1, \xi_2, \xi_3)}{\partial \xi_2} = \left[\dfrac{\partial \boldsymbol{r}(\xi_1, \xi_2, \xi_3, \xi_4)}{\partial \xi_2} - \dfrac{\partial \boldsymbol{r}(\xi_1, \xi_2, \xi_3, \xi_4)}{\partial \xi_4}\right]_{\xi_1, \xi_3 \text{ constant}} \\ \boldsymbol{\ell}^3 = \dfrac{\partial \boldsymbol{r}(\xi_1, \xi_2, \xi_3)}{\partial \xi_3} = \left[\dfrac{\partial \boldsymbol{r}(\xi_1, \xi_2, \xi_3, \xi_4)}{\partial \xi_3} - \dfrac{\partial \boldsymbol{r}(\xi_1, \xi_2, \xi_3, \xi_4)}{\partial \xi_4}\right]_{\xi_1, \xi_2 \text{ constant}} \end{cases} \tag{2.76}$$

from which the edge vectors of Figure 2.20 are found

$$\begin{cases} \boldsymbol{\ell}_{12} = \boldsymbol{\ell}^3 \\ \boldsymbol{\ell}_{13} = -\boldsymbol{\ell}^2 \\ \boldsymbol{\ell}_{14} = \boldsymbol{\ell}^2 - \boldsymbol{\ell}^3 \\ \boldsymbol{\ell}_{23} = \boldsymbol{\ell}^1 \\ \boldsymbol{\ell}_{24} = \boldsymbol{\ell}^3 - \boldsymbol{\ell}^1 \\ \boldsymbol{\ell}_{34} = \boldsymbol{\ell}^1 - \boldsymbol{\ell}^2 \end{cases} \tag{2.77}$$

The gradient vectors are determined in terms of the unitary base vectors as (see Section 3.11)

$$\begin{cases} \nabla\xi_1 = \dfrac{\boldsymbol{\ell}^2 \times \boldsymbol{\ell}^3}{\mathcal{J}} \\ \nabla\xi_2 = \dfrac{\boldsymbol{\ell}^3 \times \boldsymbol{\ell}^1}{\mathcal{J}} \\ \nabla\xi_3 = \dfrac{\boldsymbol{\ell}^1 \times \boldsymbol{\ell}^2}{\mathcal{J}} \end{cases} \tag{2.78}$$

where

$$\mathcal{J} = \boldsymbol{\ell}^1 \cdot \boldsymbol{\ell}^2 \times \boldsymbol{\ell}^3 \tag{2.79}$$

is the Jacobian of the transformation from parent to child coordinates. The remaining coordinate gradient is determined to be

$$\nabla\xi_4 = -(\nabla\xi_1 + \nabla\xi_2 + \nabla\xi_3) \tag{2.80}$$

Figure 2.20 illustrates the various edge and height vectors and normalized coordinates defined for a rectilinear tetrahedral element.

2.6.2 Lagrangian Basis Functions

2.6.2.1 *The Shape Polynomials $\alpha(p, \boldsymbol{\xi})$ to Interpolate All the Edges and Faces*

Clearly, the order of the indices (i, j, k, ℓ) used in (2.72) and (2.74) is associated with and follows the order used for the parent variables (ξ_1, ξ_2, ξ_3, ξ_4).

To fully exploit the symmetry of the tetrahedral cell, we find convenient to introduce the four *dummy* parent variables

$$\boldsymbol{\xi} = (\xi_a, \xi_b, \xi_c, \xi_d) \tag{2.81}$$

to write the scalar basis function

$$\alpha_{ijk\ell}(p, \boldsymbol{\xi}) = R_i(p, \xi_a)\, R_j(p, \xi_b)\, R_k(p, \xi_c)\, R_\ell(p, \xi_d) \tag{2.82}$$

in terms of *dummy* indices (i, j, k, ℓ) not associated anymore with the parent variables (ξ_1, ξ_2, ξ_3, ξ_4), but rather associated with the dummy variables (ξ_a, ξ_b, ξ_c, ξ_d). The dummy variables represent permutation of the parent variables and allow us to express the full collection of basis functions in terms of a minimal set. Obviously, (2.82) and (2.72) coincide for (ξ_a, ξ_b, ξ_c, ξ_d)

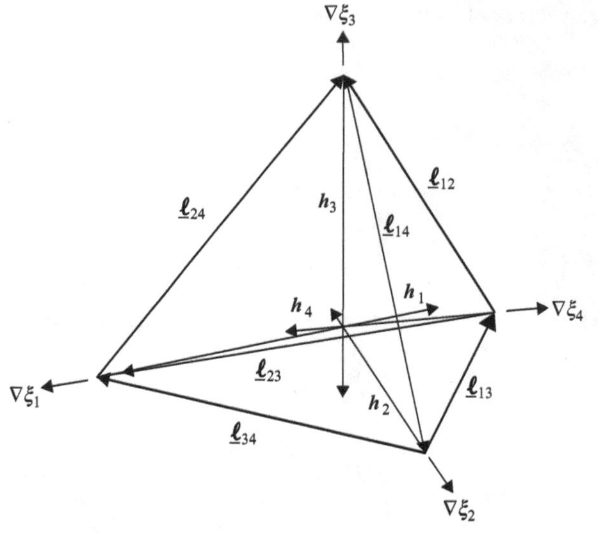

Figure 2.20 Edge, height, and gradient vectors for tetrahedral elements.

equal to $(\xi_1, \xi_2, \xi_3, \xi_4)$. The interpolation points associated with $\alpha_{ijk\ell}(p, \boldsymbol{\xi})$ for $p = 2$ are shown in Figure 2.21.

By using dummy parent variables one can show, once and for all, the conformity of the tetrahedral basis functions with the Lagrangian basis function defined for the triangular cell. In fact, on the $\xi_d = 0$ face, the polynomials (2.82) simplify into

$$\alpha_{ijk\ell}\big|_{\xi_d=0} = 0, \quad \text{for } \ell \neq 0 \tag{2.83}$$

$$\alpha_{ijk0}\big|_{\xi_d=0} = R_i(p, \xi_a)\, R_j(p, \xi_b)\, R_k(p, \xi_c), \quad \text{for } \ell = 0 \tag{2.84}$$

The latter expression obtained for $\ell = 0$ holds for $i + j + k = p$ and coincides with the expression of the basis functions α_{ijk} of a triangular cell described in terms of the three dummy parent variables (ξ_a, ξ_b, ξ_c) (see Section 2.4.2.1). Table 2.5 reports the interpolating functions $\alpha_{ijk\ell}(p, \boldsymbol{\xi})$ for tetrahedral cells, up to $p = 3$.

2.6.2.2 Lagrange Interpolation and Gradient Approximation

For a pth-order Lagrangian interpolation of a function f on the tetrahedron one has

$$\tilde{f} = \sum_{i,j,k,\ell=0}^{p} f_{ijk\ell}\, \alpha_{ijk\ell}(p, \xi_1, \xi_2, \xi_3, \xi_4) \tag{2.85}$$

$$\nabla \tilde{f} = \sum_{i,j,k,\ell=0}^{p} f_{ijk\ell}\, \nabla \alpha_{ijk\ell}(p, \xi_1, \xi_2, \xi_3, \xi_4) \tag{2.86}$$

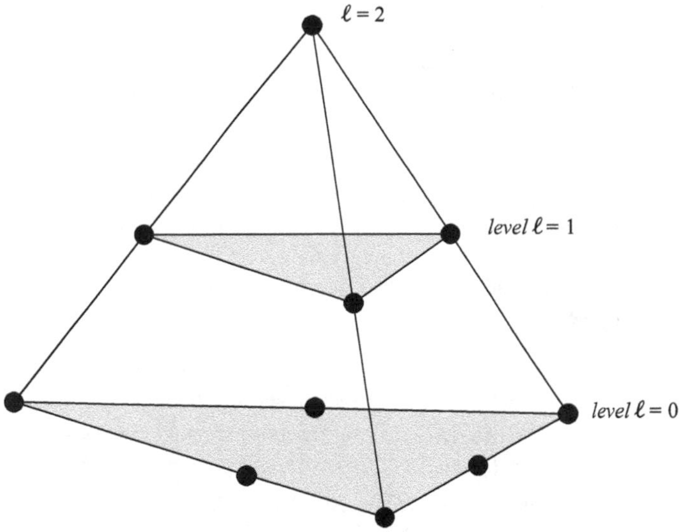

Figure 2.21 Lagrange interpolation of order $p = 2$ on a tetrahedron. The figure shows the interpolation nodes associated with the interpolatory polynomials $\alpha_{ijk\ell}(2, \boldsymbol{\xi})$. The front-left face of the tetrahedron is the $\xi_a = 0$ face, while the $\xi_d = 0$ face is the bottom face, labeled as the "*level* $\ell = 0$" triangle. The figures show the $\ell =$ constant triangular-cuts that correspond to the coordinate surfaces $\xi_d = \ell/p$, for $l = 0, 1, 2$ (recall that $p = 2$ in the figure).

$$\begin{aligned}\nabla\alpha_{ijk\ell}(p,\boldsymbol{\xi}) &= R_i'(p,\xi_1)\,R_j(p,\xi_2)\,R_k(p,\xi_3)\,R_\ell(p,\xi_4)\,\nabla\xi_1\\ &\quad + R_i(p,\xi_1)\,R_j'(p,\xi_2)\,R_k(p,\xi_3)\,R_\ell(p,\xi_4)\,\nabla\xi_2\\ &\quad + R_i(p,\xi_1)\,R_j(p,\xi_2)\,R_k'(p,\xi_3)\,R_\ell(p,\xi_4)\,\nabla\xi_3\\ &\quad + R_i(p,\xi_1)\,R_j(p,\xi_2)\,R_k(p,\xi_3)\,R_\ell'(p,\xi_4)\,\nabla\xi_4 \end{aligned} \tag{2.87}$$

with $i + j + k + \ell = p$ and where the quadruple index is used to label the interpolation points with normalized coordinates

$$(\xi_1, \xi_2, \xi_3, \xi_4) = \left(\frac{i}{p}, \frac{j}{p}, \frac{k}{p}, \frac{\ell}{p}\right) = \boldsymbol{\xi}_{ijk\ell} \tag{2.88}$$

The gradient vectors $\nabla\xi_s$ (with $s = 1, 2, 3, 4$) are given in (2.78) and (2.80).

2.7 Brick Cells

2.7.1 Cell Geometry Representation and Local Vector Bases

A brick or hexahedron is a six-faceted volumetric cell that we describe in terms of six parent variables. The local, normalized coordinates $\boldsymbol{\xi} = (\xi_1, \xi_4; \xi_2, \xi_5; \xi_3, \xi_6)$ for bricks (see Figure 2.22) are related by

Table 2.5 The first three orders of the shape polynomials interpolating all the edges and faces of a tetrahedral cell.

The $(p+1)(p+2)(p+3)/6$ polynomials $\alpha_{ijk\ell}(p,\boldsymbol{\xi}) = R_i(p,\xi_1)R_j(p,\xi_2)R_k(p,\xi_3)R_\ell(p,\xi_4)$ that interpolate a tetrahedral cell of order p are obtained for $i,j,k,\ell = 0,1,\ldots,p$, with $i+j+k+\ell = p$ and $p \geq 1$. Each of these polynomials is of order p because of the symmetry of the tetrahedral cell it is sufficient to report, for each family, only a smaller set of polynomials that is then "completed" by other polynomials, obtained from those of the smaller set by cyclic permutations of the parent variables $\{\xi_1,\xi_2,\xi_3,\xi_4\}$. The smaller set of polynomials is more conveniently written in terms of the dummy tetrahedral parent variables $\{\xi_a,\xi_b,\xi_c,\xi_d\}$. Each pth-order complete interpolatory set is then obtained by setting, in the functions $R_i(p,\xi_a)R_j(p,\xi_b)R_k(p,\xi_c)R_\ell(p,\xi_d)$:

$\{\xi_a,\xi_b,\xi_c,\xi_d\} = \{\xi_1,\xi_2,\xi_3,\xi_4\}$ first set of permuted variables;
$\{\xi_a,\xi_b,\xi_c,\xi_d\} = \{\xi_2,\xi_3,\xi_4,\xi_1\}$ second set of permuted variables;
$\{\xi_a,\xi_b,\xi_c,\xi_d\} = \{\xi_3,\xi_4,\xi_1,\xi_2\}$ third set of permuted variables;
$\{\xi_a,\xi_b,\xi_c,\xi_d\} = \{\xi_4,\xi_1,\xi_2,\xi_3\}$ fourth set of permuted variables.

Thus, all the tetrahedral polynomials are defined by using four different sets of permuted variables; notice how the number of the permuted sets equals the number of the cell corner nodes.

$p = 1$ (4 functions in total)
Dummy expressions (use set 1–4):

ξ_a (4 functions)

By permutations, the functions are:

$$\alpha_{1000} = \xi_1,\ \alpha_{0100} = \xi_2$$
$$\alpha_{0010} = \xi_3,\ \alpha_{0001} = \xi_4$$

$p = 2$ (10 functions in total)
Dummy expressions:

$\xi_a(2\xi_a - 1)$ (4 functions; use set 1–4)
$2^2\xi_a\xi_b$ (4 functions; use set 1–4)
$2^2\xi_a\xi_c$ (2 functions; use set 1 & 2)

By permutations, the functions are:

$$\alpha_{2000} = xi_1(2\xi_1 - 1)$$
$$\alpha_{0200} = \xi_2(2\xi_2 - 1)$$
$$\alpha_{0020} = \xi_3(2\xi_3 - 1)$$
$$\alpha_{0002} = \xi_4(2\xi_4 - 1)$$
$$\alpha_{1100} = 2^2\xi_1\xi_2,\quad \alpha_{0110} = 2^2\xi_2\xi_3,$$
$$\alpha_{0011} = 2^2\xi_3\xi_4,\quad \alpha_{1001} = 2^2\xi_4\xi_1,$$
$$\alpha_{1010} = 2^2\xi_1\xi_3,\quad \alpha_{0101} = 2^2\xi_2\xi_4$$

$p = 3$ (20 functions in total)
Dummy expressions (use set 1–4):

$\xi_a(3\xi_a - 1)(3\xi_a - 2)/2$ (4 functions)
$3^2\xi_a(3\xi_a - 1)\,\xi_b/2$ (4 functions)
$3^2\xi_a(3\xi_a - 1)\,\xi_c/2$ (4 functions)
$3^2\xi_a(3\xi_a - 1)\,\xi_d/2$ (4 functions)
$3^3\xi_a\xi_b\xi_c$ (4 functions)

By permutations, the functions are:

$$\alpha_{3000} = \xi_1(3\xi_1 - 1)(3\xi_1 - 2)/2$$
$$\alpha_{0300} = \xi_2(3\xi_2 - 1)(3\xi_2 - 2)/2$$
$$\alpha_{0030} = \xi_3(3\xi_3 - 1)(3\xi_3 - 2)/2$$
$$\alpha_{0003} = \xi_4(3\xi_4 - 1)(3\xi_4 - 2)/2$$
$$\alpha_{2100} = 3^2\xi_1(3\xi_1 - 1)\,\xi_2/2, \alpha_{0210} = 3^2\xi_2(3\xi_2 - 1)\,\xi_3/2,$$
$$\alpha_{0021} = 3^2\xi_3(3\xi_3 - 1)\,\xi_4/2, \alpha_{1002} = 3^2\xi_4(3\xi_4 - 1)\,\xi_1/2$$
$$\alpha_{2010} = 3^2\xi_1(3\xi_1 - 1)\,\xi_3/2, \alpha_{0201} = 3^2\xi_2(3\xi_2 - 1)\,\xi_4/2$$
$$\alpha_{1020} = 3^2\xi_3(3\xi_3 - 1)\,\xi_1/2, \alpha_{0102} = 3^2\xi_4(3\xi_4 - 1)\,\xi_2/2$$
$$\alpha_{2001} = 3^2\xi_1(3\xi_1 - 1)\,\xi_4/2, \alpha_{1200} = 3^2\xi_2(3\xi_2 - 1)\,\xi_1/2$$
$$\alpha_{0120} = 3^2\xi_3(3\xi_3 - 1)\,\xi_2/2, \alpha_{0012} = 3^2\xi_4(3\xi_4 - 1)\,\xi_3/2$$
$$\alpha_{1110} = 3^3\xi_1\xi_2\xi_3, \alpha_{0111} = 3^3\xi_2\xi_3\xi_4,$$
$$\alpha_{1011} = 3^3\xi_3\xi_4\xi_1, \alpha_{1101} = 3^3\xi_4\xi_1\xi_2$$

$$\begin{aligned}\xi_1 + \xi_4 &= 1\\ \xi_2 + \xi_5 &= 1\\ \xi_3 + \xi_6 &= 1\end{aligned} \tag{2.89}$$

where we have taken ξ_1, ξ_2, and ξ_3 as independent coordinates.

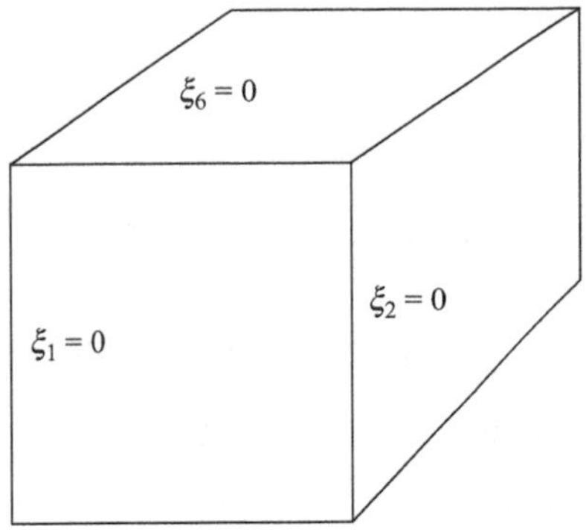

Figure 2.22 Zero-coordinate surfaces for the brick cell. The faces $\xi_i = 0$ and $\xi_{i+3} = 0$ (for $i = 1, 2, 3$) are opposite.

A pth-order complete polynomial family that (for $p \geq 1$) *interpolates* a brick cell on a *symmetric* grid of points is easily obtained by using a Cartesian product procedure, for example by multiplying the three Lagrange polynomials P_i, P_j, and P_k of the independent variables ξ_1, ξ_2, and ξ_3 as it follows

$$\begin{aligned} P_i(p;\xi_1)\,P_j(p;\xi_2)\,P_k(p;\xi_3) = {} & R_i(p,\xi_1)\,R_{p-i}(p,\xi_4) \\ & R_j(p,\xi_2)\,R_{p-j}(p,\xi_5) \\ & R_k(p,\xi_3)\,R_{p-k}(p,\xi_6) \quad \text{for } i,j,k = 0,1,\ldots,p \end{aligned} \tag{2.90}$$

Linear combinations of the polynomials (2.90) can represent any of the $(p+1)(p+2)(p+3)/6$ terms of the form $\xi_1^r\,\xi_2^s\,\xi_3^t$ obtained for $r, s, t = 0, 1, \ldots, p$; with $0 \leq r+s+t \leq p$. The set (2.90) consists of $(p+1)^3$ independent polynomials, and therefore it has $p(p+1)(5p+7)/6$ more terms than the minimum required to form a pth-order complete three-dimensional scalar polynomial family, since

$$\frac{p(p+1)(5p+7)}{6} = (p+1)^3 - \frac{(p+1)(p+2)(p+3)}{6} \tag{2.91}$$

However, as already noted for quadrilateral cells, the simple expression of the polynomials (2.90) and the symmetry of the array of the interpolation nodes are advantageous from the computational viewpoint.

Since, without justification, the expression (2.90) favors the independent parent variables ξ_1, ξ_2, and ξ_3, it is preferred to write the Lagrangian basis set in terms of the $(p+1)^3$ polynomials

$$\begin{aligned} \alpha_{i\ell;jm;kn}(p,\boldsymbol{\xi}) = {} & R_i(p,\xi_1)\,R_j(p,\xi_2)\,R_k(p,\xi_3) \\ & R_\ell(p,\xi_4)\,R_m(p,\xi_5)\,R_n(p,\xi_6) \end{aligned} \tag{2.92}$$

expressed in terms of Silvester polynomials and obtained for $i, j, k, \ell, m, n = 0, 1, \ldots, p$, with $i+\ell = j+m = k+n = p$. In (2.92), a sextuple indexing scheme is used since these polynomials interpolate the points with normalized coordinates

$$(\xi_1, \xi_4; \xi_2, \xi_5; \xi_3, \xi_6) = \left(\frac{i}{p}, \frac{\ell}{p}; \frac{j}{p}, \frac{m}{p}; \frac{k}{p}, \frac{n}{p}\right) = \boldsymbol{\xi}_{i\ell;jm;kn} \tag{2.93}$$

The Lagrangian basis functions $\alpha_{i\ell;jm;kn}$ can be used to map the parent brick into the child space. A Lagrange parameterization of order q for a curvilinear brick can for example be expressed as

$$\boldsymbol{r} = \sum_{i,j,k,\ell,m,n=0}^{q} \boldsymbol{r}_{i\ell;jm;kn}\,\alpha_{i\ell;jm;kn}(q,\xi_1,\xi_2,\xi_3,\xi_4,\xi_5,\xi_6) \qquad \text{with } i+\ell=j+m=k+n=q \tag{2.94}$$

and where the sextuple index is used to label the position vector $\boldsymbol{r}_{i\ell;jm;kn}$ interpolating the point with normalized coordinates $\boldsymbol{\xi}_{i\ell;jm;kn} = \left(\frac{i}{q},\frac{\ell}{q};\frac{j}{q},\frac{m}{q};\frac{k}{q},\frac{n}{q}\right)$.

Regardless of the geometrical parameterization used for the brick, the *unitary base vectors*, derived from the *independent* coordinates ξ_1, ξ_2, and ξ_3, are (see Section 3.14)

$$\begin{cases} \boldsymbol{\ell}^1 = \dfrac{\partial \boldsymbol{r}(\xi_1,\xi_2,\xi_3)}{\partial \xi_1} = \left[\dfrac{\partial \boldsymbol{r}(\boldsymbol{\xi})}{\partial \xi_1} - \dfrac{\partial \boldsymbol{r}(\boldsymbol{\xi})}{\partial \xi_4}\right]_{\xi_2,\xi_3 \text{ constant}} \\ \boldsymbol{\ell}^2 = \dfrac{\partial \boldsymbol{r}(\xi_1,\xi_2,\xi_3)}{\partial \xi_2} = \left[\dfrac{\partial \boldsymbol{r}(\boldsymbol{\xi})}{\partial \xi_2} - \dfrac{\partial \boldsymbol{r}(\boldsymbol{\xi})}{\partial \xi_5}\right]_{\xi_1,\xi_3 \text{ constant}} \\ \boldsymbol{\ell}^3 = \dfrac{\partial \boldsymbol{r}(\xi_1,\xi_2,\xi_3)}{\partial \xi_3} = \left[\dfrac{\partial \boldsymbol{r}(\boldsymbol{\xi})}{\partial \xi_3} - \dfrac{\partial \boldsymbol{r}(\boldsymbol{\xi})}{\partial \xi_6}\right]_{\xi_1,\xi_2 \text{ constant}} \end{cases} \tag{2.95}$$

from which the edge vectors of Figure 2.23 are found to be

$$\begin{cases} \boldsymbol{\ell}_{12} = \boldsymbol{\ell}_{24} = \boldsymbol{\ell}_{45} = -\boldsymbol{\ell}_{15} = \boldsymbol{\ell}^3 \\ \boldsymbol{\ell}_{16} = -\boldsymbol{\ell}_{13} = -\boldsymbol{\ell}_{34} = -\boldsymbol{\ell}_{46} = \boldsymbol{\ell}^2 \\ \boldsymbol{\ell}_{23} = \boldsymbol{\ell}_{35} = \boldsymbol{\ell}_{56} = -\boldsymbol{\ell}_{26} = \boldsymbol{\ell}^1 \end{cases} \tag{2.96}$$

The gradient vectors are determined in terms of the unitary base vectors as (see Section 3.14)

$$\begin{cases} \nabla\xi_1 = -\nabla\xi_4 = \dfrac{\boldsymbol{\ell}^2 \times \boldsymbol{\ell}^3}{\mathcal{J}} \\ \nabla\xi_2 = -\nabla\xi_5 = \dfrac{\boldsymbol{\ell}^3 \times \boldsymbol{\ell}^1}{\mathcal{J}} \\ \nabla\xi_3 = -\nabla\xi_6 = \dfrac{\boldsymbol{\ell}^1 \times \boldsymbol{\ell}^2}{\mathcal{J}} \end{cases} \tag{2.97}$$

where

$$\mathcal{J} = \boldsymbol{\ell}^1 \cdot \boldsymbol{\ell}^2 \times \boldsymbol{\ell}^3 \tag{2.98}$$

is the Jacobian of the transformation from parent to child coordinates. All these geometrical quantities, including the Jacobian, vary with position unless the brick is mapped into a parallelepiped.

Figure 2.23 illustrates the various edge and height vectors and normalized coordinates defined for a brick element.

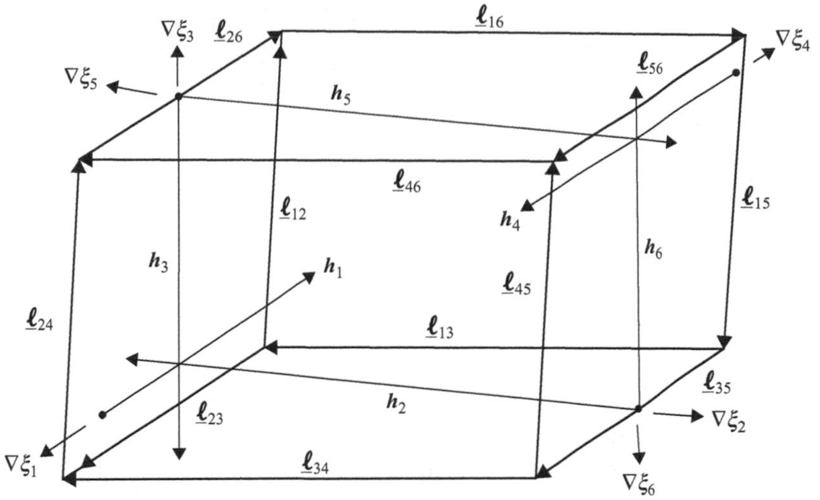

Figure 2.23 Edge, height, and gradient vectors for brick elements.

© 1997 IEEE. Reprinted with permission from R. D. Graglia, D. R. Wilton, and A. F. Peterson, "Higher order interpolatory vector bases for computational electromagnetics," special issue on "Advanced Numerical Techniques in Electromagnetics," *IEEE Trans. Antennas Propag.*, vol. 45, no. 3, pp. 329–342, Mar. 1997.

2.7.2 Lagrangian Basis Functions

2.7.2.1 The Shape Polynomials $\alpha(p, \xi)$ to Interpolate All the Edges and Faces

To fully exploit the symmetry of the brick cell, we find convenient to introduce the six *dummy* parent variables

$$\boldsymbol{\xi} = (\xi_a, \xi_d; \xi_b, \xi_e; \xi_c, \xi_f) \tag{2.99}$$

with

$$\xi_d = 1 - \xi_a, \quad \xi_e = 1 - \xi_b, \quad \xi_f = 1 - \xi_c \tag{2.100}$$

to write (for $i, j, k, \ell, m, n = 0, 1, \ldots, p$ with $i + \ell = j + m = k + n = p$) the scalar basis function

$$\begin{aligned}\alpha_{i\ell;jm;kn}(p, \boldsymbol{\xi}) = {} & R_i(p, \xi_a)\, R_\ell(p, \xi_d) \\ & R_j(p, \xi_b)\, R_m(p, \xi_e) \\ & R_k(p, \xi_c)\, R_n(p, \xi_f)\end{aligned} \tag{2.101}$$

interpolating the points with normalized coordinates

$$(\xi_a, \xi_d; \xi_b, \xi_e; \xi_c, \xi_f) = \left(\frac{i}{p}, \frac{\ell}{p}; \frac{j}{p}, \frac{m}{p}; \frac{k}{p}, \frac{n}{p}\right) = \boldsymbol{\xi}_{i\ell;jm;kn} \tag{2.102}$$

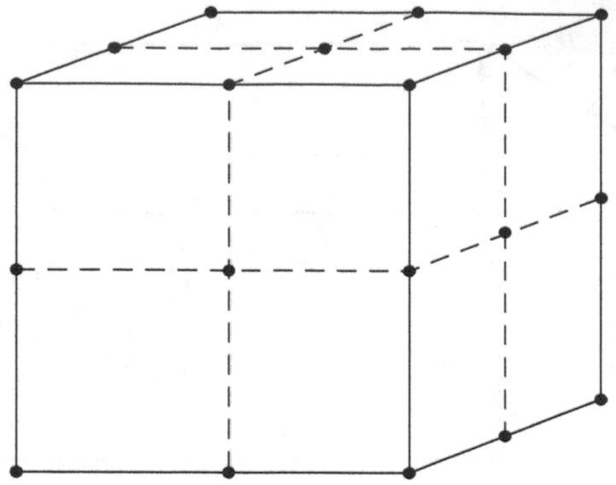

Figure 2.24 Lagrange interpolation of order $p = 2$ on a brick. The figure shows the interpolation nodes associated with the interpolatory polynomials $\alpha_{i\ell;jm;kn}(2, \boldsymbol{\xi})$ (hidden interpolation points omitted for clarity).

in terms of *dummy* indices (i, j, k, ℓ, m, n) associated with the dummy variables (ξ_a, ξ_b, ξ_c, ξ_d, ξ_e, ξ_f). These dummy variables represent permutation of the parent variables (ξ_1, ξ_2, ξ_3, ξ_4, ξ_5, ξ_6) and allow us to express the full collection of basis functions in terms of a minimal set. Obviously, (2.101) and (2.92) coincide for (ξ_a, ξ_b, ξ_c, ξ_d, ξ_e, ξ_f) equal to (ξ_1, ξ_2, ξ_3, ξ_4, ξ_5, ξ_6). The interpolation points associated with $\alpha_{i\ell;jm;kn}(p, \boldsymbol{\xi})$ for $p = 2$ are shown in Figure 2.24.

The conformity of the α basis functions of the brick on the quadrilateral face of the brick itself with the α basis functions of the same order defined for the quadrilateral cell is readily proved. Table 2.6 reports the interpolating functions $\alpha_{ijk\ell}(p, \boldsymbol{\xi})$ for bricks, up to $p = 3$.

2.7.2.2 Lagrange Interpolation and Gradient Approximation

For a pth-order Lagrangian interpolation of a function f on the brick, one has

$$\tilde{f} = \sum_{i,j,k,\ell,m,n=0}^{p} f_{i\ell;jm;kn}\, \alpha_{i\ell;jm;kn}(p, \boldsymbol{\xi}) \tag{2.103}$$

$$\nabla\tilde{f} = \sum_{i,j,k,\ell,m,n=0}^{p} f_{i\ell;jm;kn}\, \nabla\alpha_{i\ell;jm;kn}(p, \boldsymbol{\xi}) \tag{2.104}$$

where

$$\begin{aligned}\nabla\alpha_{i\ell;jm;kn}(p, \boldsymbol{\xi}) = {} & R_j(p, \xi_2)\, R_m(p, \xi_5)\, R_k(p, \xi_3)\, R_n(p, \xi_6) \left[R_i'(p, \xi_1)\, \nabla\xi_1 + R_\ell'(p, \xi_4)\, \nabla\xi_4\right] \\ & + R_i(p, \xi_1)\, R_\ell(p, \xi_4)\, R_k(p, \xi_3)\, R_n(p, \xi_6) \left[R_j'(p, \xi_2)\, \nabla\xi_2 + R_m'(p, \xi_5)\, \nabla\xi_5\right] \\ & + R_i(p, \xi_1)\, R_\ell(p, \xi_4)\, R_j(p, \xi_2)\, R_m(p, \xi_5) \left[R_k'(p, \xi_3)\, \nabla\xi_3 + R_n'(p, \xi_6)\, \nabla\xi_6\right]\end{aligned} \tag{2.105}$$

Table 2.6 The first three orders of the shape polynomials interpolating all the edges and faces of a brick cell.

The $(p+1)^3$ polynomials $\alpha_{i\ell;jm;kn}(p,\boldsymbol{\xi}) = R_i(p,\xi_1)\,R_\ell(p,\xi_4)\;R_j(p,\xi_2)\,R_m(p,\xi_5)\;R_k(p,\xi_3)\,R_n(p,\xi_6)$ that interpolate a brick cell are obtained for $i,j,k,\ell,m,n = 0,1,\ldots,p$, with $i+\ell = j+m = k+n = p$ and $p \geq 1$. Each of these polynomials is of order $3p$. Because of the symmetry of the brick cell it is sufficient to report, for each family, only a smaller set of polynomials that is then "completed" by other polynomials, obtained from those of the smaller set by permutations of the parent variables. The smaller set of polynomials is more conveniently written in terms of the dummy brick parent variables $\{\xi_a, \xi_d; \xi_b, \xi_e; \xi_c, \xi_f\}$. Each complete interpolatory set is then obtained by setting, in the polynomials reported below:

$\{\xi_a, \xi_d; \xi_b, \xi_e; \xi_c, \xi_f\} = \{\xi_1, \xi_4; \xi_2, \xi_5; \xi_3, \xi_6\}$ first set of permuted variables;
$\{\xi_a, \xi_d; \xi_b, \xi_e; \xi_c, \xi_f\} = \{\xi_1, \xi_4; \xi_2, \xi_5; \xi_6, \xi_3\}$ second set of permuted variables;
$\{\xi_a, \xi_d; \xi_b, \xi_e; \xi_c, \xi_f\} = \{\xi_1, \xi_4; \xi_5, \xi_2; \xi_3, \xi_6\}$ third set of permuted variables;
$\{\xi_a, \xi_d; \xi_b, \xi_e; \xi_c, \xi_f\} = \{\xi_1, \xi_4; \xi_5, \xi_2; \xi_6, \xi_3\}$ fourth set of permuted variables;
$\{\xi_a, \xi_d; \xi_b, \xi_e; \xi_c, \xi_f\} = \{\xi_4, \xi_1; \xi_2, \xi_5; \xi_3, \xi_6\}$ fifth set of permuted variables;
$\{\xi_a, \xi_d; \xi_b, \xi_e; \xi_c, \xi_f\} = \{\xi_4, \xi_1; \xi_2, \xi_5; \xi_6, \xi_3\}$ sixth set of permuted variables;
$\{\xi_a, \xi_d; \xi_b, \xi_e; \xi_c, \xi_f\} = \{\xi_4, \xi_1; \xi_5, \xi_2; \xi_3, \xi_6\}$ seventh set of permuted variables;
$\{\xi_a, \xi_d; \xi_b, \xi_e; \xi_c, \xi_f\} = \{\xi_4, \xi_1; \xi_5, \xi_2; \xi_6, \xi_3\}$ eighth set of permuted variables.

All the brick polynomials are thus defined by using eight different sets of permuted variables; notice how the number of the permuted sets equals the number of the cell corner nodes.

$p = 1$ (8 functions in total, use set 1–8)

$\xi_a\,\xi_b\,\xi_c$

$p = 2$ (27 functions in total)

$\xi_a(2\xi_a - 1)\,\xi_b(2\xi_b - 1)\,\xi_c(2\xi_c - 1)$	(8 functions, use set 1–8)
$2^2\,\xi_a(2\xi_a - 1)\,\xi_b(2\xi_b - 1)\,\xi_3\,\xi_6$	(4 functions, use set 1, 3, 5, & 7)
$2^2\,\xi_b(2\xi_b - 1)\,\xi_c(2\xi_c - 1)\,\xi_1\,\xi_4$	(4 functions, use set 1–4)
$2^2\,\xi_c(2\xi_c - 1)\,\xi_a(2\xi_a - 1)\,\xi_2\,\xi_5$	(4 functions, use set 1, 2, 5, & 6)
$2^4\,\xi_a(2\xi_a - 1)\,\xi_2\,\xi_5\,\xi_3\,\xi_6$	(2 functions, use set 4 & 5)
$2^4\,\xi_b(2\xi_b - 1)\,\xi_3\,\xi_6\,\xi_1\,\xi_4$	(2 functions, use set 2 & 3)
$2^4\,\xi_c(2\xi_c - 1)\,\xi_1\,\xi_4\,\xi_2\,\xi_5$	(2 functions, use set 1 & 2)
$2^6\,\xi_1\xi_4\,\xi_2\,\xi_5\,\xi_3\,\xi_6$	(1 function, set 1)

$p = 3$ (64 functions in total, use set 1–8)

$\xi_a(3\xi_a - 1)(3\xi_a - 2)\,\xi_b(3\xi_b - 1)(3\xi_b - 2)\,\xi_c(3\xi_c - 1)(3\xi_c - 2)/2^3$	(8 functions)
$3^2\,\xi_a(3\xi_a - 1)(3\xi_a - 2)\,\xi_b(3\xi_b - 1)(3\xi_b - 2)\,\xi_c(3\xi_c - 1)\xi_f/2^3$	(8 functions)
$3^2\,\xi_b(3\xi_b - 1)(3\xi_b - 2)\,\xi_c(3\xi_c - 1)(3\xi_c - 2)\,\xi_a(3\xi_a - 1)\xi_d/2^3$	(8 functions)
$3^2\,\xi_c(3\xi_c - 1)(3\xi_c - 2)\,\xi_a(3\xi_a - 1)(3\xi_a - 2)\,\xi_b(3\xi_b - 1)\xi_e/2^3$	(8 functions)
$3^4\,\xi_a(3\xi_a - 1)(3\xi_a - 2)\,\xi_b(3\xi_b - 1)\xi_e\,\xi_c(3\xi_c - 1)\xi_f/2^3$	(8 functions)
$3^4\,\xi_b(3\xi_b - 1)(3\xi_b - 2)\,\xi_c(3\xi_c - 1)\xi_f\,\xi_a(3\xi_a - 1)\xi_d/2^3$	(8 functions)
$3^4\,\xi_c(3\xi_c - 1)(3\xi_c - 2)\,\xi_a(3\xi_a - 1)\xi_d\,\xi_b(3\xi_b - 1)\xi_e/2^3$	(8 functions)
$3^6\,\xi_a(3\xi_a - 1)\xi_d\,\xi_b(3\xi_b - 1)\xi_e\,\xi_c(3\xi_c - 1)\xi_f/2^3$	(8 functions)

with $i+\ell=j+m=k+n=p$ and $p\geq 1$, and where the sextuple index is used to label the interpolation points with normalized coordinates

$$(\xi_1, \xi_4;\ \xi_2, \xi_5;\ \xi_3, \xi_6) = \left(\frac{i}{p}, \frac{\ell}{p};\ \frac{j}{p}, \frac{m}{p};\ \frac{k}{p}, \frac{n}{p}\right) = \boldsymbol{\xi}_{i\ell;jm;kn} \tag{2.106}$$

The gradient vectors $\nabla\xi_s$ (for $s=1,2,\ldots,6$) are given in (2.97).

2.8 Triangular Prism Cells

2.8.1 Cell Geometry Representation and Local Vector Bases

The local, normalized coordinates $\boldsymbol{\xi}=(\xi_1, \xi_2, \xi_3; \xi_4, \xi_5)$ for prisms (see Figure 2.25) are related by

$$\begin{aligned}\xi_1+\xi_2+\xi_3&=1\\ \xi_4+\xi_5&=1\end{aligned} \tag{2.107}$$

where we have taken ξ_1, ξ_2, and ξ_4 as independent coordinates. The prism's triangular cross section is parameterized by the "usual" dependent triangle area coordinates ξ_1, ξ_2, and ξ_3 while ξ_4 is taken as the third independent coordinate, with $\nabla\xi_4\cdot(\nabla\xi_1\times\nabla\xi_2)$ positive.

A pth-order complete set of Lagrangian scalar basis functions for the prism cell is formed by the $(p+1)^2(p+2)/2$ polynomials

$$\alpha_{ijk;\ell m}(p, \boldsymbol{\xi}) = R_i(p,\xi_1)\,R_j(p,\xi_2)\,R_k(p,\xi_3)\,R_\ell(p,\xi_4)\,R_m(p,\xi_5) \tag{2.108}$$

obtained for $i, j, k, \ell, m=0,1,\ldots,p$, with $i+j+k=\ell+m=p$ and $p\geq 1$. The index quintuplet labeling these polynomials denotes the fact that they interpolate the points with normalized coordinates

$$\boldsymbol{\xi}_{ijk;\ell m} = (\xi_1, \xi_2, \xi_3; \xi_4, \xi_5) = \left(\frac{i}{p}, \frac{j}{p}, \frac{k}{p}; \frac{\ell}{p}, \frac{m}{p}\right) \tag{2.109}$$

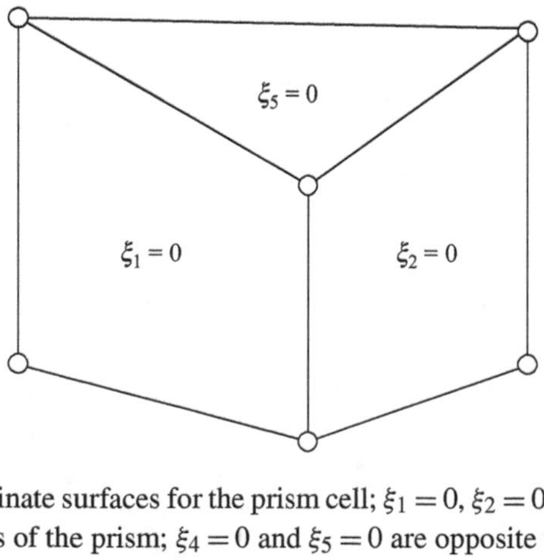

Figure 2.25 Zero-coordinate surfaces for the prism cell; $\xi_1=0$, $\xi_2=0$, and $\xi_3=0$ are the quadrilateral faces of the prism; $\xi_4=0$ and $\xi_5=0$ are opposite triangular faces.

The set (2.108) has $p(p+1)(p+2)/3$ more terms than the minimum required to form a complete three-dimensional scalar polynomial family, since

$$\frac{p(p+1)(p+2)}{3} = \frac{(p+1)^2(p+2)}{2} - \frac{(p+1)(p+2)(p+3)}{6} \tag{2.110}$$

As previously noticed, the number of Lagrangian interpolatory functions is equal to the minimum only for simplices, that is, for triangular and tetrahedral cells. For the other non-simplex cells (including the quadrilateral, the brick, and the prism), a symmetric grid of interpolation points and the simplest expression of the interpolatory polynomials is only obtained by including the "extra" polynomial terms (for the prism cell, there are $p(p+1)(p+2)/3$ extra terms).

The Lagrangian basis functions $\alpha_{ijk;\ell m}$ can be used to map the parent prism into the child space. A Lagrange parameterization of order q for a curvilinear prism can for example be expressed as

$$\boldsymbol{r} = \sum_{i,j,k,\ell,m=0}^{q} \boldsymbol{r}_{ijk;\ell m}\, \alpha_{ijk;\ell m}(q, \xi_1, \xi_2, \xi_3, \xi_4, \xi_5), \quad i+j+k = \ell + m = q \tag{2.111}$$

where the index quintuplet is again used to label the position vector $\boldsymbol{r}_{ijk;\ell m}$ interpolating the point with normalized coordinates $\boldsymbol{\xi}_{ijk;\ell m} = \left(\frac{i}{q}, \frac{j}{q}, \frac{k}{q}; \frac{\ell}{q}, \frac{m}{q}\right)$.

For any geometrical parameterization of the prism, the *unitary base vectors*, derived from the independent coordinates ξ_1, ξ_2 and ξ_4 are (see Section 3.14)

$$\begin{cases} \boldsymbol{\ell}^1 = \dfrac{\partial \boldsymbol{r}(\xi_1, \xi_2, \xi_4)}{\partial \xi_1} = \left[\dfrac{\partial \boldsymbol{r}(\boldsymbol{\xi})}{\partial \xi_1} - \dfrac{\partial \boldsymbol{r}(\boldsymbol{\xi})}{\partial \xi_3}\right]_{\xi_2, \xi_4 \text{ constant}} \\ \boldsymbol{\ell}^2 = \dfrac{\partial \boldsymbol{r}(\xi_1, \xi_2, \xi_4)}{\partial \xi_2} = \left[\dfrac{\partial \boldsymbol{r}(\boldsymbol{\xi})}{\partial \xi_2} - \dfrac{\partial \boldsymbol{r}(\boldsymbol{\xi})}{\partial \xi_3}\right]_{\xi_1, \xi_4 \text{ constant}} \\ \boldsymbol{\ell}^4 = \dfrac{\partial \boldsymbol{r}(\xi_1, \xi_2, \xi_4)}{\partial \xi_4} = \left[\dfrac{\partial \boldsymbol{r}(\boldsymbol{\xi})}{\partial \xi_4} - \dfrac{\partial \boldsymbol{r}(\boldsymbol{\xi})}{\partial \xi_5}\right]_{\xi_1, \xi_2 \text{ constant}} \end{cases} \tag{2.112}$$

from which the edge vectors of Figure 2.26 are found to be

$$\begin{cases} \boldsymbol{\ell}_{12} = -\boldsymbol{\ell}_{13} = \boldsymbol{\ell}_{23} = \boldsymbol{\ell}^4 \\ -\boldsymbol{\ell}_{14} = \boldsymbol{\ell}_{15} = \boldsymbol{\ell}^2 \\ \boldsymbol{\ell}_{24} = -\boldsymbol{\ell}_{25} = \boldsymbol{\ell}^1 \\ \boldsymbol{\ell}_{34} = -\boldsymbol{\ell}_{35} = \boldsymbol{\ell}^2 - \boldsymbol{\ell}^1 \end{cases} \tag{2.113}$$

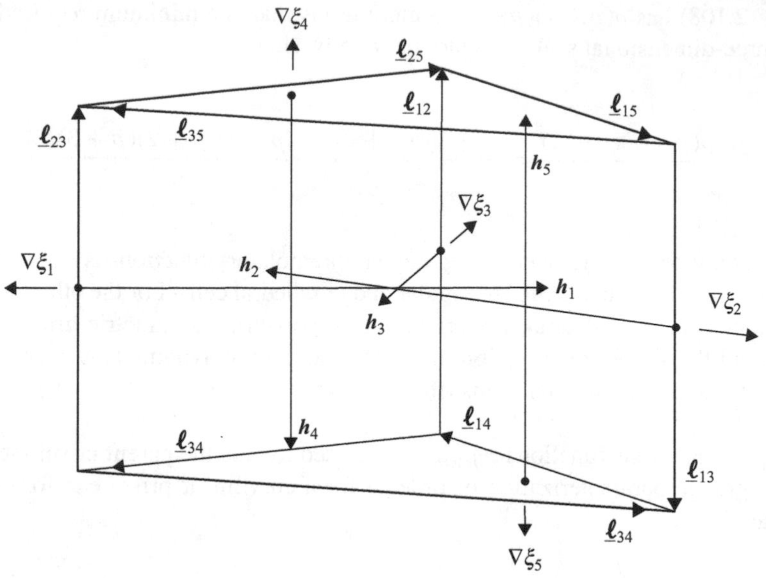

Figure 2.26 Edge, height, and gradient vectors for prism elements.

The reciprocal base or gradient vectors are determined in terms of the unitary base vectors as (see Section 3.11)

$$\begin{cases} \nabla\xi_1 = \dfrac{\boldsymbol{\ell}^2 \times \boldsymbol{\ell}^4}{\mathcal{J}} \\ \nabla\xi_2 = \dfrac{\boldsymbol{\ell}^4 \times \boldsymbol{\ell}^1}{\mathcal{J}}, \quad \nabla\xi_3 = -\nabla\xi_1 - \nabla\xi_2 \\ \nabla\xi_4 = \dfrac{\boldsymbol{\ell}^1 \times \boldsymbol{\ell}^2}{\mathcal{J}}, \quad \nabla\xi_5 = -\nabla\xi_4 \end{cases} \tag{2.114}$$

where

$$\mathcal{J} = \boldsymbol{\ell}^1 \cdot \boldsymbol{\ell}^2 \times \boldsymbol{\ell}^4 \tag{2.115}$$

is the Jacobian of the transformation from parent to child coordinates. All these geometrical quantities, including the Jacobian, vary with position unless the child-space prism is a eight-prism.

Figure 2.26 illustrates the various edge and height vectors and normalized coordinates defined for a prism element.

2.8.2 Lagrangian Basis Functions

2.8.2.1 The Shape Polynomials $\alpha(p, \xi)$ to Interpolate All the Edges and Faces

To fully exploit the symmetry of the prism cell, we find convenient to introduce the five *dummy* parent variables

$$\boldsymbol{\xi} = (\xi_a, \xi_b, \xi_c; \xi_d, \xi_e) \tag{2.116}$$

with

$$\xi_c = 1 - \xi_a - \xi_b, \quad \xi_e = 1 - \xi_d \tag{2.117}$$

to write (for $i, j, k, \ell, m = 0, 1, \ldots, p$ with $i + j + k = \ell + m = p$) the scalar basis function

$$\alpha_{ijk;\ell m}(p, \boldsymbol{\xi}) = R_i(p, \xi_a)\, R_j(p, \xi_b)\, R_k(p, \xi_c) R_\ell(p, \xi_d)\, R_m(p, \xi_e) \tag{2.118}$$

interpolating the points with normalized coordinates

$$\boldsymbol{\xi}_{ijk;\ell m} = (\xi_a, \xi_b, \xi_c; \xi_d, \xi_e) = \left(\frac{i}{p}, \frac{j}{p}, \frac{k}{p}; \frac{\ell}{p}, \frac{m}{p}\right) \tag{2.119}$$

in terms of *dummy* indices $(i, j, k; \ell, m)$ associated with the dummy variables (ξ_a, ξ_b, ξ_c; ξ_d, ξ_e). These dummy variables represent permutation of the parent variables (ξ_1, ξ_2, ξ_3; ξ_4, ξ_5) and allow us to express the full collection of basis functions in terms of a minimal set. The quadrilateral faces of the "*dummy*" prism are $\xi_a = 0$, $\xi_b = 0$, and $\xi_c = 0$, while $\xi_d = 0$ and $\xi_e = 0$ are the two opposite triangular faces. The interpolation points associated with $\alpha_{ijk;\ell m}(p, \boldsymbol{\xi})$ for $p = 2$ are shown in Figure 2.27. Table 2.7 reports the interpolating functions $\alpha_{ijk\ell}(p, \boldsymbol{\xi})$ for prisms, up to $p = 3$.

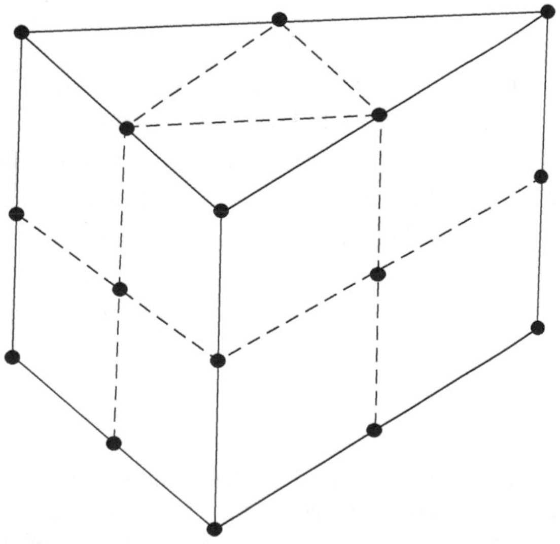

Figure 2.27 Lagrange interpolation of order $p = 2$ on a prism. The figure shows the interpolation nodes associated with the interpolatory polynomials $\alpha_{ijk;\ell m}(2, \boldsymbol{\xi})$.

Table 2.7 The first three orders of the shape polynomials interpolating all the edges and faces of a prism cell.

The $(p+1)^2(p+2)/2$ polynomials $\alpha_{ijk;\ell m}(p,\boldsymbol{\xi})=R_i(p,\xi_1)\,R_j(p,\xi_2)\,R_k(p,\xi_3)\,R_\ell(p,\xi_4)\,R_m(p,\xi_5)$ that interpolate a prism cell are obtained for $i,j,k,\ell,m=0,1,\ldots,p$, with $i+j+k=\ell+m=p$ and $p\geq 1$. Each of these polynomials is of order $2p$. In terms of dummy parent variables $\{\xi_a,\xi_b,\xi_c;\xi_d,\xi_e\}$, each polynomial is the product of a triangular function $R_i(p,\xi_a)\,R_j(p,\xi_b)\,R_k(p,\xi_c)$ with a Lagrangian function $R_\ell(p,\xi_d)\,R_m(p,\xi_e)$. Thus, each pth-order complete set can be obtained by cyclic permutations of the triangular parent variables $\{\xi_a,\xi_b,\xi_c\}$ and of the $\{\xi_d,\xi_e\}$ variables, that is by setting in the polynomials reported below

$\{\xi_a,\xi_b,\xi_c;\xi_d,\xi_e\}=\{\xi_1,\xi_2,\xi_3;\xi_4,\xi_5\}$ first set of permuted variables;
$\{\xi_a,\xi_b,\xi_c;\xi_d,\xi_e\}=\{\xi_2,\xi_3,\xi_1;\xi_4,\xi_5\}$ second set of permuted variables;
$\{\xi_a,\xi_b,\xi_c;\xi_d,\xi_e\}=\{\xi_3,\xi_1,\xi_2;\xi_4,\xi_5\}$ third set of permuted variables;
$\{\xi_a,\xi_b,\xi_c;\xi_d,\xi_e\}=\{\xi_1,\xi_2,\xi_3;\xi_5,\xi_4\}$ fourth set of permuted variables;
$\{\xi_a,\xi_b,\xi_c;\xi_d,\xi_e\}=\{\xi_2,\xi_3,\xi_1;\xi_5,\xi_4\}$ fifth set of permuted variables;
$\{\xi_a,\xi_b,\xi_c;\xi_d,\xi_e\}=\{\xi_3,\xi_1,\xi_2;\xi_5,\xi_4\}$ sixth set of permuted variables.

All the prism polynomials are thus defined by using six different sets of permuted variables; notice how the number of the permuted sets equals the number of the cell corner nodes.

$p=1$ (6 functions in total, use set 1–6)	
$\xi_a\xi_d$ (6 functions)	
$p=2$ (18 functions in total)	
$\xi_a(2\xi_a-1)\,\xi_d(2\xi_d-1)$	(6 functions, use set 1–6)
$2^2\,\xi_a\xi_b\,\xi_d(2\xi_d-1)$	(6 functions, use set 1–6)
$2^2\,\xi_a(2\xi_a-1)\,\xi_4\xi_5$	(3 functions, use set 1–3)
$2^4\,\xi_a\xi_b\,\xi_4\xi_5$	(3 functions, use set 1–3)
$p=3$ (40 functions in total)	
$\xi_a(3\xi_a-1)(3\xi_a-2)\,\xi_d(3\xi_d-1)(3\xi_d-2)/4$	(6 functions, use set 1–6)
$3^2\,\xi_a\xi_b(3\xi_b-1)\,\xi_d(3\xi_d-1)(3\xi_d-2)/4$	(6 functions, use set 1–6)
$3^2\,\xi_b\xi_a(3\xi_a-1)\,\xi_d(3\xi_d-1)(3\xi_d-2)/4$	(6 functions, use set 1–6)
$3^2\,\xi_a(3\xi_a-1)(3\xi_a-2)\,\xi_d\xi_e(3\xi_e-1)/4$	(6 functions, use set 1–6)
$3^4\,\xi_a\xi_b(3\xi_b-1)\,\xi_d\xi_e(3\xi_e-1)/4$	(6 functions, use set 1–6)
$3^4\,\xi_b\xi_a(3\xi_a-1)\,\xi_d\xi_e(3\xi_e-1)/4$	(6 functions, use set 1–6)
$3^3\,\xi_1\xi_2\xi_3\,\xi_d(3\xi_d-1)(3\xi_d-2)/2$	(2 functions, use set 1 & 4)
$3^5\,\xi_1\xi_2\xi_3\,\xi_d\xi_e(3\xi_e-1)/2$	(2 functions, use set 1 & 4)

The conformity of the α basis functions of the prism on the quadrilateral (or on the triangular) face of the prism itself with the basis functions of the same order defined for the quadrilateral (or the triangular) cell is easily demonstrated by using dummy parent variables.

2.8.2.2 Lagrange Interpolation and Gradient Approximation

For a pth-order Lagrangian interpolation of a function f on the prism (with $p \geq 1$), one has

$$\widetilde{f} = \sum_{i,j,k,\ell,m=0}^{p} f_{ijk;\ell m}\, \alpha_{ijk;\ell m}(p, \boldsymbol{\xi}) \tag{2.120}$$

$$\nabla\widetilde{f} = \sum_{i,j,k,\ell,m=0}^{p} f_{ijk;\ell m}\, \nabla\alpha_{ijk;\ell m}(p, \boldsymbol{\xi}) \tag{2.121}$$

$$\begin{aligned}\nabla\alpha_{ijk;\ell m}(p, \boldsymbol{\xi}) = {} & \left[R_i'(p, \xi_1)\, \nabla\xi_1 + R_j'(p, \xi_2)\, \nabla\xi_2 + R_k'(p, \xi_3)\, \nabla\xi_3\right] R_\ell(p, \xi_4)\, R_m(p, \xi_5) \\ & + R_i(p, \xi_1)\, R_j(p, \xi_2)\, R_k(p, \xi_3) \left[R_\ell'(p, \xi_4)\, \nabla\xi_4 + R_m'(p, \xi_5)\, \nabla\xi_5\right]\end{aligned} \tag{2.122}$$

with $i+j+k=\ell+m=p$ and where the index quintuplet $(i, j, k; \ell, m)$ is used to label the interpolation points with normalized coordinates

$$(\xi_1, \xi_2, \xi_3;\, \xi_4, \xi_5) = \left(\frac{i}{p}, \frac{j}{p}, \frac{k}{p}; \frac{\ell}{p}, \frac{m}{p}\right) = \boldsymbol{\xi}_{ijk;\ell m}. \tag{2.123}$$

The gradient vectors $\nabla\xi_s$ (for $s = 1, 2, \ldots, 5$) are given in (2.114).

2.9 Generation of Shape Functions

In this chapter, the shape polynomials for the fundamental cells for the linear, quadratic, and cubic case are presented in compact expressions in terms of dummy parent variables. In actual computer codes, these shape functions are usually computed "on the fly" and, consequently, there is no need to implement the polynomials reported in Tables 2.2 and 2.4–2.7. In fact, the (1D) Silvester polynomials used to define the shape functions (as well as their first derivatives with respect to ξ) are conveniently computed using very simple routines that implement the recurrence relations (2.4)–(2.6) given in Section 2.2. The shape polynomials are then computed from the knowledge of the ordered indexes that distinguish each interpolation point and shape polynomial.

References

[1] P. L. George, *Automatic Mesh Generation*, New York, NY: Wiley, 1991.

[2] G. F. Carey, *Computational Grids*, Washington, DC: Taylor and Francis, 1997.

[3] J. F. Thompson, B. K. Soni, and N. P. Weatherill, eds., *Handbook of Grid Generation*, Boca Raton, FL: CRC Press, 1999.

[4] P. P. Silvester and R. L. Ferrari, *Finite Elements for Electrical Engineers*, Cambridge: Cambridge Press, 1990.

[5] R. D. Graglia, D. R. Wilton, and A. F. Peterson, "Higher order interpolatory vector bases for computational electromagnetics," special issue on "Advanced Numerical Techniques in Electromagnetics," *IEEE Trans. Antennas Propag.*, vol. 45, no. 3, pp. 329–342, Mar. 1997.

[6] R. D. Graglia, D. R. Wilton, A. F. Peterson, and I.-L. Gheorma, "Higher order interpolatory vector bases on prism elements," *IEEE Trans. Antennas Propag.*, vol. 46, no. 3, pp. 442–450, Mar. 1998.

CHAPTER 3

Representation of Vector Fields in Two and Three Dimensions Using Low-Degree Polynomials

In this chapter, we consider the representation of vector functions (often referred to as "vector fields") with low-order (constant and linear) polynomial basis functions on simple cells, such as triangles or quadrilateral cells in two dimensions or tetrahedrons and bricks in three dimensions. As will soon be apparent, there are multiple ways of defining vector basis functions, and therefore the approach requires some consideration.

The proper representation of a function depends on what will be done with it—do we need to compute the curl of the function, for instance? If so, the representation might be different than if we need to compute the divergence of the function. We use the term *curl conforming* to denote the space of vector functions that maintain first-order tangential-vector continuity throughout the domain and can be differentiated via the curl operation, without producing unbounded or generalized functions (Dirac delta functions) in the process. The term *divergence conforming* is used to denote the complementary space of vector functions that maintain first-order normal-vector continuity throughout the domain and can therefore be differentiated via the divergence operation. (First order or C_0 continuity is continuity of the function itself, but not necessarily continuity of its first derivatives.) The simple low-order polynomial vector basis functions in widespread use are either curl conforming or divergence conforming; seldom we will use functions that maintain complete continuity and belong to both the curl-conforming and divergence-conforming spaces, although it is possible to define such functions.

3.1 Two-Dimensional Vector Functions on Triangles

Suppose we want to represent a vector function in two-dimensional (2D) (x, y) space. If each component is represented by a constant, such a function has the general form

$$f(x,y) = a_0\hat{\boldsymbol{x}} + b_0\hat{\boldsymbol{y}} \tag{3.1}$$

involving two DoFs. If represented to linear order, the general form is

$$f(x,y) = (a_0 + a_1x + a_2y)\hat{\boldsymbol{x}} + (b_0 + b_1x + b_2y)\hat{\boldsymbol{y}} \tag{3.2}$$

containing six DoFs. Continuing in this manner, a quadratic representation requires 12 DoFs, and so on.

Now suppose that we want to assign these DoFs to a subsectional representation on triangular-cell shapes, where each cell has a similar representation. One can associate each DoF with samples of the components of the function at some point within the triangular cell, or at some point on the triangle edges, or at one of the triangle nodes. In the constant case, there are two DoFs, which can be associated with independent vector components within the cell (the specific location does not matter since a constant only has one value, but the vector direction of each does matter and must be specified). This specification is somewhat arbitrary and may or may not be done in terms of the $\hat{\boldsymbol{x}}$ and $\hat{\boldsymbol{y}}$ components. However, if we attempt to associate those samples of the field with the values at triangle edges or nodes, we do not obtain a symmetric representation with respect to the triangle, since there are three edges and three nodes, but only two samples. Thus, it is more practical to associate the two DoFs with two independent components of the field at the center of the cell.

3.1.1 Linear Curl-Conforming Vector Basis Functions

For a polynomial expansion that is complete to linear degree, such as (3.2), the number of DoFs is a multiple of the number of nodes and the number of edges (three each), and we have several possibilities. We may assign all six DoFs in (3.2) to the cell itself, or two of the DoFs to each edge or to each node. While the assignment to nodes may appear initially to be the most obvious, we will not immediately lock into that approach. In fact, the assignment to edges is more useful, because it provides control over continuity constraints between cells. Observe when we assign a DoF to an edge, we implicitly assume that that DoF will be shared by the adjacent cell bordering that edge. When we assign a DoF to a node, it will be shared by all the other cells sharing that node. These assignments are intrinsically tied to a particular type of continuity.

Let us momentarily digress to review the continuity conditions associated with electromagnetic fields. In the general isotropic case, near a material interface, these conditions can be expressed as

$$\hat{\boldsymbol{n}} \times (\boldsymbol{E}_1 - \boldsymbol{E}_2)|_{\text{interface}} = 0 \tag{3.3}$$

$$\hat{\boldsymbol{n}} \times (\boldsymbol{H}_1 - \boldsymbol{H}_2)|_{\text{interface}} = 0 \tag{3.4}$$

$$\hat{\boldsymbol{n}} \cdot (\varepsilon_1 \boldsymbol{E}_1 - \varepsilon_2 \boldsymbol{E}_2)|_{\text{interface}} = 0 \tag{3.5}$$

$$\hat{\boldsymbol{n}} \cdot (\mu_1 \boldsymbol{H}_1 - \mu_2 \boldsymbol{H}_2)|_{\text{interface}} = 0 \tag{3.6}$$

where $\boldsymbol{E}$ denotes the electric field, $\boldsymbol{H}$ denotes the magnetic field, and $\hat{\boldsymbol{n}}$ is a unit normal vector at the interface pointing from one cell to the other. The material parameters μ and ϵ denote the permittivity and permeability, respectively. In general, the electric and magnetic fields maintain continuity of only their tangential components at the interface, while the electric and magnetic flux densities ($\boldsymbol{D} = \varepsilon \boldsymbol{E}$ and $\boldsymbol{B} = \mu \boldsymbol{H}$) maintain normal-vector continuity at such an interface. Since we may be representing either type of field, we require some flexibility in the particular scheme that we select. Let us suppose we are modeling the $\boldsymbol{E}$ or $\boldsymbol{H}$ field, which requires tangential-vector continuity across material interfaces (which for our discussion will always coincide with cell interfaces). How may we achieve that?

For a linear expansion such as (3.2), we may assign the six DoFs to represent tangential field components at two locations on each of the three cell edges. For instance, we may define one function that has unity tangential component at node 1 of edge 1, and zero tangential component at node 2 of edge 1, as well as the other two nodes of edges 2 and 3. That set of six constraints is adequate to define a single linear basis function interpolating to the tangential component at one end of edge 1. Five other functions may be defined in an analogous way, with each interpolating to a unity value at a different point and edge, to obtain a set of six. Each of the six functions interpolates to a unity value at one end of one edge of the triangular cell and has zero tangential component at the other five locations.

For a particular triangle, the set of six functions may be found explicitly in the form of the Cartesian components in (3.2). An alternative representation can be obtained in terms of the simplex or local area coordinates, in which case the six functions are given by [1,2]

$$\boldsymbol{B}_{ij} = w_{ij}\, \xi_i\, \nabla \xi_j\,, \qquad i,j = 1,2,3; \qquad i \neq j \tag{3.7}$$

where i and j denote nodes, w_{ij} is the length of the edge between nodes i and j, and the simplex coordinates are (ξ_1, ξ_2, ξ_3). Figure 3.1 shows a sketch of these basis functions. Since each function is associated with a tangential component at one end of an edge, they are particularly convenient when imposing boundary or continuity conditions on the tangential fields at a cell edge. For instance, tangential-vector continuity between two adjacent cells can be imposed by equating coefficients of the functions sharing the same interpolation point on the common edge. A Dirichlet type of boundary condition (in which the field is specified *a priori* along an edge) can be obtained by equating the coefficients of each basis function with the prescribed value. A common Dirichlet condition is that of setting the tangential field to zero on a boundary, which is easily obtained by independently setting the appropriate coefficients to zero.

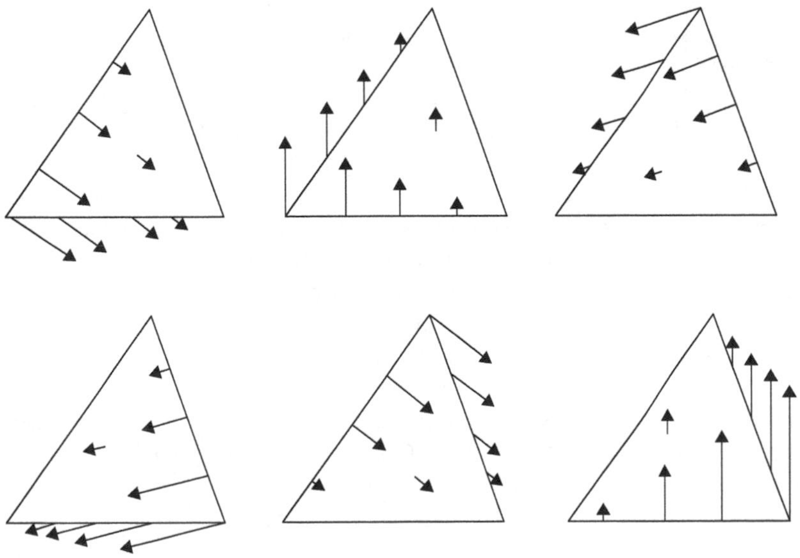

Figure 3.1 The six linear vector basis functions in (3.7).

Adapted from A. F. Peterson, S. L. Ray, and R. Mittra, *Computational Methods for Electromagnetics*, New York, NY: IEEE Press, 1998.

If the overall representation uses basis functions from (3.7) that are combined with similar functions in the adjacent cells by equating coefficients of their tangential components, the resulting representation belongs to the curl-conforming function space. These basis functions are discussed below. An expansion in these basis functions maintains tangential-vector continuity across cell boundaries. The representation does not maintain normal-vector continuity and does not belong to the divergence-conforming space.

The curl of the functions in (3.7) is

$$\nabla \times \boldsymbol{B}_{ij} = w_{ij} \nabla \xi_i \times \nabla \xi_j \tag{3.8}$$

Since the gradient of a simplex coordinate variable is a constant vector, the result in (3.8) is constant within a cell.

We contrast the preceding form of a linear representation with what we would obtain using a Cartesian definition, i.e., separating the $\hat{x}$ and $\hat{y}$ components. In that case, the obvious approach is to assign three DoFs to the $\hat{x}$ component of the vector field at the three nodes and three DoFs to the $\hat{y}$ component at those locations. While that yields an equivalent representation over the triangular cell to that provided by the set of functions in Figure 3.1, the representation differs as cell boundaries are crossed. A node-based expansion is normally shared by all the cells associated with a particular node, in such a way that it automatically imposes continuity of both field components across cell boundaries. This might appear desirable; however in the general case we observe from (3.5) and (3.6) that there may be situations where the normal component of $\boldsymbol{E}$ or $\boldsymbol{H}$ is not supposed to be continuous.

In actuality, a Cartesian node-based expansion makes it difficult to impose general boundary conditions on field components [2]. With such an expansion, the tangential component along a given triangle edge involves a linear combination of four basis function coefficients. To impose a Dirichlet boundary condition, a linear combination of the four coefficients must be set to some prescribed value, instead of a single coefficient. In most numerical solution schemes, that process leads to a set of constraint equations that must be solved simultaneously for the coefficient values and not independently assigned coefficients. This is much more cumbersome that what is proposed above, where two independent coefficients determine the tangential value! While there are other problems associated with a Cartesian representation that we will not immediately address, that one alone provides sufficient motivation to conclude that the edge-based approach has merit. It should be noted, however, that the node-based expansion tends to require the fewest global DoFs (since basis functions are shared over a broad footprint).

As we increase the polynomial degree of the representation, it happens that not all the DoFs can be uniquely associated with tangential values on cell edges. For instance, in the quadratic case, there are 12 DoFs to assign, but only three per triangle edge can be uniquely assigned to samples of the tangential field. Three of the 12 DoFs must be assigned to other fields. One sample per edge may be assigned to the normal-vector component, or to some component value at a point within the cell (away from the edges) that necessarily has a non-zero normal-vector component on the edges. This situation will be illustrated by the higher order functions developed in Chapter 4.

3.1.2 A Simpler Curl-Conforming Representation on Triangles

In the preceding section, a linear representation on triangles is achieved with six basis functions per cell, each associated with the tangential field at a unique point on the edge.

This representation is symmetric with respect to the triangle and is linear along the triangle edges and throughout the interior of the cell. We also observed that a constant representation only involves two DoFs and therefore is not symmetric with respect to the triangle. One may ask whether there is a simpler representation than the six functions that is also symmetric with respect to the triangular shape of the cell. In fact, the simplest approach that is symmetric is that where each basis function exhibits a constant tangential-vector behavior on one of the cell edges, which requires three basis functions. Such a representation can be easily obtained from a linear combination of the linear set of six, by adjusting the two coefficients on each edge to obtain a constant tangential projection, and has the simplex-coordinate form

$$\boldsymbol{B}_i = w_i(\xi_{i+1}\nabla\xi_{i-1} - \xi_{i-1}\nabla\xi_{i+1}) \tag{3.9}$$

where $(i, i+1, i-1)$ are cyclical indices representing the nodes of the triangle, and w_i is the length of the edge opposite node i. This set of three basis functions provides the constant representation of (3.1), with one additional DoF per cell in order to achieve triangular symmetry. In Cartesian coordinates, the form of each basis function is

$$\boldsymbol{B}(x,y) = \left(a_0 + \frac{a_2 - b_1}{2}y\right)\hat{\boldsymbol{x}} + \left(b_0 + \frac{b_1 - a_2}{2}x\right)\hat{\boldsymbol{y}} \tag{3.10}$$

The basis functions are sketched in Figure 3.2. While the tangential-vector component on each triangle edge is constant, the normal-vector component is linear. If the three vector basis functions are combined with analogous functions in neighboring cells and their coefficients shared to maintain tangential-vector continuity across the cell boundaries, we obtain a curl-conforming representation. This representation will be discussed below.

The curl of the basis functions in (3.9) may be obtained as

$$\nabla \times \boldsymbol{B}_i = 2w_i\nabla\xi_{i+1} \times \nabla\xi_{i-1} \tag{3.11}$$

The basis functions in (3.9) are in fact far more important than it might initially appear, and we will return to them shortly.

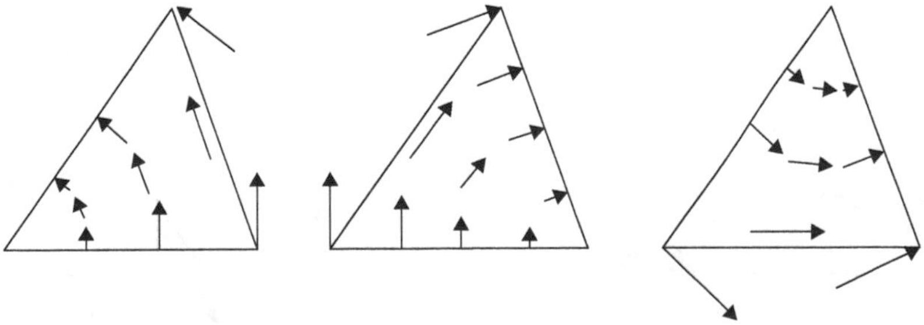

Figure 3.2 The three functions in (3.9).

Adapted from A. F. Peterson, S. L. Ray, and R. Mittra, *Computational Methods for Electromagnetics*, New York, NY: IEEE Press, 1998.

3.1.3 Alternate Approach: A Divergence-Conforming Representation on Triangles

The preceding scheme assigned DoFs associated with tangential fields to basis functions, making it simple to describe tangentially continuous quantities. The complementary approach would involve assigning DoFs to normal-vector components at cell edges, to make it easy to impose boundary conditions on the normal components, or represent quantities with normal-vector continuity. These may be the electromagnetic flux densities $\boldsymbol{D}$ or $\boldsymbol{B}$, or the current density $\boldsymbol{J}_S$.

The behavior of constant or linear vector functions that maintain normal-vector continuity is exactly complementary to the tangentially continuous functions discussed above: for the constant case, there are not enough DoFs (two) to impose continuity across three cell boundaries. For the linear case, two unknowns can be assigned to each edge to produce six normal-vector basis functions of the form

$$\boldsymbol{N}_{ij} = w_{ij}\,\xi_i(\hat{z} \times \nabla\xi_j), \qquad i,j = 1,2,3; \qquad i \neq j \tag{3.12}$$

where simplex coordinates are employed, i and j denote nodes, w_{ij} is the length of the edge between nodes i and j, and $\hat{z}$ is a unit vector perpendicular to the plane of the simplex coordinates. These functions are a 90° rotation of those depicted in Figure 3.1. The set of six functions requires six coefficients, each of which, for instance, can be associated with the normal-vector field component at one end of a cell edge.

Simpler functions with the flexibility to impose normal-vector continuity can be expressed by the three DoFs in

$$\boldsymbol{N}_i(\xi_1, \xi_2, \xi_3) = w_i\,\hat{z} \times (\xi_{i+1}\nabla\xi_{i-1} - \xi_{i-1}\nabla\xi_{i+1}) \tag{3.13}$$

where cyclical indices are employed. These functions have the Cartesian form

$$\boldsymbol{N}(x,y) = \left(a_0 + \frac{b_2 - a_1}{2}x\right)\hat{x} + \left(b_0 + \frac{b_2 - a_1}{2}y\right)\hat{y} \tag{3.14}$$

and are complementary to those in (3.9) and (3.10). The three functions are sketched in Figure 3.3.

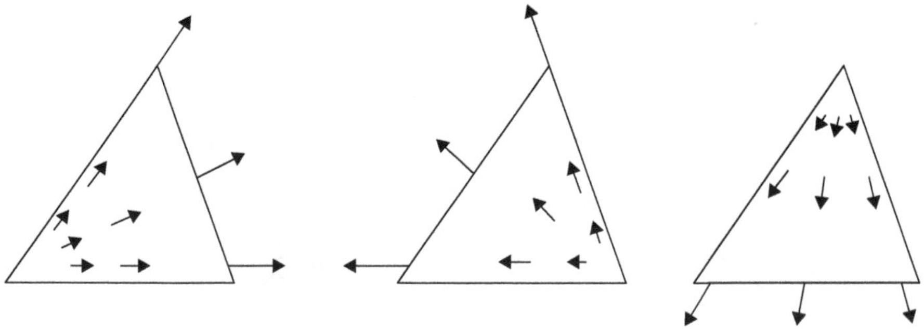

Figure 3.3 The three functions defined in (3.13).

Adapted from A. F. Peterson, S. L. Ray, and R. Mittra, *Computational Methods for Electromagnetics*, New York, NY: IEEE Press, 1998.

3.2 Tangential-Vector versus Normal-Vector Continuity: Curl-Conforming and Divergence-Conforming Bases

In the preceding sections, vector representations were introduced that impose either tangential-vector continuity or normal-vector continuity across cell edges. One or the other approach is usually sufficient for representing electromagnetic fields or currents.

Expansions that impose tangential-vector continuity are useful for problems where a curl operator is applied to the field representation. Taking the curl of a quantity that maintains tangential-vector continuity will yield a bounded result, therefore such an expansion is a *curl-conforming* representation. The simplest curl-conforming expansion on triangles is provided by the basis functions depicted in Figure 3.2, which may be joined across a common edge while maintaining tangential continuity as shown in Figure 3.4. Observe that the basis function thus derived does *not* maintain normal-vector continuity across the central edge; furthermore since the basis functions have non-zero normal components at the outer four edges of the cell pair the resulting expansion appears to exhibit large discontinuities in the normal components at all cell edges. A curl operation applied to the function in Figure 3.4 produces a constant function in each triangular cell, with the vector directed normal to the cell (in or out of the page in this case).

Furthermore, the specific form of (3.9)–(3.11) yields another interesting property: both the vector function and its curl are mathematically complete only to degree zero (constant order). Completeness of the function and its curl to the same degree is more than coincidental, and this property can be maintained even with high-order basis functions (Chapters 4 and 5). Using an expansion that is complete to the same degree as its curl helps to provide balance when numerically solving an equation with a curl operator, such as the vector Helmholtz equation (see Section 3.12 and Chapter 6).

In contrast, expansions that impose normal-vector continuity are useful for problems where a divergence operator is applied to the field or current density representation. Taking the divergence of a quantity that maintains normal-vector continuity yields a bounded result, and such an expansion is known as a *divergence-conforming* representation. The simplest divergence-conforming basis on triangles is provided by the functions depicted in Figure 3.3, which may be joined across a common edge while maintaining normal continuity as shown in Figure 3.5. Observe that the basis function thus derived does *not* maintain tangential-vector

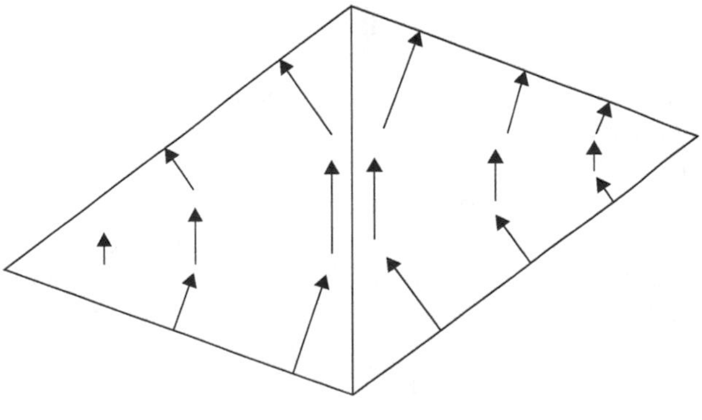

Figure 3.4 A curl-conforming basis function straddling two triangular cells.

Adapted from A. F. Peterson, S. L. Ray, and R. Mittra, *Computational Methods for Electromagnetics*, New York, NY: IEEE Press, 1998.

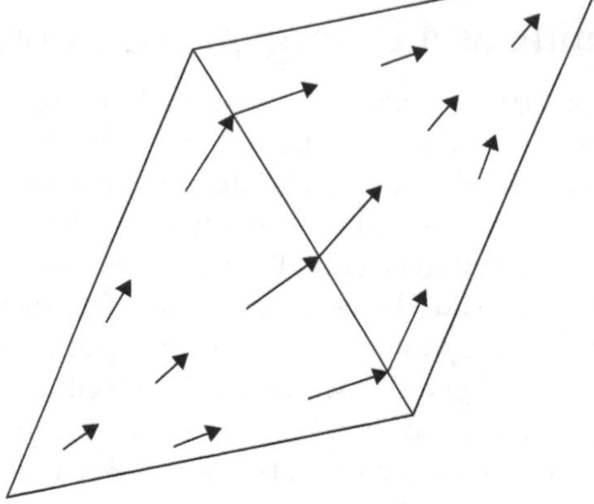

Figure 3.5 A divergence-conforming basis function straddling two triangular cells.

Adapted from A. F. Peterson, S. L. Ray, and R. Mittra, *Computational Methods for Electromagnetics*, New York, NY: IEEE Press, 1998.

continuity across the central edge; furthermore the function has a large non-zero tangential component along the four outer edges of the cell pair. Thus, the resulting expansion will generally exhibit discontinuities in the tangential components at all cell edges.

The divergence of the function in Figure 3.5 is a constant in each triangular cell. In this case, both the vector function and its divergence are complete to degree zero (constant order). This property helps to provide balance among terms when numerically solving an equation with a divergence operator, such as the electric field integral equation (Chapter 6).

An alternate notation for the divergence-conforming basis function depicted in Figure 3.5 is [3]

$$\boldsymbol{B}_n(\boldsymbol{r}) = \begin{cases} \dfrac{w_n}{2A_n^+}\boldsymbol{\rho}_n^+(\boldsymbol{r}) & \boldsymbol{r} \in T_n^+ \\ \dfrac{w_n}{2A_n^-}\boldsymbol{\rho}_n^-(\boldsymbol{r}) & \boldsymbol{r} \in T_n^- \\ \boldsymbol{0} & \text{otherwise} \end{cases} \tag{3.15}$$

where T_n^+ and T_n^- denote two triangular cells adjacent to edge n, w_n is the length of edge n, and A_n^+ and A_n^- are the areas of the two triangles. The vector $\boldsymbol{\rho}_n^+$ points away from one vertex of triangle T_n^+ toward the opposite edge (edge n), while $\boldsymbol{\rho}_n^-$ points away from edge n toward the opposite vertex of the adjacent triangular cell (T_n^-) [3]. Thus, each basis function has a constant (unit) normal component over one triangle edge and has no normal component on the other edges. The basis function has a constant divergence in each cell, given by

$$\nabla \cdot \boldsymbol{B}_n = \begin{cases} \dfrac{w_n}{A_n^+} & \boldsymbol{r} \in T_n^+ \\ -\dfrac{w_n}{A_n^-} & \boldsymbol{r} \in T_n^- \\ 0 & \text{otherwise} \end{cases} \tag{3.16}$$

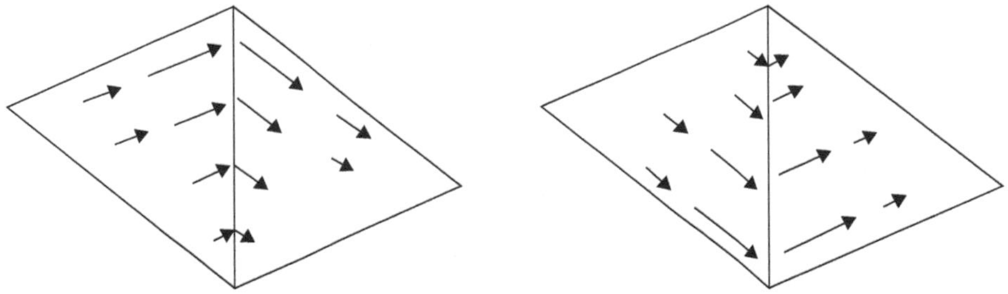

Figure 3.6 Two linear divergence-conforming basis functions straddling adjacent triangular cells. Adapted from A. F. Peterson, S. L. Ray, and R. Mittra, *Computational Methods for Electromagnetics*, New York, NY: IEEE Press, 1998.

Divergence-conforming functions of higher polynomial order may be constructed in an analogous manner. For instance, from the set of six functions in (3.7), we may also construct functions that are linear in both variables and maintain normal-vector continuity across cell edges. Figure 3.6 depicts the two functions for an edge. An inspection of the figure should make it obvious that the functions only ensure normal-vector continuity, not tangential-vector continuity. Families of higher order basis functions will be developed in Chapters 4 and 5.

3.2.1 Additional Terminology

Since there are many types of vector basis functions possible, it is convenient to introduce some additional nomenclature to more precisely describe each type of function. The simplest curl-conforming basis functions (Figure 3.4) exhibit a constant tangential-vector component and a linear normal-vector component along the cell edges. We label these "constant tangential/linear normal" or CT/LN functions. (On triangles, these are also known as "edge elements" [4–6] or "zero-order edge elements" or "Whitney elements" [7] throughout the literature.) The simplest divergence-conforming basis functions (Figure 3.5) exhibit a constant normal-vector component and a linear tangential-vector component on cell edges, thus we label them "constant normal/linear tangential" or CN/LT functions. (These are also known as "Rao Wilton Glisson" or RWG basis functions [3] and sometimes as "Raviart Thomas" functions [8].) Both types of functions are sometimes denoted as functions of "order $p = 0$" and are also sometimes denoted as functions of "order $p = 0.5$" because of their mixed-order nature.

If the set of six vector functions depicted in Figure 3.1 is used to construct a curl-conforming basis, those functions exhibit linear tangential and linear normal behavior on cell edges. They will be known as "curl-conforming LT/LN" functions. In contrast, the functions depicted in Figure 3.6, which also have linear components along cell edges, will be labeled "divergence-conforming LN/LT" functions.

3.3 Two-Dimensional Representations on Rectangular Cells

The reader might wonder why we began with a discussion of vector representations on triangles, when in fact a rectangular cell in 2D may appear to be much easier conceptually to handle. The principal reason for our approach is to motivate the reader to think beyond the Cartesian expansion, which we hope to have sufficiently motivated by the preceding discussion of triangles. Having this under our belts, we now consider the extension of the edge-based

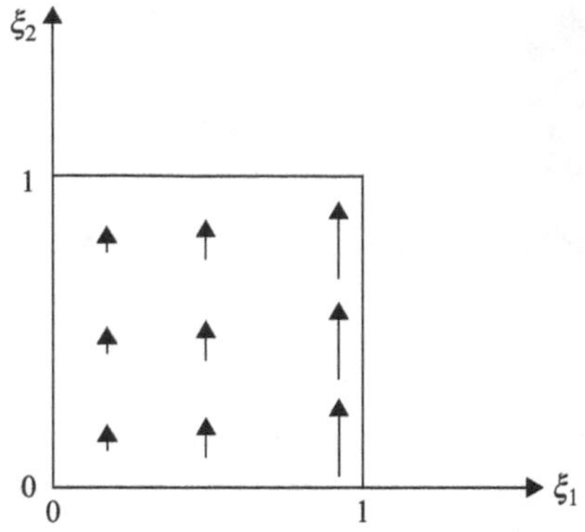

Figure 3.7 A curl-conforming CT/LN vector basis function on a rectangular cell, interpolating to the value at $\xi_1 = 1$.

representations to rectangular cells, which we will subsequently generalize to quadrilaterals by a mapping procedure. We will use a (ξ_1, ξ_2) coordinate system for the independent variables in the parent domain.

For the lowest order representation, consider a set of four basis functions that each provides a constant tangential-vector component of the field on one edge of a rectangle. Each of these functions will contribute no tangential field to any of the other three edges; thus (assuming proper cell-to-cell continuity conditions are imposed) the set can be used to provide a curl-conforming representation. This set of four basis functions exceeds the number of DoFs necessary to represent constant vector fields throughout the cell, but offers a symmetric way of providing a representation that is at least constant in each vector component. Since the functions represent the tangential component of the field on their edge, boundary and continuity conditions are easy to impose with analogous representations on adjacent cells. (In fact, the behavior along each edge is identical to that of the functions in (3.9) for triangles, thus facilitating the mixture of both cell shapes in the representation.)

The four basis functions can be succinctly expressed in terms of the local coordinates $0 \le \xi_1 \le 1, 0 \le \xi_2 \le 1$ by the expressions

$$\xi_1 \nabla \xi_2, \quad (1-\xi_1)\nabla \xi_2, \quad \xi_2 \nabla \xi_1, \quad (1-\xi_2)\nabla \xi_1 \tag{3.17}$$

Figure 3.7 depicts one of the basis functions. These are clearly "constant tangential/linear normal" basis functions. If their tangential components are adjusted in order to maintain continuity with similar functions in adjacent cells, the expansion provides a curl-conforming CT/LN representation, much like that of the triangular-cell functions depicted in Figure 3.4. The curl of these CT/LN functions is a constant throughout the cell and is therefore complete to the same degree as the CT/LN function itself.

A complementary divergence-conforming representation can be constructed from functions that each has a constant normal-vector component along one edge of a rectangle. The functions must contribute no normal component to any of the other three edges. A suitable set of functions defined on $0 \le \xi_1 \le 1, 0 \le \xi_2 \le 1$ is given by

$$\xi_1 \nabla \xi_1, \quad (1-\xi_1)\nabla \xi_1, \quad \xi_2 \nabla \xi_2, \quad (1-\xi_2)\nabla \xi_2 \tag{3.18}$$

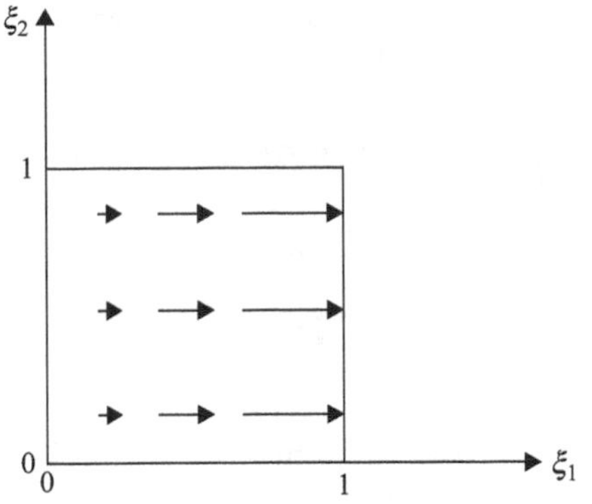

Figure 3.8 A divergence-conforming CN/LT vector basis function on a rectangular cell, interpolating to the value at $\xi_1 = 1$.

One of these is depicted in Figure 3.8. If the amplitude of each function is adjusted to maintain normal-vector continuity with a similar function in the adjacent cell, the expansion provides a divergence-conforming CN/LT representation. The divergence of these CN/LT basis functions is a constant throughout the cell and is therefore complete to the same order as the CN/LT function.

The preceding examples use square cells instead of a more general quadrilateral cell shape. However, these basis functions can be mapped from the local coordinates to more general quadrilateral cells while maintaining their curl-conforming or divergence-conforming properties. In fact, they can also be mapped to curved cells in 2D or curved surface patches in three-dimensional (3D). The coefficients associated with the tangential or normal field at some point on an edge will retain that interpretation in the general case, for curl-conforming or divergence-conforming expansions, respectively. The mathematical details associated with these transformations will be discussed below.

A consistently linear representation on rectangular cells can be developed by employing two functions per cell edge, each providing a linear tangential behavior and a linear normal behavior along that edge. This set of eight basis functions exceeds the minimum number of DoFs required for a linear expansion (six), but provides a symmetric representation on the quadrilateral cell, and a convenient means of imposing boundary and continuity conditions with other cells. A set of curl-conforming LT/LN functions may be obtained from the DoFs

$$\left\{ \begin{matrix} \xi_1\xi_2\nabla\xi_1, & (1-\xi_1)\xi_2\nabla\xi_1, & \xi_1(1-\xi_2)\nabla\xi_1, & (1-\xi_1)(1-\xi_2)\nabla\xi_1, \\ \xi_1\xi_2\nabla\xi_2, & \xi_1(1-\xi_2)\nabla\xi_2, & (1-\xi_1)\xi_2\nabla\xi_2, & (1-\xi_1)(1-\xi_2)\nabla\xi_2 \end{matrix} \right\} \tag{3.19}$$

where suitable coefficients are assigned in order to represent the tangential components at the ends of cell edges. The same group of eight functions in (3.19) may be used to define a set of divergence-conforming LN/LT basis functions, provided that coefficients are assigned in a different manner to represent normal-vector components at the ends of each edge. For both types of function, we assume that suitable continuity is maintained across cell boundaries.

As previously mentioned, more than one specific set of basis functions may be obtained with the preceding properties. Instead of associating each coefficient with the field value at the end of a cell edge, for instance, two other points may be assigned along the edge, or points off the edge may be used as an alternative. For the simple functions described in this chapter, it makes little difference. For higher order basis functions, however, the location and spacing of the interpolation points can affect the linear independence of the equations as manifested in the matrix condition numbers arising from their use [9]. Higher order functions will be developed in Chapters 4 and 5.

3.4 Quasi-Helmholtz Decomposition in 2D: Loop and Star Functions

The *Helmholtz decomposition theorem* states that any vector function $\boldsymbol{V}$ can be decomposed into a part that has zero divergence (a solenoidal function) and another part that has zero curl (an irrotational function). In equation form, $\boldsymbol{V}$ can be expressed as

$$\boldsymbol{V} = \nabla \times \boldsymbol{A} + \nabla \Psi \tag{3.20}$$

Using two mathematical identities

$$\nabla \cdot (\nabla \times \boldsymbol{A}) \equiv 0 \tag{3.21}$$

$$\nabla \times \nabla \Psi \equiv \mathbf{0} \tag{3.22}$$

the first term in (3.20) is clearly solenoidal while the second term is irrotational. This construction also demonstrates that the curl of $\boldsymbol{V}$, given by

$$\nabla \times \boldsymbol{V} = \nabla \times \nabla \times \boldsymbol{A} \tag{3.23}$$

is independent from the divergence of $\boldsymbol{V}$, given by

$$\nabla \cdot \boldsymbol{V} = \nabla \cdot \nabla \Psi \tag{3.24}$$

It is sometimes necessary to partition a numerical field representation into solenoidal and irrotational parts. Such a partitioning may provide a more physical representation in situations where the field is supposed to be solenoidal or irrotational; it may also be necessary to isolate the nullspace of the resulting matrix operator in various situations or stabilize the matrix operator at very low frequencies when other scale factors produce instabilities. Unfortunately, an exact Helmholtz decomposition cannot be obtained in terms of either curl-conforming or divergence-conforming representations alone: an irrotational field requires a curl-conforming expansion, while a solenoidal field requires a divergence-conforming expansion. However, we may obtain a quasi-Helmholtz decomposition with either curl-conforming or divergence-conforming base. In either case, one way of realizing the decomposition is in terms of *loop* and *star* vector basis functions [10], which can be directly related to the curl-conforming CT/LN and divergence-conforming CN/LT functions defined above.

Suppose that we have a representation of a vector function in terms of the curl-conforming CT/LN basis functions on triangles. The CT/LN basis functions can be exchanged for curl-conforming loop and star basis functions, as depicted in Figure 3.9. The star functions are each associated with one node of the mesh and are obtained as the superposition of the set of CT/LN functions at every edge leaving that node, with the coefficients of the individual members of the

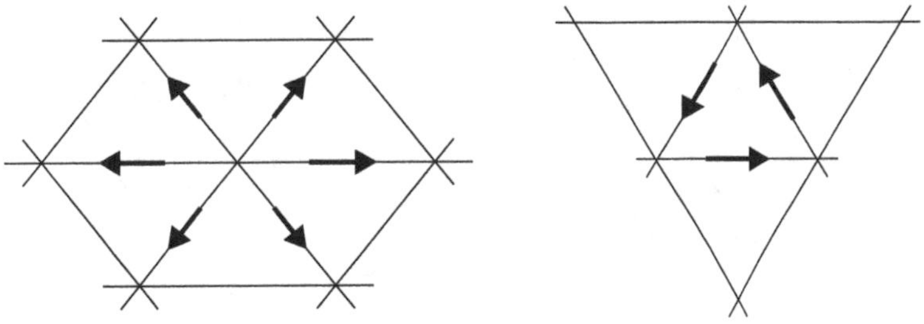

Figure 3.9 Curl-conforming star and loop basis functions.

set adjusted so that the resulting star function is globally irrotational. In this configuration, two CT/LN functions overlap each cell around the central node, with the coefficients chosen so that in each cell the curl of one CT/LN function is cancelled by the curl of the other. Each star basis function can also be thought of as the negative gradient of a linear (scalar) Lagrangian basis function centered at the appropriate node, straddling the surrounding cells. (Since the result is a gradient, the star function is automatically irrotational.) The star functions have a large global divergence. Star functions located at each node are superimposed with loop functions, which are associated with cells and are obtained as a linear combination of three CT/LN basis functions around the cell. The loop functions exhibit a relatively large global curl, but are not solenoidal (as they would be in a true Helmholtz decomposition).

The partitioning of a curl-conforming CT/LN expansion into curl-conforming loop and star functions involves trading one basis coefficient per edge (CT/LN) into one per node (star) and one per cell (loop). The number of nodes, edges, and cells are interrelated by topological properties of the structure being meshed. As an example, on a surface modeled by a simply connected 2D mesh of triangular cells, the nodes, cells, and edges are related by [11]

$$N_{\text{cells}} - N_{\text{edges}} + N_{\text{nodes}} = 1 \tag{3.25}$$

$$2N_{\text{edges}} - N_{\text{boundary edges}} = 3N_{\text{cells}} \tag{3.26}$$

Since there are N_{nodes} star functions and N_{cells} loop functions, (3.25) indicates that not all of the star and loop functions are linearly independent. In this simply-connected mesh (no holes), one of the curl-conforming star functions is linearly dependent on the other functions and must be discarded. This quasi-Helmholtz partitioning produces loop functions with relatively large global curl and star functions with relatively large global divergence, so it approximates a true Helmholtz decomposition.

In the divergence-conforming case, CN/LT basis functions can be replaced by divergence-conforming loop and star functions [10], as depicted in Figure 3.10. Each loop function is associated with one node of the mesh, and is obtained by superimposing CN/LT bases around that node with their coefficients adjusted to produce zero divergence in each adjacent cell. The loop function may also be obtained by taking the curl of a linear (scalar) Lagrangian basis function centered at the appropriate node, straddling the surrounding cells, multiplied by a unit vector normal to the plane of the mesh. The loop function constructed in this way is automatically solenoidal. Divergence-conforming star basis functions are obtained by superimposing three CN/LT functions per cell, as depicted in Figure 3.10. The loop functions exhibit a relatively large curl, while the star functions exhibit a large divergence. Since the partitioning

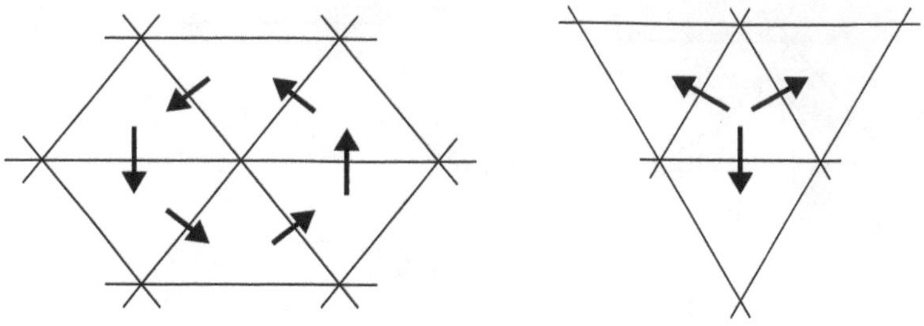

Figure 3.10 Divergence-conforming loop and star basis functions.

involved exchanging the original CN/LT basis functions (one per edge) for loop functions (one per node) and star functions (one per cell), (3.25) indicates that not all the loop and star basis functions on the global mesh are linearly independent. In a simply connected mesh, one of the loop functions must be discarded.

An alternative approach employs loop functions with *flower* functions for the same purpose [12]. In the divergence-conforming case, the flower functions are effectively node-based and may offer better conditioned Gram matrices than obtained with loop–star constructions.

The loop and star basis functions provide a conceptual means for obtaining an approximate Helmholtz decomposition. In practice, the actual partitioning is often accomplished by matrix operations acting on matrices constructed from CT/LN or CN/LT bases, rather than an explicit expansion in loop and star functions [13]. The partitioning may also be driven by an analysis of the spanning tree and co-tree of the mesh [14–17].

In situations where a purely solenoidal expansion is desired, the divergence-conforming loop functions may be directly employed as a basis for the field. Similarly, the curl-conforming star basis functions may be used to obtain an irrotational expansion.

Loop and star functions can also be defined for quadrilateral cells [18].

3.5 Projecting between Curl-Conforming and Divergence-Conforming Bases

There are occasionally situations in computational electromagnetics where it is necessary to implement both divergence and curl operators applied to the same representation of a source or field. This situation occurs with radiation boundary conditions [19], impedance boundary conditions [20–22], the combined field integral equation [23], and certain preconditioning techniques [24, 25]. In these situations, it is often necessary to compute the divergence of a curl-conforming expansion or the curl of a divergence-conforming expansion. One way to facilitate this without gross approximations is to project an expansion of one type onto an expansion of the other type; for instance curl-conforming basis functions may be projected onto divergence-conforming basis functions in order that the divergence operation may be carried out in a systematic fashion. If this is attempted with the simple vector basis functions introduced in Sections 3.1.2 and 3.1.3, a projection defined on a single mesh will usually fail due to the fact that these basis functions are almost orthogonal to each other. The most successful remedy to this situation is to employ an alternative to one type of function (divergence conforming or curl conforming) that closely approximates the other and avoids the orthogonality problem.

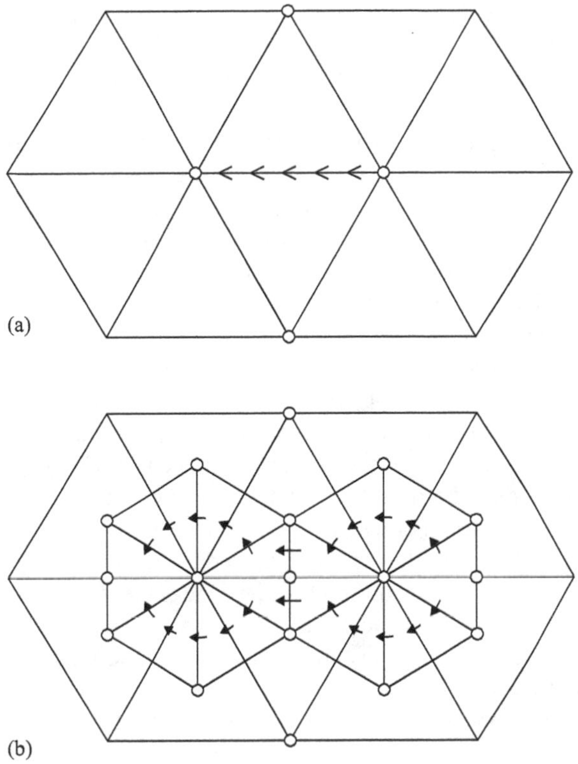

Figure 3.11 Buffa–Christiansen basis functions. (a) The marked edge indicates the location of a curl-conforming basis function on the original mesh. (b) The Buffa–Christensen approximation to that function realized by a superposition of divergence-conforming bases on a finer mesh obtained from the barycentric edges of the original mesh.

One such function is the so-called Buffa–Christiansen basis function [26] developed for triangular cells and depicted in Figure 3.11. Similar functions have been proposed for quadrilateral cells [27].

3.6 Three-Dimensional Representation on Tetrahedral Cells: Curl-Conforming Bases

The development of vector basis functions for 3D fields is analogous in most respects to those discussed above for the 2D case. Consider a tetrahedral cell, containing six edges, four faces, and four nodes. Based on the preceding discussion, we initially consider a linear representation of the form

$$\begin{aligned} f(x,y,z) &= (a_0 + a_1 x + a_2 y + a_3 z)\,\hat{x} + (b_0 + b_1 x + b_2 y + b_3 z)\,\hat{y} \\ &\quad + (c_0 + c_1 x + c_2 y + c_3 z)\,\hat{z} \end{aligned} \tag{3.27}$$

The linear representation involves 12 DoFs, which in the course of defining basis functions can be potentially associated with nodes (3 per node), edges (2 per edge), faces (3 per face), or the cell itself (all 12). In a large tetrahedral-cell mesh, the number of cells is approximately five times the number of nodes, so a cell-based representation ultimately requires the most variables ($\sim 60N$, where N is the number of nodes) of these four possibilities. The number of faces is roughly 10 times the number of nodes in a large mesh, yielding a variable count of $\sim 30N$ for a face-based representation. In contrast, an edge-based expansion requires $\sim 12N$ variables, while a node-based approach only involves $3N$. The number of variables is reduced as the global footprint of the expansion increases; a node-based approach produces the largest footprint while the cell-based expansion has the smallest.

Since a node-based expansion limits the flexibility of field continuity, we begin by considering an edge-based expansion. To obtain a curl-conforming expansion, the DoFs in (3.27) can be distributed by assigning two per edge, each associated with the tangential field at a unique point on the edge. As in the 2D case, we may define one function that has unity tangential component at node 1 of edge 1, zero tangential component at node 2 of edge 1, and zero tangential components at the two nodes of edges 2–6. That set of 12 constraints is adequate to define a single linear basis function. Eleven other functions may be defined in a similar way, with each interpolating to a unity value at a different node and edge. Each of the 12 basis functions thus defined interpolates to a unity value at one end of one edge of the tetrahedron and contributes no tangential component to the other 11 locations. This representation is symmetric with respect to the tetrahedron and is linear along the cell faces and throughout the cell interior.

A representation of these basis functions within the cell can be obtained in terms of the tetrahedral simplex coordinates (ξ_1, ξ_2, ξ_3, ξ_4), yielding the 12 functions given by

$$\boldsymbol{B}_{ij} = w_{ij}\, \xi_i\, \nabla \xi_j\,, \quad i,j = 1,2,3,4; \quad i \neq j \tag{3.28}$$

where i and j denote nodes, and where w_{ij} is the length of the edge between nodes i and j. In simplex coordinates, these basis functions are identical in form to those of the 2D (triangular) case. As in the 2D case, each 3D function is associated with a tangential component at one end of an edge, and the six functions associated with the three edges around a face are the only functions that contribute to the tangential fields on that face. By equating coefficients of the six functions associated with a face with similar functions in the adjacent cell, tangential-vector continuity can be imposed between two cells. The result is a linear tangential/linear normal (LT/LN) representation.

As in the 2D case, a simpler curl-conforming representation that is also symmetric with respect to the tetrahedron can be obtained by adjusting the two coefficients on each edge in the functions of (3.28) to obtain a constant tangent vector. This produces the six basis functions

$$\boldsymbol{B}_{ij} = w_{ij}(\xi_i \nabla \xi_j - \xi_j \nabla \xi_i) \tag{3.29}$$

While only complete to constant order, this set of six basis functions provides a symmetric representation on a tetrahedral cell, with a constant tangential-vector component along each edge, and a mixed-order (constant and linear) tangential component on the cell faces. These functions also produce a linear normal-vector component on each cell face. If the vector basis function at an edge is combined with analogous functions in neighboring cells sharing that edge, and their coefficients adjusted to maintain tangential-vector continuity across the cell boundaries, we obtain a curl-conforming representation. Based on their behavior on cell faces, these functions are "constant tangential/linear normal" or CT/LN basis functions. They are sometimes known as "edge elements," "zero-order edge elements," or "Whitney elements."

The curl of the basis functions in (3.28) is given by

$$\nabla \times \boldsymbol{B}_{ij} = w_{ij} \nabla \xi_i \times \nabla \xi_j \tag{3.30}$$

while that of the functions in (3.29) is

$$\nabla \times \boldsymbol{B}_{ij} = 2 w_{ij} \nabla \xi_i \times \nabla \xi_j \tag{3.31}$$

Both curls are constant vectors within each cell. We observe that the functions in (3.29) are complete to the same order as their curls in (3.31).

3.7 Three-Dimensional Representation on Tetrahedral Cells: Divergence-Conforming Bases

Curl-conforming expansions make it relatively easy to describe tangentially continuous quantities or to impose tangential-vector boundary conditions. To describe normally continuous fields and simplify the imposition of normal-vector boundary conditions on tetrahedral cells, we seek divergence-conforming basis functions.

In 2D, the curl-conforming and divergence-conforming basis functions are closely related, with one being a 90° rotation of the other. Such a simple relationship does not generalize to the 3D case. In fact, the number of DoFs for one type is usually different from the number of DoFs for the other.

The simplest type of divergence-conforming basis function on tetrahedral cells is a function that has a constant normal-vector component on one face, and no normal component on the other three faces of the cell. Such a function may be obtained from the linear representation in (3.27) by imposing the condition that the vector component normal to face i at each of the three nodes bordering face i has unit value (three equations), while the analogous normal components at the nodes of the other three faces be zero (nine equations). That set of 12 equations yields one function; by rotating the faces we obtain four basis functions of the divergence-conforming CN/LT type. These can be expressed in simplex coordinates by

$$\alpha_{ijk}(\xi_i \nabla \xi_j \times \nabla \xi_k + \xi_j \nabla \xi_k \times \nabla \xi_i + \xi_k \nabla \xi_i \times \nabla \xi_j), \quad i \neq j \neq k \tag{3.32}$$

where α_{ijk} is a normalization constant. Unfortunately, the simplex-coordinate description is not very helpful in visualizing the functions. If combined with the mirror-image function in the adjacent cell, with coefficients adjusted to provide a continuous normal component from one cell to the other, the resulting basis function representing the normal-vector component on face n of the mesh is sometimes written as [28]

$$\boldsymbol{N}_n(\boldsymbol{r}) = \begin{cases} \dfrac{a_n}{3V_n^+} \boldsymbol{\rho}_n^+(\boldsymbol{r}) & \boldsymbol{r} \in T_n^+ \\ \dfrac{a_n}{3V_n^-} \boldsymbol{\rho}_n^-(\boldsymbol{r}) & \boldsymbol{r} \in T_n^- \\ \boldsymbol{0} & \text{otherwise} \end{cases} \tag{3.33}$$

where T_n^+ and T_n^- denote the two cells, a_n denotes the area of face n, V_n^+ and V_n^- denote the volumes of the two tetrahedrons, the position vector ρ_n^+ points away from one vertex of tetrahedron T_n^+ toward the opposite face (face n), and the position vector ρ_n^- points away from face n toward the opposite vertex of the adjacent tetrahedral cell (T_n^-) [28].

Each basis function has a constant divergence in each cell, given by

$$\nabla \cdot N_n = \begin{cases} \dfrac{a_n}{V_n^+} & \bar{r} \in T_n^+ \\ -\dfrac{a_n}{V_n^-} & \bar{r} \in T_n^- \\ 0 & \text{otherwise} \end{cases} \tag{3.34}$$

Thus, the function and its divergence are complete to the same (constant) degree. These basis functions belong to the divergence-conforming CN/LT family. They are often known as "Schaubert–Wilton–Glisson" basis functions after the authors of Reference 28.

3.8 Three-Dimensional Expansion on Brick Cells: Curl-Conforming Case

We now consider the development of simple vector basis functions on rectangular brick cells, which we will subsequently generalize to skewed or curved hexahedral cells by a mapping procedure. We will use the (ξ_1, ξ_2, ξ_3) coordinate system, for a parent brick defined by $0 \le \xi_1 \le 1, 0 \le \xi_2 \le 1, 0 \le \xi_3 \le 1$.

A rectangular brick cell has 8 nodes, 6 faces, and 12 edges. The simplest curl-conforming CT/LN representation, with constant tangential-vector behavior along the 12 cell edges, and only one non-zero basis function per edge, can be found by inspection to consist of the 12 basis functions shown in Table 3.1.

Although these functions interpolate to 12 different components and are linearly independent, they do not realize all the DoFs in (3.27). They provide an expansion that is only complete to constant (degree 0) order, despite containing some linear DoFs. The redundant DoFs are necessary to provide a symmetric (one function per edge) representation. To build curl-conforming basis functions, their coefficients must be adjusted with those in all adjacent cells sharing the same edge to provide tangential-vector continuity between cells. Table 3.1 also shows the result of applying the curl operation to the bases. The basis functions and their curls are both complete only to constant order.

It is a straightforward matter to define a set of vector basis functions whose members are each linear along an edge, or some higher polynomial order. The development of higher order functions will be discussed in Chapters 4 and 5.

3.9 Divergence-Conforming Bases on Brick Cells

Divergence-conforming bases, defined within brick cells, can also be found by inspection and are shown in Table 3.2 for the standard unit cell defined by $0 \le \xi_1 \le 1, 0 \le \xi_2 \le 1, 0 \le \xi_3 \le 1$. Each function provides a constant normal-vector component on one face of the brick. To ensure normal-vector continuity, the coefficients of these functions must be adjusted with those in the

Table 3.1 Twelve un-normalized CT/LN basis functions, defined within a single brick cell, and their curls.

$\boldsymbol{B}$	$\nabla \times \boldsymbol{B}$
$\xi_2(1-\xi_3)\nabla\xi_1$	$-\xi_2\nabla\xi_2-(1-\xi_3)\nabla\xi_3$
$(1-\xi_2)(1-\xi_3)\nabla\xi_1$	$-(1-\xi_2)\nabla\xi_2+(1-\xi_3)\nabla\xi_3$
$\xi_2\xi_3\nabla\xi_1$	$\xi_2\nabla\xi_2-\xi_3\nabla\xi_3$
$(1-\xi_2)\xi_3\nabla\xi_1$	$(1-\xi_2)\nabla\xi_2+\xi_3\nabla\xi_3$
$\xi_1(1-\xi_3)\nabla\xi_2$	$\xi_1\nabla\xi_1+(1-\xi_3)\nabla\xi_3$
$(1-\xi_1)(1-\xi_3)\nabla\xi_2$	$(1-\xi_1)\nabla\xi_1-(1-\xi_3)\nabla\xi_3$
$\xi_1\xi_3\nabla\xi_2$	$-\xi_1\nabla\xi_1+\xi_3\nabla\xi_3$
$(1-\xi_1)\xi_3\nabla\xi_2$	$-(1-\xi_1)\nabla\xi_1-\xi_3\nabla\xi_3$
$\xi_1(1-\xi_2)\nabla\xi_3$	$-\xi_1\nabla\xi_1-(1-\xi_2)\nabla\xi_2$
$(1-\xi_1)(1-\xi_2)\nabla\xi_3$	$-(1-\xi_1)\nabla\xi_1+(1-\xi_2)\nabla\xi_2$
$\xi_1\xi_2\nabla\xi_3$	$\xi_1\nabla\xi_1-\xi_2\nabla\xi_2$
$(1-\xi_1)\xi_2\nabla\xi_3$	$(1-\xi_1)\nabla\xi_1+\xi_2\nabla\xi_2$

Table 3.2 Six un-normalized CN/LT basis functions, defined within a single brick cell, and their divergence.

$\boldsymbol{B}$	$\nabla \bullet \boldsymbol{B}$
$\xi_1\nabla\xi_1$	1
$(\xi_1-1)\nabla\xi_1$	1
$\xi_2\nabla\xi_2$	1
$(\xi_2-1)\nabla\xi_2$	1
$\xi_3\nabla\xi_3$	1
$(\xi_3-1)\nabla\xi_3$	1

adjacent cell sharing the same face. In the global hexahedral mesh, there is one basis function per face. The local divergence of the functions is also shown in Table 3.2.

Higher order bases for hexahedral cells will be described in Chapters 4 and 5.

3.10 Quasi-Helmholtz Decomposition on Tetrahedral Meshes

Consider a 3D tetrahedral mesh and a vector function $\boldsymbol{V}$ represented by the curl-conforming basis functions in (3.29). Suppose that it is necessary to partition a numerical field representation into solenoidal (zero divergence) and irrotational (zero curl) parts. While this cannot be done exactly in the context of curl-conforming functions, $\boldsymbol{V}$ can be decomposed into a part that has a relatively large curl and another part that has zero curl. The global expansion in terms of the basis functions in (3.29) can be projected onto two alternative types of functions, analogous to the loop and star functions in 2D.

For a curl-conforming expansion, the 3D loop function for each face of the mesh can be obtained by superimposing the three CT/LN basis functions on the edges of that face, while the 3D star function associated with each node is most easily constructed by taking the gradient of a linear Lagrangian basis function centered at that node. Since the star function is obtained via a gradient, it is automatically irrotational. In the global decomposition, there is one loop function per face and one star function per node. Not all of these are linearly independent, however, depending on the topology of the mesh. For a simply connected tetrahedral-cell mesh, the original expansion in terms of edge-based functions (3.29) requires N_{edges} variables, where [11]

$$N_{\text{edges}} = N_{\text{faces}} + N_{\text{nodes}} - N_{\text{cells}} - 1 \tag{3.35}$$

In that situation, the set of loop and star functions include $(N_{\text{cells}} + 1)$ dependent functions, which must be discarded to maintain linear independence.

For a divergence-conforming expansion, a quasi-Helmholtz decomposition can be realized in terms of solenoidal (zero divergence) 3D loop basis functions and 3D star functions with relatively large divergence. Each solenoidal loop function is associated with one edge of a tetrahedral mesh and is most easily obtained by taking the curl of the CT/LN basis function in (3.29), to obtain a piecewise-constant function with a vector component that circulates around the principal edge. Since this function is constructed by taking a curl, it is automatically solenoidal. The 3D star functions are constructed by superimposing the four CN/LT functions of (3.32) on a single cell (and their complementary functions in adjacent cells).

Thus, a divergence-conforming expansion in terms of the face-based functions in (3.32) may be recast into loop and star functions. As in the curl-conforming case, not all of these are linearly independent. There is one star function per cell and one loop function per edge on the global mesh. Since (3.35) holds for a simply connected mesh, in that case there are

$$N_{\text{cells}} + N_{\text{edges}} - N_{\text{faces}} = N_{\text{nodes}} - 1 \tag{3.36}$$

linearly dependent functions in the loop/star set. Algorithms have been proposed for identifying the linearly independent subspace [29, 30], and alternative algorithms can be based on a spanning tree of the mesh [16].

As in the 2D situation, the 3D loop and 3D star basis functions provide a conceptual means for obtaining an approximate Helmholtz decomposition. In situations where a purely solenoidal expansion is desired, a linearly independent subspace of the divergence-conforming solenoidal 3D loop functions may be directly employed as a basis for the field. Similarly, the curl-conforming star basis functions may be used to obtain an irrotational expansion. Both of these subspaces involve redundant DoFs, some of which must be discarded to produce a linearly independent basis.

3.11 Vector Basis Functions on Skewed Meshes or Meshes with Curved Cells

Previous sections have introduced simple vector basis functions in the local or parent coordinates. These functions may be defined on general cell shapes by a parametric mapping, as briefly discussed for scalar functions in Chapter 2. The vector nature of the functions introduces another DoF into the mapping, in that a specific mapping may be designed to preserve tangential-vector continuity (as desired for curl-conforming bases) or to preserve normal-vector continuity (as desired for divergence-conforming bases), but will not be able to do both [2].

Thus, when vector basis functions are mapped from the parent space into the child space, an appropriate mapping must be employed to preserve the desired continuity properties of the functions.

There are three situations of interest: mapping 2D bases onto general cells in 2D, mapping 2D bases onto (curved) surfaces in 3D, or mapping 3D bases onto cells in 3D. We will primarily consider the general 3D to 3D situation, then briefly discuss the 2D surface case.

3.11.1 Base and Reciprocal Base Vectors

Suppose we have the general 3D situation involving a mapping from a 3D reference space described by independent parent coordinates (ξ_1, ξ_2, ξ_3) to the child space described in terms of (x, y, z) coordinates. We assume that the mapping functions $x(\xi_1, \xi_2, \xi_3)$, $y(\xi_1, \xi_2, \xi_3)$, and $z(\xi_1, \xi_2, \xi_3)$ are given. A position vector from the origin $(0, 0, 0)$ to a point (x, y, z) in the curved cell is expressed

$$\boldsymbol{r}(\xi_1, \xi_2, \xi_3) = x(\xi_1, \xi_2, \xi_3)\hat{\boldsymbol{x}} + y(\xi_1, \xi_2, \xi_3)\hat{\boldsymbol{y}} + z(\xi_1, \xi_2, \xi_3)\hat{\boldsymbol{z}} \tag{3.37}$$

Three differential displacement vectors can be defined as

$$\boldsymbol{\ell}^i = \frac{\partial \boldsymbol{r}}{\partial \xi_i} = \frac{\partial x}{\partial \xi_i}\hat{\boldsymbol{x}} + \frac{\partial y}{\partial \xi_i}\hat{\boldsymbol{y}} + \frac{\partial z}{\partial \xi_i}\hat{\boldsymbol{z}}, \quad i = 1, 2, 3 \tag{3.38}$$

If parameters ξ_2 and ξ_3 are held constant, while ξ_1 is varied, the mapping creates a curve. The vector $\boldsymbol{\ell}^1$ is tangential to that curve. Similarly, $\boldsymbol{\ell}^2$ is tangential to a curve defined by constant values of parameters ξ_1 and ξ_3, and $\boldsymbol{\ell}^3$ is tangential to a curve defined by constant values of ξ_1 and ξ_2. These three vectors are known as *unitary base vectors*. If all three parameters are varied between constant limits to create a curvilinear cell in (x, y, z) space, two of the three base vectors are tangential to each face of that curvilinear cell. The base vectors are not necessarily mutually perpendicular at a point, nor are they unit vectors in general. For rectilinear elements, each unitary base vector coincides with (at least) one of the element's edge vectors.

Alternatively, we can define three independent vectors in terms of the gradients

$$\nabla \xi_i = \frac{\partial \xi_i}{\partial x}\hat{\boldsymbol{x}} + \frac{\partial \xi_i}{\partial y}\hat{\boldsymbol{y}} + \frac{\partial \xi_i}{\partial z}\hat{\boldsymbol{z}}, \quad i = 1, 2, 3 \tag{3.39}$$

It should be apparent from the gradient operation that the vector $\nabla \xi_i$ is normal to a surface over which ξ_i is constant. These vectors are known as *reciprocal base vectors*. The reciprocal base vectors are also not necessarily mutually perpendicular or of unit length. If parameters (ξ_1, ξ_2, ξ_3) are varied between constant limits to create a curvilinear cell in (x, y, z) space, one reciprocal base vector is normal at every point on a face of the resulting curvilinear cell.

The reciprocal base vectors can also be expressed as

$$\nabla \xi_i = -\frac{\hat{\boldsymbol{h}}_i}{h_i} \tag{3.40}$$

where $1/h_i$ is the magnitude of the gradient, and $\hat{\boldsymbol{h}}_i$ is a unit vector opposite the direction of the gradient. On the ith boundary (edge or face), $\hat{\boldsymbol{h}}_i$ is the unit outward normal to the element. The *height vector*,

$$\boldsymbol{h}_i = h_i \hat{\boldsymbol{h}}_i = -h_i^2 \nabla \xi_i \tag{3.41}$$

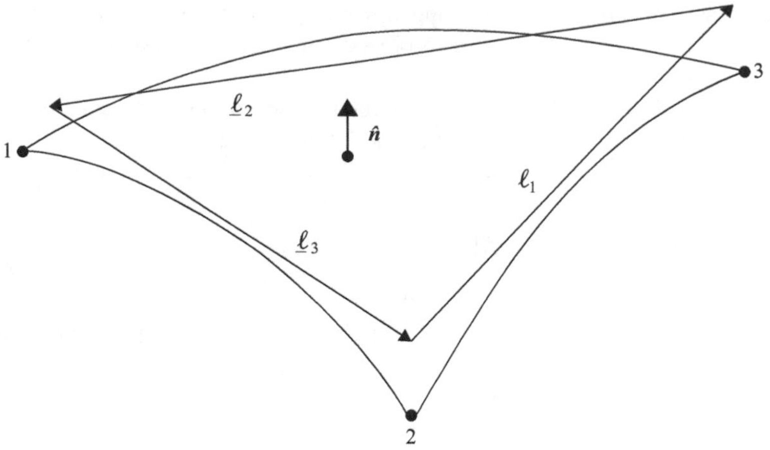

Figure 3.12 Triangle tangent to a curvilinear triangle at a point. The curvilinear and rectilinear tangent triangles have the same element coordinates, Jacobian, edge vectors, height vectors, and unit normal $\hat{n}$ at the point of tangency.

© 1997 IEEE. Reprinted with permission from R. D. Graglia, D. R. Wilton, and A. F. Peterson, "Higher order interpolatory vector bases for computational electromagnetics," special issue on "Advanced Numerical Techniques in Electromagnetics," *IEEE Trans. Antennas Propag.*, vol. 45, no. 3, pp. 329–342, Mar. 1997.

has a magnitude that measures the height of a rectilinear element perpendicular to the ith element boundary. For curvilinear elements, the edge and height vectors retain their interpretations if we apply them to a *tangent* rectilinear element that can be constructed from the local edge and height vectors at each coordinate point within the element (Figure 3.12).

In the mapping between the reference cell and the curvilinear cell, derivatives transform according to the relation

$$\begin{bmatrix} \dfrac{\partial}{\partial \xi_1} \\ \dfrac{\partial}{\partial \xi_2} \\ \dfrac{\partial}{\partial \xi_3} \end{bmatrix} = \mathbf{J} \begin{bmatrix} \dfrac{\partial}{\partial x} \\ \dfrac{\partial}{\partial y} \\ \dfrac{\partial}{\partial z} \end{bmatrix} \tag{3.42}$$

The three-by-three matrix in (3.42) is known as the Jacobian matrix

$$\mathbf{J} = \begin{bmatrix} \dfrac{\partial x}{\partial \xi_1} & \dfrac{\partial y}{\partial \xi_1} & \dfrac{\partial z}{\partial \xi_1} \\ \dfrac{\partial x}{\partial \xi_2} & \dfrac{\partial y}{\partial \xi_2} & \dfrac{\partial z}{\partial \xi_2} \\ \dfrac{\partial x}{\partial \xi_3} & \dfrac{\partial y}{\partial \xi_3} & \dfrac{\partial z}{\partial \xi_3} \end{bmatrix} \tag{3.43}$$

The differential volumes of the two spaces are related by the determinant of the Jacobian matrix

$$dx\,dy\,dz = \det[\mathbf{J}]\,d\xi_1\,d\xi_2\,d\xi_3 = \mathcal{J}\,d\xi_1\,d\xi_2\,d\xi_3 \tag{3.44}$$

It is also useful to consider the inverse relation

$$\begin{bmatrix} \dfrac{\partial}{\partial x} \\ \dfrac{\partial}{\partial y} \\ \dfrac{\partial}{\partial z} \end{bmatrix} = \mathbf{J}^{-1} \begin{bmatrix} \dfrac{\partial}{\partial \xi_1} \\ \dfrac{\partial}{\partial \xi_2} \\ \dfrac{\partial}{\partial \xi_3} \end{bmatrix} \tag{3.45}$$

where the inverse of the Jacobian matrix is given directly in the form

$$\mathbf{J}^{-1} = \begin{bmatrix} \dfrac{\partial \xi_1}{\partial x} & \dfrac{\partial \xi_2}{\partial x} & \dfrac{\partial \xi_3}{\partial x} \\ \dfrac{\partial \xi_1}{\partial y} & \dfrac{\partial \xi_2}{\partial y} & \dfrac{\partial \xi_3}{\partial y} \\ \dfrac{\partial \xi_1}{\partial z} & \dfrac{\partial \xi_2}{\partial z} & \dfrac{\partial \xi_3}{\partial z} \end{bmatrix} \tag{3.46}$$

Observe that the rows of $\mathbf{J}$ are the components of the base vectors, while the columns of $\mathbf{J}^{-1}$ are those of the reciprocal base vectors. Of course, a matrix and its inverse are also related by

$$\mathbf{J}\mathbf{J}^{-1} = \mathbf{I} \tag{3.47}$$

Equation (3.47) is equivalent to the biorthogonality relationship between $\boldsymbol{\ell}^j$ and the reciprocal basis vectors $\nabla \xi_i$

$$\boldsymbol{\ell}^j \cdot \nabla \xi_i = \delta_{ij} = \begin{cases} 1, & i = j \\ 0, & i \neq j \end{cases} \tag{3.48}$$

where δ_{ij} is the Kronecker delta. From the biorthogonality relationships, it is easily established for 3D elements that

$$\nabla \xi_i = \frac{\boldsymbol{\ell}^j \times \boldsymbol{\ell}^k}{\mathcal{J}} \tag{3.49}$$

where the Jacobian $\mathcal{J}$ can also be expressed

$$\mathcal{J} = \boldsymbol{\ell}^i \cdot \boldsymbol{\ell}^j \times \boldsymbol{\ell}^k \tag{3.50}$$

and i, j, k are taken in cyclic order. As demonstrated below, these relationships also apply in two dimensions if $\boldsymbol{\ell}^3$ and $\nabla \xi_3$ are both replaced by $\hat{\boldsymbol{n}}$.

The reciprocal base vectors cannot be computed directly, but can be found from (3.38) and (3.49). Subsequently, gradients of the *dependent* coordinates are easily found using the dependency relations. For example, since for the tetrahedron $\xi_1 + \xi_2 + \xi_3 + \xi_4 = 1$, then

$\nabla\xi_4 = -\nabla\xi_1 - \nabla\xi_2 - \nabla\xi_3$. The height vector definitions (3.40) and (3.41) also apply to the dependent coordinates. For independent coordinates, one also easily establishes that

$$\nabla\xi_j \times \nabla\xi_k = \frac{\boldsymbol{\ell}^i}{\mathcal{J}} \tag{3.51}$$

again with i, j, k in cyclic order. This relation can be generalized to apply also to dependent coordinates by introducing double subscripted quantities, $\boldsymbol{\ell}_{jk}$, defining the so-called *edge vectors* formed by the gradient cross products in (3.51)

$$\frac{\boldsymbol{\ell}_{jk}}{\mathcal{J}} = \nabla\xi_j \times \nabla\xi_k \tag{3.52}$$

Clearly, $\boldsymbol{\ell}_{jk} = -\boldsymbol{\ell}_{kj}$ and $\boldsymbol{\ell}_{jk}$ vanishes if $j = k$, or if $\nabla\xi_j$ is parallel to $\nabla\xi_k$. This extension associates a non-vanishing edge vector $\boldsymbol{\ell}_{jk}$ with *each* edge formed by intersecting constant ξ_j and ξ_k surfaces of the tangent element. Since dependent coordinates are involved, these edge vectors have linear dependencies among them and are linear combinations of the independent unitary base vectors $\boldsymbol{\ell}^i$. For triangle, quadrilateral, tetrahedral, brick, and prism elements, these relationships are found in Table 4.1 of Chapter 4. Finally, given any triplet of non-coplanar coordinate variables ξ_i, ξ_j, and ξ_k, we have

$$\nabla\xi_i \cdot (\nabla\xi_j \times \nabla\xi_k) = \frac{1}{\mathcal{J}} \tag{3.53}$$

with the indices ordered to ensure a positive Jacobian. Hence from (3.52), with the same index ordering as (3.53), we obtain the following generalization of (3.48)

$$\nabla\xi_i \cdot \boldsymbol{\ell}_{jk} = 1 \tag{3.54}$$

In two dimensions, only the case in which $\nabla\xi_k = \hat{\boldsymbol{n}}$ in (3.52) is of interest, where $\hat{\boldsymbol{n}}$ is the unit normal to the plane containing the element (or to the tangent plane at any point). In that case, the second subscript in $\boldsymbol{\ell}_{jk}$ becomes superfluous and is dropped. Thus in two dimensions, (3.52) and (3.54) become

$$\frac{\boldsymbol{\ell}_j}{\mathcal{J}} = \nabla\xi_j \times \hat{\boldsymbol{n}} \tag{3.55}$$

and

$$\nabla\xi_i \cdot \boldsymbol{\ell}_{i\pm1} = \pm 1 \tag{3.56}$$

respectively.

3.11.2 Covariant and Contravariant Projections

There are two principal ways of expressing a vector quantity in the curvilinear space [32, 33]. If represented in terms of projections onto the base vectors, known as *covariant* components, one obtains the expression

$$\boldsymbol{E} = (\boldsymbol{E} \cdot \boldsymbol{\ell}^1)\,\nabla\xi_1 + (\boldsymbol{E} \cdot \boldsymbol{\ell}^2)\,\nabla\xi_2 + (\boldsymbol{E} \cdot \boldsymbol{\ell}^3)\,\nabla\xi_3 \tag{3.57}$$

The covariant components $\boldsymbol{E} \cdot \boldsymbol{\ell}^i$ are the tangential components along the various curves defined by holding two of the three independent parameters (ξ_1, ξ_2, ξ_3) constant. When using curl-conforming basis functions, our goal is usually to maintain the tangential-vector continuity between cells and to maintain the interpolation properties, if any, of the "tangential" components of those basis functions at cell boundaries. Thus, the natural approach is to work with the covariant components of the functions and use vectors as expressed in (3.57).

Alternatively, a vector can be represented in terms of its projections onto the reciprocal base vectors, or in terms of its *contravariant* components, leading to

$$\boldsymbol{E} = (\boldsymbol{E} \cdot \nabla\xi_1)\,\boldsymbol{\ell}^1 + (\boldsymbol{E} \cdot \nabla\xi_2)\,\boldsymbol{\ell}^2 + (\boldsymbol{E} \cdot \nabla\xi_3)\,\boldsymbol{\ell}^3 \tag{3.58}$$

The contravariant components $\boldsymbol{E} \cdot \nabla\xi_i$ are the components perpendicular to the constant parameter surfaces. With divergence-conforming functions, our goal is to maintain the normal-vector continuity of the function across cell boundaries, and the interpolation properties (if any) associated with "normal" components at various locations within the cell. Since the quantities of interest are the normal-vector components, it will be convenient to work with the contravariant components as expressed in (3.58).

We now turn our attention to the mapping of the vector basis functions from the parent reference cell in (ξ_1, ξ_2, ξ_3) space to the curvilinear child cell in (x, y, z) space.[1] It is simpler to discuss the mapping of curl-conforming basis functions, so we consider those first. Curl-conforming functions maintain tangential-vector continuity across cell boundaries, and our mapping procedure must ensure that behavior. Since the principal quantities of interest are the tangential-vector components at cell boundaries, it is natural to work with the covariant components of the basis functions as expressed in (3.57).

Suppose that the basis function is defined on the parent cell by the function $\boldsymbol{R}^{\text{curl}}(\xi_1, \xi_2, \xi_3)$. In the child space (x, y, z), we define the basis function by

$$\boldsymbol{B} = R^{\text{curl}}_{\xi_1}\,\nabla\xi_1 + R^{\text{curl}}_{\xi_2}\,\nabla\xi_2 + R^{\text{curl}}_{\xi_3}\,\nabla\xi_3 \tag{3.59}$$

where

$$\boldsymbol{B} \cdot \boldsymbol{\ell}^i = R^{\text{curl}}_{\xi_i} \tag{3.60}$$

On a component-by-component level, (3.59) is equivalent to the matrix relation

$$\begin{bmatrix} B_x \\ B_y \\ B_z \end{bmatrix} = \begin{bmatrix} \dfrac{\partial\xi_1}{\partial x} & \dfrac{\partial\xi_2}{\partial x} & \dfrac{\partial\xi_3}{\partial x} \\ \dfrac{\partial\xi_1}{\partial y} & \dfrac{\partial\xi_2}{\partial y} & \dfrac{\partial\xi_3}{\partial y} \\ \dfrac{\partial\xi_1}{\partial z} & \dfrac{\partial\xi_2}{\partial z} & \dfrac{\partial\xi_3}{\partial z} \end{bmatrix} \begin{bmatrix} R^{\text{curl}}_{\xi_1} \\ R^{\text{curl}}_{\xi_2} \\ R^{\text{curl}}_{\xi_3} \end{bmatrix} = \mathbf{J}^{-1} \begin{bmatrix} R^{\text{curl}}_{\xi_1} \\ R^{\text{curl}}_{\xi_2} \\ R^{\text{curl}}_{\xi_3} \end{bmatrix} \tag{3.61}$$

There are two aspects of the basis definition to consider: (1) the continuity of tangential fields across cell boundaries and (2) the normalization of the basis functions in the curvilinear domain.

---

[1] The material in this section has been adapted from Reference 33.

As a consequence of the way that the mapping of the cell coordinates is defined, the base vectors are tangential to the boundaries of the curvilinear patch at the extreme values of the parameters ξ_1, ξ_2, and ξ_3. Within two adjacent cells sharing a boundary that coincides with constant-ξ_1 and constant-ξ_3 surfaces, and common cell endpoints in ξ_2, the base vector tangential to that boundary is defined by the derivatives of x, y, and z with respect to ξ_2, and will therefore be the same in either cell. The imposition of (3.60) with $i = 1$ and $i = 3$ in adjacent cells, for basis functions that are adjusted to maintain tangential continuity in the parent space, will ensure that the resulting vector basis functions in the (x, y, z) child space also maintain tangential continuity.

Curl-conforming basis functions can be normalized so that their tangential components have unity value at appropriate locations along a cell boundary. (This is usually desired for interpolatory bases and may or may not be necessary for hierarchical bases.) As an example, consider one such location (ξ_1, ξ_2, ξ_3) along a boundary where $\boldsymbol{\ell}^2$ is tangential and $\boldsymbol{R}^{\text{curl}}$ has a unit tangent. In the curvilinear x, y, z space, it follows that

$$\boldsymbol{B} \cdot \boldsymbol{\ell}^2 \Big|_{\xi_1, \xi_2, \xi_3} = 1 \tag{3.62}$$

However, since the vector $\boldsymbol{\ell}^2$ is not a unit vector, an additional scaling or normalization factor is necessary if the curl-conforming basis function in the curvilinear space is to exhibit a unity tangential value at that location. This scale factor is the magnitude of $\boldsymbol{\ell}^2$ at that point, which can be determined from the entries of the Jacobian matrix.

Now, let us turn our attention to divergence-conforming basis functions. Divergence-conforming functions maintain normal-vector continuity across cell boundaries, and our mapping procedure must ensure that behavior in the x, y, z space. It also may be desirable for those functions to interpolate to normal-vector components at a specific location in the cell. Since the principal quantities of interest are the normal-vector components on cell boundaries, it will be necessary to work with the contravariant components of the basis functions as expressed in (3.58).

The mapping functions $x(\xi_1, \xi_2, \xi_3)$, $y(\xi_1, \xi_2, \xi_3)$, and $z(\xi_1, \xi_2, \xi_3)$ in (3.37) used to create skewed or curved cells do not ensure the continuity of the reciprocal base vectors across a cell boundary. From (3.49), and using the continuity properties of the base vectors, it can be shown that it is sufficient to scale by the determinant $\mathcal{J}$ to obtain reciprocal base vectors of the same length on either side of a cell boundary where they are normal. Thus, to define basis functions in the child (x, y, z) space with normal-vector continuity, we use the contravariant components, scaled by the function $\mathcal{J}(\xi_1, \xi_2, \xi_3)$, and obtain

$$\boldsymbol{B} = \frac{1}{\mathcal{J}} R^{\text{div}}_{\xi_1} \boldsymbol{\ell}^1 + \frac{1}{\mathcal{J}} R^{\text{div}}_{\xi_2} \boldsymbol{\ell}^2 + \frac{1}{\mathcal{J}} R^{\text{div}}_{\xi_3} \boldsymbol{\ell}^3 \tag{3.63}$$

where $\boldsymbol{R}^{\text{div}}(\xi_1, \xi_2, \xi_3)$ is the basis function in the parent coordinates, and where the contravariant components are

$$\boldsymbol{B} \cdot \nabla \xi_i = \frac{1}{\mathcal{J}} R^{\text{div}}_{\xi_i} \tag{3.64}$$

Equivalently, the basis function in the (x, y, z) space can be written as

$$\begin{bmatrix} B_x \\ B_y \\ B_z \end{bmatrix} = \frac{1}{\mathcal{J}} \begin{bmatrix} \frac{\partial x}{\partial \xi_1} & \frac{\partial x}{\partial \xi_2} & \frac{\partial x}{\partial \xi_3} \\ \frac{\partial y}{\partial \xi_1} & \frac{\partial y}{\partial \xi_2} & \frac{\partial y}{\partial \xi_3} \\ \frac{\partial z}{\partial \xi_1} & \frac{\partial z}{\partial \xi_2} & \frac{\partial z}{\partial \xi_3} \end{bmatrix} \begin{bmatrix} R_{\xi_1}^{\text{div}} \\ R_{\xi_2}^{\text{div}} \\ R_{\xi_3}^{\text{div}} \end{bmatrix} = \frac{1}{\mathcal{J}} \mathbf{J}^T \begin{bmatrix} R_{\xi_1}^{\text{div}} \\ R_{\xi_2}^{\text{div}} \\ R_{\xi_3}^{\text{div}} \end{bmatrix} \tag{3.65}$$

where $\mathbf{J}^T$ is the transpose of the Jacobian matrix.

The divergence-conforming basis functions may be scaled so that their normal components have unity value at appropriate locations along a cell boundary. The appropriate normalization factor is the magnitude of the base vector at the interpolation point.

To summarize, curl-conforming functions are best represented by their covariant projections onto the base vectors, resulting in the definition (3.61) of the function in (x, y, z) space. In contrast, divergence-conforming functions are best represented by their contravariant projections onto the reciprocal base vectors, leading to the child-space definition in (3.65). In either case, an additional normalization factor is necessary to properly scale the tangential or normal component in the child space.

3.11.3 Derivatives in the Parent Space

An intrinsic feature of the type of analysis under consideration is that all operations involving the basis functions on the curvilinear cell (child space) can be transferred to the reference cell (parent space). First, we note that the gradient operator may be expressed as

$$\nabla f = \frac{\partial f}{\partial \xi_1} \nabla \xi_1 + \frac{\partial f}{\partial \xi_2} \nabla \xi_2 + \frac{\partial f}{\partial \xi_3} \nabla \xi_3 \tag{3.66}$$

which is also equivalent to

$$\nabla f = -\sum_{i=1}^{N} \frac{\partial f}{\partial \xi_i} \frac{\hat{\boldsymbol{h}}_i}{h_i} \tag{3.67}$$

Consider a curl-conforming basis function defined by

$$\boldsymbol{B} = R_{\xi_1}^{\text{curl}} \nabla \xi_1 + R_{\xi_2}^{\text{curl}} \nabla \xi_2 + R_{\xi_3}^{\text{curl}} \nabla \xi_3 \tag{3.68}$$

Using the standard vector identity

$$\nabla \times (f \nabla \xi_i) = \nabla f \times \nabla \xi_i \tag{3.69}$$

the curl of $\boldsymbol{B}$ can be expanded to produce

$$\begin{aligned} \nabla \times \boldsymbol{B} &= \nabla \times (R_{\xi_1}^{\text{curl}} \nabla \xi_1) + \nabla \times (R_{\xi_2}^{\text{curl}} \nabla \xi_2) + \nabla \times (R_{\xi_3}^{\text{curl}} \nabla \xi_3) \\ &= \nabla (R_{\xi_1}^{\text{curl}}) \times \nabla \xi_1 + \nabla (R_{\xi_2}^{\text{curl}}) \times \nabla \xi_2 + \nabla (R_{\xi_3}^{\text{curl}}) \times \nabla \xi_3 \end{aligned} \tag{3.70}$$

From (3.66), the gradients in (3.70) can be expressed as

$$\nabla(R_{\xi_i}^{\text{curl}}) = \frac{\partial R_{\xi_i}^{\text{curl}}}{\partial \xi_1}\nabla\xi_1 + \frac{\partial R_{\xi_i}^{\text{curl}}}{\partial \xi_2}\nabla\xi_2 + \frac{\partial R_{\xi_i}^{\text{curl}}}{\partial \xi_3}\nabla\xi_3 \tag{3.71}$$

Thus, (3.70) can be written as

$$\begin{aligned}\nabla \times \boldsymbol{B} = &\frac{\partial R_{\xi_1}^{\text{curl}}}{\partial \xi_1}\nabla\xi_1 \times \nabla\xi_1 + \frac{\partial R_{\xi_1}^{\text{curl}}}{\partial \xi_2}\nabla\xi_2 \times \nabla\xi_1 + \frac{\partial R_{\xi_1}^{\text{curl}}}{\partial \xi_3}\nabla\xi_3 \times \nabla\xi_1 \\ &+ \frac{\partial R_{\xi_2}^{\text{curl}}}{\partial \xi_1}\nabla\xi_1 \times \nabla\xi_2 + \frac{\partial R_{\xi_2}^{\text{curl}}}{\partial \xi_2}\nabla\xi_2 \times \nabla\xi_2 + \frac{\partial R_{\xi_2}^{\text{curl}}}{\partial \xi_3}\nabla\xi_3 \times \nabla\xi_2 \\ &+ \frac{\partial R_{\xi_3}^{\text{curl}}}{\partial \xi_1}\nabla\xi_1 \times \nabla\xi_3 + \frac{\partial R_{\xi_3}^{\text{curl}}}{\partial \xi_2}\nabla\xi_2 \times \nabla\xi_3 + \frac{\partial R_{\xi_3}^{\text{curl}}}{\partial \xi_3}\nabla\xi_3 \times \nabla\xi_3\end{aligned} \tag{3.72}$$

Three terms in (3.72) vanish, and the other terms can be simplified using (3.52) to produce

$$\nabla \times \boldsymbol{B} = \frac{1}{\mathcal{J}}\left\{\left(\frac{\partial R_{\xi_3}^{\text{curl}}}{\partial \xi_2} - \frac{\partial R_{\xi_2}^{\text{curl}}}{\partial \xi_3}\right)\boldsymbol{\ell}^1 + \left(\frac{\partial R_{\xi_1}^{\text{curl}}}{\partial \xi_3} - \frac{\partial R_{\xi_3}^{\text{curl}}}{\partial \xi_1}\right)\boldsymbol{\ell}^2 + \left(\frac{\partial R_{\xi_2}^{\text{curl}}}{\partial \xi_1} - \frac{\partial R_{\xi_1}^{\text{curl}}}{\partial \xi_2}\right)\boldsymbol{\ell}^3\right\} \tag{3.73}$$

Thus, the curl in the child space is just the curl in the parent space, scaled by the Jacobian, as expressed

$$\nabla \times \boldsymbol{B}|_{x,y,z} = \frac{1}{\mathcal{J}}\,\nabla \times \boldsymbol{R}^{\text{curl}}\Big|_{\xi_1,\xi_2,\xi_3} \tag{3.74}$$

Consider a divergence-conforming basis function defined by

$$\begin{aligned}\boldsymbol{B} &= \frac{1}{\mathcal{J}}R_{\xi_1}^{\text{div}}\,\boldsymbol{\ell}^1 + \frac{1}{\mathcal{J}}R_{\xi_2}^{\text{div}}\,\boldsymbol{\ell}^2 + \frac{1}{\mathcal{J}}R_{\xi_3}^{\text{div}}\,\boldsymbol{\ell}^3 \\ &= R_{\xi_1}^{\text{div}}(\nabla\xi_2 \times \nabla\xi_3) + R_{\xi_2}^{\text{div}}(\nabla\xi_3 \times \nabla\xi_1) + R_{\xi_3}^{\text{div}}(\nabla\xi_1 \times \nabla\xi_2)\end{aligned} \tag{3.75}$$

where we have used (3.51). The vector identities

$$\nabla \cdot (\boldsymbol{A} \times \boldsymbol{B}) = \boldsymbol{B} \cdot \nabla \times \boldsymbol{A} - \boldsymbol{A} \cdot \nabla \times \boldsymbol{B} \tag{3.76}$$

and

$$\nabla \times \nabla f = 0 \tag{3.77}$$

can be used to show that

$$\nabla \cdot (\nabla\xi_i \times \nabla\xi_j) = 0 \tag{3.78}$$

It follows from the additional vector identity

$$\nabla \cdot (f\,\boldsymbol{V}) = f\,\nabla \cdot \boldsymbol{V} + \nabla f \cdot \boldsymbol{V} \tag{3.79}$$

that the divergence of $\boldsymbol{B}$ is given by

$$\begin{aligned}
\nabla\cdot\boldsymbol{B} &= \nabla(R^{\text{div}}_{\xi_1})\cdot(\nabla\xi_2\times\nabla\xi_3)+\nabla(R^{\text{div}}_{\xi_2})\cdot(\nabla\xi_3\times\nabla\xi_1)+\nabla(R^{\text{div}}_{\xi_3})\cdot(\nabla\xi_1\times\nabla\xi_2)\\
&= \nabla(R^{\text{div}}_{\xi_1})\cdot\frac{1}{\mathcal{J}}\boldsymbol{\ell}^1+\nabla(R^{\text{div}}_{\xi_2})\cdot\frac{1}{\mathcal{J}}\boldsymbol{\ell}^2+\nabla(R^{\text{div}}_{\xi_3})\cdot\frac{1}{\mathcal{J}}\boldsymbol{\ell}^3\\
&= \frac{1}{\mathcal{J}}\left\{\left[\frac{\partial R^{\text{div}}_{\xi_1}}{\partial\xi_1}\nabla\xi_1+\frac{\partial R^{\text{div}}_{\xi_1}}{\partial\xi_2}\nabla\xi_2+\frac{\partial R^{\text{div}}_{\xi_1}}{\partial\xi_3}\nabla\xi_3\right]\cdot\boldsymbol{\ell}^1\right.\\
&\quad+\left[\frac{\partial R^{\text{div}}_{\xi_2}}{\partial\xi_1}\nabla\xi_1+\frac{\partial R^{\text{div}}_{\xi_2}}{\partial\xi_2}\nabla\xi_2+\frac{\partial R^{\text{div}}_{\xi_2}}{\partial\xi_3}\nabla\xi_3\right]\cdot\boldsymbol{\ell}^2\\
&\quad\left.+\left[\frac{\partial R^{\text{div}}_{\xi_3}}{\partial\xi_1}\nabla\xi_1+\frac{\partial R^{\text{div}}_{\xi_3}}{\partial\xi_2}\nabla\xi_2+\frac{\partial R^{\text{div}}_{\xi_3}}{\partial\xi_3}\nabla\xi_3\right]\cdot\boldsymbol{\ell}^3\right\}\\
&= \frac{1}{\mathcal{J}}\left[\frac{\partial R^{\text{div}}_{\xi_1}}{\partial\xi_1}+\frac{\partial R^{\text{div}}_{\xi_2}}{\partial\xi_2}+\frac{\partial R^{\text{div}}_{\xi_3}}{\partial\xi_3}\right]
\end{aligned} \tag{3.80}$$

Therefore, the divergence of (3.75) in x–y–z space can be calculated directly from the application of a divergence operator in the parent coordinates, scaled by the determinant, to produce

$$\nabla\cdot\boldsymbol{B}|_{x,y,z}=\frac{1}{\mathcal{J}}\,\nabla\cdot\boldsymbol{R}^{\text{div}}\Big|_{\xi_1,\xi_2,\xi_3} \tag{3.81}$$

3.11.4 Restriction to Surfaces

When mapping from a square or triangular planar reference cell (2D) to a curved surface in a 3D space, the third variable (ξ_3) is not involved. The mapping functions x, y, and z are only functions of ξ_1 and ξ_2. In common with our previous approach, the position vector from the origin to a point (x, y, z) is given by

$$\boldsymbol{r}(\xi_1,\xi_2)=x(\xi_1,\xi_2)\hat{\boldsymbol{x}}+y(\xi_1,\xi_2)\hat{\boldsymbol{y}}+z(\xi_1,\xi_2)\hat{\boldsymbol{z}} \tag{3.82}$$

The base vectors for this cell are

$$\boldsymbol{\ell}^i=\frac{\partial\boldsymbol{r}}{\partial\xi_i}=\frac{\partial x}{\partial\xi_i}\hat{\boldsymbol{x}}+\frac{\partial y}{\partial\xi_i}\hat{\boldsymbol{y}}+\frac{\partial z}{\partial\xi_i}\hat{\boldsymbol{z}},\quad i=1,2 \tag{3.83}$$

while the reciprocal base vectors are

$$\nabla\xi_i=\frac{\partial\xi_i}{\partial x}\hat{\boldsymbol{x}}+\frac{\partial\xi_i}{\partial y}\hat{\boldsymbol{y}}+\frac{\partial\xi_i}{\partial z}\hat{\boldsymbol{z}},\quad i=1,2 \tag{3.84}$$

The base vectors are tangential to the surface defined by (3.82), while the reciprocal base vectors are normal to constant-ξ_1 or constant-ξ_2 surfaces, respectively.

Derivatives transform according to

$$\begin{bmatrix} \dfrac{\partial}{\partial \xi_1} \\ \dfrac{\partial}{\partial \xi_2} \end{bmatrix} = \begin{bmatrix} \dfrac{\partial x}{\partial \xi_1} & \dfrac{\partial y}{\partial \xi_1} & \dfrac{\partial z}{\partial \xi_1} \\ \dfrac{\partial x}{\partial \xi_2} & \dfrac{\partial y}{\partial \xi_2} & \dfrac{\partial z}{\partial \xi_2} \end{bmatrix} \begin{bmatrix} \dfrac{\partial}{\partial x} \\ \dfrac{\partial}{\partial y} \\ \dfrac{\partial}{\partial z} \end{bmatrix} \tag{3.85}$$

and

$$\begin{bmatrix} \dfrac{\partial}{\partial x} \\ \dfrac{\partial}{\partial y} \\ \dfrac{\partial}{\partial z} \end{bmatrix} = \begin{bmatrix} \dfrac{\partial \xi_1}{\partial x} & \dfrac{\partial \xi_2}{\partial x} \\ \dfrac{\partial \xi_1}{\partial y} & \dfrac{\partial \xi_2}{\partial y} \\ \dfrac{\partial \xi_1}{\partial z} & \dfrac{\partial \xi_2}{\partial z} \end{bmatrix} \begin{bmatrix} \dfrac{\partial}{\partial \xi_1} \\ \dfrac{\partial}{\partial \xi_2} \end{bmatrix} \tag{3.86}$$

The Jacobian matrix in (3.85) has dimension 2×3 in accordance with the dimensionality of the mapping, while the inverse Jacobian matrix in (3.86) has dimension 3×2.

Since the functions x, y, and z are defined explicitly, the entries of the Jacobian matrix in (3.85) are readily available. However, the matrix entries in (3.86) are not. In situations where the inverse Jacobian entries are required (to be encountered below), it is necessary to be able to invert the Jacobian matrix numerically, which is not possible if it is not a square matrix! Thus, it is desirable to introduce a dummy parameter ξ_3, to fill the equations in (3.85) and (3.86) out to 3×3 systems. Since the third variable is arbitrary, the dummy parameter may be constrained so that [33]

$$\nabla \xi_3 = \frac{\boldsymbol{\ell}^1 \times \boldsymbol{\ell}^2}{|\boldsymbol{\ell}^1 \times \boldsymbol{\ell}^2|} \tag{3.87}$$

which, by virtue of its definition, is a unit vector $\hat{\boldsymbol{n}}$. (The only assumption being made here is that ξ_3 can be chosen to make $\nabla \xi_3$ a unit vector.) From (3.83),

$$\left|\boldsymbol{\ell}^1 \times \boldsymbol{\ell}^2\right| = \sqrt{\begin{aligned} &\left(\frac{\partial y}{\partial \xi_1}\frac{\partial z}{\partial \xi_2} - \frac{\partial z}{\partial \xi_1}\frac{\partial y}{\partial \xi_2}\right)^2 + \left(\frac{\partial z}{\partial \xi_1}\frac{\partial x}{\partial \xi_2} - \frac{\partial x}{\partial \xi_1}\frac{\partial z}{\partial \xi_2}\right)^2 \\ &+ \left(\frac{\partial x}{\partial \xi_1}\frac{\partial y}{\partial \xi_2} - \frac{\partial y}{\partial \xi_1}\frac{\partial x}{\partial \xi_2}\right)^2 \end{aligned}} \tag{3.88}$$

However, (3.49) still holds in this situation, implying that (3.88) is equivalent to the determinant

$$\det[\mathbf{J}] = \mathcal{J}(\xi_1, \xi_2) = |\boldsymbol{\ell}^1 \times \boldsymbol{\ell}^2| \tag{3.89}$$

One way by which (3.89) can be true is if

$$\boldsymbol{\ell}^3 = \frac{\partial \boldsymbol{r}}{\partial \xi_3} = \frac{1}{\mathcal{J}}\left(\frac{\partial y}{\partial \xi_1}\frac{\partial z}{\partial \xi_2} - \frac{\partial z}{\partial \xi_1}\frac{\partial y}{\partial \xi_2}\right)\hat{\boldsymbol{x}} + \frac{1}{\mathcal{J}}\left(\frac{\partial z}{\partial \xi_1}\frac{\partial x}{\partial \xi_2} - \frac{\partial x}{\partial \xi_1}\frac{\partial z}{\partial \xi_2}\right)\hat{\boldsymbol{y}}$$
$$+\frac{1}{\mathcal{J}}\left(\frac{\partial x}{\partial \xi_1}\frac{\partial y}{\partial \xi_2} - \frac{\partial y}{\partial \xi_1}\frac{\partial x}{\partial \xi_2}\right)\hat{\boldsymbol{z}} \tag{3.90}$$

which is the same as

$$\boldsymbol{\ell}^3 = \frac{1}{\mathcal{J}}\boldsymbol{\ell}^1 \times \boldsymbol{\ell}^2 = \nabla \xi_3 \tag{3.91}$$

The implications of this choice for ξ_3 are summarized as follows:

1. $\boldsymbol{\ell}^3 = \nabla\xi_3 = \hat{\boldsymbol{n}}$, where both are unit vectors normal to the surface of constant ξ_3.
2. Vectors $\boldsymbol{\ell}^1$ and $\boldsymbol{\ell}^2$ are tangential to the surface of constant ξ_3, by definition. Since $\boldsymbol{\ell}^3$ is perpendicular to the surface (by our choice of ξ_3), the reciprocal base vectors $\nabla\xi_1$ and $\nabla\xi_2$ are also tangential to the surface of constant ξ_3. It follows that

$$\nabla\xi_1 = \frac{1}{\mathcal{J}}\boldsymbol{\ell}^2 \times \boldsymbol{\ell}^3 \tag{3.92}$$

$$\nabla\xi_2 = \frac{1}{\mathcal{J}}\boldsymbol{\ell}^3 \times \boldsymbol{\ell}^1 \tag{3.93}$$

3. Since (a) $\boldsymbol{\ell}^3$ is a unit vector, (b) $\boldsymbol{\ell}^3$ is perpendicular to $\boldsymbol{\ell}^2$, and (c) $\boldsymbol{\ell}^3$ is perpendicular to $\boldsymbol{\ell}^1$, the magnitudes of the reciprocal base vectors are related to those of the base vectors by

$$|\nabla\xi_1| = \left|\frac{1}{\mathcal{J}}\boldsymbol{\ell}^2 \times \boldsymbol{\ell}^3\right| = \frac{|\boldsymbol{\ell}^2|}{\mathcal{J}} \tag{3.94}$$

$$|\nabla\xi_2| = \left|\frac{1}{\mathcal{J}}\boldsymbol{\ell}^3 \times \boldsymbol{\ell}^1\right| = \frac{|\boldsymbol{\ell}^1|}{\mathcal{J}} \tag{3.95}$$

4. The explicit form of the 3×3 Jacobian matrix is given by

$$\mathbf{J} = \begin{bmatrix} \dfrac{\partial x}{\partial \xi_1} & \dfrac{\partial y}{\partial \xi_1} & \dfrac{\partial z}{\partial \xi_1} \\ \dfrac{\partial x}{\partial \xi_2} & \dfrac{\partial y}{\partial \xi_2} & \dfrac{\partial z}{\partial \xi_2} \\ \dfrac{1}{\mathcal{J}}\left(\dfrac{\partial y}{\partial \xi_1}\dfrac{\partial z}{\partial \xi_2} - \dfrac{\partial z}{\partial \xi_1}\dfrac{\partial y}{\partial \xi_2}\right) & \dfrac{1}{\mathcal{J}}\left(\dfrac{\partial z}{\partial \xi_1}\dfrac{\partial x}{\partial \xi_2} - \dfrac{\partial x}{\partial \xi_1}\dfrac{\partial z}{\partial \xi_2}\right) & \dfrac{1}{\mathcal{J}}\left(\dfrac{\partial x}{\partial \xi_1}\dfrac{\partial y}{\partial \xi_2} - \dfrac{\partial y}{\partial \xi_1}\dfrac{\partial x}{\partial \xi_2}\right) \end{bmatrix} \tag{3.96}$$

In summary, when working with the base and reciprocal base vectors on a 2D surface in 3D space, it will be convenient to invoke the assumption in (3.87). This ensures that base vectors $\boldsymbol{\ell}^1$ and $\boldsymbol{\ell}^2$, and reciprocal base vectors $\nabla\xi_1$ and $\nabla\xi_2$, are tangential to the surface and permits the computation of all needed parameters.

3.11.5 Example: Quadrilateral Cells

To illustrate the preceding development, consider the mapping of the 2D CT/LN curl-conforming bases defined for square cells in Section 3.3 to quadrilateral cells defined by the four corner nodes (x_1, y_1) to (x_4, y_4). A mapping can be constructed using the $p = 1$ (linear) shape functions from Section 2.5 to interpolate between these nodes, to yield the functions

$$x(\xi_1,\xi_2) = x_1(1-\xi_1)(1-\xi_2) + x_2\xi_1(1-\xi_2) + x_3(1-\xi_1)\xi_2 + x_4\xi_1\xi_2 \tag{3.97}$$

$$y(\xi_1,\xi_2) = y_1(1-\xi_1)(1-\xi_2) + y_2\xi_1(1-\xi_2) + y_3(1-\xi_1)\xi_2 + y_4\xi_1\xi_2 \tag{3.98}$$

From these equations, the entries of the Jacobian matrix are obtained as

$$\frac{\partial x}{\partial \xi_1} = -x_1(1-\xi_2) + x_2(1-\xi_2) - x_3\xi_2 + x_4\xi_2 \tag{3.99}$$

$$\frac{\partial x}{\partial \xi_2} = -x_1(1-\xi_1) - x_2\xi_1 + x_3(1-\xi_1) + x_4\xi_1 \tag{3.100}$$

with two analogous equations for the derivatives with respect to y.

Basis functions can be defined on the skewed quadrilateral cell according to the 2D version of (3.61), given by

$$\begin{bmatrix} B_x \\ B_y \end{bmatrix} = \mathbf{J}^{-1} \begin{bmatrix} R^{\text{curl}}_{\xi_1} \\ R^{\text{curl}}_{\xi_2} \end{bmatrix} \tag{3.101}$$

where $R^{\text{curl}}_{\xi_1}$ and $R^{\text{curl}}_{\xi_2}$ are the basis functions defined in the (square) parent domain in Section 3.3. Note that this definition requires the inverse of the Jacobian matrix, which is a function of position and must be inverted explicitly at each required value of (ξ_1,ξ_2). Thus, (3.101) provides the definition of the basis functions for use within various element matrix integrals in differential equation or integral equation formulations. As these integrals are evaluated by quadrature, the Jacobian, inverse Jacobian, and other quantities may be computed *at those quadrature points* during the course of the evaluation. Furthermore, to ensure tangential-vector continuity at the appropriate cell boundaries, each basis function must be normalized as described in Section 3.11.2.

To extend the procedure to curved cells, the linear mapping in (3.97) and (3.98) can be replaced by a quadratic, cubic, or other mapping defined by higher order shape functions (See Table 2.4 in Chapter 2). These mappings will involve additional nodes to define the cell shapes.

3.12 The Mixed-Order Nédélec Spaces

In a 1980 paper, Jean-Claude Nédélec proposed mixed-order spaces of vector functions for use in representing electromagnetic quantities [34]. The idea of discarding some of the DoFs contained in a "complete" polynomial space may not seem intuitive. However, this approach has been widely used, and the motivation for it will be illustrated by an example.

Consider the linear expansion from (3.2)

$$f(x,y) = (a_0 + a_1 x + a_2 y)\hat{x} + (b_0 + b_1 x + b_2 y)\hat{y} \tag{3.102}$$

Suppose that a set of curl-conforming expansion functions, such as those in (3.7), are to be used within a numerical solution of the vector Helmholtz equation,

$$\nabla \times \left(\frac{1}{\mu_r} \nabla \times \boldsymbol{E} \right) - k^2 \varepsilon_r \boldsymbol{E} = 0 \tag{3.103}$$

to find the electric field $\boldsymbol{E}$ in a source-free region with relative permittivity ε_r and relative permeability μ_r. The parameter k is the wavenumber

$$k = \frac{\omega}{c_0} \sqrt{\mu_r \varepsilon_r} \tag{3.104}$$

where c_0 is the speed of light in a vacuum. This vector Helmholtz operator has an interesting property, namely that it admits two families of solutions. One family consists of the desired electromagnetic fields, which satisfy Maxwell's equations, and therefore that family has the properties

$$\nabla \times \boldsymbol{E} \rightarrow \text{non-zero} \tag{3.105}$$

$$\nabla \cdot (\varepsilon_r \boldsymbol{E}) = 0 \tag{3.106}$$

The other family has the form

$$\boldsymbol{E} = \nabla \Psi \tag{3.107}$$

and therefore exhibits the properties

$$\nabla \times \boldsymbol{E} = 0 \tag{3.108}$$

$$\nabla \cdot (\varepsilon_r \boldsymbol{E}) \rightarrow \text{non-zero} \tag{3.109}$$

Thus, the second family does not satisfy Maxwell's equations, although it satisfies the Helmholtz equation for $k = 0$. This family of solutions constitutes the *nullspace* of the vector Helmholtz operator.

Numerical solutions of (3.103) obtained by traditional techniques will generally contain solutions from both families. As a specific example, consider an air-filled 2D rectangular cavity with perfect electric walls. The resonant modes of that cavity may be obtained by treating (3.103) as an eigenvalue equation and solving for the resonant modes and their associated wavenumbers. (The details of this numerical solution procedure will be deferred until Chapter 6.) Table 3.3 shows numerical values for the lowest resonant wavenumbers of a rectangular cavity, bounded by perfect electric walls, compared to the exact analytical results. The set of curl-conforming vector basis functions from (3.7) is used, which requires two unknown coefficients per edge of the mesh. A triangular-cell mesh with 144 cells is used to model the rectangular domain. For the transverse magnetic (TM) modes, obtained by applying the vector Helmholtz operator to the $\boldsymbol{H}$ field and imposing the Neumann type of boundary condition

$$\hat{\boldsymbol{n}} \times \nabla \times \boldsymbol{H} = 0 \tag{3.110}$$

on the cavity walls, the system of equations has order 468. (No equations are eliminated by the boundary condition since it is a "natural" condition.) For the transverse electric (TE) modes, obtained by treating the transverse vector as the $\boldsymbol{E}$ field and imposing the Dirichlet type of boundary condition

$$\hat{\boldsymbol{n}} \times \boldsymbol{E} = 0 \tag{3.111}$$

on the cavity walls, the system has order 396. (For the TE boundary condition, two equations are eliminated from the system per boundary edge.) Once the system of equations is constructed, a matrix eigenvalue algorithm is employed to solve the generalized matrix eigenvalue problem.

From Table 3.3, we observe that the non-zero numerical results exhibit good correlation with the exact wavenumbers for the rectangular cavity. However, many of the numerical wavenumbers are zero (324 for the TM case, 253 for the TE case). These solutions have been drawn from the nullspace. Since the nullspace solutions are non-physical, and more than half of the results are fictitious, a substantial percentage of the computational effort to generate the numerical solutions is being wasted on nullspace solutions. Therefore, it is reasonable to ask if there is some solution process that would eliminate the nullspace from the computational solution space.

Table 3.3 Numerical results for the resonant wavenumbers of a rectangular cavity of dimension 1.0 by 0.5, obtained from the vector Helmholtz equation using LT/LN curl conforming vector basis and testing functions of the form in (3.7). The zero and smallest non-zero results are shown and compared to the exact results obtained from $k = \sqrt{\left(\frac{m\pi}{a}\right)^2 + \left(\frac{n\pi}{b}\right)^2}$.

Mode	TM	TE	Exact
	0.000 (324)	0.000 (253)	
10		3.148	3.1416
01		6.331	6.2832
20		6.331	6.2832
11	7.091	7.092	7.0248
21	9.019	9.016	8.8858
30		9.588	9.4248
31	11.601	11.595	11.3272
02		12.944	12.5664
40		12.950	12.5664
12	13.367	13.363	12.9531

The Nédélec curl-conforming spaces are designed to accomplish this, at least in part. Since the nullspace solutions have the form in (3.107), one avenue of approach would be to try to suppress solutions of that form from the expansion set. In fact, the linear representation

$$f(x,y) = (a_0 + a_1 x + a_2 y)\hat{x} + (b_0 + b_1 x + b_2 y)\hat{y} \tag{3.112}$$

can be decomposed into two parts

$$f_{\text{curl}}(x,y) = \left(a_0 + \frac{a_2 - b_1}{2}y\right)\hat{x} + \left(b_0 + \frac{b_1 - a_2}{2}x\right)\hat{y} \tag{3.113}$$

and

$$f_{\text{grad}}(x,y) = \left(a_1 x + \frac{a_2 + b_1}{2}y\right)\hat{x} + \left(\frac{b_1 + a_2}{2}x + b_2 y\right)\hat{y} \tag{3.114}$$

where $\boldsymbol{f}_{\text{curl}} + \boldsymbol{f}_{\text{grad}} = \boldsymbol{f}$ [35]. Observe that

$$f_{\text{grad}} = \nabla \left\{ \frac{a_1}{2}x^2 + \frac{a_2 + b_1}{2}xy + \frac{b_2}{2}y^2 \right\} \tag{3.115}$$

In this case, the three DoFs associated with f_{grad} will only be able to represent nullspace solutions since these functions have identically zero curl. By employing a representation with the form of (3.113), we omit those DoFs. It happens that the mixed-order basis functions in (3.9), and depicted in Figure 3.4, have the Cartesian form in (3.113). Those basis functions only involve one unknown per edge of the mesh, and if they are used instead of the full linear set in (3.7) they reduce the order of the system (for the TM case) from 468 to 234, and the TE system from 396 to 198. Using this reduced set, we obtain the numerical results depicted in Table 3.4.

Table 3.4 Numerical results for the resonant wavenumbers of a rectangular cavity of dimension 1.0 by 0.5, obtained from the vector Helmholtz equation using CT/LN curl-conforming vector basis and testing functions of the form in (3.9). The zero and smallest non-zero results are shown, and compared to the exact results obtained from $k = \sqrt{\left(\frac{m\pi}{a}\right)^2 + \left(\frac{n\pi}{b}\right)^2}$.

Mode	TM	TE	Exact
	0.000 (90)	0.000 (55)	
10		3.139	3.1416
01		6.259	6.2832
20		6.259	6.2832
11	7.023	7.024	7.0248
21	8.919	8.915	8.8858
30		9.347	9.4248
31	11.362	11.356	11.3272
02		12.374	12.5664
40		12.380	12.5664
12	12.813	12.811	12.9531

From Table 3.4, we observe that the number of nullspace solutions has been reduced by 234 for the TM case and by 198 in the TE case. The number of non-zero results remains the same as in Table 3.3. In other words, every DoF discarded by moving from the LT/LN basis to the CT/LN basis was originally used to generate additional nullspace solutions! Furthermore, the CT/LN eigenvalues appear to be slightly more accurate than those produced by the LT/LN expansion.

The mixed-order space of basis functions in (3.113) constitutes the lowest order ($p = 0.5$) member of the curl-conforming Nédélec spaces and omits the DoFs in (3.114) associated with the gradient of a quadratic polynomial. The resulting space does not omit all the possible gradients in the solution space; in fact there remains some nullspace solutions in Table 3.4. However, the Nédélec spaces eliminate all the gradient DoFs that can be eliminated locally, within the definition of the basis functions in the parent space. In a sense, these are the "easy" ones that can be discarded. The remaining gradient DoFs can only be eliminated globally, by constraining the linear system of equations after they have been defined in the child space.

The Nédélec spaces can be extended to any order. The $p = 1.5$ Nédélec space is a linear and quadratic vector space that omits certain DoFs associated with the gradient of a cubic polynomial. The DoFs associated with a vector quadratic space may be expressed as

$$\begin{aligned} \boldsymbol{f} &= (a_0 + a_1 x + a_2 y + a_3 x^2 + a_4 xy + a_5 y^2)\hat{\boldsymbol{x}} \\ &\quad + (b_0 + b_1 x + b_2 y + b_3 x^2 + b_4 xy + b_5 y^2)\hat{\boldsymbol{y}} \end{aligned} \tag{3.116}$$

and divided into the two subsets [35]

$$\begin{aligned} \boldsymbol{f}_{\text{curl}} &= \left(a_0 + a_1 x + a_2 y + \frac{a_4 - 2b_3}{3} xy + \frac{2a_5 - b_4}{3} y^2\right)\hat{\boldsymbol{x}} \\ &\quad + \left(b_0 + b_1 x + b_2 y + \frac{2b_3 - a_4}{3} x^2 + \frac{b_4 - 2a_5}{3} xy\right)\hat{\boldsymbol{y}} \end{aligned} \tag{3.117}$$

$$\begin{aligned} \boldsymbol{f}_{\text{grad}} &= \left(a_3 x^2 + \frac{2(a_4 + b_3)}{3} xy + \frac{(a_5 + b_4)}{3} y^2\right)\hat{\boldsymbol{x}} \\ &\quad + \left(\frac{(b_3 + a_4)}{3} x^2 + \frac{2(b_4 + a_5)}{3} xy + b_5 y^2\right)\hat{\boldsymbol{y}} \end{aligned} \tag{3.118}$$

In this case, the four DoFs in (3.118) can only be used to represent the gradient of a cubic polynomial. By eliminating these DoFs, we should improve the solution process. A new set of curl-conforming basis functions, which we denote linear tangential/quadratic normal (LT/QN), can be designed to represent the DoFs in (3.117). It is interesting that these functions involve eight DoFs, and distributing these on triangular cells seems very non-intuitive at first. Two DoFs will be assigned to tangential fields on each cell edge, and those DoFs will be shared by the adjacent cell to impose tangential-vector continuity. The remaining two DoFs in (3.117) will be assigned to the cell itself, and in such a way that they do not contribute to the tangential fields on cell edges. Specific forms of these functions will be presented in Chapters 4 and 5. We observe that the LT/QN functions can be thought of as starting with the set of six LT/LN functions and augmenting them with two additional cell-based functions per cell that provide a quadratic normal component along the cell edges.

Table 3.5 LT/QN results for rectangular cavity.

Mode	TM	TE	Exact
	0.000 (324)	0.000 (253)	
10		3.1416	3.1416
01		6.2832	6.2832
20		6.2832	6.2832
11	7.0250	7.0250	7.0248
21	8.8866	8.8866	8.8858
30		9.4249	9.4248
31	11.3295		11.3272
02		12.5668	12.5664
40		12.5668	12.5664
12	12.9547	12.9546	12.9531

To conclude the previous example, Table 3.5 presents analogous results for resonant wavenumbers for a rectangular cavity, obtained by solving the vector Helmholtz equation with the LT/QN basis and testing functions. With two unknowns per edge and two unknowns per cell, the system of equations for the TM case has order 756. For the TE case, the imposition of boundary conditions reduces the system order to 684. The results in Table 3.5 improve in accuracy over the LT/LN and CT/LN results and contain exactly the same number of nullspace eigenvalues as the LT/LN results. In essence, the two "extra" DoFs assigned to the LT/QN basis space substantially improve the accuracy without adding any additional nullspace solutions. In fact, these are the minimum additional DoFs required to complete the representation to the $p = 1.5$ order.

Nédélec mixed-order curl-conforming spaces for the 3D, tetrahedral cell situation are also available [34]. In simplex coordinates, the 3D basis functions for tetrahedral cells are similar to the 2D basis functions for triangular cells. Specific basis functions from these spaces will be developed in Chapters 4 and 5. In addition, mixed-order spaces for quadrilateral and hexahedral cells are also defined in Reference 34. These spaces involve additional terms, needed to create a symmetric set of basis functions. Thus, four $p = 0.5$ functions (instead of three) are used with quadrilateral cells, while twelve $p = 0.5$ functions are used on bricks or hexahedral cells. For the $p = 1.5$ functions, 2 per edge (shared by the adjacent cell) and 4 per cell (entirely local) are used with quadrilateral cells, for a total of 12, while 54 functions of order $p = 1.5$ are necessary to define an LT/QN representation on a brick cell.

The Nédélec curl-conforming spaces eliminate some of the DoFs associated with the nullspace of the vector Helmholtz operator. They also improve the balance of terms in the numerical discretization of that operator, due to the fact that the curl of the basis functions is complete to the same polynomial order as the basis function itself. Thus, even if the operator had no nullspace, the mixed-order basis functions are well suited for use with an operator containing a leading-order curl operation.

Mixed-order divergence-conforming spaces are also of interest for treating the electric field integral equation or other equations involving a divergence operator. Nédélec also described divergence-conforming spaces in his 1980 paper. In 2D, the divergence-conforming spaces are closely related to the curl-conforming spaces (the basis functions are a 90° rotation of one another). While this simple relation does not apply in 3D, the 3D divergence-conforming spaces eliminate some of the DoFs associated with the curl of a higher order (vector) polynomial. As in the curl-conforming case, there are many different basis sets that can be developed from these spaces. Two specific families will be developed in Chapters 4 and 5.

3.13 The De Rham Complex

The De Rham complex is a mathematical structure used to characterize functions belonging to different spaces related by differential operators. As used in electromagnetics, the "complex" consists of three differential operators and four vector spaces obtained from two copies of L^2 (that is the space of the square integrable scalar functions) and two copies of $\boldsymbol{L}^2$ (the space of the square integrable vector functions). Various electromagnetic quantities belong to different spaces as dictated by the type of operators (Maxwell's equations) that act upon them. For example, voltage potential fields reside in a space of scalar functions. Scalar functions (0-forms) can be denoted by the generic function ϕ that belongs to the function space H^1 (functions that satisfy an order-one Sobolev norm, so that the function and its gradient are square integrable) [36]. In contrast, the electric and magnetic fields belong to a space that must be curl conforming. These 1-forms belong to the function space $\boldsymbol{H}(\text{curl})$ and will be denoted by the vector function $\boldsymbol{v}$. Electric and magnetic flux densities are 2-forms that belong to a divergence-conforming space $\boldsymbol{H}(\text{div})$ and will be denoted by the generic vector function $\boldsymbol{f}$. Finally, there are 3-forms such as the charge density, which will be denoted by the generic square integrable function ρ.

The typical De Rham complex is defined in terms of the gradient ($\text{grad}_{\mathcal{B}}$), the curl ($\text{curl}_{\mathcal{B}}$), and the divergence ($\text{div}_{\mathcal{B}}$) operators on a simply connected domain $\mathcal{D}$ (with no holes) bounded by a simply connected boundary $\mathcal{B}$ on which boundary conditions are set. For illustration, we restrict the complex to functions that satisfy the homogeneous boundary conditions

$$\begin{aligned} \phi &= 0 \quad \text{on } \mathcal{B} \\ \hat{n} \times \boldsymbol{v} &= 0 \quad \text{on } \mathcal{B} \\ \hat{n} \cdot \boldsymbol{f} &= 0 \quad \text{on } \mathcal{B} \end{aligned} \tag{3.119}$$

The De Rham diagram reported at top of Figure 3.13 represents the complex in the "continuous" case and relates the vector spaces H_0^1, $\boldsymbol{H}_0(\text{curl})$, $\boldsymbol{H}_0(\text{div})$, and L_0^2; zero-subscripts are used for H_0^1, $\boldsymbol{H}_0(\text{curl})$, and $\boldsymbol{H}_0(\text{div})$ to indicate that the elements of these spaces satisfy homogenous boundary conditions, while L_0^2 stands for L^2-functions with zero average.

From the relations in the diagram, we observe that ϕ must be continuous across medium interfaces in order to have $\phi \in H_0^1$; similarly, the tangential component of $\boldsymbol{v}$ and the normal component of $\boldsymbol{f}$ must be continuous across medium interfaces for $\boldsymbol{v} \in \boldsymbol{H}_0(\text{curl})$ and $\boldsymbol{f} \in \boldsymbol{H}_0(\text{div})$.

Furthermore, it is of importance to understand that the range of each operator shown in Figure 3.13 coincides with the nullspace of the next operator in the sequence. For example, by recalling that the Helmholtz decomposition theorem states that any vector function $\boldsymbol{v}$ can be decomposed into a solenoidal part (zero divergence) plus another part that has zero curl (irrotational), the space of the vectors $\boldsymbol{v} \in \boldsymbol{H}_0(\text{curl})$ is made by the subspace formed by all the irrotational vectors that are gradient of a scalar $\phi \in H_0^1$ that vanish on $\mathcal{B}$ plus the subspace of vectors orthogonal to gradients, that is, divergence-free fields whose tangent component

$$
\begin{array}{ccccccc}
H_0^1 & \xrightarrow{\nabla} & \boldsymbol{H}_0(\text{curl}) & \xrightarrow{\nabla\times} & \boldsymbol{H}_0(\text{div}) & \xrightarrow{\nabla\bullet} & L_0^2 \\
\downarrow\Pi & & \downarrow\Pi^{\text{curl}} & & \downarrow\Pi^{\text{div}} & & \downarrow P_L \\
W^0 & \xrightarrow{\nabla} & \boldsymbol{W}^1(\text{curl}) & \xrightarrow{\nabla\times} & \boldsymbol{W}^2(\text{div}) & \xrightarrow{\nabla\bullet} & W^3
\end{array}
$$

Figure 3.13 The De Rham's complex in the *continuous* (top row) and *discrete* case (bottom row). The discretization process "projects" the spaces of the continuous case onto finite-dimensional spaces; this involves some projection (or interpolation) operators Π, Π^{curl}, Π^{div}, and P_L (shown in the middle row of the figure) which may or may not be defined explicitly.

vanishes on $\mathcal{B}$. Thus, the subspace formed by all gradients of $\phi \in H_0^1$ is mapped by the operator $\nabla\times$ into $\boldsymbol{f} = 0$ (that is the nullspace of $\boldsymbol{H}_0(\text{div})$).

In performing numerical computations to solve differential or integral equations (using the Finite Element Method (FEM) or the Method of Moments (MoM)), one starts by choosing appropriate finite-dimensional subspaces for the unknown quantities (this is done when choosing the basis functions). In light of the De Rham diagram, it seems beneficial to insist that the discretization scheme also complies with the discrete version of the De Rham complex shown in Figure 3.13, thereby requiring the *conformity* of the basis functions, given by

$$
\begin{aligned}
W^0 &\subset \text{dom}\,(\text{grad}_{\mathcal{B}}) \\
\boldsymbol{W}^1(\text{curl}) &\subset \text{dom}\,(\text{curl}_{\mathcal{B}}) \\
\boldsymbol{W}^2(\text{div}) &\subset \text{dom}\,(\text{div}_{\mathcal{B}})
\end{aligned}
\tag{3.120}
$$

where W^0, $\boldsymbol{W}^1(\text{curl})$, and $\boldsymbol{W}^2(\text{div})$ indicate the linear spaces generated by all the possible linear combinations of the basis functions.

In the discretized case, the "conformity" of the basis functions requires the continuity across the edges and faces of the mesh of the basis functions used to discretize ϕ and of the tangential and normal component of the vector basis functions used to discretize the functions $\boldsymbol{v}$ and $\boldsymbol{f}$, respectively. The discretized sequence of Figure 3.13 is "exact" if

$$
\begin{aligned}
\nabla W^0 &\quad \text{is the kernel of } \text{curl}_{\mathcal{B}} \text{ in } \boldsymbol{W}^1 \\
\nabla \times \boldsymbol{W}^1 &\quad \text{is the kernel of } \text{div}_{\mathcal{B}} \text{ in } \boldsymbol{W}^2
\end{aligned}
$$

Furthermore, we say that the *commuting diagram property* is fulfilled if (1) the top and bottom rows of the diagram of Figure 3.13 are exact and (2) the whole diagram of Figure 3.13 commutes. The latter requirement means, for example, that if $\phi \in H_0^1$, $\boldsymbol{v} \in \boldsymbol{H}_0(\text{curl})$ with $\boldsymbol{v} = \nabla\phi$ while $\boldsymbol{v}_d$, ϕ_d are the corresponding "projected" (discretized) quantities, then one has $\boldsymbol{v}_d = \nabla\phi_d$. The commuting diagram property is not met by all the known families of finite element spaces. It is possible to prove (under certain restricting conditions) that finite element families that fulfill the commuting diagram property also fulfill the so-called *discrete compactness property* [37, 38]. In addition to other properties, basis families that satisfy this property are known to exhibit h-convergence of the numerical solution to the analytical solution as the cell size h shrinks to zero.

The curl- and divergence-conforming basis functions discussed in this book span the Nédélec mixed-order spaces and are associated with a De Rham exact discretized sequence.

3.14 Conclusion

This chapter considered the representation of vector fields or currents by low-order polynomial functions. Several different types of vector basis functions have been introduced, with emphasis on curl-conforming and divergence-conforming bases that span the Nédélec mixed-order spaces. The mapping of these basis functions to curved cells has also been described in detail. Subsequent chapters will present interpolatory and hierarchical vector basis families for the most common cell shapes.

References

[1] D. R. Tanner and A. F. Peterson, "Vector expansion functions for the numerical solution of Maxwell's equations," *Microwave Opt. Technol. Lett.*, vol. 2, pp. 331–334, Sept. 1989.

[2] A. F. Peterson, S. L. Ray, and R. Mittra, *Computational Methods for Electromagnetics*, New York, NY: IEEE Press, 1998.

[3] S. M. Rao, D. R. Wilton, and A. W. Glisson, "Electromagnetic scattering by surfaces of arbitrary shape," *IEEE Trans. Antennas Propag.*, vol. AP-30, pp. 409–418, May 1982.

[4] A. Bossavit, "A rationale for edge-elements in 3-D fields computations," *IEEE Trans. Magn.*, vol. 24, no. 1, pp. 74–79, Jan. 1988.

[5] J. P. Webb, "Edge elements and what they can do for you," *IEEE Trans. Magn.*, vol. 29, pp. 1460–1465, Mar. 1993.

[6] G. Mur, "Edge elements, their advantages and their disadvantages," *IEEE Trans. Magn.*, vol. 30, pp. 3552–3557, Sept. 1994.

[7] A. Bossavit, "Whitney forms: a class of finite elements for three-dimensional computations in electromagnetics," *IEE Proc.*, vol. 135, pt. A, no. 8, pp. 493–500, 1988.

[8] P. A. Raviart and J. M. Thomas, "A mixed finite element method for 2nd order elliptic problems," in *Mathematical Aspects of Finite Element Methods*, A. Dold and B. Eckmann, eds., New York, NY: Springer-Verlag, pp. 292–315, 1977.

[9] R. N. Rieben, D. A. White, and G. H. Rodrigue, "Improved conditioning of finite element matrices using new high-order interpolatory bases," *IEEE Trans. Antennas Propag.*, vol. 52, pp. 2675–2683, Oct. 2004.

[10] D. R. Wilton, "Topological considerations in surface patch and volume cell modeling of electromagnetic scatterers," *Proceedings of the URSI Symposium on Electromagnetic Theory*, Santiago de Compostela (Spain), pp. 65–68, Aug. 23–26, 1983.

[11] G. F. Carey, *Computational Grids*, Bristol: Taylor and Francis, 1997.

[12] G. Xiao, "Applying loop-flower basis functions to analyze electromagnetic scattering problems of PEC scatterers," *Int. J. Antennas Propag.*, vol. 2014, Article 905935, 2014.

[13] F. P. Andriulli, "Loop–star and loop-tree decompositions: analysis and efficient algorithms," *IEEE Trans. Antennas Propag.*, vol. 60, pp. 2347–2356, May 2012.

[14] R. Albanese and G. Rubinacci, "Integral formulation for 3D eddy-current computation using edge elements," *IEE Proc. A*, vol. 135, pp. 457–462, Sept. 1988.

[15] R. Albanese and G. Rubinacci, "Analysis of three-dimensional electromagnetic fields using edge elements," *J. Comp. Phys.*, vol. 108, pp. 236–245, Oct. 1993.

[16] J. B. Manges and Z. J. Cendes, "A generalized tree–cotree gauge for magnetic field computations," *IEEE Trans. Magn.*, vol. 31, pp. 1342–1347, May 1995.

[17] L. Bai, *An Efficient Algorithm for Finding the Minimal Loop Basis of a Graph and its Application in Computational Electromagnetics*, M. S. Thesis, University of Illinois, 2000.

[18] G. Vecchi, "Loop star decomposition of basis functions in the discretization of the EFIE," *IEEE Trans. Antennas Propag.*, vol. 47, pp. 339–346, Feb. 1999.

[19] A. F. Peterson, "Absorbing boundary conditions for the vector wave equation," *Microwave Opt. Technol. Lett.*, vol. 1, pp. 62–64, April 1988.

[20] F. Collino, F. Millot, and S. Pernet, "Boundary-integral methods for iterative solution of scattering problems with variable impedance surface conditions," *Progr. Electromagn. Res.*, vol. PIER 80, pp. 1–28, 2008.

[21] P. Yla-Oijala, S. P. Kiminki, and S. Javenpaa, "Solving IBC-CFIE with dual basis functions," *IEEE Trans. Antennas Propag.*, vol. 58, pp. 3997–4004, Dec. 2010.

[22] W.-D. Li, W. Hong, H.-X. Zhou, and Z. Song, "Novel Buffa–Christiansen functions for improving CFIE with impedance boundary condition," *IEEE Trans. Antennas Propag.*, vol. 60, pp. 3763–3771, Aug. 2012.

[23] M. H. Smith and A. F. Peterson, "Numerical solution of the CFIE using vector bases and dual interlocking meshes," *IEEE Trans. Antennas Propag.*, vol. 53, pp. 3334–3339, Oct. 2005.

[24] F. P. Andriulli, K. Cools, H. Bagci, F. Olyslager, A. Buffa, S. H. Christiansen, and E. Michielssen, "A multiplicative Calderon preconditioner for the electric field integral equation," *IEEE Trans. Antennas Propag.*, vol. 56, pp. 2398–2412, Aug. 2008.

[25] M. B. Stephanson and J.-F. Lee, "Preconditioned electric field integral equation using Calderon identities and dual loop/star basis functions," *IEEE Trans. Antennas Propag.*, vol. 57, pp. 1274–1279, Apr. 2009.

[26] A. Buffa and S. H. Christiansen, "A dual finite element complex on the barycentric refinement," *Math. Comput.*, vol. 76, pp. 1743–1769, 2007.

[27] R. Chang and V. Lomakin, "Quadrilateral barycentric basis functions for surface integral equations," *IEEE Trans. Antennas Propag.*, vol. 61, pp. 6039–6050, Dec. 2013.

[28] D. H. Schaubert, D. R. Wilton, and A. W. Glisson, "A tetrahedral modeling method for electromagnetic scattering by arbitrarily shaped inhomogeneous dielectric bodies," *IEEE Trans. Antennas Propag.*, vol. AP-32, pp. 77–85, Jan. 1984.

[29] R. A. Lemdiasov and R. Ludwig, "The determination of linearly independent rotational basis functions in volumetric electric field integral equations," *IEEE Trans. Antennas Propag.*, vol. 54, pp. 2166–2169, July 2006.

[30] A. Obi, R. Lemdiasov, and R. Ludwig, "Minimizing the 3D solenoidal basis set in method of moments based volume integral equation," *ACES J.*, vol. 28, pp. 903–908, Oct. 2013.

[31] R. D. Graglia, D. R. Wilton, and A. F. Peterson, "Higher order interpolatory vector bases for computational electromagnetics," special issue on "Advanced Numerical Techniques in Electromagnetics," *IEEE Trans. Antennas Propag.*, vol. 45, no. 3, pp. 329–342, Mar. 1997.

[32] C. W. Crowley, *Mixed-order Covariant Projection Finite Elements for Vector Fields*, Ph.D. Dissertation, McGill University, Montreal, Quebec, Feb. 1988.

[33] A. F. Peterson, *Mapped Vector Basis Functions for Electromagnetic Integral Equations*, San Rafael, CA: Morgan & Claypool Synthesis Lectures, 2006.

[34] J. C. Nédélec, "Mixed finite elements in R3," *Numer. Math.,* vol. 35, pp. 315–341, 1980.

[35] A. F. Peterson and D. R. Wilton, "Curl-conforming mixed-order edge elements for discretizing the 2D and 3D vector Helmholtz equation," in *Finite Element Software for Microwave Engineering*, T. Itoh, G. Pelosi, and P. P. Silvester, eds., New York, NY: Wiley, pp. 101–125, 1996.

[36] L. Demkowicz, *Computing with hp-Adaptive Finite Elements, Volume 1*. Boca Raton, FL: Chapman and Hall, 2007.

[37] P. Monk and L. Demkowicz, "Discrete compactness and the approximation of Maxwell's equations in R^3," *Math. Comput.,* vol. 70, no. 234, pp. 507–523, 2000.

[38] D. Boffi, "A note on the De Rham complex and a discrete compactness property," *Appl. Math. Lett.,* vol. 14, pp. 33–38, 2001.

CHAPTER 4

Interpolatory Vector Bases of Arbitrary Order

Chapter 3 introduced several types of low-order vector basis functions for representing vector fields on conforming meshes. Here, interpolatory vector bases of arbitrary polynomial order are proposed. As in previous chapters, we focus on the primary canonical cell shapes, namely the triangle and quadrilateral for two-dimensional domains and the tetrahedron, the brick, and the triangular prism for three-dimensional domains. As discussed in Chapters 2 and 3, each cell of the meshed $x-y-z$ child space is defined by a different mapping $\boldsymbol{r}=\boldsymbol{r}(\boldsymbol{\xi})$ from its parent ξ-space to the child space $\boldsymbol{r}$, for example by using the interpolatory shape functions described in Chapter 2. The material in this chapter expands and systematically arranges the results available in References 1 and 2 published by the authors of this book together with other co-authors.

4.1 Development of Vector Bases

There is an extensive literature on vector basis functions, beginning with papers by Raviart and Thomas [3] on divergence-conforming bases and Nédélec [4] on curl-conforming bases. Throughout the 1980s and 1990s, there were many papers proposing vector basis functions for applications in electromagnetic scattering, eddy current analysis, and waveguide analysis, including References 5–28. These papers proposed bases for triangular and quadrilateral cell shapes in 2D and tetrahedral and brick cells in 3D. Some of these papers illustrated the mapping of vector basis functions to curved cells [14, 16, 26, 27]. Early attempts to solve the vector Helmholtz equation for waveguides were complicated by the appearance of spurious modes, and it was shown that the use of curl-conforming vector bases eliminated the spurious solutions [21–24]. Vector bases were also extended to triangular prisms in References 2, 29, and 30.

Most of the early papers proposed low-order functions, while higher order functions were gradually developed over time.

4.2 The Construction of Vector Bases

Vector bases[1] of higher polynomial order p are constructed by taking the product of *zeroth-order* basis vectors $\boldsymbol{B}$ with a set of *scalar* polynomials complete to the pth order. The scalar

[1]Recall from Chapter 3 that the most useful vector bases are the curl- and divergence-conforming ones, which have continuous tangential and normal components, respectively, across adjacent elements.

polynomials, as well as the zeroth-order basis vectors, used by this construction technique are conveniently defined in terms of the ξ-parent variables.

The high-order bases obtained by this multiplicative construction process easily inherit important properties of the zeroth-order base vectors $\boldsymbol{B}$. For example:

1. High-order bases with continuous tangential or normal components across adjacent elements are obtained by multiplying the zeroth-order base vectors $\boldsymbol{B}$ (that already fulfill this property on adjacent elements) with high-order *scalar* functions that are *continuous* across adjacent elements.
2. Furthermore, by multiplying a high-order scalar *polynomial* belonging to a scalar base of order p with a zeroth-order vector $\boldsymbol{B}$ of the Nédélec type one always gets a vector function that satisfies the pth-order Nédélec constraint equations. (These are constraints that eliminate some DoFs from the representation, as discussed in Section 3.12.) This follows from the fact that the Nédélec constraint equations impose the vanishing of appropriate linear combinations of the pth-order partial derivatives of the vector components.[2]

Among the polynomial-complete bases one may construct, the Nédélec bases are of major importance because of their curl- or divergence-conforming nature (see point 1 above) and because they have the following equivalent DoF minimizing properties (see Section 3.12):

- For conforming bases of fixed polynomial-complete order, the Nédélec functions minimize the number of nullspace DoFs associated with a curl or divergence operator, in the sense that they eliminate all the nullspace DoFs that can be eliminated *locally*, within the definition of the basis functions in the parent space.
- For conforming bases of fixed polynomial-complete order, the Nédélec functions achieve the next higher order of approximation with the smallest number of additional DoFs.

Thus, in this chapter, we consider only high-order bases constructed by taking the product of *zeroth-order* Nédélec vector basis functions $\boldsymbol{B}$ with *polynomial* families that are interpolatory. A similar approach is used to construct hierarchical bases in Chapter 5. Here, we derive interpolatory polynomial vector bases of the form

$$I_{\text{interp}}(\boldsymbol{\xi})\,\boldsymbol{B} \tag{4.1}$$

of arbitrarily high order, simply by defining appropriate interpolatory scalar polynomials $I_{\text{interp}}(\boldsymbol{\xi})$ that are continuous across adjacent differently shaped elements. In numerical applications, the curl and divergence of the high-order bases functions (4.1) are conveniently computed from knowledge of the gradient of $I_{\text{interp}}(\boldsymbol{\xi})$, since

$$\nabla \times \left[I_{\text{interp}}(\boldsymbol{\xi})\,\boldsymbol{B}\right] = \nabla I_{\text{interp}}(\boldsymbol{\xi}) \times \boldsymbol{B} + I_{\text{interp}}(\boldsymbol{\xi})\,\nabla \times \boldsymbol{B} \tag{4.2}$$

$$\nabla \cdot \left[I_{\text{interp}}(\boldsymbol{\xi})\,\boldsymbol{B}\right] = \nabla I_{\text{interp}}(\boldsymbol{\xi}) \cdot \boldsymbol{B} + I_{\text{interp}}(\boldsymbol{\xi})\,\nabla \cdot \boldsymbol{B} \tag{4.3}$$

[2]For additional information about the Nédélec constraint equations, the interested reader is referred to References 4 and 11.

Furthermore, all the interpolatory vector bases considered in this chapter have interpolation points located on regular interpolation grids.[3] For each canonical cell, the regular distribution of the interpolation points is obtained by extensive use of Lagrange interpolation polynomials defined in terms of the parent variables $\boldsymbol{\xi}$. These polynomials are discussed in detail in the rest of this chapter, for each canonical cell. This chapter also provides a brief review of the zeroth-order basis vectors of each canonical cell (which were discussed for several types of cells in Chapter 3).

The vector basis functions, regardless of their polynomial order, are expressed in terms of the edge and gradient base vectors introduced in Chapter 2 and discussed in detail in Chapter 3. The number of the edge vectors is equal to the number of the cell edges, while the number of the gradient vectors ($\nabla\xi_i$) equals the number of the cell parent variables (this latter number equals the number of edges for 2D cells, or the number of faces for 3D cells). Table 4.1 provides a comprehensive framework of all the edge and gradient vectors, together with the parent-variable dependency relations that hold for each canonical cell. Note that all the tables discussed in this chapter are reported in Section 4.8, for convenience of reference.

As discussed in Chapter 2, a 2D cell is described by two independent parent variables plus $(n_e - 2)$ dependent parent variables, where n_e is the number of the edges of the cell. Similarly, a 3D cell is described by three independent parent variables plus $(n_f - 3)$ dependent parent variables, where n_f is the number of the faces of the cell. The number of dependency relations reported in the third column of Table 4.1 is equal to $(n_e - 2)$ and $(n_f - 3)$ for 2D and 3D cells, respectively.

In this connection, notice also that in Table 4.1, we use only one subscript to label the edge vectors ($\boldsymbol{\ell}_i$) of a two-dimensional cell and two subscripts to label the edge vectors ($\boldsymbol{\ell}_{ij} = -\boldsymbol{\ell}_{ji}$) of a three-dimensional cell. This is done with no ambiguity because each edge vector is tangent to one edge of the cell that, in turn, is identified by one or two subscripts in the 2D and 3D cases, respectively.[4]

In other words, the subscripts labeling each edge vector are those that label the cell edge to which the given edge vector is tangent to.

4.3 Zeroth-Order Vector Bases for the Canonical 2D Cells

The zeroth-order Nédélec curl- (and divergence-)conforming bases for triangular and quadrilateral cells were defined in Chapter 3 such that they interpolate a vector component tangential (or normal) to the ith edge of the element at its *midpoint*. These bases are reported in Tables 4.2 and 4.3 together with their major properties. Observe that the bases of Tables 4.2 and 4.3 are un-normalized; their normalized form[5] is obtained by setting the order p to zero in the expression for the higher order forms presented in Section 4.6 (see also Tables 4.9 and 4.10).

---

[3] The reader should however keep in mind that the interpolation grids could actually be chosen *at pleasure*, provided one does not violate the tangential or normal continuity of the vector bases across adjacent elements. See References 31 and 32.

[4] Recall from Chapter 2 that the ith edge of a 2D cell is along the coordinate-line $\xi_i = 0$, while edge ij of a 3D cell lies along the line in common to the coordinate surfaces $\xi_i = 0$, $\xi_j = 0$.

[5] To avoid gross blunders in numerical applications, the zeroth-order bases *must* always be normalized since the edges of a practical mesh are *always* of unequal length.

The curl-conforming function $\mathbf{\Omega}_i(\boldsymbol{r})$ of Table 4.2 interpolates a tangential-vector component on the ith edge of the cell, while its tangential component vanishes along the remaining edges of the cell; on rectilinear elements these bases have constant tangential and linear normal (CT/LN) components along the element edges. Similarly, the divergence-conforming function $\mathbf{\Lambda}_i(\boldsymbol{r})$ of Table 4.3 interpolates a normal-vector component on the ith edge of the cell, while its normal component vanishes along the remaining edges of the cell; on rectilinear elements these bases have constant normal and linear tangential (CN/LT) components along the element edges. These bases are often called "edge bases" [22] because in a numerical solution the basis coefficient can be interpreted as the (un-normalized) tangential or normal component along the edge for the vector it represents.

As seen in Chapter 3, the divergence-conforming bases of a two-dimensional element (given in Table 4.3) may be obtained as the cross product of the associated curl-conforming bases with the unit vector $\hat{\boldsymbol{n}}$ normal to the element. In Tables 4.2 and 4.3, $\mathcal{J}$ indicates the Jacobian of the transformation from parent to child-space coordinates.

Despite the appearance of linear terms in the bases of Tables 4.2 and 4.3, these basis sets are clearly *not* complete to first order since six DoFs are required to model linear variations in two independent vector components on a surface. For example, the curl-conforming triangular bases are missing the curl-free combinations $\xi_1 \nabla \xi_1$, $\xi_2 \nabla \xi_2$, and $\xi_1 \nabla \xi_2 + \xi_2 \nabla \xi_1$ which would complete the bases to first order. Similarly, the divergence-conforming triangular bases are missing the divergence-free combinations $\xi_1 \boldsymbol{\ell}_1/\mathcal{J}$, $\xi_2 \boldsymbol{\ell}_2/\mathcal{J}$, and $(\xi_1 \boldsymbol{\ell}_2 + \xi_2 \boldsymbol{\ell}_1)/\mathcal{J}$. Completeness to zeroth order is, however, assured by noticing that there are two linear combinations (out of those reported in the second row of Tables 4.2 and 4.3) that are able to represent two independent basis vectors on a 2D element.

As a result of the Nédélec conditions, the curl-conforming bases $\mathbf{\Omega}_i$ (and the divergence-conforming bases $\mathbf{\Lambda}_i$) are complete in the curl (or divergence) to the same order as the bases. This follows from the results reported in the third row of Table 4.2 (or 4.3).

4.4 Zeroth-Order Vector Bases for the Canonical 3D Cells

The zeroth-order curl- and divergence-conforming bases for the canonical volumetric cells (tetrahedrons, bricks, and prisms) are reported in Tables 4.4 and 4.5, respectively, together with their major properties. Although the zeroth-order bases for tetrahedrons and bricks were discussed in Chapter 3, the reader will find more details on these bases in Section 4.7. In particular, the bases of Tables 4.4 and 4.5 are un-normalized; their normalized form[6] is obtained by setting the order p of the higher order forms presented in the rest of this chapter to zero.

The curl-conforming function $\mathbf{\Omega}_{ij}(\boldsymbol{r})$ of Table 4.4 interpolates a tangential-vector component on the edge in common to the ith and jth faces of the cell, its tangential component vanishes along the remaining edges of the cell; on rectilinear elements these bases have CT components along the element edges. The curl-conforming bases for the rectilinear tetrahedron have linear normal components along the element edges, that is, they are of CT/LN form.

Conversely, the divergence-conforming function $\mathbf{\Lambda}_i(\boldsymbol{r})$ of Table 4.5 interpolates a normal-vector component on the ith face of the cell, its normal component vanishes along the remaining

---

[6]Once again recall that in numerical applications the zeroth-order bases *must* always be normalized before using them because the edges and faces of a practical mesh are *always* of unequal size.

faces of the cell; on rectilinear elements these bases have CN components along the element faces.[7] The divergence-conforming bases for the rectilinear tetrahedron have also linear tangential components along the element faces, that is, they are of CN/LT form.

Despite the appearance of linear terms in the bases of Tables 4.4 and 4.5, these basis sets are clearly not complete to first order since, as noted above, they are of CT or of CN form in the curl- and divergence-conforming cases, respectively. For example, the curl-conforming bases cannot represent the curl-free vectors $\xi_i \nabla \xi_i$, where $i = 1, 2, 3, 4$ for the tetrahedron, $i = 1, 2, 3, 4, 5, 6$ for the brick, and $i = 1, 2, 4$, for the prism. In fact, the curl-conforming bases (apart those of the brick) and the divergence-conforming bases are formed by less than nine functions, while nine DoFs are required to model linear variations in three independent vector components on a volume. Completeness to zeroth order is, however, assured by noticing that there are three linear combinations (reported in the third row of Table 4.4 and in the second row of Table 4.5) that are able to represent three independent basis vectors on a volumetric element.

The curl-conforming bases $\mathbf{\Omega}_{ij}$ (and the divergence-conforming bases $\mathbf{\Lambda}_i$) are complete to the same degree in the curl (or divergence) as the bases themselves. This follows from the results reported in the third row of Table 4.4 (or 4.5).

4.5 The High-Order Vector Basis Construction Method

A simple approach to the construction of higher order curl- and divergence-conforming bases follows from the observation that the product of zeroth-order conforming bases and a complete set of polynomials of order p yields conforming bases complete to order p. Complete sets of scalar polynomials of different forms are merely linear combinations of one another and hence the form may be chosen for convenience.

For example, for the *triangular* cell described by the three parent variables (ξ_1, ξ_2, ξ_3), the polynomials may be

- of *homogeneous* form

$$\xi_1^\alpha \xi_2^\beta \xi_3^\gamma, \quad \alpha + \beta + \gamma = p$$

- of *inhomogeneous* form

$$\xi_i^\alpha \xi_j^\beta, \quad 0 \leq \alpha + \beta \leq p, \quad i \neq j$$

- of *interpolatory form*, given in terms of the Silvester polynomials $R_i(p, \xi)$

$$R_\alpha(p, \xi_1) R_\beta(p, \xi_2) R_\gamma(p, \xi_3), \quad \alpha + \beta + \gamma = p$$

- or of similar form employing the shifted interpolatory polynomials $\hat{R}_i(p, \xi)$ discussed in Chapter 2, or other polynomials (for example those belonging to a hierarchical family)[8].

---

[7]The interested reader can also verify that $\mathbf{\Omega}_{ij}(\mathbf{r})$ is proportional to the cross product $\mathbf{\Lambda}_i(\mathbf{r}) \times \mathbf{\Lambda}_j(\mathbf{r})$.

[8]Indeed, in the next chapter we employ hierarchical polynomials to construct hierarchical high-order vector bases.

Similarly, for the *tetrahedral* cell described by the four parent variables $(\xi_1, \xi_2, \xi_3, \xi_4)$, the multiplicative polynomials may be

- of *homogeneous* form

$$\xi_1^\alpha \xi_2^\beta \xi_3^\gamma \xi_4^\delta, \quad \alpha+\beta+\gamma+\delta=p$$

- of *inhomogeneous* form

$$\xi_i^\alpha \xi_j^\beta \xi_k^\gamma, \quad 0\leq\alpha+\beta+\gamma\leq p, \quad i\neq j\neq k$$

- of *interpolatory form*

$$R_\alpha(p,\xi_1)R_\beta(p,\xi_2)R_\gamma(p,\xi_3)R_\delta(p,\xi_4), \quad \alpha+\beta+\gamma+\delta=p$$

- or of similar form employing the shifted interpolatory polynomials $\hat{R}_i(p,\xi)$ discussed in Chapter 2, or hierarchical polynomials (such as those used in Chapter 5).

4.5.1 Completeness of the High-Order Vector Bases for 2D Cells

If the *homogeneous* forms are used as the multiplicative polynomials, it follows from the second row of Table 4.2 that the following linear combinations of the *triangular* bases yield complete vector polynomials in two independent directions:

$$\xi_1^\alpha \xi_2^\beta \xi_3^\gamma [\boldsymbol{\Omega}_2(\boldsymbol{r}) - \boldsymbol{\Omega}_3(\boldsymbol{r})] = \xi_1^\alpha \xi_2^\beta \xi_3^\gamma \nabla \xi_1 \tag{4.4}$$

$$\xi_1^\alpha \xi_2^\beta \xi_3^\gamma [\boldsymbol{\Omega}_3(\boldsymbol{r}) - \boldsymbol{\Omega}_1(\boldsymbol{r})] = \xi_1^\alpha \xi_2^\beta \xi_3^\gamma \nabla \xi_2 \tag{4.5}$$

The curl of the resulting bases are polynomials with only a component normal to the triangle, but these polynomials are also complete to the same order. This can be shown most easily by choosing the *inhomogeneous* forms for the multiplying polynomials. Completeness of the curl then follows from the result that the curl of the product generates terms of like order:

$$\nabla \times [\xi_i^\alpha \xi_j^\beta \boldsymbol{\Omega}_k(\boldsymbol{r})] = \frac{\alpha+\beta+2}{\mathcal{J}} \xi_i^\alpha \xi_j^\beta \hat{\boldsymbol{n}}, \quad i\neq j\neq k \tag{4.6}$$

where, once again, $\mathcal{J}$ is the Jacobian (a constant for rectilinear triangles). The procedure used to show the completeness of the curl-conforming quadrilateral bases is similar to that used above for the triangular cell; Table 4.6 summarizes the formulas to prove the completeness of the higher order curl-conforming bases of the canonical 2D cells obtained by this multiplicative process.

We omit a table similar to Table 4.6 for the divergence-conforming bases since the divergence-conforming bases of 2D cells are obtained by taking the cross product of the associated curl-conforming bases with the unit vector $\hat{\boldsymbol{n}}$ normal to the element. In other words, for 2D cells, the completeness of the higher order divergence-conforming bases follows from the completeness of the higher order curl-conforming bases.

4.5.2 Completeness of the High-Order Vector Bases for 3D Cells

If the *homogeneous* forms are used as the multiplicative polynomials, it follows from the third row of Table 4.4 that the following linear combinations of the curl-conforming bases yield complete vector polynomials of degree p in three independent directions:

- *For the tetrahedral bases*:

$$\begin{aligned}\xi_1^\alpha\xi_2^\beta\xi_3^\gamma\xi_4^\delta\,[\boldsymbol{\Omega}_{23}(\boldsymbol{r})-\boldsymbol{\Omega}_{24}(\boldsymbol{r})+\boldsymbol{\Omega}_{34}(\boldsymbol{r})]&=\xi_1^\alpha\xi_2^\beta\xi_3^\gamma\xi_4^\delta\nabla\xi_1,\\ \xi_1^\alpha\xi_2^\beta\xi_3^\gamma\xi_4^\delta\,[\boldsymbol{\Omega}_{14}(\boldsymbol{r})-\boldsymbol{\Omega}_{34}(\boldsymbol{r})-\boldsymbol{\Omega}_{13}(\boldsymbol{r})]&=\xi_1^\alpha\xi_2^\beta\xi_3^\gamma\xi_4^\delta\nabla\xi_2,\\ \xi_1^\alpha\xi_2^\beta\xi_3^\gamma\xi_4^\delta\,[\boldsymbol{\Omega}_{12}(\boldsymbol{r})-\boldsymbol{\Omega}_{14}(\boldsymbol{r})+\boldsymbol{\Omega}_{24}(\boldsymbol{r})]&=\xi_1^\alpha\xi_2^\beta\xi_3^\gamma\xi_4^\delta\nabla\xi_3,\\ \text{with } \alpha+\beta+\gamma+\delta&=p\end{aligned}\tag{4.7}$$

- *For the brick bases*:

$$\begin{aligned}\xi_1^\alpha\xi_2^\beta\xi_3^\gamma\xi_4^\delta\xi_5^\epsilon\xi_6^\zeta\,[\boldsymbol{\Omega}_{23}(\boldsymbol{r})-\boldsymbol{\Omega}_{26}(\boldsymbol{r})+\boldsymbol{\Omega}_{35}(\boldsymbol{r})+\boldsymbol{\Omega}_{56}(\boldsymbol{r})]&=\xi_1^\alpha\xi_2^\beta\xi_3^\gamma\xi_4^\delta\xi_5^\epsilon\xi_6^\zeta\nabla\xi_1,\\ \xi_1^\alpha\xi_2^\beta\xi_3^\gamma\xi_4^\delta\xi_5^\epsilon\xi_6^\zeta\,[\boldsymbol{\Omega}_{16}(\boldsymbol{r})-\boldsymbol{\Omega}_{13}(\boldsymbol{r})-\boldsymbol{\Omega}_{34}(\boldsymbol{r})-\boldsymbol{\Omega}_{46}(\boldsymbol{r})]&=\xi_1^\alpha\xi_2^\beta\xi_3^\gamma\xi_4^\delta\xi_5^\epsilon\xi_6^\zeta\nabla\xi_2,\\ \xi_1^\alpha\xi_2^\beta\xi_3^\gamma\xi_4^\delta\xi_5^\epsilon\xi_6^\zeta\,[\boldsymbol{\Omega}_{12}(\boldsymbol{r})-\boldsymbol{\Omega}_{15}(\boldsymbol{r})+\boldsymbol{\Omega}_{24}(\boldsymbol{r})+\boldsymbol{\Omega}_{45}(\boldsymbol{r})]&=\xi_1^\alpha\xi_2^\beta\xi_3^\gamma\xi_4^\delta\xi_5^\epsilon\xi_6^\zeta\nabla\xi_3,\\ \text{with } \alpha+\beta+\gamma+\delta+\epsilon+\zeta&=p\end{aligned}\tag{4.8}$$

- *For the prism bases*:

$$\begin{aligned}\xi_1^\alpha\xi_2^\beta\xi_3^\gamma\xi_4^\delta\xi_5^\epsilon\,[\boldsymbol{\Omega}_{24}(\boldsymbol{r})-\boldsymbol{\Omega}_{34}(\boldsymbol{r})-\boldsymbol{\Omega}_{25}(\boldsymbol{r})+\boldsymbol{\Omega}_{35}(\boldsymbol{r})]&=\xi_1^\alpha\xi_2^\beta\xi_3^\gamma\xi_4^\delta\xi_5^\epsilon\nabla\xi_1,\\ \xi_1^\alpha\xi_2^\beta\xi_3^\gamma\xi_4^\delta\xi_5^\epsilon\,[\boldsymbol{\Omega}_{34}(\boldsymbol{r})-\boldsymbol{\Omega}_{14}(\boldsymbol{r})-\boldsymbol{\Omega}_{35}(\boldsymbol{r})+\boldsymbol{\Omega}_{15}(\boldsymbol{r})]&=\xi_1^\alpha\xi_2^\beta\xi_3^\gamma\xi_4^\delta\xi_5^\epsilon\nabla\xi_2,\\ \xi_1^\alpha\xi_2^\beta\xi_3^\gamma\xi_4^\delta\xi_5^\epsilon\,[\boldsymbol{\Omega}_{12}(\boldsymbol{r})+\boldsymbol{\Omega}_{23}(\boldsymbol{r})-\boldsymbol{\Omega}_{13}(\boldsymbol{r})]&=\xi_1^\alpha\xi_2^\beta\xi_3^\gamma\xi_4^\delta\xi_5^\epsilon\nabla\xi_4,\\ \text{with } \alpha+\beta+\gamma+\delta+\epsilon&=p\end{aligned}\tag{4.9}$$

where $\nabla\xi_1$, $\nabla\xi_2$, and $\nabla\xi_3$ are the gradient vectors of the independent parent variables used for the tetrahedron and the brick, while $\nabla\xi_1$, $\nabla\xi_2$, and $\nabla\xi_4$ are the gradient vectors of the independent parent variables used for the prism (see Table 4.1).

Conversely, completeness in the curl is proved by using polynomials of *inhomogeneous* form. In this connection, we first observe that any polynomial vector of order $(p+1=\alpha+\beta+\gamma)$ can be expressed as the sum of a curl-free vector of order $(p+1)$ plus a vector that can be represented in terms of curl-conforming functions of order p:

$$\begin{aligned}(p+2)\,\xi_1^\alpha\,\xi_2^\beta\,\xi_3^\gamma\,\nabla\xi_1&=\nabla\left(\xi_1^{\alpha+1}\,\xi_2^\beta\,\xi_3^\gamma\right)+\beta\,\xi_1^\alpha\,\xi_2^{\beta-1}\,\xi_3^\gamma\,\boldsymbol{\Upsilon}_2-\gamma\,\xi_1^\alpha\,\xi_2^\beta\,\xi_3^{\gamma-1}\,\boldsymbol{\Upsilon}_1\\ (p+2)\,\xi_1^\alpha\,\xi_2^\beta\,\xi_3^\gamma\,\nabla\xi_2&=\nabla\left(\xi_1^\alpha\,\xi_2^{\beta+1}\,\xi_3^\gamma\right)+\gamma\,\xi_1^\alpha\,\xi_2^\beta\,\xi_3^{\gamma-1}\,\boldsymbol{\Upsilon}_3-\alpha\,\xi_1^{\alpha-1}\,\xi_2^\beta\,\xi_3^\gamma\,\boldsymbol{\Upsilon}_2\\ (p+2)\,\xi_1^\alpha\,\xi_2^\beta\,\xi_3^\gamma\,\nabla\xi_3&=\nabla\left(\xi_1^\alpha\,\xi_2^\beta\,\xi_3^{\gamma+1}\right)+\alpha\,\xi_1^{\alpha-1}\,\xi_2^\beta\,\xi_3^\gamma\,\boldsymbol{\Upsilon}_1-\beta\,\xi_1^\alpha\,\xi_2^{\beta-1}\,\xi_3^\gamma\,\boldsymbol{\Upsilon}_3\end{aligned}\tag{4.10}$$

with

$$\begin{aligned}&p=\alpha+\beta+\gamma-1\geq 0\\ &\alpha,\beta,\gamma\geq 0\end{aligned}\tag{4.11}$$

and where the vectors

$$\begin{aligned} \boldsymbol{\Upsilon}_1(\boldsymbol{r}) &= \xi_1 \nabla \xi_3 - \xi_3 \nabla \xi_1 \\ \boldsymbol{\Upsilon}_2(\boldsymbol{r}) &= \xi_2 \nabla \xi_1 - \xi_1 \nabla \xi_2 \\ \boldsymbol{\Upsilon}_3(\boldsymbol{r}) &= \xi_3 \nabla \xi_2 - \xi_2 \nabla \xi_3 \end{aligned} \tag{4.12}$$

are linear combinations of curl-conforming bases,[9] as shown in Table 4.4. Notice that the vectors $\nabla\left(\xi_1^{\alpha+1}\,\xi_2^{\beta}\,\xi_3^{\gamma}\right)$, $\nabla\left(\xi_1^{\alpha}\,\xi_2^{\beta+1}\,\xi_3^{\gamma}\right)$, and $\nabla\left(\xi_1^{\alpha}\,\xi_2^{\beta}\,\xi_3^{\gamma+1}\right)$ of order $(p+1)$ are the gradients of inhomogeneous polynomials of order $(p+2)$ and, because they are gradients, are curl free. Taking the curl of both sides of (4.10), one finds that the curl of any vector of order $(p+1)$ (yielding a vector of order p) can always be expressed as a linear combination of the curl of curl-conforming bases of order p. Hence the curl of curl-conforming bases of order p are complete to order p within the space of vectors derivable from the curl of vectors of order $p+1$. These bases appear in inhomogeneous polynomial form in (4.10), but they are, of course, linear combinations of the interpolatory polynomial bases defined in Section 4.7.

The completeness of the high-order divergence-conforming bases is readily proved by multiplying the linear combinations reported in the second row of Table 4.5 with polynomials of *homogeneous* forms, to get complete vector polynomials in the three independent directions $\boldsymbol{\ell}^1$, $\boldsymbol{\ell}^2$, and $\boldsymbol{\ell}^3$ (or $\boldsymbol{\ell}^4$, for prism cells). Similarly, as shown in the third row of Table 4.5, completeness in the divergence is easily proved by using polynomials of *inhomogeneous* form. Notice that for the divergence-conforming bases, the completeness is with respect to $1/\mathcal{J}$ as a weighting factor.

4.5.3 Use of Shifted Silvester Polynomials to Move the Interpolation Points to the Element Interior

To obtain interpolatory properties for the high-order conforming bases, we now employ Lagrange interpolating polynomials as multiplying polynomials. However, in the vector case, we cannot use the shape polynomials formed in Chapter 2 by the product of Silvester polynomials, because these have interpolation points on the cell's boundaries. These are not appropriate when building high-order interpolatory *vector* functions with a vanishing tangential or normal component along some of the cell boundaries. For these functions, we need to shift the interpolation points of the shape polynomials away from the edges and faces along which the tangential (in the curl-conforming case) or the normal components (in the divergence-conforming case) of the zeroth-order bases vanish. Since the number of cell boundaries (that is, the number of edges and faces) varies with the cell, the procedure to shift the interpolation points away from these boundaries is different for each canonical cell, as discussed in detail in the rest of this chapter.

The easiest way to shift the interpolation points away from an edge or a face is to substitute a shifted Silvester polynomial for the appropriate (unshifted) Silvester polynomial appearing in the multiplicative expression of the shape function given in Section 2.2 of Chapter 2.

---

[9] When dealing with prism-element functions, one must replace ξ_3 by ξ_4 and $\boldsymbol{\Upsilon}_3$ by $\boldsymbol{\Upsilon}_4 = \xi_4 \nabla \xi_2 - \xi_2 \nabla \xi_4$ in (4.10) and (4.12).

Obviously, the order of the polynomials before and after the substitution must remain equal. As we will see shortly, this implies that we must increase the number of uniform subintervals into which the ξ-interval [0, 1] of each parent variable is divided. To form the Lagrangian shape functions of Chapter 2 we used p subintervals for each (unshifted) Silvester polynomial forming the shape function. The number of subintervals of each shifted or unshifted polynomial forming the new interpolating polynomial is however higher than p and depends on the symmetry of the parent cell at issue although, as said, the total order of the shape and of the new interpolatory polynomial remains the same.

4.6 Vector Bases for the Canonical 2D Cells

In this section, we first define, in Sections 4.6.1 and 4.6.2, the Lagrange interpolating polynomials used to form the vector bases. The high-order bases for the triangular and quadrilateral cells are then defined in Sections 4.6.3 and 4.6.4, respectively.

4.6.1 The $\hat{\alpha}(p, \xi)$ Polynomials with Edge-Interpolation Points on Only One Edge of a Triangular Cell

The function

$$\begin{aligned} &\hat{\alpha}^{a}_{IJK}(p,\xi_a,\xi_b,\xi_c) = R_I(p+2,\xi_a)\,\hat{R}_J(p+2,\xi_b)\,\hat{R}_K(p+2,\xi_c) \\ &\quad \text{with } I=0,1,\ldots,p; \quad J,K=1,2,\ldots,p+1; \\ &\quad \text{where } I+J+K=p+2 \end{aligned} \tag{4.13}$$

is a pth-order Lagrange polynomial interpolating the points

$$(\xi_a,\xi_b,\xi_c)_{IJK} = \left(\frac{I}{p+2},\frac{J}{p+2},\frac{K}{p+2}\right) = \boldsymbol{\xi}_{IJK} \tag{4.14}$$

on the interior of a triangular cell and on the $\xi_a = 0$ edge, with no interpolation points on the other two edges $\xi_b = 0$, $\xi_c = 0$. $R(p+2,\xi)$ denotes a Silvester polynomial, and $\hat{R}(p+2,\xi)$ denotes a shifted Silvester polynomial, as defined in Section 2.2.

The interpolation points associated with $\hat{\alpha}^{a}_{IJK}(p,\xi_a,\xi_b,\xi_c)$ for $p = 3$ are shown in Figure 4.1. The interpolation points have the same Pascal triangle arrangement valid for the shape polynomial $\alpha_{IJK}(p,\xi_a,\xi_b,\xi_c)$ of the same order (see (2.34) and Figure 2.9), except that the nodes along the two edges $\xi_b = 0$, $\xi_c = 0$ have been shifted to the triangle's interior. For notational purposes, the superscript a in the expression of $\hat{\alpha}^{a}_{IJK}$ also indicates that the Silvester polynomials of the ξ_a dummy variable that appear in (4.13) are unshifted, while the other Silvester polynomials forming the product (4.13) are "careted" (i.e., shifted).

To express $\hat{\alpha}^{a}_{IJK}$, we have used *dummy* normalized coordinates (ξ_a,ξ_b,ξ_c) since the array of the interpolation points in the dummy space does not change by equating ξ_a with any of

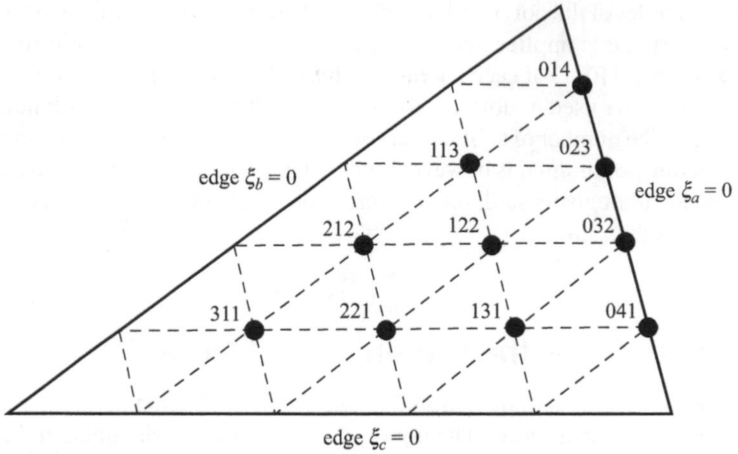

Figure 4.1 Interpolation nodes on triangular cells for a Lagrange interpolation of order $p=3$ obtained by using Silvester and shifted Silvester polynomials.

the three normalized coordinates ξ_1, ξ_2, or ξ_3. The interpolation grid for each of these cases is obtained by either permuting the edge numbers, thereby setting

$$\begin{aligned} &(\xi_a, \xi_b, \xi_c) = (\xi_1, \xi_2, \xi_3), \\ &(\xi_a, \xi_b, \xi_c) = (\xi_2, \xi_3, \xi_1), \\ \text{or} \quad &(\xi_a, \xi_b, \xi_c) = (\xi_3, \xi_1, \xi_2) \end{aligned} \tag{4.15}$$

or by rotating the *dummy* pattern; for example, the pattern shown in Figure 4.1 for $p=3$. Note that the interior node patterns remain the same after permutation and, most importantly, that none of the interpolation points is on a corner node of the triangle. Again with reference to Figure 4.1, notice that the interpolation point is interior to the triangular cell if none of the three ordered indices labeling the interpolation point is equal to zero. Only one of these indices is equal to zero for interpolation points located on the triangle edge $\xi_a = 0$.

In this connection it is of importance to stress the fact that the indices (I, J, K) used in (4.13) are dummy indices associated with the dummy variables (ξ_a, ξ_b, ξ_c). Therefore, the coordinates (4.14) of the interpolation point are not given in the true parent space (ξ_1, ξ_2, ξ_3). In the true parent space, the coordinates of the interpolation points in terms of the *true* indices (i, j, k) are

$$(\xi_1, \xi_2, \xi_3) = \left(\frac{i}{p+2}, \frac{j}{p+2}, \frac{k}{p+2}\right) = \boldsymbol{\xi}_{ijk} \tag{4.16}$$

with

$$(i, j, k) = \begin{cases} (I, J, K) & \text{if } (\xi_a, \xi_b, \xi_c) = (\xi_1, \xi_2, \xi_3) \\ (K, I, J) & \text{if } (\xi_a, \xi_b, \xi_c) = (\xi_2, \xi_3, \xi_1) \\ (J, K, I) & \text{if } (\xi_a, \xi_b, \xi_c) = (\xi_3, \xi_1, \xi_2) \end{cases} \tag{4.17}$$

depending on which permutation (4.15) one is using. Only the true indices (i, j, k) label the three polynomial bases $\hat{\alpha}^1_{ijk}$, $\hat{\alpha}^2_{ijk}$, $\hat{\alpha}^3_{ijk}$, and their interpolation points in a unique and consistent manner.

In the rest of the chapter, we use dummy indices and variables to write the bases of the canonical cells in compact form. However, in the following, dummy indices are denoted with lower case letters $(i, j, k, \ell, m$, etc.) just as "true" indices. This eases the notation (since capital-letter subscripts are avoided when reporting dummy expressions), while the relation between dummy and true indices is straightforward and can easily be obtained by the reader.

The $\hat{\alpha}^a_{ijk}(p, \boldsymbol{\xi})$ polynomials of the triangular cell are reported in Table 4.7 in terms of dummy variables and indices (with dummy indices denoted by lower case letters).

As an example, consider the polynomial $\hat{\alpha}^a_{032}$ that interpolates at node 032 in Figure 4.1. This function can be expressed as

$$\hat{\alpha}^a_{032} = R_0(5, \xi_a)\,\hat{R}_3(5, \xi_b)\,\hat{R}_2(5, \xi_c) = \frac{1}{2}(5\xi_b - 1)(5\xi_b - 2)(5\xi_c - 1) \tag{4.18}$$

One should observe that the factor $(5\xi_b - 1)$ zeros the polynomial at $\xi_b = 1/5$, or specifically at nodes 311, 212, 113, and 014 in Figure 4.1. Similarly, the factor $(5\xi_b - 2)$ zeros the polynomial at nodes 221, 122, and 023. The factor $(5\xi_c - 1)$ zeros the polynomial at nodes 311, 221, 131, and 041. Thus, the interpolation polynomial vanishes at all nodes identified in Figure 4.1 except for node 032, where it has value $\hat{\alpha}^a_{032} = 1$. It should be clear that the construction of the interpolation polynomial is primarily driven by the need to zero the function at all nodes other than that at which it interpolates.

4.6.2 The $\hat{\alpha}(p, \xi)$ Polynomials with Edge-Interpolation Points on Only One Edge of a Quadrilateral Cell

To build high-order interpolatory *vector* functions with a vanishing tangential or normal component along three edges of the quadrilateral cell, we move the interpolation points of the shape polynomials lying on the cell's boundary to the cell's interior by using the following polynomials of order $2p$ (dummy notation)

$$\begin{aligned} &\hat{\alpha}^a_{ik;\,j\ell}(p, \xi_a, \xi_c; \xi_b, \xi_d) = R_i(p+1, \xi_a)\hat{R}_k(p+1, \xi_c)\hat{R}_j(p+2, \xi_b)\hat{R}_\ell(p+2, \xi_d),\\ &\quad i = 0, 1, \ldots, p; \quad j, k, \ell = 1, 2, \ldots, p+1, \quad i + k = p+1; \quad j + \ell = p+2 \end{aligned} \tag{4.19}$$

written in terms of shifted Silvester polynomials and interpolating the point with (dummy) normalized coordinates

$$(\xi_a, \xi_c; \xi_b, \xi_d)_{ik;\,j\ell} = \left(\frac{i}{p+1}, \frac{k}{p+1}; \frac{j}{p+2}, \frac{\ell}{p+2}\right) = \boldsymbol{\xi}_{ik;\,j\ell} \tag{4.20}$$

with

$$\begin{aligned} \xi_a + \xi_c &= 1\\ \xi_b + \xi_d &= 1 \end{aligned} \tag{4.21}$$

In common with the triangular cell, the polynomials (4.19) are conveniently written in terms of *dummy* parent variables and indices since the dummy array of the interpolation points remains unchanged for ξ_a equal to one of the four normalized coordinates ξ_1, ξ_2, ξ_3, or ξ_4. These polynomials interpolate points on the interior of a quadrilateral cell and on the $\xi_a = 0$ edge, with no interpolation points on the other three edges $\xi_b = 0$, $\xi_c = 0$, and $\xi_d = 0$. The interpolation points associated with $\hat{\alpha}^a_{ik;j\ell}(p, \xi_a, \xi_c; \xi_b, \xi_d)$ for $p = 2$ are shown in Figure 4.2. The interpolation points have the same arrangement valid for the *shape* polynomial $\alpha_{ik;j\ell}(p, \xi_a, \xi_c; \xi_b, \xi_d)$ of the same order (see (2.66) and Figure 2.18), except that the nodes along the three edges $\xi_b = 0$, $\xi_c = 0$, and $\xi_d = 0$ have been shifted to the cell's interior. As a matter of fact, the superscript a in the expression of $\hat{\alpha}^a_{ik;\,j\ell}$ also indicates that the Silvester polynomial of the ξ_a dummy variable that appears in its product expression is unshifted, while the other Silvester polynomials in the product are "careted" (i.e., shifted).

Notice that none of the interpolation points of $\hat{\alpha}^a_{ik;j\ell}$ is on a corner of the quadrilateral and that the *interior* node patterns remain the same on "double" permutation of the parent variables, that is, for example, with $(\xi_a, \xi_c; \xi_b, \xi_d) = (\xi_1, \xi_3; \xi_2, \xi_4)$ and $(\xi_a, \xi_c; \xi_b, \xi_d) = (\xi_3, \xi_1; \xi_4, \xi_2)$. However, the node pattern obtained by setting $(\xi_a, \xi_c; \xi_b, \xi_d) = (\xi_2, \xi_4; \xi_1, \xi_3)$ is different from that obtained by setting $(\xi_a, \xi_c; \xi_b, \xi_d) = (\xi_1, \xi_3; \xi_2, \xi_4)$ by a 90° rotation.

The $\hat{\alpha}^a_{ij;k\ell}(p, \boldsymbol{\xi})$ polynomials of the quadrilateral cell are reported in Table 4.8 in terms of dummy variables and indices.

As an example, consider the polynomial $\hat{\alpha}^a_{12;22}$ that interpolates at node 12;22 in Figure 4.2. This function can be expressed as

$$\hat{\alpha}^a_{12;22} = R_1(3, \xi_a)\,\hat{R}_2(3, \xi_c)\,\hat{R}_2(3, \xi_b)\,\hat{R}_2(3, \xi_d) = (3\xi_a)(3\xi_c - 1)(3\xi_b - 1)(3\xi_d - 1) \quad (4.22)$$

Observe that the factor $3\xi_a$ zeros the polynomial at $\xi_a = 0$, or specifically at nodes 03;31, 03;22, and 03;13 in Figure 4.2. Similarly, the factor $(3\xi_c - 1)$ zeros the polynomial at nodes 21;31, 21;22, and 21:13. The factor $(3\xi_b - 1)$ zeros the polynomial at nodes 03;13, 12;13, and 21;13.

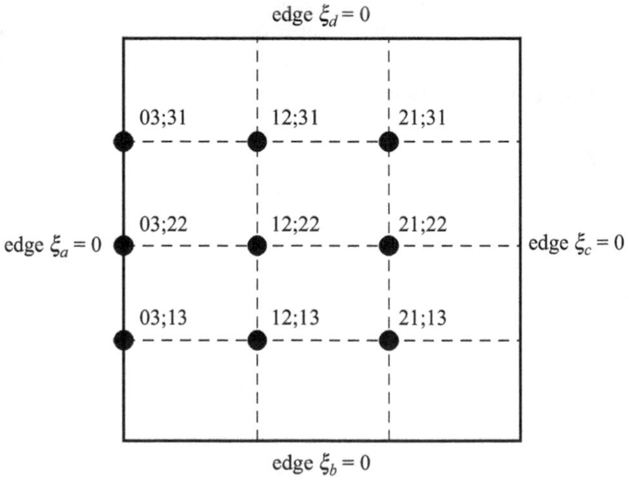

Figure 4.2 Interpolation nodes on quadrilateral cells for a Lagrange interpolation of order $p = 2$ obtained by using Silvester and shifted Silvester polynomials.

Finally, $(3\xi_d - 1)$ zeros the polynomial at 03;31, 12;31, and 21;31. Thus, the interpolation polynomial vanishes at all nodes identified in Figure 4.2 except for node 12;22, where it has value $\hat{\alpha}^a_{12;22} = 1$.

4.6.3 Order p Vector Bases for Triangular Cells

Table 4.9 reports, in terms of the parent variables $\{\xi_1, \xi_2, \xi_3\}$ and (true) indices (i, j, k), the curl- and divergence-conforming bases of arbitrary polynomial order p for the triangular cell. These bases are obtained by multiplying the un-normalized zeroth-order bases of Tables 4.2 and 4.3 with the $\hat{\alpha}$ polynomials described in Section 4.6.1.

4.6.3.1 Curl-Conforming Bases on Triangles

The high-order curl-conforming bases of Table 4.9 are obtained by index and subscript permutation in the compact dummy expression (see Section 4.6.1)

$$\boldsymbol{\Omega}^a_{ijk}(\boldsymbol{r}) = N^a_{ijk}\,\hat{\alpha}^a_{ijk}(p, \xi_a, \xi_b, \xi_c)\,\boldsymbol{\Omega}_a(\boldsymbol{r}) = N^a_{ijk}\,R_i(p+2, \xi_a)\hat{R}_j(p+2, \xi_b)\hat{R}_k(p+2, \xi_c)\,\boldsymbol{\Omega}_a(\boldsymbol{r}),$$
$$i = 0, 1, \ldots, p; \quad j, k = 1, 2, \ldots, p+1, \quad i + j + k = p + 2 \tag{4.23}$$

Completeness of both the bases and their curls to order p follows from the observation that the interpolation polynomials are simply linear combinations of the polynomial forms used in the complete arguments of Section 4.5. Clearly, the curl of (4.23)

$$\nabla \times \boldsymbol{\Omega}^a_{ijk}(\boldsymbol{r}) = N^a_{ijk}\left[\nabla\hat{\alpha}^a_{ijk} \times \boldsymbol{\Omega}_a(\boldsymbol{r}) + \hat{\alpha}^a_{ijk}\nabla \times \boldsymbol{\Omega}_a(\boldsymbol{r})\right] \tag{4.24}$$

is evaluated by computing the gradient of $\hat{\alpha}^a_{ijk}$ and by using the un-normalized expression of $\boldsymbol{\Omega}_a(\boldsymbol{r})$ and $\nabla \times \boldsymbol{\Omega}_a(\boldsymbol{r})$ reported in Table 4.2.

Normalization Coefficients

The normalization coefficients (dummy expression)

$$N^a_{ijk} = \left.\frac{(p+2)}{(p+2-i)}\ell_a\right|_{\boldsymbol{\xi}_{ijk}} \tag{4.25}$$

in the bases above are chosen to ensure that the component of $\boldsymbol{\Omega}^a_{ijk}(\boldsymbol{r})$ along $\hat{\boldsymbol{\ell}}_a$ at the interpolation point

$$\boldsymbol{\xi}_{ijk} = (\xi_a, \xi_b, \xi_c)_{ijk} = \left(\frac{i}{p+2}, \frac{j}{p+2}, \frac{k}{p+2}\right) \tag{4.26}$$

is unity. The normalization coefficients in terms of the parent variables $\{\xi_1, \xi_2, \xi_3\}$ and of the *true* indices (i, j, k) are given in Table 4.9.

To illustrate using the previous example, consider the basis function that interpolates at node 032 in Figure 4.1. This function can be expressed as

$$\boldsymbol{\Omega}^a_{032} = N^a_{032}\frac{1}{2}(5\xi_b - 1)(5\xi_b - 2)(5\xi_c - 1)\boldsymbol{\Omega}_a(\boldsymbol{r}) \tag{4.27}$$

As previously discussed, the polynomial factors zero the basis function at all nodes in Figure 4.1 except for node 032. The vector dependence is provided by the base vector $\boldsymbol{\Omega}_a(\boldsymbol{r}) = \xi_b \nabla \xi_c - \xi_c \nabla \xi_b$.

Dependency Relations at Interior Nodes

Though both the vector bases and their curls are complete to order p, it happens that dependencies in the basis set exist for $p \geq 1$, and these must be eliminated before using them. We note that the bases (4.23) interpolate (for $p = 3$) the points shown in Figure 4.1. Since the interpolation points for the remaining bases may be obtained by permuting the edge numbers in Figure 4.1, it is clear that three different components are interpolated at each interior node of the triangle. Since only *two* vector components can be independent on a two-dimensional element, there must exist a dependency among the bases interpolating the interior points. This dependency is formally revealed by observing that there always exists a linear combination of the three bases interpolating an interior point which contains the factor

$$\xi_1 \boldsymbol{\Omega}_1(\boldsymbol{r}) + \xi_2 \boldsymbol{\Omega}_2(\boldsymbol{r}) + \xi_3 \boldsymbol{\Omega}_3(\boldsymbol{r}) = \boldsymbol{0} \tag{4.28}$$

(This dependency identity is reported in the third row of Table 4.6.) In fact, for $i, j, k \neq 0$ (i.e., at interior nodes), (4.28) yields

$$\frac{i\boldsymbol{\Omega}^1_{ijk}(\boldsymbol{r})}{N^1_{ijk}} + \frac{j\boldsymbol{\Omega}^2_{ijk}(\boldsymbol{r})}{N^2_{ijk}} + \frac{k\boldsymbol{\Omega}^3_{ijk}(\boldsymbol{r})}{N^3_{ijk}} = \boldsymbol{0}$$

Hence in using the bases of Table 4.9, we must discard one member from the set of three basis functions interpolating a given interior node. Also, reversal of a basis function's sign may be necessary in order to maintain tangential continuity if it is defined at an edge-interpolating node common to more than a single element. One possible scheme for determining this is to select reference directions for edge-interpolating bases such that their interpolated components are directed from lower to higher (global) vertex numbers on the edge. Note that basis functions interpolating interior nodes do not straddle multiple cells and therefore require no sign reversal.

Number of DoFs

The total number of DoFs for curl-conforming basis functions of order p on a triangle may be determined as follows (it could be helpful to see Figure 4.1):

- 1 component $\times$ $(p+1)$ DoFs $\times$ 3 edges $= 3(p+1)$ edge DoFs,
- 2 components $\times$ $\frac{p(p+1)}{2}$ DoFs $\times$ 1 face $= p(p+1)$ triangle interior DoFs,

for a total of $(p+1)(p+3)$ DoFs per triangle. This agrees with the number of DoFs previously determined by Nédélec [4], who presented explicitly only $p = 0, 1$ order, non-interpolatory curl-conforming forms. These and the bases presented in Reference 28 for $p = 0, 1, 2$ are linear combinations of the bases given here.

An alternative method of counting DoFs recognizes that the pth-order complete basis is actually of mixed order and contains terms of order $p+1$. If complete to order $p+1$, there would be $(p+2)(p+3)$ such terms, but the highest order terms with vanishing curls, which have the form $\xi_1^{p+1}\nabla\xi_1$, $\xi_2^{p+1}\nabla\xi_2$, and $\xi_1^{\alpha}\xi_2^{\beta}[(\alpha+1)\xi_2\nabla\xi_1 + (\beta+1)\xi_1\nabla\xi_2]$, $0 \leq \alpha+\beta \leq p$, are missing. There are a total of $p+3$ such terms, leaving $(p+1)(p+3)$ terms as previously determined.

The interpolatory form of the curl-conforming Nédélec bases has several advantages over other forms. Using Silvester's form of the Lagrange interpolating polynomials, for example, they may be written in closed form for any order. Since they are interpolatory, the bases are unisolvent (i.e., their linear independence is guaranteed), and the basis coefficients in a field representation have a physical significance as the (covariant) component of the vector field at the interpolation point. Interpolatory bases also have a greater degree of independence than so-called nodeless bases (such as the hierarchical bases presented in Chapter 5), thus reducing one source of ill-conditioning in high-order schemes. Since the bases are normalized and interpolate tangential components, it is also trivial to enforce tangential continuity of fields across element boundaries. And finally, the generalization of Silvester's multiple indexing scheme provides a unified approach to labeling both the bases and their interpolation points. The closed-form representation and the unified indexing properties significantly simplify implementation of computer codes for the generation of arbitrary order vector bases.

4.6.3.2 Divergence-Conforming Bases on Triangles

Divergence-conforming bases on a triangle may be obtained as the cross product of the associated curl-conforming bases with the unit vector $\hat{\boldsymbol{n}}$ normal to the element. The pth-order interpolatory divergence-conforming bases are given in Table 4.9. The normalizing coefficients ensure that the component of $\boldsymbol{\Lambda}^a_{ijk}(\boldsymbol{r})$ along $\hat{\boldsymbol{h}}_a$ at the interpolation point is unity. The bases for $p=0$ are identical to those described by Rao et al. [7] and have CN/LT components at element edges. All other properties of the divergence-conforming bases follow from analogous properties of the curl-conforming bases. For functions that interpolate at an edge, the sign of a divergence-conforming basis function must be chosen to maintain normal continuity with the complementary function in the adjacent cell. One possible scheme for determining this is to select reference directions for these bases such that their interpolated components are directed from the element having the lowest (global) element number to those having higher element numbers at the edge.

Clearly, the divergence of the bases (dummy expression)

$$\nabla \cdot \boldsymbol{\Lambda}^a_{ijk}(\boldsymbol{r}) = N^a_{ijk}\left[\nabla \hat{\alpha}^a_{ijk} \cdot \boldsymbol{\Lambda}_a(\boldsymbol{r}) + \hat{\alpha}^a_{ijk} \nabla \cdot \boldsymbol{\Lambda}_a(\boldsymbol{r})\right] \tag{4.29}$$

may be evaluated by computing the gradient of the (dummy) polynomials $\hat{\alpha}^a_{ijk}$ and by using the un-normalized expression of $\boldsymbol{\Lambda}_a(\boldsymbol{r})$ and $\nabla \cdot \boldsymbol{\Lambda}_a(\boldsymbol{r})$ reported in Table 4.3.

4.6.4 Order p Vector Bases for Quadrilateral Cells

Table 4.10 reports the curl- and divergence-conforming bases of arbitrary polynomial order p for the quadrilateral cell. The bases are obtained by multiplying the un-normalized zeroth-order bases of Tables 4.2 and 4.3 with the $\hat{\alpha}^a$ polynomials described in Section 4.6.2.

4.6.4.1 Curl-Conforming Bases on Quadrilaterals

Curl-conforming bases complete to the pth order and which interpolate a vector function on a quadrilateral are given in Table 4.10. The bases are the product of the zeroth-order bases given in the right-hand column of Table 4.2 with the interpolatory polynomials $\hat{\alpha}^a$ described in Section 4.6.2. In terms of dummy parent variables $\{\xi_a, \xi_c; \xi_b, \xi_d\}$ and indices they read

$$\begin{aligned}\mathbf{\Omega}^a_{ik;\,j\ell}(\boldsymbol{r}) = N^a_{ik;j\ell}\,\hat{\alpha}^a_{i,j;k,\ell}(p, \xi_a, \xi_c; \xi_b, \xi_d)\,\mathbf{\Omega}_a(\boldsymbol{r}),&\\ i=0,1,\ldots,p;\quad j,k,\ell=1,2,\ldots,p+1,&\\ i+k=p+1;\quad j+\ell=p+2&\end{aligned} \tag{4.30}$$

The full set of Table 4.10 is then simply obtained from the dummy expression (4.30) by "double" permutation of the edge numbers and indices, thereby setting

$$\begin{aligned}\{\xi_a, \xi_c; \xi_b, \xi_d\} &= \{\xi_1, \xi_3; \xi_2, \xi_4\}\\ \{\xi_a, \xi_c; \xi_b, \xi_d\} &= \{\xi_3, \xi_1; \xi_4, \xi_2\}\\ \{\xi_a, \xi_c; \xi_b, \xi_d\} &= \{\xi_2, \xi_4; \xi_1, \xi_3\}\\ \{\xi_a, \xi_c; \xi_b, \xi_d\} &= \{\xi_4, \xi_2; \xi_3, \xi_1\}\end{aligned} \tag{4.31}$$

The interpolation points for bases of the form $\mathbf{\Omega}^a_{ik;\,j\ell}$ are shown in Figure 4.2 (for $p=2$). The arrangement of interpolation points is the same as that of the *shape* polynomials of the same order on a quadrilateral, except that the pattern contracts away from the edges where the tangential components of the associated zeroth-order bases vanish. Only edge a is interpolated by the basis subset illustrated in the figure; the remaining three basis subsets provide interpolants for the remaining edges. The arrangement of interpolation points for the latter bases may be determined from Figure 4.2 by permutation of the edge numbers.

Completeness of both the bases and their curls to order p follows from the observation that the interpolation polynomials $\hat{\alpha}^a(p, \boldsymbol{\xi})$ are simply linear combinations of the polynomial forms used in the complete arguments of Section 4.5.

Normalization Coefficients

The normalization coefficients (dummy notation)

$$N^a_{ik;\,j\ell} = \frac{p+1}{p+1-i}\,\frac{\mathcal{J}}{h_a}\bigg|_{\boldsymbol{\xi}^a_{ik;\,j\ell}} \tag{4.32}$$

are chosen to ensure that the component of $\mathbf{\Omega}^a_{ik;\,j\ell}(\boldsymbol{r})$ along $\hat{\boldsymbol{\ell}}_a$ is unity. The normalization coefficients in terms of the parent variables $\{\xi_1, \xi_3; \xi_2, \xi_4\}$ and true indices $(i, k;\, j, \ell)$ are given in Table 4.10.

As an example, consider the basis function that interpolates at node 12;22 in Figure 4.2. This function can be expressed as

$$\mathbf{\Omega}^a_{12;22} = N^a_{12;22}(3\xi_a)(3\xi_c - 1)(3\xi_b - 1)(3\xi_d - 1)\mathbf{\Omega}_a(\boldsymbol{r}) \tag{4.33}$$

As previously discussed, the polynomial factors zero the interpolation function at all nodes other than 12;22 in Figure 4.2. Similarly, the factor $(3\xi_c - 1)$ zeros the polynomial at nodes 21;31, 21;22, and 21:13. The vector direction is provided by the base vector $\mathbf{\Omega}_a = \xi_c \nabla \xi_d$.

Dependency Relations at Interior Nodes

At each interior node, either $\boldsymbol{\Omega}^1_{ik;\,j\ell}$ and $\boldsymbol{\Omega}^3_{ik;\,j\ell}$ both interpolate the $\hat{\boldsymbol{\ell}}_1$ $(=-\hat{\boldsymbol{\ell}}_3)$ component or $\boldsymbol{\Omega}^2_{ik;\,j\ell}$ and $\boldsymbol{\Omega}^4_{ik;\,j\ell}$ both interpolate the $\hat{\boldsymbol{\ell}}_2$ $(=-\hat{\boldsymbol{\ell}}_4)$ component; in either case, one of the dependent pair of bases should be discarded to produce a linearly independent set. The dependency between bases also follows from the fact that linear combinations of $\boldsymbol{\Omega}^1_{ik;\,j\ell}$ and $\boldsymbol{\Omega}^3_{ik;\,j\ell}$ which interpolate an interior point can always be found containing the factor

$$\xi_1 \boldsymbol{\Omega}_1(\boldsymbol{r}) + \xi_3 \boldsymbol{\Omega}_3(\boldsymbol{r}) = \mathbf{0}$$

while similar combinations of $\boldsymbol{\Omega}^2_{ik;\,j\ell}$ and $\boldsymbol{\Omega}^4_{ik;\,j\ell}$ contain the factor

$$\xi_2 \boldsymbol{\Omega}_2(\boldsymbol{r}) + \xi_4 \boldsymbol{\Omega}_4(\boldsymbol{r}) = \mathbf{0}$$

(These dependency identities are reported in the third row, right-hand column of Table 4.6.) At interior nodes, the previous identities yield

$$\frac{i\,\boldsymbol{\Omega}^1_{ik;\,j\ell}(\boldsymbol{r})}{N^1_{ik;\,j\ell}} + \frac{k\,\boldsymbol{\Omega}^3_{ik;\,j\ell}(\boldsymbol{r})}{N^3_{ik;\,j\ell}} = \mathbf{0} \quad \text{for } i, k \neq 0$$

$$\frac{j\,\boldsymbol{\Omega}^2_{ik;\,j\ell}(\boldsymbol{r})}{N^2_{ik;\,j\ell}} + \frac{\ell\,\boldsymbol{\Omega}^4_{ik;\,j\ell}(\boldsymbol{r})}{N^4_{ik;\,j\ell}} = \mathbf{0} \quad \text{for } j, \ell \neq 0$$

As with the curl-conforming bases on triangles, sign reversals may be required to maintain a basis function's tangential continuity from element to element at edge-interpolating nodes. Functions that interpolate at interior nodes are confined to one cell and do not require sign adjustments.

Number of DoFs

The number of DoFs for curl-conforming basis functions of order p on a quadrilateral may be determined as follows (it could be helpful to see Figure 4.2):

- 1 component $\times$ $(p+1)$ DoFs $\times$ 4 edges $= 4(p+1)$ edge DoFs,
- 2 components $\times$ $p(p+1)$ DoFs $= 2p(p+1)$ quadrilateral interior DoFs,

for a total of $2(p+1)(p+2)$ DoFs per quadrilateral. We note that for $p=1$ these bases are the same in number and are linear combinations of the so-called covariant projection elements of Crowley et al. [14] having LT and quadratic normal (LT/QN) variation.

Note that the term of highest degree appearing in the Cartesian product representation used to construct the quadrilateral bases is $2p$, while it is p for the triangular bases. As with scalar bases, by abandoning the Cartesian product scheme, one may construct pth-order curl-conforming quadrilateral bases with fewer DoFs. However, there is little or no advantage in using other parameterizations since the bases obtained in those cases do not have simple, straightforward expressions as the bases reported above do.

4.6.4.2 Divergence-Conforming Vector Bases on Quadrilaterals

The divergence-conforming bases on quadrilaterals reported in Table 4.10 may be obtained from the cross product of the associated curl-conforming bases with the unit vector $\hat{\boldsymbol{n}}$ normal to the element. The normalization constants for the divergence-conforming bases on quadrilaterals ensure that the component of $\mathbf{\Lambda}^{a}_{ik;j\ell}(\boldsymbol{r})$ along $\hat{\boldsymbol{h}}_a$ is unity at the interpolation point. All other properties and dependencies of the divergence-conforming bases follow from analogous properties of the curl-conforming bases.

4.7 Vector Bases for the Canonical 3D Cells

In this section we define, for each canonical 3D cell, the Lagrange interpolating polynomials used to form the vector bases before providing the high-order expressions of the basis vectors. The polynomials and the basis vectors are then summarized into tables reported in the last section of this chapter.

4.7.1 Tetrahedral Cells

4.7.1.1 The $\hat{\alpha}(p,\xi)$ Functions, with Edge-Interpolation Points on Only One Edge

The convenience in using dummy parent variables becomes more clear when the interpolation points of the *shape* polynomials of Chapter 2 are moved to the interior of a volumetric cell, as is required to build high-order interpolatory vector functions. In fact, in terms of dummy variables and indices, the pth-order complete set of the $(p+1)(p+2)(p+3)/6$ polynomials

$$\hat{\alpha}^{ad}_{ijk\ell}(p,\boldsymbol{\xi}) = R_i(p+2,\xi_a)\,\hat{R}_j(p+2,\xi_b)\,\hat{R}_k(p+2,\xi_c)\,R_\ell(p+2,\xi_d)$$
$$\text{obtained for } i,\ell=0,1,\ldots,p;\quad j,k=1,2,\ldots,p+1; \qquad (4.34)$$
$$\text{with } i+j+k+\ell=p+2$$

interpolates (for $p\geq 0$) the points

$$(\xi_a,\xi_b,\xi_c,\xi_d) = \left(\frac{i}{p+2},\frac{j}{p+2},\frac{k}{p+2},\frac{\ell}{p+2}\right) = \boldsymbol{\xi}_{ijk\ell} \qquad (4.35)$$

located on the interior of the tetrahedral cell and on the ξ_a and $\xi_d=0$ face, with no interpolation points on the vertices and on the edges of the tetrahedron, except for the edge in common to the $\xi_a=0$ and the $\xi_d=0$ face, and with no interpolation points on the ξ_b or the $\xi_c=0$ face of the tetrahedron (since j and k are never equal to zero in (4.34)). As a matter of fact, the superscript *ad* in the expression of $\hat{\alpha}^{ad}_{ijk\ell}$ is also used to indicate that the Silvester polynomials of the ξ_a and ξ_d dummy variables that appear in (4.34) are unshifted, while the other Silvester polynomials forming the product (4.34) are "careted" (i.e., shifted).

The interpolation points associated with $\hat{\alpha}^{ad}_{ijk\ell}(p,\boldsymbol{\xi})$ for $p=2$ are shown in Figures 4.3b and 4.4b. On the $\xi_d=0$ face, the polynomials (4.34) simplify into

$$\hat{\alpha}^{ad}_{ijk\ell}\Big|_{\xi_d=0} = 0, \quad \text{for } \ell\neq 0 \qquad (4.36)$$

$$\hat{\alpha}^{ad}_{ijk0}\Big|_{\xi_d=0} = R_i(p+2,\xi_a)\,\hat{R}_j(p+2,\xi_b)\,\hat{R}_k(p+2,\xi_c), \quad \text{for } \ell=0 \qquad (4.37)$$

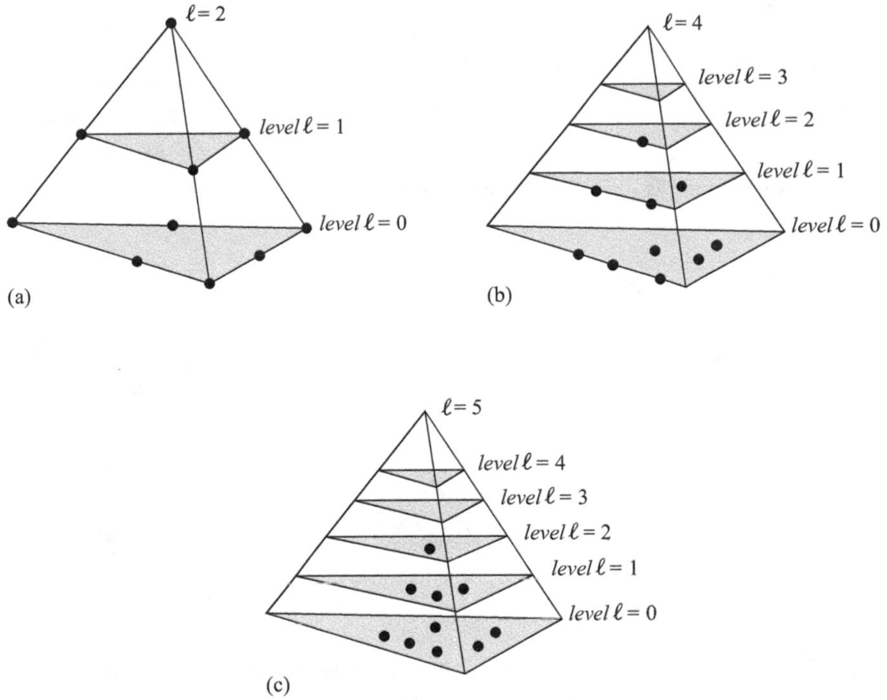

Figure 4.3 Lagrange interpolation of order $p=2$ on a tetrahedron. The interpolation nodes associated with the interpolatory polynomials $\alpha_{ijk\ell}(2,\boldsymbol{\xi})$, $\hat{\alpha}^{ad}_{ijk\ell}(2,\boldsymbol{\xi})$, and $\hat{\beta}^{d}_{ijk\ell}(2,\boldsymbol{\xi})$ are shown in (a), (b), and (c), respectively. For each tetrahedron shown in the figure, the front-left face is the $\xi_a=0$ face, while the $\xi_d=0$ face is the bottom face, labeled as the "*level* $\ell=0$" triangle. The figures show the $\ell=$ constant triangular-cuts that correspond to the coordinate surfaces $\xi_d=\ell/s$, where s is equal to $p, p+2$, and $p+3$ in (a), (b), and (c), respectively (recall that $p=2$ in the figure). The interpolation points of (b) have the same Pascal tetrahedron arrangement valid for the shape polynomial $\alpha_{ijk\ell}(p,\boldsymbol{\xi})$ of the same order shown in (a), except that the nodes along the two faces $\xi_b=0$ and $\xi_c=0$ have been shifted to the tetrahedron's interior. The interpolation points of (c) have the same Pascal tetrahedron arrangement of (a), except that the nodes along the three faces $\xi_a=0$, $\xi_b=0$, and $\xi_c=0$ have been shifted to the tetrahedron's interior. (Nodes are sketched in approximate positions.)

The latter expression obtained for $\ell=0$ holds for $i+j+k=p+2$ and coincides with the expression (4.13) of the basis functions $\hat{\alpha}^{a}_{ijk}$ valid for a triangular cell described in terms of the three dummy parent variables (ξ_a,ξ_b,ξ_c). The $\hat{\alpha}^{ad}_{ijk\ell}$ polynomials are used to build curl-conforming vector functions with a vanishing *tangential* component along the $\xi_b=0$ and the $\xi_c=0$ faces and along all the edges of the tetrahedron with the exception of the $\xi_a=\xi_d=0$ edge.

The dummy arrays of the interpolation points of $\hat{\alpha}^{ad}_{ijk\ell}(p,\xi_a,\xi_b,\xi_c,\xi_d)$ can be used unchanged for ξ_a, ξ_d each equal to one of the four normalized coordinates ξ_1, ξ_2, ξ_3, or ξ_4, with $\xi_a\neq\xi_d$.

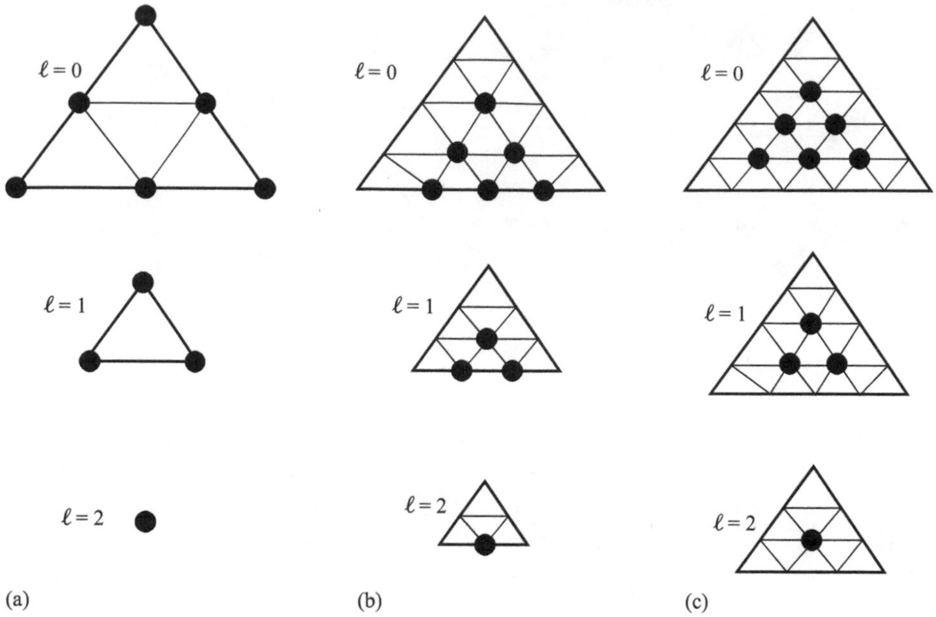

Figure 4.4 Arrangement of the interpolation nodes for Lagrangian interpolations of order $p=2$ on a tetrahedron. The interpolation nodes associated with the polynomials $\alpha_{ijk\ell}(2,\boldsymbol{\xi})$, $\hat{\alpha}^{ad}_{ijk\ell}(2,\boldsymbol{\xi})$, and $\hat{\beta}^{d}_{ijk\ell}(2,\boldsymbol{\xi})$ are shown in columns (a), (b), and (c), respectively. The nodes are staggered on the triangular layers obtained by cutting the tetrahedron along the coordinate surfaces $\xi_d=\ell/s$ for $\ell=0,1,2$, and with $s=p,p+2,p+3$ in (a), (b), and (c), respectively (recall that $p=2$ in the figure). The bottom edge of each triangle in the figure lies on the $\xi_a=0$ face of the tetrahedron. At each level $\ell=$ constant, the interpolation points of (b) and (c) have the same Pascal triangle arrangement valid for the shape polynomial $\alpha_{ijk\ell}(p,\boldsymbol{\xi})$ of the same order shown in (a), except that the nodes have been shifted to the cell's interior.

4.7.1.2 The $\hat{\beta}(p,\xi)$ Functions, with Face-Interpolation Points on One Face Only

To build divergence-conforming vector functions that have a vanishing *normal* component along three cell's faces we also need the following complete set of $(p+1)(p+2)(p+3)/6$ polynomials

$$\begin{aligned}&\hat{\beta}^{d}_{ijk\ell}(p,\boldsymbol{\xi})=\hat{R}_i(p+3,\xi_a)\,\hat{R}_j(p+3,\xi_b)\,\hat{R}_k(p+3,\xi_c)\,R_\ell(p+3,\xi_d)\\&\quad\text{obtained for } \ell=0,1,\ldots,p;\quad i,j,k=1,2,\ldots,p+1;\\&\quad\text{with } i+j+k+\ell=p+3\end{aligned}\tag{4.38}$$

interpolating (for $p\geq 0$) the points

$$(\xi_a,\xi_b,\xi_c,\xi_d)_{ijk\ell}=\left(\frac{i}{p+3},\frac{j}{p+3},\frac{k}{p+3},\frac{\ell}{p+3}\right)=\boldsymbol{\xi}_{ijk\ell}\tag{4.39}$$

located on the interior of the tetrahedral cell and on the $\xi_d = 0$ face, with no interpolation points on the vertices, edges, or faces of the tetrahedron with the exception of the $\xi_d = 0$ face (since i, j, and k are never equal to zero in (4.38)). As a matter of fact, the superscript d in the expression of $\hat{\beta}^d_{ijk\ell}$ is also used to indicate that the Silvester polynomial of the ξ_d dummy variable that appears in (4.38) is unshifted, while the other Silvester polynomials forming the product (4.38) are "careted" (i.e., shifted). The interpolation points associated with $\hat{\beta}^d_{ijk\ell}(p, \boldsymbol{\xi})$ for $p = 2$ are shown in Figures 4.3c and 4.4c. On the $\xi_d = 0$ face, the polynomials (4.38) simplify into

$$\hat{\beta}^d_{ijk\ell}\Big|_{\xi_d=0} = 0, \quad \text{for } \ell \neq 0 \tag{4.40}$$

$$\hat{\beta}^d_{ijk0}\Big|_{\xi_d=0} = \hat{R}_i(p+3, \xi_a)\,\hat{R}_j(p+3, \xi_b)\,\hat{R}_k(p+3, \xi_c), \quad \text{for } \ell = 0 \tag{4.41}$$

The dummy arrays of the interpolation points of $\hat{\beta}^d_{ijk\ell}(p, \xi_a, \xi_b, \xi_c, \xi_d)$ can be used unchanged for ξ_d equal to one of the four normalized coordinates ξ_1, ξ_2, ξ_3, or ξ_4.

4.7.1.3 Curl-Conforming Vector Bases of Order p for Tetrahedral Cells

By using dummy indices and the interpolatory polynomials $\hat{\alpha}^{ad}_{ijk\ell}(p, \boldsymbol{\xi})$ of Section 4.7.1.1, the curl-conforming bases complete to pth order that interpolate a vector on a tetrahedron can be written as (Table 4.11)

$$\boldsymbol{\Omega}^{ad}_{ijk\ell}(\boldsymbol{r}) = N^{ad}_{ijk\ell}\,\hat{\alpha}^{ad}_{ijk\ell}(p, \boldsymbol{\xi})\,\boldsymbol{\Omega}_{ad}(\boldsymbol{r}), \quad i, \ell = 0, 1, \ldots, p; \quad j, k = 1, 2, \ldots, p+1 \tag{4.42}$$

with

$$i + j + k + \ell = p + 2 \tag{4.43}$$

$$N^{ad}_{ijk\ell} = \frac{p+2}{p+2-i-\ell}\,\ell_{ad}\,\big|_{\boldsymbol{\xi}^{ad}_{ijk\ell}}, \qquad \boldsymbol{\xi}^{ad}_{ijk\ell} = \left(\frac{i}{p+2}, \frac{j}{p+2}, \frac{k}{p+2}, \frac{\ell}{p+2}\right) \tag{4.44}$$

and where $\boldsymbol{\Omega}_{ad}$ indicates the generic un-normalized zeroth-order basis vector of Table 4.4 (left-hand column). The expressions of the curl-conforming bases in the parent space $(\xi_1, \xi_2, \xi_3, \xi_4)$ reported in Table 4.12 are obtained by permutation of the indices and subscripts in the above given dummy expressions.

The arrangement of interpolation nodes is similar to that of scalar Lagrange bases of the same order on a tetrahedron, except that the pattern is contracted away from the two faces where tangential components of the zeroth-order bases vanish; there are thus no vertex interpolation nodes and only a single basis function interpolates components along a given edge. However, each interior interpolation node of the tetrahedron is interpolated by six bases; since only three can be independent, half of these must be discarded. To be independent, those retained cannot all have zeroth-order basis factors associated with edges bounding the same face. Similarly, each node interior to a face has tangential components interpolated by three bases functions; since only two of them can be independent, one must be discarded. More precisely, for nodes interior to a face one has to discard one of the bases:

- $\boldsymbol{\Omega}^{12}_{ijk\ell}$, $\boldsymbol{\Omega}^{13}_{ijk\ell}$, or $\boldsymbol{\Omega}^{14}_{ijk\ell}$ for $i = 0$ and $j, k, \ell \neq 0$ ($\xi_1 = 0$ face);
- $\boldsymbol{\Omega}^{12}_{ijk\ell}$, $\boldsymbol{\Omega}^{23}_{ijk\ell}$, or $\boldsymbol{\Omega}^{24}_{ijk\ell}$ for $j = 0$ and $i, k, \ell \neq 0$ ($\xi_2 = 0$ face);

- $\boldsymbol{\Omega}_{ijk\ell}^{13}$, $\boldsymbol{\Omega}_{ijk\ell}^{23}$, or $\boldsymbol{\Omega}_{ijk\ell}^{34}$ for $k=0$ and $i,j,\ell \neq 0$ ($\xi_3=0$ face);
- $\boldsymbol{\Omega}_{ijk\ell}^{14}$, $\boldsymbol{\Omega}_{ijk\ell}^{24}$, or $\boldsymbol{\Omega}_{ijk\ell}^{34}$ for $\ell=0$ and $i,j,k \neq 0$ ($\xi_4=0$ face);

whereas, for nodes interior to the tetrahedron (obtained for i, j, k, and $\ell \neq 0$) one can for example retain only the bases $\boldsymbol{\Omega}_{ijk\ell}^{12}$, $\boldsymbol{\Omega}_{ijk\ell}^{13}$, and $\boldsymbol{\Omega}_{ijk\ell}^{24}$.

With reference to the third row of Table 4.4 (left-hand column), notice that *at the interpolation nodes* one has

$$
\begin{aligned}
&\frac{\boldsymbol{\Omega}_{ijk\ell}^{23}(\boldsymbol{r})}{N_{ijk\ell}^{23}} - \frac{\boldsymbol{\Omega}_{ijk\ell}^{24}(\boldsymbol{r})}{N_{ijk\ell}^{24}} + \frac{\boldsymbol{\Omega}_{ijk\ell}^{34}(\boldsymbol{r})}{N_{ijk\ell}^{34}} = \nabla\xi_1 \\
&\frac{\boldsymbol{\Omega}_{ijk\ell}^{14}(\boldsymbol{r})}{N_{ijk\ell}^{14}} - \frac{\boldsymbol{\Omega}_{ijk\ell}^{34}(\boldsymbol{r})}{N_{ijk\ell}^{34}} - \frac{\boldsymbol{\Omega}_{ijk\ell}^{13}(\boldsymbol{r})}{N_{ijk\ell}^{13}} = \nabla\xi_2 \\
&\frac{\boldsymbol{\Omega}_{ijk\ell}^{12}(\boldsymbol{r})}{N_{ijk\ell}^{12}} - \frac{\boldsymbol{\Omega}_{ijk\ell}^{14}(\boldsymbol{r})}{N_{ijk\ell}^{14}} + \frac{\boldsymbol{\Omega}_{ijk\ell}^{24}(\boldsymbol{r})}{N_{ijk\ell}^{24}} = \nabla\xi_3 \\
&\frac{\boldsymbol{\Omega}_{ijk\ell}^{13}(\boldsymbol{r})}{N_{ijk\ell}^{13}} - \frac{\boldsymbol{\Omega}_{ijk\ell}^{12}(\boldsymbol{r})}{N_{ijk\ell}^{12}} - \frac{\boldsymbol{\Omega}_{ijk\ell}^{23}(\boldsymbol{r})}{N_{ijk\ell}^{23}} = \nabla\xi_4
\end{aligned}
\tag{4.45}
$$

Therefore, as a matter of practical implementation, any three of the above four linear combinations, suitably renormalized, provide convenient alternative bases for interpolating interior points (that is those obtained for i, j, k, and $\ell \neq 0$). On face i, the tangential components of $\nabla\xi_i$ and, hence, those of the corresponding linear combination above vanish at the interpolation point; two of the three remaining combinations, however, can serve as convenient bases for interpolating nodes on face i. With (4.45) interpolating interior and face nodes and with (4.42) interpolating edge nodes, the DoFs simply become the covariant vector components at each interpolation point. The dependency relations for faces and interior nodes are detailed below.

Normalization Constants

The normalization constants $N_{ijk\ell}^{ad}$ are chosen to ensure that the component of $\boldsymbol{\Omega}_{ijk\ell}^{ad}(\boldsymbol{r})$ along $\boldsymbol{\ell}_{ad}$ at the interpolation point is unity. The normalization coefficients in terms of the parent variables $(\xi_1, \xi_2, \xi_3, \xi_4)$ and of the true indices (i, j, k, ℓ) are given in Table 4.12.

Dependency Relations at Face and Interior Nodes

As discussed following (4.44), only two of the three pth-order bases that are non-vanishing at an interpolation point on a face are independent. Similarly, only three of the six bases that interpolate an interior point of the tetrahedron are independent. The dependencies arise from linear combinations of the bases that contain one of the following identities as a factor:

$$
\begin{aligned}
&\xi_2\boldsymbol{\Omega}_{12}(\boldsymbol{r}) + \xi_3\boldsymbol{\Omega}_{13}(\boldsymbol{r}) + \xi_4\boldsymbol{\Omega}_{14}(\boldsymbol{r}) = \mathbf{0} \\
&\xi_3\boldsymbol{\Omega}_{23}(\boldsymbol{r}) + \xi_4\boldsymbol{\Omega}_{24}(\boldsymbol{r}) - \xi_1\boldsymbol{\Omega}_{12}(\boldsymbol{r}) = \mathbf{0} \\
&\xi_4\boldsymbol{\Omega}_{34}(\boldsymbol{r}) - \xi_1\boldsymbol{\Omega}_{13}(\boldsymbol{r}) - \xi_2\boldsymbol{\Omega}_{23}(\boldsymbol{r}) = \mathbf{0} \\
&\xi_1\boldsymbol{\Omega}_{14}(\boldsymbol{r}) + \xi_2\boldsymbol{\Omega}_{24}(\boldsymbol{r}) + \xi_3\boldsymbol{\Omega}_{34}(\boldsymbol{r}) = \mathbf{0}
\end{aligned}
\tag{4.46}
$$

These can be written more succinctly as

$$\xi_{i+1}\boldsymbol{\Omega}_{i,i+1}(\boldsymbol{r}) + \xi_{i+2}\boldsymbol{\Omega}_{i,i+2}(\boldsymbol{r}) + \xi_{i+3}\boldsymbol{\Omega}_{i,i+3}(\boldsymbol{r}) = \boldsymbol{0} \tag{4.47}$$

where $\boldsymbol{\Omega}_{i,j}(\boldsymbol{r}) = -\boldsymbol{\Omega}_{j,i}(\boldsymbol{r})$ and $i = 1, 2, 3, 4$ with index arithmetic performed modulo 4. For nodes not lying on edges, the previous identities yield

$$\begin{aligned}
&\frac{j\,\boldsymbol{\Omega}^{12}_{ijk\ell}(\boldsymbol{r})}{N^{12}_{ijk\ell}} + \frac{k\,\boldsymbol{\Omega}^{13}_{ijk\ell}(\boldsymbol{r})}{N^{13}_{ijk\ell}} + \frac{\ell\,\boldsymbol{\Omega}^{14}_{ijk\ell}(\boldsymbol{r})}{N^{14}_{ijk\ell}} = \boldsymbol{0}, \quad \text{for } j,k,\ell \neq 0\\
&\frac{k\,\boldsymbol{\Omega}^{23}_{ijk\ell}(\boldsymbol{r})}{N^{23}_{ijk\ell}} + \frac{\ell\,\boldsymbol{\Omega}^{24}_{ijk\ell}(\boldsymbol{r})}{N^{24}_{ijk\ell}} - \frac{i\,\boldsymbol{\Omega}^{12}_{ijk\ell}(\boldsymbol{r})}{N^{12}_{ijk\ell}} = \boldsymbol{0}, \quad \text{for } i,k,\ell \neq 0;\\
&\frac{\ell\,\boldsymbol{\Omega}^{34}_{ijk\ell}(\boldsymbol{r})}{N^{34}_{ijk\ell}} - \frac{i\,\boldsymbol{\Omega}^{13}_{ijk\ell}(\boldsymbol{r})}{N^{13}_{ijk\ell}} - \frac{j\,\boldsymbol{\Omega}^{23}_{ijk\ell}(\boldsymbol{r})}{N^{23}_{ijk\ell}} = \boldsymbol{0}, \quad \text{for } i,j,\ell \neq 0\\
&\frac{i\,\boldsymbol{\Omega}^{14}_{ijk\ell}(\boldsymbol{r})}{N^{14}_{ijk\ell}} + \frac{j\,\boldsymbol{\Omega}^{24}_{ijk\ell}(\boldsymbol{r})}{N^{24}_{ijk\ell}} + \frac{k\,\boldsymbol{\Omega}^{34}_{ijk\ell}(\boldsymbol{r})}{N^{34}_{ijk\ell}} = \boldsymbol{0}, \quad \text{for } i,j,k \neq 0
\end{aligned} \tag{4.48}$$

Number of DoFs

The number of DoFs for curl-conforming bases of order p on a tetrahedron may be determined as follows (it could be helpful to see Figure 4.3b):

- 1 component $\times$ $(p+1)$ DoFs $\times$ 6 edges $= 6(p+1)$ edge DoFs,
- 2 components $\times$ $\frac{p(p+1)}{2}$ DoFs $\times$ 4 faces $= 4p(p+1)$ face DoFs,
- 3 components $\times$ $\frac{p(p^2-1)}{6}$ DoFs $\times$ 1 cell $= \frac{p(p^2-1)}{2}$ DoFs interior to a tetrahedron,

for a total of $\frac{(p+1)(p+3)(p+4)}{2}$ DoFs per tetrahedron. This agrees with the result previously obtained by Nédélec [4], who presented explicitly only the $p=0$ and $p=1$ order, non-interpolatory bases. These and the bases presented in Reference 28 for $p = 0, 1, 2$ are linear combinations of the bases provided here.

4.7.1.4 Divergence-Conforming Vector Bases on Tetrahedrons

By using dummy indices and the interpolatory polynomials $\hat{\beta}^d_{ijk\ell}(p, \boldsymbol{\xi})$ of Section 4.7.1.2, the divergence-conforming bases complete to pth order and which interpolate a vector function on a tetrahedron can be expressed as

$$\boldsymbol{\Lambda}^d_{ijk\ell}(\boldsymbol{r}) = N^d_{ijk\ell}\,\hat{\beta}^d_{ijk\ell}(p, \xi_a, \xi_b, \xi_c, \xi_d)\,\boldsymbol{\Lambda}_d(\boldsymbol{r}), \quad \ell = 0, 1, \ldots, p; \quad i,j,k = 1, 2, \ldots, p+1 \tag{4.49}$$

with

$$i + j + k + \ell = p + 3 \tag{4.50}$$

$$N^d_{ijk\ell} = \frac{p+3}{p+3-\ell}\,\frac{\mathcal{J}}{h_d}\bigg|_{\boldsymbol{\xi}^d_{ijk\ell}}, \qquad \boldsymbol{\xi}^d_{ijk\ell} = \left(\frac{i}{p+3}, \frac{j}{p+3}, \frac{k}{p+3}, \frac{\ell}{p+3}\right) \tag{4.51}$$

and where $\mathbf{\Lambda}_d(\boldsymbol{r})$ indicates the generic un-normalized zeroth-order basis vector of Table 4.5 (left-hand column).

In terms of the true indices, the above expression in the parent space $(\xi_1, \xi_2, \xi_3, \xi_4)$ reads as reported in Table 4.13.

The arrangement of interpolation points is similar to that of scalar Lagrange bases of the same order on a tetrahedron, except that the pattern contracts away from the three faces where normal components of the zeroth-order bases vanish. Note that no interpolation points lie at vertices or along edges of the tetrahedron and that a single basis function interpolates the normal at a point on a given face. This is not the case at interior points, however, where four bases contribute component at each interpolation point. Clearly, only three can be independent, and hence one interpolating basis must be eliminated for each interior point. The dependency relations that hold at interior nodes are given below.

For $p=0$, the normalized forms are identical to the basis functions presented in Reference 8.

Normalization Constants

The normalization constants of Table 4.13 are chosen such that the components of the bases $\mathbf{\Lambda}^1_{ijk\ell}$, $\mathbf{\Lambda}^2_{ijk\ell}$, $\mathbf{\Lambda}^3_{ijk\ell}$, and $\mathbf{\Lambda}^4_{ijk\ell}$ along $\hat{\boldsymbol{h}}_1, \hat{\boldsymbol{h}}_2, \hat{\boldsymbol{h}}_3$, and $\hat{\boldsymbol{h}}_4$, respectively, are unity at the interpolation points $\boldsymbol{\xi}^a_{ijk\ell} = (\frac{i}{p+3}, \frac{j}{p+3}, \frac{k}{p+3}, \frac{\ell}{p+3})$ (with $a=1,2,3$, or 4).

Dependency Relations at Interior Nodes

As discussed following (4.51), only three of the four pth-order divergence-conforming bases at an interior point of the tetrahedron are independent. The dependencies arise from linear combinations of the bases that contain the identity

$$\xi_1\mathbf{\Lambda}_1 + \xi_2\mathbf{\Lambda}_2 + \xi_3\mathbf{\Lambda}_3 + \xi_4\mathbf{\Lambda}_4 = \mathbf{0}$$

as a factor. Indeed, for $i, j, k, \ell \neq 0$, one obtains

$$\frac{i\,\mathbf{\Lambda}^1_{ijk\ell}(\boldsymbol{r})}{N^1_{ijk\ell}} + \frac{j\,\mathbf{\Lambda}^2_{ijk\ell}(\boldsymbol{r})}{N^2_{ijk\ell}} + \frac{k\,\mathbf{\Lambda}^3_{ijk\ell}(\boldsymbol{r})}{N^3_{ijk\ell}} + \frac{\ell\,\mathbf{\Lambda}^4_{ijk\ell}(\boldsymbol{r})}{N^4_{ijk\ell}} = \mathbf{0}$$

Number of DoFs

The number of DoFs for divergence-conforming bases of order p on a tetrahedron may be determined as follows (refer to Figure 4.3c):

- 1 component $\times \frac{(p+1)(p+2)}{2}$ DoFs $\times$ 4 faces $= 2(p+1)(p+2)$ face DoFs,
- 3 components $\times \frac{p(p+1)(p+2)}{6}$ DoFs $\times$ 1 cell $= \frac{p(p+1)(p+2)}{2}$ cell interior DoFs,

for a total of $\frac{(p+1)(p+2)(p+4)}{2}$ DoFs per tetrahedron. This agrees with the result previously determined by Nédélec [4].

4.7.2 Brick Cells

4.7.2.1 The $\hat{\alpha}(p, \xi)$ Functions, with Edge-Interpolation Points on Only One Edge

In terms of dummy variables and indices, the pth-order complete set of the $(p+1)^3$ polynomials

$$
\begin{aligned}
\hat{\alpha}^{ab}_{i\ell;\,jm;kn}(p, \boldsymbol{\xi}) = {} & R_i(p+1, \xi_a)\, \hat{R}_\ell(p+1, \xi_d) \\
& R_j(p+1, \xi_b)\, \hat{R}_m(p+1, \xi_e) \\
& \hat{R}_k(p+2, \xi_c)\, \hat{R}_n(p+2, \xi_f) \\
\text{for} \quad & i, j = 0, 1, \ldots, p; \\
& k, \ell, m, n = 1, 2, \ldots, p+1; \\
\text{with} \quad & i+\ell = j+m = p+1; \quad k+n = p+2
\end{aligned}
\tag{4.52}
$$

interpolates (for $p \geq 0$) the points

$$
(\xi_a, \xi_d; \xi_b, \xi_e; \xi_c, \xi_f)_{i\ell;jm;kn} = \boldsymbol{\xi}^{ab}_{i\ell;jm;kn} = \left(\frac{i}{p+1}, \frac{\ell}{p+1}; \frac{j}{p+1}, \frac{m}{p+1}; \frac{k}{p+2}, \frac{n}{p+2} \right)
\tag{4.53}
$$

located on the interior of the brick cell or on the ξ_a and $\xi_b = 0$ face, with no interpolation points on the other faces of the brick (since k, ℓ, m, and n are never equal to zero), and no interpolation points on the vertices and on the edges of the brick, except for the edge in common to the $\xi_a = 0$ and the $\xi_b = 0$ faces. The interpolation points associated with $\hat{\alpha}^{ab}_{i\ell;jm;kn}(p, \boldsymbol{\xi})$ for $p = 2$ are shown in Figure 4.5b. The arrangement of interpolation nodes is similar to that of scalar Lagrange bases $\alpha_{i\ell;jm;kn}(p, \boldsymbol{\xi})$ of the same order on a brick element, except that the pattern is contracted away from the four faces $\xi_c, \xi_d, \xi_e, \xi_f = 0$.

The arrangement of the interpolation nodes for the remaining bases interpolating an edge different from the $\xi_a = \xi_b = 0$ edge may be determined from the figure by rotating the pattern to put the edge-interpolation nodes along the new edge. As a matter of fact, the superscript ab in the expression of $\hat{\alpha}^{ab}_{i\ell;\,jm;kn}$ is also used to indicate that the Silvester polynomials of the ξ_a and ξ_b dummy variables that appear in (4.52) are unshifted, while the other Silvester polynomials forming the product (4.52) are "careted" (i.e., shifted).

On a given quadrilateral face of the brick, the conformity of the $\hat{\alpha}$ basis functions of order p defined for the brick with those defined for a quadrilateral cell (in Section 4.6.2) is readily proved, to the point that the details can be left to the reader. For example, by recalling that we are using "*dummy*" parent variables and indices, it is enough to compare (4.19) with the expression (4.52) on the $\xi_a = 0$ face

$$
\begin{aligned}
&\hat{\alpha}^{ab}_{i=0,\ell=p+1;jm;kn}(p, \boldsymbol{\xi}) = R_j(p+1, \xi_b)\hat{R}_m(p+1, \xi_e)\hat{R}_k(p+2, \xi_c)\hat{R}_n(p+2, \xi_f), \\
&j = 0, 1, \ldots, p; \quad k, m, n = 1, 2, \ldots, p+1, \quad j+m = p+1; \quad k+n = p+2
\end{aligned}
\tag{4.54}
$$

as it is simplified for $i = 0$, $\ell = p+1$ (or, on the $\xi_b = 0$ face, for $j = 0$, $m = p+1$). Alternatively, one can just compare the interpolation grids defined on the two quadrilateral faces at issue (for example, for $p = 2$, compare Figure 4.5b with Figure 4.2). This latter result on conformity is of importance since it proves that the interpolatory polynomials $\hat{\alpha}$ of the brick are conforming

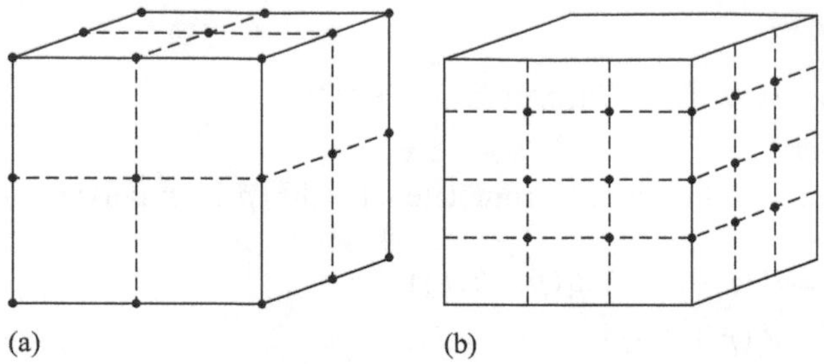

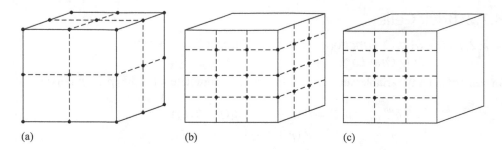

Figure 4.5 Lagrange interpolation of order $p=2$ on a brick. The $\xi_d=0$, $\xi_e=0$, and $\xi_f=0$ faces are opposite to the $\xi_a=0$, $\xi_b=0$, and $\xi_c=0$ faces, respectively. For each brick shown in the figure, the $\xi_a=0$ and the $\xi_c=0$ faces are the front and bottom faces, respectively; the right-hand face is the $\xi_b=0$ face. The interpolation nodes associated with the interpolatory polynomials $\alpha_{i\ell;\,jm;kn}(2,\boldsymbol{\xi})$, $\hat{\alpha}^{ab}_{i\ell;\,jm;kn}(2,\boldsymbol{\xi})$, and $\hat{\beta}^{a}_{i\ell;\,jm;kn}(2,\boldsymbol{\xi})$ are shown in (a), (b), and (c), respectively (with interior interpolation points omitted for clarity). The interpolation points of (b) have the same arrangement valid for the shape polynomial $\alpha_{ijk\ell}(p,\boldsymbol{\xi})$ of the same order shown in (a), except that the pattern is contracted away from the four faces $\xi_c, \xi_d, \xi_e, \xi_f=0$. In this manner, only one edge (the $\xi_a=\xi_b=0$ edge) is interpolated by the polynomials $\hat{\alpha}^{ab}_{i\ell;\,jm;kn}$. The interpolation points of (c) have the same arrangement of (a), except that the nodes along the five faces $\xi_b, \xi_c, \xi_d, \xi_e, \xi_f=0$ have been shifted to the brick's interior. In this manner, only one face (the $\xi_a=0$ face) is interpolated by the polynomials $\hat{\beta}^{a}_{i\ell;\,jm;kn}$.

with those defined on adjacent cells having a quadrilateral face in common to the brick, where the adjacent cells can be either bricks or triangular prisms (discussed in Section 4.7.3).

4.7.2.2 The $\hat{\beta}(p,\xi)$ Functions, with Face-Interpolation Points on Only One Face

Vector functions that have a vanishing *normal* component along five brick faces are built by using the following complete set of $(p+1)^3$ polynomials

$$
\begin{aligned}
\hat{\beta}^{a}_{i\ell;jm;kn}(p,\boldsymbol{\xi}) = {} & R_i(p+1,\xi_a)\,\hat{R}_\ell(p+1,\xi_d)\\
& \hat{R}_j(p+2,\xi_b)\,\hat{R}_m(p+2,\xi_e)\\
& \hat{R}_k(p+2,\xi_c)\,\hat{R}_n(p+2,\xi_f)
\end{aligned}
\tag{4.55}
$$

$$
\begin{aligned}
&\text{obtained for} \quad i=0,1,\ldots,p;\ \ j,k,\ell,m,n=1,2,\ldots,p+1\\
&\text{with} \quad i+\ell=p+1, j+m=k+n=p+2
\end{aligned}
$$

interpolating (for $p\geq 0$) the points

$$
\boldsymbol{\xi}^{a}_{i\ell;\,jm;kn} = \left(\frac{i}{p+1},\frac{\ell}{p+1};\frac{j}{p+2},\frac{m}{p+2};\frac{k}{p+2},\frac{n}{p+2}\right) \tag{4.56}
$$

located on the interior of the brick cell or on the $\xi_a=0$ face, with no interpolation points on the vertices, edges, or faces of the brick with the exception of the $\xi_a=0$ face. In fact, the superscript a in the expression of $\hat{\beta}^{a}_{i\ell;\,jm;kn}$ is also used to indicate that the Silvester polynomial of the ξ_a

dummy variable that appears in (4.55) is unshifted, while the other Silvester polynomials forming the product (4.55) are "careted" (i.e., shifted).

On the $\xi_a = 0$ face, the polynomials (4.55) simplify into

$$\hat{\beta}^a_{i\ell;jm;kn}\Big|_{\xi_a=0} = 0, \quad \text{for } i \neq 0 \tag{4.57}$$

$$\hat{\beta}^a_{i\ell;jm;kn}\Big|_{\xi_a=0} = \hat{R}_j(p+2,\xi_b)\,\hat{R}_m(p+2,\xi_e) \\ \hat{R}_k(p+2,\xi_c)\,\hat{R}_n(p+2,\xi_f), \quad \text{for } i=0 \tag{4.58}$$

The interpolation points associated with $\hat{\beta}^a_{i\ell;\,jm;kn}(p,\boldsymbol{\xi})$ for $p=2$ are shown in Figure 4.5c. The dummy arrays of the interpolation points of $\hat{\beta}^a_{i\ell;\,jm;kn}$ can be used unchanged for ξ_a equal to one of the six normalized coordinates ξ_1, ξ_2, ξ_3, ξ_4, ξ_5, or ξ_6 (Table 4.14).

4.7.2.3 Curl-Conforming Vector Bases of Order p for Brick Cells

By using dummy indices and the interpolatory polynomials $\hat{\alpha}^{ab}_{i\ell;jm;kn}(p,\boldsymbol{\xi})$ of Section 4.7.2.1, the curl-conforming bases complete to pth order that interpolate a vector on a brick can be written

$$\boldsymbol{\Omega}^{ab}_{i\ell;\,jm;kn}(\boldsymbol{r}) = N^{ab}_{i\ell;\,jm;kn}\,\hat{\alpha}^{ab}_{i\ell;jm;kn}(p,\boldsymbol{\xi})\,\boldsymbol{\Omega}_{ab}(\boldsymbol{r}), \\ \text{for} \quad i,j=0,1,\ldots,p; \quad k,\ell,m,n=1,2,\ldots,p+1 \tag{4.59}$$

with

$$i+\ell = j+m = p+1; \quad k+n = p+2 \tag{4.60}$$

$$N^{ab}_{i\ell;jm;kn} = \frac{(p+1)^2}{(p+1-i)(p+1-j)}\,\ell_{ab}\Big|_{\boldsymbol{\xi}^{ab}_{i\ell;jm;kn}} \tag{4.61}$$

$$\boldsymbol{\xi}^{ab}_{i\ell;jm;kn} = \left(\frac{i}{p+1},\frac{\ell}{p+1};\frac{j}{p+1},\frac{m}{p+1};\frac{k}{p+2},\frac{n}{p+2}\right) \tag{4.62}$$

and where $\boldsymbol{\Omega}_{ab}(\boldsymbol{r})$ indicates the generic un-normalized zeroth-order basis vector of Table 4.4 (right-hand column).

In terms of the true indices, the above expressions in the parent space $(\xi_1,\xi_4;\xi_2,\xi_5;\xi_3,\xi_6)$ read as reported in Table 4.15, with normalization constants given in Table 4.16.

The arrangement of interpolation points is similar to that of scalar Lagrange bases of the same order on a brick, except that the pattern contracts away from the four faces where tangential components of the zeroth-order bases vanish. Note that no vertices of the brick element are interpolated and only a single basis function interpolates a component tangential to a given edge. The tangential components at each interpolation point on a face are interpolated by the bases containing as factors the zeroth-order bases associated with the four edges bounding that face. But on a face, only two tangential components can be independent. Hence only two of these basis functions should be retained. Similarly, on the interior, only three of the twelve bases that interpolate each interior point should be retained to provide interpolation of the three independent components. These three should contain zeroth-order basis factors associated with edges having independent edge vectors. The dependency relations for face and interior nodes are given below.

Normalization Constants

The normalization constants of Table 4.16 are chosen to ensure that the component of $\boldsymbol{\Omega}^{ab}_{i\ell;\, jm;kn}(\boldsymbol{r})$ along $\boldsymbol{\ell}_{ab}$ at the given interpolation point is unity.

Dependency Relations at Face and Interior Nodes

As discussed following (4.62), only two of the four pth-order bases that are non-vanishing at an interpolation point on a face are independent. Similarly, only three of the twelve bases that interpolate an interior point of the brick are independent. The dependencies arise from linear combinations of the bases that contain the following identities as factors:

$$\xi_j \boldsymbol{\Omega}_{ij}(\boldsymbol{r}) + \xi_{j+3}\boldsymbol{\Omega}_{i,j+3}(\boldsymbol{r}) = \boldsymbol{0}, \quad i = 1, 2, \ldots, 6; \quad j = i+1, i+2 \tag{4.63}$$

with $\boldsymbol{\Omega}_{ij}(\boldsymbol{r}) = -\boldsymbol{\Omega}_{ji}(\boldsymbol{r})$. Indeed, *at face and interior nodes*, the zeroth-order basis definitions immediately yield

$$\frac{j\,\boldsymbol{\Omega}^{12}_{i\ell;jm;kn}(\boldsymbol{r})}{N^{12}_{i\ell;jm;kn}} + \frac{m\,\boldsymbol{\Omega}^{15}_{i\ell;jm;kn}(\boldsymbol{r})}{N^{15}_{i\ell;jm;kn}} = \boldsymbol{0}, \qquad \frac{m\,\boldsymbol{\Omega}^{45}_{i\ell;jm;kn}(\boldsymbol{r})}{N^{45}_{i\ell;jm;kn}} - \frac{j\,\boldsymbol{\Omega}^{24}_{i\ell;jm;kn}(\boldsymbol{r})}{N^{24}_{i\ell;jm;kn}} = \boldsymbol{0}$$

$$\frac{k\,\boldsymbol{\Omega}^{13}_{i\ell;jm;kn}(\boldsymbol{r})}{N^{13}_{i\ell;jm;kn}} + \frac{n\,\boldsymbol{\Omega}^{16}_{i\ell;jm;kn}(\boldsymbol{r})}{N^{16}_{i\ell;jm;kn}} = \boldsymbol{0}, \qquad \frac{n\,\boldsymbol{\Omega}^{46}_{i\ell;jm;kn}(\boldsymbol{r})}{N^{46}_{i\ell;jm;kn}} - \frac{k\,\boldsymbol{\Omega}^{34}_{i\ell;jm;kn}(\boldsymbol{r})}{N^{34}_{i\ell;jm;kn}} = \boldsymbol{0}$$

$$\frac{k\,\boldsymbol{\Omega}^{23}_{i\ell;jm;kn}(\boldsymbol{r})}{N^{23}_{i\ell;jm;kn}} + \frac{n\,\boldsymbol{\Omega}^{26}_{i\ell;jm;kn}(\boldsymbol{r})}{N^{26}_{i\ell;jm;kn}} = \boldsymbol{0}, \qquad \frac{n\,\boldsymbol{\Omega}^{56}_{i\ell;jm;kn}(\boldsymbol{r})}{N^{56}_{i\ell;jm;kn}} - \frac{k\,\boldsymbol{\Omega}^{35}_{i\ell;jm;kn}(\boldsymbol{r})}{N^{35}_{i\ell;jm;kn}} = \boldsymbol{0}$$

$$\frac{\ell\,\boldsymbol{\Omega}^{24}_{i\ell;jm;kn}(\boldsymbol{r})}{N^{24}_{i\ell;jm;kn}} - \frac{i\,\boldsymbol{\Omega}^{12}_{i\ell;jm;kn}(\boldsymbol{r})}{N^{12}_{i\ell;jm;kn}} = \boldsymbol{0}, \qquad \frac{i\,\boldsymbol{\Omega}^{15}_{i\ell;jm;kn}(\boldsymbol{r})}{N^{15}_{i\ell;jm;kn}} + \frac{\ell\,\boldsymbol{\Omega}^{45}_{i\ell;jm;kn}(\boldsymbol{r})}{N^{45}_{i\ell;jm;kn}} = \boldsymbol{0}$$

$$\frac{\ell\,\boldsymbol{\Omega}^{34}_{i\ell;jm;kn}(\boldsymbol{r})}{N^{34}_{i\ell;jm;kn}} - \frac{i\,\boldsymbol{\Omega}^{13}_{i\ell;jm;kn}(\boldsymbol{r})}{N^{13}_{i\ell;jm;kn}} = \boldsymbol{0}, \qquad \frac{i\,\boldsymbol{\Omega}^{16}_{i\ell;jm;kn}(\boldsymbol{r})}{N^{16}_{i\ell;jm;kn}} + \frac{\ell\,\boldsymbol{\Omega}^{46}_{i\ell;jm;kn}(\boldsymbol{r})}{N^{46}_{i\ell;jm;kn}} = \boldsymbol{0}$$

$$\frac{m\,\boldsymbol{\Omega}^{35}_{i\ell;jm;kn}(\boldsymbol{r})}{N^{35}_{i\ell;jm;kn}} - \frac{j\,\boldsymbol{\Omega}^{23}_{i\ell;jm;kn}(\boldsymbol{r})}{N^{23}_{i\ell;jm;kn}} = \boldsymbol{0}, \qquad \frac{j\,\boldsymbol{\Omega}^{26}_{i\ell;jm;kn}(\boldsymbol{r})}{N^{26}_{i\ell;jm;kn}} + \frac{m\,\boldsymbol{\Omega}^{56}_{i\ell;jm;kn}(\boldsymbol{r})}{N^{56}_{i\ell;jm;kn}} = \boldsymbol{0}$$

Number of DoFs

The number of DoFs for curl-conforming bases of order p on a brick element may be determined as follows (it could be helpful to see Figure 4.5b):

- 1 component $\times\ (p+1)$ DoFs $\times$ 12 edges $= 12(p+1)$ edge DoFs,
- 2 components $\times\ p(p+1)$ DoFs $\times$ 6 faces $= 12p(p+1)$ face DoFs,
- 3 components $\times\ p^2(p+1)$ DoFs $= 3p^2(p+1)$ brick-interior DoFs,

for a total of $3(p+1)(p+2)^2$ DoFs per brick element. This total agrees with the result previously determined by Nédélec [4].

4.7.2.4 Divergence-Conforming Vector Bases on Bricks

By using dummy indices and the interpolatory polynomials $\hat{\beta}^a_{i\ell;jm;kn}$ of Section 4.7.2.2, the divergence-conforming bases complete to pth order and which interpolate a vector function on a brick may be written as

$$\mathbf{\Lambda}^a_{i\ell;jm;kn}(\boldsymbol{r}) = N^a_{i\ell;jm;kn}\,\hat{\beta}^a_{i\ell;jm;kn}(p,\boldsymbol{\xi})\,\mathbf{\Lambda}_a(\boldsymbol{r}), \quad i=0,1,\ldots,p; \quad j,k,\ell,m,n=1,2,\ldots,p+1 \tag{4.64}$$

with

$$i+\ell=p+1; \quad j+m=k+n=p+2 \tag{4.65}$$

$$N^a_{i\ell;jm;kn} = \frac{p+1}{p+1-i}\,\frac{\mathcal{J}}{h_a}\bigg|_{\boldsymbol{\xi}^a_{i\ell;jm;kn}} \tag{4.66}$$

$$\boldsymbol{\xi}^a_{i\ell;jm;kn} = \left(\frac{i}{p+1},\frac{\ell}{p+1};\frac{j}{p+2},\frac{m}{p+2};\frac{k}{p+2},\frac{n}{p+2}\right) \tag{4.67}$$

and where $\mathbf{\Lambda}_a(\boldsymbol{r})$ indicates the generic un-normalized zeroth-order basis vector of Table 4.5 (right-hand column).

In terms of the true indices, the above expressions in the parent space $(\xi_1,\xi_4;\xi_2,\xi_5;\xi_3,\xi_6)$ read as reported in Table 4.17.

The arrangement of interpolation points is similar to that of scalar Lagrange bases of the same order on a brick, except that the pattern contracts away from the five faces where normal components of the zeroth-order bases vanish. Notice that no points at vertices or edges of the brick element are interpolated and that only a single basis function interpolates a component normal to a given face. On the interior, only three of the six bases that interpolate each interior point should be retained to maintain independence of the bases. These three should contain zeroth-order basis factors associated with faces that are not opposite one another. The dependency relations for interior nodes are given below.

Normalization Constants

The normalization constants $N^a_{i\ell;jm;kn}$ are chosen to ensure that the component of $\mathbf{\Lambda}^a_{i\ell;jm;kn}(\boldsymbol{r})$ along $\hat{\boldsymbol{h}}_a$ at the given interpolation point $\boldsymbol{\xi}^a_{i\ell;jm;kn}$ is unity. The normalization constants and the interpolation points are given in Table 4.17.

Dependency Relations at Interior Nodes

As discussed following (4.67), only three of the six pth-order divergence-conforming bases at an interior point of the brick element are independent. The dependencies arise from linear combinations of the bases that contain the identities

$$\xi_1\mathbf{\Lambda}_1(\boldsymbol{r}) + \xi_4\mathbf{\Lambda}_4(\boldsymbol{r}) = \mathbf{0}$$
$$\xi_2\mathbf{\Lambda}_2(\boldsymbol{r}) + \xi_5\mathbf{\Lambda}_5(\boldsymbol{r}) = \mathbf{0}$$
$$\xi_3\mathbf{\Lambda}_3(\boldsymbol{r}) + \xi_6\mathbf{\Lambda}_6(\boldsymbol{r}) = \mathbf{0}$$

Indeed, *at interior nodes*, the previous identities immediately yield

$$\frac{i\,\mathbf{\Lambda}^1_{i\ell;jm;kn}(\boldsymbol{r})}{N^1_{i\ell;\,jm;kn}} + \frac{\ell\,\mathbf{\Lambda}^4_{i\ell;jm;kn}(\boldsymbol{r})}{N^4_{i\ell;\,jm;kn}} = \mathbf{0}$$

$$\frac{j\,\mathbf{\Lambda}^2_{i\ell;jm;kn}(\boldsymbol{r})}{N^2_{i\ell;\,jm;kn}} + \frac{m\,\mathbf{\Lambda}^5_{i\ell;jm;kn}(\boldsymbol{r})}{N^5_{i\ell;\,jm;kn}} = \mathbf{0}$$

$$\frac{k\,\mathbf{\Lambda}^3_{i\ell;jm;kn}(\boldsymbol{r})}{N^3_{i\ell;\,jm;kn}} + \frac{n\,\mathbf{\Lambda}^6_{i\ell;jm;kn}(\boldsymbol{r})}{N^6_{i\ell;\,jm;kn}} = \mathbf{0}$$

Number of DoFs

The number of DoFs for divergence-conforming bases of order p on a brick element may be determined as follows (it could be helpful to see Figure 4.5c):

- 1 component $\times\,(p+1)^2$ DoFs $\times$ 6 faces $=6(p+1)^2$ face DoFs,
- 3 components $\times\, p(p+1)^2$ DoFs $=3p(p+1)^2$ brick-interior DoFs,

for a total of $3(p+2)(p+1)^2$ DoFs per brick element. This result agrees with that found by Nédélec [4].

4.7.3 Triangular Prism Cells

In the $(\xi_1, \xi_2, \xi_3; \xi_4, \xi_5)$ parent space, the quadrilateral faces of the prism are the $\xi_1=0$, the $\xi_2=0$, and the $\xi_3=0$ faces, while the $\xi_4=0$ and the $\xi_5=0$ faces of the prism are triangular. In this subsection, however, we often use also a dummy notation and indicate two generic quadrilateral faces of the prism as the $\xi_a=0$ and $\xi_b=0$ faces, with an edge in common along the line $\xi_a=\xi_b=0$. That is, in the following, the *dummy* couple (ξ_a, ξ_b) stays for the parent-variable couple (ξ_1, ξ_2), (ξ_2, ξ_3), or (ξ_3, ξ_1). Similarly, the dummy triangular face $\xi_d=0$ indicates the triangular face $\xi_4=0$ or the $\xi_5=0$ triangular face. The dummy faces $\xi_a=0$ and $\xi_d=0$ have a common edge along the line $\xi_a=\xi_d=0$, and the *dummy* couple (ξ_a, ξ_d) indicates the parent-variable couples (ξ_1, ξ_4), (ξ_2, ξ_4), (ξ_3, ξ_4), (ξ_1, ξ_5), (ξ_2, ξ_5), or (ξ_3, ξ_5).

4.7.3.1 *The $\hat{\alpha}(p, \xi)$ Functions, with Edge-Interpolation Points on Only One Edge*

To build high-order interpolatory vector functions with a tangential component that vanishes along all the edges of the prism but one we need to move the interpolation points of the prism shape function to the prism's interior. Once again, these interpolatory functions are conveniently written in terms of dummy parent variables. In fact, in terms of dummy variables and indices, the pth-order complete set of the $(p+1)^2(p+2)/2$ polynomials

$$\begin{aligned}
\hat{\alpha}^{ad}_{ijk;\ell m}(p,\boldsymbol{\xi}) &= R_i(p+2,\xi_a)\,\hat{R}_j(p+2,\xi_b)\,\hat{R}_k(p+2,\xi_c) \\
&\quad R_\ell(p+1,\xi_d)\,\hat{R}_m(p+1,\xi_e) \\
\text{for} \quad & i,\ell = 0,1,\ldots,p; \\
& j,k,m = 1,2,\ldots,p+1; \\
\text{with} \quad & i+j+k = p+2; \quad \ell+m = p+1
\end{aligned} \tag{4.68}$$

interpolates (for $p \geq 0$) the points

$$(\xi_a,\xi_b,\xi_c;\xi_d,\xi_e)_{ijk;\ell m} = \boldsymbol{\xi}^{ad}_{ijk;\ell m} = \left(\frac{i}{p+2},\frac{j}{p+2},\frac{k}{p+2};\frac{\ell}{p+1},\frac{m}{p+1}\right) \tag{4.69}$$

located on the interior of the prism cell or on the ξ_a and $\xi_d = 0$ faces, with no interpolation points on the other faces of the prism (since j, k, and m are never equal to zero) and no interpolation points on the vertices and on the edges of the prism, except for the edge in common to the quadrilateral $\xi_a = 0$ face and the triangular $\xi_d = 0$ face. As a matter of fact, the superscript ad in the expression of $\hat{\alpha}^{ad}_{ijk;\ell m}$ is also used to indicate that the Silvester polynomials of the ξ_a and ξ_d dummy variable that appear in (4.68) are unshifted, while the other Silvester polynomials forming the product (4.68) are "careted" (i.e., shifted).

By using dummy parent variables, the conformity of the $\hat{\alpha}^{ad}$ basis functions of the prism on the triangular (or on the quadrilateral) face of the prism itself with the Lagrangian basis functions $\hat{\alpha}$ of the same order defined for the triangular (or the quadrilateral) cell can readily be proved by the reader (see (4.13) and (4.19)).

For example, the interpolation points associated with $\hat{\alpha}^{ad}_{ijk;\ell m}(p,\boldsymbol{\xi})$ for $p=2$ are shown in Figure 4.6a. The arrangement of interpolation nodes is similar to that of scalar Lagrange bases $\alpha_{ijk;\ell m}(p,\boldsymbol{\xi})$ of the same order on a prism element (for $p=2$, see Figure 2.27), except that the pattern is contracted away from the three faces $\xi_b,\xi_c,\xi_e = 0$. The arrangement of the interpolation nodes for the remaining bases interpolating the other five edges where triangular and quadrilateral faces intersect may be determined from Figure 4.6a by rotating the pattern to put the edge-interpolation nodes along the new edge. Note that on rotation, the pattern of the interpolation nodes on the triangular face and in the prism interior remains the same.

Similarly, in terms of dummy variables and indices, the pth-order complete set of the $(p+1)^2(p+2)/2$ polynomials

$$\begin{aligned}
\hat{\alpha}^{ab}_{ijk;\ell m}(p,\boldsymbol{\xi}) &= R_i(p+1,\xi_a)\,R_j(p+1,\xi_b)\,\hat{R}_k(p+1,\xi_c) \\
&\quad \hat{R}_\ell(p+2,\xi_d)\,\hat{R}_m(p+2,\xi_e) \\
\text{for} \quad & i,j = 0,1,\ldots,p; \\
& k,\ell,m = 1,2,\ldots,p+1; \\
\text{with} \quad & i+j+k = p+1; \quad \ell+m = p+2
\end{aligned} \tag{4.70}$$

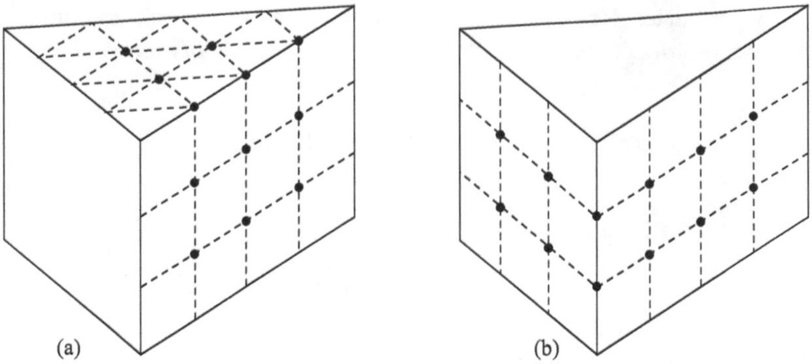

Figure 4.6 Interpolation points for curl-conforming interpolatory bases of order $p=2$ on prism elements (interior interpolation points omitted for clarity). (a) Nodes in basis subset $\hat{\alpha}^{ad}_{ijk;\ell m}(p,\boldsymbol{\xi})$. (b) Nodes in basis subset $\hat{\alpha}^{ab}_{ijk;\ell m}(p,\boldsymbol{\xi})$. The interpolation points have the same arrangement valid for the shape polynomial $\alpha_{ijk;\ell m}(p,\boldsymbol{\xi})$ of the same order shown in Figure 2.27, except that the nodes along three (out of five) faces have been shifted to the prism's interior. The $\xi_a=0$ and the $\xi_b=0$ face of the two prisms in the figure are the front-right and the front-left faces, respectively; the top-face of the prisms is the $\xi_d=0$ face.

interpolates (for $p \geq 0$) the points

$$(\xi_a,\xi_b,\xi_c;\xi_d,\xi_e)_{ijk;\ell m}=\boldsymbol{\xi}^{ab}_{ijk;\ell m}=\left(\frac{i}{p+1},\frac{j}{p+1},\frac{k}{p+1};\frac{\ell}{p+2},\frac{m}{p+2}\right) \tag{4.71}$$

located on the interior of the prism cell or on the ξ_a and $\xi_b=0$ faces, with no interpolation points on the other faces of the prism (since k, ℓ, and m are never equal to zero), and no interpolation points on the vertices and on the edges of the prism, except for the edge in common to the two quadrilateral faces $\xi_a=0$ and $\xi_b=0$. As a matter of fact, the superscript ab in the expression of $\hat{\alpha}^{ab}_{ijk;\ell m}$ is also used to indicate that the Silvester polynomials of the ξ_a and ξ_b dummy variables that appear in (4.70) are unshifted, while the other Silvester polynomials forming the product (4.70) are "careted" (i.e., shifted).

The interpolation points associated with $\hat{\alpha}^{ab}_{ijk;\ell m}(p,\boldsymbol{\xi})$ for $p=2$ are shown in Figure 4.6b. The arrangement of interpolation nodes is similar to that of scalar Lagrange bases $\alpha_{ijk;\ell m}(p,\boldsymbol{\xi})$ of the same order on a prism element (for example, for $p=2$, see Figure 2.27, except that the pattern is contracted away from the four faces $\xi_c,\xi_d,\xi_e,\xi_f=0$. The arrangement of the interpolation nodes for the remaining bases interpolating the other two edges $\xi_b=\xi_c=0$ and $\xi_c=\xi_a=0$ in common with the other two quadrilateral faces may be determined from Figure 4.6b by rotating the pattern to put the edge-interpolation nodes along the new edge. Despite rotation, the pattern of the interpolation nodes in the interior of the quadrilateral faces and in the prism interior remains the same.

The conformity of the prism basis functions $\hat{\alpha}^{ab}_{ijk;\ell m}(p,\boldsymbol{\xi})$ with the Lagrangian basis function $\hat{\alpha}(p,\boldsymbol{\xi})$ defined for the quadrilateral cell can readily be proved by the reader (see (4.19)).

4.7.3.2 The $\hat{\beta}(p,\xi)$ Functions, with Face-Interpolation Points on One Face Only

To build vector functions that have a vanishing *normal* component along five prism faces, we need the following complete set of $(p+1)^2(p+2)/2$ polynomials (Table 4.18)

$$\begin{aligned}\hat{\beta}^{a}_{ijk;\ell m}(p,\boldsymbol{\xi}) &= R_i(p+2,\xi_a)\,\hat{R}_j(p+2,\xi_b)\,\hat{R}_k(p+2,\xi_c)\\ &\quad \hat{R}_\ell(p+2,\xi_d)\,\hat{R}_m(p+2,\xi_e)\\ \text{obtained for}\quad & i=0,1,\ldots,p;\quad j,k,\ell,m=1,2,\ldots,p+1\\ \text{with}\quad & i+j+k=\ell+m=p+2\end{aligned} \tag{4.72}$$

interpolating (for $p \geq 0$) the points

$$\boldsymbol{\xi}^{a}_{ijk;\ell m} = (\xi_a,\xi_b,\xi_c;\xi_d,\xi_e)_{ijk;\ell m} = \left(\frac{i}{p+2},\frac{j}{p+2},\frac{k}{p+2};\frac{\ell}{p+2},\frac{m}{p+2}\right) \tag{4.73}$$

located on the interior of the prism cell or on the $\xi_a = 0$ *quadrilateral* face, with no interpolation points on the vertices, edges, or faces of the prism with the exception of the $\xi_a = 0$ face.

Similarly, the $(p+1)^2(p+2)/2$ polynomials

$$\begin{aligned}\hat{\beta}^{d}_{ijk;\ell m}(p,\boldsymbol{\xi}) &= \hat{R}_i(p+3,\xi_a)\,\hat{R}_j(p+3,\xi_b)\,\hat{R}_k(p+3,\xi_c)\\ &\quad R_\ell(p+1,\xi_d)\,\hat{R}_m(p+1,\xi_e)\\ \text{obtained for}\quad & \ell=0,1,\ldots,p;\quad i,j,k,m=1,2,\ldots,p+1\\ \text{with}\quad & i+j+k=p+3;\quad \ell+m=p+1\end{aligned} \tag{4.74}$$

interpolate (for $p \geq 0$) the points

$$\boldsymbol{\xi}^{d}_{ijk;\ell m} = (\xi_a,\xi_b,\xi_c;\xi_d,\xi_e)_{ijk;\ell m} = \left(\frac{i}{p+3},\frac{j}{p+3},\frac{k}{p+3};\frac{\ell}{p+1},\frac{m}{p+1}\right) \tag{4.75}$$

located on the interior of the prism cell or on the $\xi_d = 0$ *triangular* face, with no interpolation points on the vertices, edges, or faces of the prism with the exception of the $\xi_d = 0$ face.

As a matter of fact, the superscript a (or d) in the expression of $\hat{\beta}^{a}_{ijk;\ell m}$ (or $\hat{\beta}^{d}_{ijk;\ell m}$) is also used to indicate that the Silvester polynomial of the ξ_a (or ξ_d) dummy variable that appears in (4.72) (or in (4.74)) is unshifted, while the other Silvester polynomials forming the product (4.72) (or (4.74)) are "careted" (i.e., shifted).

The dummy arrays of the interpolation points of $\hat{\beta}^{a}_{ijk;\ell m}$ can be used unchanged for ξ_a equal to one of the three normalized coordinates ξ_1, ξ_2, or ξ_3 (with the same interpolation nodes in the interior of the prism). Similarly, the dummy arrays of the interpolation points of $\hat{\beta}^{d}_{ijk;\ell m}$ can be used unchanged for ξ_d equal to one of the two normalized coordinates ξ_4 or ξ_5 (with the same interpolation nodes in the interior of the prism).

The reader can readily prove that the prism functions $\hat{\beta}^{a}_{ijk;\ell m}(p,\boldsymbol{\xi})$ and the brick functions $\hat{\beta}^{a}_{ik;j\ell;mn}(p,\boldsymbol{\xi})$ are conforming on a ($\xi_a = 0$) quadrilateral face shared by both cells (a prism attached to a brick) (see (4.58)). Similarly, the prism functions $\hat{\beta}^{d}_{ijk;\ell m}(p,\boldsymbol{\xi})$ and the tetrahedral functions $\hat{\beta}^{d}_{ijk\ell}(p,\boldsymbol{\xi})$ are conforming on the ($\xi_d = 0$) triangular face shared by both cells (see (4.38), (4.40), and (4.41)).

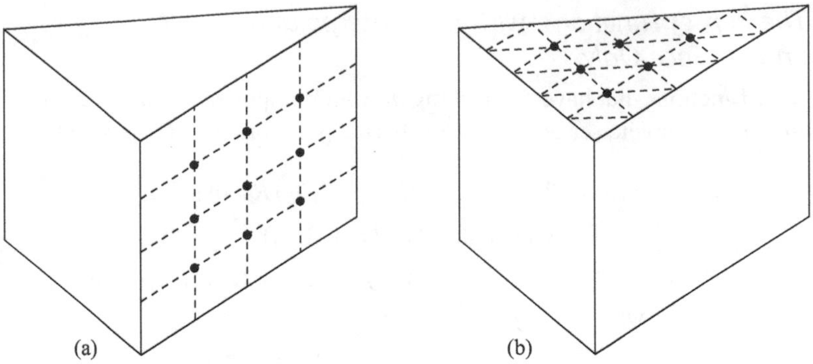

Figure 4.7 Interpolation points for divergence-conforming interpolatory bases of order $p=2$ on prism elements (interior interpolation points omitted for clarity). The front-left face is the $\xi_a=0$ face; the top-face is the $\xi_d=0$ face. (a) Nodes in basis subset $\hat{\beta}^a_{ijk;\ell m}(p,\boldsymbol{\xi})$. (b) Nodes in basis subset $\hat{\beta}^d_{ijk;\ell m}(p,\boldsymbol{\xi})$. The interpolation points have the same arrangement valid for the shape polynomial $\alpha_{ijk\ell}(p,\boldsymbol{\xi})$ of the same order shown in Figure 2.27, except that the nodes along four (out of five) faces have been shifted to the prism's interior.

© 1998 IEEE. Reprinted with permission from R. D. Graglia, D. R. Wilton, A. F. Peterson, and I.-L. Gheorma, "Higher order interpolatory vector bases on prism elements," *IEEE Trans. Antennas Propag.*, vol. 46, no. 3, pp. 442–450, Mar. 1998.

For $p=2$, the interpolation points associated with $\hat{\beta}^a_{ijk;\ell m}(p,\boldsymbol{\xi})$ and $\hat{\beta}^e_{ijk;\ell m}(p,\boldsymbol{\xi})$ are shown in Figure 4.7a and b, respectively.

4.7.3.3 Curl-Conforming Bases of Order p for Prism Cells

By using dummy indices and the interpolatory polynomials $\hat{\alpha}^{ad}_{ijk;\ell m}(p,\boldsymbol{\xi})$ and $\hat{\alpha}^{ab}_{ijk;\ell m}(p,\boldsymbol{\xi})$ of Section 4.7.3.1, the curl-conforming bases complete to pth order that interpolate a vector on a prism can be written

$$\begin{aligned}\boldsymbol{\Omega}^{ad}_{ijk;\ell m}(\boldsymbol{r}) &= N^{ad}_{ijk;\ell m}\, R_i(p+2,\xi_a)\,\hat{R}_j(p+2,\xi_b)\,\hat{R}_k(p+2,\xi_c)\\ &\quad R_\ell(p+1,\xi_d)\,\hat{R}_m(p+1,\xi_e)\,\boldsymbol{\Omega}_{ad}(\boldsymbol{r})\\ \text{for}\quad & i,\ell=0,1,\ldots,p;\\ & j,k,m=1,2,\ldots,p+1;\\ \text{with}\quad & i+j+k=p+2;\quad \ell+m=p+1\end{aligned} \tag{4.76}$$

$$\begin{aligned}\boldsymbol{\Omega}^{ab}_{ijk;\ell m}(\boldsymbol{r}) &= N^{ab}_{ijk;\ell m}\, R_i(p+1,\xi_a)\,R_j(p+1,\xi_b)\,\hat{R}_k(p+1,\xi_c)\\ &\quad \hat{R}_\ell(p+2,\xi_d)\,\hat{R}_m(p+2,\xi_e)\,\boldsymbol{\Omega}_{ab}(\boldsymbol{r})\\ \text{for}\quad & i,j=0,1,\ldots,p;\\ & k,\ell,m=1,2,\ldots,p+1;\\ \text{with}\quad & i+j+k=p+1;\quad \ell+m=p+2\end{aligned} \tag{4.77}$$

with

$$N^{ad}_{ijk;\ell m} = \frac{(p+2)(p+1)}{(p+2-i)(p+1-\ell)} \,\ell_{ad}\Big|_{\boldsymbol{\xi}^{ad}_{ijk;\ell m}}, \qquad \boldsymbol{\xi}^{ad}_{ijk;\ell m} = \left(\tfrac{i}{p+2}, \tfrac{j}{p+2}, \tfrac{k}{p+2}; \tfrac{\ell}{p+1}, \tfrac{m}{p+1}\right)$$

$$N^{ab}_{ijk;\ell m} = \frac{p+1}{p+1-i-j} \,\ell_{ab}\Big|_{\boldsymbol{\xi}^{ab}_{ijk;\ell m}}, \qquad \boldsymbol{\xi}^{ab}_{ijk;\ell m} = \left(\tfrac{i}{p+1}, \tfrac{j}{p+1}, \tfrac{k}{p+1}; \tfrac{\ell}{p+2}, \tfrac{m}{p+2}\right) \tag{4.78}$$

and where $\boldsymbol{\Omega}_{ad}(\boldsymbol{r})$ and $\boldsymbol{\Omega}_{ab}(\boldsymbol{r})$ indicate one of the un-normalized zeroth-order basis vectors of Table 4.4 (central column). In terms of the true indices, the above expression in the parent space $(\xi_1, \xi_2, \xi_3; \xi_4, \xi_5)$ reads as reported in Table 4.19, with normalization constants given in Table 4.20.

The arrangement of interpolation points is similar to that of scalar Lagrange bases of the same order on a prism, except that the pattern contracts away from the three faces where tangential components of the zeroth-order bases vanish. Note that no vertices of the prism element are interpolated and only a single basis function interpolates a component tangential to a given edge. The tangential components at each interpolation point on a face are interpolated by the bases containing as factors zeroth-order basis functions that are associated with the edges bounding that face. But on a face, only two of these tangential components can be independent. Hence two basis functions on rectangular faces and one basis function on triangular faces at each interpolation point must be eliminated. For interpolation points on rectangular faces, only pairs of basis functions with zeroth-order basis factors associated with edges bounding the face and having a common vertex should be eliminated. Similarly, on the interior, only three of the nine bases that interpolate each interior point should be retained to provide interpolation of the three independent components. One of these should have a zeroth-order basis factor associated with an edge formed by intersecting rectangular faces; the remaining two should have zeroth-order basis factors associated with any two edges of, say, one of the triangular faces. The dependency relations for face and interior nodes are given below.

Normalization Constants

The normalization constants above are chosen to ensure that the component of $\boldsymbol{\Omega}^{ad}_{ijk;\ell m}(\boldsymbol{r})$ along $\boldsymbol{\ell}_{ad}$ (and the component of $\boldsymbol{\Omega}^{ab}_{ijk;\ell m}(\boldsymbol{r})$ along $\boldsymbol{\ell}_{ab}$) at the given interpolation point $\boldsymbol{\xi}^{ad}_{ijk;\ell m}$ (or $\boldsymbol{\xi}^{ab}_{ijk;\ell m}$) is unity. The normalization constants and the interpolation points are given in Table 4.20.

Dependency Relations at Face and Interior Nodes

As discussed following (4.78), only two of the three or four pth-order bases for triangular or rectangular faces, respectively, which are non-vanishing at an interpolation point on a face are independent. Similarly, only three of the nine bases that interpolate an interior point of the prism are independent. The dependencies arise from linear combinations of the bases that contain one of the following eight identities as factors:

$$
\begin{aligned}
\xi_1 \boldsymbol{\Omega}_{1j}(\boldsymbol{r}) + \xi_2 \boldsymbol{\Omega}_{2j}(\boldsymbol{r}) + \xi_3 \boldsymbol{\Omega}_{3j}(\boldsymbol{r}) &= \mathbf{0}, \quad \text{for } j = 4, 5 \\
\xi_4 \boldsymbol{\Omega}_{i4}(\boldsymbol{r}) + \xi_5 \boldsymbol{\Omega}_{i5}(\boldsymbol{r}) &= \mathbf{0}, \quad \text{for } i = 1, 2, 3 \\
\xi_1 \boldsymbol{\Omega}_{12}(\boldsymbol{r}) - \xi_3 \boldsymbol{\Omega}_{23}(\boldsymbol{r}) &= \mathbf{0} \\
\xi_1 \boldsymbol{\Omega}_{13}(\boldsymbol{r}) + \xi_2 \boldsymbol{\Omega}_{23}(\boldsymbol{r}) &= \mathbf{0} \\
\xi_2 \boldsymbol{\Omega}_{12}(\boldsymbol{r}) + \xi_3 \boldsymbol{\Omega}_{13}(\boldsymbol{r}) &= \mathbf{0}
\end{aligned}
$$

Indeed, at face and interior nodes, the previous identities immediately yield

$$
\left.
\begin{aligned}
\frac{i\,\boldsymbol{\Omega}^{14}_{ijk;\ell m}(\boldsymbol{r})}{N^{14}_{ijk;\ell m}} + \frac{j\,\boldsymbol{\Omega}^{24}_{ijk;\ell m}(\boldsymbol{r})}{N^{24}_{ijk;\ell m}} + \frac{k\,\boldsymbol{\Omega}^{34}_{ijk;\ell m}(\boldsymbol{r})}{N^{34}_{ijk;\ell m}} &= \mathbf{0}, \\
\frac{i\,\boldsymbol{\Omega}^{15}_{ijk;\ell m}(\boldsymbol{r})}{N^{15}_{ijk;\ell m}} + \frac{j\,\boldsymbol{\Omega}^{25}_{ijk;\ell m}(\boldsymbol{r})}{N^{25}_{ijk;\ell m}} + \frac{k\,\boldsymbol{\Omega}^{35}_{ijk;\ell m}(\boldsymbol{r})}{N^{35}_{ijk;\ell m}} &= \mathbf{0},
\end{aligned}
\right\} \quad \text{for } i, j, k \neq 0
$$

$$
\left.
\begin{aligned}
\frac{\ell\,\boldsymbol{\Omega}^{14}_{ijk;\ell m}(\boldsymbol{r})}{N^{14}_{ijk;\ell m}} + \frac{m\,\boldsymbol{\Omega}^{15}_{ijk;\ell m}(\boldsymbol{r})}{N^{15}_{ijk;\ell m}} &= \mathbf{0}, \\
\frac{\ell\,\boldsymbol{\Omega}^{24}_{ijk;\ell m}(\boldsymbol{r})}{N^{24}_{ijk;\ell m}} + \frac{m\,\boldsymbol{\Omega}^{25}_{ijk;\ell m}(\boldsymbol{r})}{N^{25}_{ijk;\ell m}} &= \mathbf{0}, \\
\frac{\ell\,\boldsymbol{\Omega}^{34}_{ijk;\ell m}(\boldsymbol{r})}{N^{34}_{ijk;\ell m}} + \frac{m\,\boldsymbol{\Omega}^{35}_{ijk;\ell m}(\boldsymbol{r})}{N^{35}_{ijk;\ell m}} &= \mathbf{0}
\end{aligned}
\right\} \quad \text{for } \ell, m \neq 0
$$

$$
\frac{i\,\boldsymbol{\Omega}^{12}_{ijk;\ell m}(\boldsymbol{r})}{N^{12}_{ijk;\ell m}} - \frac{k\,\boldsymbol{\Omega}^{23}_{ijk;\ell m}(\boldsymbol{r})}{N^{23}_{ijk;\ell m}} = \mathbf{0} \quad \text{for } i, k \neq 0
$$

$$
\frac{i\,\boldsymbol{\Omega}^{13}_{ijk;\ell m}(\boldsymbol{r})}{N^{13}_{ijk;\ell m}} + \frac{j\,\boldsymbol{\Omega}^{23}_{ijk;\ell m}(\boldsymbol{r})}{N^{23}_{ijk;\ell m}} = \mathbf{0} \quad \text{for } i, j \neq 0
$$

$$
\frac{j\,\boldsymbol{\Omega}^{12}_{ijk;\ell m}(\boldsymbol{r})}{N^{12}_{ijk;\ell m}} + \frac{k\,\boldsymbol{\Omega}^{13}_{ijk;\ell m}(\boldsymbol{r})}{N^{13}_{ijk;\ell m}} = \mathbf{0} \quad \text{for } j, k \neq 0
$$

Number of DoFs

The number of DoFs for curl-conforming bases of order p on a prism element may be determined as follows (it could be helpful to see Figure 4.6):

- 1 component $\times\ (p+1)$ DoFs $\times$ 9 edges $= 9(p+1)$ edge DoFs,
- 2 components $\times\ \frac{p(p+1)}{2}$ DoFs $\times$ 2 triangular faces $= 2p(p+1)$ face DoFs,
- 2 components $\times\ p(p+1)$ DoFs $\times$ 3 rectangular faces $= 6p(p+1)$ face DoFs,

- 2 components × $\frac{p^2(p+1)}{2}$ interior DoFs $= p^2(p+1)$ prism-interior DoFs,
- 1 component × $\frac{p(p-1)(p+1)}{2}$ interior DoFs $= \frac{p(p-1)(p+1)}{2}$ prism-interior DoFs,

for a total of $\frac{3(p+1)(p+2)(p+3)}{2}$ DoFs per prism element.

4.7.3.4 Divergence-Conforming Bases on Prisms

By using dummy indices and the interpolatory polynomials $\hat{\beta}^a_{ijk;\ell m}(p,\boldsymbol{\xi})$ and $\hat{\beta}^d_{ijk;\ell m}(p,\boldsymbol{\xi})$ of Section 4.7.3.2, the divergence-conforming bases complete to pth order that interpolate a vector on a prism can be expressed

$$\begin{aligned}\mathbf{\Lambda}^a_{ijk;\ell m}(\boldsymbol{r}) &= N^a_{ijk;\ell m}\,\hat{\beta}^a_{ijk;\ell m}(p,\boldsymbol{\xi})\,\mathbf{\Lambda}_a(\boldsymbol{r}),\\ &\quad i=0,1,\ldots,p;\quad j,k,\ell,m=1,2,\ldots,p+1,\\ &\quad i+j+k=p+2;\quad \ell+m=p+2\end{aligned} \tag{4.79}$$

$$\begin{aligned}\mathbf{\Lambda}^d_{ijk;\ell m}(\boldsymbol{r}) &= N^d_{ijk;\ell m}\,\hat{\beta}^d_{ijk;\ell m}(p,\boldsymbol{\xi})\,\mathbf{\Lambda}_d(\boldsymbol{r}),\\ &\quad \ell=0,1,\ldots,p;\quad i,j,k,m=1,2,\ldots,p+1,\\ &\quad i+j+k=p+3;\quad \ell+m=p+1\end{aligned} \tag{4.80}$$

where $\mathbf{\Lambda}_a(\boldsymbol{r})$ (with $a=1,2,3$) and $\mathbf{\Lambda}_d(\boldsymbol{r})$ (with $d=4,5$) indicate an un-normalized zeroth-order basis vector of Table 4.5 (central column). In the dummy parent space, the interpolation points of (4.79) and (4.80)

$$\boldsymbol{\xi}^a_{ijk;\ell m} = \left(\frac{i}{p+2},\frac{j}{p+2},\frac{k}{p+2};\frac{\ell}{p+2},\frac{m}{p+2}\right) \tag{4.81}$$

$$\boldsymbol{\xi}^d_{ijk;\ell m} = \left(\frac{i}{p+3},\frac{j}{p+3},\frac{k}{p+3};\frac{\ell}{p+1},\frac{m}{p+1}\right) \tag{4.82}$$

define two different grids. In fact, (4.79) and (4.80) interpolate the points (4.81) and (4.82), respectively. The $\xi_a=0$ and $\xi_d=0$ faces are interpolated only by (4.79) and (4.80), respectively. All the other interpolation points of these functions are interior to the element, that is, these functions do not interpolate the vertices, the edges, and the remaining (four) faces of the prism. The dummy expressions of the normalization constants appearing in (4.79) and (4.80) are

$$\begin{aligned}N^a_{ijk;\ell m} &= \left.\frac{p+2}{p+2-i}\,\frac{\mathcal{J}}{h_a}\right|_{\boldsymbol{\xi}^a_{ijk;\ell m}}\\ N^d_{ijk;\ell m} &= \left.\frac{p+1}{p+1-\ell}\,\frac{\mathcal{J}}{h_d}\right|_{\boldsymbol{\xi}^d_{ijk;\ell m}}\end{aligned} \tag{4.83}$$

In terms of the true indices, the bases (4.79) and (4.80) in the parent space $(\xi_1, \xi_2, \xi_3; \xi_4, \xi_5)$ read as reported in Table 4.21. The interpolation points are arranged similar to those of scalar Lagrange bases of the same order on a prism, except that the pattern contracts away from the four faces where normal components of the zeroth-order bases must vanish. Recall that no vertex or edge points of the prism element are interpolated and that only a single basis function interpolates a component normal to a given face. At each interior point, three bases interpolate vector components in the plane of the triangle cross section, and two bases interpolate the orthogonal component. One basis should be discarded in each case to guarantee independence. The dependency relations for interior nodes are given below.

Normalization Constants

Table 4.21 reports the expressions of the normalization constants (4.83) and of the interpolation points in the parent space $(\xi_1, \xi_2, \xi_3; \xi_4, \xi_5)$, in terms of the true indices. The normalization constants are chosen to ensure that the component of $\mathbf{\Lambda}^a_{ijk;\ell m}(\boldsymbol{r})$ along $\hat{\boldsymbol{h}}_a$, and $\mathbf{\Lambda}^d_{ijk;\ell m}(\boldsymbol{r})$ along $\hat{\boldsymbol{h}}_d$, at the given interpolation point is unity.

Dependency Relations at Interior Nodes

As discussed following (4.83), only three of the five pth-order bases that are non-vanishing at an interior interpolation point are independent. The dependencies arise from linear combinations of the bases that contain the following identities as factors:

$$\xi_1 \mathbf{\Lambda}_1(\boldsymbol{r}) + \xi_2 \mathbf{\Lambda}_2(\boldsymbol{r}) + \xi_3 \mathbf{\Lambda}_3(\boldsymbol{r}) = \mathbf{0}$$
$$\xi_4 \mathbf{\Lambda}_4(\boldsymbol{r}) + \xi_5 \mathbf{\Lambda}_5(\boldsymbol{r}) = \mathbf{0}$$

Indeed, at interior nodes, the previous identities immediately yield

$$\frac{i\,\mathbf{\Lambda}^1_{ijk;\ell m}(\boldsymbol{r})}{N^1_{ijk;\ell m}} + \frac{j\,\mathbf{\Lambda}^2_{ijk;\ell m}(\boldsymbol{r})}{N^2_{ijk;\ell m}} + \frac{k\,\mathbf{\Lambda}^3_{ijk;\ell m}(\boldsymbol{r})}{N^3_{ijk;\ell m}} = \mathbf{0} \quad \text{for } i, j, k \neq 0$$

$$\frac{\ell\,\mathbf{\Lambda}^4_{ijk;\ell m}(\boldsymbol{r})}{N^4_{ijk;\ell m}} + \frac{m\,\mathbf{\Lambda}^5_{ijk;\ell m}(\boldsymbol{r})}{N^5_{ijk;\ell m}} = \mathbf{0} \quad \text{for } \ell, m \neq 0$$

Number of DoFs

The number of DoFs for divergence-conforming bases of order p on a prism element may be determined as follows (refer to Figure 4.7):

- 1 component $\times$ $\frac{(p+1)(p+2)}{2}$ DoFs $\times$ 2 faces $+$ 1 component $\times$ $(p+1)^2$ DoFs $\times$ 3 faces $=$ $(p+1)(4p+5)$ face DoFs,
- 2 components $\times$ $\frac{p(p+1)^2}{2}$ $+$ 1 component $\times$ $\frac{p(p+1)(p+2)}{2}$ interior DoFs $=$ $\frac{p(p+1)(3p+4)}{2}$ prism-interior DoFs,

for a total of $\frac{(p+1)(3p^2+12p+10)}{2}$ DoFs.

4.8 Synoptic Tables

Table 4.1 Coordinate, edge vector, and gradient vector relations for the canonical elements.

Element type	Independent coordinates	Coordinate dependency relations	Edge vectors in terms of unitary base vectors, $\boldsymbol{\ell}^i = \frac{\partial \mathbf{r}}{\partial \xi_i}$, ξ_i independent	Gradient vectors $\nabla \xi_i = -\frac{\hat{\mathbf{h}}_i}{h_i}$ in terms of unitary base vectors
Triangle	ξ_1, ξ_2	$\xi_1 + \xi_2 + \xi_3 = 1$	$\boldsymbol{\ell}_1 = -\boldsymbol{\ell}^2$, $\boldsymbol{\ell}_2 = \boldsymbol{\ell}^1$, $\boldsymbol{\ell}_3 = \boldsymbol{\ell}^2 - \boldsymbol{\ell}^1$	$\nabla\xi_1 = -\frac{\hat{\mathbf{n}} \times \boldsymbol{\ell}^2}{\mathcal{J}}$, $\nabla\xi_2 = \frac{\hat{\mathbf{n}} \times \boldsymbol{\ell}^1}{\mathcal{J}}$, $\nabla\xi_3 = -\nabla\xi_1 - \nabla\xi_2$
Quadrilateral	ξ_1, ξ_2	$\xi_1 + \xi_3 = 1$, $\xi_2 + \xi_4 = 1$	$\boldsymbol{\ell}_3 = -\boldsymbol{\ell}_1 = \boldsymbol{\ell}^2$, $\boldsymbol{\ell}_2 = -\boldsymbol{\ell}_4 = \boldsymbol{\ell}^1$	$\nabla\xi_1 = -\frac{\hat{\mathbf{n}} \times \boldsymbol{\ell}^2}{\mathcal{J}}$, $\nabla\xi_2 = \frac{\hat{\mathbf{n}} \times \boldsymbol{\ell}^1}{\mathcal{J}}$, $\nabla\xi_3 = -\nabla\xi_1$, $\nabla\xi_4 = -\nabla\xi_2$
Tetrahedron	ξ_1, ξ_2, ξ_3	$\xi_1 + \xi_2 + \xi_3 + \xi_4 = 1$	$\boldsymbol{\ell}_{12} = \boldsymbol{\ell}^3$, $\boldsymbol{\ell}_{13} = -\boldsymbol{\ell}^2$, $\boldsymbol{\ell}_{14} = \boldsymbol{\ell}^2 - \boldsymbol{\ell}^3$, $\boldsymbol{\ell}_{23} = \boldsymbol{\ell}^1$, $\boldsymbol{\ell}_{24} = \boldsymbol{\ell}^3 - \boldsymbol{\ell}^1$, $\boldsymbol{\ell}_{34} = \boldsymbol{\ell}^1 - \boldsymbol{\ell}^2$	$\nabla\xi_1 = \frac{\boldsymbol{\ell}^2 \times \boldsymbol{\ell}^3}{\mathcal{J}}$, $\nabla\xi_2 = \frac{\boldsymbol{\ell}^3 \times \boldsymbol{\ell}^1}{\mathcal{J}}$, $\nabla\xi_3 = \frac{\boldsymbol{\ell}^1 \times \boldsymbol{\ell}^2}{\mathcal{J}}$, $\nabla\xi_4 = -\nabla\xi_1 - \nabla\xi_2 - \nabla\xi_3$
Brick	ξ_1, ξ_2, ξ_3	$\xi_1 + \xi_4 = 1$, $\xi_2 + \xi_5 = 1$, $\xi_3 + \xi_6 = 1$	$\boldsymbol{\ell}_{12} = \boldsymbol{\ell}_{24} = \boldsymbol{\ell}_{45} = -\boldsymbol{\ell}_{15} = \boldsymbol{\ell}^3$, $\boldsymbol{\ell}_{16} = -\boldsymbol{\ell}_{13} = -\boldsymbol{\ell}_{34} = -\boldsymbol{\ell}_{46} = \boldsymbol{\ell}^2$, $\boldsymbol{\ell}_{23} = \boldsymbol{\ell}_{35} = \boldsymbol{\ell}_{56} = -\boldsymbol{\ell}_{26} = \boldsymbol{\ell}^1$	$\nabla\xi_1 = -\nabla\xi_4 = \frac{\boldsymbol{\ell}^2 \times \boldsymbol{\ell}^3}{\mathcal{J}}$, $\nabla\xi_2 = -\nabla\xi_5 = \frac{\boldsymbol{\ell}^3 \times \boldsymbol{\ell}^1}{\mathcal{J}}$, $\nabla\xi_3 = -\nabla\xi_6 = \frac{\boldsymbol{\ell}^1 \times \boldsymbol{\ell}^2}{\mathcal{J}}$
Prism	ξ_1, ξ_2, ξ_4	$\xi_1 + \xi_2 + \xi_3 = 1$, $\xi_4 + \xi_5 = 1$	$\boldsymbol{\ell}_{12} = -\boldsymbol{\ell}_{13} = \boldsymbol{\ell}_{23} = \boldsymbol{\ell}^4$, $-\boldsymbol{\ell}_{14} = \boldsymbol{\ell}_{15} = \boldsymbol{\ell}^2$, $\boldsymbol{\ell}_{24} = -\boldsymbol{\ell}_{25} = \boldsymbol{\ell}^1$, $\boldsymbol{\ell}_{34} = -\boldsymbol{\ell}_{35} = \boldsymbol{\ell}^2 - \boldsymbol{\ell}^1$	$\nabla\xi_1 = \frac{\boldsymbol{\ell}^2 \times \boldsymbol{\ell}^4}{\mathcal{J}}$, $\nabla\xi_2 = \frac{\boldsymbol{\ell}^4 \times \boldsymbol{\ell}^1}{\mathcal{J}}$, $\nabla\xi_3 = -\nabla\xi_1 - \nabla\xi_2$, $\nabla\xi_4 = \frac{\boldsymbol{\ell}^1 \times \boldsymbol{\ell}^2}{\mathcal{J}}$, $\nabla\xi_5 = -\nabla\xi_4$

In the table, the subscript i denoting the parent coordinate ξ_i is the "local" number introduced to distinguish the various coordinates of each parent element; similarly, for each element, the edge and gradient vectors are "locally" numbered.

Table 4.2 Zeroth-order curl-conforming Nédélec bases for 2D cells (un-normalized form).

	Triangle	Quadrilateral
Zeroth-order bases	$\mathbf{\Omega}_1(\boldsymbol{r}) = \xi_2\nabla\xi_3 - \xi_3\nabla\xi_2$ $\mathbf{\Omega}_2(\boldsymbol{r}) = \xi_3\nabla\xi_1 - \xi_1\nabla\xi_3$ $\mathbf{\Omega}_3(\boldsymbol{r}) = \xi_1\nabla\xi_2 - \xi_2\nabla\xi_1$	$\mathbf{\Omega}_1(\boldsymbol{r}) = \xi_3\nabla\xi_4$ $\mathbf{\Omega}_2(\boldsymbol{r}) = \xi_4\nabla\xi_1$ $\mathbf{\Omega}_3(\boldsymbol{r}) = \xi_1\nabla\xi_2$ $\mathbf{\Omega}_4(\boldsymbol{r}) = \xi_2\nabla\xi_3$
Completeness of zeroth-order bases	$\mathbf{\Omega}_2(\boldsymbol{r}) - \mathbf{\Omega}_3(\boldsymbol{r}) = \nabla\xi_1$ $\mathbf{\Omega}_3(\boldsymbol{r}) - \mathbf{\Omega}_1(\boldsymbol{r}) = \nabla\xi_2$ $\mathbf{\Omega}_1(\boldsymbol{r}) - \mathbf{\Omega}_2(\boldsymbol{r}) = \nabla\xi_3$	$\mathbf{\Omega}_2(\boldsymbol{r}) - \mathbf{\Omega}_4(\boldsymbol{r}) = \nabla\xi_1 = -\nabla\xi_3$ $\mathbf{\Omega}_3(\boldsymbol{r}) - \mathbf{\Omega}_1(\boldsymbol{r}) = \nabla\xi_2 = -\nabla\xi_4$
Curl completeness of zeroth-order bases	$\nabla \times \mathbf{\Omega}_i(\boldsymbol{r}) = \frac{2}{\mathcal{J}}\hat{\boldsymbol{n}}, \quad i = 1, 2, 3$	$\nabla \times \mathbf{\Omega}_i(\boldsymbol{r}) = \frac{\hat{\boldsymbol{n}}}{\mathcal{J}}, \quad i = 1, 2, 3, 4$

The expressions of the gradient vectors $\nabla\xi_i$ and of the edge vectors $\boldsymbol{\ell}_i$ in terms of the three unitary base vectors are reported in Table 4.1; $\hat{\boldsymbol{n}}$ is the element unit normal. The first row reports the (un-normalized) zeroth-order vector basis functions; their normalized form is obtained by setting $p = 0$ in the bases of Tables 4.9 and 4.10. On rectilinear elements, these bases have CT/LN components along the element edges. While the basis function $\mathbf{\Omega}_i(\boldsymbol{r})$ interpolates a tangential-vector component on the ith edge of the cell, its tangential component vanishes along the remaining edges, with

$$\mathbf{\Omega}_i(\boldsymbol{r}) \cdot \boldsymbol{\ell}_i|_{\xi_i=0} = 1, \qquad \mathbf{\Omega}_i(\boldsymbol{r}) \cdot \boldsymbol{\ell}_k|_{\xi_k=0} = 0 \quad \text{for } k \neq i.$$

Although the zeroth-order bases contain linear terms, they are not complete to first order. Completeness to zeroth order follows from the first two linear combinations reported in the second row of the table yielding two independent vectors, which are constant in case of rectilinear cells. The relationships reported in the third row of the table readily prove the completeness to zeroth order in the curl with respect to $1/\mathcal{J}$ as a weighting factor.

Table 4.3 Zeroth-order divergence-conforming Nédélec bases for 2D cells (un-normalized form).

	Triangle	Quadrilateral
Zeroth-order bases	$\boldsymbol{\Lambda}_1(\boldsymbol{r}) = \frac{1}{\mathcal{J}}(\xi_2\boldsymbol{\ell}_3 - \xi_3\boldsymbol{\ell}_2)$ $\boldsymbol{\Lambda}_2(\boldsymbol{r}) = \frac{1}{\mathcal{J}}(\xi_3\boldsymbol{\ell}_1 - \xi_1\boldsymbol{\ell}_3)$ $\boldsymbol{\Lambda}_3(\boldsymbol{r}) = \frac{1}{\mathcal{J}}(\xi_1\boldsymbol{\ell}_2 - \xi_2\boldsymbol{\ell}_1)$	$\boldsymbol{\Lambda}_1(\boldsymbol{r}) = \frac{\xi_3\boldsymbol{\ell}_4}{\mathcal{J}},\ \boldsymbol{\Lambda}_2(\boldsymbol{r}) = \frac{\xi_4\boldsymbol{\ell}_1}{\mathcal{J}},$ $\boldsymbol{\Lambda}_3(\boldsymbol{r}) = \frac{\xi_1\boldsymbol{\ell}_2}{\mathcal{J}},\ \boldsymbol{\Lambda}_4(\boldsymbol{r}) = \frac{\xi_2\boldsymbol{\ell}_3}{\mathcal{J}}$
Completeness of zeroth-order bases	$\boldsymbol{\Lambda}_2(\boldsymbol{r}) - \boldsymbol{\Lambda}_3(\boldsymbol{r}) = \frac{\boldsymbol{\ell}_1}{\mathcal{J}}$ $\boldsymbol{\Lambda}_3(\boldsymbol{r}) - \boldsymbol{\Lambda}_1(\boldsymbol{r}) = \frac{\boldsymbol{\ell}_2}{\mathcal{J}}$ $\boldsymbol{\Lambda}_1(\boldsymbol{r}) - \boldsymbol{\Lambda}_2(\boldsymbol{r}) = \frac{\boldsymbol{\ell}_3}{\mathcal{J}}$	$\boldsymbol{\Lambda}_2(\boldsymbol{r}) - \boldsymbol{\Lambda}_4(\boldsymbol{r}) = \frac{\boldsymbol{\ell}_1}{\mathcal{J}} = -\frac{\boldsymbol{\ell}_3}{\mathcal{J}}$ $\boldsymbol{\Lambda}_3(\boldsymbol{r}) - \boldsymbol{\Lambda}_1(\boldsymbol{r}) = \frac{\boldsymbol{\ell}_2}{\mathcal{J}} = -\frac{\boldsymbol{\ell}_4}{\mathcal{J}}$
Divergence completeness of zeroth-order bases	$\nabla \cdot \boldsymbol{\Lambda}_i(\boldsymbol{r}) = \frac{2}{\mathcal{J}},\quad i = 1, 2, 3$	$\nabla \cdot \boldsymbol{\Lambda}_i(\boldsymbol{r}) = \frac{1}{\mathcal{J}},\quad i = 1, 2, 3, 4$

The expressions of the edge vectors $\boldsymbol{\ell}_i$ in terms of the three unitary base vectors are reported in Table 4.1; $\hat{\boldsymbol{n}}$ is the element unit normal. The first row reports the (un-normalized) zeroth-order vector basis functions; their normalized form is obtained by setting $p=0$ in the bases of Tables 4.9 and 4.10. On rectilinear elements, these bases have CN/LT components along the element edges. While the basis function $\boldsymbol{\Lambda}_i(\boldsymbol{r}) = \boldsymbol{\Omega}_i(\boldsymbol{r}) \times \hat{\boldsymbol{n}}$ interpolates a normal-vector component on the ith edge of the cell, its normal component vanishes along the remaining edges. Although the zeroth-order bases contain linear terms, they are not complete to first order. Completeness to zeroth order follows from the first two linear combinations reported in the second row of the table yielding two independent vectors, which are constant in case of rectilinear cells. The relationships reported in the third row of the table readily prove the completeness to zeroth order in the divergence with respect to $1/\mathcal{J}$ as a weighting factor.

Table 4.4 Curl-conforming bases for volumetric cells (un-normalized form, from References 1 and 2).

	Tetrahedron	Prism	Brick
Zeroth-order bases	$\boldsymbol{\Omega}_{12}(\boldsymbol{r})=\xi_4\nabla\xi_3-\xi_3\nabla\xi_4,$ $\boldsymbol{\Omega}_{13}(\boldsymbol{r})=\xi_2\nabla\xi_4-\xi_4\nabla\xi_2,$ $\boldsymbol{\Omega}_{14}(\boldsymbol{r})=\xi_3\nabla\xi_2-\xi_2\nabla\xi_3,$ $\boldsymbol{\Omega}_{23}(\boldsymbol{r})=\xi_4\nabla\xi_1-\xi_1\nabla\xi_4,$ $\boldsymbol{\Omega}_{24}(\boldsymbol{r})=\xi_1\nabla\xi_3-\xi_3\nabla\xi_1,$ $\boldsymbol{\Omega}_{34}(\boldsymbol{r})=\xi_2\nabla\xi_1-\xi_1\nabla\xi_2.$	$\boldsymbol{\Omega}_{14}(\boldsymbol{r})=\xi_5(\xi_2\nabla\xi_3-\xi_3\nabla\xi_2),$ $\boldsymbol{\Omega}_{15}(\boldsymbol{r})=\xi_4(\xi_3\nabla\xi_2-\xi_2\nabla\xi_3),$ $\boldsymbol{\Omega}_{24}(\boldsymbol{r})=\xi_5(\xi_3\nabla\xi_1-\xi_1\nabla\xi_3),$ $\boldsymbol{\Omega}_{25}(\boldsymbol{r})=\xi_4(\xi_1\nabla\xi_3-\xi_3\nabla\xi_1),$ $\boldsymbol{\Omega}_{34}(\boldsymbol{r})=\xi_5(\xi_1\nabla\xi_2-\xi_2\nabla\xi_1),$ $\boldsymbol{\Omega}_{35}(\boldsymbol{r})=\xi_4(\xi_2\nabla\xi_1-\xi_1\nabla\xi_2),$ $\boldsymbol{\Omega}_{13}(\boldsymbol{r})=\xi_2\nabla\xi_5,\ \boldsymbol{\Omega}_{12}(\boldsymbol{r})=\xi_3\nabla\xi_4,$ $\boldsymbol{\Omega}_{23}(\boldsymbol{r})=\xi_1\nabla\xi_4.$	$\boldsymbol{\Omega}_{12}(\boldsymbol{r})=\xi_4\xi_5\nabla\xi_3,\ \boldsymbol{\Omega}_{13}(\boldsymbol{r})=\xi_4\xi_6\nabla\xi_5,$ $\boldsymbol{\Omega}_{15}(\boldsymbol{r})=\xi_2\xi_4\nabla\xi_6,\ \boldsymbol{\Omega}_{16}(\boldsymbol{r})=\xi_3\xi_4\nabla\xi_2,$ $\boldsymbol{\Omega}_{23}(\boldsymbol{r})=\xi_5\xi_6\nabla\xi_1,\ \boldsymbol{\Omega}_{24}(\boldsymbol{r})=\xi_1\xi_5\nabla\xi_3,$ $\boldsymbol{\Omega}_{26}(\boldsymbol{r})=\xi_3\xi_5\nabla\xi_4,\ \boldsymbol{\Omega}_{34}(\boldsymbol{r})=\xi_1\xi_6\nabla\xi_5,$ $\boldsymbol{\Omega}_{35}(\boldsymbol{r})=\xi_2\xi_6\nabla\xi_1,\ \boldsymbol{\Omega}_{45}(\boldsymbol{r})=\xi_1\xi_2\nabla\xi_3,$ $\boldsymbol{\Omega}_{46}(\boldsymbol{r})=\xi_1\xi_3\nabla\xi_5,\ \boldsymbol{\Omega}_{56}(\boldsymbol{r})=\xi_2\xi_3\nabla\xi_1.$
Curl of the zeroth-order bases in terms of the edge vectors	$\nabla\times\boldsymbol{\Omega}_{12}(\boldsymbol{r})=-2\boldsymbol{\ell}_{34}/\mathcal{J},$ $\nabla\times\boldsymbol{\Omega}_{13}(\boldsymbol{r})=+2\boldsymbol{\ell}_{24}/\mathcal{J},$ $\nabla\times\boldsymbol{\Omega}_{14}(\boldsymbol{r})=-2\boldsymbol{\ell}_{23}/\mathcal{J},$ $\nabla\times\boldsymbol{\Omega}_{23}(\boldsymbol{r})=-2\boldsymbol{\ell}_{14}/\mathcal{J},$ $\nabla\times\boldsymbol{\Omega}_{24}(\boldsymbol{r})=+2\boldsymbol{\ell}_{13}/\mathcal{J},$ $\nabla\times\boldsymbol{\Omega}_{34}(\boldsymbol{r})=-2\boldsymbol{\ell}_{12}/\mathcal{J}.$	$\nabla\times\boldsymbol{\Omega}_{14}(\boldsymbol{r})=[+2\xi_5\boldsymbol{\ell}_{23}-\xi_2\boldsymbol{\ell}_{35}+\xi_3\boldsymbol{\ell}_{25}]/\mathcal{J},$ $\nabla\times\boldsymbol{\Omega}_{15}(\boldsymbol{r})=[-2\xi_4\boldsymbol{\ell}_{23}+\xi_2\boldsymbol{\ell}_{34}-\xi_3\boldsymbol{\ell}_{24}]/\mathcal{J},$ $\nabla\times\boldsymbol{\Omega}_{24}(\boldsymbol{r})=[-2\xi_5\boldsymbol{\ell}_{13}+\xi_1\boldsymbol{\ell}_{35}-\xi_3\boldsymbol{\ell}_{15}]/\mathcal{J},$ $\nabla\times\boldsymbol{\Omega}_{25}(\boldsymbol{r})=[+2\xi_4\boldsymbol{\ell}_{13}-\xi_1\boldsymbol{\ell}_{34}+\xi_3\boldsymbol{\ell}_{14}]/\mathcal{J},$ $\nabla\times\boldsymbol{\Omega}_{34}(\boldsymbol{r})=[+2\xi_5\boldsymbol{\ell}_{12}-\xi_1\boldsymbol{\ell}_{25}+\xi_2\boldsymbol{\ell}_{15}]/\mathcal{J},$ $\nabla\times\boldsymbol{\Omega}_{35}(\boldsymbol{r})=[-2\xi_4\boldsymbol{\ell}_{12}+\xi_1\boldsymbol{\ell}_{24}-\xi_2\boldsymbol{\ell}_{14}]/\mathcal{J},$ $\nabla\times\boldsymbol{\Omega}_{13}(\boldsymbol{r})=\boldsymbol{\ell}_{25}/\mathcal{J},$ $\nabla\times\boldsymbol{\Omega}_{12}(\boldsymbol{r})=\boldsymbol{\ell}_{34}/\mathcal{J},$ $\nabla\times\boldsymbol{\Omega}_{23}(\boldsymbol{r})=\boldsymbol{\ell}_{14}/\mathcal{J}.$	$\nabla\times\boldsymbol{\Omega}_{12}(\boldsymbol{r})=[+\xi_4\boldsymbol{\ell}_{35}-\xi_5\boldsymbol{\ell}_{34}]/\mathcal{J},$ $\nabla\times\boldsymbol{\Omega}_{13}(\boldsymbol{r})=[-\xi_4\boldsymbol{\ell}_{56}+\xi_6\boldsymbol{\ell}_{45}]/\mathcal{J},$ $\nabla\times\boldsymbol{\Omega}_{15}(\boldsymbol{r})=[+\xi_2\boldsymbol{\ell}_{46}+\xi_4\boldsymbol{\ell}_{26}]/\mathcal{J},$ $\nabla\times\boldsymbol{\Omega}_{16}(\boldsymbol{r})=[-\xi_3\boldsymbol{\ell}_{24}-\xi_4\boldsymbol{\ell}_{23}]/\mathcal{J},$ $\nabla\times\boldsymbol{\Omega}_{23}(\boldsymbol{r})=[-\xi_5\boldsymbol{\ell}_{16}-\xi_6\boldsymbol{\ell}_{15}]/\mathcal{J},$ $\nabla\times\boldsymbol{\Omega}_{24}(\boldsymbol{r})=[-\xi_1\boldsymbol{\ell}_{35}+\xi_5\boldsymbol{\ell}_{13}]/\mathcal{J},$ $\nabla\times\boldsymbol{\Omega}_{26}(\boldsymbol{r})=[-\xi_3\boldsymbol{\ell}_{45}+\xi_5\boldsymbol{\ell}_{34}]/\mathcal{J},$ $\nabla\times\boldsymbol{\Omega}_{34}(\boldsymbol{r})=[-\xi_1\boldsymbol{\ell}_{56}+\xi_6\boldsymbol{\ell}_{15}]/\mathcal{J},$ $\nabla\times\boldsymbol{\Omega}_{35}(\boldsymbol{r})=[-\xi_2\boldsymbol{\ell}_{16}-\xi_6\boldsymbol{\ell}_{12}]/\mathcal{J},$ $\nabla\times\boldsymbol{\Omega}_{45}(\boldsymbol{r})=[+\xi_1\boldsymbol{\ell}_{23}+\xi_2\boldsymbol{\ell}_{13}]/\mathcal{J},$ $\nabla\times\boldsymbol{\Omega}_{46}(\boldsymbol{r})=[+\xi_1\boldsymbol{\ell}_{35}+\xi_3\boldsymbol{\ell}_{15}]/\mathcal{J},$ $\nabla\times\boldsymbol{\Omega}_{56}(\boldsymbol{r})=[-\xi_2\boldsymbol{\ell}_{13}-\xi_3\boldsymbol{\ell}_{12}]/\mathcal{J}.$
Completeness of the zeroth-order bases	$\boldsymbol{\Omega}_{23}(\boldsymbol{r})-\boldsymbol{\Omega}_{24}(\boldsymbol{r})+\boldsymbol{\Omega}_{34}(\boldsymbol{r})=\nabla\xi_1,$ $\boldsymbol{\Omega}_{14}(\boldsymbol{r})-\boldsymbol{\Omega}_{34}(\boldsymbol{r})-\boldsymbol{\Omega}_{13}(\boldsymbol{r})=\nabla\xi_2,$ $\boldsymbol{\Omega}_{12}(\boldsymbol{r})-\boldsymbol{\Omega}_{14}(\boldsymbol{r})+\boldsymbol{\Omega}_{24}(\boldsymbol{r})=\nabla\xi_3.$	$\boldsymbol{\Omega}_{24}(\boldsymbol{r})-\boldsymbol{\Omega}_{34}(\boldsymbol{r})-\boldsymbol{\Omega}_{25}(\boldsymbol{r})+\boldsymbol{\Omega}_{35}(\boldsymbol{r})=\nabla\xi_1,$ $\boldsymbol{\Omega}_{34}(\boldsymbol{r})-\boldsymbol{\Omega}_{14}(\boldsymbol{r})-\boldsymbol{\Omega}_{35}(\boldsymbol{r})+\boldsymbol{\Omega}_{15}(\boldsymbol{r})=\nabla\xi_2,$ $\boldsymbol{\Omega}_{12}(\boldsymbol{r})+\boldsymbol{\Omega}_{23}(\boldsymbol{r})-\boldsymbol{\Omega}_{13}(\boldsymbol{r})=\nabla\xi_4.$	$\boldsymbol{\Omega}_{23}(\boldsymbol{r})-\boldsymbol{\Omega}_{26}(\boldsymbol{r})+\boldsymbol{\Omega}_{35}(\boldsymbol{r})+\boldsymbol{\Omega}_{56}(\boldsymbol{r})=\nabla\xi_1,$ $\boldsymbol{\Omega}_{16}(\boldsymbol{r})-\boldsymbol{\Omega}_{13}(\boldsymbol{r})-\boldsymbol{\Omega}_{34}(\boldsymbol{r})-\boldsymbol{\Omega}_{46}(\boldsymbol{r})=\nabla\xi_2,$ $\boldsymbol{\Omega}_{12}(\boldsymbol{r})-\boldsymbol{\Omega}_{15}(\boldsymbol{r})+\boldsymbol{\Omega}_{24}(\boldsymbol{r})+\boldsymbol{\Omega}_{45}(\boldsymbol{r})=\nabla\xi_3.$
	$\nabla\times\boldsymbol{\Omega}_{14}(\boldsymbol{r})=-2\boldsymbol{\ell}^1/\mathcal{J},$ $\nabla\times\boldsymbol{\Omega}_{24}(\boldsymbol{r})=-2\boldsymbol{\ell}^2/\mathcal{J},$ $\nabla\times\boldsymbol{\Omega}_{34}(\boldsymbol{r})=-2\boldsymbol{\ell}^3/\mathcal{J}.$	$\nabla\times\boldsymbol{\Omega}_{13}(\boldsymbol{r})=-\boldsymbol{\ell}^1/\mathcal{J},$ $\nabla\times\boldsymbol{\Omega}_{23}(\boldsymbol{r})=-\boldsymbol{\ell}^2/\mathcal{J},$ $\nabla\times[\boldsymbol{\Omega}_{34}(\boldsymbol{r})-\boldsymbol{\Omega}_{35}(\boldsymbol{r})]=+2\boldsymbol{\ell}^4/\mathcal{J}.$	$\nabla\times[\boldsymbol{\Omega}_{45}(\boldsymbol{r})-\boldsymbol{\Omega}_{15}(\boldsymbol{r})]=\boldsymbol{\ell}^1/\mathcal{J},$ $\nabla\times[\boldsymbol{\Omega}_{26}(\boldsymbol{r})-\boldsymbol{\Omega}_{56}(\boldsymbol{r})]=-\boldsymbol{\ell}^2/\mathcal{J},$ $\nabla\times[\boldsymbol{\Omega}_{13}(\boldsymbol{r})-\boldsymbol{\Omega}_{16}(\boldsymbol{r})]=\boldsymbol{\ell}^3/\mathcal{J}.$
Completeness to order p in the curl	$\boldsymbol{\Upsilon}_1(\boldsymbol{r})=\xi_1\nabla\xi_3-\xi_3\nabla\xi_1=\boldsymbol{\Omega}_{24}(\boldsymbol{r})$ $\boldsymbol{\Upsilon}_2(\boldsymbol{r})=\xi_2\nabla\xi_1-\xi_1\nabla\xi_2=\boldsymbol{\Omega}_{34}(\boldsymbol{r})$ $\boldsymbol{\Upsilon}_3(\boldsymbol{r})=\xi_3\nabla\xi_2-\xi_2\nabla\xi_3=\boldsymbol{\Omega}_{14}(\boldsymbol{r})$	$\boldsymbol{\Upsilon}_1(\boldsymbol{r})=\xi_1\nabla\xi_4-\xi_4\nabla\xi_1$ $=\boldsymbol{\Omega}_{23}(\boldsymbol{r})+\boldsymbol{\Omega}_{25}(\boldsymbol{r})-\boldsymbol{\Omega}_{35}(\boldsymbol{r})$ $\boldsymbol{\Upsilon}_2(\boldsymbol{r})=\xi_2\nabla\xi_1-\xi_1\nabla\xi_2$ $=\boldsymbol{\Omega}_{35}(\boldsymbol{r})-\boldsymbol{\Omega}_{34}(\boldsymbol{r})$ $\boldsymbol{\Upsilon}_4(\boldsymbol{r})=\xi_4\nabla\xi_2-\xi_2\nabla\xi_4$ $=\boldsymbol{\Omega}_{13}(\boldsymbol{r})+\boldsymbol{\Omega}_{15}(\boldsymbol{r})-\boldsymbol{\Omega}_{35}(\boldsymbol{r})$	$\boldsymbol{\Upsilon}_1(\boldsymbol{r})=\xi_1\nabla\xi_3-\xi_3\nabla\xi_1$ $=\boldsymbol{\Omega}_{24}(\boldsymbol{r})+\boldsymbol{\Omega}_{26}(\boldsymbol{r})+\boldsymbol{\Omega}_{45}(\boldsymbol{r})-\boldsymbol{\Omega}_{56}(\boldsymbol{r})$ $\boldsymbol{\Upsilon}_2(\boldsymbol{r})=\xi_2\nabla\xi_1-\xi_1\nabla\xi_2$ $=\boldsymbol{\Omega}_{34}(\boldsymbol{r})+\boldsymbol{\Omega}_{35}(\boldsymbol{r})+\boldsymbol{\Omega}_{46}(\boldsymbol{r})+\boldsymbol{\Omega}_{56}(\boldsymbol{r})$ $\boldsymbol{\Upsilon}_3(\boldsymbol{r})=\xi_3\nabla\xi_2-\xi_2\nabla\xi_3$ $=\boldsymbol{\Omega}_{15}(\boldsymbol{r})+\boldsymbol{\Omega}_{16}(\boldsymbol{r})-\boldsymbol{\Omega}_{45}(\boldsymbol{r})-\boldsymbol{\Omega}_{46}(\boldsymbol{r})$

In the table, the subscript i denoting the parent coordinate ξ_i and gradient vector $\nabla\xi_i$ is the "local" number introduced to distinguish the various coordinates of each parent element. The expressions of the gradient vectors $\nabla\xi_i$ and of the edge vectors $\boldsymbol{\ell}_{ij}$ in terms of the three unitary base vectors are reported in Table 4.1.

The first row reports the (un-normalized) zeroth-order curl-conforming vector basis functions; their normalized form is obtained by setting $p=0$ in the bases of Tables 4.12, 4.15, and 4.19. The basis function $\boldsymbol{\Omega}_{ij}(\boldsymbol{r})$ interpolates the vector component tangent to the midpoint of the edge formed by the intersection of faces i and j. The bases for the tetrahedral cell contain linear terms and are of CT/LN form. The bases for the prism and the brick also contain quadratic terms and are of CT form. In spite of the presence of first- and second-order terms, the bases are not complete to first order since they cannot represent, for example, vectors of the form $\xi_i\nabla\xi_i$ (with $i=1,2,3$ for tetrahedrons and bricks, and $i=1,2,4$ for prisms). The zeroth-order completeness of the bases and of their curl follows from the linear combinations reported in the third row of the table yielding three independent vectors. The high-order completeness of the bases is readily proved by multiplying the first three linear combinations reported in the third row of the table with *homogeneous* polynomials. Conversely, to prove the completeness to order p in the curl for all the higher order polynomial vector sets obtained by multiplying the zeroth-order vector bases with a pth-order complete polynomial set one has to use the relationships reported in the fourth row of the table. Completeness in the curl is proved as a consequence of the fact that the zeroth-order curl-conforming functions contain linear terms able to model the linear vectors $\boldsymbol{\Upsilon}_1(\boldsymbol{r})$, $\boldsymbol{\Upsilon}_2(\boldsymbol{r})$, and $\boldsymbol{\Upsilon}_3(\boldsymbol{r})$ (or $\boldsymbol{\Upsilon}_4(\boldsymbol{r})$ for the prism) reported in the fourth row of the table.

Table 4.5 Divergence-conforming bases for volumetric cells (un-normalized form, from References 1 and 2).

	Tetrahedron	Prism	Brick
Zeroth-order bases	$\boldsymbol{\Lambda}_1(\boldsymbol{r}) = -(\xi_2\boldsymbol{\ell}_{34} - \xi_3\boldsymbol{\ell}_{24} + \xi_4\boldsymbol{\ell}_{23})/\mathcal{J}$ $\boldsymbol{\Lambda}_2(\boldsymbol{r}) = (\xi_3\boldsymbol{\ell}_{41} - \xi_4\boldsymbol{\ell}_{31} + \xi_1\boldsymbol{\ell}_{34})/\mathcal{J}$ $\boldsymbol{\Lambda}_3(\boldsymbol{r}) = -(\xi_4\boldsymbol{\ell}_{12} - \xi_1\boldsymbol{\ell}_{42} + \xi_2\boldsymbol{\ell}_{41})/\mathcal{J}$ $\boldsymbol{\Lambda}_4(\boldsymbol{r}) = (\xi_1\boldsymbol{\ell}_{23} - \xi_2\boldsymbol{\ell}_{13} + \xi_3\boldsymbol{\ell}_{12})/\mathcal{J}$	$\boldsymbol{\Lambda}_1(\boldsymbol{r}) = (\xi_2\boldsymbol{\ell}_{34} - \xi_3\boldsymbol{\ell}_{24})/\mathcal{J}$ $\boldsymbol{\Lambda}_2(\boldsymbol{r}) = (\xi_3\boldsymbol{\ell}_{14} - \xi_1\boldsymbol{\ell}_{34})/\mathcal{J}$ $\boldsymbol{\Lambda}_3(\boldsymbol{r}) = (\xi_1\boldsymbol{\ell}_{24} - \xi_2\boldsymbol{\ell}_{14})/\mathcal{J}$ $\boldsymbol{\Lambda}_4(\boldsymbol{r}) = \xi_5\boldsymbol{\ell}_{13}/\mathcal{J}$ $\boldsymbol{\Lambda}_5(\boldsymbol{r}) = \xi_4\boldsymbol{\ell}_{12}/\mathcal{J}$	$\boldsymbol{\Lambda}_1(\boldsymbol{r}) = \xi_4\boldsymbol{\ell}_{26}/\mathcal{J}$ $\boldsymbol{\Lambda}_2(\boldsymbol{r}) = \xi_5\boldsymbol{\ell}_{13}/\mathcal{J}$ $\boldsymbol{\Lambda}_3(\boldsymbol{r}) = \xi_6\boldsymbol{\ell}_{15}/\mathcal{J}$ $\boldsymbol{\Lambda}_4(\boldsymbol{r}) = \xi_1\boldsymbol{\ell}_{23}/\mathcal{J}$ $\boldsymbol{\Lambda}_5(\boldsymbol{r}) = \xi_2\boldsymbol{\ell}_{16}/\mathcal{J}$ $\boldsymbol{\Lambda}_6(\boldsymbol{r}) = \xi_3\boldsymbol{\ell}_{12}/\mathcal{J}$
Completeness of zeroth-order bases	$\boldsymbol{\Lambda}_4(\boldsymbol{r}) - \boldsymbol{\Lambda}_1(\boldsymbol{r}) = \boldsymbol{\ell}_{23}/\mathcal{J} = \boldsymbol{\ell}^1/\mathcal{J}$ $\boldsymbol{\Lambda}_4(\boldsymbol{r}) - \boldsymbol{\Lambda}_2(\boldsymbol{r}) = \boldsymbol{\ell}_{31}/\mathcal{J} = \boldsymbol{\ell}^2/\mathcal{J}$ $\boldsymbol{\Lambda}_4(\boldsymbol{r}) - \boldsymbol{\Lambda}_3(\boldsymbol{r}) = \boldsymbol{\ell}_{12}/\mathcal{J} = \boldsymbol{\ell}^3/\mathcal{J}$	$\boldsymbol{\Lambda}_3(\boldsymbol{r}) - \boldsymbol{\Lambda}_1(\boldsymbol{r}) = \boldsymbol{\ell}_{24}/\mathcal{J} = \boldsymbol{\ell}^1/\mathcal{J}$ $\boldsymbol{\Lambda}_3(\boldsymbol{r}) - \boldsymbol{\Lambda}_2(\boldsymbol{r}) = \boldsymbol{\ell}_{41}/\mathcal{J} = \boldsymbol{\ell}^2/\mathcal{J}$ $\boldsymbol{\Lambda}_5(\boldsymbol{r}) - \boldsymbol{\Lambda}_4(\boldsymbol{r}) = \boldsymbol{\ell}_{12}/\mathcal{J} = \boldsymbol{\ell}^4/\mathcal{J}$	$\boldsymbol{\Lambda}_4(\boldsymbol{r}) - \boldsymbol{\Lambda}_1(\boldsymbol{r}) = \boldsymbol{\ell}_{23}/\mathcal{J} = \boldsymbol{\ell}^1/\mathcal{J}$ $\boldsymbol{\Lambda}_5(\boldsymbol{r}) - \boldsymbol{\Lambda}_2(\boldsymbol{r}) = \boldsymbol{\ell}_{16}/\mathcal{J} = \boldsymbol{\ell}^2/\mathcal{J}$ $\boldsymbol{\Lambda}_6(\boldsymbol{r}) - \boldsymbol{\Lambda}_3(\boldsymbol{r}) = \boldsymbol{\ell}_{12}/\mathcal{J} = \boldsymbol{\ell}^3/\mathcal{J}$
Completeness to order p in the divergence (α, β, γ, and $\delta \geq 0$)	$\nabla \cdot \left[\xi_i^\alpha \xi_j^\beta \xi_k^\gamma \boldsymbol{\Lambda}_\ell(\boldsymbol{r})\right] = \left(\xi_i^\alpha \xi_j^\beta \xi_k^\gamma\right) \dfrac{(3+\alpha+\beta+\gamma)}{\mathcal{J}}$ for i, j, k and $\ell = 1, 2, 3,$ or 4 and $0 \leq \alpha + \beta + \gamma \leq p$	$\nabla \cdot \left[\xi_1^\alpha \xi_2^\beta \xi_3^\gamma \xi_\ell^\delta \boldsymbol{\Lambda}_m(\boldsymbol{r})\right] = \left(\xi_1^\alpha \xi_2^\beta \xi_3^\gamma \xi_\ell^\delta\right)(1+\delta)/\mathcal{J}$ $\nabla \cdot \left[\xi_i^\alpha \xi_j^\beta \xi_\ell^\delta \boldsymbol{\Lambda}_k(\boldsymbol{r})\right] = \left(\xi_i^\alpha \xi_j^\beta \xi_\ell^\delta\right)(2+\alpha+\beta)/\mathcal{J}$ for $i, j, k = 1, 2,$ or 3; $\ell, m = 4$ or 5 and $0 \leq \alpha + \beta + \gamma + \delta \leq p$	$\nabla \cdot \left[\xi_{i+3}^\alpha \xi_{i+2}^\beta \xi_{i+1}^\gamma \boldsymbol{\Lambda}_i(\boldsymbol{r})\right] = \left(\xi_{i+3}^\alpha \xi_{i+2}^\beta \xi_{i+1}^\gamma\right) \dfrac{(1+\alpha)}{\mathcal{J}}$ for $i = 1, 2, \ldots, 6$ (with index arithmetic computed modulo 6) and $0 \leq \alpha + \beta + \gamma \leq p$

The expressions of the edge-vectors $\boldsymbol{\ell}_{ij}$ in terms of the three unitary base vectors are reported in Table 4.1.

The first row reports the (un-normalized) zeroth order vector basis functions; their normalized form is obtained by setting $p = 0$ in the bases of Tables 4.13, 4.17, and 4.21. The basis function $\boldsymbol{\Lambda}_i(\boldsymbol{r})$ interpolates the vector component normal to the centroid of face i, and each basis has CN (constant normal) variation on a face. Although the zeroth order bases contain linear terms, they are not complete to first order. Completeness to zeroth order with respect to $1/\mathcal{J}$ as a weighting factor follows from the three linear combinations reported in the second row of the table yielding three independent vectors.

The relationships reported in the third row of the table readily prove the completeness to order p in the divergence (with respect to $1/\mathcal{J}$ as a weighting factor) for all the higher-order polynomial vector sets obtained by multiplying the zeroth-order vector bases with a p-th order complete polynomial set. The set may take one of several different forms (homogeneous, inhomogeneous, interpolatory or hierarchical) chosen for convenience; since all are complete, they span the same space and are linear combinations of one another. As a matter of fact, the completeness in the divergence is most easily shown by using an inhomogeneous polynomial of order p (such as those used in the third row of the table). Since the right-hand side of each equation in the third row of the table is a complete polynomial of order p, the divergence of the higher-order vector set is complete to the same order. (It is understood that the values of the subscripts $(i, j, k, \ell,$ etc.) appearing in the expressions of the third row are all different.) The completeness of the higher-order polynomial vector sets stems from the completeness of the zeroth-order bases shown in the second row of the Table.

Notice that the divergence of the zeroth-order bases is obtained from the expressions reported in the third row of the table simply by setting α, β, γ, and $\delta = 0$.

Table 4.6 Formulas to prove the completeness of high-order curl-conforming bases for 2D cells.

	Triangle	Quadrilateral
Completeness of the bases	$\xi_1^\alpha\xi_2^\beta\xi_3^\gamma[\boldsymbol{\Omega}_2(\boldsymbol{r})-\boldsymbol{\Omega}_3(\boldsymbol{r})]=\xi_1^\alpha\xi_2^\beta\xi_3^\gamma\nabla\xi_1$ $\xi_1^\alpha\xi_2^\beta\xi_3^\gamma[\boldsymbol{\Omega}_3(\boldsymbol{r})-\boldsymbol{\Omega}_1(\boldsymbol{r})]=\xi_1^\alpha\xi_2^\beta\xi_3^\gamma\nabla\xi_2$ for α, β, and $\gamma\geq 0$ with $\alpha+\beta+\gamma=p$	$\xi_1^\alpha\,\xi_2^\beta\,\xi_3^\gamma\,\xi_4^\delta\,[\boldsymbol{\Omega}_2(\boldsymbol{r})-\boldsymbol{\Omega}_4(\boldsymbol{r})]=\xi_1^\alpha\,\xi_2^\beta\,\xi_3^\gamma\,\xi_4^\delta\nabla\xi_1$ $\xi_1^\alpha\,\xi_2^\beta\,\xi_3^\gamma\,\xi_4^\delta\,[\boldsymbol{\Omega}_3(\boldsymbol{r})-\boldsymbol{\Omega}_1(\boldsymbol{r})]=\xi_1^\alpha\,\xi_2^\beta\,\xi_3^\gamma\,\xi_4^\delta\nabla\xi_2$ for α, β, γ, and $\delta\geq 0$ with $\alpha+\beta+\gamma+\delta=p$
Completeness in the curl	$\nabla\times[\,\xi_i^\alpha\,\xi_j^\beta\,\boldsymbol{\Omega}_k(\boldsymbol{r})\,]=\frac{\alpha+\beta+2}{\mathcal{J}}\,\xi_i^\alpha\,\xi_j^\beta\,\hat{\boldsymbol{n}}$ for $i\neq j\neq k$ and $0\leq\alpha,\beta\leq p$	$\nabla\times[\,\xi_3^\alpha\xi_4^\beta\boldsymbol{\Omega}_1(\boldsymbol{r})\,]=\frac{(\alpha+1)\xi_3^\alpha\xi_4^\beta}{\mathcal{J}}\hat{\boldsymbol{n}}$ $\nabla\times[\,\xi_4^\alpha\xi_1^\beta\boldsymbol{\Omega}_2(\boldsymbol{r})\,]=\frac{(\alpha+1)\xi_4^\alpha\xi_1^\beta}{\mathcal{J}}\hat{\boldsymbol{n}}$ $\nabla\times[\,\xi_1^\alpha\xi_2^\beta\boldsymbol{\Omega}_3(\boldsymbol{r})\,]=\frac{(\alpha+1)\xi_1^\alpha\xi_2^\beta}{\mathcal{J}}\hat{\boldsymbol{n}}$ $\nabla\times[\,\xi_2^\alpha\xi_3^\beta\boldsymbol{\Omega}_4(\boldsymbol{r})\,]=\frac{(\alpha+1)\xi_2^\alpha\xi_3^\beta}{\mathcal{J}}\hat{\boldsymbol{n}}$ for $0\leq\alpha,\beta\leq p$
Dependency relations for higher orders	$\xi_1\,\boldsymbol{\Omega}_1(\boldsymbol{r})+\xi_2\,\boldsymbol{\Omega}_2(\boldsymbol{r})+\xi_3\,\boldsymbol{\Omega}_3(\boldsymbol{r})=\boldsymbol{0}$	$\xi_1\,\boldsymbol{\Omega}_1(\boldsymbol{r})+\xi_3\,\boldsymbol{\Omega}_3(\boldsymbol{r})=\boldsymbol{0}$ $\xi_2\,\boldsymbol{\Omega}_2(\boldsymbol{r})+\xi_4\,\boldsymbol{\Omega}_4(\boldsymbol{r})=\boldsymbol{0}$

1. The completeness of high-order bases is readily proved by using multiplicative polynomials of *homogeneous* forms. The first row of the table reports two linear combinations of the bases that yield complete vector polynomials in two independent directions.
2. The curl of the resulting high-order bases are polynomials with only a component normal to the element. The completeness in the curl can be shown most easily by choosing multiplicative polynomials of *inhomogeneous* forms. Completeness of the curl then follows from the result that the curl of the product generates terms of like order, as shown in the second row of the table.
3. The dependency relations reported in the third row of the table follow trivially from the bases expressions given in the first row of Table 4.2. These expressions are useful to eliminate dependencies in the higher order basis sets, since there always exist a linear combination of the high-order bases interpolating an interior point that contains the factor(s) reported in the third row of this table. (The dependency relations useful to form higher order divergence-conforming bases are obtained by substituting $\boldsymbol{\Lambda}_i$ for $\boldsymbol{\Omega}_i$ in the third row expressions.)

Table 4.7 Interpolatory polynomials for the triangular cell (dummy expressions).

Polynomial	**Index rule**	**Interpolation point** $\boldsymbol{\xi}_{ijk} = (\xi_a, \xi_b, \xi_c)$
$\alpha_{ijk}(p, \boldsymbol{\xi}) =$ $R_i(p, \xi_a) R_j(p, \xi_b) R_k(p, \xi_c)$	$i, j, k = 0, 1, \ldots, p$; with: $i + j + k = p$	$\boldsymbol{\xi}_{ijk} = \left(\frac{i}{p}, \frac{j}{p}, \frac{k}{p}\right)$
$\hat{\alpha}^a_{ijk}(p, \boldsymbol{\xi}) =$ $R_i(p+2, \xi_a)\hat{R}_j(p+2, \xi_b)\hat{R}_k(p+2, \xi_c)$	$i = 0, 1, \ldots, p$; $j, k = 1, 2, \ldots, p+1$; with: $i + j + k = p + 2$	$\boldsymbol{\xi}^a_{ijk} = \left(\frac{i}{p+2}, \frac{j}{p+2}, \frac{k}{p+2}\right)$

The left-hand column of the table reports the dummy expressions of the pth-order complete polynomial sets formed by $(p+1)(p+2)/2$ terms that individuate a regular grid of $(p+1)(p+2)/2$ interpolation points. These polynomials attain a unit value at the given interpolation point (reported in the third column) while they are equal to zero at the other interpolation points. The polynomial expressions are reported in terms of "dummy" variables and indices. In terms of the *local* numbers used to distinguish the coordinates of the parent triangle, ξ_a can be equal to ξ_1, ξ_2, or ξ_3.

- The *shape* polynomials $\alpha_{ijk}(p, \boldsymbol{\xi})$ are used for *scalar* interpolation or for geometry description; these polynomials interpolate all the edges of the triangle.
- The polynomials $\hat{\alpha}^a_{ijk}(p, \boldsymbol{\xi})$ are used to build the curl-conforming and the divergence-conforming *vector* bases; these polynomials have $(p+1)$ edge-interpolation points only on the $\xi_a = 0$ edge, and $p(p+1)/2$ interior interpolation points.

Table 4.8 Interpolatory polynomials for the quadrilateral cell (dummy expressions).

Polynomial	**Index rule**	**Interpolation point** $\boldsymbol{\xi}_{ik;j\ell} = (\xi_a, \xi_c; \xi_b, \xi_d)$
$\alpha_{ik;j\ell}(p, \boldsymbol{\xi}) =$ $R_i(p, \xi_a) R_k(p, \xi_c)$ $R_j(p, \xi_b) R_\ell(p, \xi_d)$	$i, j = 0, 1, \ldots, p$; with: $i + k = j + \ell = p$	$\boldsymbol{\xi}_{ik;j\ell} = \left(\frac{i}{p}, \frac{k}{p}; \frac{j}{p}, \frac{\ell}{p}\right)$
$\hat{\alpha}^a_{ik;j\ell}(p, \boldsymbol{\xi}) =$ $R_i(p+1, \xi_a)\hat{R}_k(p+1, \xi_c)$ $\hat{R}_j(p+2, \xi_b)\hat{R}_\ell(p+2, \xi_d)$	$i = 0, 1, \ldots, p$; $j, k, \ell = 1, 2, \ldots, p+1$, with: $i + k = p + 1; j + \ell = p + 2$	$\boldsymbol{\xi}^a_{ik;j\ell} = \left(\frac{i}{p+1}, \frac{k}{p+1}; \frac{j}{p+2}, \frac{\ell}{p+2}\right)$

The left-hand column of the table reports pth-order complete polynomial sets formed by $(p+1)^2$ terms that individuate a regular grid of $(p+1)^2$ interpolation points. These polynomials attain a unit value at the given interpolation point (reported in the third column) while they are equal to zero at the other interpolation points. The polynomial expressions are reported in terms of "dummy" variables. The "dummy" edges $\xi_a = 0$ and $\xi_b = 0$ are opposite to the edges $\xi_c = 0$ and $\xi_d = 0$, respectively. In terms of the *local* numbers used to distinguish the coordinates of the parent quadrilateral, ξ_a can be equal to ξ_1, ξ_2, ξ_3, or ξ_4.

- The *shape* polynomials $\alpha_{ik;j\ell}(p, \boldsymbol{\xi})$ are used for *scalar* interpolation or for geometry description; these polynomials interpolate all the edges of the quadrilateral.
- The polynomials $\hat{\alpha}^a_{ik;j\ell}(p, \boldsymbol{\xi})$ are used to build the curl-conforming and the divergence-conforming *vector* bases; these polynomials have $(p+1)$ edge-interpolation points only on the $\xi_a = 0$ edge, and $p(p+1)$ interior interpolation points.

Table 4.9 Interpolatory vector bases of order p for triangular cells.

$\boldsymbol{\Omega}^1_{ijk}(\boldsymbol{r}) = N^1_{ijk}\, R_i(p+2,\xi_1)\hat{R}_j(p+2,\xi_2)\hat{R}_k(p+2,\xi_3)\, \boldsymbol{\Omega}_1(\boldsymbol{r})$ $i=0,1,\ldots,p;\quad j,k=1,2,\ldots,p+1$ $\boldsymbol{\Omega}^2_{ijk}(\boldsymbol{r}) = N^2_{ijk}\, \hat{R}_i(p+2,\xi_1)R_j(p+2,\xi_2)\hat{R}_k(p+2,\xi_3)\, \boldsymbol{\Omega}_2(\boldsymbol{r})$ $j=0,1,\ldots,p;\quad i,k=1,2,\ldots,p+1$ $\boldsymbol{\Omega}^3_{ijk}(\boldsymbol{r}) = N^3_{ijk}\, \hat{R}_i(p+2,\xi_1)\hat{R}_j(p+2,\xi_2)R_k(p+2,\xi_3)\, \boldsymbol{\Omega}_3(\boldsymbol{r})$ $k=0,1,\ldots,p;\quad i,j=1,2,\ldots,p+1$ with $i+j+k=p+2$	$N^1_{ijk} = \frac{(p+2)}{(p+2-i)}\ell_1\Big\vert_{\boldsymbol{\xi}_{ijk}}$ $N^2_{ijk} = \frac{(p+2)}{(p+2-j)}\ell_2\Big\vert_{\boldsymbol{\xi}_{ijk}}$ $N^3_{ijk} = \frac{(p+2)}{(p+2-k)}\ell_3\Big\vert_{\boldsymbol{\xi}_{ijk}}$ $\boldsymbol{\xi}_{ijk} = (\xi_1,\xi_2,\xi_3)_{ijk}$ $= \left(\frac{i}{p+2}, \frac{j}{p+2}, \frac{k}{p+2}\right)$
$\boldsymbol{\Lambda}^1_{ijk}(\boldsymbol{r}) = N^1_{ijk} R_i(p+2,\xi_1)\hat{R}_j(p+2,\xi_2)\hat{R}_k(p+2,\xi_3)\boldsymbol{\Lambda}_1(\boldsymbol{r})$ $i=0,1,\ldots,p;\; j,k=1,2,\ldots,p+1$ $\boldsymbol{\Lambda}^2_{ijk}(\boldsymbol{r}) = N^2_{ijk}\hat{R}_i(p+2,\xi_1)R_j(p+2,\xi_2)\hat{R}_k(p+2,\xi_3)\boldsymbol{\Lambda}_2(\boldsymbol{r})$ $j=0,1,\ldots,p;\; i,k=1,2,\ldots,p+1$ $\boldsymbol{\Lambda}^3_{ijk}(\boldsymbol{r}) = N^3_{ijk}\hat{R}_i(p+2,\xi_1)\hat{R}_j(p+2,\xi_2)R_k(p+2,\xi_3)\boldsymbol{\Lambda}_3(\boldsymbol{r})$ $k=0,1,\ldots,p;\; i,j=1,2,\ldots,p+1$ with $i+j+k=p+2$	$N^1_{011} = \ell_1\vert_{\boldsymbol{\xi}_{011}}$ $N^2_{101} = \ell_2\vert_{\boldsymbol{\xi}_{101}}$ $N^3_{110} = \ell_3\vert_{\boldsymbol{\xi}_{110}}$ $\boldsymbol{\xi}_{011} = (0, \frac{1}{2}, \frac{1}{2})$ $\boldsymbol{\xi}_{101} = (\frac{1}{2}, 0, \frac{1}{2})$ $\boldsymbol{\xi}_{110} = (\frac{1}{2}, \frac{1}{2}, 0)$

The table reports the curl- ($\boldsymbol{\Omega}^a_{ijk}$) and the divergence-conforming ($\boldsymbol{\Lambda}^a_{ijk}$) basis functions of order p on a triangle. These functions are obtained by multiplying the un-normalized zeroth-order functions $\boldsymbol{\Omega}_a$ and $\boldsymbol{\Lambda}_a$ of Tables 4.2 and 4.3 with the interpolatory polynomials $\hat{\alpha}^a$ described in Section 4.6.1 and reported in Table 4.7. The normalization constant N^a_{ijk} is defined in terms of the magnitude of the edge vector (ℓ_a) evaluated at the given interpolation point ($\boldsymbol{\xi}_{ijk}$) and is chosen to ensure that the component of the vector functions $\boldsymbol{\Omega}^a_{ijk}$ (or $\boldsymbol{\Lambda}^a_{ijk}$) along the edge (or the height) vector at the interpolation point is unity. The normalization constants and the interpolation points for the $p=0$ case (zeroth-order bases) are reported in the bottom right-hand corner of the table.

Though both the vector bases and their curls are complete to order p, it happens that dependencies in the basis set exist for $p \geq 1$. Redundancy is easily eliminated by discarding one member from the set of the three basis functions interpolating interior nodes. For example, the third one, formed by all the functions $\boldsymbol{\Omega}^3_{ijk}$ and $\boldsymbol{\Lambda}^3_{ijk}$ obtained with $k \neq 0$. Hence, the total number of DoFs for the conforming basis functions of order p on a triangle is $(p+1)(p+3)$, with $3(p+1)$ edge DoF and $p(p+1)$ triangle interior DoF.

Table 4.10 Interpolatory vector bases of order p for quadrilateral cells.

$\boldsymbol{\Omega}^1_{ik;\,j\ell}(\boldsymbol{r}) = N^1_{ik;\,j\ell}\, R_i(p+1,\xi_1)\hat{R}_j(p+2,\xi_2)\hat{R}_k(p+1,\xi_3)\hat{R}_\ell(p+2,\xi_4)\,\boldsymbol{\Omega}_1(\boldsymbol{r}),$ $i=0,1,\ldots,p;\; j,k,\ell=1,2,\ldots,p+1,$ $i+k=p+1;\; j+\ell=p+2.$ $\boldsymbol{\Omega}^3_{ik;\,j\ell}(\boldsymbol{r}) = N^3_{ik;\,j\ell}\, \hat{R}_i(p+1,\xi_1)\hat{R}_j(p+2,\xi_2)R_k(p+1,\xi_3)\hat{R}_\ell(p+2,\xi_4)\,\boldsymbol{\Omega}_3(\boldsymbol{r}),$ $k=0,1,\ldots,p;\; i,j,\ell=1,2,\ldots,p+1,$ $i+k=p+1;\; j+\ell=p+2.$ $\boldsymbol{\Omega}^2_{ik;\,j\ell}(\boldsymbol{r}) = N^2_{ik;\,j\ell}\, \hat{R}_i(p+2,\xi_1)R_j(p+1,\xi_2)\hat{R}_k(p+2,\xi_3)\hat{R}_\ell(p+1,\xi_4)\,\boldsymbol{\Omega}_2(\boldsymbol{r}),$ $j=0,1,\ldots,p;\; i,k,\ell=1,2,\ldots,p+1,$ $i+k=p+2;\; j+\ell=p+1.$ $\boldsymbol{\Omega}^4_{ik;\,j\ell}(\boldsymbol{r}) = N^4_{ik;\,j\ell}\, \hat{R}_i(p+2,\xi_1)\hat{R}_j(p+1,\xi_2)\hat{R}_k(p+2,\xi_3)R_\ell(p+1,\xi_4)\,\boldsymbol{\Omega}_4(\boldsymbol{r}),$ $\ell=0,1,\ldots,p;\; i,j,k=1,2,\ldots,p+1,$ $i+k=p+2;\; j+\ell=p+1.$	$N^1_{ik;\,j\ell} = \frac{p+1}{p+1-i}\,\frac{\mathcal{J}}{h_1}\Big\vert_{\boldsymbol{\xi}^1_{ik;\,j\ell}},$ $N^3_{ik;\,j\ell} = \frac{p+1}{p+1-k}\,\frac{\mathcal{J}}{h_3}\Big\vert_{\boldsymbol{\xi}^3_{ik;\,j\ell}},$ $N^2_{ik;\,j\ell} = \frac{p+1}{p+1-j}\,\frac{\mathcal{J}}{h_2}\Big\vert_{\boldsymbol{\xi}^2_{ik;\,j\ell}},$ $N^4_{ik;\,j\ell} = \frac{p+1}{p+1-\ell}\,\frac{\mathcal{J}}{h_4}\Big\vert_{\boldsymbol{\xi}^4_{ik;\,j\ell}}.$ $\boldsymbol{\xi}^1_{ik;\,j\ell} = \left(\frac{i}{p+1}, \frac{k}{p+1}; \frac{j}{p+2}, \frac{\ell}{p+2}\right),$ $\boldsymbol{\xi}^3_{ik;\,j\ell} = \left(\frac{i}{p+1}, \frac{k}{p+1}; \frac{j}{p+2}, \frac{\ell}{p+2}\right),$ $\boldsymbol{\xi}^2_{ik;\,j\ell} = \left(\frac{i}{p+2}, \frac{k}{p+2}; \frac{j}{p+1}, \frac{\ell}{p+1}\right),$ $\boldsymbol{\xi}^4_{ik;\,j\ell} = \left(\frac{i}{p+2}, \frac{k}{p+2}; \frac{j}{p+1}, \frac{\ell}{p+1}\right),$ where $\boldsymbol{\xi} = (\xi_1, \xi_3; \xi_2, \xi_4)$.
$\boldsymbol{\Lambda}^1_{ik;\,j\ell}(\boldsymbol{r}) = N^1_{ik;\,j\ell} R_i(p+1,\xi_1)\hat{R}_j(p+2,\xi_2)\hat{R}_k(p+1,\xi_3)\hat{R}_\ell(p+2,\xi_4)\boldsymbol{\Lambda}_1(\boldsymbol{r}),$ $i=0,1,\ldots,p;\; j,k,\ell=1,2,\ldots,p+1,$ $i+k=p+1;\; j+\ell=p+2.$ $\boldsymbol{\Lambda}^3_{ik;\,j\ell}(\boldsymbol{r}) = N^3_{ik;\,j\ell} \hat{R}_i(p+1,\xi_1)\hat{R}_j(p+2,\xi_2)R_k(p+1,\xi_3)\hat{R}_\ell(p+2,\xi_4)\boldsymbol{\Lambda}_3(\boldsymbol{r}),$ $k=0,1,\ldots,p;\; i,j,\ell=1,2,\ldots,p+1,$ $i+k=p+1;\; j+\ell=p+2.$ $\boldsymbol{\Lambda}^2_{ik;\,j\ell}(\boldsymbol{r}) = N^2_{ik;\,j\ell} \hat{R}_i(p+2,\xi_1)R_j(p+1,\xi_2)\hat{R}_k(p+2,\xi_3)\hat{R}_\ell(p+1,\xi_4)\boldsymbol{\Lambda}_2(\boldsymbol{r}),$ $j=0,1,\ldots,p;\; i,k,\ell=1,2,\ldots,p+1,$ $i+k=p+2;\; j+\ell=p+1.$ $\boldsymbol{\Lambda}^4_{ik;\,j\ell}(\boldsymbol{r}) = N^4_{ik;\,j\ell} \hat{R}_i(p+2,\xi_1)\hat{R}_j(p+1,\xi_2)\hat{R}_k(p+2,\xi_3)R_\ell(p+1,\xi_4)\boldsymbol{\Lambda}_4(\boldsymbol{r}),$ $\ell=0,1,\ldots,p;\; i,j,k=1,2,\ldots,p+1,$ $i+k=p+2;\; j+\ell=p+1.$	$N^1_{01;11} = \frac{\mathcal{J}}{h_1}\Big\vert_{\boldsymbol{\xi}^1_{01;11}},$ $N^3_{10;11} = \frac{\mathcal{J}}{h_3}\Big\vert_{\boldsymbol{\xi}^3_{10;11}},$ $N^2_{11;01} = \frac{\mathcal{J}}{h_2}\Big\vert_{\boldsymbol{\xi}^2_{11;01}},$ $N^4_{11;10} = \frac{\mathcal{J}}{h_4}\Big\vert_{\boldsymbol{\xi}^4_{11;10}}.$ $\boldsymbol{\xi}^1_{01;11} = (0, 1; \tfrac{1}{2}, \tfrac{1}{2}),$ $\boldsymbol{\xi}^3_{10;11} = (1, 0; \tfrac{1}{2}, \tfrac{1}{2}),$ $\boldsymbol{\xi}^2_{11;01} = (\tfrac{1}{2}, \tfrac{1}{2}; 0, 1),$ $\boldsymbol{\xi}^4_{11;10} = (\tfrac{1}{2}, \tfrac{1}{2}; 1, 0).$

The table reports the curl- ($\boldsymbol{\Omega}^a_{ik;\,j\ell}$) and the divergence-conforming ($\boldsymbol{\Lambda}^a_{ik;\,j\ell}$) basis functions of order p on a quadrilateral. These functions are obtained by multiplying the un-normalized zeroth-order functions $\boldsymbol{\Omega}_a$ and $\boldsymbol{\Lambda}_a$ of Tables 4.2 and 4.3 with the interpolatory polynomials $\hat{\alpha}^a$ described in Section 4.6.2 and reported in Table 4.8. The normalization constant $N^a_{ik;\,j\ell}$ is chosen to ensure that the component of the vector functions $\boldsymbol{\Omega}^a_{ik;\,j\ell}$ (or $\boldsymbol{\Lambda}^a_{ik;\,j\ell}$) along the edge (or the height) vector at the interpolation point $\boldsymbol{\xi}^a_{ik;\,j\ell}$ is unity. The normalization constants and the interpolation points for the $p=0$ case (zeroth-order bases) are reported in the bottom right-hand corner of the table. Though both the vector bases and their curls are complete to order p, it happens that dependencies in the basis set exist for $p \geq 1$. At each interior node (that is for i, j, k, and $\ell \neq 0$), one of the dependent bases $\boldsymbol{\Omega}^1_{ik;\,j\ell}$, $\boldsymbol{\Omega}^3_{ik;\,j\ell}$ (or $\boldsymbol{\Lambda}^1_{ik;\,j\ell}$, $\boldsymbol{\Lambda}^3_{ik;\,j\ell}$) and one of the dependent bases $\boldsymbol{\Omega}^2_{ik;\,j\ell}$, $\boldsymbol{\Omega}^4_{ik;\,j\ell}$ (or $\boldsymbol{\Lambda}^2_{ik;\,j\ell}$, $\boldsymbol{\Lambda}^4_{ik;\,j\ell}$) should be discarded to produce a linearly independent set. Hence, the total number of DoFs for the conforming basis functions of order p on a quadrilateral is $2(p+1)(p+2)$, with $4(p+1)$ edge DoF and $2p(p+1)$ quadrilateral interior DoF.

Table 4.11 Interpolatory polynomials for the tetrahedral cell (dummy expressions).

Polynomial	Index rule	Interpolation point $\boldsymbol{\xi}_{ijk\ell} = (\xi_a, \xi_c; \xi_b, \xi_d)_{ijk\ell}$
$\alpha_{ijk\ell}(p, \boldsymbol{\xi}) = R_i(p, \xi_a)\, R_j(p, \xi_b)\, R_k(p, \xi_c)\, R_\ell(p, \xi_d)$	$i, j, k, \ell = 0, 1, \ldots, p;$ with: $i + j + k + \ell = p$	$\boldsymbol{\xi}_{ijk\ell} = \left(\frac{i}{p}, \frac{j}{p}, \frac{k}{p}, \frac{\ell}{p}\right)$
$\hat{\alpha}^{ad}_{ijk\ell}(p, \boldsymbol{\xi}) = R_i(p+2, \xi_a)\, \hat{R}_j(p+2, \xi_b)\, \hat{R}_k(p+2, \xi_c)\, R_\ell(p+2, \xi_d)$	$i, \ell = 0, 1, \ldots, p;$ $j, k = 1, 2, \ldots, p+1;$ with: $i + j + k + \ell = p + 2$	$\boldsymbol{\xi}^{ad}_{ijk\ell} = \left(\frac{i}{p+2}, \frac{j}{p+2}, \frac{k}{p+2}, \frac{\ell}{p+2}\right)$
$\hat{\beta}^{d}_{ijk\ell}(p, \boldsymbol{\xi}) = \hat{R}_i(p+3, \xi_a)\, \hat{R}_j(p+3, \xi_b)\, \hat{R}_k(p+3, \xi_c)\, R_\ell(p+3, \xi_d)$	$\ell = 0, 1, \ldots, p;$ $i, j, k = 1, 2, \ldots, p+1;$ with: $i + j + k + \ell = p + 3$	$\boldsymbol{\xi}^{d}_{ijk\ell} = \left(\frac{i}{p+3}, \frac{j}{p+3}, \frac{k}{p+3}, \frac{\ell}{p+3}\right)$

The left-hand column of the table reports pth-order complete polynomial sets formed by $(p+1)(p+2)(p+3)/6$ terms that individuate a regular grid of $(p+1)(p+2)(p+3)/6$ interpolation points. These polynomials attain a unit value at the given interpolation point (reported in the third column) while they are equal to zero at the other interpolation points. The polynomial expressions are reported in terms of "dummy" variables. In terms of the *local* numbers used to distinguish the coordinates of the parent tetrahedron, ξ_a (and $\xi_d \neq \xi_a$) can be equal to ξ_1, ξ_2, ξ_3, or ξ_4.

- The shape polynomials $\alpha_{ijk\ell}(p, \boldsymbol{\xi})$ are used for *scalar* interpolation or for geometry description; these polynomials interpolate all the edges and faces of the tetrahedron.
- The polynomials $\hat{\alpha}^{ad}_{ijk\ell}(p, \boldsymbol{\xi})$ are used to build the curl-conforming *vector* bases; these polynomials have $(p+1)$ edge-interpolation points only on the $\xi_a = \xi_d = 0$ edge, $p(p+1)/2$ face-interpolation point on the $\xi_a = 0$ and on the $\xi_d = 0$ face, and $p(p^2 - 1)/6$ interior interpolation points.
- The polynomials $\hat{\beta}^{d}_{ijk\ell}(p, \boldsymbol{\xi})$ are used to build the divergence-conforming *vector* bases; these polynomials have $(p+1)(p+2)/2$ face-interpolation points only on the $\xi_d = 0$ face, and $p(p+1)(p+2)/6$ interior interpolation points.

Table 4.12 Interpolatory curl-conforming bases of order p for tetrahedral cells.

$$\boldsymbol{\Omega}^{12}_{ijk\ell}(\boldsymbol{r}) = N^{12}_{ijk\ell}\, R_i(p+2,\xi_1)R_j(p+2,\xi_2)\hat{R}_k(p+2,\xi_3)\hat{R}_\ell(p+2,\xi_4)\,\boldsymbol{\Omega}_{12}(\boldsymbol{r}),$$
$$i,j=0,1,\ldots,p;\;\; k,\ell=1,2,\ldots,p+1,$$

$$\boldsymbol{\Omega}^{13}_{ijk\ell}(\boldsymbol{r}) = N^{13}_{ijk\ell}\, R_i(p+2,\xi_1)\hat{R}_j(p+2,\xi_2)R_k(p+2,\xi_3)\hat{R}_\ell(p+2,\xi_4)\,\boldsymbol{\Omega}_{13}(\boldsymbol{r}),$$
$$i,k=0,1,\ldots,p;\;\; j,\ell=1,2,\ldots,p+1,$$

$$\boldsymbol{\Omega}^{14}_{ijk\ell}(\boldsymbol{r}) = N^{14}_{ijk\ell}\, R_i(p+2,\xi_1)\hat{R}_j(p+2,\xi_2)\hat{R}_k(p+2,\xi_3)R_\ell(p+2,\xi_4)\,\boldsymbol{\Omega}_{14}(\boldsymbol{r}),$$
$$i,\ell=0,1,\ldots,p;\;\; j,k=1,2,\ldots,p+1,$$

$$\boldsymbol{\Omega}^{23}_{ijk\ell}(\boldsymbol{r}) = N^{23}_{ijk\ell}\, \hat{R}_i(p+2,\xi_1)R_j(p+2,\xi_2)R_k(p+2,\xi_3)\hat{R}_\ell(p+2,\xi_4)\,\boldsymbol{\Omega}_{23}(\boldsymbol{r}),$$
$$j,k=0,1,\ldots,p;\;\; i,\ell=1,2,\ldots,p+1,$$

$$\boldsymbol{\Omega}^{24}_{ijk\ell}(\boldsymbol{r}) = N^{24}_{ijk\ell}\, \hat{R}_i(p+2,\xi_1)R_j(p+2,\xi_2)\hat{R}_k(p+2,\xi_3)R_\ell(p+2,\xi_4)\,\boldsymbol{\Omega}_{24}(\boldsymbol{r}),$$
$$j,\ell=0,1,\ldots,p;\;\; i,k=1,2,\ldots,p+1,$$

$$\boldsymbol{\Omega}^{34}_{ijk\ell}(\boldsymbol{r}) = N^{34}_{ijk\ell}\, \hat{R}_i(p+2,\xi_1)\hat{R}_j(p+2,\xi_2)R_k(p+2,\xi_3)R_\ell(p+2,\xi_4)\,\boldsymbol{\Omega}_{34}(\boldsymbol{r}),$$
$$k,\ell=0,1,\ldots,p;\;\; i,j=1,2,\ldots,p+1,$$

where $i+j+k+\ell=p+2$.

$N^{12}_{ijk\ell} = \dfrac{p+2}{p+2-i-j}\,\ell_{12}\,\vert_{\boldsymbol{\xi}_{ijk\ell}}$, $\quad N^{23}_{ijk\ell} = \dfrac{p+2}{p+2-j-k}\,\ell_{23}\,\vert_{\boldsymbol{\xi}_{ijk\ell}}$,	$N^{12}_{0011} = \ell_{12}\,\vert_{\boldsymbol{\xi}_{0011}}$, $N^{13}_{0101} = \ell_{13}\,\vert_{\boldsymbol{\xi}_{0101}}$,
$N^{13}_{ijk\ell} = \dfrac{p+2}{p+2-i-k}\,\ell_{13}\,\vert_{\boldsymbol{\xi}_{ijk\ell}}$, $\quad N^{24}_{ijk\ell} = \dfrac{p+2}{p+2-j-\ell}\,\ell_{24}\,\vert_{\boldsymbol{\xi}_{ijk\ell}}$,	$N^{14}_{0110} = \ell_{14}\,\vert_{\boldsymbol{\xi}_{0110}}$,
$N^{14}_{ijk\ell} = \dfrac{p+2}{p+2-i-\ell}\,\ell_{14}\,\vert_{\boldsymbol{\xi}_{ijk\ell}}$, $\quad N^{34}_{ijk\ell} = \dfrac{p+2}{p+2-k-\ell}\,\ell_{34}\,\vert_{\boldsymbol{\xi}_{ijk\ell}}$,	$N^{23}_{1001} = \ell_{23}\,\vert_{\boldsymbol{\xi}_{1001}}$,
$\boldsymbol{\xi}_{ijk\ell} = (\xi_1,\xi_2,\xi_3,\xi_4)_{ijk\ell} = \left(\frac{i}{p+2},\frac{j}{p+2},\frac{k}{p+2},\frac{\ell}{p+2}\right)$.	$N^{24}_{1010} = \ell_{24}\,\vert_{\boldsymbol{\xi}_{1010}}$, $N^{34}_{1100} = \ell_{34}\,\vert_{\boldsymbol{\xi}_{1100}}$.

The table reports the curl-conforming basis functions $\boldsymbol{\Omega}^{ad}_{ijk\ell}$ of order p on a tetrahedron. These functions are obtained by multiplying the un-normalized zeroth-order functions $\boldsymbol{\Omega}_{ad}$ of Table 4.4 (left-hand column) with the interpolatory polynomials $\hat{\alpha}^{ad}$ described in Section 4.7.1.1, and reported in Table 4.11. The normalization constant $N^{ad}_{ijk\ell}$ is defined in terms of the magnitude of the edge vector (ℓ_{ad}) evaluated at the given interpolation point $\boldsymbol{\xi}_{ijk\ell}$ and is chosen to ensure that the component of the vector functions $\boldsymbol{\Omega}^{ad}_{ijk\ell}$ along the edge vector at the interpolation point is unity. The normalization constants for the $p=0$ case (zeroth-order bases) are reported in the bottom right-hand corner of the table. Although both the vector bases and their curls are complete to order p, it happens that dependencies in the basis set exist for $p \geq 1$. Redundancy is easily eliminated by discarding three members from the set of the six basis functions interpolating interior nodes, and one member from the set of the three basis functions interpolating face-interior nodes, as explained in Section 4.7.1.3. The total number of DoFs for the curl-conforming basis functions of order p on a tetrahedron is $(p+1)(p+3)(p+4)/2$.

Table 4.13 Interpolatory divergence-conforming bases of order p for tetrahedral cells.

$$\mathbf{\Lambda}^1_{ijk\ell}(\boldsymbol{r}) = N^1_{ijk\ell}\, R_i(p+3,\xi_1)\hat{R}_j(p+3,\xi_2)\hat{R}_k(p+3,\xi_3)\hat{R}_\ell(p+3,\xi_4)\, \mathbf{\Lambda}_1(\boldsymbol{r}),$$
$$i=0,1,\ldots,p;\; j,k,\ell=1,2,\ldots,p+1,$$

$$\mathbf{\Lambda}^2_{ijk\ell}(\boldsymbol{r}) = N^2_{ijk\ell}\, \hat{R}_i(p+3,\xi_1)R_j(p+3,\xi_2)\hat{R}_k(p+3,\xi_3)\hat{R}_\ell(p+3,\xi_4)\, \mathbf{\Lambda}_2(\boldsymbol{r}),$$
$$j=0,1,\ldots,p;\; i,k,\ell=1,2,\ldots,p+1,$$

$$\mathbf{\Lambda}^3_{ijk\ell}(\boldsymbol{r}) = N^3_{ijk\ell}\, \hat{R}_i(p+3,\xi_1)\hat{R}_j(p+3,\xi_2)R_k(p+3,\xi_3)\hat{R}_\ell(p+3,\xi_4)\, \mathbf{\Lambda}_3(\boldsymbol{r}),$$
$$k=0,1,\ldots,p;\; i,j,\ell=1,2,\ldots,p+1,$$

$$\mathbf{\Lambda}^4_{ijk\ell}(\boldsymbol{r}) = N^4_{ijk\ell}\, \hat{R}_i(p+3,\xi_1)\hat{R}_j(p+3,\xi_2)\hat{R}_k(p+3,\xi_3)R_\ell(p+3,\xi_4)\, \mathbf{\Lambda}_4(\boldsymbol{r}),$$
$$\ell=0,1,\ldots,p;\; i,j,k=1,2,\ldots,p+1,$$
$$\text{where } i+j+k+\ell=p+3.$$

$N^1_{ijk\ell} = \dfrac{p+3}{p+3-i}\dfrac{\mathcal{J}}{h_1}\bigg\vert_{\boldsymbol{\xi}_{ijk\ell}}$,	$N^1_{0111} = \dfrac{\mathcal{J}}{h_1}\bigg\vert_{\boldsymbol{\xi}_{0111}}$,
$N^2_{ijk\ell} = \dfrac{p+3}{p+3-j}\dfrac{\mathcal{J}}{h_2}\bigg\vert_{\boldsymbol{\xi}_{ijk\ell}}$,	$N^2_{1011} = \dfrac{\mathcal{J}}{h_2}\bigg\vert_{\boldsymbol{\xi}_{1011}}$,
$N^3_{ijk\ell} = \dfrac{p+3}{p+3-k}\dfrac{\mathcal{J}}{h_3}\bigg\vert_{\boldsymbol{\xi}_{ijk\ell}}$,	$N^3_{1101} = \dfrac{\mathcal{J}}{h_3}\bigg\vert_{\boldsymbol{\xi}_{1101}}$,
$N^4_{ijk\ell} = \dfrac{p+3}{p+3-\ell}\dfrac{\mathcal{J}}{h_4}\bigg\vert_{\boldsymbol{\xi}_{ijk\ell}}$,	$N^4_{1110} = \dfrac{\mathcal{J}}{h_4}\bigg\vert_{\boldsymbol{\xi}_{1110}}$.
$\boldsymbol{\xi}_{ijk\ell} = (\xi_1,\xi_2,\xi_3,\xi_4)_{ijk\ell} = \left(\dfrac{i}{p+3},\dfrac{j}{p+3},\dfrac{k}{p+3},\dfrac{\ell}{p+3}\right)$.	

The table reports the divergence-conforming basis functions $\mathbf{\Lambda}^d_{ijk\ell}$ of order p on a tetrahedron. These functions are obtained by multiplying the un-normalized zeroth-order functions $\mathbf{\Lambda}_d$ of Table 4.5 (left-hand column) with the interpolatory polynomials $\hat{\beta}^d$ described in Section 4.7.1.2, and reported in Table 4.11. The normalization constant $N^d_{ijk\ell}$ is chosen such that the components of the bases along $\hat{\boldsymbol{h}}_d$ is unity at the given interpolation point $\boldsymbol{\xi}_{ijk\ell}$. The normalization constants for the $p=0$ case (zeroth-order bases) are reported in the bottom right-hand corner of the table. Although both the vector bases and their divergence are complete to order p, it happens that dependencies in the basis set exist for $p \geq 1$. Redundancy is easily eliminated by discarding one member from the set of the four basis functions interpolating interior nodes. The total number of DoFs for the divergence-conforming basis functions of order p on a tetrahedron is $(p+1)(p+2)(p+4)/2$.

Table 4.14 Interpolatory polynomials for the brick cell.

Polynomial	Index rule	Interpolation point $\boldsymbol{\xi}_{i\ell;jm;kn} = (\xi_a, \xi_d; \xi_b, \xi_e; \xi_c, \xi_f)$
$\alpha_{i\ell;\,jm;kn}(p,\boldsymbol{\xi}) =$ $R_i(p,\xi_a)\,R_\ell(p,\xi_d)$ $R_j(p,\xi_b)\,R_m(p,\xi_e)$ $R_k(p,\xi_c)\,R_n(p,\xi_f)$	$i,j,k,\ell,m,n=0,1,\ldots,p$; with: $i+\ell=j+m=k+n=p$	$\boldsymbol{\xi}_{i\ell;\,jm;kn} = \left(\frac{i}{p},\frac{\ell}{p};\frac{j}{p},\frac{m}{p};\frac{k}{p},\frac{n}{p}\right)$
$\hat{\alpha}^{ab}_{i\ell;\,jm;kn}(p,\boldsymbol{\xi}) =$ $R_i(p+1,\xi_a)\,\hat{R}_\ell(p+1,\xi_d)$ $R_j(p+1,\xi_b)\,\hat{R}_m(p+1,\xi_e)$ $\hat{R}_k(p+2,\xi_c)\,\hat{R}_n(p+2,\xi_f)$	$i,j=0,1,\ldots,p$; $k,\ell,m,n=1,2,\ldots,p+1$; with: $i+\ell=j+m=p+1$; $k+n=p+2$	$\boldsymbol{\xi}^{ab}_{i\ell;\,jm;kn} = \left(\frac{i}{p+1},\frac{\ell}{p+1};\frac{j}{p+1},\frac{m}{p+1};\frac{k}{p+2},\frac{n}{p+2}\right)$
$\hat{\beta}^{a}_{i\ell;\,jm;kn}(p,\boldsymbol{\xi}) =$ $R_i(p+1,\xi_a)\,\hat{R}_\ell(p+1,\xi_d)$ $\hat{R}_j(p+2,\xi_b)\,\hat{R}_m(p+2,\xi_e)$ $\hat{R}_k(p+2,\xi_c)\,\hat{R}_n(p+2,\xi_f)$	$i=0,1,\ldots,p$; $j,k,\ell,m,n=1,2,\ldots,p+1$ with: $i+\ell=p+1$, $j+m=k+n=p+2$	$\boldsymbol{\xi}^{a}_{i\ell;\,jm;kn} = \left(\frac{i}{p+1},\frac{\ell}{p+1};\frac{j}{p+2},\frac{m}{p+2};\frac{k}{p+2},\frac{n}{p+2}\right)$

The left-hand column of the table reports pth-order complete polynomial sets formed by $(p+1)^3$ terms that individuate a regular grid of $(p+1)^3$ interpolation points. These polynomials attain a unit value at the given interpolation point (reported in the third column) while they are equal to zero at the other interpolation points. The polynomial expressions are reported in terms of "dummy" variables. The "dummy" faces $\xi_a=0$, $\xi_b=0$, and $\xi_c=0$ are opposite to the faces $\xi_d=0$, $\xi_e=0$, and $\xi_f=0$, respectively, while the "dummy" faces $\xi_a=0$ and $\xi_b=0$ have a common edge (the $\xi_a=\xi_b=0$ edge). In terms of the *local* numbers used to distinguish the coordinates of the parent brick, ξ_a (or ξ_b) can be equal to ξ_1, ξ_2, ξ_3, ξ_4, ξ_5, or ξ_6.

- The shape polynomials $\alpha_{i\ell;\,jm;kn}(p,\boldsymbol{\xi})$ are used for *scalar* interpolation or for geometry description; these polynomials interpolate all the edges and faces of the brick.
- The polynomials $\hat{\alpha}^{ab}_{i\ell;\,jm;kn}(p,\boldsymbol{\xi})$ are used to build the curl-conforming *vector* bases; these polynomials have $(p+1)$ edge-interpolation points only on the $\xi_a=\xi_b=0$ edge, $p(p+1)$ face-interior interpolation points on the $\xi_a=0$ and on the $\xi_b=0$ face, and $p^2(p+1)$ element-interior interpolation points.
- The polynomials $\hat{\beta}^{a}_{i\ell;\,jm;kn}(p,\boldsymbol{\xi})$ are used to build the divergence-conforming *vector* bases; these polynomials have $(p+1)^2$ face-interpolation points only on the $\xi_a=0$ face, and $p(p+1)^2$ brick-interior interpolation point.

Table 4.15 Interpolatory curl-conforming bases of order p for brick cells.

$\mathbf{\Omega}^{12}_{i\ell;\,jm;kn}(\boldsymbol{r}) =$ $N^{12}_{i\ell;\,jm;kn} R_i(p+1,\xi_1) R_j(p+1,\xi_2) \hat{R}_k(p+2,\xi_3)$ $\cdot \hat{R}_\ell(p+1,\xi_4) \hat{R}_m(p+1,\xi_5) \hat{R}_n(p+2,\xi_6) \mathbf{\Omega}_{12}(\boldsymbol{r}),$ $i,j=0,1,\ldots,p;\; k,\ell,m,n=1,2,\ldots,p+1;$ $i+\ell=j+m=p+1;\; k+n=p+2,$	$\mathbf{\Omega}^{26}_{i\ell;\,jm;kn}(\boldsymbol{r}) =$ $N^{26}_{i\ell;\,jm;kn} \hat{R}_i(p+2,\xi_1) R_j(p+1,\xi_2) \hat{R}_k(p+1,\xi_3)$ $\cdot \hat{R}_\ell(p+2,\xi_4) \hat{R}_m(p+1,\xi_5) R_n(p+1,\xi_6) \mathbf{\Omega}_{26}(\boldsymbol{r}),$ $j,n=0,1,\ldots,p;\; i,k,\ell,m=1,2,\ldots,p+1;$ $j+m=k+n=p+1;\; i+\ell=p+2,$
$\mathbf{\Omega}^{13}_{i\ell;\,jm;kn}(\boldsymbol{r}) =$ $N^{13}_{i\ell;\,jm;kn} R_i(p+1,\xi_1) \hat{R}_j(p+2,\xi_2) R_k(p+1,\xi_3)$ $\cdot \hat{R}_\ell(p+1,\xi_4) \hat{R}_m(p+2,\xi_5) \hat{R}_n(p+1,\xi_6) \mathbf{\Omega}_{13}(\boldsymbol{r}),$ $i,k=0,1,\ldots,p;\; j,\ell,m,n=1,2,\ldots,p+1;$ $i+\ell=k+n=p+1;\; j+m=p+2,$	$\mathbf{\Omega}^{34}_{i\ell;\,jm;kn}(\boldsymbol{r}) =$ $N^{34}_{i\ell;\,jm;kn} \hat{R}_i(p+1,\xi_1) \hat{R}_j(p+2,\xi_2) R_k(p+1,\xi_3)$ $\cdot R_\ell(p+1,\xi_4) \hat{R}_m(p+2,\xi_5) \hat{R}_n(p+1,\xi_6) \mathbf{\Omega}_{34}(\boldsymbol{r}),$ $k,\ell=0,1,\ldots,p;\; i,j,m,n=1,2,\ldots,p+1;$ $i+\ell=k+n=p+1;\; j+m=p+2,$
$\mathbf{\Omega}^{15}_{i\ell;\,jm;kn}(\boldsymbol{r}) =$ $N^{15}_{i\ell;\,jm;kn} R_i(p+1,\xi_1) \hat{R}_j(p+1,\xi_2) \hat{R}_k(p+2,\xi_3)$ $\cdot \hat{R}_\ell(p+1,\xi_4) R_m(p+1,\xi_5) \hat{R}_n(p+2,\xi_6) \mathbf{\Omega}_{15}(\boldsymbol{r}),$ $i,m=0,1,\ldots,p;\; j,k,\ell,n=1,2,\ldots,p+1;$ $i+\ell=j+m=p+1;\; k+n=p+2,$	$\mathbf{\Omega}^{35}_{i\ell;\,jm;kn}(\boldsymbol{r}) =$ $N^{35}_{i\ell;\,jm;kn} \hat{R}_i(p+2,\xi_1) \hat{R}_j(p+1,\xi_2) R_k(p+1,\xi_3)$ $\cdot \hat{R}_\ell(p+2,\xi_4) R_m(p+1,\xi_5) \hat{R}_n(p+1,\xi_6) \mathbf{\Omega}_{35}(\boldsymbol{r}),$ $k,m=0,1,\ldots,p;\; i,j,\ell,n=1,2,\ldots,p+1;$ $j+m=k+n=p+1;\; i+\ell=p+2,$
$\mathbf{\Omega}^{16}_{i\ell;\,jm;kn}(\boldsymbol{r}) =$ $N^{16}_{i\ell;\,jm;kn} R_i(p+1,\xi_1) \hat{R}_j(p+2,\xi_2) \hat{R}_k(p+1,\xi_3)$ $\cdot \hat{R}_\ell(p+1,\xi_4) \hat{R}_m(p+2,\xi_5) R_n(p+1,\xi_6) \mathbf{\Omega}_{16}(\boldsymbol{r}),$ $i,n=0,1,\ldots,p;\; j,k,\ell,m=1,2,\ldots,p+1;$ $i+\ell=k+n=p+1;\; j+m=p+2,$	$\mathbf{\Omega}^{45}_{i\ell;\,jm;kn}(\boldsymbol{r}) =$ $N^{45}_{i\ell;\,jm;kn} \hat{R}_i(p+1,\xi_1) \hat{R}_j(p+1,\xi_2) \hat{R}_k(p+2,\xi_3)$ $\cdot R_\ell(p+1,\xi_4) R_m(p+1,\xi_5) \hat{R}_n(p+2,\xi_6) \mathbf{\Omega}_{45}(\boldsymbol{r}),$ $\ell,m=0,1,\ldots,p;\; i,j,k,n=1,2,\ldots,p+1;$ $i+\ell=j+m=p+1;\; k+n=p+2,$
$\mathbf{\Omega}^{23}_{i\ell;\,jm;kn}(\boldsymbol{r}) =$ $N^{23}_{i\ell;\,jm;kn} \hat{R}_i(p+2,\xi_1) R_j(p+1,\xi_2) R_k(p+1,\xi_3)$ $\cdot \hat{R}_\ell(p+2,\xi_4) \hat{R}_m(p+1,\xi_5) \hat{R}_n(p+1,\xi_6) \mathbf{\Omega}_{23}(\boldsymbol{r}),$ $j,k=0,1,\ldots,p;\; i,\ell,m,n=1,2,\ldots,p+1;$ $j+m=k+n=p+1;\; i+\ell=p+2,$	$\mathbf{\Omega}^{46}_{i\ell;\,jm;kn}(\boldsymbol{r}) =$ $N^{46}_{i\ell;\,jm;kn} \hat{R}_i(p+1,\xi_1) \hat{R}_j(p+2,\xi_2) \hat{R}_k(p+1,\xi_3)$ $\cdot R_\ell(p+1,\xi_4) \hat{R}_m(p+2,\xi_5) R_n(p+1,\xi_6) \mathbf{\Omega}_{46}(\boldsymbol{r}),$ $\ell,n=0,1,\ldots,p;\; i,j,k,m=1,2,\ldots,p+1;$ $i+\ell=k+n=p+1;\; j+m=p+2,$
$\mathbf{\Omega}^{24}_{i\ell;\,jm;kn}(\boldsymbol{r}) =$ $N^{24}_{i\ell;\,jm;kn} \hat{R}_i(p+1,\xi_1) R_j(p+1,\xi_2) \hat{R}_k(p+2,\xi_3)$ $\cdot R_\ell(p+1,\xi_4) \hat{R}_m(p+1,\xi_5) \hat{R}_n(p+2,\xi_6) \mathbf{\Omega}_{24}(\boldsymbol{r}),$ $j,\ell=0,1,\ldots,p;\; i,k,m,n=1,2,\ldots,p+1;$ $i+\ell=j+m=p+1;\; k+n=p+2,$	$\mathbf{\Omega}^{56}_{i\ell;\,jm;kn}(\boldsymbol{r}) =$ $N^{56}_{i\ell;\,jm;kn} \hat{R}_i(p+2,\xi_1) \hat{R}_j(p+1,\xi_2) \hat{R}_k(p+1,\xi_3)$ $\cdot \hat{R}_\ell(p+2,\xi_4) R_m(p+1,\xi_5) R_n(p+1,\xi_6) \mathbf{\Omega}_{56}(\boldsymbol{r}),$ $m,n=0,1,\ldots,p;\; i,j,k,\ell=1,2,\ldots,p+1;$ $j+m=k+n=p+1;\; i+\ell=p+2,$

The table reports the curl-conforming basis functions $\mathbf{\Omega}^{ab}_{i\ell;\,jm;kn}$ of order p on a brick. These functions are obtained by multiplying the un-normalized zeroth-order functions $\mathbf{\Omega}_{ab}$ of Table 4.4 (right-hand column) with the interpolatory polynomials $\hat{\alpha}^{ab}$ described in Section 4.7.2.1, and reported in Table 4.14. The normalization constant $N^{ab}_{i\ell;\,jm;kn}$ is chosen such that the components of the bases along $\boldsymbol{\ell}_{ab}$ at the interpolation point $\boldsymbol{\xi}^{ab}_{i\ell;\,jm;kn}$ are unity. The normalization constants and the interpolation points are given in Table 4.16. Although both the vector bases and their curl are complete to order p, it happens that dependencies in the basis set exist for $p \geq 1$. Redundancy is easily eliminated as explained in Section 4.7.2.3. The total number of DoFs for the curl-conforming basis functions of order p on a brick is $3(p+1)(p+2)^2$.

Table 4.16 Normalization constants and interpolatory points for brick curl-conforming bases.

$N^{12}_{i\ell;\,jm;kn} = \dfrac{(p+1)^2}{(p+1-i)(p+1-j)}\,\ell_{12}\Big\vert_{\xi^{12}_{i\ell;\,jm;kn}}$,	$\xi^{12}_{i\ell;\,jm;kn} = \left(\dfrac{i}{p+1}, \dfrac{\ell}{p+1}; \dfrac{j}{p+1}, \dfrac{m}{p+1}; \dfrac{k}{p+2}, \dfrac{n}{p+2}\right)$,
$N^{13}_{i\ell;\,jm;kn} = \dfrac{(p+1)^2}{(p+1-i)(p+1-k)}\,\ell_{13}\Big\vert_{\xi^{13}_{i\ell;\,jm;kn}}$,	$\xi^{13}_{i\ell;\,jm;kn} = \left(\dfrac{i}{p+1}, \dfrac{\ell}{p+1}; \dfrac{j}{p+2}, \dfrac{m}{p+2}; \dfrac{k}{p+1}, \dfrac{n}{p+1}\right)$,
$N^{15}_{i\ell;\,jm;kn} = \dfrac{(p+1)^2}{(p+1-i)(p+1-m)}\,\ell_{15}\Big\vert_{\xi^{15}_{i\ell;\,jm;kn}}$,	$\xi^{15}_{i\ell;\,jm;kn} = \left(\dfrac{i}{p+1}, \dfrac{\ell}{p+1}; \dfrac{j}{p+1}, \dfrac{m}{p+1}; \dfrac{k}{p+2}, \dfrac{n}{p+2}\right)$,
$N^{16}_{i\ell;\,jm;kn} = \dfrac{(p+1)^2}{(p+1-i)(p+1-n)}\,\ell_{16}\Big\vert_{\xi^{16}_{i\ell;\,jm;kn}}$,	$\xi^{16}_{i\ell;\,jm;kn} = \left(\dfrac{i}{p+1}, \dfrac{\ell}{p+1}; \dfrac{j}{p+2}, \dfrac{m}{p+2}; \dfrac{k}{p+1}, \dfrac{n}{p+1}\right)$,
$N^{23}_{i\ell;\,jm;kn} = \dfrac{(p+1)^2}{(p+1-j)(p+1-k)}\,\ell_{23}\Big\vert_{\xi^{23}_{i\ell;\,jm;kn}}$,	$\xi^{23}_{i\ell;\,jm;kn} = \left(\dfrac{i}{p+2}, \dfrac{\ell}{p+2}; \dfrac{j}{p+1}, \dfrac{m}{p+1}; \dfrac{k}{p+1}, \dfrac{n}{p+1}\right)$,
$N^{24}_{i\ell;\,jm;kn} = \dfrac{(p+1)^2}{(p+1-j)(p+1-\ell)}\,\ell_{24}\Big\vert_{\xi^{24}_{i\ell;\,jm;kn}}$,	$\xi^{24}_{i\ell;\,jm;kn} = \left(\dfrac{i}{p+1}, \dfrac{\ell}{p+1}; \dfrac{j}{p+1}, \dfrac{m}{p+1}; \dfrac{k}{p+2}, \dfrac{n}{p+2}\right)$,
$N^{26}_{i\ell;\,jm;kn} = \dfrac{(p+1)^2}{(p+1-j)(p+1-n)}\,\ell_{26}\Big\vert_{\xi^{26}_{i\ell;\,jm;kn}}$,	$\xi^{26}_{i\ell;\,jm;kn} = \left(\dfrac{i}{p+2}, \dfrac{\ell}{p+2}; \dfrac{j}{p+1}, \dfrac{m}{p+1}; \dfrac{k}{p+1}, \dfrac{n}{p+1}\right)$,
$N^{34}_{i\ell;\,jm;kn} = \dfrac{(p+1)^2}{(p+1-k)(p+1-\ell)}\,\ell_{34}\Big\vert_{\xi^{34}_{i\ell;\,jm;kn}}$,	$\xi^{34}_{i\ell;\,jm;kn} = \left(\dfrac{i}{p+1}, \dfrac{\ell}{p+1}; \dfrac{j}{p+2}, \dfrac{m}{p+2}; \dfrac{k}{p+1}, \dfrac{n}{p+1}\right)$,
$N^{35}_{i\ell;\,jm;kn} = \dfrac{(p+1)^2}{(p+1-k)(p+1-m)}\,\ell_{35}\Big\vert_{\xi^{35}_{i\ell;\,jm;kn}}$,	$\xi^{35}_{i\ell;\,jm;kn} = \left(\dfrac{i}{p+2}, \dfrac{\ell}{p+2}; \dfrac{j}{p+1}, \dfrac{m}{p+1}; \dfrac{k}{p+1}, \dfrac{n}{p+1}\right)$,
$N^{45}_{i\ell;\,jm;kn} = \dfrac{(p+1)^2}{(p+1-\ell)(p+1-m)}\,\ell_{45}\Big\vert_{\xi^{45}_{i\ell;\,jm;kn}}$,	$\xi^{45}_{i\ell;\,jm;kn} = \left(\dfrac{i}{p+1}, \dfrac{\ell}{p+1}; \dfrac{j}{p+1}, \dfrac{m}{p+1}; \dfrac{k}{p+2}, \dfrac{n}{p+2}\right)$,
$N^{46}_{i\ell;\,jm;kn} = \dfrac{(p+1)^2}{(p+1-\ell)(p+1-n)}\,\ell_{46}\Big\vert_{\xi^{46}_{i\ell;\,jm;kn}}$,	$\xi^{46}_{i\ell;\,jm;kn} = \left(\dfrac{i}{p+1}, \dfrac{\ell}{p+1}; \dfrac{j}{p+2}, \dfrac{m}{p+2}; \dfrac{k}{p+1}, \dfrac{n}{p+1}\right)$,
$N^{56}_{i\ell;\,jm;kn} = \dfrac{(p+1)^2}{(p+1-m)(p+1-n)}\,\ell_{56}\Big\vert_{\xi^{56}_{i\ell;\,jm;kn}}$.	$\xi^{56}_{i\ell;\,jm;kn} = \left(\dfrac{i}{p+2}, \dfrac{\ell}{p+2}; \dfrac{j}{p+1}, \dfrac{m}{p+1}; \dfrac{k}{p+1}, \dfrac{n}{p+1}\right)$.
$N^{12}_{01;01;11} = \ell_{12}\vert_{\xi^{12}_{01;01;11}}$, $N^{26}_{11;01;10} = \ell_{26}\vert_{\xi^{26}_{11;01;10}}$,	$\xi^{12}_{01;01;11} = (0,1;0,1;\tfrac{1}{2},\tfrac{1}{2})$, $\xi^{26}_{11;01;10} = (\tfrac{1}{2},\tfrac{1}{2};0,1;1,0)$,
$N^{13}_{01;11;01} = \ell_{13}\vert_{\xi^{13}_{01;11;01}}$, $N^{34}_{10;11;01} = \ell_{34}\vert_{\xi^{34}_{10;11;01}}$,	$\xi^{13}_{01;11;01} = (0,1;\tfrac{1}{2},\tfrac{1}{2};0,1)$, $\xi^{34}_{10;11;01} = (1,0;\tfrac{1}{2},\tfrac{1}{2};0,1)$,
$N^{15}_{01;10;11} = \ell_{15}\vert_{\xi^{15}_{01;10;11}}$, $N^{35}_{11;10;01} = \ell_{35}\vert_{\xi^{35}_{11;10;01}}$,	$\xi^{15}_{01;10;11} = (0,1;1,0;\tfrac{1}{2},\tfrac{1}{2})$, $\xi^{35}_{11;10;01} = (\tfrac{1}{2},\tfrac{1}{2};1,0;0,1)$,
$N^{16}_{01;11;10} = \ell_{16}\vert_{\xi^{16}_{01;11;10}}$, $N^{45}_{10;10;11} = \ell_{45}\vert_{\xi^{45}_{10;10;11}}$,	$\xi^{16}_{01;11;10} = (0,1;\tfrac{1}{2},\tfrac{1}{2};1,0)$, $\xi^{45}_{10;10;11} = (1,0;1,0;\tfrac{1}{2},\tfrac{1}{2})$,
$N^{23}_{11;01;01} = \ell_{23}\vert_{\xi^{23}_{11;01;01}}$, $N^{46}_{10;11;10} = \ell_{46}\vert_{\xi^{46}_{10;11;10}}$,	$\xi^{23}_{11;01;01} = (\tfrac{1}{2},\tfrac{1}{2};0,1;0,1)$, $\xi^{46}_{10;11;10} = (1,0;\tfrac{1}{2},\tfrac{1}{2};1,0)$,
$N^{24}_{10;01;11} = \ell_{24}\vert_{\xi^{24}_{10;01;11}}$, $N^{56}_{11;10;10} = \ell_{56}\vert_{\xi^{56}_{11;10;10}}$.	$\xi^{24}_{10;01;11} = (1,0;0,1;\tfrac{1}{2},\tfrac{1}{2})$, $\xi^{56}_{11;10;10} = (\tfrac{1}{2},\tfrac{1}{2};1,0;1,0)$.

The table reports the normalization constants $N^{ab}_{i\ell;\,jm;kn}$ and the interpolation points $\xi^{ab}_{i\ell;\,jm;kn}$ of the brick curl-conforming bases of Table 4.15. The results for the $p=0$ case (zeroth-order bases) are reported in the bottom row of the table.

Table 4.17 Interpolatory divergence-conforming bases of order p for brick cells.

$$\mathbf{\Lambda}^{1}_{i\ell;\,jm;kn}(\mathbf{r}) = N^{1}_{i\ell;\,jm;kn} R_i(p+1,\xi_1)\hat{R}_j(p+2,\xi_2)\hat{R}_k(p+2,\xi_3) \cdot \hat{R}_\ell(p+1,\xi_4)\hat{R}_m(p+2,\xi_5)\hat{R}_n(p+2,\xi_6)\mathbf{\Lambda}_1(\mathbf{r}),$$
$$i=0,1,\ldots,p;\; j,k,\ell,m,n=1,2,\ldots,p+1,\quad i+\ell=p+1;\; j+m=k+n=p+2.$$

$$\mathbf{\Lambda}^{2}_{i\ell;\,jm;kn}(\mathbf{r}) = N^{2}_{i\ell;\,jm;kn} \hat{R}_i(p+2,\xi_1)R_j(p+1,\xi_2)\hat{R}_k(p+2,\xi_3) \cdot \hat{R}_\ell(p+2,\xi_4)\hat{R}_m(p+1,\xi_5)\hat{R}_n(p+2,\xi_6)\mathbf{\Lambda}_2(\mathbf{r}),$$
$$j=0,1,\ldots,p;\; i,k,\ell,m,n=1,2,\ldots,p+1,\quad j+m=p+1;\; i+\ell=k+n=p+2.$$

$$\mathbf{\Lambda}^{3}_{i\ell;\,jm;kn}(\mathbf{r}) = N^{3}_{i\ell;\,jm;kn} \hat{R}_i(p+2,\xi_1)\hat{R}_j(p+2,\xi_2)R_k(p+1,\xi_3) \cdot \hat{R}_\ell(p+2,\xi_4)\hat{R}_m(p+2,\xi_5)\hat{R}_n(p+1,\xi_6)\mathbf{\Lambda}_3(\mathbf{r}),$$
$$k=0,1,\ldots,p;\; i,j,\ell,m,n=1,2,\ldots,p+1,\quad k+n=p+1;\; i+\ell=j+m=p+2.$$

$$\mathbf{\Lambda}^{4}_{i\ell;\,jm;kn}(\mathbf{r}) = N^{4}_{i\ell;\,jm;kn} \hat{R}_i(p+1,\xi_1)\hat{R}_j(p+2,\xi_2)\hat{R}_k(p+2,\xi_3) \cdot R_\ell(p+1,\xi_4)\hat{R}_m(p+2,\xi_5)\hat{R}_n(p+2,\xi_6)\mathbf{\Lambda}_4(\mathbf{r}),$$
$$\ell=0,1,\ldots,p;\; i,j,k,m,n=1,2,\ldots,p+1,\quad i+\ell=p+1;\; j+m=k+n=p+2.$$

$$\mathbf{\Lambda}^{5}_{i\ell;\,jm;kn}(\mathbf{r}) = N^{5}_{i\ell;\,jm;kn} \hat{R}_i(p+2,\xi_1)\hat{R}_j(p+1,\xi_2)\hat{R}_k(p+2,\xi_3) \cdot \hat{R}_\ell(p+2,\xi_4)R_m(p+1,\xi_5)\hat{R}_n(p+2,\xi_6)\mathbf{\Lambda}_5(\mathbf{r}),$$
$$m=0,1,\ldots,p;\; i,j,k,\ell,n=1,2,\ldots,p+1,\quad j+m=p+1;\; i+\ell=k+n=p+2.$$

$$\mathbf{\Lambda}^{6}_{i\ell;\,jm;kn}(\mathbf{r}) = N^{6}_{i\ell;\,jm;kn} \hat{R}_i(p+2,\xi_1)\hat{R}_j(p+2,\xi_2)\hat{R}_k(p+1,\xi_3) \cdot \hat{R}_\ell(p+2,\xi_4)\hat{R}_m(p+2,\xi_5)R_n(p+1,\xi_6)\mathbf{\Lambda}_6(\mathbf{r}),$$
$$n=0,1,\ldots,p;\; i,j,k,\ell,m=1,2,\ldots,p+1,\quad k+n=p+1;\; i+\ell=j+m=p+2.$$

$$N^{1}_{i\ell;\,jm;kn} = \frac{p+1}{p+1-i}\left.\frac{\mathcal{J}}{h_1}\right|_{\boldsymbol{\xi}^{1}_{i\ell;\,jm;kn}},\quad N^{2}_{i\ell;\,jm;kn} = \frac{p+1}{p+1-j}\left.\frac{\mathcal{J}}{h_2}\right|_{\boldsymbol{\xi}^{2}_{i\ell;\,jm;kn}},\quad N^{3}_{i\ell;\,jm;kn} = \frac{p+1}{p+1-k}\left.\frac{\mathcal{J}}{h_3}\right|_{\boldsymbol{\xi}^{3}_{i\ell;\,jm;kn}},$$
$$N^{4}_{i\ell;\,jm;kn} = \frac{p+1}{p+1-\ell}\left.\frac{\mathcal{J}}{h_4}\right|_{\boldsymbol{\xi}^{4}_{i\ell;\,jm;kn}},\quad N^{5}_{i\ell;\,jm;kn} = \frac{p+1}{p+1-m}\left.\frac{\mathcal{J}}{h_5}\right|_{\boldsymbol{\xi}^{5}_{i\ell;\,jm;kn}},\quad N^{6}_{i\ell;\,jm;kn} = \frac{p+1}{p+1-n}\left.\frac{\mathcal{J}}{h_6}\right|_{\boldsymbol{\xi}^{6}_{i\ell;\,jm;kn}}.$$

$$\boldsymbol{\xi}^{1}_{i\ell;\,jm;kn} = \left(\frac{i}{p+1},\frac{\ell}{p+1};\frac{j}{p+2},\frac{m}{p+2};\frac{k}{p+2},\frac{n}{p+2}\right),$$
$$\boldsymbol{\xi}^{2}_{i\ell;\,jm;kn} = \left(\frac{i}{p+2},\frac{\ell}{p+2};\frac{j}{p+1},\frac{m}{p+1};\frac{k}{p+2},\frac{n}{p+2}\right),$$
$$\boldsymbol{\xi}^{3}_{i\ell;\,jm;kn} = \left(\frac{i}{p+2},\frac{\ell}{p+2};\frac{j}{p+2},\frac{m}{p+2};\frac{k}{p+1},\frac{n}{p+1}\right),$$
$$\boldsymbol{\xi}^{4}_{i\ell;\,jm;kn} = \left(\frac{i}{p+1},\frac{\ell}{p+1};\frac{j}{p+2},\frac{m}{p+2};\frac{k}{p+2},\frac{n}{p+2}\right),$$
$$\boldsymbol{\xi}^{5}_{i\ell;\,jm;kn} = \left(\frac{i}{p+2},\frac{\ell}{p+2};\frac{j}{p+1},\frac{m}{p+1};\frac{k}{p+2},\frac{n}{p+2}\right),$$
$$\boldsymbol{\xi}^{6}_{i\ell;\,jm;kn} = \left(\frac{i}{p+2},\frac{\ell}{p+2};\frac{j}{p+2},\frac{m}{p+2};\frac{k}{p+1},\frac{n}{p+1}\right).$$

$$N^{1}_{01;11;11} = \left.\frac{\mathcal{J}}{h_1}\right|_{\boldsymbol{\xi}^{1}_{01;11;11}},\quad N^{4}_{10;11;11} = \left.\frac{\mathcal{J}}{h_4}\right|_{\boldsymbol{\xi}^{4}_{10;11;11}},$$
$$N^{2}_{11;01;11} = \left.\frac{\mathcal{J}}{h_2}\right|_{\boldsymbol{\xi}^{2}_{11;01;11}},\quad N^{5}_{11;10;11} = \left.\frac{\mathcal{J}}{h_5}\right|_{\boldsymbol{\xi}^{5}_{11;10;11}},$$
$$N^{3}_{11;11;01} = \left.\frac{\mathcal{J}}{h_3}\right|_{\boldsymbol{\xi}^{3}_{11;11;01}},\quad N^{6}_{11;11;10} = \left.\frac{\mathcal{J}}{h_6}\right|_{\boldsymbol{\xi}^{6}_{11;11;10}}.$$

$$\boldsymbol{\xi}^{1}_{01;11;11} = (0,1;\tfrac{1}{2},\tfrac{1}{2};\tfrac{1}{2},\tfrac{1}{2}),$$
$$\boldsymbol{\xi}^{2}_{11;01;11} = (\tfrac{1}{2},\tfrac{1}{2};0,1;\tfrac{1}{2},\tfrac{1}{2}),$$
$$\boldsymbol{\xi}^{3}_{11;11;01} = (\tfrac{1}{2},\tfrac{1}{2};\tfrac{1}{2},\tfrac{1}{2};0,1),$$
$$\boldsymbol{\xi}^{4}_{10;11;11} = (1,0;\tfrac{1}{2},\tfrac{1}{2};\tfrac{1}{2},\tfrac{1}{2}),$$
$$\boldsymbol{\xi}^{5}_{11;10;11} = (\tfrac{1}{2},\tfrac{1}{2};1,0;\tfrac{1}{2},\tfrac{1}{2}),$$
$$\boldsymbol{\xi}^{6}_{11;11;10} = (\tfrac{1}{2},\tfrac{1}{2};\tfrac{1}{2},\tfrac{1}{2};1,0).$$

The table reports the divergence-conforming basis functions $\mathbf{\Lambda}^{a}_{i\ell;\,jm;kn}$ of order p on a brick. These functions are obtained by multiplying the un-normalized zeroth-order functions $\mathbf{\Lambda}_a$ of Table 4.5 (right-hand column) with the interpolatory polynomials $\hat{\beta}^a$ described in Section 4.7.2.2, and reported in Table 4.14. The normalization constants $N^{a}_{i\ell;\,jm;kn}$ are chosen to ensure that the component of $\mathbf{\Lambda}^{a}_{i\ell;\,jm;kn}(\mathbf{r})$ along $\hat{\boldsymbol{h}}_a$ at the interpolation point $\boldsymbol{\xi}^{a}_{i\ell;\,jm;kn}$ is unity. The normalization constants for the $p=0$ case (zeroth-order bases) are reported in the bottom right-hand corner of the table. Though both the vector bases and their divergence are complete to order p, it happens that dependencies in the basis set exist for $p \geq 1$. Redundancy is easily eliminated as explained in Section 4.7.2.4. The total number of DoFs for the divergence-conforming basis functions of order p on a brick is $3(p+2)(p+1)^2$.

Table 4.18 Interpolatory polynomials for the triangular prism cell.

Polynomial	Index rule	Interpolation point $\xi_{ijk;\ell m} = (\xi_a, \xi_b, \xi_c; \xi_d, \xi_e)$
$\alpha_{ijk;\ell m}(p, \boldsymbol{\xi}) =$ $R_i(p, \xi_a)\, R_j(p, \xi_b)\, R_k(p, \xi_c)$ $R_\ell(p, \xi_d)\, R_m(p, \xi_e)$	$i, j, k, \ell, m = 0, 1, \ldots, p;$ with: $i + j + k = \ell + m = p$	$\xi_{ijk;\ell m} = \left(\frac{i}{p}, \frac{j}{p}, \frac{k}{p}; \frac{\ell}{p}, \frac{m}{p}\right)$
$\hat{\alpha}^{ad}_{ijk;\ell m}(p, \boldsymbol{\xi}) =$ $R_i(p+2, \xi_a)\, \hat{R}_j(p+2, \xi_b)\, \hat{R}_k(p+2, \xi_c)$ $R_\ell(p+1, \xi_d)\, \hat{R}_m(p+1, \xi_e)$	$i, \ell = 0, 1, \ldots, p;$ $j, k, m = 1, 2, \ldots, p+1;$ with: $i + j + k = p + 2;$ $\ell + m = p + 1$	$\xi^{ad}_{ijk;\ell m} = \left(\frac{i}{p+2}, \frac{j}{p+2}, \frac{k}{p+2}; \frac{\ell}{p+1}, \frac{m}{p+1}\right)$
$\hat{\alpha}^{ab}_{ijk;\ell m}(p, \boldsymbol{\xi}) =$ $R_i(p+1, \xi_a)\, R_j(p+1, \xi_b)\, \hat{R}_k(p+1, \xi_c)$ $\hat{R}_\ell(p+2, \xi_d)\, \hat{R}_m(p+2, \xi_e)$	$i, j = 0, 1, \ldots, p;$ $k, \ell, m = 1, 2, \ldots, p+1;$ with: $i + j + k = p + 1;$ $\ell + m = p + 2$	$\xi^{ab}_{ijk;\ell m} = \left(\frac{i}{p+1}, \frac{j}{p+1}, \frac{k}{p+1}; \frac{\ell}{p+2}, \frac{m}{p+2}\right)$
$\hat{\beta}^{a}_{ijk;\ell m}(p, \boldsymbol{\xi}) =$ $R_i(p+2, \xi_a)\, \hat{R}_j(p+2, \xi_b)\, \hat{R}_k(p+2, \xi_c)$ $\hat{R}_\ell(p+2, \xi_d)\, \hat{R}_m(p+2, \xi_e)$	$i = 0, 1, \ldots, p;$ $j, k, \ell, m = 1, 2, \ldots, p+1;$ with: $i + j + k = \ell + m = p + 2$	$\xi^{a}_{ijk;\ell m} = \left(\frac{i}{p+2}, \frac{j}{p+2}, \frac{k}{p+2}; \frac{\ell}{p+2}, \frac{m}{p+2}\right)$
$\hat{\beta}^{d}_{ijk;\ell m}(p, \boldsymbol{\xi}) =$ $\hat{R}_i(p+3, \xi_a)\, \hat{R}_j(p+3, \xi_b)\, \hat{R}_k(p+3, \xi_c)$ $R_\ell(p+1, \xi_d)\, \hat{R}_m(p+1, \xi_e)$	$\ell = 0, 1, \ldots, p;$ $i, j, k, m = 1, 2, \ldots, p+1;$ with: $i + j + k = p + 3;$ $\ell + m = p + 1$	$\xi^{d}_{ijk;\ell m} = \left(\frac{i}{p+3}, \frac{j}{p+3}, \frac{k}{p+3}; \frac{\ell}{p+1}, \frac{m}{p+1}\right)$

The left-hand column of the table reports pth-order complete polynomial sets formed by $(p+1)^2(p+2)/2$ terms that individuate a regular grid of $(p+1)^2(p+2)/2$ interpolation points. These polynomials attain a unit value at the given interpolation point (reported in the third column) while they are equal to zero at the other interpolation points. The polynomial expressions are reported in terms of "dummy" variables. The "dummy" quadrilateral faces of the prism are the $\xi_a = 0$, $\xi_b = 0$, and $\xi_c = 0$ faces. Hence, in terms of the *local* numbers used to distinguish the coordinates of the parent prism, ξ_a and $\xi_b (\neq \xi_a)$ can be equal to ξ_1, ξ_2, or ξ_3. The "dummy" triangular faces of the prism are the $\xi_d = 0$, and $\xi_e = 0$ faces. In terms of *local* numbers, ξ_d and $\xi_e (\neq \xi_d)$ can then be equal to ξ_4 or ξ_5.

- The shape polynomials $\alpha_{ijk;\ell m}(p, \boldsymbol{\xi})$ are used for *scalar* interpolation or for geometry description; these polynomials interpolate all the edges and faces of the prism.
- The polynomials $\hat{\alpha}^{ad}_{ijk;\ell m}(p, \boldsymbol{\xi})$ are used to build the curl-conforming *vector* bases; these polynomials have $(p+1)$ edge-interpolation points only on the $\xi_a = \xi_d = 0$ edge, $p(p+1)$ face-interior interpolation points on the $\xi_a = 0$ quadrilateral face, $p(p+1)/2$ face-interior interpolation points on the $\xi_d = 0$ triangular face, and $p^2(p+1)/2$ prism-interior interpolation points.
- The polynomials $\hat{\alpha}^{ab}_{ijk;\ell m}(p, \boldsymbol{\xi})$ are used to build the curl-conforming *vector* bases; these polynomials have $(p+1)$ edge-interpolation points only on the $\xi_a = \xi_b = 0$ edge, $p(p+1)$ face-interior interpolation points on the $\xi_a = 0$ and on the $\xi_b = 0$ quadrilateral face, and $p(p-1)(p+1)/2$ prism-interior interpolation points.
- The polynomials $\hat{\beta}^{a}_{ijk;\ell m}(p, \boldsymbol{\xi})$ are used to build the divergence-conforming *vector* bases; these polynomials have $(p+1)^2$ face-interior interpolation points only on the quadrilateral $\xi_a = 0$ face, and $p(p+1)^2/2$ prism-interior interpolation point.
- The polynomials $\hat{\beta}^{d}_{ijk;\ell m}(p, \boldsymbol{\xi})$ are used to build the divergence-conforming *vector* bases; these polynomials have $(p+1)(p+2)/2$ face-interior interpolation points only on the triangular $\xi_d = 0$ face, and $p(p+1)(p+2)/2$ prism-interior interpolation point.

Table 4.19 Interpolatory curl-conforming bases of order p for prisms.

$$\mathbf{\Omega}^{14}_{ijk;\ell m}(\mathbf{r}) = N^{14}_{ijk;\ell m} R_i(p+2,\xi_1)\hat{R}_j(p+2,\xi_2)\hat{R}_k(p+2,\xi_3)R_\ell(p+1,\xi_4)\hat{R}_m(p+1,\xi_5)\mathbf{\Omega}_{14}(\mathbf{r}),$$
$$i,\ell=0,1,\ldots,p;\; j,k,m=1,2,\ldots,p+1;\text{ with } i+j+k=p+2;\; \ell+m=p+1,$$

$$\mathbf{\Omega}^{15}_{ijk;\ell m}(\mathbf{r}) = N^{15}_{ijk;\ell m} R_i(p+2,\xi_1)\hat{R}_j(p+2,\xi_2)\hat{R}_k(p+2,\xi_3)\hat{R}_\ell(p+1,\xi_4)R_m(p+1,\xi_5)\mathbf{\Omega}_{15}(\mathbf{r}),$$
$$i,m=0,1,\ldots,p;\; j,k,\ell=1,2,\ldots,p+1;\text{ with } i+j+k=p+2;\; \ell+m=p+1,$$

$$\mathbf{\Omega}^{24}_{ijk;\ell m}(\mathbf{r}) = N^{24}_{ijk;\ell m} \hat{R}_i(p+2,\xi_1)R_j(p+2,\xi_2)\hat{R}_k(p+2,\xi_3)R_\ell(p+1,\xi_4)\hat{R}_m(p+1,\xi_5)\mathbf{\Omega}_{24}(\mathbf{r}),$$
$$j,\ell=0,1,\ldots,p;\; i,k,m=1,2,\ldots,p+1;\text{ with } i+j+k=p+2;\; \ell+m=p+1,$$

$$\mathbf{\Omega}^{25}_{ijk;\ell m}(\mathbf{r}) = N^{25}_{ijk;\ell m} \hat{R}_i(p+2,\xi_1)R_j(p+2,\xi_2)\hat{R}_k(p+2,\xi_3)\hat{R}_\ell(p+1,\xi_4)R_m(p+1,\xi_5)\mathbf{\Omega}_{25}(\mathbf{r}),$$
$$j,m=0,1,\ldots,p;\; i,k,\ell=1,2,\ldots,p+1;\text{ with } i+j+k=p+2;\; \ell+m=p+1,$$

$$\mathbf{\Omega}^{34}_{ijk;\ell m}(\mathbf{r}) = N^{34}_{ijk;\ell m} \hat{R}_i(p+2,\xi_1)\hat{R}_j(p+2,\xi_2)R_k(p+2,\xi_3)R_\ell(p+1,\xi_4)\hat{R}_m(p+1,\xi_5)\mathbf{\Omega}_{34}(\mathbf{r}),$$
$$k,\ell=0,1,\ldots,p;\; i,j,m=1,2,\ldots,p+1;\text{ with } i+j+k=p+2;\; \ell+m=p+1,$$

$$\mathbf{\Omega}^{35}_{ijk;\ell m}(\mathbf{r}) = N^{35}_{ijk;\ell m} \hat{R}_i(p+2,\xi_1)\hat{R}_j(p+2,\xi_2)R_k(p+2,\xi_3)\hat{R}_\ell(p+1,\xi_4)R_m(p+1,\xi_5)\mathbf{\Omega}_{35}(\mathbf{r}),$$
$$k,m=0,1,\ldots,p;\; i,j,\ell=1,2,\ldots,p+1;\text{ with } i+j+k=p+2;\; \ell+m=p+1,$$

$$\mathbf{\Omega}^{13}_{ijk;\ell m}(\mathbf{r}) = N^{13}_{ijk;\ell m} R_i(p+1,\xi_1)\hat{R}_j(p+1,\xi_2)R_k(p+1,\xi_3)\hat{R}_\ell(p+2,\xi_4)\hat{R}_m(p+2,\xi_5)\mathbf{\Omega}_{13}(\mathbf{r}),$$
$$i,k=0,1,\ldots,p;\; j,\ell,m=1,2,\ldots,p+1;\text{ with } i+j+k=p+1;\; \ell+m=p+2,$$

$$\mathbf{\Omega}^{12}_{ijk;\ell m}(\mathbf{r}) = N^{12}_{ijk;\ell m} R_i(p+1,\xi_1)R_j(p+1,\xi_2)\hat{R}_k(p+1,\xi_3)\hat{R}_\ell(p+2,\xi_4)\hat{R}_m(p+2,\xi_5)\mathbf{\Omega}_{12}(\mathbf{r}),$$
$$i,j=0,1,\ldots,p;\; k,\ell,m=1,2,\ldots,p+1;\text{ with } i+j+k=p+1;\; \ell+m=p+2,$$

$$\mathbf{\Omega}^{23}_{ijk;\ell m}(\mathbf{r}) = N^{23}_{ijk;\ell m} \hat{R}_i(p+1,\xi_1)R_j(p+1,\xi_2)R_k(p+1,\xi_3)\hat{R}_\ell(p+2,\xi_4)\hat{R}_m(p+2,\xi_5)\mathbf{\Omega}_{23}(\mathbf{r}),$$
$$j,k=0,1,\ldots,p;\; i,\ell,m=1,2,\ldots,p+1;\text{ with } i+j+k=p+1;\; \ell+m=p+2.$$

The table reports the curl-conforming basis functions of order p on a prism. The quadrilateral faces of the prism are the $\xi_1=0$, the $\xi_2=0$, and the $\xi_3=0$ faces, while the $\xi_4=0$ and the $\xi_5=0$ faces of the prism are triangular. By using the dummy notation of Section 4.7.3.1, the curl-conforming bases for the prism are the $\mathbf{\Omega}^{ad}_{ijk;\ell m}$ and $\mathbf{\Omega}^{ab}_{ijk;\ell m}$ bases, where it is understood that $\xi_a=0$ and $\xi_b=0$ indicate two quadrilateral faces of the prism with a common edge (the $\xi_a=\xi_b=0$ edge), while $\xi_d=0$ is a triangular face with an edge in common to the $\xi_a=0$ quadrilateral face.

The functions $\mathbf{\Omega}^{ad}_{ijk;\ell m}$ (and $\mathbf{\Omega}^{ab}_{ijk;\ell m}$) are obtained by multiplying the un-normalized zeroth-order functions $\mathbf{\Omega}_{ad}$ ($\mathbf{\Omega}_{ab}$) of Table 4.4 (central column) with the interpolatory polynomials $\hat{\alpha}^{ad}$ ($\hat{\alpha}^{ab}$) described in Section 4.7.3.1, and reported in Table 4.18. The normalization constant $N^{ab}_{i\ell;\,jm;kn}$ (and $N^{ad}_{i\ell;\,jm;kn}$) is chosen such that the components of the bases along $\boldsymbol{\ell}_{ab}$ ($\boldsymbol{\ell}_{ad}$) at the interpolation point $\boldsymbol{\xi}^{ab}_{i\ell;\,jm;kn}$ ($\boldsymbol{\xi}^{ad}_{i\ell;\,jm;kn}$) are unity. The normalization constants and the interpolation points are given in Table 4.20. Though both the vector bases and their curl are complete to order p, it happens that dependencies in the basis set exist for $p \geq 1$. Redundancy is easily eliminated as explained in Section 4.7.3.3. The total number of DoFs for the curl-conforming basis functions of order p on a prism is $3(p+1)(p+2)(p+3)/2$.

Table 4.20 Normalization constants and interpolatory points for prism curl-conforming bases.

$N^{14}_{ijk;\ell m} = \dfrac{(p+2)(p+1)}{(p+2-i)(p+1-\ell)}\, \ell_{14}\Big\|_{\xi^{14}_{ijk;\ell m}},$	$\xi^{14}_{ijk;\ell m} = \left(\dfrac{i}{p+2}, \dfrac{j}{p+2}, \dfrac{k}{p+2}; \dfrac{\ell}{p+1}, \dfrac{m}{p+1}\right),$
$N^{15}_{ijk;\ell m} = \dfrac{(p+2)(p+1)}{(p+2-i)(p+1-m)}\, \ell_{15}\Big\|_{\xi^{15}_{ijk;\ell m}},$	$\xi^{15}_{ijk;\ell m} = \left(\dfrac{i}{p+2}, \dfrac{j}{p+2}, \dfrac{k}{p+2}; \dfrac{\ell}{p+1}, \dfrac{m}{p+1}\right),$
$N^{24}_{ijk;\ell m} = \dfrac{(p+2)(p+1)}{(p+2-j)(p+1-\ell)}\, \ell_{24}\Big\|_{\xi^{24}_{ijk;\ell m}},$	$\xi^{24}_{ijk;\ell m} = \left(\dfrac{i}{p+2}, \dfrac{j}{p+2}, \dfrac{k}{p+2}; \dfrac{\ell}{p+1}, \dfrac{m}{p+1}\right),$
$N^{25}_{ijk;\ell m} = \dfrac{(p+2)(p+1)}{(p+2-j)(p+1-m)}\, \ell_{25}\Big\|_{\xi^{25}_{ijk;\ell m}},$	$\xi^{25}_{ijk;\ell m} = \left(\dfrac{i}{p+2}, \dfrac{j}{p+2}, \dfrac{k}{p+2}; \dfrac{\ell}{p+1}, \dfrac{m}{p+1}\right),$
$N^{34}_{ijk;\ell m} = \dfrac{(p+2)(p+1)}{(p+2-k)(p+1-\ell)}\, \ell_{34}\Big\|_{\xi^{34}_{ijk;\ell m}},$	$\xi^{34}_{ijk;\ell m} = \left(\dfrac{i}{p+2}, \dfrac{j}{p+2}, \dfrac{k}{p+2}; \dfrac{\ell}{p+1}, \dfrac{m}{p+1}\right),$
$N^{35}_{ijk;\ell m} = \dfrac{(p+2)(p+1)}{(p+2-k)(p+1-m)}\, \ell_{35}\Big\|_{\xi^{35}_{ijk;\ell m}},$	$\xi^{35}_{ijk;\ell m} = \left(\dfrac{i}{p+2}, \dfrac{j}{p+2}, \dfrac{k}{p+2}; \dfrac{\ell}{p+1}, \dfrac{m}{p+1}\right),$
$N^{13}_{ijk;\ell m} = \dfrac{p+1}{p+1-i-k}\, \ell_{13}\Big\|_{\xi^{13}_{ijk;\ell m}},$	$\xi^{13}_{ijk;\ell m} = \left(\dfrac{i}{p+1}, \dfrac{j}{p+1}, \dfrac{k}{p+1}; \dfrac{\ell}{p+2}, \dfrac{m}{p+2}\right),$
$N^{12}_{ijk;\ell m} = \dfrac{p+1}{p+1-i-j}\, \ell_{12}\Big\|_{\xi^{12}_{ijk;\ell m}},$	$\xi^{12}_{ijk;\ell m} = \left(\dfrac{i}{p+1}, \dfrac{j}{p+1}, \dfrac{k}{p+1}; \dfrac{\ell}{p+2}, \dfrac{m}{p+2}\right),$
$N^{23}_{ijk;\ell m} = \dfrac{p+1}{p+1-j-k}\, \ell_{23}\Big\|_{\xi^{23}_{ijk;\ell m}}.$	$\xi^{23}_{ijk;\ell m} = \left(\dfrac{i}{p+1}, \dfrac{j}{p+1}, \dfrac{k}{p+1}; \dfrac{\ell}{p+2}, \dfrac{m}{p+2}\right).$
$N^{14}_{011;01} = \ell_{14}\big\|_{\xi^{14}_{011;01}},$	$\xi^{14}_{011;01} = \left(0, \tfrac{1}{2}, \tfrac{1}{2}; 0, 1\right),$
$N^{15}_{011;10} = \ell_{15}\big\|_{\xi^{15}_{011;10}},$	$\xi^{15}_{011;10} = \left(0, \tfrac{1}{2}, \tfrac{1}{2}; 1, 0\right),$
$N^{24}_{101;01} = \ell_{24}\big\|_{\xi^{24}_{101;01}},$	$\xi^{24}_{101;01} = \left(\tfrac{1}{2}, 0, \tfrac{1}{2}; 0, 1\right),$
$N^{25}_{101;10} = \ell_{25}\big\|_{\xi^{25}_{101;10}},$	$\xi^{25}_{101;10} = \left(\tfrac{1}{2}, 0, \tfrac{1}{2}; 1, 0\right),$
$N^{34}_{110;01} = \ell_{34}\big\|_{\xi^{34}_{110;01}},$	$\xi^{34}_{110;01} = \left(\tfrac{1}{2}, \tfrac{1}{2}, 0; 0, 1\right),$
$N^{35}_{110;10} = \ell_{35}\big\|_{\xi^{35}_{110;10}},$	$\xi^{35}_{110;10} = \left(\tfrac{1}{2}, \tfrac{1}{2}, 0; 1, 0\right),$
$N^{13}_{010;11} = \ell_{13}\big\|_{\xi^{13}_{010;11}},$	$\xi^{13}_{010;11} = \left(0, 1, 0; \tfrac{1}{2}, \tfrac{1}{2}\right),$
$N^{12}_{001;11} = \ell_{12}\big\|_{\xi^{12}_{001;11}},$	$\xi^{12}_{001;11} = \left(0, 0, 1; \tfrac{1}{2}, \tfrac{1}{2}\right),$
$N^{23}_{100;11} = \ell_{23}\big\|_{\xi^{23}_{100;11}}.$	$\xi^{23}_{100;11} = \left(1, 0, 0; \tfrac{1}{2}, \tfrac{1}{2}\right).$

The table reports the normalization constants $N^{ab}_{i\ell;\, jm;kn}$, $N^{ad}_{i\ell;\, jm;kn}$, and the interpolation points $\xi^{ab}_{i\ell;\, jm;kn}$, $\xi^{ad}_{i\ell;\, jm;kn}$ of the prism curl-conforming bases of Table 4.19. The results for the $p=0$ case (zeroth-order bases) are reported in the bottom row of the table.

Table 4.21 Interpolatory divergence-conforming bases of order p for prism cells.

$$\begin{aligned}
\mathbf{\Lambda}^1_{ijk;\ell m}(\boldsymbol{r}) &= N^1_{ijk;\ell m} R_i(p+2,\xi_1)\hat{R}_j(p+2,\xi_2)\hat{R}_k(p+2,\xi_3)\hat{R}_\ell(p+2,\xi_4)\hat{R}_m(p+2,\xi_5)\mathbf{\Lambda}_1(\boldsymbol{r}),\\
&\qquad i=0,1,\ldots,p;\ j,k,\ell,m=1,2,\ldots,p+1,\ \text{with}: i+j+k=p+2;\ \ell+m=p+2,
\end{aligned}$$

$$\begin{aligned}
\mathbf{\Lambda}^2_{ijk;\ell m}(\boldsymbol{r}) &= N^2_{ijk;\ell m} \hat{R}_i(p+2,\xi_1)R_j(p+2,\xi_2)\hat{R}_k(p+2,\xi_3)\hat{R}_\ell(p+2,\xi_4)\hat{R}_m(p+2,\xi_5)\mathbf{\Lambda}_2(\boldsymbol{r}),\\
&\qquad j=0,1,\ldots,p;\ i,k,\ell,m=1,2,\ldots,p+1,\ \text{with}: i+j+k=p+2;\ \ell+m=p+2,
\end{aligned}$$

$$\begin{aligned}
\mathbf{\Lambda}^3_{ijk;\ell m}(\boldsymbol{r}) &= N^3_{ijk;\ell m} \hat{R}_i(p+2,\xi_1)\hat{R}_j(p+2,\xi_2)R_k(p+2,\xi_3)\hat{R}_\ell(p+2,\xi_4)\hat{R}_m(p+2,\xi_5)\mathbf{\Lambda}_3(\boldsymbol{r}),\\
&\qquad k=0,1,\ldots,p;\ i,j,\ell,m=1,2,\ldots,p+1,\ \text{with}: i+j+k=p+2;\ \ell+m=p+2,
\end{aligned}$$

$$\begin{aligned}
\mathbf{\Lambda}^4_{ijk;\ell m}(\boldsymbol{r}) &= N^4_{ijk;\ell m} \hat{R}_i(p+3,\xi_1)\hat{R}_j(p+3,\xi_2)\hat{R}_k(p+3,\xi_3)R_\ell(p+1,\xi_4)\hat{R}_m(p+1,\xi_5)\mathbf{\Lambda}_4(\boldsymbol{r}),\\
&\qquad \ell=0,1,\ldots,p;\ i,j,k,m=1,2,\ldots,p+1,\ \text{with}: i+j+k=p+3;\ \ell+m=p+1,
\end{aligned}$$

$$\begin{aligned}
\mathbf{\Lambda}^5_{ijk;\ell m}(\boldsymbol{r}) &= N^5_{ijk;\ell m} \hat{R}_i(p+3,\xi_1)\hat{R}_j(p+3,\xi_2)\hat{R}_k(p+3,\xi_3)\hat{R}_\ell(p+1,\xi_4)R_m(p+1,\xi_5)\mathbf{\Lambda}_5(\boldsymbol{r}).\\
&\qquad m=0,1,\ldots,p;\ i,j,k,\ell=1,2,\ldots,p+1,\ \text{with}: i+j+k=p+3;\ \ell+m=p+1.
\end{aligned}$$

$$\begin{aligned}
N^1_{ijk;\ell m} &= \frac{p+2}{p+2-i}\,\frac{\mathcal{J}}{h_1}\bigg|_{\boldsymbol{\xi}^1_{ijk;\ell m}}, & \boldsymbol{\xi}^1_{ijk;\ell m} &= \left(\tfrac{i}{p+2},\tfrac{j}{p+2},\tfrac{k}{p+2};\tfrac{\ell}{p+2},\tfrac{m}{p+2}\right),\\
N^2_{ijk;\ell m} &= \frac{p+2}{p+2-j}\,\frac{\mathcal{J}}{h_2}\bigg|_{\boldsymbol{\xi}^2_{ijk;\ell m}}, & \boldsymbol{\xi}^2_{ijk;\ell m} &= \left(\tfrac{i}{p+2},\tfrac{j}{p+2},\tfrac{k}{p+2};\tfrac{\ell}{p+2},\tfrac{m}{p+2}\right),\\
N^3_{ijk;\ell m} &= \frac{p+2}{p+2-k}\,\frac{\mathcal{J}}{h_3}\bigg|_{\boldsymbol{\xi}^3_{ijk;\ell m}}, & \boldsymbol{\xi}^3_{ijk;\ell m} &= \left(\tfrac{i}{p+2},\tfrac{j}{p+2},\tfrac{k}{p+2};\tfrac{\ell}{p+2},\tfrac{m}{p+2}\right),\\
N^4_{ijk;\ell m} &= \frac{p+1}{p+1-\ell}\,\frac{\mathcal{J}}{h_4}\bigg|_{\boldsymbol{\xi}^4_{ijk;\ell m}}, & \boldsymbol{\xi}^4_{ijk;\ell m} &= \left(\tfrac{i}{p+3},\tfrac{j}{p+3},\tfrac{k}{p+3};\tfrac{\ell}{p+1},\tfrac{m}{p+1}\right),\\
N^5_{ijk;\ell m} &= \frac{p+1}{p+1-m}\,\frac{\mathcal{J}}{h_5}\bigg|_{\boldsymbol{\xi}^5_{ijk;\ell m}}, & \boldsymbol{\xi}^5_{ijk;\ell m} &= \left(\tfrac{i}{p+3},\tfrac{j}{p+3},\tfrac{k}{p+3};\tfrac{\ell}{p+1},\tfrac{m}{p+1}\right).
\end{aligned}$$

$$\begin{aligned}
N^1_{011;11} &= \frac{\mathcal{J}}{h_1}\bigg|_{\boldsymbol{\xi}^1_{011;11}}, & \boldsymbol{\xi}^1_{011;11} &= \left(0,\tfrac{1}{2},\tfrac{1}{2};\tfrac{1}{2},\tfrac{1}{2}\right),\\
N^2_{101;11} &= \frac{\mathcal{J}}{h_2}\bigg|_{\boldsymbol{\xi}^2_{101;11}}, & \boldsymbol{\xi}^2_{101;11} &= \left(\tfrac{1}{2},0,\tfrac{1}{2};\tfrac{1}{2},\tfrac{1}{2}\right),\\
N^3_{110;11} &= \frac{\mathcal{J}}{h_3}\bigg|_{\boldsymbol{\xi}^3_{110;11}}, & \boldsymbol{\xi}^3_{110;11} &= \left(\tfrac{1}{2},\tfrac{1}{2},0;\tfrac{1}{2},\tfrac{1}{2}\right),\\
N^4_{111;01} &= \frac{\mathcal{J}}{h_4}\bigg|_{\boldsymbol{\xi}^4_{111;01}}, & \boldsymbol{\xi}^4_{111;01} &= \left(\tfrac{1}{3},\tfrac{1}{3},\tfrac{1}{3};0,1\right),\\
N^5_{111;10} &= \frac{\mathcal{J}}{h_5}\bigg|_{\boldsymbol{\xi}^5_{111;10}}, & \boldsymbol{\xi}^5_{111;10} &= \left(\tfrac{1}{3},\tfrac{1}{3},\tfrac{1}{3};1,0\right).
\end{aligned}$$

The table reports the divergence-conforming basis functions $\mathbf{\Lambda}^a_{ijk;\ell m}(\boldsymbol{r})$ and $\mathbf{\Lambda}^d_{ijk;\ell m}(\boldsymbol{r})$ of order p on a prism. These functions are obtained by multiplying the interpolatory polynomials $\hat{\beta}^a$ and $\hat{\beta}^b$ (described in Section 4.7.3.2 and reported in Table 4.18) with the un-normalized zeroth-order functions $\mathbf{\Lambda}_a$ and $\mathbf{\Lambda}_d$ (given in the central column of Table 4.5), respectively. The normalization constants $N^a_{ijk;\ell m}$ (or $N^d_{ijk;\ell m}$) are chosen to ensure that the component of $\mathbf{\Lambda}^a_{ijk;\ell m}(\boldsymbol{r})$ (or $\mathbf{\Lambda}^d_{ijk;\ell m}(\boldsymbol{r})$) along $\hat{\boldsymbol{h}}_a$ ($\hat{\boldsymbol{h}}_d$) at the given interpolation point $\boldsymbol{\xi}^a_{ijk;\ell m}$ ($\boldsymbol{\xi}^d_{ijk;\ell m}$) is unity. The normalization constants for the $p=0$ case (zeroth-order bases) are reported in the bottom row of the table. Though both the vector bases and their divergence are complete to order p, it happens that dependencies in the basis set exist for $p \geq 1$. Redundancy is easily eliminated as explained in Section 4.7.3.4. The total number of DoFs for the divergence-conforming basis functions of order p on a prism is $(p+1)(3p^2+12p+10)/2$.

References

[1] R. D. Graglia, D. R. Wilton, and A. F. Peterson, "Higher order interpolatory vector bases for computational electromagnetics," special issue on "Advanced Numerical Techniques in Electromagnetics," *IEEE Trans. Antennas Propag.*, vol. 45, no. 3, pp. 329–342, Mar. 1997.

[2] R. D. Graglia, D. R. Wilton, A. F. Peterson, and I.-L. Gheorma, "Higher order interpolatory vector bases on prism elements," *IEEE Trans. Antennas Propag.*, vol. 46, no. 3, pp. 442–450, Mar. 1998.

[3] P. A. Raviart and J. M. Thomas, "A mixed finite element method for 2nd order elliptic problems," in *Mathematical Aspects of Finite Element Methods*, A. Dold and B. Eckmann, eds., New York, NY: Springer-Verlag, pp. 292–315, 1977.

[4] J. C. Nédélec, "Mixed finite elements in R^3," *Numer. Math.*, vol. 35, pp. 315–341, 1980.

[5] A. W. Glisson and D. R. Wilton, "Simple and efficient numerical methods for problems of electromagnetic radiation and scattering from surfaces," *IEEE Trans. Antennas Propag.*, AP-28, no. 5, pp. 593–603, Sept. 1980.

[6] A. Bossavit and J.-C. Vérité, "A mixed FEM-BIEM method to solve 3-D eddy current problems," *IEEE Trans. Magn.*, MAG-18, no. 2, pp. 431–435, Mar. 1982.

[7] S. M. Rao, D. R. Wilton, and A. W. Glisson, "Electromagnetic scattering by surfaces of arbitrary shape," *IEEE Trans. Antennas Propag.*, AP-30, no. 3, pp. 409–418, May 1982.

[8] D. H. Schaubert, D. R. Wilton, and A. W. Glisson, "A tetrahedral modeling method for electromagnetic scattering by arbitrarily shaped inhomogeneous dielectric bodies," *IEEE Trans. Antennas Propag.*, AP-32, no. 1, pp. 77–85, Jan. 1984.

[9] M. Hano, "Finite-element analysis of dielectric-loaded waveguides," *IEEE Trans. Microwave Theory Tech.*, MTT-32, no. 10, pp. 1275–1279, Oct. 1984.

[10] J. S. van Welij, "Calculation of eddy currents in terms of H on hexahedra," *IEEE Trans. Magn.*, MAG-21, no. 6, pp. 2239–2241, Nov. 1985.

[11] J. C. Nédélec, "A new family of mixed finite elements in R3," *Numer. Math.*, vol. 50, pp. 57–81, 1986.

[12] M. L. Barton and Z. J. Cendes, "New vector elements for three-dimensional magnetic field computation," *J. Appl. Phys.*, vol. 61, no. 8, pp. 3919–3921, Apr. 1987.

[13] Z. J. Cendes, "Overview of CAE/CAD/AI electromagnetic field computation," in *Proceedings of the Second IEEE Conference on Electromagnetic Field Computation*, A. Konrad, ed., Schenectady, NY, 1987.

[14] C. W. Crowley, P. P. Silvester, and H. Hurwitz, "Covariant projection elements for 3D vector field problems," *IEEE Trans. Magn.*, MAG-24, no. 1, pp. 397–400, Jan. 1988.

[15] A. Bossavit, "Mixed finite elements and the complex of Whitney forms," in *The Mathematics of Finite Elements and Applications VI*, J. R. Whiteman, ed., pp. 137–144, London: Academic Press, 1988.

[16] R. D. Graglia, "The use of parametric elements in the moment method solution of static and dynamic volume integral equations", *IEEE Trans. Antennas Propag.*, vol. AP-36, pp. 636–646, May 1988.

[17] A. Bossavit, "Whitney forms: a class of finite elements for three-dimensional computations in electromagnetics," *IEE Proc.*, vol. 135, pt. A, no. 8, pp. 493–500, 1988.

[18] A. Bossavit and I. Mayergoyz, "Edge-elements for scattering problems," *IEEE Trans. Magn.*, MAG-25, no. 4, pp. 2816–2821, July 1989.

[19] R. D. Graglia, P. L. E. Uslenghi, and R. S. Zich, "Moment method with isoparametric elements for three-dimensional anisotropic scatterers", (invited paper), *"Proceedings of the IEEE"* – Special Issue on *"Radar Cross Sections of Complex Objects"*, vol. 77, no. 5, pp. 750–760, May 1989. Also available in the *IEEE* book, *"Radar Cross Sections of Complex Objects"*, pp. 206–216, 1990.

[20] Z.J. Cendes, "Vector finite elements for electromagnetic field computation," *IEEE Trans. Magn.*, MAG-27, no. 5, pp. 3958–3966, Sept. 1991.

[21] J.-F. Lee, D.-K. Sun, and Z. J. Cendes, "Full-wave analysis of dielectric waveguides using tangential vector finite elements," *IEEE Trans. Microwave Theory Tech.*, MTT-39, no. 8, pp. 1262–1271, Aug. 1991.

[22] J. P. Webb, "Edge elements and what they can do for you," *IEEE Trans. Magn.*, MAG-29, no. 2, pp. 1460–1465, Mar. 1993.

[23] W. Schroeder and I. Wolff, "The origin of spurious modes in numerical solutions of electromagnetic field eigenvalue problems," *IEEE Trans. Microwave Theory Tech.*, MTT-42, no. 4, pp. 644–653, Apr. 1994.

[24] D. Sun, J. Manges, X. Yuan, and Z. Cendes, "Spurious modes in finite element methods," *IEEE Antennas Propag. Mag.*, vol. 37, no. 5, pp. 12–24, Oct. 1995.

[25] P. P. Silvester and R. L. Ferrari, *Finite Elements for Electrical Engineers*, 3rd ed., Cambridge: Cambridge Press, 1996.

[26] D. R. Wilton and W. J. Brown, "Higher order modeling using curvilinear elements and singular bases," 1996 Radio Science Meeting, Boulder, CO, Jan, 1996.

[27] J. S. Savage and A. F. Peterson, "Extension of higher-order 3-D vector finite elements to curved cells and open-region problems," *Proceedings of the 12th Annual Review of Progress in Applied Computational Electromagnetics*, Monterey, CA, pp. 988–994, Mar. 1996.

[28] A. F. Peterson and D. R. Wilton, "Curl-conforming mixed-order edge elements for discretizing the 2D and 3D vector Helmholtz equation," in *Finite Element Software for Microwave Engineering*, T. Itoh, P. Silvester, and G. Pelosi, eds., New York, NY: Wiley, pp. 101–126, 1996.

[29] T. Özdemir and J. L. Volakis, "Triangular prisms for edge-based vector finite element analysis of conformal antennas," *IEEE Trans. Antennas Propag.*, AP-45, no. 5, pp. 788–797, May 1997.

[30] I.-L. Gheorma and R. D. Graglia, "Higher order vectorial modeling using curvilinear prism elements," Proceedings of the fifth *International Conference on Electromagnetics in Aerospace Applications (ICEAA)*, pp. 179–182, Torino, Italy, Sept. 15–18, 1997.

[31] L. E. Garcia-Castillo and M. Salazar-Palma, "Second order Nédélec tetrahedral element for computational electromagnetics," *Int. J. Numer. Modell.*, vol. 13, pp. 261–287, 2000.

[32] L. E. Garcia-Castillo, A. J. Ruiz-Gonoves, M. Salazar-Palma, and T. K. Sarkar, "Third order Nédélec curl-conforming finite element," *IEEE Trans. Magn.*, vol. 38, no. 5, pp. 2370–2372, Sept. 2002.

CHAPTER 5

Hierarchical Bases

High-order bases can be categorized as interpolatory or hierarchical. Interpolatory bases interpolate the value of a field quantity at a number of interpolation points such that only one basis function is non-zero at a given interpolation point. As seen in Chapter 4, this property allows a direct physical interpretation of the expansion coefficients, but does not facilitate mixing different orders throughout the computational domain. Conversely a base of order n is said to be hierarchical if it includes, as a subset, all the basis functions forming the $(n-1)$th-order base. This property prevents one from associating the coefficient of a hierarchical basis function with the solution at a given point; in other words, a hierarchical base is by definition *nodeless*. However, being nodeless, hierarchical bases allow for a selective expansion using different orders in different regions of the computational domain. For example, lowest order hierarchical bases can be employed in sub-domains where the to-be-approximated quantity is expected to vary slowly whereas higher order bases can be employed in sub-domains where a rapid variation of the quantity is anticipated. In the numerical solution of a field problem, a selective choice of the order to be used on each sub-domain can lead to a memory and CPU-time reduction as well as improved accuracy. It goes without saying that hierarchical bases are not intended for interpolation problems, where the values of the field quantities at certain points have to be re-constructed out of the knowledge of the quantities on a certain number of isolated points. Hierarchical bases are of interest in numerical problems where the field quantities are completely unknown throughout the entire computational domain.

In this connection, we ought to observe that the full potential of numerical analysis is expected to be realized through the use of *adaptive* refinement (possibly a mixture of h-refinement and p-refinement) to produce a desired accuracy with minimal computational costs [1–3]. Adaptive h-refinement involves adjusting cell sizes, while adaptive p-refinement varies the polynomial order, in an attempt to produce an optimal non-uniform representation throughout the computational domain. Note that to implement any h-refinement procedure,[1] it is necessary to have full control over the mesh-generating process, while this is not required by a p-refinement process. Adaptation also requires an error estimator to drive the process without user intervention. Adaptive p-refinement methods are best built around hierarchical bases since, unlike interpolatory bases, they facilitate mixing different orders throughout the domain.

---

[1] The commercially available mesh-generators typically do not permit *normal* users full control of the mesh-generating process.

This chapter describes both scalar and vector *hierarchical* bases that can be used together in the same mesh. Both types of basis functions may find use, for example, when dealing with inhomogeneous waveguiding structures where one has to model both the transverse vector field and the longitudinal field component (the latter being a scalar quantity), or for magneto-quasistatic problems, for which a vector is needed in the conducting regions, but a scalar is sufficient elsewhere. The hierarchical vector bases discussed here are either tangentially or normally continuous (i.e. curl or divergence conforming) to facilitate the imposition of the boundary conditions on interfaces of abrupt material discontinuity.

5.1 The Ill-Conditioning Issue

The principal drawback arising from using hierarchical or, in general, *nodeless* bases is the fact that they typically produce system matrices with a worse condition number than those obtained from interpolatory polynomial bases of the same order. In fact, for polynomial orders lower than the sixth, the interpolatory basis functions have a higher degree of built-in linear independence (or orthogonality) since, as seen in the previous chapter

1. Each interpolatory basis function is properly *normalized* to a given value at one node of the interpolatory grid while it vanishes at all the other nodes of the grid.
2. There is only one non-zero basis function at each node of the interpolatory grid.
3. The interpolatory grid of each subdomain is *regular*, with well-separated nodes located only inside or on the border of the subdomain.

The degree of orthogonality (or of linear independence) of a *scalar* basis function set is typically measured by the condition number (CN) of the local Gram matrix **G** with entries

$$g_{mn} = \int_{\mathcal{D}} B_m B_n \, \mathrm{d}\mathcal{D} \tag{5.1}$$

where B_n is a basis function of the scalar set, and with integrals computed over the domain $\mathcal{D}$ on which the basis functions are defined.[2,3] In the vector case, the entries of the local Gram matrix **G** used to measure the degree of orthogonality of a *vector* basis function set are

$$g_{mn} = \int_{\mathcal{D}} \boldsymbol{B}_m \cdot \boldsymbol{B}_n \, \mathrm{d}\mathcal{D} \tag{5.2}$$

where $\boldsymbol{B}_n$ is a basis function of the vector set while, once again, the integrals are computed over the domain $\mathcal{D}$ on which the basis functions are defined.

[2]The degree of the linear independence of the basis functions can be measured by the condition number of the matrix **G** only if all the basis functions in the set are square integrable.

[3]In principle, the Gram matrix and the integrals (5.1) and (5.2) should be computed in the child space. However, in the following, the orthogonality of the scalar polynomials that define the hierarchical basis functions will be enforced in the parent space by computing integrals of the form (5.1) on the parent domain. The convenience of that will become apparent to the reader later on. Therefore, in the following, the space where the integrals (5.1) or (5.2) are computed is decided every time upon convenience although the integration domain, either in the parent or in the child space, will be indicated by the same symbol $\mathcal{D}$ with no confusion.

To construct hierarchical bases that lead to lower system matrix condition numbers, the optimum approach would be to use mutually *orthogonal* functions that are *normalized* in a convenient manner. To reduce the computational burden, it is also convenient to define these functions in terms of normalized orthogonal functions at the onset. There are several advantages in doing this that should be immediately apparent by comparing the Lagrangian interpolatory polynomial bases of Chapter 2 for scalar one-dimensional problem with the corresponding conveniently normalized hierarchical bases obtained by following the construction technique outlined above.

Recall from Chapter 2 the convenience arising from one-dimensional scalar basis functions defined directly in the parent domain $\{0 \leq \xi \leq 1\}$, that are eventually mapped onto each different child domain. The basis functions are then functions of the parent variable ξ (although, as seen in Chapter 2, it is convenient to define them by introducing two parent variables, $\xi_1 = \xi$ and $\xi_2 = 1 - \xi$, related by the dependency relation $\xi_1 + \xi_2 = 1$). For example, one may use the pth-order interpolatory base formed by the following $(p+1)$ Lagrangian interpolatory polynomials of order p

$$\alpha_{ij}(p, \xi_1, \xi_2) = R_i(p, \xi_1) R_j(p, \xi_2), \qquad i, j = 0, 1, \ldots, p, \quad \text{with } i + j = p \tag{5.3}$$

The basis set (5.3) is formed by $(p-1)$ *bubble* functions[4] that vanish at $\xi = 0$ and at $\xi = 1$ because of the presence of a common $\xi_1\,\xi_2$ factor, plus two *vertex* basis functions interpolating the scalar quantity at the two extremes of the interval, that is at the $\xi = 0$ and $\xi = 1$ nodes. The vertex functions of the set (5.3) are obtained for $(i = 0, j = p)$ and $(i = p, j = 0)$. For example, the interpolatory vertex functions are ξ_1 and ξ_2 if $p = 1$, while they are $\xi_1(2\xi_1 - 1)$ and $\xi_2(2\xi_2 - 1)$ in the $p = 2$ case. The interpolatory vertex functions change with the base order p because they have to interpolate $(p+1)$ nodes. At $p = 1$, the basis set (5.3) contains only the two vertex functions, while bubble functions *appear* only for bases of order higher than the first; for example, for $p = 2$, there is only one bubble function $(4\,\xi_1\,\xi_2)$ while the number of the bubble functions linearly increases with the order. However, in the interpolatory case, the expressions of each bubble function are again order-dependent because the interpolatory grid does change with the order p of the base. Thus, the entire set of interpolator functions changes as the order p changes.

Conversely, in the hierarchical case, the *triangular* vertex functions ξ_1 and ξ_2 defined for $p = 1$ remain the same for any order, and cannot be changed to improve the linear independence of higher order basis sets. The two hierarchical vertex functions (independent of base order) are given by the dummy expression

$$V_1 = \xi_a \tag{5.4}$$

with ξ_a equal to ξ_1 or ξ_2, and where the subscript "1" used for V indicates that the vertex functions are linear (that is of first order).

---

[4] Any function that vanishes on the cell borders AND that is *nodeless* is a bubble function. The $(p-1)$ *interpolatory* functions (5.3) obtained for $i \neq 0$ and $j \neq 0$ vanish at the extreme nodes of the interval but are node based. Hence, strictly speaking, they are not bubble functions, although here we call them so because, with this *extended* definition, the hierarchical and the interpolatory bases discussed here are formed by two (node based) *vertex* functions plus $(p-1)$ bubble functions.

To improve the linear independence of the hierarchical sets, we may however define the bubble functions by using *normalized orthogonal* polynomials. Since the bubble functions contain $\xi_1\,\xi_2$ as a common factor, they can be written as

$$\begin{aligned} B_{m2}(\xi) &= \xi_1\,\xi_2\,\mathcal{U}_m(\xi_1-\xi_2) \\ &= \frac{(1-\xi_{12}^2)}{4}\,\mathcal{U}_m(\xi_{12}) = \frac{(1-\xi_{21}^2)}{4}\,\mathcal{U}_m(-\xi_{21}) \end{aligned} \tag{5.5}$$

where $\mathcal{U}_m(\xi_{12})=\mathcal{U}_m(-\xi_{21})$ is an appropriate polynomial of order m given in terms of the new *auxiliary* variable

$$\xi_{12}=(\xi_1-\xi_2)=(2\xi-1) \tag{5.6}$$

or

$$\xi_{21}=-\xi_{12}=(\xi_2-\xi_1) \tag{5.7}$$

The order of the function B_{m2} is given by the sum of its subscripts $(m+2)$. The first bubble function B_{02} introduced to form the $p=2$ base is proportional to $\xi_1\,\xi_2$ and must remain the same for all the other higher order basis sets. The second bubble function B_{12} is introduced to complete the $p=2$ base to the third ($p=3$) order, and therefore it is of the form $\xi_1\,\xi_2\,\mathcal{U}_1(\xi_{12})$, $\mathcal{U}_1(\xi_{12})$ being a first-order polynomial of the auxiliary ξ_{12} variable. In the hierarchical case, the second bubble function must remain the same for all the other higher orders. It is clear that, for $p\geq 2$, the higher order hierarchical base of order p is obtained by adding to the basis set of order $(p-1)$ only one bubble function of the form $\xi_1\,\xi_2\,\mathcal{U}_{p-2}(\xi_{12})$. To reduce the condition number of the local Gram matrix, it is therefore convenient to form the bubble functions by using *normalized orthogonal* polynomials $\mathcal{U}_q(\xi_{12})$ of the auxiliary ξ_{12} variable. For the entries (5.1) of the local Gram matrix $\mathbf{G}$ obtained by considering the generic bubble functions B_{m2} and B_{n2}, this is done by setting

$$\begin{aligned} g_{m+2,n+2} &= \int_0^1 B_{m2}(\xi)\,B_{n2}(\xi)\,\mathrm{d}\xi = \int_{-1}^{+1} \xi_1^2\,\xi_2^2\,\mathcal{U}_m(\xi_{12})\,\mathcal{U}_n(\xi_{12})\,\mathrm{d}\xi_{12} \\ &= \frac{1}{4}\int_{-1}^{+1} (1+\xi_{12})^2(1-\xi_{12})^2\,\mathcal{U}_m(\xi_{12})\,\mathcal{U}_n(\xi_{12})\,\mathrm{d}\xi_{12} = \delta_{mn} \end{aligned} \tag{5.8}$$

where

$$\delta_{mn} = \begin{cases} 0 & \text{for } m\neq n \\ 1 & \text{for } m=n \end{cases} \tag{5.9}$$

is the Kronecker delta. The orthogonal polynomials $\mathcal{U}_q(\xi_{12})$ that satisfy (5.8) are, by definition, the following rescaled version of the Jacobi orthogonal polynomials $P_q^{(2,2)}(\xi_{12})$

$$\mathcal{U}_q(\xi_{12}) = \sqrt{\frac{(3+q)\,(4+q)\,(5+2\,q)}{(1+q)\,(2+q)}}\;P_q^{(2,2)}(\xi_{12}) \tag{5.10}$$

obtained by the recurrence relation

$$a_{q1}\,\mathcal{U}_{q+1}(\xi_{12}) = a_{q2}\,\xi_{12}\,\mathcal{U}_q(\xi_{12}) - a_{q3}\,\mathcal{U}_{q-1}(\xi_{12}) \tag{5.11}$$

with

$$\begin{cases} a_{q1} = \sqrt{(1+q)\;(5+q)\,(3+2\,q)} \\ a_{q2} = \sqrt{(3+2\,q)(5+2\,q)(7+2\,q)} \\ a_{q3} = \sqrt{q\;(4+q)\,(7+2\,q)} \end{cases} \tag{5.12}$$

The series begins with

$$\mathcal{U}_0(\xi_{12})=\sqrt{30}, \quad \mathcal{U}_1(\xi_{12})=\sqrt{210}\,\xi_{12} \tag{5.13}$$

Notice that $\mathcal{U}_q(\xi_{12})$ is either symmetric (for even or zero q values) or antisymmetric (for odd q values) in ξ_1 and ξ_2, with $\mathcal{U}_q(\xi_{21})=\mathcal{U}_q(-\xi_{12})=(-1)^q\,\mathcal{U}_q(\xi_{12})$. The symmetry or the antisymmetry of the basis functions is an issue of major importance when building the hierarchical basis functions for the canonical two- and three-dimensional cells. Here, we just observe that if we superimpose two one-dimensional elements, the *local* variable ξ_1 used for the first can be equal to either the *local* variable ξ_1 or $\xi_2(=1-\xi_1)$ of the second element. In this event, it is clear that to guarantee we are using the same bubble function on the two superimposed elements there is only the need to adjust the sign of the bubble function on the two elements (this is required only if $\mathcal{U}_q$ is antisymmetric in ξ_1 and ξ_2) by selecting a reference direction, for example, from the lower to the higher *global* vertex number of the two superimposed elements.

Figure 5.1 compares the CNs of the local Gram matrix obtained using the interpolatory bases (5.3) and the hierarchical bases formed by the vertex functions (5.4) and the normalized orthogonal bubble functions discussed above. Although the local Gram matrix condition number growth rate is exponential for the interpolatory bases while it is only polynomial for the hierarchical bases, the degree of linear independence of the basis functions of the interpolatory bases is higher than that of the hierarchical bases for orders lower than 7.5. As discussed, this happens mainly because the triangular *vertex* functions of the hierarchical bases do not change with the order, while in the interpolatory case the *vertex* functions change with the order. Notice that vertex functions are necessary to form scalar bases while they are not used to form *vector* bases on two- and three-dimensional elements; this simplifies the vector bases conditioning problem.

In Figure 5.1, the order of the base is denoted by a half integer as sometimes done in the FEM literature. For example, we use "order 0.5" to denote the $p=1$ base. This is done here simply to recall that set formed by the first-order derivatives of the basis functions of the pth-order set is complete to the $(p-1)$ order. The quantity approximated by using the one-dimensional scalar base could be, for example, the longitudinal component of the field in an inhomogeneous cylindrical waveguide; in this case, the transverse vector field should then be expanded by using a vector base of the same (half integer) order. When discussing the conforming vector bases used to represent a transverse fields on a surface elements, we use "order 0.5" to denote the $p=0$ vector basis functions which must be used together with the $p=1$ scalar functions of "order 0.5" to achieve the same degree of completeness in the transverse and longitudinal directions. Recall however that, for $p=1$, (5.4) (and (5.3)) is a complete polynomial set of first order able to represent polynomials of degree one.

Figure 5.1 clarifies the *ill-conditioning* issue associated with the use of hierarchical bases. In fact the figure shows that the CN of the second-order hierarchical base is comparable to that of the fifth-order interpolatory base. Note that bases of order higher than the sixth are seldom used in numerical applications (typical high-order applications involve bases of order from the second up to the fourth).

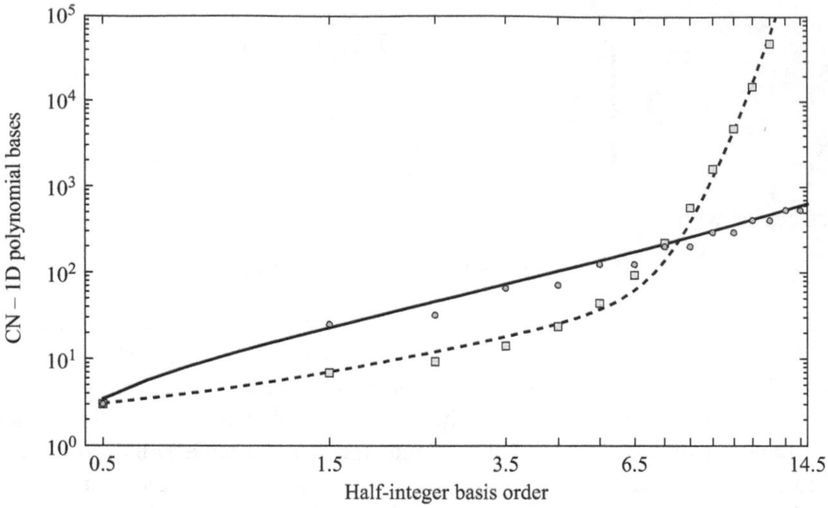

Figure 5.1 Local Gram matrix condition numbers for the scalar, one-dimensional hierarchical (CNH) and interpolatory (CNI) polynomial bases up to order 14.5. The interpolatory normalized bases are given in (5.3). The figure compares (using logarithmic scales) the condition numbers with reference growth-rate lines. The hierarchical condition numbers, CNH, are reported by circles; the interpolatory condition numbers, CNI, are reported by squares. The solid-line depicts a polynomial growth rate $g_1 = 1.8 \times \text{order}^2 + 16 \times \text{order} - 5$; the dashed-line represents an exponential growth rate $g_2 = 3.65^{(\text{order}-4)} + 3.1 \times \text{order}^{1.3} + 1.8$. The condition number growth rate is exponential for the interpolatory bases and polynomial for the hierarchical bases.

5.2 Hierarchical Scalar Bases

5.2.1 Tetrahedral and Triangular Bases

As seen in Chapter 2, the polynomials (5.14) and (5.15) reported below define the interpolatory scalar base of order $p(\geq 1)$ for the tetrahedral and the triangular cell, respectively:

$$\begin{aligned}\alpha_{ijk\ell}(p, \xi_1, \xi_2, \xi_3, \xi_4) &= R_i(p, \xi_1)\, R_j(p, \xi_2)\, R_k(p, \xi_3)\, R_\ell(p, \xi_4) \\ & \quad i, j, k, \ell = 0, 1, \ldots, p, \\ & \quad i + j + k + \ell = p\end{aligned} \tag{5.14}$$

$$\begin{aligned}\alpha_{ijk}(p, \xi_1, \xi_2, \xi_3) &= R_i(p, \xi_1) R_j(p, \xi_2) R_k(p, \xi_3) \\ & \quad i, j, k = 0, 1, \ldots, p, \\ & \quad i + j + k = p\end{aligned} \tag{5.15}$$

where $R_n(p, \xi)$ indicates the Silvester polynomial of degree n, with ξ in the interval [0, 1], and where the parameter p is the number of uniform subintervals into which this interval is divided. The tetrahedral base (5.14) contains $(p+1)(p+2)(p+3)/6$ terms while the triangular base (5.15) is formed by $(p+1)(p+2)/2$ polynomials.

The global polynomial order or degree of (5.14) and (5.15) is p, while $(p-1)$ is the degree of the first derivatives of (5.14) and (5.15) with respect to any of the parent variables. The half-integer order of these bases is then equal to $(p-0.5)$ because the basis functions can represent a scalar quantity and its first derivatives up to the same complete $(p-1)$ order within a triangle or tetrahedron.

The tetrahedral base (5.14) contains[5]

$\forall p \geq 1$: **four vertex polynomials** obtained by setting three entries of the index-list (i, j, k, ℓ) to zero with the fourth remaining index (i, j, k, or ℓ) set equal to p;
$\forall p \geq 2$: **$(p-1)$ edge-based polynomials per edge** obtained by setting two entries of the index-list (i, j, k, ℓ) to zero and the two remaining indexes to a value different from zero (the tetrahedron has in fact six edges corresponding to the $(i=j=0)$, $(i=k=0)$, $(i=\ell=0)$, $(j=k=0)$, $(j=\ell=0)$, and $(k=\ell=0)$ case);
$\forall p \geq 3$: **$(p-1)(p-2)/2$ face-based polynomials per face** obtained by setting three entries of the index-list (i, j, k, ℓ) to a value different from zero with the other fourth remaining index set to zero (the tetrahedron has four faces corresponding to the $i=0, j=0, k=0$, and $\ell=0$ case);
$\forall p \geq 4$: **$(p-1)(p-2)(p-3)/6$ volume-based polynomials** obtained by setting $i, j, k, \ell \neq 0$.

Similarly, the triangular basis function set (5.15) is formed by

$\forall p \geq 1$: **three vertex functions** obtained by setting two entries of the index-list (i, j, k) to zero with the third remaining index (i, j, or k) set equal to p;
$\forall p \geq 2$: **$(p-1)$ edge-based functions per edge** obtained by setting one entry of the index-list (i, j, k) to zero and the two remaining indexes to a value different from zero (the triangle has in fact three edges corresponding to the i, j, and $k=0$ case);
$\forall p \geq 3$: **$(p-1)(p-2)/2$ face-based functions** obtained by setting $i, j, k \neq 0$.

Although the definition of vertex-, edge-, face-, and volume-based polynomials is momentarily deferred, we observe that hierarchical scalar bases are linear combinations of the polynomials (5.14) and (5.15), and can therefore be formed by the given number of vertex-, edge-, face-, and volume-based polynomials. However, umpteen alternatives exist to form hierarchical bases by linearly combining the polynomials (5.14) and (5.15), and a clever strategy is needed to avoid hierarchical basis functions defined by lengthy complex expressions. The hierarchical polynomial sets presented in this chapter are obtained by applying a Gram–Schmidt orthogonalization process and, in principle, one could do that by linearly combining the *interpolatory* polynomials of the pth-order base with the *hierarchical* ones previously derived for

[5]To avoid any confusion, in this connection it is important to point out immediately that the lowest integer order of a scalar polynomial base is $p=1$, with half-integer order equal to 0.5; conversely, the vector bases of order 0.5 are built by multiplying the zeroth-order vector bases of Chapter 4 with polynomials of integer order $p=0$. In problems that require a scalar base together with a vector base, for example to model a given (longitudinal) field component with the scalar base and a (transverse) vector field with the vector base, the half-integer order of the scalar and vector bases must be equal. That is, if one uses a vector base of integer order p (and half-integer order $p+0.5$), the scalar base must also be of integer order $p+1$ (and half-integer order $p+0.5$).

all the lower order bases; unfortunately, a blind application of this procedure yields polynomials with complex expressions that are not suitable for implementation into a numerical code. As we will see, the best approach to form the hierarchical bases is to use orthogonal polynomials from the very beginning and exploit the symmetries of the cell as much as possible.[6]

Contrary to intuition, as we will see shortly, it is also more convenient to start by defining the hierarchical bases of the tetrahedral cell before defining those of the triangular cell. Furthermore, for reasons that will become apparent during the definition process, it is more convenient to start by defining the *volume-based* functions of the tetrahedral cell.

To avoid reporting at length the expression of the hierarchical polynomials, in the following we exploit the symmetry of the cells and denote the normalized coordinates by dummy parent variables ξ_a, ξ_b, ξ_c, and ξ_d, with $\xi_a + \xi_b + \xi_c + \xi_d = 1$. The tetrahedral cell is described by all four variables while the triangular cell is described only by the three variables (ξ_a, ξ_b, ξ_c); in this case, $\xi_d = 0$ is understood. The variable ξ_a can then be equal to ξ_1, ξ_2, ξ_3, or ξ_4, as convenient.

Furthermore, to minimize the expressions of the hierarchical polynomials that follow, it is also convenient to exploit the symmetries of the cell thereby introducing the following new variables

$$\begin{aligned} \xi_{ab} &= \xi_a - \xi_b \\ \xi_{cd} &= \xi_c - \xi_d \\ \chi_{ab} &= \xi_a + \xi_b \\ \chi_{cd} &= \xi_c + \xi_d \end{aligned} \tag{5.16}$$

with dependency relation

$$\chi_{ab} + \chi_{cd} = 1 \tag{5.17}$$

The expressions of the four parent variables

$$\begin{aligned} \xi_a &= (\chi_{ab} + \xi_{ab})/2 \\ \xi_b &= (\chi_{ab} - \xi_{ab})/2 \\ \xi_c &= (\chi_{cd} + \xi_{cd})/2 \\ \xi_d &= (\chi_{cd} - \xi_{cd})/2 \end{aligned} \tag{5.18}$$

in terms of the new dependent variables are straightforwardly obtained by inversion of (5.16).

5.2.1.1 Volume-Based Polynomials

For $p \geq 4$, we define $(p-1)(p-2)(p-3)/6$ volume-based (or bubble) hierarchical polynomials

$$V_{\ell mn4}(\boldsymbol{\xi}) = \xi_a\, \xi_b\, \xi_c\, \xi_d\, \mathcal{U}_{\ell mn}(\boldsymbol{\xi}) \tag{5.19}$$

---

[6]The Gram–Schmidt orthogonalization process discussed here can be effectively applied only by running powerful algebraic manipulators on modern computers. These were not available to the great mathematicians of the eighteenth and nineteenth centuries, otherwise those "giants" would have already derived the hierarchical orthogonal polynomials for the two- and three-dimensional cells considered in this book.

of order

$$g = \ell + m + n + 4, \quad \text{with } 4 \leq g \leq p \tag{5.20}$$

and where $\boldsymbol{\xi} = \{\xi_a, \xi_b, \xi_c, \xi_d\}$ could be any permutation of the parent variables $\{\xi_1, \xi_2, \xi_3, \xi_4\}$. These polynomials vanish along the four faces of the tetrahedral cell. They are obtained by Gram–Schmidt orthogonalizing over the tetrahedral simplex T^3 the polynomials $\xi_a\,\xi_b\,\xi_c\,\xi_d\,\chi_{ab}^{\ell}\,P_m(\xi_{ab})\,P_n(\xi_{cd})$ using a triple nested loop where

- $g = 4, 5, \ldots, p$ (outer loop on the global order g)
- $n = 0, 1, \ldots, g-4$ (middle loop)
- $m = 0, 1, \ldots, g-4-n$ (inner loop)

with $\ell = (g-4-n-m)$ fixed in the inner loop, and where $P_q(z)$ indicates the qth degree Legendre polynomial. The $V_{\ell mn4}$ polynomials are mutually orthogonal over the tetrahedral simplex and normalized so that

$$\iint_{T^3} V_g^2(\boldsymbol{\xi})\,\mathrm{d}T^3 = \frac{1}{(2g+2)(2g+3)} \tag{5.21}$$

where g indicates the global polynomial order of $V_{\ell mn4}$, equal to the sum of its subscripts. Table 5.1 reports the polynomials $\mathcal{U}_{\ell mn}(\boldsymbol{\xi})$ (obtained as explained above) that permit one to construct the volume-based functions for the tetrahedron up to the order $p = 7$.

Observe the symmetries of the compact expression of the polynomials of Table 5.1; these symmetries would not be present if we orthogonalize a different polynomial set. For example, by applying the Gram–Schmidt procedure to the polynomials $\xi_a\,\xi_b\,\xi_c\,\xi_d\,P_\ell(\xi_{ab})\,P_m(2\xi_c-1)\,P_n(2\xi_d-1)$ one obtains the non-symmetric expressions

$$\mathcal{U}_{000} = 18\sqrt{70} \tag{5.22}$$

$$\begin{cases} \mathcal{U}_{100} = 30\sqrt{462}\xi_{ab} \\ \mathcal{U}_{010} = 30\sqrt{77}\,(\chi_{ab} - \chi_{cd} - 2\xi_{cd}) \\ \mathcal{U}_{001} = 30\sqrt{154}\,(\chi_{ab} - \chi_{cd} + \xi_{cd}) \end{cases} \tag{5.23}$$

$$\begin{cases} \mathcal{U}_{200} = 60\sqrt{3}\,(91\xi_{ab}^2 - 13\chi_{ab} + 3)\,/\sqrt{19} \\ \mathcal{U}_{110} = 60\sqrt{39}\xi_{ab}(14\xi_c - 3) \\ \mathcal{U}_{020} = 30\sqrt{78}\,(14\xi_{ab}^2 - 1463\xi_c^2 - 2\chi_{ab} + 836\xi_c - 96)\,/\sqrt{4,579} \\ \mathcal{U}_{101} = 30\sqrt{273}\xi_{ab}(8\xi_d - 3\chi_{ab}) \\ \mathcal{U}_{011} = 10\sqrt{\dfrac{858}{241}}\left(5\xi_{ab}^2 + 321\xi_c^2 - 390\xi_c + 2\chi_{ab}(241\xi_c - 52) + 69\right) \\ \mathcal{U}_{002} = 5\sqrt{429}\,(24\xi_d^2 - 32\chi_{ab}\xi_d + 7\chi_{ab}^2 - \xi_{ab}^2) \end{cases} \tag{5.24}$$

The expressions in Table 5.1 are far more compact than these.

Table 5.1 Scalar bases – hierarchical volume-based functions for the tetrahedral cell.

For $p \geq 4$, there are $(p-1)(p-2)(p-3)/6$ hierarchical bubble functions of the form $V_{\ell mn4}(\boldsymbol{\xi}) = \xi_a\, \xi_b\, \xi_c\, \xi_d\, \mathcal{U}_{\ell mn}(\boldsymbol{\xi})$ that vanish along the four faces. The $V_{\ell mn4}$ polynomials are mutually orthogonal over the tetrahedral simplex and normalized so that

$$\iint_{T^3} V_g^2(\boldsymbol{\xi})\, dT^3 = \frac{1}{(2g+2)(2g+3)}$$

where g indicates the global polynomial order of $V_{\ell mn4}$, equal to the sum of its subscripts. The table reports the polynomials $\mathcal{U}_{\ell mn}(\boldsymbol{\xi})$ used to construct the volume-based functions up to the order $p = 7$.

$\mathcal{U}_{000} = 18\sqrt{70}$	$\mathcal{U}_{300} = 45\sqrt{2{,}002}\,(\chi_{ab} - \chi_{cd})(1 - 5\chi_{ab}\chi_{cd})$
$\mathcal{U}_{100} = 30\sqrt{231}\,(\chi_{ab} - \chi_{cd})$	$\mathcal{U}_{030} = 105\sqrt{429}\,\xi_{ab}\,(\chi_{ab}^2 - 3\xi_{ab}^2)$
$\mathcal{U}_{010} = 30\sqrt{462}\,\xi_{ab}$	$\mathcal{U}_{003} = 105\sqrt{429}\,\xi_{cd}\,(\chi_{cd}^2 - 3\xi_{cd}^2)$
$\mathcal{U}_{001} = 30\sqrt{462}\,\xi_{cd}$	$\mathcal{U}_{210} = 15\sqrt{3{,}003}\,\xi_{ab}\,(40\chi_{cd}^2 - 35\chi_{cd} + 7)/\sqrt{2}$
$\mathcal{U}_{200} = 60\sqrt{33}\,(3 - 13\chi_{ab}\chi_{cd})$	$\mathcal{U}_{201} = 15\sqrt{3{,}003}\,\xi_{cd}\,(40\chi_{ab}^2 - 35\chi_{ab} + 7)/\sqrt{2}$
$\mathcal{U}_{020} = 15\sqrt{429}\,(\chi_{ab}^2 - 7\xi_{ab}^2)$	$\mathcal{U}_{120} = 15\sqrt{3{,}003}\,(8\chi_{cd} - 3)(\chi_{ab}^2 - 7\xi_{ab}^2)/\sqrt{8}$
$\mathcal{U}_{002} = 15\sqrt{429}\,(\chi_{cd}^2 - 7\xi_{cd}^2)$	$\mathcal{U}_{102} = 15\sqrt{3{,}003}\,(8\chi_{ab} - 3)(\chi_{cd}^2 - 7\xi_{cd}^2)/\sqrt{8}$
$\mathcal{U}_{110} = 30\sqrt{429}\,\xi_{ab}\,(7\chi_{ab} - 4)$	$\mathcal{U}_{111} = 315\sqrt{1{,}430}\,\xi_{ab}\,\xi_{cd}\,(\chi_{ab} - \chi_{cd})$
$\mathcal{U}_{101} = 30\sqrt{429}\,\xi_{cd}\,(7\chi_{cd} - 4)$	$\mathcal{U}_{021} = 15\sqrt{15{,}015}\,\xi_{cd}\,(\chi_{ab}^2 - 7\xi_{ab}^2)$
$\mathcal{U}_{011} = 60\sqrt{3{,}003}\,\xi_{ab}\,\xi_{cd}$	$\mathcal{U}_{012} = 15\sqrt{15{,}015}\,\xi_{ab}\,(\chi_{cd}^2 - 7\xi_{cd}^2)$

With $\xi_{ab} = \xi_a - \xi_b$, $\chi_{ab} = \xi_a + \xi_b$, $\xi_{cd} = \xi_c - \xi_d$, $\chi_{cd} = \xi_c + \xi_d$, $\boldsymbol{\xi} = (\xi_a, \xi_b, \xi_c, \xi_d)$, and $\xi_a + \xi_b + \xi_c + \xi_d = 1$.

5.2.1.2 Face-Based Polynomials

For $p \geq 3$, we can easily define $(p-1)(p-2)/2$ face-based hierarchical polynomials

$$\mathcal{F}_{mn3}(\boldsymbol{\xi}) = \xi_a\, \xi_b\, \xi_c\, \mathcal{P}_{mn}(\xi_{ab}, \chi_{ab}) \tag{5.25}$$

obtained by Gram–Schmidt orthogonalizing over the triangular simplex $T^2 \equiv \{0 \leq \xi_a, \xi_b, \xi_c \leq 1 \colon \xi_a + \xi_b + \xi_c = 1\}$ the polynomials

$$\xi_a\, \xi_b\, \xi_c\, P_m(\xi_a - \xi_b)\, P_n(\xi_a + \xi_b) = \xi_a\, \xi_b\, \xi_c\, P_m(\xi_{ab})\, P_n(\chi_{ab}) \tag{5.26}$$

where P_q indicates the qth-order Legendre polynomial. (For the triangular cell, we assumed $\xi_d = 0$; in this case, $\boldsymbol{\xi} = \{\xi_a, \xi_b, \xi_c\}$ could be any permutation of the parent variables $\{\xi_1, \xi_2, \xi_3\}$.)

The global polynomial order (or degree) of $\mathcal{F}_{mn3}$ is

$$g = m + n + 3 = t + 3, \quad \text{with } 3 \leq g \leq p,\ 0 \leq t \leq p - 3 \tag{5.27}$$

and the polynomials are orthogonalized by using a double-nested loop where

- $t = 0, 1, \ldots, p-3$ (outer loop)
- $m = 0, 1, \ldots, t$ (inner loop)

with $n = (t-m)$ fixed in the inner loop. The polynomials (5.25) vanish along the three coordinate surfaces $\xi_a, \xi_b, \xi_c = 0$ (i.e. along the three edges of the triangular cell T^2 or along three faces of the tetrahedral cell T^3); they are mutually orthogonal over the triangular simplex and normalized so that

$$\iint_{T^2} \mathcal{F}_g^2(\boldsymbol{\xi})\,\mathrm{d}T^2 = \frac{1}{2g+2} \tag{5.28}$$

where g indicates the global polynomial order of $\mathcal{F}_{mn3}$, equal to the sum of its subscripts. The first term of this face-based family

$$\mathcal{F}_{003}(\boldsymbol{\xi}) = N_{003}\,\xi_a\,\xi_b\,\xi_c = 3\sqrt{70}\,\xi_a\,\xi_b\,\xi_c \tag{5.29}$$

is obtained by setting $N_{003}^2 = 630$ as per the equation

$$\iint_{T^2} \mathcal{F}_{003}^2(\boldsymbol{\xi})\,\mathrm{d}T^2 = N_{003}^2 \iint_{T^2} (\xi_a\,\xi_b\,\xi_c)^2\,\mathrm{d}T^2 = \frac{1}{8} \tag{5.30}$$

Without modifications, the polynomials $\mathcal{F}_{mn3}(\xi_a, \xi_b, \xi_c)$ could be the polynomials based on the $\xi_d = 0$ face of the tetrahedral cell of (dummy) parent variables $\{\xi_a, \xi_b, \xi_c, \xi_d\}$ and also used to form the triangular-face-based functions of the triangular prism of parent variables $\{\xi_a, \xi_b, \xi_c; \xi_d, \xi_e\}$. To get the latter, it is just sufficient to multiply $\mathcal{F}_{mn3}(\xi_a, \xi_b, \xi_c)$ with the fourth (ξ_d) or fifth ($\xi_e = 1 - \xi_d$) parent variable of the prism, as we will see in Section 5.2.4. The functions obtained in this manner would then be mutually orthogonal on the volume of the prism element, but not on the volume of the tetrahedral cell because of the differences in the dependency relations

$$\xi_a + \xi_b + \xi_c = \chi_{ab} + \xi_c = \begin{cases} 1 - \xi_d & \text{for the tetrahedron} \\ 1 & \text{for the prism} \end{cases} \tag{5.31}$$

(Obviously, on their associated triangular face, the face-based polynomials of the tetrahedral cell must be equal to the face-based polynomials of the prism cell, apart from a possible sign adjustment, to permit one to use meshes that contain both tetrahedral and triangular prism cells.)

In order to obtain triangular face-based functions $F_{mn3}(\xi_a, \xi_b, \xi_c)$ that are mutually orthogonal also on the volume of the tetrahedron $\{0 \le \xi_a, \xi_b, \xi_c, \xi_d \le 1;\ \xi_a + \xi_b + \xi_c + \xi_d = 1\}$, we need to add to $\mathcal{F}_{mn3}(\xi_a, \xi_b, \xi_c)$ an appropriate polynomial of the same global order of $\mathcal{F}_{mn3}$ that vanishes for $\xi_d = 0$, as well as for $\xi_a, \xi_b, \xi_c = 0$. That polynomial is a linear combination of the volume-based polynomials discussed in Section 5.2.1.1. To get the triangular-face based polynomials

$$F_{mn3}(\boldsymbol{\xi}) = \xi_a\,\xi_b\,\xi_c\,\mathcal{U}_{mn}(\xi_a, \xi_b, \xi_c) \tag{5.32}$$

we then only need to substitute $(1 - \chi_{ab} - \xi_c)$ for ξ_d after the addition is done.

Although for the sake of brevity, we omit the details of this procedure, we notice that the polynomials $\mathcal{F}_{mn3}(\xi_a, \xi_b, \xi_c)$ are uniquely determined by the double-loop procedure discussed above while, in general, there is more than one linear combination of $\mathcal{F}_{mn3}(\xi_a, \xi_b, \xi_c)$

with volume-based polynomials that renders the functions $F_{mn3}(\xi_a, \xi_b, \xi_c)$ mutually orthogonal on the triangular face and on the tetrahedron volume. The triangular-face-based functions F_{mn3} reported in Table 5.2 were obtained by adding to $\mathcal{F}_{mn3}$, the linear combination of the tetrahedron-volume-based functions that most simplify the expression of F_{mn3}. The functions F_{mn3} of Table 5.2 are mutually orthogonal over the triangular simplex T^2 and over the tetrahedral simplex T^3, and normalized so that

$$\iint_{T^2} F_g^2(\boldsymbol{\xi})\,\mathrm{d}T^2 = \frac{1}{2g+2}, \quad \iiint_{T^3} F_g^2(\boldsymbol{\xi})\,\mathrm{d}T^3 = \frac{1}{(2g+2)(2g+3)} \tag{5.33}$$

where g indicates the global polynomial order of F_{mn3}, equal to the sum of its subscripts. Notice also that F_{mn3} is either symmetric (for even or zero m values) or antisymmetric (for odd m values) in ξ_a and ξ_b.

5.2.1.3 Edge-Based Polynomials

For $p \geq 2$, we associate with the $\xi_c = \xi_d = 0$ edge $(p-1)$ hierarchical polynomials

$$E_{m2}(\boldsymbol{\xi}) = \xi_a\,\xi_b\,\mathcal{U}_m(\boldsymbol{\xi}) \tag{5.34}$$

of order $g = m + 2$, with $2 \leq g \leq p$. These polynomials vanish for $\xi_a = 0$ and $\xi_b = 0$ and are mutually orthogonal on the $\xi_c = \xi_d = 0$ edge. They are obtained by orthogonalizing over the edge $\xi_c = \xi_d = 0$ the polynomials $\xi_a\,\xi_b\,P_m(\xi_a - \xi_b)$ for $m = 0, 1, \ldots, p-2$, and then by adding, to each obtained polynomial, an appropriate linear combination of the face-based polynomials $F_g(\boldsymbol{\xi})$ of order $g \leq m + 2$ to render the polynomials E_{m2} mutually orthogonal on the triangular simplex T^2. The polynomials E_{m2} associated with the $\xi_c = \xi_d = 0$ edge are normalized so that

$$\int_0^1 E_g^2\Big|_{\xi_c=0}\,\mathrm{d}\xi_a = 1, \quad \iint_{T^2} E_g^2(\boldsymbol{\xi})\,\mathrm{d}T^2 = \frac{1}{2g+2} \tag{5.35}$$

where $g = m + 2$ indicates the global polynomial order of E_{m2}, equal to the sum of its subscripts. The polynomials E_{m2} obtained with the above procedure are also mutually orthogonal over the tetrahedral simplex T^3, with

$$\iiint_{T^3} E_g^2(\boldsymbol{\xi})\,\mathrm{d}T^3 = \frac{1}{(2g+2)(2g+3)} \tag{5.36}$$

With reference to Table 5.2, it is of importance to observe that, on their associated $\xi_c = \xi_d = 0$ edge, the edge-based functions (5.34) simplify into the one-dimensional edge-based functions (5.5), where $\chi_{ab} = \xi_a + \xi_b = 1$ for $\xi_c = \xi_d = 0$. In fact, for $\chi_{ab} = 1$ (i.e. for $\xi_c = \xi_d = 0$),

Table 5.2 Hierarchical scalar bases for the triangular and tetrahedral cell

VERTEX FUNCTIONS

For $p \geq 1$, there are three (for the triangle) or four (for the tetrahedron) first-order vertex functions of the form $V_1 = \xi_a$. The function associated with the ith vertex is obtained by setting $\xi_a = \xi_i$. V_1 vanishes on the $\xi_a = 0$ edge (or face) and attains a value of unity at the vertex $\xi_a = 1$, where $\xi_b = \xi_c\,(=\xi_d) = 0$.

EDGE-BASED FUNCTIONS

For $p \geq 2$, there are $(p-1)$ edge-based functions per edge. The edge-based functions associated with the $\xi_c = 0$ edge ($\xi_d = 0$ is understood) vanish on the $\xi_a = 0$ and the $\xi_b = 0$ edge and are of the form $E_{m2} = \xi_a\,\xi_b\,\mathcal{U}_m(\boldsymbol{\xi})$, for $m = 0, 1, \ldots, p-2$. The functions associated with the $\xi_i = 0$ edge are obtained by setting $\xi_c = \xi_i$, $\xi_a = \xi_{i+1}$, $\xi_b = \xi_{i-1}$.

$\mathcal{U}_0 = \sqrt{30}$	$\mathcal{U}_2 = 3\sqrt{5}\,(7\,\xi_{ab}^2 - \chi_{ab}^2)\,/\sqrt{2}$	$\mathcal{U}_4 = \sqrt{1{,}365}\,(33\,\xi_{ab}^4 - 18\,\xi_{ab}^2\,\chi_{ab}^2 + \chi_{ab}^4)\,/8$
$\mathcal{U}_1 = \sqrt{210}\,\xi_{ab}$	$\mathcal{U}_3 = \sqrt{1{,}155}\,\xi_{ab}\,(3\,\xi_{ab}^2 - \chi_{ab}^2)\,/\sqrt{2}$	$\mathcal{U}_5 = 3\sqrt{35}\,\xi_{ab}\,(143\,\xi_{ab}^4 - 110\,\xi_{ab}^2\,\chi_{ab}^2 + 15\,\chi_{ab}^4)\,/8$

FACE-BASED FUNCTIONS

For $p \geq 3$, there are $(p-1)(p-2)/2$ face-based functions of the form $F_{mn3} = \xi_a\,\xi_b\,\xi_c\,\mathcal{U}_{mn}(\boldsymbol{\xi})$ that vanish along the three edges of the $\xi_d = 0$ face. The bubble functions are obtained by setting $\boldsymbol{\xi} = \{\xi_a, \xi_b, \xi_c\} = \{\xi_i, \xi_{i+1}, \xi_{i-1}\}$, with $m, n = 0, 1, \ldots, p-3$ and $0 \leq m + n \leq p-3$.

$\mathcal{U}_{00} = 3\sqrt{70}$	$\mathcal{U}_{01} = 6\sqrt{70}\,(\chi_{ab} - 2\,\xi_c)$	$\mathcal{U}_{10} = 6\sqrt{210}\,\xi_{ab}$
$\mathcal{U}_{02} = 6\sqrt{5}\,(6\,\chi_{ab}^2 - 28\,\xi_c\,\chi_{ab} + 21\,\xi_c^2)$ $\mathcal{U}_{11} = 15\sqrt{14}\,\xi_{ab}\,(3\chi_{ab} - 8\,\xi_c)$ $\mathcal{U}_{20} = 15\sqrt{\frac{11}{2}}\,(\chi_{ab}^2 - 7\,\xi_{ab}^2)$	$\mathcal{U}_{03} = 6\sqrt{15}\,(5\,\chi_{ab}^3 - 40\,\xi_c\,\chi_{ab}^2 + 70\,\xi_c^2\,\chi_{ab} - 28\,\xi_c^3)$ $\mathcal{U}_{12} = 30\sqrt{77}\,\xi_{ab}\,(\chi_{ab}^2 - 6\,\xi_c\,\chi_{ab} + 6\,\xi_c^2)$ $\mathcal{U}_{21} = 3\sqrt{55}\,(\chi_{ab}^2 - 7\xi_{ab}^2)\,(3\,\chi_{ab} - 10\,\xi_c)$ $\mathcal{U}_{30} = 3\sqrt{5{,}005}\,\xi_{ab}\,(\chi_{ab}^2 - 3\,\xi_{ab}^2)$	$\mathcal{U}_{04} = 3\sqrt{110}\,(5\,\chi_{ab}^4 - 60\,\xi_c\,\chi_{ab}^3 + 180\,\xi_c^2\,\chi_{ab}^2 - 168\,\xi_c^3\,\chi_{ab} + 42\,\xi_c^4)$ $\mathcal{U}_{13} = 15\sqrt{154}\,\xi_{ab}\,(2\chi_{ab}^3 - 20\,\xi_c\,\chi_{ab}^2 + 45\,\xi_c^2\,\chi_{ab} - 24\,\xi_c^3)$ $\mathcal{U}_{22} = 3\sqrt{65}\,(\chi_{ab}^2 - 7\,\xi_{ab}^2)\,(6\,\chi_{ab}^2 - 44\,\xi_c\,\chi_{ab} + 55\,\xi_c^2)\,/\sqrt{2}$ $\mathcal{U}_{31} = 21\sqrt{715}\,\xi_{ab}\,(\chi_{ab}^2 - 3\,\xi_{ab}^2)\,(\chi_{ab} - 4\,\xi_c)/\sqrt{2}$ $\mathcal{U}_{40} = 105\sqrt{13}\,(\chi_{ab}^4 - 18\,\xi_{ab}^2\,\chi_{ab}^2 + 33\,\xi_{ab}^4)\,/8$

Above, $\xi_{ab} = \xi_a - \xi_b$, $\chi_{ab} = \xi_a + \xi_b$, $\boldsymbol{\xi} = (\xi_a, \xi_b, \xi_c)$, and $\xi_a + \xi_b + \xi_c = 1$ (for $\xi_d = 0$). The reported polynomials are normalized as follows

$$\int_0^1 E_g^2\big|_{\xi_c=0}\,d\xi_a = 1; \quad \iint_{T^2} E_g^2(\boldsymbol{\xi})\,dT^2 = \frac{1}{2g+2}; \quad \iint_{T^2} F_g^2(\boldsymbol{\xi})\,dT^2 = \frac{1}{2g+2}$$

where T^2 is the triangular simplex while g indicates the global order of the polynomial, equal to the sum of its subscripts.

the entries of the second row of Table 5.2 simplify into the rescaled Jacobi polynomial family (5.10), whose first eight terms are

$$\begin{aligned}
\mathcal{U}_0(\xi_{ab}) &= \sqrt{30} \\
\mathcal{U}_1(\xi_{ab}) &= \sqrt{210}\,\xi_{ab} \\
\mathcal{U}_2(\xi_{ab}) &= 3\sqrt{5}\left(7\xi_{ab}^2 - 1\right)/\sqrt{2} \\
\mathcal{U}_3(\xi_{ab}) &= \sqrt{1,155}\,\xi_{ab}\left(3\xi_{ab}^2 - 1\right)/\sqrt{2} \\
\mathcal{U}_4(\xi_{ab}) &= \sqrt{1,365}\left(33\xi_{ab}^4 - 18\xi_{ab}^2 + 1\right)/8 \\
\mathcal{U}_5(\xi_{ab}) &= 3\sqrt{35}\,\xi_{ab}\left(143\xi_{ab}^4 - 110\xi_{ab}^2 + 15\right)/8 \\
\mathcal{U}_6(\xi_{ab}) &= 3\sqrt{595}\left(143\xi_{ab}^6 - 143\xi_{ab}^4 + 33\xi_{ab}^2 - 1\right)/16 \\
\mathcal{U}_7(\xi_{ab}) &= 3\sqrt{1,045}\,\xi_{ab}\left(221\xi_{ab}^6 - 273\xi_{ab}^4 + 91\xi_{ab}^2 - 7\right)/16 \\
&\text{with } \xi_{ab} = \xi_a - \xi_b
\end{aligned} \tag{5.37}$$

At this point, it should be clear that the orthogonal polynomials $\mathcal{U}_q(\xi_{ab}, \chi_{ab})$ in the second row of Table 5.2 are a two-dimensional (properly normalized) extension of the Jacobi polynomials $P_q^{(2,2)}(\xi_{ab})$, with $\mathcal{U}_q(-\xi_{ab}, \chi_{ab}) = (-1)^q\, \mathcal{U}_q(\xi_{ab}, \chi_{ab})$. The polynomials $\mathcal{U}_q(\xi_{ab}, \chi_{ab})$ are in fact orthogonal with respect to the weight $\xi_a^2\, \xi_b^2$ on the triangular simplex $T^2 \equiv \{0 \le \xi_a, \xi_b, \xi_c \le 1 : \xi_a + \xi_b + \xi_c = 1\}$, with $\xi_{ab} = \xi_a - \xi_b$, $\chi_{ab} = \xi_a + \xi_b$.

5.2.1.4 Vertex Polynomials

The lowest order interpolatory set is obtained by setting $p = 1$ in (5.14) and (5.15) and is defined by the linear vertex functions

$$V_1 = \xi_a \tag{5.38}$$

where the subscript 1 denotes the order of the polynomial function. The vertex functions are obtained by setting $\xi_a = \xi_1$, $\xi_a = \xi_2$, $\xi_a = \xi_3$ (and also $\xi_a = \xi_4$ for the tetrahedral cell). The functions (5.38) are vertex functions because they vanish on the $\xi_a = 0$ face and attain a value of unity on the vertex $\xi_a = 1$ opposite that face. The functions (5.38) define the first-order hierarchical sets.

The polynomial sets (5.32), (5.34), and (5.38) are hierarchical in the sense that the $(p+1)$th-order set contains all the functions of the pth-order set. The hierarchical edge and face-based polynomials up to global order 7 are reported in Table 5.2.

In terms of the *true* local variables ξ_1, ξ_2, ξ_3 (and by assuming $\xi_4 = 0$), the *dummy* couple (ξ_a, ξ_b) associated with the $\xi_1 = 0$ edge could either be (ξ_2, ξ_3) or (ξ_3, ξ_2), but the sign of the auxiliary variable $\xi_{ab} = \xi_a - \xi_b$ is obviously different in the two cases. One possible scheme to avoid any sign ambiguity is to select a reference direction on each edge from lower to higher global vertex numbers, and order the local variables of the (ξ_a, ξ_b) couples according to the edge's reference direction. In other words, because of the symmetry or antisymmetry of the Jacobi polynomials, a reversal of an edge-based basis function's sign may be necessary in order to maintain the continuity of the scalar quantity between adjacent elements with an edge in common. The way to choose the reference directions for the face-based polynomials of the tetrahedral cell is discussed below when dealing with hierarchical vector bases (see Section 5.4.1).

5.2.1.5 Condition Number Comparison

The principal concern of hierarchical bases tends to be the matrix conditioning arising from their use. To establish that the rate of growth in condition number for the hierarchical bases of this section is not substantially worse than that of the interpolatory bases of (5.14) and (5.15), Figure 5.2 reports results for the matrix condition numbers of the local Gram matrix **G**, computed with integrals over the triangular simplex T^2 (at top) or the tetrahedral simplex T^3 (at bottom) (see (5.1)). Table 5.3 explicitly reports the matrix condition numbers of the local Gram matrix **G** for rectilinear triangular cells. The individual element mass-matrix condition numbers reported in the table and in the figure are valid for rectilinear (triangular or tetrahedral) cells of any shape since the Jacobian of the transformation from parent to child space is constant if the cell is rectilinear in the child-domain. As sometimes done in the vector FEM literature,

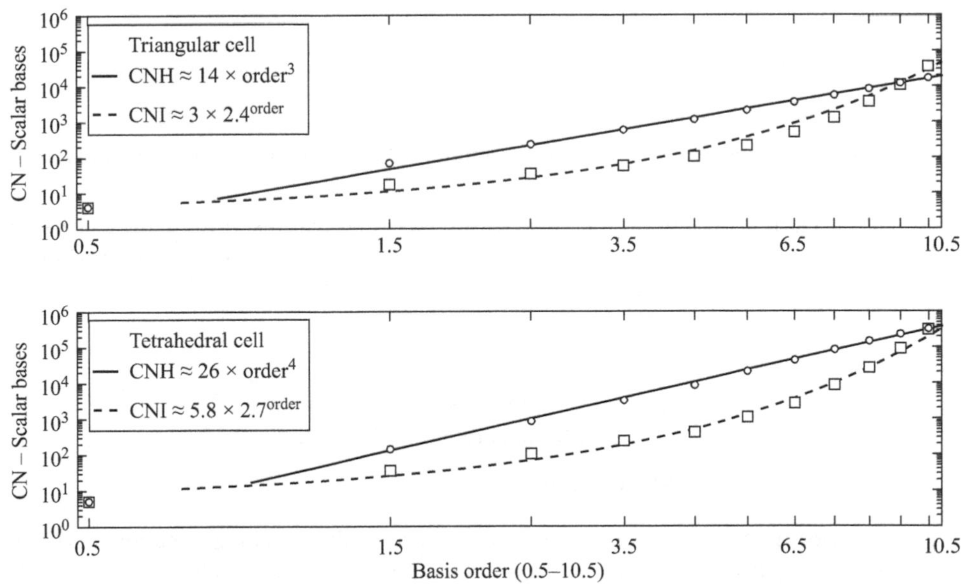

Figure 5.2 Individual element mass-matrix condition numbers for the hierarchical (CNH) and the interpolatory (CNI) scalar polynomial bases up to order 10.5, for triangular (at top) and tetrahedral (at bottom) rectilinear cells of any shape. The figure compares (using logarithmic scales) the condition numbers with reference growth-rate lines. The hierarchical condition numbers, CNH, are reported by circles; the interpolatory condition numbers, CNI, are reported by squares. The solid-lines depict a growth rate $g_H = 14 \times (\text{order})^3$, $g_H = 26 \times (\text{order})^4$ for the triangular and tetrahedral cells, respectively. The dashed-lines represent an exponential growth rate $g_I = 3 \times 2.4^{\text{order}}$, $g_I = 5.8 \times 2.7^{\text{order}}$ for the triangular and tetrahedral cells, respectively. The condition number growth rate is exponential for the interpolatory bases and polynomial for the hierarchical bases.

Table 5.3 Individual element mass-matrix condition numbers for the triangular scalar bases.

Basis order	p	G-matrix order	CNI	CNH	CN for the bases of [4]
0.5	1	3	4	4	4
1.5	2	6	17.21	69.99	271.1
2.5	3	10	33.97	225.8	1.563×10^4
3.5	4	15	57.74	576.9	1.035×10^6
4.5	5	21	103.4	1,123	8.815×10^7
5.5	6	28	214.9	2,101	8.093×10^9
6.5	7	36	494.6	3,441	7.624×10^{11}
7.5	8	45	1,269	5,525	7.267×10^{13}
8.5	9	55	3,594	8,193	6.939×10^{15}
9.5	10	66	10,850	11,997	1.143×10^{18}
10.5	11	78	34,799	16,661	6.369×10^{19}

Individual element mass-matrix condition numbers for the hierarchical (CNH) and interpolatory (CNI) scalar polynomial bases up to order 10.5, for rectilinear triangular cells of any shape. The last column on the right-hand side of the table reports the condition numbers obtained using the *unscaled* hierarchical bases given in [4] that do not employ orthogonal polynomials.

the basis order in the table and in the figure is denoted by a half integer: for example, "order 0.5" denotes the $p = 1$ scalar bases. The results show that

- the condition number growth rate is exponential for the interpolatory bases and polynomial for the hierarchical bases of this section;
- the condition number (CNH) of the hierarchical base of order p is of the same order of the condition number (CNI) of the interpolatory base of order $p + 3$, for $2 \le p \le 5$; while, for bases of the same order p, one gets

$$\begin{aligned} &\mathrm{CNH} = \mathrm{CNI} && \text{for } p = 1 \text{ (at order 0.5)} \\ &\mathrm{CNH} < \mathrm{CNI} && \begin{cases} \text{for triangular scalar bases and } p \ge 10 \\ \text{for tetrahedral scalar bases and } p \ge 11 \end{cases} \\ &\mathrm{CNH} > \mathrm{CNI} && \text{otherwise} \end{aligned}$$

At any rate bases of order higher than the sixth are very rarely used in applications. With this in mind, we can say that, in practice, the condition number of the hierarchical bases is higher than the condition number of the interpolatory bases of the same order by a factor from 4 to 10 for triangular cells, or from 4 to 20 for tetrahedral cells, respectively (see Figure 5.3). For these orders, the hierarchical condition numbers are higher than the interpolatory ones primarily because the vertex functions of the hierarchical bases do not change with the basis order, while those of the interpolatory bases do change with the basis order.

To prove how necessary orthogonalization is in forming hierarchical bases, Table 5.3 also reports (in the right-hand column) the condition numbers obtained from the unscaled, non-orthogonal hierarchical bases given in [4]; the poor conditioning associated with those bases can be slightly improved by appropriate scaling or, as specified in [4], by preconditioning algorithms that, more likely, are able to compensate for poor scale factors rather than for an inherent lack of linear independence.

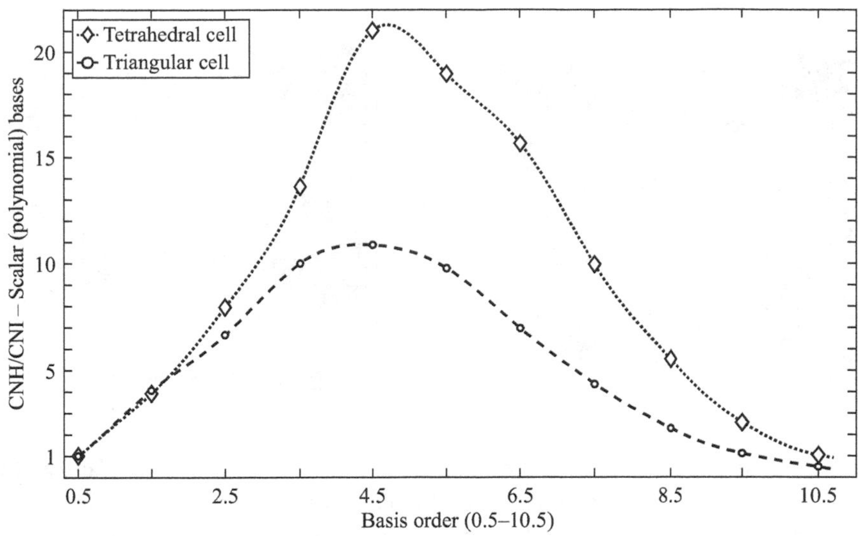

Figure 5.3 The ratio CNH/CNI of the condition number of the local Gram matrix obtained with hierarchical bases (CNH) to that obtained with interpolatory bases (CNI), versus the basis order for the triangular and the tetrahedral rectilinear cells.

Although the hierarchical bases are constructed with an orthogonalization process, the *individual* Gram matrix is not diagonal because orthogonality is enforced only between the edge-based functions associated with the same edge. That is, the edge-based functions of a given edge are not orthogonal to those of a different edge, nor they are orthogonal to the face-based or volume-based functions. Similarly, the face-based functions of a given face are mutually orthogonal but they are not orthogonal to those of a different face, neither they are orthogonal to the volume-based functions. The volume-based functions are mutually orthogonal, but they are not orthogonal to any edge- or face-based function. For example, Figure 5.4 graphically displays the structure of the local Gram matrix obtained for hierarchical scalar polynomials of order $p = 7$ defined on a rectilinear tetrahedron. In Figure 5.4, the zero entries are shown by white cells and the non-zero entries by black cells. In general, the *individual* Gram matrices obtained with interpolatory bases have no zero entries.

The condition number of the individual Gram matrices is studied to assess the degree of linear independence of the basis functions. In finite element applications, it is normally expected that the condition number of the global system matrix is lowered by using bases associated with individual Gram matrix having a lower condition number. At any rate, the condition number of the global system matrix obtained with any finite method can be improved by appropriate conditioning algorithms.

In other words, the hierarchical polynomial scalar bases for the triangle and the tetrahedron were derived in this subsection by using a procedure aimed at enhancing the linear independence of the basis function set as measured by the conditioning of the *mass* matrix (5.1). Different hierarchical polynomial scalar bases and orthogonalization techniques to improve the matrix conditioning of the *stiffness* matrix are available in the literature; those alternative bases are useful when dealing with more specific problems such as Poisson problems for which the mass matrix does not arise.

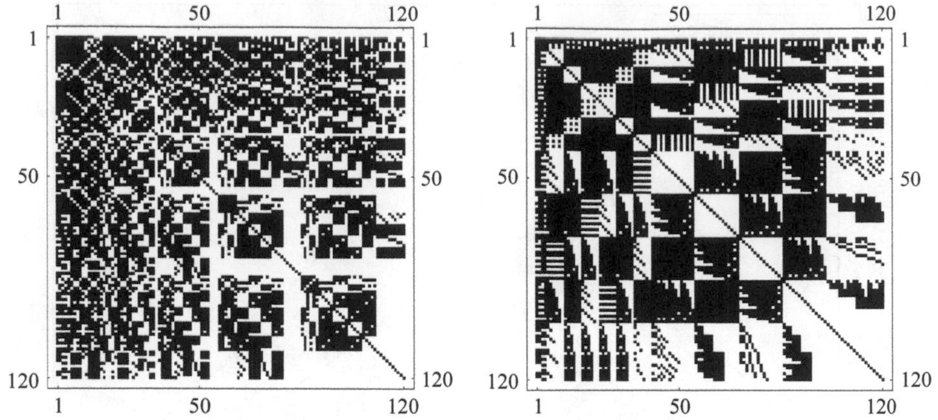

Figure 5.4 Local Gram matrix structure obtained for the $p = 7$ hierarchical scalar base (of half-integer order 6.5) on a rectilinear tetrahedron. The zero entries of the matrix are shown by white cells; the non-zero entries by black cells. The size of the local Gram matrix is 120×120. The matrix at left is obtained by ordering the basis functions in a hierarchical fashion, and for increasing polynomial order, by starting with the four vertex functions, the edge-based functions of order $p = 2$ and $p = 3$, the face-based functions of order $p = 3$ and so on, according to the scheme edge based + face based + volume based of order n followed by edge based + face based + volume based of order $n + 1$, etc.

The matrix at right is obtained by ordering the basis functions by group to highlight the mutual orthogonality of the functions belonging to the same group. The first group for the four vertex functions produces the sub-matrix of order four at the top-left corner. The groups that follow are associated with the edge-based functions, where the six edges of the tetrahedron produce the six diagonal sub-matrices (located along the main diagonal of the matrix) of order $(p - 1) = 6$. Those groups are then followed by the face-based function groups where the four tetrahedron faces produce the four diagonal sub-matrices of order $(p - 1)(p - 2)/2 = 15$. The last group for the volume-based functions produces the diagonal sub-matrix at the bottom-right corner of order $(p - 1)(p - 2)(p - 3)/6 = 20$.

5.2.2 Quadrilateral Bases

The one-dimensional scalar bases of Section 5.1 can be used to expand functions on a two- or three-dimensional space along each of its independent coordinates. The hierarchical scalar bases of the quadrilateral and the brick element can then be simply obtained by forming the Cartesian products of the one-dimensional bases of the appropriate variables. This construction procedure is straightforward, and we only need to separate the basis functions into different groups formed by the vertex, the edge, the face and, for the brick element, by the volume-based (or bubble) functions.

For the quadrilateral element of normalized coordinates $\{\xi_1, \xi_3; \xi_2, \xi_4\}$ (discussed in Section 2.5), it is convenient to write, as usual, the basis functions in terms of the dummy coordinates $\{\xi_a, \xi_c; \xi_b, \xi_d\}$ with

$$\begin{aligned} \xi_a + \xi_c &= 1 \\ \xi_b + \xi_d &= 1 \end{aligned} \tag{5.39}$$

For the $p=1$ order, there are four quadratic vertex functions

$$V_2 = \xi_a \, \xi_b \tag{5.40}$$

obtained for $\xi_a = \xi_1$, ξ_3, and $\xi_b = \xi_2$, ξ_4, that attain a unit value on the $\xi_a = \xi_b = 1$ vertex and vanish on the other three vertices.

As the order of the polynomial set increases, for $p \geq 2$, the set contains $(p-1)$ edge-based functions per edge. Those associated with the dummy $\xi_c = 0$ (or $\xi_a = 1$) edge vanish along the other three edges $\xi_a, \xi_b, \xi_d = 0$ and are expressed by

$$E_{m3} = \xi_a \, \xi_b \, \xi_d \, \mathcal{U}_m(\xi_{bd}), \quad m = 0, 1, \ldots, p-2 \tag{5.41}$$

with $\mathcal{U}_m(\xi_{bd})$ given in (5.10) and

$$\xi_{bd} = \xi_b - \xi_d \tag{5.42}$$

The four edges are spanned by setting $\xi_c = \xi_1, \xi_2, \xi_3$, and ξ_4, with $\xi_a = \xi_3, \xi_4, \xi_1, \xi_2$, respectively.

Finally, for $p \geq 2$, there are $(p-1)^2$ face-based (bubble) functions of the form

$$F_{mn4} = \xi_a \, \xi_c \, \mathcal{U}_m(\xi_{ac}) \, \xi_b \, \xi_d \, \mathcal{U}_n(\xi_{bd}), \quad m, n = 0, 1, \ldots, p-2 \tag{5.43}$$

that vanish along the four quadrilateral edges.

The order of V_2, E_{m3}, and F_{mn4} is given by the sum of the subscripts, that is, 2 for V_2, $(m+3)$ for E_{m3}, and $(m+n+4)$ for F_{mn4}. Clearly, this hierarchical base is formed by a total of $(p+1)^2$ basis functions since

$$(p+1)^2 = 4 + 4(p-1) + (p-1)^2$$

The maximum polynomial order is $2p$, in common with the interpolatory base

$$\begin{aligned} \alpha_{ik;j\ell}(p, \xi_a, \xi_c; \xi_b, \xi_d) = {} & R_i(p, \xi_a) \, R_k(p, \xi_c) \, R_j(p, \xi_b) \, R_\ell(p, \xi_d) \\ & i, j = 0, 1, \ldots, p, \\ & \text{with } i + k = j + \ell = p \end{aligned}$$

given in (2.68) and discussed in Section 2.5.2.

5.2.3 Brick Bases

Similarly, the Cartesian product of the one-dimensional hierarchical scalar bases of the previous Section 5.1 yields the hierarchical scalar bases for the brick or hexahedral element of normalized coordinates $\{\xi_1, \xi_4; \xi_2, \xi_5; \xi_3, \xi_6\}$ (discussed in Section 2.7). The basis functions are once again given in terms of the dummy coordinates $\{\xi_a, \xi_d; \xi_b, \xi_e; \xi_c, \xi_f\}$ with

$$\begin{aligned} \xi_a + \xi_d &= 1 \\ \xi_b + \xi_e &= 1 \\ \xi_c + \xi_f &= 1 \end{aligned} \tag{5.44}$$

For the $p=1$ order, the basis set consists entirely of eight *cubic* vertex functions of the form

$$V_2 = \xi_a \, \xi_b \, \xi_c \tag{5.45}$$

that attain a unit value on the $\xi_a = \xi_b = \xi_c = 1$ vertex and vanish on the other seven vertices. In terms of the true parent variables $\{\xi_1, \xi_4; \xi_2, \xi_5; \xi_3, \xi_6\}$ the vertex functions are

$$\begin{Bmatrix} \xi_1 \, \xi_2 \, \xi_3, \; \xi_1 \, \xi_5 \, \xi_3, \; \xi_1 \, \xi_2 \, \xi_6, \; \xi_1 \, \xi_5 \, \xi_6, \\ \xi_4 \, \xi_2 \, \xi_3, \; \xi_4 \, \xi_5 \, \xi_3, \; \xi_4 \, \xi_2 \, \xi_6, \; \xi_4 \, \xi_5 \, \xi_6 \end{Bmatrix}$$

For $p \geq 2$, the set is augmented with $(p-1)$ edge-based functions per edge. Those associated with the dummy $\xi_e = \xi_f = 0$ edge vanish along the other 11 edges and are expressed by

$$E_{m4} = \xi_a \, \xi_d \, \mathcal{U}_m(\xi_{ad}) \, \xi_b \, \xi_c, \quad m = 0, 1, \ldots, p-2 \tag{5.46}$$

with $\mathcal{U}_m(\xi_{ad})$ given in (5.10). The edge-based functions in terms of the true parent variables are obtained by setting in (5.46)

$$(\xi_a, \xi_d) = (\xi_1, \xi_4), \; (\xi_b, \xi_c) = \begin{cases} (\xi_2, \xi_3), \; (\xi_2, \xi_6), \\ (\xi_5, \xi_3), \; (\xi_5, \xi_6) \end{cases} \tag{5.47}$$

$$(\xi_a, \xi_d) = (\xi_2, \xi_5), \; (\xi_b, \xi_c) = \begin{cases} (\xi_1, \xi_3), \; (\xi_1, \xi_6), \\ (\xi_4, \xi_3), \; (\xi_4, \xi_6) \end{cases} \tag{5.48}$$

$$(\xi_a, \xi_d) = (\xi_3, \xi_6), \; (\xi_b, \xi_c) = \begin{cases} (\xi_1, \xi_2), \; (\xi_1, \xi_5), \\ (\xi_4, \xi_2), \; (\xi_4, \xi_5) \end{cases} \tag{5.49}$$

For $p \geq 2$, the set is also augmented with $(p-1)^2$ face-based functions per face. Those associated with the dummy $\xi_f = 0$ (or $\xi_c = 1$) face have the form

$$F_{mn5} = \xi_a \, \xi_d \, \mathcal{U}_m(\xi_{ad}) \, \xi_b \, \xi_e \, \mathcal{U}_n(\xi_{be}) \, \xi_c, \quad m, n = 0, 1, \ldots, p-2 \tag{5.50}$$

and vanish along the other five brick faces $\xi_a, \xi_b, \xi_c, \xi_d, \xi_e = 0$. The face-based functions in terms of the true parent variables are obtained by setting in (5.50)

$$(\xi_a, \xi_d) = (\xi_2, \xi_5), \; (\xi_b, \xi_e) = (\xi_3, \xi_6), \; \xi_c = \xi_1, \xi_4 \tag{5.51}$$

$$(\xi_a, \xi_d) = (\xi_1, \xi_4), \; (\xi_b, \xi_e) = (\xi_3, \xi_6), \; \xi_c = \xi_2, \xi_5 \tag{5.52}$$

$$(\xi_a, \xi_d) = (\xi_1, \xi_4), \; (\xi_b, \xi_e) = (\xi_2, \xi_5), \; \xi_c = \xi_3, \xi_6 \tag{5.53}$$

Finally, for $p \geq 2$, the set also contains $(p-1)^3$ volume-based (bubble) functions of the form

$$\begin{aligned} V_{\ell mn6} = \; & \xi_a \, \xi_d \, \mathcal{U}_\ell(\xi_{ad}) \, \xi_b \, \xi_e \, \mathcal{U}_m(\xi_{be}) \, \xi_c \, \xi_f \, \mathcal{U}_n(\xi_{cf}) \\ & \ell, m, n = 0, 1, \ldots, p-2 \end{aligned} \tag{5.54}$$

with $\{\xi_a, \xi_d; \xi_b, \xi_e; \xi_c, \xi_f\} = \{\xi_1, \xi_4; \xi_2, \xi_5; \xi_3, \xi_6\}$. The bubble functions vanish along all the edges and faces of the brick.

The order of V_3, E_{m4}, F_{mn5}, and $V_{\ell mn6}$ is given by the sum of the subscripts, that is, 3 for V_3, $(m+4)$ for E_{m4}, $(m+n+5)$ for F_{mn5}, and $(\ell+m+n+6)$ for $V_{\ell mn6}$. Clearly, this hierarchical base is formed by a total of $(p+1)^3$ basis functions since

$$(p+1)^3 = 8 + 12(p-1) + 6(p-1)^2 + (p-1)^3$$

The maximum polynomial order is $3p$, in common with the interpolatory base

$$\begin{aligned}\alpha_{i\ell;jm;kn}(p,\boldsymbol{\xi}) &= R_i(p,\xi_a)\,R_\ell(p,\xi_d)\,R_j(p,\xi_b)\,R_m(p,\xi_e)\,R_k(p,\xi_c)\,R_n(p,\xi_f)\\ &\quad i,j,k,\ell,m,n = 0,1,\ldots,p,\\ &\quad \text{with } i+\ell = j+m = k+n = p\end{aligned}$$

given in (2.114) and discussed in Section 2.7.2.

5.2.4 Prism Bases

The Cartesian product of the one-dimensional scalar bases of Section 5.1 with the triangular scalar bases of Section 5.2.1 yields the *normalized* hierarchical scalar bases for the prism element of normalized coordinates $\{\xi_1, \xi_2, \xi_3; \xi_4, \xi_5\}$ (discussed in Section 2.8). The basis functions are once again given in terms of the dummy coordinates $\{\xi_a, \xi_b, \xi_c; \xi_d, \xi_e\}$ with

$$\begin{aligned}\xi_a + \xi_b + \xi_c &= 1\\ \xi_d + \xi_e &= 1\end{aligned} \tag{5.55}$$

For the $p=1$ order, the basis set consists entirely of the six *quadratic* vertex functions of the form

$$V_2 = \xi_a\,\xi_d \tag{5.56}$$

that attain a unit value on the $\xi_a = \xi_d = 1$ vertex and vanish on the other five vertices. In terms of the true parent variables $\{\xi_1, \xi_2, \xi_3; \xi_4, \xi_5\}$, the vertex functions are

$$\begin{Bmatrix}\xi_1\,\xi_4, & \xi_2\,\xi_4, & \xi_3\,\xi_4,\\ \xi_1\,\xi_5, & \xi_2\,\xi_5, & \xi_3\,\xi_5\end{Bmatrix}$$

For $p \geq 2$, the set is augmented with $(p-1)$ edge-based functions per edge. Those associated with the dummy $\xi_c = 0$ edge of the $\xi_d = 0$ triangular face are expressed by

$$E_{m3}(\boldsymbol{\xi}) = E_{m2}(\xi_{ab}, \chi_{ab})\,\xi_e = \xi_a\,\xi_b\,\mathcal{U}_m(\xi_{ab}, \chi_{ab})\,\xi_e, \quad m = 0,1,\ldots,p-2 \tag{5.57}$$

with E_{m2} given in (5.34). The order of these normalized polynomials is $g = m+3$; they vanish along the $\xi_a, \xi_b, \xi_e = 0$ faces and are mutually orthogonal on the $\xi_c = \xi_d = 0$ edge, as well as on the prism volume (because orthogonal on the $\xi_d = 0$ triangular face). Similarly, the $(p-1)$ edge-based polynomials associated with the dummy edge in common to the quadrilateral faces $\xi_b, \xi_c = 0$ are

$$E_{m3} = \xi_a\,\xi_d\,\xi_e\,\mathcal{U}_m(\xi_{de}), \quad m = 0,1,\ldots,p-2 \tag{5.58}$$

with $\mathcal{U}_m(\xi_{bd})$ given in (5.10) and $\xi_{de} = \xi_d - \xi_e$.

For $p \geq 2$, there are $(p-1)^2$ face-based functions associated with each quadrilateral face. Those associated with the $\xi_a = 0$ face are

$$F_{mn4} = \xi_b\, \xi_c\, \mathcal{U}_m(\xi_{bc})\, \xi_d\, \xi_e\, \mathcal{U}_n(\xi_{de}), \quad m, n = 0, 1, \ldots, p-2 \tag{5.59}$$

and vanish along the $\xi_b, \xi_c = 0$ quadrilateral faces, as well as on the $\xi_d, \xi_e = 0$ triangular faces.

For $p \geq 3$, we define $(p-1)(p-2)/2$ face-based polynomials on each triangular face ($\xi_4 = 0$ or $\xi_5 = 0$) as

$$\begin{aligned} F_{mn4}(\boldsymbol{\xi}) &= \xi_d\, F_{mn3}(\xi_a, \xi_b, \xi_c) = \xi_d\, \xi_a\, \xi_b\, \xi_c\, \mathcal{U}_{mn}(\xi_a, \xi_b, \xi_c), \\ & m, n = 0, 1, \ldots, p-3, \\ & 0 \leq m + n \leq p-3 \end{aligned} \tag{5.60}$$

where $\xi_d = \xi_4, \xi_5$, and F_{mn3} is given in (5.32). These functions have order $m + n + 4$.

Finally, for $p \geq 3$, we define $(p-1)^2(p-2)/2$ volume-based (bubble) polynomials

$$\begin{aligned} V_{\ell mn5}(\boldsymbol{\xi}) &= \xi_d\, \xi_e\, \mathcal{U}_\ell(\xi_{de})\, F_{mn3}(\xi_a, \xi_b, \xi_c) \\ &= \xi_d\, \xi_e\, \mathcal{U}_\ell(\xi_{de})\, \xi_a\, \xi_b\, \xi_c\, \mathcal{U}_{mn}(\xi_a, \xi_b, \xi_c), \\ & \ell = 0, 1, \ldots, p-2, \quad m, n = 0, 1, \ldots, p-3, \\ & 0 \leq m + n \leq p-3 \end{aligned} \tag{5.61}$$

of order $\ell + m + n + 5$, with $\mathcal{U}_\ell(\xi_{de})$ and F_{mn3} given in (5.10) and (5.32), respectively.

Clearly, this hierarchical base is formed by a total of $(p+1)^2(p+2)/2$ basis functions since

$$\frac{(p+1)^2(p+2)}{2} = 6 + 9(p-1) + 3(p-1)^2 + 2\frac{(p-1)(p-2)}{2} + \frac{(p-1)^2(p-2)}{2}$$

The maximum polynomial order is $2p$, in common with the interpolatory base

$$\begin{aligned} \alpha_{ijk;\ell m}(p, \boldsymbol{\xi}) &= R_i(p, \xi_1)\, R_j(p, \xi_2)\, R_k(p, \xi_3)\, R_\ell(p, \xi_4)\, R_m(p, \xi_5) \\ & i, j, k, \ell, m = 0, 1, \ldots, p, \quad i + j + k = \ell + m = p \end{aligned}$$

given in (2.108) and discussed in Section 2.8.1.

5.3 Hierarchical Curl-Conforming Vector Bases

Vector basis functions are widely used in electromagnetics for volumetric discretizations of the vector Helmholtz equation in 2D and 3D and surface and volume discretizations of the electric and magnetic field integral equations in 3D. Interpolatory functions of this type have been introduced in Chapters 3 and 4. In this section, we introduce hierarchical vector bases in order to facilitate adaptive refinement procedures. Since hierarchical bases often exhibit poor linear independence as the order of the representation is increased, we attempt to develop vector bases that alleviate the loss of linear independence.

Table 5.4 Classification of hierarchical curl-conforming vector bases available in the literature.

Bases given in [ref.], year	Group	Element shapes		General formulas available	Explicit bases presented to degree
		2D	3D		
[6], 1993	A	None			2
[7], 1997	A	△	None	Yes	
[8], 1997	A	△	None	Yes	
[9], 1998 [10], 1999	B	△			2.5
[11], 1999	A	△			3
[12], 2001	A	None			3
[13], 2001	A	△ □	None	Yes	
[14], 2003	A	None		Yes	
[15], 2003	C	△	None		4.5
[16], 2004	C	△	None		4.5
[17], 2006	C	None			4.5
[18], 2005 [19], 2006	A	△ □		Yes	
This book	B	△ □	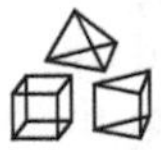		6.5

The published hierarchical basis functions here are classified into three groups: (A) those that span complete polynomial vector spaces, (B) those that span the mixed-order spaces of Nédélec [5], and (C) those with subsets that exactly span both types of spaces. The procedure to obtain some of the bases considered in the table is quite complex; for this reason, the table also shows the maximum polynomial degree of the bases explicitly reported in the relevant publication.

Papers proposing hierarchical vector basis functions began appearing in the electromagnetics literature in the early 1990s. Most of the proposed basis functions are of the curl-conforming variety, which on triangles or quadrilaterals are easily converted into divergence-conforming functions. Table 5.4 summarizes some of the existing curl-conforming hierarchical vector bases suitable for two- and/or three-dimensional cells.

The published basis functions can be classified into three groups[7]: (A) those that span complete polynomial vector spaces, (B) those that span the mixed-order spaces of Nédélec [5] (sometimes known as *reduced-gradient* spaces for curl-conforming functions), and (C) those with subsets that exactly span both types of spaces. As an example, the interpolatory vector basis functions discussed in Chapter 4 fall into group B, since those basis functions span the mixed-order spaces of Nédélec but do not contain subsets that exactly span polynomial-complete spaces. The hierarchical bases presented in the following subsections also belong to group B.

The hierarchical curl-conforming vector bases described in this section were originally developed and discussed by the authors of this book in a long series of papers [20–28]. Two references that have appeared since the publication of these papers propose new bases for triangles [29] and tetrahedrons [30] and use an orthogonalization procedure somewhat similar to that employed here. Our bases have four distinguishing features: (a) the vector basis functions are subdivided from the outset into three different groups of edge-, face-, and volume-based functions; (b) each basis function is obtained by using one *generating* edge-, face-, or volume-based polynomial whose analytical expression involves all the dependent parent variables that describe the cell; (c) in each group, all the generating polynomials are mutually orthogonal independent of the definition domain of the inner product, i.e. either the volume, the face, or the edge of the cell; (d) the hierarchical vector functions are either symmetric or antisymmetric with respect to the parent variables that describe each edge and face of the cell.

The four features outlined above yield the following outcomes, respectively: (a) different individual polynomial orders can be used on each edge, face, and volumetric element of a given mesh, thereby facilitating the use of vector bases of different orders together in the same mesh (p-adaption); (b) the generating polynomials for the edge-, the face-, and the volume-based vector functions can be implemented in routines which can be used without modification to build the vector basis functions for three- or two-dimensional cells; this greatly simplifies the implementation of the numerical codes; (c) our higher order bases maintain excellent linear independence because they are derived after an *analytical* orthogonalization of the generating scalar polynomials, which is done in the element parent domain; and (d) the procedure to enforce the conformity of the approximation across element interfaces is drastically simplified.

The outcome (c) is of importance because hierarchical bases are typically ill conditioned at high orders and usually necessitate a cumbersome (partial) orthogonalization process to improve system conditioning. As illustrated by Abdul-Rahman and Kasper [31], considerable effort is required to directly orthogonalize the vector functions. In contrast, our bases are defined from orthogonal generating scalar polynomials, to enhance the conditioning of the system matrices. The outcome (d) is also of importance because enforcement of the continuity of the tangential component across adjacent elements (for the curl-conforming case) can be difficult [14, 32]; the proposed basis functions reduce this problem to one of determining the correct sign of each basis function with respect to an arbitrarily selected reference direction along adjacent elements.

---

[7] Although several families of bases appearing in Table 5.4 are considered to be "Nédélec" bases, here we classify them as "type A" because they do not contain subspaces that properly span the reduced-gradient spaces of [5] on triangles or tetrahedra. For instance, neither the face-based R_{30} functions of [11] nor the R^3 functions of [12] properly span the Nédélec space of order 2.5; the "type 2" element-based functions of [18] do not properly span the Nédélec spaces of order 1.5 or higher. Thus, these functions are listed here as belonging to type A.

5.3.1 Tetrahedral and Triangular Bases

The hierarchical vector functions are obtained by a three-step process. First, we orthogonalize on a given parent element appropriate linear combinations of the interpolatory scalar polynomials given in Chapter 4, to obtain hierarchical scalar polynomials. These polynomials are then multiplied by the zeroth-order vector functions of the element under consideration to obtain a set of vector functions. Finally, using a procedure similar to the one given in Chapter 4, any redundant basis function is eliminated from the resulting vector set.

As previously observed in connection with the hierarchical scalar bases of Section 5.2.1, it is convenient to introduce dummy parent variables. The bases for the tetrahedral and the triangular cells are then derived at the same time by simply considering the triangular cell described by the three parent variables (ξ_a, ξ_b, ξ_c) as the bounding $(\xi_d = 0)$ face of the tetrahedral cell described by the four parent variables $\boldsymbol{\xi} = \{\xi_a, \xi_b, \xi_c, \xi_d\}$. Let us consider a tetrahedral element whose faces are labeled by these four parent variables and, at the same time, the triangular element defined by the $\xi_d = 0$ face of this tetrahedron (see Figure 5.5). In terms of these parent variables, the zeroth-order curl-conforming vector function *associated* with the

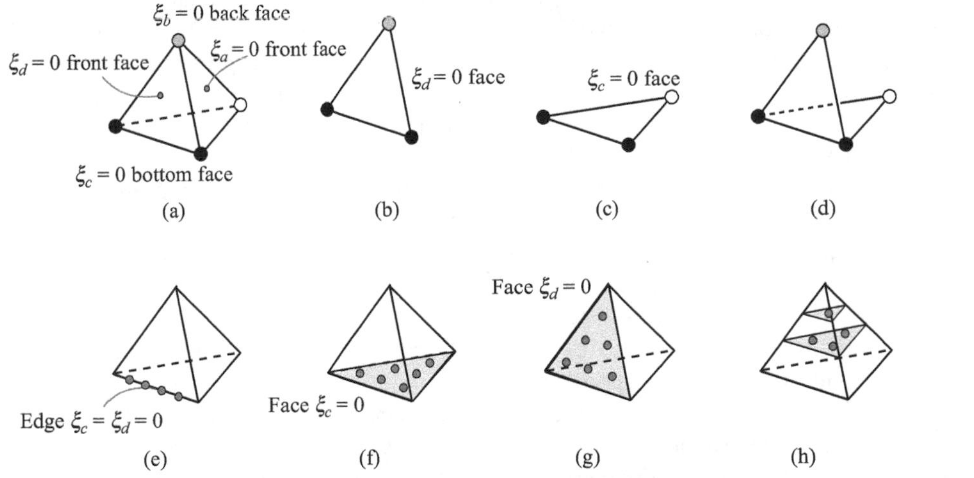

Figure 5.5 The hierarchial polynomials that generate the curl-conforming functions associated with the zeroth-order vector function relative to the edge formed by the intersection of the $\xi_c = 0$ and $\xi_d = 0$ faces have features similar to those of the *equivalent* interpolatory polynomials of Chapter 4. Vanishing regions: (a) the edge-based polynomials are different from zero on the cell boundaries; (b) all the polynomials based on the $\xi_c = 0$ face vanish on the face $\xi_d = 0$; (c) all the polynomials based on the $\xi_d = 0$ face vanish on the face $\xi_c = 0$; and (d) the volume-based polynomials vanish on both the $\xi_c = 0$ and $\xi_d = 0$ faces. The number of hierarchical and interpolatory edge-, face-, and volume-based polynomials of order p is the same. There are: (e) $(p+1)$ edge-based polynomials; (f) and (g) $p(p+1)/2$ polynomials based on the $\xi_c = 0$ and the $\xi_d = 0$ faces; and (h) $p(p^2 - 1)/6$ volume-based polynomials. The figures from (e) to (h) show the interpolation nodes of the interpolatory polynomials of Chapter 4 for $p = 3$.

© 2011 IEEE. Reprinted, with permission, from R. D. Graglia, A. F. Peterson, and F. P. Andriulli, "Curl-conforming hierarchical vector bases for triangles and tetrahedra," *IEEE Trans. Antennas Propag.*, vol. 59, no. 3, pp. 950–959, Mar. 2011.

edge at the intersection of the $\xi_c = 0$ and $\xi_d = 0$ face reads $\mathbf{\Omega}_{cd}(r) = (\xi_b \nabla \xi_a - \xi_a \nabla \xi_b)$ and $\mathbf{\Omega}_c(r) = -(\xi_b \nabla \xi_a - \xi_a \nabla \xi_b)$ for the tetrahedral and the triangular elements, respectively (see Tables 4.2 and 4.4 in Chapter 4). Despite of the sign difference in the previous two expressions, which can be eventually eliminated by reorienting the triangle unit normal, both functions turn out to be antisymmetric with respect to the two parent variables ξ_a and ξ_b, since $\mathbf{\Omega}_{cd}(\xi_a, \xi_b) = -\mathbf{\Omega}_{cd}(\xi_b, \xi_a)$ and $\mathbf{\Omega}_c(\xi_a, \xi_b) = -\mathbf{\Omega}_c(\xi_b, \xi_a)$. Because of this property, the enforcement of the tangential continuity of the field across element boundaries is greatly simplified. The continuity of the tangential component is ensured by adjusting the basis function sign to correspond to an arbitrarily selected reference direction along the adjacent elements.

Higher order interpolatory bases are constructed in Chapter 4 by multiplying the zeroth-order vector functions with Silvester–Lagrange interpolatory polynomials. Here, linear combinations of those interpolatory polynomials are used to obtain symmetric or antisymmetric hierarchical scalar polynomials which, within each group, are constructed *a priori* to be mutually orthogonal.

The hierarchical vector functions are constructed using the same technique given in Chapter 4, where we simply *substitute* the new scalar hierarchical polynomials for the interpolatory ones of Chapter 4.

With reference to Figure 5.5, the interpolatory polynomials associated with the edge at the intersection of the $\xi_c = 0$ and $\xi_d = 0$ faces are subdivided into four different groups. The first group is formed by all the polynomials interpolating (up to a given order) at the edge (Figure 5.5e) and that (in general) do not vanish on the other tetrahedral boundaries (Figure 5.5a); a second group is formed by the polynomials that interpolate (up to a given order) at the $\xi_c = 0$ face (Figure 5.5f) and that vanish on the face $\xi_d = 0$ (Figure 5.5b); a third group is given by polynomials which interpolate at the $\xi_d = 0$ face (Figure 5.5g) and vanish on $\xi_c = 0$ (Figure 5.5c); the last group is given by the remaining interpolating polynomials which vanish on both the $\xi_c = 0$ and $\xi_d = 0$ faces (Figure 5.5d and h). Appropriate linear combinations of these interpolatory polynomials, together with extensive symmetry considerations, provide four groups of *normalized* orthogonal hierarchical polynomials. These polynomials are explicitly reported in Tables 5.5 and 5.6 up to the sixth order. We have obtained hierarchical families up to eleventh order with this approach.

In Tables 5.5 and 5.6, the first superscript s or a labels symmetric or antisymmetric polynomials of the ξ_a and ξ_b variables, respectively; similarly, the second superscript (used only for the edge- and the volume-based polynomials) labels symmetric and antisymmetric polynomials of the ξ_c and ξ_d variables.

All the edge-based hierarchical polynomials E_p of Table 5.5 are symmetric in the ξ_c and ξ_d variables. In Table 5.5, $P_p(\xi_{ab})$ indicates the Legendre polynomial of order p, with $\xi_{ab} = \xi_a - \xi_b$. The polynomials based on the edge $\xi_c = 0$ of the T^2 simplex (triangular element, with $\xi_a + \xi_b + \xi_c = 1$) are obtained by setting $\xi_d = 0$. For the T^3 simplex (tetrahedral element, with $\xi_a + \xi_b + \xi_c + \xi_d = 1$), the polynomials reported are those based on the edge $\xi_c = 0$, $\xi_d = 0$. Along its associated edge, E_p behaves as the Legendre polynomial $P_p(\xi_{ab})$.

For the face-based polynomials of Table 5.5, one has to set $\xi_d = 0$ while dealing with the T^2 simplex; for the T^3 simplex, the polynomials reported are those associated with the $\xi_d = 0$ face.

The edge-based hierarchical polynomials of Table 5.5 are orthogonal on the T^1, T^2, and T^3 simplices, while the face-based polynomials are orthogonal on both the T^2 and T^3 simplices.

The number of DoFs for curl-conforming bases of order p on a tetrahedron is $(p+1)(p+3)$ $(p+4)/2$; the number of DoFs for curl- and divergence-conforming bases of order p on a triangle is $(p+1)(p+3)$ (see Chapter 4). By following the same procedure reported in

Table 5.5 Vector bases on triangular and tetrahedral cells – edge- and face-based hierarchical polynomials up to the sixth order.

Edge-based polynomials $E_p(\boldsymbol{\xi})$ of global order p, mutually orthogonal on the T^1, T^2, and T^3 simplices:

$E_0^{ss} = P_0(\xi_{ab}) = 1$	$E_3^{as} = \sqrt{7}\,\{P_3(\xi_{ab}) - 3\,\chi_{cd}\,(\chi_{cd} - 2)\,\xi_{ab}/2\}$
$E_1^{as} = \sqrt{3}\;P_1(\xi_{ab}) = \sqrt{3}\,\xi_{ab}$	$E_4^{ss} = \sqrt{9}\left\{P_4(\xi_{ab}) + 3\,\chi_{cd}\,(\chi_{cd} - 2)\left(1 + 40\,\xi_a\,\xi_b - 9\,\chi_{ab}^2\right)/8\right\}$
$E_2^{ss} = \sqrt{5}\,\{P_2(\xi_{ab}) - \chi_{cd}\,(\chi_{cd} - 2)/2\}$	$E_5^{as} = \sqrt{11}\left\{P_5(\xi_{ab}) + 5\,\chi_{cd}\,(\chi_{cd} - 2)\,\xi_{ab}\left[3 + 56\,\xi_a\,\xi_b - 11\,\chi_{ab}^2\right]/8\right\}$

$$E_6^{ss} = \sqrt{13}\left\{P_6(\xi_{ab}) - 5\,\chi_{cd}\,(\chi_{cd} - 2)\left[1 + 43\,\chi_{ab}^4 + 84\,\xi_a\,\xi_b\,(1 + 12\,\xi_a\,\xi_b) - 20\,\chi_{ab}^2\,(1 + 21\,\xi_a\,\xi_b)\right]/16\right\}$$

Face-based polynomials $F_{mn}(\boldsymbol{\xi})$ of global order $p = (m + n)$, mutually orthogonal on the T^2 and T^3 simplices:

$$F_{01}^s = 2\sqrt{3}\,\xi_c$$

$$F_{02}^s = 2\sqrt{3}\,\xi_c\,(5\,\xi_c - 3 + 3\,\xi_d)$$

$$F_{11}^a = 6\sqrt{5}\,\xi_c\,\xi_{ab}$$

$$F_{03}^s = 2\sqrt{30}\,\xi_c\left[7\,\xi_c^2 - 8\,\xi_c + 2 + 2\,\xi_d\,(4\,\xi_c - 2 + \xi_d)\right]$$

$$F_{12}^a = 2\sqrt{30}\,\xi_c\,\xi_{ab}\,(7\,\xi_c - 3 + 3\,\xi_d)$$

$$F_{21}^s = 2\sqrt{210}\,\xi_c\,\chi_2$$

$$F_{04}^s = 2\sqrt{15}\,\xi_c\left\{42\,\xi_c^3 - 70\,\xi_c^2 + 35\,\xi_c - 5 + 5\,\xi_d\,[1 + \chi_{ab} + \chi_{ab}^2 - \xi_c\;(6 + 5\,\chi_{ab} - 8\,\xi_c)]\right\}$$

$$F_{13}^a = 2\sqrt{105}\,\xi_c\,\xi_{ab}\left[18\,\xi_c^2 - 16\,\xi_c + 3 - \xi_d\,(3 + 3\,\chi_{ab} - 13\,\xi_c)\right]$$

$$F_{22}^s = 10\sqrt{42}\,\xi_c\,\chi_2\,(3\,\xi_c - 1 + \xi_d)$$

$$F_{31}^a = 6\sqrt{70}\,\xi_c\,\chi_3$$

$$F_{05}^s = 2\sqrt{105}\,\xi_c\,\{66\,\xi_c^4 - 144\,\xi_c^3 + 108\,\xi_c^2 - 32\,\xi_c + 3 - \xi_d\left[3\,(1 + \chi_{ab}) \times (1 + \chi_{ab}^2) - \xi_c\,[29 + \chi_{ab}\,(26 + 23\,\chi_{ab})] + \xi_c^2\,(79 + 53\,\chi_{ab}) - 65\,\xi_c^3\right]\}$$

$$F_{14}^a = 6\sqrt{70}\,\xi_c\,\xi_{ab}\left\{33\,\xi_c^3 - 45\,\xi_c^2 + 18\,\xi_c - 2 + \xi_d\left[2\,(1 + \chi_{ab}^2 + \chi_{ab}\,(1 - 7\,\xi_c)) - 16\,\xi_c + 29\,\xi_c^2\right]\right\}$$

$$F_{23}^s = 6\sqrt{10}\,\xi_c\,\chi_2\left[55\,\xi_c^2 - 40\,\xi_c + 6 + 2\xi_d\,(3\xi_d + 20\xi_c - 6)\right]$$

$$F_{32}^a = 6\sqrt{35}\,\xi_c\,\chi_3\,(11\,\xi_c - 3 + 3\xi_d)$$

$$F_{41}^s = 6\sqrt{165}\,\xi_c\,\chi_4$$

$$F_{06}^s = 2\sqrt{42}\,\xi_c\left\{429\,\xi_c^5 - 1155\,\xi_c^4 + 1155\,\xi_c^3 - 525\,\xi_c^2 + 105\,\xi_c - 7 + 7\,\xi_d\left[1 + \chi_{ab} + \chi_{ab}^2 + \chi_{ab}^3 + \chi_{ab}^4\;-(14 + 13\,\chi_{ab} + 12\,\chi_{ab}^2 + 11\,\chi_{ab}^3)\,\xi_c + (61 + 48\,\chi_{ab} + 36\,\chi_{ab}^2)\,\xi_c^2 - 8\,(13 + 7\,\chi_{ab})\xi_c^3 + 61\,\xi_c^4\right]\right\}$$

$$F_{15}^a = 6\sqrt{70}\,\xi_c\,\xi_{ab}\left\{143\,\xi_c^4 - 264\,\xi_c^3 + 165\,\xi_c^2 - 40\,\xi_c + 3 + \xi_d\left[136\,\xi_c^3 - 3 \times (1 + \chi_{ab} + \chi_{ab}^2 + \chi_{ab}^3) + (37 + 34\,\chi_{ab} + 31\,\chi_{ab}^2)\,\xi_c - 2\,(64 + 47\,\chi_{ab})\,\xi_c^2\right]\right\}$$

$$F_{24}^s = 6\sqrt{35}\,\xi_c\,\chi_2\left\{143\,\xi_c^3 - 165\,\xi_c^2 + 55\,\xi_c - 5 + 5\,\xi_d\left[1 + \chi_{ab} + \chi_{ab}^2\;- (10 + 9\,\chi_{ab})\,\xi_c + 23\,\xi_c{}^2\right]\right\}$$

$$F_{33}^a = 14\sqrt{165}\,\xi_c\,\chi_3\left[13\,\xi_c{}^2 - 8\,\xi_c + 1 - \xi_d\,(2 - 8\,\xi_c - \xi_d)\right]$$

$$F_{42}^s = 6\sqrt{77}\,\xi_c\,\chi_4\;(13\,\xi_c - 3 + 3\,\xi_d)$$

$$F_{51}^a = 2\sqrt{3{,}003}\,\xi_c\,\chi_5$$

With $\xi_{ab} = \xi_a - \xi_b$, $\chi_{ab} = \xi_a + \xi_b$, and $\chi_{cd} = \xi_c + \xi_d$. Several equivalent polynomial expressions are obtained by using the dependency relation $\xi_a + \xi_b + \xi_c + \xi_d = 1$. The reported expressions were indeed made more compact by using the dependency relation and by setting:

$$\chi_2 = \xi_a{}^2 - 4\,\xi_a\,\xi_b + \xi_b{}^2; \qquad \chi_4 = \xi_a{}^4 - 16\,\xi_a{}^3\,\xi_b + 36\,\xi_a{}^2\,\xi_b{}^2 - 16\,\xi_a\,\xi_b{}^3 + \xi_b{}^4;$$

$$\chi_3 = \xi_{ab}\,(\xi_a{}^2 - 8\,\xi_a\,\xi_b + \xi_b{}^2);\; \chi_5 = \xi_{ab}\,(\xi_a{}^4 - 24\,\xi_a{}^3\,\xi_b + 76\,\xi_a{}^2\,\xi_b{}^2 - 24\,\xi_a\,\xi_b{}^3 + \xi_b{}^4).$$

The first superscripts s and a label symmetric and antisymmetric polynomials of the ξ_a and ξ_b variables, respectively; the second superscript s, used only for E_p, indicates that the edge-based polynomials are symmetric in the ξ_c and ξ_d variables. The table reports only the edge- and the face-based polynomials associated with the $\xi_c = \xi_d = 0$ edge and with the $\xi_d = 0$ face, respectively. The face-based polynomials associated with the $\xi_c = 0$ face of the tetrahedral element T^3 are obtained by interchanging ξ_c with ξ_d in the reported expressions. The polynomials for the triangular element T^2 are obtained by setting $\xi_d = 0$. Along its associated edge, E_p behaves as the Legendre polynomial $P_p(\xi_{ab})$ of order p. These polynomials are normalized as follows (where p indicates the global order of the polynomial, equal to the sum of its subscripts):

$$\int_{T^1} E_p^2(\boldsymbol{\xi})\,\mathrm{d}T^1 = 1;\; \iint_{T^2} E_p^2(\boldsymbol{\xi})\,\mathrm{d}T^2 = \frac{1}{2p+2};\; \iiint_{T^3} E_p^2(\boldsymbol{\xi})\,\mathrm{d}T^3 = \frac{1}{(2p+2)(2p+3)}$$

$$\iint_{T^2} F_p^2(\boldsymbol{\xi})\,\mathrm{d}T^2 = 1; \qquad \iiint_{T^3} F_p^2(\boldsymbol{\xi})\,\mathrm{d}T^3 = \frac{1}{2p+3}$$

Adapted from R. D. Graglia, A. F. Peterson, and F. P. Andriulli, "Curl-conforming hierarchical vector bases for triangles and tetrahedra," *IEEE Trans. Antennas Propag.*, vol. 59, no. 3, pp. 950–959, Mar. 2011.

Table 5.6 Curl-conforming vector bases on tetrahedral cells – volume-based hierarchical polynomials $V_{ijk} = \xi_c\, \xi_d\, \mathcal{U}_{i-2,j,k}$ up to the sixth order.

$\mathcal{U}^{ss}_{000} = 6\sqrt{35}$	$\mathcal{U}^{ss}_{100} = 6\sqrt{105}\,(4\,\chi_{ab} - 1)$ $\mathcal{U}^{as}_{010} = 36\sqrt{35}\,\xi_{ab}$ $\mathcal{U}^{sa}_{001} = 12\sqrt{105}\,\xi_{cd}$
$\mathcal{U}^{ss}_{200} = 12\sqrt{55}\,(1 + 3\,\chi_{ab}\,(5\,\chi_{ab} - 3))$ $\mathcal{U}^{as}_{110} = 6\sqrt{2,310}\,\xi_{ab}\,(5\,\chi_{ab} - 2)$ $\mathcal{U}^{ss}_{020} = 15\sqrt{462}\,(3\,\xi^2_{ab} - \chi^2_{ab})$ $\mathcal{U}^{sa}_{101} = 6\sqrt{1,155}\,\xi_{cd}\,(5\,\chi_{ab} - 1)$ $\mathcal{U}^{aa}_{011} = 60\sqrt{231}\,\xi_{ab}\,\xi_{cd}$ $\mathcal{U}^{ss}_{002} = 15\sqrt{11}\,(7\,\xi^2_{cd} - \chi^2_{cd})$	$\mathcal{U}^{ss}_{400} = 30\,(5 + 11\,\chi_{ab}\,(-10 + \chi_{ab}\,(60 + 13\,\chi_{ab}\,(7\,\chi_{ab} - 10))))$ $\mathcal{U}^{as}_{310} = 90\sqrt{22}\,\xi_{ab}\,(-5 + \chi_{ab}\,(45 + 13\,\chi_{ab}\,(7\,\chi_{ab} - 9)))$ $\mathcal{U}^{ss}_{220} = 30\sqrt{1,155}\,(3 - 13\,\chi_{cd}\,\chi_{ab})\,(3\,\xi^2_{ab} - \chi^2_{ab})$ $\mathcal{U}^{as}_{130} = 15\sqrt{3,003}\,\xi_{ab}\,(7\,\chi_{ab} - 4)\,(5\,\xi^2_{ab} - 3\,\chi^2_{ab})$ $\mathcal{U}^{ss}_{040} = 45\sqrt{1,001}\,(3\,\chi^4_{ab} - 30\,\chi^2_{ab}\,\xi^2_{ab} + 35\,\xi^4_{ab})/4$ $\mathcal{U}^{sa}_{301} = 30\sqrt{154}\,\xi_{cd}\,(-1 + \chi_{ab}\,(18 + 13\,\chi_{ab}\,(7\,\chi_{ab} - 6)))$ $\mathcal{U}^{aa}_{211} = 30\sqrt{231}\,\xi_{ab}\,\xi_{cd}\,(10 + 13\,\chi_{ab}\,(7\,\chi_{ab} - 5))$ $\mathcal{U}^{sa}_{121} = 15\sqrt{15,015}\,\xi_{cd}\,(7\,\chi_{ab} - 3)\,(3\,\xi^2_{ab} - \chi^2_{ab})$ $\mathcal{U}^{aa}_{031} = 210\sqrt{429}\,\xi_{ab}\,\xi_{cd}\,(5\,\xi^2_{ab} - 3\,\chi^2_{ab})$ $\mathcal{U}^{ss}_{202} = 15\sqrt{6}\,(3 + 13\,\chi_{ab}\,(7\,\chi_{ab} - 3))\,(7\,\xi^2_{cd} - \chi^2_{cd})$ $\mathcal{U}^{as}_{112} = 45\sqrt{143}\,\xi_{ab}\,(7\,\chi_{ab} - 2)\,(7\,\xi^2_{cd} - \chi^2_{cd})$ $\mathcal{U}^{ss}_{022} = 15\sqrt{15,015}\,(\chi^2_{ab} - 3\,\xi^2_{ab})\,(\chi^2_{cd} - 7\,\xi^2_{cd})/2$ $\mathcal{U}^{sa}_{103} = 15\sqrt{1,001}\,\xi_{cd}\,(7\,\chi_{ab} - 1)\,(3\,\xi^2_{cd} - \chi^2_{cd})$ $\mathcal{U}^{aa}_{013} = 105\sqrt{858}\,\xi_{ab}\,\xi_{cd}\,(3\,\xi^2_{cd} - \chi^2_{cd})$ $\mathcal{U}^{ss}_{004} = 105\sqrt{13}\,(\chi^4_{cd} - 18\,\chi^2_{cd}\,\xi^2_{cd} + 33\,\xi^4_{cd})/(4\sqrt{2})$
$\mathcal{U}^{ss}_{300} = 6\sqrt{390}\,(5\,\chi_{ab}\,(3 - 11\,\chi_{cd}\,\chi_{ab}) - 1)$ $\mathcal{U}^{as}_{210} = 6\sqrt{390}\,\xi_{ab}\,(10 + 11\,\chi_{ab}\,(6\,\chi_{ab} - 5))$ $\mathcal{U}^{ss}_{120} = 15\sqrt{6,006}\,(2\,\chi_{ab} - 1)\,(3\,\xi^2_{ab} - \chi^2_{ab})$ $\mathcal{U}^{as}_{030} = 6\sqrt{15,015}\,\xi_{ab}\,(5\,\xi^2_{ab} - 3\,\chi^2_{ab})$ $\mathcal{U}^{sa}_{201} = 30\sqrt{91}\,\xi_{cd}\,(1 + 11\,\chi_{ab}\,(2\,\chi_{ab} - 1))$ $\mathcal{U}^{aa}_{111} = 30\sqrt{6,006}\,\xi_{ab}\,\xi_{cd}\,(3\,\chi_{ab} - 1)$ $\mathcal{U}^{sa}_{021} = 30\sqrt{3,003}\,\xi_{cd}\,(3\,\xi^2_{ab} - \chi^2_{ab})$ $\mathcal{U}^{ss}_{102} = 3\sqrt{715}\,(6\,\chi_{ab} - 1)\,(7\,\xi^2_{cd} - \chi^2_{cd})$ $\mathcal{U}^{as}_{012} = 15\sqrt{858}\,\xi_{ab}\,(7\,\xi^2_{cd} - \chi^2_{cd})$ $\mathcal{U}^{sa}_{003} = 3\sqrt{10,010}\,\xi_{cd}\,(3\,\xi^2_{cd} - \chi^2_{cd})$	

The table reports the functions $\mathcal{U}_{\ell mn}$ up to the order $(\ell + m + n) = 4$. These functions form the set of the volume-based polynomials $V_{ijk} = \xi_c\, \xi_d\, \mathcal{U}_{i-2,j,k}$ up to the sixth order. It is understood that $\xi_a + \xi_b + \xi_c + \xi_d = 1$; $\xi_{ab} = \xi_a - \xi_b$; $\xi_{cd} = \xi_c - \xi_d$; $\chi_{ab} = \xi_a + \xi_b$; and $\chi_{cd} = \xi_c + \xi_d$. All the volume-based polynomials are mutually orthogonal on the T^3 simplex and are normalized so that $\iiint_{T^3} V_p^2(\boldsymbol{\xi})\, dT^3 = 1$, where p is the global order of the volume polynomial, equal to the sum of its subscripts.

Adapted from R. D. Graglia, A. F. Peterson, and F. P. Andriulli, "Curl-conforming hierarchical vector bases for triangles and tetrahedra," *IEEE Trans. Antennas Propag.*, vol. 59, no. 3, pp. 950–959, Mar. 2011.

Chapter 4, the $p(p^2 - 1)/2$ elements of an order p hierarchical vector base associated with DoFs internal to the tetrahedral element $\mathcal{T}^3(\xi_a, \xi_b, \xi_c, \xi_d)$ are obtained by forming the product of the $p(p^2 - 1)/6$ volume-based hierarchical polynomials of Table 5.6 with three different zeroth-order curl-conforming functions. To guarantee basis function independence, the chosen zeroth-order basis factors cannot be associated with edges bounding the same face (see Chapter 4). Similarly, the $p(p + 1)$ elements of a pth-order hierarchical base associated with DoFs

Table 5.7 Vector bases on triangular and tetrahedral cells – correspondence between dummy and parent variables.

Zeroth-order basis factor	$\{\xi_a, \xi_b, \xi_c, \xi_d\}$	Zeroth-order basis factor	$\{\xi_a, \xi_b, \xi_c, \xi_d\}$
$\mathbf{\Omega}_1$, $\mathbf{\Lambda}_1$	$\{\xi_2, \xi_3, \xi_1, 0\}$	$\mathbf{\Omega}_2$, $\mathbf{\Lambda}_2$	$\{\xi_3, \xi_1, \xi_2, 0\}$
$\mathbf{\Omega}_3$, $\mathbf{\Lambda}_3$	$\{\xi_1, \xi_2, \xi_3, 0\}$		
	Based on face #		Based on face #
$\mathbf{\Omega}_{12}$	$\{\xi_3, \xi_4, \xi_1, \xi_2\}$ 2 $\{\xi_3, \xi_4, \xi_2, \xi_1\}$ 1	$\mathbf{\Omega}_{23}$	$\{\xi_1, \xi_4, \xi_2, \xi_3\}$ 3 $\{\xi_1, \xi_4, \xi_3, \xi_2\}$ 2
	Based on face #		Based on face #
$\mathbf{\Omega}_{13}$	$\{\xi_4, \xi_2, \xi_1, \xi_3\}$ 3 $\{\xi_4, \xi_2, \xi_3, \xi_1\}$ 1	$\mathbf{\Omega}_{24}$	$\{\xi_3, \xi_1, \xi_2, \xi_4\}$ 4 $\{\xi_3, \xi_1, \xi_4, \xi_2\}$ 2
	Based on face #		Based on face #
$\mathbf{\Omega}_{14}$	$\{\xi_2, \xi_3, \xi_1, \xi_4\}$ 4 $\{\xi_2, \xi_3, \xi_4, \xi_1\}$ 1	$\mathbf{\Omega}_{34}$	$\{\xi_1, \xi_2, \xi_3, \xi_4\}$ 4 $\{\xi_1, \xi_2, \xi_4, \xi_3\}$ 3

Hierarchical basis functions for the triangular and tetrahedral cells are the product of the polynomials of Tables 5.5 and 5.6 with the zeroth-order vector functions of Chapter 4 listed here in the left-hand columns. In forming the products, the dummy parent variables $\{\xi_a, \xi_b, \xi_c, \xi_d\}$ appearing in the polynomials reported in Tables 5.5 and 5.6 are replaced by the parent variables reported in the right-hand columns. The first two rows consider both the curl- ($\mathbf{\Omega}$) and the divergence-conforming ($\mathbf{\Lambda}$) triangular bases; in this case, one has to set $\xi_d = 0$ in Table 5.5. The last three rows hold for the curl-conforming bases of the tetrahedral element; in this case, when forming the face-based basis functions, one has to choose the parent variables reported in the appropriate row.

internal to the triangular face $\mathcal{T}^2(\xi_a, \xi_b, \xi_c)$ (possibly, the $\xi_d = 0$ face bounding the tetrahedral element $\mathcal{T}^3$) are obtained by forming the product of the $p(p+1)/2$ face-based hierarchical polynomials of Table 5.5 with two zeroth-order curl-conforming functions associated with two edges bounding $\mathcal{T}^2$. Finally, the $(p+1)$ elements of a pth-order hierarchical base associated with edge DoFs relative to the edge $\mathcal{T}^1(\xi_a, \xi_b)$ (possibly, the $\xi_c = \xi_d = 0$ edge bounding the tetrahedral element $\mathcal{T}^3$, or the $\xi_c = 0$ edge bounding the triangular element $\mathcal{T}^2$) are obtained by forming the product of the $(p+1)$ edge-based hierarchical polynomials of Table 5.5 with the zeroth-order curl-conforming function along that edge. In this construction process, the dummy parent variables $\{\xi_a, \xi_b, \xi_c, \xi_d\}$ appearing in the polynomial expressions reported in Tables 5.5 and 5.6 are replaced by the permutation of $\{\xi_1, \xi_2, \xi_3, \xi_4\}$ that corresponds with the appropriate zeroth-order basis factor shown in Table 5.7.

In general, a generating hierarchical scalar polynomial of one group (either the volume-, the face-, or the edge-based group) is not orthogonal to a hierarchical polynomial of a different group, but all the polynomials within each group are mutually orthogonal independent of the definition domain of the Legendre inner product (i.e. either the volume T^3, the face T^2, or the edge T^1 of the element). This is evident in Figure 5.6 that shows the *patterns* of the non-zero entries of the symmetric *Gram* matrix $\mathbf{G}_6$ having coefficient g_{ij} equal to the Legendre inner product of the ith and jth polynomials of the sixth-order complete triangular (at left) and tetrahedral families (at right) reported in Tables 5.5 and 5.6. The generating hierarchical scalar polynomial sets that build the vector bases do not contain any vertex-associated polynomial; conversely, the sets that build the scalar bases (Tables 5.1 and 5.2) contain vertex-associated

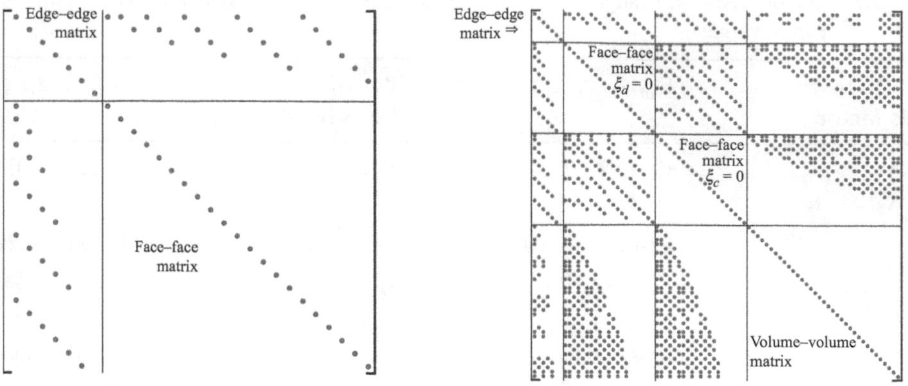

Figure 5.6 Non-zero entries of the Gram matrix $\mathbf{G}_6$ formed with all the scalar polynomials (up to the sixth order) given in Tables 5.5 and 5.6. These polynomials are those able to form the hierarchical vector bases of half-integer order 6.5 on a triangle (left matrix) or on a tetrahedron (right matrix). The triangular-family result reported at left is obtained with inner product performed on T^2 (the $\xi_d = 0$ tetrahedral face). The tetrahedral-family result reported at right is with inner product on T^3. Notice that the edge- and face-based hierarchical polynomials are orthogonal on both T^2 and T^3 simplcies, while the volume-based polynomials are orthogonal on the T^3 simplex. (Please refer to footnote 5 on page 189 before comparing these results with the one shown on the right-hand side of Figure 5.4.)

polynomials. The effect of the vertex-associated polynomials on the sparsity of the Gram matrix is readily appreciated by comparing Figure 5.6 with the right-hand side of Figure 5.4. (Please read footnote 5 on page 189 before comparing the two figures.)

Furthermore, for $p \geq 3$, the hierarchical generating scalar polynomials used to form the vector bases have a better degree of linear independence (measured with the condition number of the Gram matrix $\mathbf{G}_p$) than that of the corresponding generating interpolatory scalar polynomials of Chapter 4 (see for example [21, Figure 2]).

5.3.1.1 Auxiliary Polynomials

Let us begin by introducing some auxiliary polynomials needed to construct hierarchical polynomial bases with terms subdivided into volume-, face-, and edge-based polynomials. In the following, the symbol used for these polynomials is related to their further use; that is, $\mathcal{V}$, $\mathcal{F}$, and $\mathcal{E}$ are the auxiliary polynomials used to construct the volume-, the face-, and the edge-based polynomials, respectively. Then, the following three subsubsections show how to use the auxiliary polynomials to construct the hierarchical volume-, face-, and edge-based polynomials of Tables 5.5 and 5.6, which are labeled V, F, and E, respectively. In each group, the polynomials are mutually orthogonal independent of the definition domain of the inner product, i.e. either the volume, the face, or the edge of the element. This is a very important feature of our hierarchical bases because one can use the same polynomial bases either on tetrahedral, triangular or line elements, with no need to modify their expressions. The edge-, face-, and volume-based polynomials are normalized as reported at the bottom of Tables 5.5 and 5.6; in numerical applications, these polynomials can be normalized differently whenever convenient. To clarify to the reader that the polynomials of Table 5.5 and 5.6 are linear combinations of the interpolatory ones discussed in Chapter 4, below we report the *historical* construction technique used in [25]. This technique can be simplified by starting with appropriate orthogonal polynomials instead of by starting with the interpolatory ones of Chapter 4, as we may do later on for convenience to derive the bases for the other (quadrilateral, brick, and triangular prism) cells. As a matter of fact, it is useful to show, at least once, that the polynomials generating the vector functions are always linear combinations of those of Chapter 4.

To fully exploit the cell symmetries and produce compact expressions, and in common with Section 5.2.1, we introduce variables (see (5.16))

$$\begin{aligned} \xi_{ab} &= \xi_a - \xi_b \\ \xi_{cd} &= \xi_c - \xi_d \\ \chi_{ab} &= \xi_a + \xi_b \\ \chi_{cd} &= \xi_c + \xi_d \end{aligned} \tag{5.62}$$

with dependency relation

$$\chi_{ab} + \chi_{cd} = 1 \tag{5.63}$$

The expressions of the four parent variables $\{\xi_a, \xi_b, \xi_c, \xi_d\}$ in terms of the new dependent variables are straightforwardly obtained by inversion of (5.62).

Polynomials with particular symmetry properties in the dependent parent variables $\boldsymbol{\xi} = \{\xi_a, \xi_b, \xi_c, \xi_d\}$ used to describe the $T^1(\xi_a, \xi_b)$, the $T^2(\xi_a, \xi_b, \xi_c)$, and the $T^3(\xi_a, \xi_b, \xi_c, \xi_d)$ simplices are obtained by linear combinations of the interpolatory polynomials of Chapter 4 expressed in terms of the Silvester (R_k) and the shifted Silvester ($\hat{R}_k$) polynomials. For example, for $p \geq 0$, the pth-order polynomials

$$\mathcal{E}_p(\xi_a, \xi_b) = \sum_{i=1}^{i_{\max}} K_{p,i} \left[\hat{R}_{q-i}(q, \xi_a)\, \hat{R}_i(q, \xi_b) + (-1)^p \hat{R}_i(q, \xi_a)\, \hat{R}_{q-i}(q, \xi_b) \right] \tag{5.64}$$

with

$$q = p + 2, \quad i_{\max} = \begin{cases} q/2, & q \text{ even} \\ (q-1)/2, & q \text{ odd} \end{cases} \tag{5.65}$$

are either symmetric (for even or zero p values) or antisymmetric (for odd p values) in ξ_a and ξ_b, with $\mathcal{E}_p(\xi_b, \xi_a) = (-1)^p\, \mathcal{E}_p(\xi_a, \xi_b)$.

The coefficients $K_{p,i}$ in (5.64) are determined to define hierarchical polynomials associated with the edge $\xi_a + \xi_b = 1$. They are obtained by imposing, along the edge at issue (i.e. the $T^1(\xi_a, \xi_b)$ simplex), the orthogonality of each $\mathcal{E}_p(\xi_a, \xi_b)$ with respect to all the polynomials $\mathcal{E}_m(\xi_a, \xi_b)$ of lower order m

$$\int_0^1 \mathcal{E}_p(\xi_a, 1-\xi_a)\, \mathcal{E}_m(\xi_a, 1-\xi_a)\, d\xi_a = \frac{\delta_{mp}}{2p+1} \tag{5.66}$$

where δ_{mp} is the Kronecker delta. The normalization (5.66) involves a unit constant weight function and makes $\mathcal{E}_p(\xi_a, 1-\xi_a)$ equal to the shifted Legendre polynomial $P_p^*(\xi_a)$. (In principle, other normalizations with different weight functions are possible to make $\mathcal{E}_p$, for example, equal to Chebyshev or Jacobi polynomial.) Convenient expressions for the $\mathcal{E}_p$ defined in (5.64) can be given in terms of the two new dependent variables ξ_{ab} and χ_{cd} given in (5.62).

All the polynomials $\mathcal{E}_p$ are implicitly symmetric in the ξ_c and ξ_d variables since $\chi_{cd}(\xi_c, \xi_d) = \chi_{cd}(\xi_d, \xi_c) = \xi_c + \xi_d$.

To construct the face-based polynomials of Section 5.3.1.3, we also need the polynomials $\mathcal{F}_n(\xi_c)$ obtained by orthogonalizing the Silvester polynomials $R_n(p+2, \xi_c)$, for $n = 1, 2, \ldots, p$. All these polynomials contain a common factor ξ_c and do not have any symmetry or antisymmetry property in ξ_c, ξ_d because they are independent of ξ_d; they are normalized by setting

$$\iint_{T^2} \mathcal{F}_n(\xi_c)\, \mathcal{F}_m(\xi_c)\, \mathrm{d}T^2 = \frac{\delta_{mn}}{2n(n+1)(n+2)} \tag{5.67}$$

which yields $\mathcal{F}_n(1) = 1$ and $|\mathcal{F}_n(\xi_c)| \leq 1$ for $\{0 \leq \xi_c \leq 1\}$. These polynomials are obtained from the lowest order functions

$$\begin{cases} \mathcal{F}_1(\xi_c) = \xi_c \\ \mathcal{F}_2(\xi_c) = \xi_c\,(5\,\xi_c - 3)/2 \end{cases} \tag{5.68}$$

by using the following recurrence relation with respect to the degree n

$$a_{1n}\mathcal{F}_{n+1}(\xi_c) = (a_{2n}\xi_c - a_{3n})\mathcal{F}_n(\xi_c) - a_{4n}\mathcal{F}_{n-1}(\xi_c) \tag{5.69}$$

with

$$\begin{cases} a_{1n} = (n+3)(2n+1) \\ a_{2n} = 2(2n+1)(2n+3) \\ a_{3n} = 2(2n^2+4n+3) \\ a_{4n} = (n-1)(2n+3) \end{cases} \tag{5.70}$$

Orthogonal volume-based polynomials symmetric or antisymmetric in the (ξ_a, ξ_b) and the (ξ_c, ξ_d) variables can be obtained by linearly combining the Silvester and shifted Silvester interpolatory polynomials in a way similar to that used to produce (5.64). To obtain these polynomials, however, it is much more convenient to start with expressions that involve the appropriate Legendre polynomials from the beginning. In fact, in Section 5.3.1.2, we use the following volume-based linearly independent polynomials of order $p = 2 + \ell + m + n$

$$\mathcal{V}_{\ell+2,m,n}(\xi_a, \xi_b, \xi_c, \xi_d) = \xi_c\, \xi_d\, \chi_{ab}^{\ell}\, P_m(\xi_{ab})\, P_n(\xi_{cd}) \tag{5.71}$$

which are symmetric in (ξ_a, ξ_b) and (ξ_c, ξ_d) for even values of m and n, respectively (antisymmetric otherwise), with

$$\begin{aligned} \mathcal{V}_{imn}(\xi_b, \xi_a, \xi_c, \xi_d) &= (-1)^m\, \mathcal{V}_{imn}(\xi_a, \xi_b, \xi_c, \xi_d) \\ \mathcal{V}_{imn}(\xi_a, \xi_b, \xi_d, \xi_c) &= (-1)^n\, \mathcal{V}_{imn}(\xi_a, \xi_b, \xi_c, \xi_d) \\ \mathcal{V}_{imn}(\xi_b, \xi_a, \xi_d, \xi_c) &= (-1)^{m+n}\, \mathcal{V}_{imn}(\xi_a, \xi_b, \xi_c, \xi_d) \end{aligned} \tag{5.72}$$

The integral of the product of a symmetric and an antisymmetric polynomial of the kind given in (5.64) automatically vanishes over the T^1, T^2, and T^3 simplices; similarly, the product of a symmetric and an antisymmetric polynomial of the kind given in (5.71) has a vanishing integral over T^3.

5.3.1.2 Volume-Based Polynomials

In terms of the (dummy) parent variables, the $p(p^2-1)/6$ polynomials which in Chapter 4 interpolate internal points of the T^3 simplex and vanish on the $\xi_c=0$ and $\xi_d=0$ faces because of the presence of a common $\xi_c\,\xi_d$ factor read as follows

$$\begin{aligned}\hat{\alpha}^{cd}_{ijk\ell}(p,\boldsymbol{\xi}) &= \hat{R}_i(p+2,\xi_a)\,\hat{R}_j(p+2,\xi_b)\,R_k(p+2,\xi_c)\,R_\ell(p+2,\xi_d)\\ &\text{for } i,j=1,2,\ldots,p+1;\ \ k,\ell=1,2,\ldots,p;\\ &i+j+k+\ell=p+2\end{aligned} \tag{5.73}$$

According to our definition, the above polynomials are volume based. An equivalent pth-order hierarchical family consisting of $p(p^2-1)/6$ volume-based polynomials is obtained by applying the Gram–Schmidt orthogonalization process to the polynomial set (5.71), performed by using the Legendre inner product over the T^3 simplex. The hierarchical volume-based polynomials are thus obtained by orthogonalizing in order (from the first to the last polynomial) the list of the polynomials (5.71) provided by running a three-nested loop: for $g=2,3,\ldots,p$ (outer loop on the *global* order $g=\ell+2+m+n$); $n=0,1,\ldots,g-2$, and $m=0,1,\ldots,g-2-n$ (inner loop), with $\ell=(g-2-m-n)$ (fixed in the inner loop).

The hierarchical polynomials $V_{ijk}(\boldsymbol{\xi})$ obtained in this manner are reported in Table 5.6 up to the sixth order. The global order of these polynomials is equal to $p=i+j+k$ and equivalently to the sum of the subscripts appearing in their expressions.

5.3.1.3 Face-Based Polynomials

In terms of the (dummy) parent variables, the $p(p+1)/2$ polynomials which in Chapter 4 interpolate the points located inside the triangular face described by the three parent variables ξ_a, ξ_b, and ξ_c read as follows

$$\begin{aligned}\hat{\alpha}^{c}_{ijk}(p,\boldsymbol{\xi}) &= \hat{R}_i(p+2,\xi_a)\hat{R}_j(p+2,\xi_b)\,R_k(p+2,\xi_c)\\ &\text{for } i,j=1,2,\ldots,p+1;\ \ k=1,2,\ldots,p;\\ &\text{and } i+j+k=p+2\end{aligned} \tag{5.74}$$

The $k=0$ case is excluded here since it yields edge-based functions that interpolate the $\xi_c=0$ edge of the triangle; those functions are considered in the following subsubsection.

The face-based interpolatory polynomials (5.74) are replaced by hierarchical polynomials obtained by orthogonalizing in order over T^2, within a two-nested loop for $k=1,2,\ldots,p$ (outer loop) and $n=k,k-1,\ldots,1$ (inner loop; see Figure 5.7), the kth-order polynomials

$$\mathcal{P}_{mn}(\xi_a,\xi_b,\xi_c)=\mathcal{E}_m(\xi_a,\xi_b)\,\mathcal{F}_n(\xi_c) \tag{5.75}$$

where $k=m+n$ and with $\mathcal{E}_m$, $\mathcal{F}_n$ given in Section 5.3.1.1.

This orthogonalization process yields a polynomial set $\mathcal{Q}_{mn}(\xi_a,\xi_b,\xi_c)$ that contains all the *normalized* orthogonal polynomials $\mathcal{F}_n(\xi_c)$, with $\mathcal{Q}_{0n}(\xi_a,\xi_b,\xi_c)=\mathcal{F}_n(\xi_c)$, and $\mathcal{Q}_{0n}=1$ at $\xi_c=1$. The order of $\mathcal{Q}_{mn}$ is $(m+n)$ and coincides with the sum of the subscripts m and n of the $\mathcal{E}_m$ and $\mathcal{F}_n$ functions in (5.75). Furthermore, $\mathcal{Q}_{mn}$ is symmetric in ξ_a, ξ_b if m is even or equal to zero, whereas $\mathcal{Q}_{mn}$ is antisymmetric in ξ_a, ξ_b if m is odd.

The hierarchical polynomials $\mathcal{Q}_{mn}(\xi_a,\xi_b,\xi_c)$ and $\mathcal{Q}_{mn}(\xi_a,\xi_b,\xi_d)$ obtained by this procedure are mutually orthogonal *only* over the triangular simplices $T^2(\xi_a,\xi_b,\xi_c)$ and $T^2(\xi_a,\xi_b,\xi_d)$, respectively. The $\mathcal{Q}_{mn}(\xi_a,\xi_b,\xi_c)$ are face-based and non-zero on the triangular face $\xi_d=0$, and

Order

$k = 1$ $\mathcal{E}_0(\xi_a, \xi_b)\,\mathcal{F}_1(\xi_c) = \mathcal{Q}_{01}(\xi_a, \xi_b, \xi_c)$

$k = 2$ $\mathcal{E}_0(\xi_a, \xi_b)\,\mathcal{F}_2(\xi_c) \to \mathcal{E}_1(\xi_a, \xi_b)\,\mathcal{F}_1(\xi_c)$

$k = 3$ $\mathcal{E}_0(\xi_a, \xi_b)\,\mathcal{F}_3(\xi_c) \to \mathcal{E}_1(\xi_a, \xi_b)\,\mathcal{F}_2(\xi_c) \to \mathcal{E}_2(\xi_a, \xi_b)\,\mathcal{F}_1(\xi_c)$

$k = 4$ $\mathcal{E}_0(\xi_a, \xi_b)\,\mathcal{F}_4(\xi_c) \to \mathcal{E}_1(\xi_a, \xi_b)\,\mathcal{F}_3(\xi_c) \to \mathcal{E}_2(\xi_a, \xi_b)\,\mathcal{F}_2(\xi_c) \to \mathcal{E}_3(\xi_a, \xi_b)\,\mathcal{F}_1(\xi_c)$

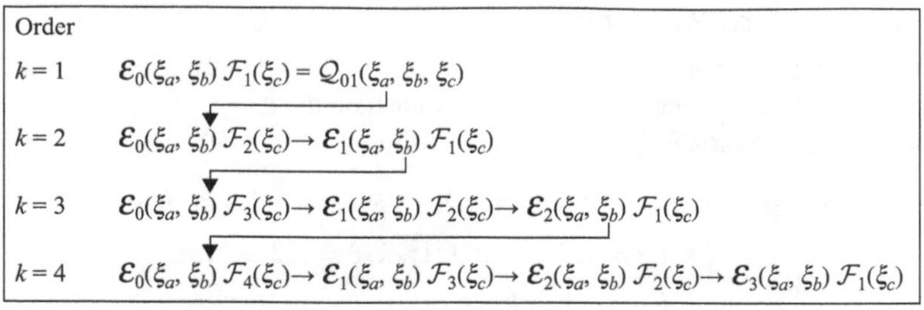

Figure 5.7 Orthogonalization path followed by the two-nested loop used to build the face-based hierarchical polynomials $\mathcal{Q}_{mn}$. The outer loop is for $k = 1, 2, \ldots, p$; the inner loop is for $n = k, k-1, \ldots, 1$. Thus, the kth-order polynomial $\mathcal{Q}_{mn}$ (with $k = m + n$) is obtained by orthogonalizing $\mathcal{E}_m\,\mathcal{F}_n$ with respect to all the face-based orthogonal polynomials $\mathcal{Q}_{j\ell}$ previously obtained up to the order $j + \ell = k - 1$ as well as, in the case of $m \geq 1$, by orthogonalizing $\mathcal{E}_m\,\mathcal{F}_n$ with respect to the polynomials $\mathcal{Q}_{j\ell}$ previously obtained for $j = 0, 1, \ldots, m - 1$ with $\ell = k - j$. The figure just shows the path from $k = 1$ up to $k = 4$. The procedure starts with $k = 1$, where $\mathcal{Q}_{01}$ is known. For $k = 2$, we first build $\mathcal{Q}_{02}$ by orthogonalizing $\mathcal{E}_0\,\mathcal{F}_2$ with respect to $\mathcal{Q}_{01}$, then we build $\mathcal{Q}_{11}$ by orthogonalizing $\mathcal{E}_1\,\mathcal{F}_1$ with respect to $\mathcal{Q}_{01}$ and $\mathcal{Q}_{02}$. For $k = 3$, we first build $\mathcal{Q}_{03}$ by orthogonalizing $\mathcal{E}_0\,\mathcal{F}_3$ with respect to $\mathcal{Q}_{01}$, $\mathcal{Q}_{02}$, and $\mathcal{Q}_{11}$; then we build, in order, $\mathcal{Q}_{12}$ and $\mathcal{Q}_{21}$, etc.

© 2011 IEEE. Reprinted with permission from R. D. Graglia, A. F. Peterson, and F. P. Andriulli, "Curl-conforming hierarchical vector bases for triangles and tetrahedra," *IEEE Trans. Antennas Propag.*, vol. 59, no. 3, pp. 950–959, Mar. 2011.

equal to zero at $\xi_c = 0$ because of the presence of a common ξ_c factor. In order to obtain a hierarchical family of face-based polynomials $F_{mn}(\boldsymbol{\xi})$ mutually orthogonal on both the $T^2(\xi_a, \xi_b, \xi_c)$ and the T^3 simplices, it is sufficient to add to the polynomials $\mathcal{Q}_{mn}(\xi_a, \xi_b, \xi_c)$ an appropriate linear combination of the volume-based polynomials $V_p(\boldsymbol{\xi})$ (derived in Section 5.3.1.2) of global order $p \leq (m + n)$, and which share the same symmetry properties of $\mathcal{Q}_{mn}$ with respect to the ξ_a and ξ_b variables. The $F_{mn}(\boldsymbol{\xi})$ polynomials define the following polynomials of global order $(m + n)$

$$\widetilde{\mathcal{E}}_{m+n}^{s|a,\,s}(\boldsymbol{\xi}) = F_{mn}^{s|a}(\xi_a, \xi_b, \xi_c, \xi_d) + F_{mn}^{s|a}(\xi_a, \xi_b, \xi_d, \xi_c) \tag{5.76}$$

used in Section 5.3.1.4 to construct edge-based polynomials symmetric in the ξ_c and ξ_d variables, and orthogonal on the $T^2(\xi_a, \xi_b, \xi_c)$, the $T^2(\xi_a, \xi_b, \xi_d)$, and the T^3 simplices.

Table 5.5 reports, up to the sixth order, the normalized face-based hierarchical polynomials $F_{mn}(\boldsymbol{\xi})$ for the T^2 and T^3 simplices.

5.3.1.4 Edge-Based Polynomials

In terms of the (dummy) parent variables, the $(p + 1)$ polynomials which in Chapter 4 interpolate the edge described by the two parent variables ξ_a, ξ_b read as follows

$$\hat{\alpha}_{ij}(p, \xi_a, \xi_b) = \hat{R}_i(p + 2, \xi_a)\,\hat{R}_j(p + 2, \xi_b)$$
$$\text{for } i, j = 1, 2, \ldots, p + 1; \tag{5.77}$$
$$\text{and } i + j = p + 2$$

and could be replaced by the hierarchical polynomials $\mathcal{E}_p(\xi_a, \xi_b) = \mathcal{E}_p(\xi_{ab}, \chi_{cd})$ of Section 5.3.1.1.

The hierarchical polynomials for the line element (i.e. the T^1 simplex) are obtained from $\mathcal{E}_p(\xi_{ab}, \chi_{cd})$ by setting $\chi_{cd} = 0$, which is equivalent to $\xi_a + \xi_b = 1$. In the line-element case, the functions $\mathcal{E}_p$ are easily obtained by the recurrence relation available in [33], since $\mathcal{E}_p(\chi_{cd} = 0)$ coincides by construction with the Legendre polynomial $P_p(\xi_{ab})$ or, equivalently, with the shifted Legendre polynomial $P_p^*(\xi_a)$.

However, unfortunately, the edge-based hierarchical polynomials $\mathcal{E}_p$ are mutually orthogonal *only* over the simplex T^1. In order to obtain a hierarchical family of edge-based polynomials symmetric in the ξ_c and ξ_d variables, and also mutually orthogonal on the $T^3(\xi_a, \xi_b, \xi_c, \xi_d)$ simplex, the $T^2(\xi_a, \xi_b, \xi_c)$ simplex (where $\xi_d = 0$), and the $T^2(\xi_a, \xi_b, \xi_d)$ simplex (where $\xi_c = 0$), it is sufficient to add to the pth-order polynomial $\mathcal{E}_p$ an appropriate linear combination of the polynomials $\widetilde{\mathcal{E}}_m$ of Section 5.3.1.3 (given in (5.76)), and of the volume-based polynomials $V_m(\boldsymbol{\xi})$ of Section 5.3.1.2, which share the same symmetry properties of $\mathcal{E}_p$ with respect to the four parent variables. All the polynomials involved in this combination are of order $m \leq p$. The linear combination at issue here involves only those volume-based polynomials that are symmetric with respect to the ξ_c and ξ_d variables.

The hierarchical polynomials E_p obtained in this manner are reported in Table 5.5, up to the sixth order.

5.3.2 Quadrilateral and Brick Bases

This subsection presents hierarchical vector bases for quadrilateral and hexahedral cells. These functions span the reduced-gradient curl-conforming spaces of Nédélec [5], which implies that the number of independent vector-valued polynomials that form a pth-order complete hierarchical base on a quadrilateral and hexahedral cell is equal to $2(p+1)(p+2)$ and $3(p+1)(p+2)^2$, respectively.

Several basis families with similar properties have been proposed [13, 19, 34–36]. However, hierarchical bases often exhibit poor linear independence as the order of the representation is increased, resulting in an ill-conditioned system of equations. The basis functions reported here are constructed from orthogonal polynomials, by a process similar to that used to generate vector bases for triangles and tetrahedra in Section 5.3.1, and are shown to alleviate the loss of linear independence. They were first reported in [22, 23] and [26].

In the following, we describe the development of these basis functions, and provide general expressions that can be used to obtain basis functions of any polynomial degree. Results demonstrate that the bases exhibit a very slow growth in matrix condition number, indicating that they are able to maintain excellent linear independence as their polynomial order increases.

In Chapter 4, interpolatory high-order vector bases on brick and quadrilateral cells are constructed by multiplying the zeroth-order curl-conforming vector functions with Silvester–Lagrange interpolatory polynomials. Linear combinations of those interpolatory polynomials are used here to obtain *generating* hierarchical polynomials which are subdivided from the outset into three different groups of edge- (E), face- (F), and volume-based (V) functions. In each group, all the hierarchical polynomials are constructed *a priori* to be mutually orthogonal independent of the definition domain of the relevant inner product, that is, either the edge, the face, or the volume of the brick. The hierarchical vector bases are then constructed by using the same technique given in Chapter 4, where we simply *substitute* the new scalar hierarchical polynomials for the interpolatory ones of Chapter 4. Thus, for any given polynomial order, the number of the new hierarchical vector basis functions is equal to the number of the interpolatory basis set given in Chapter 4, and the space spanned by these two bases is exactly the same.

In this subsection, the quadrilateral and the brick vector bases are derived at the same time by considering the quadrilateral cell as the $\xi_b = 0$ bounding face of a brick cell described by its six dummy parent variables $\boldsymbol{\xi} = \{\xi_a, \xi_d; \xi_b, \xi_e; \xi_c, \xi_f\}$ (see Chapter 4). The variables ξ_a, ξ_b, and ξ_c are assumed to be independent, with

$$\begin{aligned} \xi_d &= 1 - \xi_a \\ \xi_e &= 1 - \xi_b \\ \xi_f &= 1 - \xi_c \end{aligned} \tag{5.78}$$

and the parent brick cell Q^3 of the parent space (ξ_a, ξ_b, ξ_c) is the cube $\{0 \le \xi_a, \xi_b, \xi_c \le 1\}$ of unitary edge-length. In the following, Q^1 indicates the parent edge formed by the intersection of faces $\xi_a = 0$ and $\xi_b = 0$, while Q_a^2 and Q_b^2 denote the $\xi_a = 0$ and the $\xi_b = 0$ parent faces, respectively. Similarly, a quadrilateral cell Q^2 is described by four *dummy* parent variables $\boldsymbol{\xi} = \{\xi_a, \xi_d; \xi_c, \xi_f\}$ where, once again, we assume ξ_a, ξ_c to be independent, with ξ_d and ξ_f given as in (5.78); recall that the quadrilateral cell Q^2 is the $\xi_b = 0$ face of the brick (i.e. $Q^2 = Q_b^2$). In the (ξ_a, ξ_c) parent space, Q^2 is the square $\{0 \le \xi_a, \xi_c \le 1\}$ of unitary edge-length, and Q^1 is the parent edge $\xi_a = 0$.

5.3.2.1 Generating Polynomials

For the brick element, *interpolatory* curl-conforming bases complete to pth order were obtained in Chapter 4 by multiplying the zeroth-order curl-conforming vector function $\boldsymbol{\Omega}_{ab} = \xi_d \xi_e \nabla \xi_c$, associated with the edge formed by the intersection of faces $\xi_a = 0$, $\xi_b = 0$, with the $(p+1)^3$ Silvester–Lagrange interpolatory polynomials defined by

$$\begin{aligned} \hat{\alpha}^{ab}_{i\ell;jm;kn}(\boldsymbol{\xi}) = {} & R_i(p+1, \xi_a)\, \hat{R}_\ell(p+1, \xi_d)\, R_j(p+1, \xi_b)\, \hat{R}_m(p+1, \xi_e) \\ & \hat{R}_k(p+2, \xi_c)\, \hat{R}_n(p+2, \xi_f) \\ & \text{for } i, j = 0, 1, \ldots, p;\;\; k, \ell, m, n = 1, 2, \ldots, p+1; \\ & \text{with } i + \ell = j + m = p + 1;\;\; k + n = p + 2 \end{aligned} \tag{5.79}$$

For the quadrilateral cell, the zeroth-order vector factor associated with the $\xi_a = 0$ edge is $\boldsymbol{\Omega}_a = \xi_d \nabla \xi_c$ while the $(p+1)^2$ Silvester–Lagrange interpolatory polynomials used to build the pth-order complete bases are obtained from (5.79) by setting $\xi_b = 0$ and $j = 0$, which yield $\xi_e = 1$, $m = p+1$, and $R_j(p+1, \xi_b)\hat{R}_m(p+1, \xi_e) = 1$.

The number of vector functions provided by this multiplicative process for the quadrilateral and brick cells is $4(p+1)^2$ and $12(p+1)^3$, respectively. Some of these vector functions are linearly dependent on the others. Using a procedure similar to that discussed in Chapter 4 for the interpolatory case, the dependent face and volume-based polynomials from the generating scalar set must be eliminated. For the quadrilateral set, $2p(p+1)$ face-based vector functions are linearly dependent. For the brick set, $12p(p+1)$ face-based vector functions and $9p^2(p+1)$ volume-based vector functions are dependent and must be eliminated. After removing the redundant functions, the total number of vector bases obtained is $2(p+1)(p+2)$ and $3(p+1)(p+2)^2$ for the quadrilateral and the brick cells, respectively; these are the DoFs necessary to span the Nédélec mixed-order spaces [5, 37].

Table 5.8 Vector bases on quadrilateral and brick cells – correspondence between dummy and parent variables.

Function	$\xi_a, \xi_d; \xi_b, \xi_e; \xi_c, \xi_f$	Function	$\xi_a, \xi_d; \xi_b, \xi_e; \xi_c, \xi_f$
$\mathbf{\Omega}_1, \mathbf{\Lambda}_1$	$\xi_1, \xi_3;\ \bullet, \bullet;\ \xi_4, \xi_2$	$\mathbf{\Omega}_3, \mathbf{\Lambda}_3$	$\xi_3, \xi_1;\ \bullet, \bullet;\ \xi_2, \xi_4$
$\mathbf{\Omega}_2, \mathbf{\Lambda}_2$	$\xi_2, \xi_4;\ \bullet, \bullet;\ \xi_1, \xi_3$	$\mathbf{\Omega}_4, \mathbf{\Lambda}_4$	$\xi_4, \xi_2;\ \bullet, \bullet;\ \xi_3, \xi_1$
$\mathbf{\Omega}_{12}$	$\xi_1, \xi_4; \xi_2, \xi_5; \xi_3, \xi_6$	$\mathbf{\Omega}_{26}$	$\xi_2, \xi_5; \xi_6, \xi_3; \xi_4, \xi_1$
$\mathbf{\Omega}_{13}$	$\xi_1, \xi_4; \xi_3, \xi_6; \xi_5, \xi_2$	$\mathbf{\Omega}_{34}$	$\xi_3, \xi_6; \xi_4, \xi_1; \xi_5, \xi_2$
$\mathbf{\Omega}_{15}$	$\xi_1, \xi_4; \xi_5, \xi_2; \xi_6, \xi_3$	$\mathbf{\Omega}_{35}$	$\xi_3, \xi_6; \xi_5, \xi_2; \xi_1, \xi_4$
$\mathbf{\Omega}_{16}$	$\xi_1, \xi_4; \xi_6, \xi_3; \xi_2, \xi_5$	$\mathbf{\Omega}_{45}$	$\xi_4, \xi_1; \xi_5, \xi_2; \xi_3, \xi_6$
$\mathbf{\Omega}_{23}$	$\xi_2, \xi_5; \xi_3, \xi_6; \xi_1, \xi_4$	$\mathbf{\Omega}_{46}$	$\xi_4, \xi_1; \xi_6, \xi_3; \xi_5, \xi_2$
$\mathbf{\Omega}_{24}$	$\xi_2, \xi_5; \xi_4, \xi_1; \xi_3, \xi_6$	$\mathbf{\Omega}_{56}$	$\xi_5, \xi_2; \xi_6, \xi_3; \xi_1, \xi_4$

Hierarchical vector basis functions are the product of the polynomials of Table 5.10 with the zeroth-order vector functions of Chapter 4 listed here in the left-hand columns. In forming the products, the dummy parent variables $\{\xi_a, \xi_d; \xi_b, \xi_e; \xi_c, \xi_f\}$ appearing in the polynomials reported in Table 5.10 are replaced by the parent variables reported here in the right-hand columns. The first two rows consider both the curl- ($\mathbf{\Omega}$) and the divergence-conforming ($\mathbf{\Lambda}$) quadrilateral bases, with zeroth-order basis factors $\mathbf{\Omega}_a = \xi_d \nabla \xi_c$, $\mathbf{\Lambda}_a = \mathbf{\Omega}_a \times \hat{n}$, respectively. In this case, one has to set $\xi_b = 0$, $\xi_e = 1$ in Table 5.10. The last six rows hold for the curl-conforming bases of brick elements, with zeroth-order basis factor $\mathbf{\Omega}_{ab} = \xi_d \xi_e \nabla \xi_c$; in this case, when forming the face-based basis functions, recall that the parent variables reported in the right-hand columns hold for the functions based on the $\xi_b = 0$ face (see Table 5.10).

The dummy parent variable set $\boldsymbol{\xi} = \{\xi_a, \xi_d;\ \xi_b, \xi_e;\ \xi_c, \xi_f\}$ used to form the polynomials (5.79) is an appropriate permutation of the brick parent variables $\{\xi_1, \xi_4;\ \xi_2, \xi_5;\ \xi_3, \xi_6\}$ that depends on the associated zeroth-order edge factor $\mathbf{\Omega}_{ab}$ (or $\mathbf{\Omega}_a$ for the quadrilateral cell), as reported in Table 5.8. Notice that $\mathbf{\Omega}_{ab} = \xi_d \xi_e \nabla \xi_c$ vanishes on the two brick faces $\xi_d = 0$, $\xi_e = 0$, while $\mathbf{\Omega}_a = \xi_d \nabla \xi_c$ vanishes for $\xi_d = 0$, although the polynomials (5.79) do not vanish for those values.

Set (5.79) was first introduced in [37] because of its interpolatory properties and of its symmetry relations which involve $(p+1)^3$ polynomials, that is, a number of independent polynomials higher than what is sufficient to build pth-order complete *scalar* bases. (In fact, a minimum of $(p+1)(p+2)/2$ and $(p+1)(p+2)(p+3)/6$ polynomials is required for the 2D and the 3D cases, respectively.) The total order of each polynomial in set (5.79) is equal to $3p$ and $2p$ for the brick and the quadrilateral cells, respectively (see Chapter 4 and [37]). By using the dependency relations (5.78) to rewrite (5.79) in terms of the independent parent variables only, and by considering the index summation rules reported at bottom of (5.79), one immediately recognizes that the order of these polynomials is p in the ξ_a, the ξ_b, and the ξ_c independent variables.

Various pth-order complete *non-interpolatory* sets are easily obtained by linearly combining the $(p+1)^3$ polynomials of (5.79). Similarly, a pth-order complete set formed by $(p+1)^3$ linearly independent *non-interpolatory* polynomials is *equivalent* to (5.79) if the total order of each of its terms is lower or equal to $3p$, and if linear combinations of its terms can produce an interpolatory polynomial set on the same interpolation grid of set (5.79). More precisely, a polynomial set is equivalent to (5.79) if its $(p+1)^3$ linearly independent polynomials comprise:

- $p^2(p+1)$ volume-based polynomials V_{ijk} that completely vanish on the $\xi_a = 0$ and the $\xi_b = 0$ faces (these polynomials are equivalent to the interpolating ones listed in (5.79) for $i, j, k \neq 0$);
- $p(p+1)$ face-based polynomials F_{0jk} with DoFs associated with the $\xi_a = 0$ face, and that vanish for $\xi_b = 0$ (these polynomials are equivalent to the ones listed in (5.79) for $j = 0$; $i = 1, 2, \ldots, p$; $k = 1, 2, \ldots, p+1$);
- $p(p+1)$ face-based polynomials F_{i0k} based on the $\xi_b = 0$ face and that vanish at $\xi_a = 0$; that is, with DoFs associated with the $\xi_b = 0$ face (these are equivalent to those listed in (5.79) for $i = 0$; $j = 1, 2, \ldots, p$; $k = 1, 2, \ldots, p+1$);
- $(p+1)$ edge-based polynomials E_k based on the $\xi_a = \xi_b = 0$ edge (these are equivalent to the polynomials listed in (5.79) that interpolate, for $i = j = 0$ and $\ell, m = p+1$, the edge formed by the intersection of faces $\xi_a = 0$, $\xi_b = 0$).

and if the order of the equivalent polynomials, when written in terms of their independent parent variables only, is lower or equal to p in the ξ_a, the ξ_b, and the ξ_c variables; this implies that the total order of each polynomial of the equivalent set cannot be higher than $3p$ (or $2p$ for the quadrilateral cell). Each equivalent set, including (5.79), takes the form

$$\begin{cases} H_{00k}(\boldsymbol{\xi}) = E_k(\boldsymbol{\xi}) \\ H_{i0k}(\boldsymbol{\xi}) = F_{i0k}(\boldsymbol{\xi}) \\ H_{0jk}(\boldsymbol{\xi}) = F_{0jk}(\boldsymbol{\xi}) \\ H_{ijk}(\boldsymbol{\xi}) = V_{ijk}(\boldsymbol{\xi}) \end{cases} \tag{5.80}$$

$$\text{for } i, j = 1, 2, \ldots, p; \quad k = 0, 1, \ldots, p$$

Note that the indices i, j, and k appearing in (5.80) are different from and do not play the same role as the indices i, j, and k in (5.79). In general, the polynomials of the equivalent sets (5.80) are not interpolatory and usually have inhomogeneous form [37]. In fact, we are only interested in equivalent hierarchical polynomial sets. Set (5.80) is hierarchical if:

- for $p = 0$ (zeroth order), the set contains only the edge-based polynomial $H_{000}(\boldsymbol{\xi})$;
- for $p \geq 1$, the pth-order complete set is obtained by incrementing the set of order $(p-1)$ with the $(3p^2 + 3p + 1)$ polynomials $H_{ijk}(\boldsymbol{\xi})$ with subscripts i, j, and k given by the following three double-nested loops

$$i = p; \quad j = 0, 1, \ldots, p; \quad k = 0, 1, \ldots, p-1 \tag{5.81}$$

$$j = p; \quad i, k = 0, 1, \ldots, p-1 \tag{5.82}$$

$$k = p; \quad i, j = 0, 1, \ldots, p \tag{5.83}$$

In the following, the hierarchical polynomials $H_{ijk}(\boldsymbol{\xi})$ are conveniently written in terms of the variables

$$\begin{cases} \xi_{ad} = \xi_a - \xi_d \\ \xi_{be} = \xi_b - \xi_e \\ \xi_{cf} = \xi_c - \xi_f \end{cases} \tag{5.84}$$

which, because of (5.78), are equivalent to

$$\begin{cases} \xi_{ad} = 2\xi_a - 1 \\ \xi_{be} = 2\xi_b - 1 \\ \xi_{cf} = 2\xi_c - 1 \end{cases} \tag{5.85}$$

5.3.2.2 Volume-, Face-, and Edge-Based Polynomials

We now form each group of volume-, face-, and edge-based hierarchical polynomials with mutually orthogonal polynomials. Obviously, different choices of the weight function used to define the orthogonality will produce different polynomial sets. For the brick cell, all the higher order vector functions obtained in this manner take the form

$$\begin{aligned} \mathbf{\Omega}_{ijk}^{ab} &= H_{ijk}(\boldsymbol{\xi})\, \mathbf{\Omega}_{ab} = H_{ijk}(\boldsymbol{\xi})\, \xi_d\, \xi_e\, \nabla \xi_c \\ &= H_{ijk}(\boldsymbol{\xi})\, \sqrt{w}\, \nabla \xi_c \end{aligned} \tag{5.86}$$

and are therefore conveniently obtained by using scalar *generating* polynomials $H_{ijk}(\boldsymbol{\xi})$ orthogonal with respect to the weight function

$$w = \xi_d^2\, \xi_e^2 = (1 - \xi_a)^2 (1 - \xi_b)^2 \tag{5.87}$$

which, on Q^1, Q^2 $(=Q_b^2)$, and Q_a^2 simplifies as follows

$$\begin{aligned} w_1 &= w|_{\xi_a = \xi_b = 0} = 1 \\ w_2 &= w|_{\xi_b = 0} = (1 - \xi_a)^2 \\ w_{2a} &= w|_{\xi_a = 0} = (1 - \xi_b)^2 \end{aligned} \tag{5.88}$$

For these weight functions, the *generating* hierarchical polynomials turn out to be naturally expressed in terms of the Jacobi polynomials $P_n^{(\alpha,\alpha)}(z)$, with $\alpha = 0$ or 2, and where n is the polynomial degree. These Jacobi polynomials are either even or odd functions of z, with $P_n^{(\alpha,\alpha)}(-z) = (-1)^n P_n^{(\alpha,\alpha)}(z)$. In particular, the polynomials $P_n^{(0,0)}(z)$ associated with the simplified weight function $w_1 = 1$ are equal to the Legendre polynomials $P_n(z)$. For $\alpha = 0$, 2, the Jacobi polynomials are obtained from the lowest order functions

$$\begin{cases} P_0(z) = 1 \\ P_1(z) = z \end{cases} \tag{5.89}$$

$$\begin{cases} P_0^{(2,2)}(z) = 1 \\ P_1^{(2,2)}(z) = 3z \end{cases} \tag{5.90}$$

by using the following recurrence relations with respect to the degree n

$$(n+1)P_{n+1}(z) = (2n+1)zP_n(z) - nP_{n-1}(z) \tag{5.91}$$

$$(n+1)(n+5)P_{n+1}^{(2,2)}(z) = (n+3)\left[(2n+5)zP_n^{(2,2)}(z) - (n+2)P_{n-1}^{(2,2)}(z)\right] \tag{5.92}$$

The *normalized* hierarchical polynomial set *equivalent* to set (5.79) obtained in this manner consists of the following $(p+1)^3$ polynomials

$$\begin{cases} H_{00k} = E_k = \sqrt{2k+1}\, P_k(\xi_{cf}) \\ H_{i0k} = F_{i0k} = \xi_a f_{i-1}(\xi_{ad})\, E_k(\xi_{cf}) \\ H_{0jk} = F_{0jk} = \xi_b f_{j-1}(\xi_{be})\, E_k(\xi_{cf}) \\ H_{ijk} = V_{ijk} = \xi_a\, \xi_b f_{i-1}(\xi_{ad}) f_{j-1}(\xi_{be})\, E_k(\xi_{cf}) \end{cases} \tag{5.93}$$

for $i, j = 1, 2, \ldots, p$; $k = 0, 1, \ldots, p$; with ξ_{ad}, ξ_{be}, ξ_{cf} given in (5.84), and where

$$f_n(z) = \sqrt{\frac{(2n+5)(n+3)(n+4)}{3(n+1)(n+2)}}\, P_n^{(2,2)}(z) \tag{5.94}$$

indicates the re-scaled Jacobi polynomial of order n. The polynomials $H_{0jk} = F_{0jk}$ associated with the $\xi_a = 0$ face (that vanish on the $\xi_b = 0$ face) are simply obtained from the polynomials F_{i0k} associated with the $\xi_b = 0$ face by changing ξ_b to ξ_a, ξ_{be} to ξ_{ad}, and j to i.

The re-scaled Jacobi polynomials appearing in the expressions (5.93) satisfy the following orthogonality relations, obtained from [33] after setting $z = 2x - 1$ (see (5.85))

$$\int_0^1 E_m(2x-1)\, E_p(2x-1)\, \mathrm{d}x = \delta_{mp} \tag{5.95}$$

$$\int_0^1 x^2(1-x)^2 f_m(2x-1) f_p(2x-1)\, \mathrm{d}x = \frac{\delta_{mp}}{3} \tag{5.96}$$

where δ_{mp} is the Kronecker delta. Equations (5.95) and (5.96) together with (5.85) yield the inner product results of Table 5.9.

As far as the quadrilateral element is concerned, recall that the zeroth-order vector factor associated with the $\xi_a = 0$ edge is $\mathbf{\Omega}_a = \xi_d \nabla \xi_c$. In this case, it is convenient to make the polynomials mutually orthogonal with respect to the weight function $w_2 = \xi_d^2$; it is then readily proved that the $(p+1)^2$ polynomials to be used for the quadrilateral element are the polynomials H_{00k} and H_{i0k} already given in (5.93).

Note also that the order of the hierarchical polynomials H_{ijk} in (5.93) equals the sum of the subscripts (i, j, k) used to denote them. To facilitate numerical implementations and code debugging, the edge-, face-, and volume-based polynomials up to the third order are explicitly reported in Table 5.10.

Equation (5.85), together with (5.95) and (5.96), readily prove that, within each group, the polynomials are mutually orthogonal with respect to the weight functions (5.87). In particular:

- the volume-based polynomials are orthogonal over the brick parent cell Q^3;
- the face-based polynomials are orthogonal over Q^3, as well as over the two quadrilateral faces Q_a^2 and Q^2 $(=Q_b^2)$;
- the edge-based polynomials are orthogonal over Q^3, over both the faces Q_a^2 and Q^2 attached to the $\xi_a = \xi_b = 0$ edge, and also over the $\xi_a = \xi_b = 0$ edge Q^1.

In general, however, a polynomial of one group is not orthogonal to those of a different group. Table 5.9 reports the inner products of the hierarchical polynomials in the matrix form; the normalization used for the polynomials of each group is reported along the main diagonal of the matrices of Table 5.9.

Table 5.9 Vector bases on quadrilateral and brick cells – inner products of the hierarchical polynomials of Table 5.10.

Inner products over Q^3

	E_n	$F_{\ell 0n}$	F_{0mn}	$V_{\ell mn}$
E_k	$\delta_{kn}/9$	$\eta_\ell \delta_{kn}/9$	$\eta_m \delta_{kn}/9$	$\eta_\ell \eta_m \delta_{kn}/9$
F_{i0k}	$\eta_i \delta_{kn}/9$	$\delta_{i\ell}\delta_{kn}/9$	$\eta_i \eta_m \delta_{kn}/9$	$\eta_m \delta_{i\ell}\delta_{kn}/9$
F_{0jk}	$\eta_j \delta_{kn}/9$	$\eta_j \eta_\ell \delta_{kn}/9$	$\delta_{jm}\delta_{kn}/9$	$\eta_\ell \delta_{jm}\delta_{kn}/9$
V_{ijk}	$\eta_i \eta_j \delta_{kn}/9$	$\eta_j \delta_{i\ell}\delta_{kn}/9$	$\eta_i \delta_{jm}\delta_{kn}/9$	$\delta_{i\ell}\delta_{jm}\delta_{kn}/9$

Inner products over:

Q_a^2

	E_n	F_{0mn}
E_k	$\delta_{kn}/3$	$\eta_m \delta_{kn}/3$
F_{0jk}	$\eta_j \delta_{kn}/3$	$\delta_{jm}\delta_{kn}/3$

Q_b^2

	E_n	$F_{\ell 0n}$
E_k	$\delta_{kn}/3$	$\eta_\ell \delta_{kn}/3$
F_{i0k}	$\eta_i \delta_{kn}/3$	$\delta_{i\ell}\delta_{kn}/3$

Q^1

	E_n
E_k	δ_{kn}

Above, δ_{rs} is the Kronecker delta, $\eta_q = (-1)^{(q+1)} \sqrt{\dfrac{3(2q+3)}{q(q+1)(q+2)(q+3)}}$,

and $-\eta_q f_{q-1}(z) = (-1)^q \dfrac{2q+3}{q(q+1)} P_{q-1}^{(2,2)}(z)$.

This table reports only the non-zero values of the inner products

$$
\begin{aligned}
\langle H_{ijk}, H_{\ell mn}\rangle_{Q^3} &= \int_{Q^3} w\, H_{ijk}\, H_{\ell mn}\, \mathrm{d}Q^3 \\
\langle H_{ijk}, H_{\ell mn}\rangle_{Q_a^2} &= \int_{Q_a^2} w\, H_{ijk}\, H_{\ell mn}\, \mathrm{d}Q_a^2 \\
\langle H_{ijk}, H_{\ell mn}\rangle_{Q_b^2} &= \int_{Q_b^2} w\, H_{ijk}\, H_{\ell mn}\, \mathrm{d}Q_b^2 \\
\langle H_{ijk}, H_{\ell mn}\rangle_{Q^1} &= \int_{Q^1} w\, H_{ijk}\, H_{\ell mn}\, \mathrm{d}Q^1
\end{aligned}
$$

for $i, \ell, j, m = 1, 2, \ldots, p$; $k, n = 0, 1, \ldots, p$, and w given in (5.87).

5.3.2.3 Symmetry Relations of the Hierarchical Bases Used to Guarantee Tangential Continuity

For brick cells, the high-order vector functions associated with the face $\xi_b = 0$ ($\xi_e = 1$) oriented in the $\nabla\xi_c$ direction are (see (5.86) and (5.93))

$$
\begin{aligned}
\mathbf{\Omega}_{i0k}^{ab} &= \xi_a f_{i-1}(\xi_{ad})\, E_k(\xi_{cf})\, [\xi_d\, \xi_e\, \nabla\xi_c] \\
\mathbf{\Omega}_{i0k}^{db} &= \xi_d f_{i-1}(\xi_{da})\, E_k(\xi_{cf})\, [\xi_a\, \xi_e\, \nabla\xi_c] \\
&\text{for } i = 1, 2, \ldots, p; \quad k = 0, 1, \ldots, p
\end{aligned}
\tag{5.97}
$$

Table 5.10 Vector bases on quadrilateral and brick cells – edge-, face-, and volume-based hierarchical polynomials.

EDGE-BASED

$p=0$ order:	$E_0(\xi_{cf})=1$
$p=1$ order: increment the $p=0$ set with	$E_1(\xi_{cf})=\sqrt{3}\,\xi_{cf}$
$p=2$ order: increment the $p=1$ set with	$E_2(\xi_{cf})=\sqrt{5}\,(3\xi_{cf}^2-1)/2$
$p=3$ order: increment the $p=2$ set with	$E_3(\xi_{cf})=\sqrt{7}\,\xi_{cf}\,(5\,\xi_{cf}^2-3)/2$

Rule for $p \geq 1$ order: increment the $(p-1)$ set with $E_p(\xi_{cf})=\sqrt{2p+1}\,P_p(\xi_{cf})$ (see (5.93))

FACE-BASED

$p=0$ order; no function

$p=1$ order:

$$F_{100}(\boldsymbol{\xi}) = \sqrt{10}\,\xi_a$$
$$F_{101}(\boldsymbol{\xi}) = \sqrt{30}\,\xi_a\xi_{cf}$$

$p=2$ order: increment the $p=1$ set with

$$F_{200}(\boldsymbol{\xi}) = \sqrt{70}\,\xi_a\,\xi_{ad}$$
$$F_{201}(\boldsymbol{\xi}) = \sqrt{210}\,\xi_a\,\xi_{ad}\xi_{cf}$$
$$F_{102}(\boldsymbol{\xi}) = 5\,\xi_a\,(3\xi_{cf}^2-1)/\sqrt{2}$$
$$F_{202}(\boldsymbol{\xi}) = 5\sqrt{7}\,\xi_a\,\xi_{ad}\,(3\xi_{cf}^2-1)/\sqrt{2}$$

$p=3$ order: increment the $p=2$ set with

$$F_{300}(\boldsymbol{\xi}) = \sqrt{30}\,\xi_a\,(7\,\xi_{ad}^2-1)/2$$
$$F_{301}(\boldsymbol{\xi}) = 3\sqrt{10}\,\xi_a\,(7\,\xi_{ad}^2-1)\,\xi_{cf}/2$$
$$F_{302}(\boldsymbol{\xi}) = 5\sqrt{3}\,\xi_a\,(7\,\xi_{ad}^2-1)\,(3\,\xi_{cf}^2-1)/2\sqrt{2}$$
$$F_{103}(\boldsymbol{\xi}) = \sqrt{35}\,\xi_a\,\xi_{cf}\,(5\,\xi_{cf}^2-3)/\sqrt{2}$$
$$F_{203}(\boldsymbol{\xi}) = 7\sqrt{5}\,\xi_a\,\xi_{ad}\,\xi_{cf}\,(5\,\xi_{cf}^2-3)/\sqrt{2}$$
$$F_{303}(\boldsymbol{\xi}) = \sqrt{105}\,\xi_a\,(7\,\xi_{ad}^2-1)\,\xi_{cf}\,(5\,\xi_{cf}^2-3)/2\sqrt{2}$$

Rule for $p \geq 1$ order: increment the $(p-1)$ set with $F_{i0k}=\xi_a f_{i-1}(\xi_{ad})E_k(\xi_{cf})$ (see (5.93) and (5.94)) for $i=p$; $k=0,1,\ldots,p-1$; and for $k=p$; $i=1,2\ldots,p$

VOLUME-BASED

$p=0$ order $\Rightarrow$ no function

$p=1$ order $\Rightarrow$ two functions:

$$V_{110}=10\xi_a\,\xi_b$$
$$V_{111}=10\sqrt{3}\xi_a\,\xi_b\,\xi_{cf}$$

$p=2$ order $\Rightarrow$ increment the $p=1$ set with ten functions:

$V_{210}=10\sqrt{7}\xi_a\,\xi_b\,\xi_{ad}$	$V_{112}=5\sqrt{5}\xi_a\,\xi_b\,\chi_{cf}^2$
$V_{220}=70\xi_a\,\xi_b\,\xi_{ad}\,\xi_{be}$	$V_{212}=5\sqrt{35}\xi_a\,\xi_b\,\xi_{ad}\,\chi_{cf}^2$
$V_{211}=10\sqrt{21}\xi_a\,\xi_b\,\xi_{ad}\xi_{cf}$	$V_{222}=35\sqrt{5}\xi_a\,\xi_b\,\xi_{ad}\,\xi_{be}\,\chi_{cf}^2$
$V_{221}=70\sqrt{3}\xi_a\,\xi_b\,\xi_{ad}\,\xi_{be}\xi_{cf}$	$V_{120},\ V_{121},\ V_{122}$

$p=3$ order $\Rightarrow$ increment the $p=2$ set with twenty four functions:

$V_{310} = 5\sqrt{3}\,\xi_a\,\xi_b\,\chi_{ad}^2,$	$V_{311} = 15\,\xi_a\,\xi_b\,\chi_{ad}^2\,\xi_{cf},$	$V_{312} = 5\sqrt{15}\,\xi_a\,\xi_b\,\chi_{ad}^2\,\chi_{cf}^2/2,$
$V_{320} = 5\sqrt{21}\,\xi_a\,\xi_b\,\chi_{ad}^2\,\xi_{be},$	$V_{321} = 15\sqrt{7}\,\xi_a\,\xi_b\,\chi_{ad}^2\,\xi_{be}\,\xi_{cf},$	$V_{322} = 5\sqrt{105}\,\xi_a\,\xi_b\,\chi_{ad}^2\,\xi_{be}\,\chi_{cf}^2/2,$
$V_{330} = 15\,\xi_a\,\xi_b\,\chi_{ad}^2\,\chi_{be}^2/2,$	$V_{331} = 15\sqrt{3}\,\xi_a\,\xi_b\,\chi_{ad}^2\,\chi_{be}^2\,\xi_{cf}/2,$	$V_{332} = 15\sqrt{5}\,\xi_a\,\xi_b\,\chi_{ad}^2\,\chi_{be}^2\,\chi_{cf}^2/4,$
$V_{130},\ V_{230},\ V_{131},$	$V_{313} = 5\sqrt{21}\,\xi_a\,\xi_b\,\chi_{ad}^2\,\chi_{cf}^3/2,$	$V_{113} = 5\sqrt{7}\,\xi_a\,\xi_b\,\chi_{cf}^3,$
$V_{231},\ V_{132},\ V_{232},$	$V_{323} = 35\sqrt{3}\,\xi_a\,\xi_b\,\chi_{ad}^2\,\xi_{be}\,\chi_{cf}^3/2,$	$V_{213} = 35\,\xi_a\,\xi_b\,\xi_{ad}\,\chi_{cf}^3,$
$V_{133},\ V_{233},\ V_{123},$	$V_{333} = 15\sqrt{7}\,\xi_a\,\xi_b\,\chi_{ad}^2\,\chi_{be}^2\,\chi_{cf}^3/4,$	$V_{223} = 35\sqrt{7}\,\xi_a\,\xi_b\,\xi_{ad}\,\xi_{be}\,\chi_{cf}^3$

Rule for $p \geq 2$ order $\Rightarrow$ increment the $(p-1)$ set with $p(3p-1)$ functions (see (5.93) and (5.94)):

$V_{ijk}=\xi_a\xi_b f_{i-1}(\xi_{ad})f_{j-1}(\xi_{be})E_k(\xi_{cf})$ for $i=p$; $j=1,2,\ldots,p$; $k=0,1,\ldots,p-1$
for $j=p$; $i=1,2,\ldots,p-1$; $k=0,1,\ldots,p-1$
for $k=p$; $i=1,2,\ldots,p$; $j=1,2\ldots,p$

Table 5.10 (*Continued*).

With $\xi_a + \xi_d = 1;\ \ \xi_{ad} = \xi_a - \xi_d;\ \ \chi^2_{ad} = (7\,\xi^2_{ad} - 1);$

$\xi_b + \xi_e = 1;\ \ \xi_{be} = \xi_b - \xi_e;\ \ \chi^2_{be} = (7\,\xi^2_{be} - 1);$

$\xi_c + \xi_f = 1;\ \ \xi_{cf} = \xi_c - \xi_f;\ \ \chi^2_{cf} = \left(3\,\xi^2_{cf} - 1\right);\ \ \chi^3_{cf} = \xi_{cf}\left(5\,\xi^2_{cf} - 3\right)$

The edge-based polynomials are mutually orthogonal on Q^1, Q^2_a, Q^2, and Q^3; the face-based polynomials are mutually orthogonal on Q^2_a, Q^2, and Q^3; the volume-based polynomials are mutually orthogonal on Q^3. These polynomials are normalized as shown in Table 5.9. The reported polynomials F_{i0k} are those associated with the $\xi_b = 0$ face, which all vanish for $\xi_a = 0$. The polynomials F_{0ik} based on the $\xi_a = 0$ face must be included to form a complete polynomial base on Q^3. An expression for these polynomials is obtained by replacing ξ_b with ξ_a, ξ_a with ξ_b, ξ_{be} with ξ_{ad}, and ξ_{ad} with ξ_{be} in the expression for $F_{i0k}(\xi_a, \xi_b, \xi_{ad}, \xi_{be})$ reported above, as follows:

$$F_{0ik}(\xi_a, \xi_b, \xi_{ad}, \xi_{be}) = F_{i0k}(\xi_b, \xi_a, \xi_{be}, \xi_{ad})$$

An expression for the polynomial V_{jik} (for $j \neq i$) not explicitly reported is obtained by substituting ξ_b with ξ_a, ξ_a with ξ_b, ξ_{be} with ξ_{ad}, and ξ_{ad} with ξ_{be} in the reported expression for V_{ijk}.

with

$$f_{i-1}(\xi_{ad}) = (-1)^{(i+1)} f_{i-1}(\xi_{da}) \tag{5.98}$$

$$\mathbf{\Omega}^{ab}_{i0k} = (-1)^{(i+1)}\, \mathbf{\Omega}^{db}_{i0k} \tag{5.99}$$

In (5.97), the zeroth-order factors used to construct the higher order vector functions are reported in square brackets.

Clearly, (5.99) shows that to secure basis independence one has to discard either $\mathbf{\Omega}^{db}_{i0k}$ or $\mathbf{\Omega}^{ab}_{i0k}$ from the basis set. Furthermore, since $\nabla\xi_c = -\nabla\xi_f$ and $E_k(\xi_{cf}) = (-1)^k\, E_k(\xi_{fc})$, one has

$$\mathbf{\Omega}^{ab}_{i0k}(\xi_c, \xi_f) = (-1)^{(k+1)}\, \mathbf{\Omega}^{ab}_{i0k}(\xi_f, \xi_c) \tag{5.100}$$

Equations (5.99) and (5.100) show that to impose tangential continuity across adjacent elements that share the same quadrilateral face, it is sufficient to adjust the sign of the face-based vector functions in use in one of the two adjacent cells.

Similarly, for the $(p+1)$ higher order vector functions associated with the edge formed by the intersection of the brick faces $\xi_a = 0$, $\xi_b = 0$ ($\mathbf{\Omega}^{ab}_{00k}$), or associated with the $\xi_a = 0$ edge of a quadrilateral cell ($\mathbf{\Omega}^{a}_{00k}$) one has

$$\mathbf{\Omega}^{ab}_{00k}(\xi_c, \xi_f) = (-1)^{(k+1)}\, \mathbf{\Omega}^{ab}_{00k}(\xi_f, \xi_c) \tag{5.101}$$

$$\begin{aligned}\mathbf{\Omega}^{a}_{00k}(\xi_c, \xi_f) &= (-1)^{(k+1)}\, \mathbf{\Omega}^{a}_{00k}(\xi_f, \xi_c)\\ &\text{for } k = 0, 1, \ldots, p\end{aligned} \tag{5.102}$$

Once again, (5.101) and (5.102) show that to secure tangential continuity across adjacent elements that share the same edge, it is sufficient to adjust the sign of the edge-based vector functions in use in all but one of the adjacent cells.

In connection with the tangential continuity issue, it is important to note that the hierarchical edge-based and face-based functions provided here match those of the other cells discussed in this section and, along a given common edge or face, are normalized in the same manner. This permits one to mesh two-dimensional domains with a mixture of triangular and quadrilateral cells, and to use meshes with a mixture of brick, prism, and tetrahedral cells in volumetric regions.

5.3.2.4 Equivalent Bases with Improved Linear Independence

In general, as discussed in Section 5.3.2.2 and apparent from the results of Table 5.9, a polynomial of the edge-, face-, or volume-based group is not orthogonal to the polynomials of a different group. However, appropriate linear combinations of the volume-based polynomials can be added to any given face-based polynomial to obtain a new face-based polynomial orthogonal to the volume-based ones. The same holds for any given edge-based polynomial, which can be linearly combined with its non-orthogonal face and volume-based polynomials to get a new edge-based polynomial orthogonal to the face- and volume-based ones. Linear combinations as such are driven by the analytical results of Table 5.9.

Thus, for $p \geq 1$, one can obtain a pth-order complete set by incrementing the set of order $(p-1)$ with $(3p^2+3p+1)$ *mutually orthogonal* polynomials $\widetilde{H}_{ijk}(\boldsymbol{\xi})$ different from those of Table 5.10, but with subscripts i, j, and k again given by (5.81)–(5.83). Although the elements of the new hierarchical family obtained in this manner for the $(p-1)$ order are not completely orthogonal to the $(3p^2+3p+1)$ elements added to complete this base to pth order, the linear independence of the elements of the new pth-order basis is improved relative to the bases given in Table 5.10. This further orthogonalization process produces the equivalent polynomial set

$$\begin{aligned} \widetilde{E}_k(\boldsymbol{\xi}) &= \Upsilon_k(\xi_a)\,\Upsilon_k(\xi_b)\,E_k(\xi_{cf}) \\ \widetilde{F}_{i0k}(\boldsymbol{\xi}) &= \Upsilon_\ell(\xi_b)\,F_{i0k}(\boldsymbol{\xi}) \\ \widetilde{F}_{0ik}(\boldsymbol{\xi}) &= \Upsilon_\ell(\xi_a)\,F_{0ik}(\boldsymbol{\xi}) \\ \widetilde{V}_{ijk}(\boldsymbol{\xi}) &= \tfrac{3}{4}\,V_{ijk}(\boldsymbol{\xi}) \\ &\text{for } i,j=1,2,\ldots,p;\ k=0,1,\ldots,p; \\ &\ell=\max\,[i,k] \end{aligned} \tag{5.103}$$

that is also given in terms of the polynomials of Table 5.10, corrected by the factors that involve the re-scaled Jacobi polynomial

$$\Upsilon_n(z)=\frac{(-1)^n}{(n+1)}\,P_n^{(2,1)}(2z-1) \tag{5.104}$$

of order n, with

$$\Upsilon_n(0)=1 \tag{5.105}$$

The re-scaled Jacobi polynomials $\Upsilon_n(z)$ are easily obtained from the lowest order functions

$$\begin{cases} \Upsilon_0(z) = 1 \\ \Upsilon_1(z) = (2-5z)/2 \end{cases} \tag{5.106}$$

by using the following recurrence relations with respect to the degree n

$$a_{1n}\Upsilon_{n+1}(z)=(a_{2n}-a_{3n}z)\Upsilon_n(z)-a_{4n}\Upsilon_{n-1}(z) \tag{5.107}$$

with

$$\begin{cases} a_{1n} = (n+4)(2n+3) \\ a_{2n} = 4(n+1)(n+3) \\ a_{3n} = 2(2n+3)(2n+5) \\ a_{4n} = n(2n+5) \end{cases} \tag{5.108}$$

The volume-based polynomials $\widetilde{V}_{ijk}$ are obtained by multiplying those of Table 5.10 by 3/4, while in general, the order of the edge- and face-based polynomials in (5.103) is higher than the sum of the relevant i, j, and k subscripts. The normalization constants used for (5.103) were chosen to improve the linear independence of the polynomials, established by studying (and lowering) the condition numbers of the Gram matrices having coefficients $G_{r,s}$ equal to the inner product (over Q^2 and Q^3) of the rth and sth polynomials of the new pth-order complete family. These polynomials are thus normalized as follows

$$\begin{aligned} &\int_{Q^1} w\, \widetilde{E}_k^2(\boldsymbol{\xi})\, \mathrm{d}Q^1 = 1 \\ &\int_{Q_b^2} w\, \widetilde{F}_{i0k}^2(\boldsymbol{\xi})\, \mathrm{d}Q_b^2 = \int_{Q_a^2} w\, \widetilde{F}_{0ik}^2(\boldsymbol{\xi})\, \mathrm{d}Q_a^2 = \frac{1}{3} \\ &\int_{Q^3} w\, \widetilde{V}_{ijk}^2(\boldsymbol{\xi})\, \mathrm{d}Q^3 = 1/16 \end{aligned} \tag{5.109}$$

that is, the new edge-based polynomials remain normalized to a unitary inner product over their primary edge, and the new face-based polynomials remain normalized to an inner product equal to 1/3 over their associated face, as do the polynomials of Table 5.10.

5.3.3 Prism Bases

Finite element analysis usually involves the discretization of a problem domain into tetrahedral or hexahedral cells. The triangular prism is a third cell shape that enables the transition between the other two cell types. Prisms may also provide an efficient means of modeling a problem domain that is thin in one dimension or has some other feature that renders the field variation in that dimension small (or large) relative to the other dimensions.

As in the previous subsections, we consider vector bases from the reduced-gradient curl-conforming spaces proposed by Nédélec in 1986 [38] for prisms. The lowest order base vectors for prisms were first used in finite element analysis by Dular et al. [39], Sacks and Lee [40], and Özdemir and Volakis [41]. The prism-interpolatory basis functions of arbitrary polynomial order reported in Chapter 4 were originally developed in [42]. A few ad hoc basis functions of higher order have also been developed for specialized situations [43–45], including that when the field variation in one direction is much different than in the others. A set of hierarchical basis functions of arbitrary order for prisms was proposed by Zaglmayr in 2006 [19]; these functions span polynomial-complete spaces but do not span the reduced-gradient Nédélec spaces.

In the following, we present the hierarchical basis functions of arbitrary order for the prism that were originally developed in [28]. These curl-conforming functions are compatible with the similar basis functions for tetrahedral and hexahedral cells discussed before in this

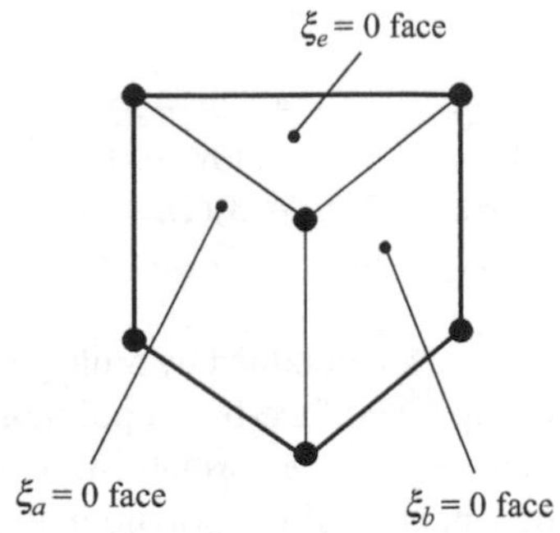

Figure 5.8 The parent prism Q^3 is the flat-faceted rectilinear prism $\{0 \leq \xi_a, \xi_b, \xi_c \leq 1; 0 \leq \xi_d, \xi_e \leq 1\}$ of unit height and with two equal and parallel triangular bases. The quadrilateral faces are the ξ_a, ξ_b, and ξ_c zero-coordinate surfaces; the two opposite triangular faces are the ξ_d and ξ_e zero-coordinate surfaces.

section, in the sense that the prism bases maintain tangential-vector continuity across edges and faces with the tetrahedra functions and with the hexahedra functions. The prism functions are constructed from the product of the prism base vectors with orthogonal scalar polynomials; such an approach facilitates good linear independence (as measured by matrix condition numbers) as the order grows and ensures that the functions properly span the reduced-gradient Nédélec spaces. General expressions are provided that can be used to obtain basis functions of any polynomial degree.

As discussed in Chapter 4, a *local* system of normalized parent coordinates $\boldsymbol{\xi} = (\xi_1, \xi_2, \ldots, \xi_5)$ is associated with each prism cell. The ith face is the zero-coordinate surface for the normalized coordinate ξ_i, and each edge is the locus of points where two *local* parent coordinates vanish. The coordinate ξ_i varies linearly across the prism, attaining a value of unity at the face, edge, or vertex opposite the zero-coordinate surface. The coordinate gradient $\nabla\xi_i$ is therefore perpendicular to the ith face and, on this face, it points toward the element's interior. The prism can be curved or distorted.

In order to avoid reporting lengthy expressions for all the polynomial bases, we exploit the symmetry of the prism and denote its normalized coordinates by *dummy* parent variables, distinguished by a literal (i.e. not numerical) subscript. Thus, as shown in Figure 5.8, the $\xi_a = 0$, $\xi_b = 0$, and $\xi_c = 0$ faces of the prism are quadrilateral while its two opposite triangular faces are defined by the zero-coordinate surfaces $\xi_d = 0$ and $\xi_e = 0$. The variables ξ_a, ξ_b, and ξ_d are assumed to be independent with

$$\xi_c = 1 - \xi_a - \xi_b \tag{5.110}$$

$$\xi_e = 1 - \xi_d \tag{5.111}$$

Occasionally, in the following, we indicate the variable subset $\{\xi_a, \xi_b, \xi_c\}$ with the symbol $\boldsymbol{\xi}_t$ to write, for example

$$\begin{aligned} \boldsymbol{\xi}_t &= \{\xi_a, \xi_b, \xi_c\} \\ \boldsymbol{\xi} &= \{\boldsymbol{\xi}_t; \xi_d, \xi_e\} \end{aligned} \tag{5.112}$$

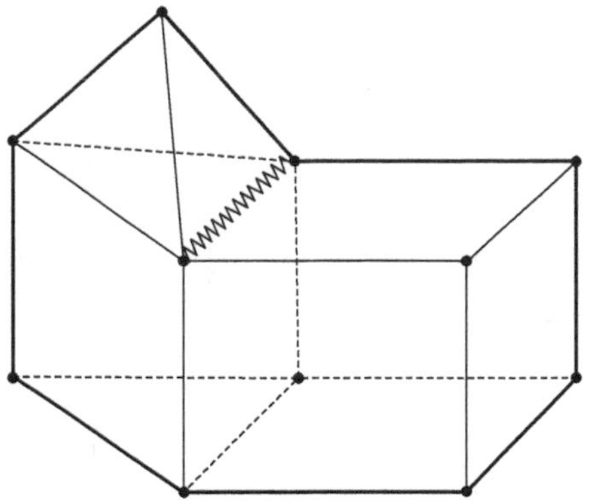

Figure 5.9 The *dummy* edge Q^1_{cd} of a prism (denoted by a snaked line) is shared by the tetrahedral and brick cells.

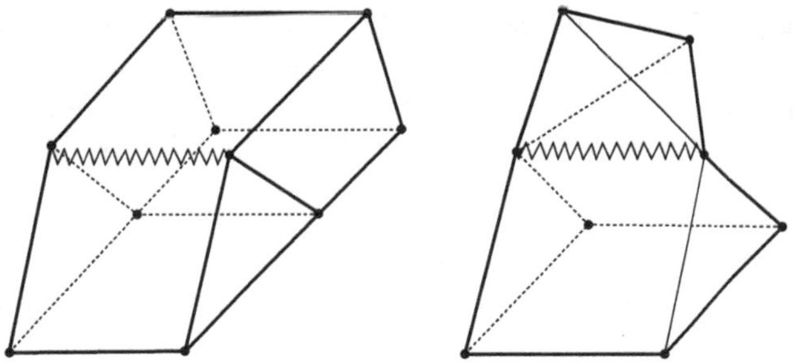

Figure 5.10 The *dummy* edge Q^1_{ab} of a prism is in common with a brick cell in figure (a) and a tetrahedral cell in figure (b). The edge at issue is denoted by a snaked line in the figures.

Furthermore, Q^2_α indicates the face $\xi_\alpha = 0$ of the prism, while $Q^1_{\alpha\beta}$ indicates the edge formed by the intersection of the faces $\xi_\alpha = 0$, $\xi_\beta = 0$.

The introduction of dummy parent variables is convenient because it permits one to discuss only the vector functions associated with the edges Q^1_{ab} and Q^1_{cd}. In fact, the *dummy* edge Q^1_{cd} typically represents the intersection of a quadrilateral ($\xi_c = 0$) face with a triangular ($\xi_d = 0$) face of the prism while Q^1_{ab} represents the *typical* intersection of two quadrilateral prism faces. Obviously, the hierarchical vector functions associated with these edges must conform with the hierarchical tetrahedral and brick vector functions already reported in Sections 5.3.1 and 5.3.2 since these edges could be shared by differently shaped cells, as shown in Figures 5.9 and 5.10. To further clarify how conformity between differently shaped adjacent elements is ensured, let us observe that each triangular (or quadrilateral) face of any given three-dimensional cell is always mapped by the same triangular (or square) cell of a two-dimensional parent space

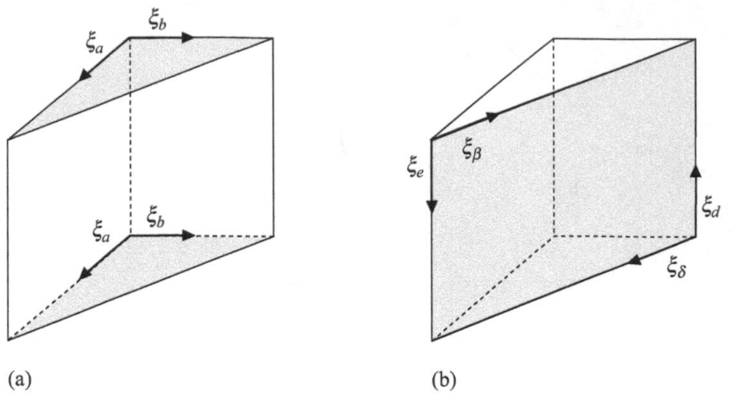

Figure 5.11 The triangular or quadrilateral face of any given cell is mapped by the same fundamental two-dimensional parent cell. (a) For the prism, the triangular faces $\xi_d = 0$ and $\xi_e = 0$ are described by the same parent triangle $\{0 \le \xi_a, \xi_b, \xi_c \le 1\}$, with $\xi_c = 1 - \xi_a - \xi_b$. (b) Each quadrilateral face of a prism is described by the same parent square $\{0 \le \xi_\beta, \xi_\delta \le 1; 0 \le \xi_d, \xi_e \le 1\}$, with $\xi_\delta = 1 - \xi_\beta$ and $\xi_e = 1 - \xi_d$. For the $\xi_c = 0$ quadrilateral face shown in figure (b), one has $\xi_\beta = \xi_b$ and $\xi_\delta = \xi_a$. Similarly, on the $\xi_a = 0$ face of a prism one sets $\xi_\beta = \xi_c$ and $\xi_\delta = \xi_b$, while the $\xi_b = 0$ face of a prism is described by setting $\xi_\beta = \xi_a$ and $\xi_\delta = \xi_c$.

© 2012 IEEE. Reprinted with permission from R. D. Graglia and A. F. Peterson, "Hierarchical curl-conforming Nédélec elements for triangular-prism cells," *IEEE Trans. Antennas Propag.*, vol. 60, no. 7, pp. 3314–3324, Jul. 2012.

described in terms of three (or four) *dummy* parent variables. For example, with reference to Figure 5.11b, any quadrilateral face is mapped by the same parent square cell $\{0 \le \xi_\alpha, \xi_\gamma \le 1; 0 \le \xi_\beta, \xi_\delta \le 1\}$, with $\xi_\gamma = 1 - \xi_\alpha$ and $\xi_\delta = 1 - \xi_\beta$.

In terms of the dummy parent variables, each of the nine zeroth-order curl-conforming vector functions of the prism element takes one of the following forms (see Chapter 4)

$$\mathbf{\Omega}_{ab} = \xi_c \nabla \xi_d \tag{5.113}$$

$$\mathbf{\Omega}_{cd} = \xi_e \left(\xi_a \nabla \xi_b - \xi_b \nabla \xi_a\right) \tag{5.114}$$

where it is understood that $\mathbf{\Omega}_{ab}$ is associated with the edge formed by the intersection of two quadrilateral faces ($\xi_a = 0$, $\xi_b = 0$), whereas $\mathbf{\Omega}_{cd}$ is associated with the edge formed by the intersection of a quadrilateral face ($\xi_c = 0$) with a triangular face ($\xi_d = 0$). Notice here that $\mathbf{\Omega}_{ab}$ vanishes on the quadrilateral face $\xi_c = 0$, while $\mathbf{\Omega}_{cd}$ vanishes on the triangular face $\xi_e = 0$.

Apart from a sign adjustment, the edge functions of (5.113) and (5.114) match with the functions defined on the elements attached to the prism, that could have tetrahedral or hexahedral shape, as well as prismatic shape. In fact, with reference to the zeroth-order vector functions discussed in Chapter 4, $\mathbf{\Omega}_{cd}$ simplifies into $(\xi_a \nabla \xi_b - \xi_b \nabla \xi_a)$ on the triangular face $\xi_d = 0$ (where $\xi_e = 1$), to coincide with one of the three zeroth-order curl-conforming vector functions of the triangular element defined by the $\xi_d = 0$ face of the prism. Similarly, on the $\xi_c = 0$ quadrilateral face, the component $\xi_e \nabla \xi_b$ $(= -\xi_e \nabla \xi_a)$ of $\mathbf{\Omega}_{cd}$ tangent to the $\xi_c = 0$ surface coincides with one of the four zeroth-order curl-conforming vector functions of a quadrilateral element, that is, the surface element defined by the $\xi_c = 0$ face of the prism. Finally, on the $\xi_a = 0$ (and on the $\xi_b = 0$) quadrilateral face, the edge function $\mathbf{\Omega}_{ab}$ naturally coincides with one of the four zeroth-order curl-conforming vector functions of a quadrilateral element, that is, the surface element defined by the $\xi_a = 0$ (or the $\xi_b = 0$) face of the prism.

The prism interpolatory higher order vector functions of Chapter 4

$$\boldsymbol{\Omega}^{ab}_{\text{interp.}} = \hat{\alpha}^{ab}_{ijk;\ell m}(\boldsymbol{\xi})\, \boldsymbol{\Omega}_{ab} \tag{5.115}$$

$$\boldsymbol{\Omega}^{cd}_{\text{interp.}} = \hat{\alpha}^{cd}_{ijk;\ell m}(\boldsymbol{\xi})\, \boldsymbol{\Omega}_{cd} \tag{5.116}$$

are generated by multiplying the zeroth-order vector functions $\boldsymbol{\Omega}_{ab}$ and $\boldsymbol{\Omega}_{cd}$ given in (5.113) and (5.114) with the following interpolatory polynomials

$$\begin{aligned} \hat{\alpha}^{ab}_{ijk;\ell m}(\boldsymbol{\xi}) = {} & R_i(p+1,\xi_a)\, R_j(p+1,\xi_b)\, \hat{R}_k(p+1,\xi_c)\, \hat{R}_\ell(p+2,\xi_d)\, \hat{R}_m(p+2,\xi_e) \\ & \text{for} \quad i,j=0,1,\ldots,p; \quad k,\ell,m=1,2,\ldots,p+1; \\ & \text{with } i+j+k=p+1; \quad \ell+m=p+2 \end{aligned} \tag{5.117}$$

$$\begin{aligned} \hat{\alpha}^{cd}_{ijk;\ell m}(\boldsymbol{\xi}) = {} & \hat{R}_i(p+2,\xi_a)\, \hat{R}_j(p+2,\xi_b)\, R_k(p+2,\xi_c)\, R_\ell(p+1,\xi_d)\, \hat{R}_m(p+1,\xi_e) \\ & \text{for} \quad k,\ell=0,1,\ldots,p; \quad i,j,m=1,2,\ldots,p+1; \\ & \text{with } i+j+k=p+2; \quad \ell+m=p+1 \end{aligned} \tag{5.118}$$

where $R_n(q,\xi)$ and $\hat{R}_n(q,\xi)$ are the Silvester and the *shifted* Silvester polynomial, respectively; ξ is in the interval [0, 1] and the parameter q is the number of uniform subintervals into which this interval is divided. The degrees of $R_n(q,\xi)$ and $\hat{R}_n(q,\xi)$ are n and $n-1$, respectively. Once again notice that, apart from a sign adjustment, the high-order functions of (5.115)–(5.118) match with the functions defined on the elements attached to the prism, that could have tetrahedral or hexahedral shape, as well as prismatic shape (see Figures 5.9 and 5.10).

The dummy parent variable set $\boldsymbol{\xi}$ used to form the interpolatory polynomials is an appropriate permutation of the prism parent variables $\boldsymbol{\xi} = \{\xi_1, \xi_2, \xi_3; \xi_4, \xi_5\}$ defined in Chapter 4 that depend on the associated zeroth-order edge factor $\boldsymbol{\Omega}_{ab}$ or $\boldsymbol{\Omega}_{cd}$, as reported in Table 5.11.

As summarized in Table 5.12, (5.117) and (5.118) are easily subdivided into volume-, face-, or edge-based polynomial according to the common factors that they contain. The polynomials (5.117) used to build the higher order vectors $\boldsymbol{\Omega}^{ab}$ are volume based if they contain a common $\xi_a\xi_b$ factor (since they vanish on both the $\xi_a = 0$ and $\xi_b = 0$ faces); they are based on the $\xi_b = 0$ (or on the $\xi_a = 0$) quadrilateral face if they contain a common ξ_a (or ξ_b) factor and, finally, they are based on the $\xi_a = \xi_b = 0$ edge if they do not vanish for $\xi_a = \xi_b = 0$. A similar subdivision holds for the polynomials (5.118) which construct $\boldsymbol{\Omega}^{cd}$; they are volume based if they contain a common $\xi_c\xi_d$ factor; they are based on the $\xi_d = 0$ triangular face (or on the $\xi_c = 0$ quadrilateral face) if they contain a common ξ_c (or ξ_d) factor and, finally, they are based on the $\xi_c = \xi_d = 0$ edge if they do not vanish for $\xi_c = \xi_d = 0$.

Sets (5.117) and (5.118) were first introduced in [42] because of their interpolatory properties and of their symmetry relations which involve $(p+1)^2(p+2)/2$ polynomials, which is a number of independent polynomials higher than what is sufficient to build a pth-order complete *scalar* basis. The total order of each polynomial in (5.117) and (5.118) is in fact equal to $2p$. As a matter of fact, a minimum of $(p+1)(p+2)(p+3)/6$ polynomials is required to form a pth-order complete 3D scalar bases; that is, sets (5.117) and (5.118) both have $p(p+1)(p+2)/3$ more polynomials than needed. By using the dependency relations (5.110) and (5.111) to rewrite (5.117) and (5.118) in terms of the independent parent variables only, and

Table 5.11 Curl-conforming vector bases on prism cells – correspondence between dummy and parent variables.

Function	$\{\xi_a, \xi_b, \xi_c, \xi_d, \xi_e\}$	Function	$\{\xi_a, \xi_b, \xi_c, \xi_d, \xi_e\}$
$\mathbf{\Omega}_{12}$	$\{\xi_1, \xi_2, \xi_3; \xi_4, \xi_5\}$	$\mathbf{\Omega}_{13}$	$\{\xi_3, \xi_1, \xi_2; \xi_5, \xi_4\}$
$\mathbf{\Omega}_{23}$	$\{\xi_2, \xi_3, \xi_1; \xi_4, \xi_5\}$		
$\mathbf{\Omega}_{34}$	$\{\xi_1, \xi_2, \xi_3; \xi_4, \xi_5\}$	$\mathbf{\Omega}_{35}$	$\{\xi_2, \xi_1, \xi_3; \xi_5, \xi_4\}$
$\mathbf{\Omega}_{14}$	$\{\xi_2, \xi_3, \xi_1; \xi_4, \xi_5\}$	$\mathbf{\Omega}_{15}$	$\{\xi_3, \xi_2, \xi_1; \xi_5, \xi_4\}$
$\mathbf{\Omega}_{24}$	$\{\xi_3, \xi_1, \xi_2; \xi_4, \xi_5\}$	$\mathbf{\Omega}_{25}$	$\{\xi_1, \xi_3, \xi_2; \xi_5, \xi_4\}$

Hierarchical (or interpolatory) vector basis functions on a prism element are the product of hierarchical (interpolatory) polynomials with the zeroth-order vector functions of Chapter 4 listed here in the left-hand columns. In forming the products, the dummy parent variables $\{\xi_a, \xi_b, \xi_c; \xi_d, \xi_e\}$ appearing in the hierarchical polynomials (5.133) and (5.139) (or in the interpolatory polynomials (5.117) and (5.118)) are replaced by the parent variables reported here in the right-hand columns. The first two rows hold for the curl-conforming functions associated with zeroth-order basis factors $\mathbf{\Omega}_{ab} = \xi_c \nabla \xi_d$. In this case, the variables ξ_a and ξ_b needed to form the hierarchical face-based functions are decided as reported in the second and third row of (5.133); that is, the right-hand columns show only one of the two needed sets. The last three rows in the table hold for the curl-conforming functions associated with zeroth-order basis factor $\mathbf{\Omega}_{cd} = \xi_e\,(\xi_a \nabla \xi_b - \xi_b \nabla \xi_a)$.

Adapted from R. D. Graglia and A. F. Peterson, "Hierarchical curl-conforming Nédélec elements for triangular-prism cells," *IEEE Trans. Antennas Propag.*, vol. 60, no. 7, pp. 3314–3324, Jul. 2012.

Table 5.12 Curl-conforming vector bases on prism cells – number of volume-, face-, and edge-based generating polynomials used to form $\mathbf{\Omega}^{ab}$ and $\mathbf{\Omega}^{cd}$.

	Vector functions $\mathbf{\Omega}^{ab}$ of pth order			Vector functions $\mathbf{\Omega}^{cd}$ of pth order		
	Common factor	Number of polynomials	i, j indices used in (5.117)	Common factor	Number of polynomials	k, ℓ indices used in (5.118)
Volume-based	$\xi_a\xi_b$	$\frac{p(p-1)(p+1)}{2}$	$i,j \neq 0$	$\xi_c\xi_d$	$\frac{p^2(p+1)}{2}$	$k,\ell \neq 0$
Triangular face-based				ξ_c	$p(p+1)/2$	$\ell = 0; k \neq 0$
Quadrilateral face-based	ξ_a ξ_b	$p(p+1)$ $p(p+1)$	$j=0; i \neq 0$ $i=0; j \neq 0$	ξ_d	$p(p+1)$	$k=0; \ell \neq 0$
Edge-based	1	$(p+1)$	$i,j=0$	1	$(p+1)$	$k,\ell = 0$

The interpolatory (5.117) and (5.118) as well as the hierarchical (5.133) and (5.139) polynomials are subdivided from the outset into volume-, face-, and edge-based polynomials according to the common factor that they contain. The table reports the number of these polynomials for the arbitrary pth order, together with their associated common factor.

Adapted from R. D. Graglia and A. F. Peterson, "Hierarchical curl-conforming Nédélec elements for triangular-prism cells," *IEEE Trans. Antennas Propag.*, vol. 60, no. 7, pp. 3314–3324, Jul. 2012.

by considering the index rules reported at the bottom of (5.117) and (5.118), one immediately recognizes that these polynomials are the product of two polynomials of order p: one in the ξ_a and the ξ_b independent variables multiplied with a second one in the ξ_d independent variable.

Various pth-order complete *non-interpolatory* sets are easily obtained by linearly combining the polynomials of (5.117), or those of (5.118). Similarly, a pth-order complete set formed by $(p+1)^2(p+2)/2$ linearly independent *non-interpolatory* polynomials is *equivalent* to (5.117) (or to (5.118)) if the total order of each of its terms is lower or equal to $2p$, and if linear combinations of its terms can produce an interpolatory polynomial set on the same interpolation grid of set (5.117) (or (5.118)). More precisely, a polynomial set is equivalent to (5.117) (or to (5.118)) if its $(p+1)^2(p+2)/2$ linearly *independent* polynomials comprise:

- volume-based polynomials in the same number and with the same common factor of Table 5.12;
- face-based polynomials in the same number and with the same common factor of Table 5.12;
- $(p+1)$ edge-based polynomials;

and if each equivalent polynomial is the product of two polynomials of order lower or equal to p: one in the ξ_a and the ξ_b independent variables multiplied with a second one in the ξ_d independent variable; this implies that the total order of each equivalent polynomial cannot bc higher than $2p$. In general, the polynomials of equivalent sets are not interpolatory and usually have inhomogeneous form.

In the following subsubsections, hierarchical *generating* polynomials are obtained by subdividing from the outset linear combinations of the interpolatory polynomials (5.117) and (5.118) into three different groups of edge- (E), face- (F), and volume-based (V) functions. In each group, all the hierarchical polynomials are constructed *a priori* to be mutually orthogonal independent of the definition domain of the relevant inner product, that is, either the edge, the face, or the volume of the prism. The hierarchical vector bases are then constructed by using the same technique used before, where we simply *substitute* the new scalar hierarchical polynomials for the interpolatory ones of Chapter 4.

5.3.3.1 Edge-, Face-, and Volume-Based Bases

Hierarchical high-order vector functions

$$\boldsymbol{\Omega}^{ab} = H_{ijk}^{ab}(\boldsymbol{\xi})\,\boldsymbol{\Omega}_{ab} \tag{5.119}$$

$$\boldsymbol{\Omega}^{cd} = H_{ijk}^{cd}(\boldsymbol{\xi})\,\boldsymbol{\Omega}_{cd} \tag{5.120}$$

are expressed in terms of the two hierarchical polynomial sets $H_{ijk}^{ab}(\boldsymbol{\xi})$ and $H_{ijk}^{cd}(\boldsymbol{\xi})$ which, in the following, are written in terms of the variables

$$\begin{cases} \xi_{ab} = \xi_a - \xi_b \\ \xi_{ac} = \xi_a - \xi_c \\ \xi_{de} = \xi_d - \xi_e \end{cases} \tag{5.121}$$

with

$$\xi_{de} = 2\xi_d - 1 \tag{5.122}$$

because of (5.111). The total order of each polynomial H_{ijk} in (5.119) and (5.120) equals the sum of the subscripts (i, j, k) used to denote it, and is always less than or equal to $2p$. Obviously, the indices i, j, and k appearing in (5.119) and (5.120) are different from and do not play the same role as the indices i, j, k, ℓ, and m used in (5.115)–(5.118).

Each group of volume-, face-, and edge-based hierarchical polynomials is formed by mutually orthogonal polynomials obtained by extensive use of the nth-order Jacobi polynomials $P_n^{(0,\alpha)}(z)$ and $P_n^{(\alpha,\alpha)}(z)$, with $\alpha = 0$ or 2. Jacobi polynomials are naturally brought forth by the weight functions used to define our orthogonal hierarchical families, given that the weight functions are suitably chosen to maintain conformity with the differently shaped cells possibly attached to the prism. More specifically, the Jacobi polynomials $P_n^{(\alpha,\alpha)}(z)$ are either even or odd functions of z, with $P_n^{(\alpha,\alpha)}(-z) = (-1)^n P_n^{(\alpha,\alpha)}(z)$ while, in particular, the polynomials $P_n^{(0,0)}(z)$ are equal to the Legendre polynomials $P_n(z)$. The recurrence relations with respect to the degree n to construct these polynomials from the lowest order functions

$$P_0(z) = P_0^{(2,2)}(z) = P_0^{(0,2)}(z) = 1 \tag{5.123}$$

$$P_1(z) = z \tag{5.124}$$

$$P_1^{(2,2)}(z) = 3\,z \tag{5.125}$$

$$P_1^{(0,2)}(z) = 2\,z - 1 \tag{5.126}$$

are reported below for the sake of completeness

$$(n+1)P_{n+1}(z) = (2n+1)zP_n(z) - nP_{n-1}(z) \tag{5.127}$$

$$(n+1)(n+5)P_{n+1}^{(2,2)}(z) = (n+3)\left[(2n+5)\,z\,P_n^{(2,2)}(z) - (n+2)P_{n-1}^{(2,2)}(z)\right] \tag{5.128}$$

$$(n+1)^2(n+3)P_{n+1}^{(0,2)}(z) = (2n+3)\left[(n+1)(n+2)\,z - 1\right]P_n^{(0,2)}(z) - n(n+2)^2\,P_{n-1}^{(0,2)}(z) \tag{5.129}$$

In the subsubsections that follow, the *normalized* hierarchical *vector* functions are derived for arbitrary pth order.

5.3.3.2 The Hierarchical Vector Family $\boldsymbol{\Omega}^{ab}_{hier.}$

The higher order hierarchical vector functions that build upon a base vector located at the intersection of two quadrilateral faces of the prism have the form

$$\begin{aligned}\boldsymbol{\Omega}_{ijk}^{ab} &= H_{ijk}(\boldsymbol{\xi})\,\boldsymbol{\Omega}_{ab} = H_{ijk}(\boldsymbol{\xi})\,\xi_c \nabla \xi_d \\ &= H_{ijk}(\boldsymbol{\xi})\sqrt{w^{ab}}\,\nabla \xi_d\end{aligned} \tag{5.130}$$

These are conveniently defined in terms of scalar *generating* polynomials $H_{ijk}(\boldsymbol{\xi})$ orthogonal on the parent prism cell Q^3 with respect to the weight function

$$w^{ab} = \xi_c^2 = (1 - \xi_a - \xi_b)^2 \tag{5.131}$$

which, for integrals on Q^1_{ab}, Q^2_a, and Q^2_b, simplifies as follows

$$\begin{aligned} w_1 &= w^{ab}\big|_{\xi_a=\xi_b=0} &&= 1 \\ w_{2a} &= w^{ab}\big|_{\xi_a=0} &&= (1-\xi_b)^2 \\ w_{2b} &= w^{ab}\big|_{\xi_b=0} &&= (1-\xi_a)^2 \end{aligned} \tag{5.132}$$

A hierarchical pth-order complete polynomial set *equivalent* to (5.117) consists of the following $(p+1)^2(p+2)/2$ polynomials

$$\begin{cases} H_{00k}^{ab} = E_k^{ab} = E_k(\xi_{de}) \\ H_{i0k}^{ab} = F_{i0k}^{ab} = \xi_a \left[f_{i-1}(2\xi_a - 1) + \xi_b \mathcal{A}_{i-2}(\xi_a, \xi_c)\right] E_k^{ab}(\xi_{de}) \\ H_{0jk}^{ab} = F_{0jk}^{ab} = \xi_b \left[f_{j-1}(2\xi_b - 1) + \xi_a \mathcal{A}_{j-2}(\xi_b, \xi_c)\right] E_k^{ab}(\xi_{de}) \\ H_{ijk}^{ab} = V_{ijk}^{ab} = \xi_a \xi_b \mathcal{C}_{i-1,j-1}(\xi_{ab}, \xi_c) E_k^{ab}(\xi_{de}) \end{cases} \tag{5.133}$$

obtained for $k=0,1,\ldots,p$; $i,j=1,2,\ldots,p$, with $(i+j)\le p$, and (ξ_{ab}, ξ_{de}) given in (5.121). Set (5.133) is hierarchically constructed as follows:

- for $p=0$ (i.e. at zeroth order), the set consists of only one edge-based polynomial, that is, H_{000}^{ab};
- for $p\ge 1$, one has to increment the set of order $(p-1)$ with the edge-based function H_{00p}^{ab} plus $2(p+1)$ face-based functions H_{p0k}^{ab}, H_{0pk}^{ab} obtained for $k=0,1,\ldots,p$;
- for $p\ge 2$, one has to further add to the set under construction $2(p-1)$ face-based functions H_{n0p}^{ab}, H_{0np}^{ab}, obtained for $n=1,2,\ldots,p-1$; plus $p(p-1)$ volume-based functions $H_{n,p-n,k}^{ab}$ obtained for $n=1,2,\ldots,p-1$ and $k=0,1,\ldots,p-1$; plus $p(p-1)/2$ volume-based functions H_{ijp}^{ab} obtained for $i,j=1,2,\ldots,p-1$ with $(i+j)\le p$.

The auxiliary polynomials $\mathcal{A}_m$ and $\mathcal{C}_{mn}$ appearing in (5.133) are reported in the top part of Table 5.13 up to $m,n=4$ (these polynomials form all the sets up to the complete sixth order); whereas E_k^{ab} is expressed in terms of the re-scaled Legendre polynomial

$$E_n(z) = \sqrt{2n+1}\, P_n(z) \tag{5.134}$$

already appearing in the first of (5.93) while f_n indicates the re-scaled Jacobi polynomial of order n given in (5.94).

Before dealing with the technique to construct the polynomials (5.133), or better, the $\mathcal{A}_m$ and $\mathcal{C}_{mn}$ polynomials, we immediately observe that conformity between adjacent, differently shaped vector elements is easily ensured by adjusting the sign of the polynomials (5.133). In fact, the expression of the hierarchical polynomials $E_k^{ab}(\xi_{de})$ is identical to the expression of the edge-based polynomials of the tetrahedral element (given in Table 5.5) along its $\xi_a=\xi_b=0$ edge; at the same time, $E_k^{ab}(\xi_{de})$ also equals the expression of the edge-based polynomials of the brick element given in (5.93), over the $\xi_a=0$ or $\xi_b=0$ brick face. Furthermore, on their associated (ξ_b or $\xi_a=0$) quadrilateral face, the face-based functions in (5.133) simplify into the following polynomials

$$\begin{aligned} F_{i0k}(\boldsymbol{\xi}) &= \xi_a f_{i-1}(2\xi_a - 1)\, E_k^{ab}(\xi_{de}) \\ F_{0jk}(\boldsymbol{\xi}) &= \xi_b f_{j-1}(2\xi_b - 1)\, E_k^{ab}(\xi_{de}) \\ &\text{for } i,j=1,2,\ldots,p;\ k=0,1,\ldots,p \end{aligned} \tag{5.135}$$

that exactly match the face-based polynomials of the brick element (on that same ξ_b or $\xi_a=0$ face possibly in common to the prism) given in (5.93). In this connection, notice that the polynomials F_{0jk}^{ab} associated with the $\xi_a=0$ face (that vanish on the $\xi_b=0$ face) are simply obtained from the polynomials F_{i0k}^{ab} associated with the $\xi_b = 0$ face by changing ξ_b to ξ_a, ξ_a to ξ_b, and j to i.

We can now proceed to describe the technique used to build (5.133). The $p(p-1)/2$ hierarchical polynomial factors $\xi_a\, \xi_b\, \mathcal{C}_{i-1,j-1}(\xi_{ab}, \xi_c)$ used to express the volume-based functions V_{ijk}^{ab} are uniquely determined by applying the Gram–Schmidt orthogonalization process to the *ordered* list of polynomials $\mathcal{P}_{ij} = \xi_a\, \xi_b\, P_{i-1}^{(2,2)}(\xi_{ab})\, P_{j-1}^{(0,2)}(2\xi_c - 1)$ as created by the nested loops:

- for $s = 2, 3, \ldots, p$ (external loop), for increasing total order s;
- for $j = 1, 2, \ldots, s-1$ (inner loop), with $i = s - j$ fixed in the inner loop.

The *inner product* used for this orthogonalization process is defined by the 2D integral over the Q_d^2 triangular face of the product of the two polynomials $(\mathcal{P}_{ij} \times \mathcal{P}_{\ell m})$ and the weight function w^{ab} given in (5.131). The final set of functions V_{ijk}^{ab} obviously depends on the order in which the orthogonalization of $\mathcal{P}_{ij}$ is performed; this order is specified by the loop definitions above. Note also that (5.94) and (5.134) are uniquely defined, while the polynomials (5.135) are unambiguously defined in Section 5.3.2.2.

Each group of the face-based polynomials F_{i0k}^{ab}, F_{0jk}^{ab} given in (5.133) is formed by polynomials mutually orthogonal on the prism volume and on their associated quadrilateral face, for the relevant weights given in (5.132). These polynomials are obtained by imposing the orthogonality on the prism volume of linear combinations of the volume-based functions $V_{\ell mk}^{ab}$ with each face-based brick polynomial (5.135) where, in the combination, we use the same value of k while keeping $\ell + m = i$ or j. This latter orthogonalization process yields the auxiliary polynomial functions $\mathcal{A}_{i-2}$ that can be written in compact (and elegant) expressions by replacing ξ_b with $1 - (\xi_a + \xi_c)$, as per the dependency relation (5.110). Notice also that the *ordering* of the functions does not affect the auxiliary polynomials $\mathcal{A}_m$, since each of these polynomials is obtained by linearly combining *one* face-based brick polynomial (5.135) with volume-based polynomials $V_{\ell mk}^{ab}$ that are already orthogonal with respect to each other.

The above-mentioned linear combinations to form F^{ab} are performed because the functions (5.135) are not orthogonal on the prism volume with respect to the weight w^{ab}, but only on their associated $\xi_a = 0$ or $\xi_b = 0$ face (for the weight function w_{2b} or w_{2a}, respectively). As an alternative, the face-based functions (5.135) may be used on a prism cell, whenever orthogonality on the prism volume is not required.

The normalization of the polynomials (5.133) is discussed in Section 5.3.3.4.

5.3.3.3 The Hierarchical Vector Family $\mathbf{\Omega}_{hier.}^{cd}$

Hierarchical higher order vector functions that build upon a base vector located at the intersection of a triangular face and a quadrilateral face of the prism have the form

$$\begin{aligned}\mathbf{\Omega}_{ijk}^{cd} &= H_{ijk}(\boldsymbol{\xi})\, \mathbf{\Omega}_{cd} \\ &= H_{ijk}(\boldsymbol{\xi})\, \xi_e\, (\xi_a \nabla \xi_b - \xi_b \nabla \xi_a) \\ &= H_{ijk}(\boldsymbol{\xi})\, \sqrt{w^{cd}}\, (\xi_a \nabla \xi_b - \xi_b \nabla \xi_a) \end{aligned} \tag{5.136}$$

These are conveniently expressed in terms of scalar *generating* polynomials $H_{ijk}(\boldsymbol{\xi})$ orthogonal on Q^3 with respect to the weight function

$$w^{cd} = \xi_e^2 = (1 - \xi_d)^2 \tag{5.137}$$

which on Q_c^2, Q_d^2, and Q_{cd}^1 (where $\xi_c = 0$, $\xi_d = 0$, and $\xi_c = \xi_d = 0$, respectively) simplifies as follows

$$\begin{aligned} w_{2c} &= w^{cd} \\ w_{2d} &= w_1 = 1 \end{aligned} \tag{5.138}$$

In this connection, (5.132) and (5.138) clearly show that for both the $\mathbf{\Omega}^{ab}$ and $\mathbf{\Omega}^{cd}$ families:

- the weight function w_1 that defines the orthogonality between edge-based polynomials (with inner product given by the one-dimensional integral along the edge at issue) is always equal to unity;
- the weight function w_2 that defines the orthogonality between polynomials based on a given quadrilateral face (with inner product given by the two-dimensional integral along the face at issue) is always of the form $w_2 = (1 - \xi^2)$;
- the weight function w_2 that defines the orthogonality between polynomials based on a given triangular face (with inner product given by the two-dimensional integral along the face at issue) is always equal to unity.

The above results are intentional and indeed essential to guarantee the conformity between differently shaped adjacent elements.

A hierarchical pth-order complete polynomial set *equivalent* to (5.118) consists of the following $(p+1)^2(p+2)/2$ polynomials

$$\begin{cases} H_{i00}^{cd} = E_i^{cd} = E_i(\xi_{ab}) + \xi_c \mathcal{R}_{i-1}(\boldsymbol{\xi}_t) \\ H_{i0k}^{cd} = F_{i0k}^{cd} = \xi_d f_{k-1}(\xi_{de}) E_i^{cd}(\boldsymbol{\xi}_t) \\ H_{ij0}^{cd} = F_{ij0}^{cd} = \xi_c \mathcal{U}_{i,j-1}(\boldsymbol{\xi}_t) \\ H_{ijk}^{cd} = V_{ijk}^{cd} = \xi_c \xi_d f_{k-1}(\xi_{de}) \sqrt{3}\, \mathcal{U}_{i,j-1}(\boldsymbol{\xi}_t) \end{cases} \tag{5.139}$$

obtained for $i = 0, 1, \ldots, p$; $j, k = 1, 2, \ldots, p$, with $(i+j) \le p$, and (ξ_{ab}, ξ_{de}) given in (5.121). (Note that the rule to run the indices (i, j, k) in (5.139) is different from that used in (5.133).) Set (5.139) is hierarchically constructed as follows:

- for $p = 0$ (i.e. at zeroth order), the set consists of only one edge-based polynomial, that is, H_{000}^{cd};
- for $p \ge 1$, one has to increment the set of order $(p-1)$ with the edge-based function H_{p00}^{cd} plus $2p$ functions H_{p0k}^{cd}, H_{i0p}^{cd} (based on the quadrilateral face) obtained for $i = 0, 1, \ldots, p-1$; $k = 1, 2, \ldots, p$; plus p functions $H_{p-j,j,0}^{cd}$ (based on the triangular face) obtained for $j = 1, 2, \ldots, p$; plus $p(p+1)/2$ volume-based functions H_{ijp}^{cd} obtained for $i = 0, 1, \ldots, p-1; j = 1, 2, \ldots, p$, with $(i+j) \le p$;
- for $p \ge 2$, one has to further add to the set under construction $p(p-1)$ volume-based functions $H_{i,p-i,k}^{cd}$ obtained for $i = 0, 1, \ldots, p-1$; $k = 1, 2, \ldots, p-1$.

The auxiliary polynomials $\mathcal{R}_m$ and $\mathcal{U}_{mn}$ appearing in (5.139) are reported up to $m = 5$, $n = 4$ in the bottom part of Table 5.13 (these polynomials form all the sets up to the complete sixth order), while $E_i(\xi_{ab})$ and $f_{k-1}(\xi_{de})$ are given in (5.134) and (5.94), respectively.

Once again, conformity between adjacent, differently shaped vector elements is easily ensured by adjusting the sign of the polynomials (5.139). In fact, the $(p+1)$ edge-based

Table 5.13 Curl-conforming vector bases on prism cells – polynomials used to build the hierarchical vector prism functions up to the sixth order.

Polynomials $\mathcal{A}_m(\xi_a, \xi_c)$ of global order $\max[m, 0]$;
they are used to build the functions F^{ab}_{i0k} based on the $\xi_b = 0$ face (for $i = m+2$ and arbitrary k value):

$\mathcal{A}_{-1} = 0$	$\mathcal{A}^a_1 = 7\sqrt{30}\,\xi_{ac}$	$\mathcal{A}^a_3 = 12\sqrt{455}\,\xi_{ac}\,(\xi_a^2 - 4\,\xi_a\,\xi_c + \xi_c^2)$
$\mathcal{A}^s_0 = \sqrt{70}$	$\mathcal{A}^s_2 = 4\sqrt{770}\,(3\,\xi_a^2 - 8\,\xi_a\,\xi_c + 3\,\xi_c^2)/3$	$\mathcal{A}^s_4 = 50\sqrt{105}\,(\xi_a^4 - 8\,\xi_a^3\,\xi_c + 15\,\xi_a^2\,\xi_c^2 - 8\,\xi_a\,\xi_c^3 + \xi_c^4)$

Polynomials $\mathcal{C}_{mn}(\xi_{ab}, \xi_c)$ of global order $m+n$ (normalized so that $\iint \xi_c^2\,\left[\xi_a\,\xi_b\,\mathcal{C}_{mn}(\boldsymbol{\xi}_t)\right]^2 \mathrm{d}\boldsymbol{\xi}_t = 1$);
they are used to build the volume-based functions V^{ab}_{ijk} (for $i = m+1$; $j = n+1$, and arbitrary k value):

$\mathcal{C}^s_{00} = 12\sqrt{35}$	$\mathcal{C}^a_{30} = 210\sqrt{\frac{22}{29}}\,\xi_{ab}\,(-4 + 13\,\xi_{ab}^2 + 6\,\xi_c)$
$\mathcal{C}^a_{10} = 60\sqrt{21}\,\xi_{ab}$ $\mathcal{C}^s_{01} = 60\sqrt{7}\,(3\,\xi_c - 1)$	$\mathcal{C}^s_{21} = 42\sqrt{\frac{30}{19}}\,(4 - 33\,\xi_{ab}^2 - 22\,\xi_c + 143\,\xi_{ab}^2\,\xi_c + 22\,\xi_c^2)$
$\mathcal{C}^s_{20} = 4\sqrt{15}\,(-7 + 55\,\xi_{ab}^2 + 10\,\xi_c)$ $\mathcal{C}^a_{11} = 30\sqrt{42}\,\xi_{ab}\,(11\,\xi_c - 3)$ $\mathcal{C}^s_{02} = 5\sqrt{66}\,[7 - \xi_{ab}^2 - (46 - 63\,\xi_c)\,\xi_c]$	$\mathcal{C}^a_{12} = 21\sqrt{\frac{1430}{29}}\,\xi_{ab}\,(7 - \xi_{ab}^2 - 54\,\xi_c + 87\,\xi_c^2)$ $\mathcal{C}^s_{03} = -21\sqrt{\frac{110}{19}}\,(9 - 3\,\xi_{ab}^2 - 97\,\xi_c + 13\,\xi_{ab}^2\,\xi_c + 287\,\xi_c^2 - 247\,\xi_c^3)$

$\mathcal{C}^s_{40} = 4\sqrt{2{,}310}\,\left[12 - 36\,\xi_c + 13\,(\xi_{ab}^2\,(-18 + 35\,\xi_{ab}^2) + 28\,\xi_{ab}^2\,\xi_c + 2\,\xi_c^2)\right]/\sqrt{197}$

$\mathcal{C}^a_{31} = 12\sqrt{770}\,\xi_{ab}\,\left[27 - 169\,\xi_c + 91\,(-\xi_{ab}^2 + 5\,\xi_{ab}^2\,\xi_c + 2\,\xi_c^2)\right]/\sqrt{41}$

$\mathcal{C}^s_{22} = 6\sqrt{10{,}010}\,\left[2{,}758\,\xi_c^3 - 141 - \xi_{ab}^2\,(175\,\xi_{ab}^2 - 1{,}272) + 1{,}408\,\xi_c - 11{,}172\,\xi_{ab}^2\,\xi_c + 3\,(6{,}895\,\xi_{ab}^2 - 1{,}251)\,\xi_c^2\right]/\sqrt{61{,}661}$

$\mathcal{C}^a_{13} = 210\sqrt{286}\,\xi_{ab}\,(123\,\xi_c^3 - 3 + \xi_{ab}^2 + 37\,\xi_c - 5\,\xi_{ab}^2\,\xi_c - 125\,\xi_c^2)/\sqrt{41}$

$\mathcal{C}^s_{04} = 35\sqrt{429}\,\left[33 + \xi_{ab}^2\,(\xi_{ab}^2 - 18) - 516\,\xi_c + 164\,\xi_{ab}^2\,\xi_c - 2\,(153\,\xi_{ab}^2 - 1{,}235)\,\xi_c^2 - 4{,}548\,\xi_c^3 + 2{,}817\,\xi_c^4\right]/(2\sqrt{313})$

Polynomials $\mathcal{R}_m(\boldsymbol{\xi}_t)$ of global order $\max[m, 0]$; they are used to build
– the functions E^{cd}_{i00} based on the $\xi_c = \xi_d = 0$ edge (for $i = m+1$)
– the functions F^{cd}_{i0k} based on the $\xi_c = 0$ quadrilateral face (for $i = m+1$ and arbitrary k value):

$\mathcal{R}^s_{-1} = \mathcal{R}^a_0 = 0$ $\mathcal{R}^s_1 = \sqrt{5}\,(2-\xi_c)/2$	$\mathcal{R}^s_3 = 9\,(\xi_c - 2) \times \left(1 + 40\,\xi_a\,\xi_b - 9\,\chi^2_{ab}\right)/\,8$	$\mathcal{R}^a_4 = 5\,\sqrt{11}\,\xi_{ab}\,(\xi_c - 2) \times \left(3 + 56\,\xi_a\,\xi_b - 11\,\chi^2_{ab}\right)/\,8$
$\mathcal{R}^a_2 = 3\,\sqrt{7}\,\xi_{ab}\,(2-\xi_c)/2$	$\mathcal{R}^s_5 = 5\,\sqrt{13}\,(2-\xi_c) \times \left[1 + 43\,\chi^4_{ab} + 84\,\xi_a\,\xi_b\,(1 + 12\,\xi_a\,\xi_b) - 20\,\chi^2_{ab}\,(1 + 21\,\xi_a\,\xi_b)\right]/16$	

Polynomials $\mathcal{U}_{mn}(\boldsymbol{\xi}_t)$ of global order $m+n$ (normalized so that $\iint \left[\xi_c\,\mathcal{U}_{mn}(\boldsymbol{\xi}_t)\right]^2 \mathrm{d}\boldsymbol{\xi}_t = 1$);
they are used to build the face (F^{cd}_{ij0}) and volume-based functions V^{cd}_{ijk} (for $i = m; j = n+1$, and arbitrary k value):

$\mathcal{U}^s_{00} = 2\,\sqrt{3}$	$\mathcal{U}^s_{04} = 2\,\sqrt{105}\,(3 - 32\,\xi_c + 108\,\xi_c^2 - 144\,\xi_c^3 + 66\,\xi_c^4)$
$\mathcal{U}^s_{01} = 2\,\sqrt{3}\,(-3 + 5\,\xi_c)$ $\mathcal{U}^a_{10} = 6\,\sqrt{5}\,\xi_{ab}$	$\mathcal{U}^a_{13} = 6\,\sqrt{70}\,\xi_{ab}\,(-2 + 18\,\xi_c - 45\,\xi_c{}^2 + 33\,\xi_c{}^3)$ $\mathcal{U}^s_{22} = 6\,\sqrt{10}\,\chi_2\,(6 - 40\,\xi_c + 55\,\xi_c{}^2)$
$\mathcal{U}^s_{02} = 2\,\sqrt{30}\,(2 - 8\,\xi_c + 7\,\xi_c^2)$ $\mathcal{U}^a_{11} = 2\,\sqrt{30}\,\xi_{ab}\,(-3 + 7\,\xi_c)$ $\mathcal{U}^s_{20} = 2\,\sqrt{210}\,\chi_2$	$\mathcal{U}^a_{31} = 6\,\sqrt{35}\,\chi_3\,(-3 + 11\,\xi_c)$ $\mathcal{U}^s_{40} = 6\,\sqrt{165}\,\chi_4$ $\mathcal{U}^s_{05} = 2\,\sqrt{42}\left[-7 + 105\,\xi_c\,(1 - 5\,\xi_c + 11\,\xi_c^2 - 11\xi_c^3) + 429\,\xi_c^5\right]$
$\mathcal{U}^s_{03} = 2\,\sqrt{15}\,(-5 + 35\,\xi_c - 70\,\xi_c{}^2 + 42\,\xi_c{}^3)$ $\mathcal{U}^a_{12} = 2\,\sqrt{105}\,\xi_{ab}\,(3 - 16\,\xi_c + 18\,\xi_c{}^2)$ $\mathcal{U}^s_{21} = 10\,\sqrt{42}\,\chi_2\,(-1 + 3\,\xi_c)$ $\mathcal{U}^a_{30} = 6\,\sqrt{70}\,\chi_3$	$\mathcal{U}^a_{14} = 6\,\sqrt{70}\,\xi_{ab}\,(3 - 40\,\xi_c + 165\,\xi_c^2 - 264\,\xi_c^3 + 143\,\xi_c^4)$ $\mathcal{U}^s_{23} = 6\,\sqrt{35}\,\chi_2\,(-5 + 55\,\xi_c - 165\,\xi_c^2 + 143\,\xi_c^3)$ $\mathcal{U}^a_{32} = 14\,\sqrt{165}\,\chi_3\,(1 - 8\,\xi_c + 13\,\xi_c{}^2)$ $\mathcal{U}^s_{41} = 6\,\sqrt{77}\,\chi_4\,(-3 + 13\,\xi_c)$ $\mathcal{U}^a_{50} = 2\,\sqrt{3,003}\,\chi_5$

With $\xi_a + \xi_b + \xi_c = 1$; $\xi_{ab} = \xi_a - \xi_b$; $\xi_{ac} = \xi_a - \xi_c$; $\chi_{ab} = \xi_a + \xi_b$;

$\chi_2 = \xi_a{}^2 - 4\,\xi_a\,\xi_b + \xi_b{}^2$; $\chi_4 = \xi_a{}^4 - 16\,\xi_a{}^3\,\xi_b + 36\,\xi_a{}^2\,\xi_b{}^2 - 16\,\xi_a\,\xi_b{}^3 + \xi_b{}^4$;

$\chi_3 = \xi_{ab}\,\left(\xi_a{}^2 - 8\,\xi_a\,\xi_b + \xi_b{}^2\right)$; $\chi_5 = \xi_{ab}\,\left(\xi_a{}^4 - 24\,\xi_a{}^3\,\xi_b + 76\,\xi_a{}^2\,\xi_b{}^2 - 24\,\xi_a\,\xi_b{}^3 + \xi_b{}^4\right)$.

The superscripts s and a used for $\mathcal{A}_m$ label symmetric and antisymmetric polynomials of the ξ_a and ξ_c variables, respectively; whereas, for $\mathcal{C}_{mn}$, $\mathcal{R}_m$, and $\mathcal{S}_{mn}$, they label symmetric and antisymmetric polynomials of the ξ_a and ξ_b variables, respectively.

Adapted from R. D. Graglia and A. F. Peterson, "Hierarchical curl-conforming Nédélec elements for triangular-prism cells," *IEEE Trans. Antennas Propag.*, vol. 60, no. 7, pp. 3314–3324, Jul. 2012.

polynomials $E_i^{cd}(\boldsymbol{\xi}_t)$ are nothing else than the *edge*-based generating functions already reported in Section 5.3.1.4 for the $\xi_c = 0$ edge of the $\xi_d = 0$ *triangular* face of a tetrahedron; thesc polynomials, for $\xi_c = 0$, coincide with the edge-based polynomials of the brick element given in Section 5.3.2.2. The polynomials E_i^{cd} are mutually orthogonal on the volume, on the $\xi_d = 0$ and the $\xi_c = 0$ faces, and on the $\xi_c = \xi_d = 0$ edge of the prism. Similarly, the $p(p+1)/2$ face-based polynomials $F_{ij0}^{cd}(\boldsymbol{\xi}_t) = \xi_c\, \mathcal{U}_{i,j-1}(\boldsymbol{\xi}_t)$ appearing in (5.139) are orthogonal on the prism volume and on their associated $\xi_d = 0$ face; they simply coincide with the face-based generating functions reported in Section 5.3.1.3 for the $\xi_d = 0$ *triangular* element. The polynomials E_i^{cd} and F_{ij0}^{cd} are thus obtained and uniquely defined as explained in Section 5.3.1, where they are explicitly reported for $1 \leq (i+j) \leq 6$ (i.e. all the E_i^{cd} and F_{ij0}^{cd} for $(i+j) \leq 6$ can be obtained by setting $\xi_d = 0$ in Table 5.5).

Finally, the $p(p+1)$ face-based functions F_{i0k}^{cd} are orthogonal on the prism volume and exactly match, on the associated $\xi_c = 0$ quadrilateral face, the face-based brick polynomials of Section 5.3.2.2

$$F_{i0k}(\boldsymbol{\xi}) = \xi_d\, f_{k-1}(\xi_{de})\, E_i(\xi_{ab}) \qquad \text{for } i = 0, 1, \ldots, p;\ k = 1, 2, \ldots, p \tag{5.140}$$

which are already mutually orthogonal over the $\xi_c = 0$ prism face.

The normalization of the polynomials (5.139) is discussed in the following subsubsection.

The conformity between our tetrahedral, hexahedral, and prism hierarchical bases highlighted above permits one to mesh volumetric regions with a mixture of brick, tetrahedral, and prism cells. However, in order to achieve maximum flexibility, one still needs to develop hierarchical vector bases for pyramids [46]; these bases will be considered in future editions.

5.3.3.4 Normalization of the Polynomial Families

Within each group, the hierarchical volume-, face-, and edge-based polynomials are mutually orthogonal, with orthogonality defined by the vanishing (for $i \neq \ell$, or $j \neq m$, or $k \neq n$) of the following inner products performed on the volume (for all the polynomials of the same group), on the associated face (for all the face- or edge-based polynomials of the same group) and on the associated edge (for all the edge-based polynomials)

$$\begin{aligned}
\left\langle H_{ijk}^{\alpha\beta}, H_{\ell mn}^{\alpha\beta} \right\rangle_{Q^3} &= \int_{Q^3} w^{\alpha\beta}\, H_{ijk}^{\alpha\beta}\, H_{\ell mn}^{\alpha\beta}\, \mathrm{d}Q^3 \\
\left\langle H_{ijk}^{\alpha\beta}, H_{\ell mn}^{\alpha\beta} \right\rangle_{Q_\alpha^2} &= \int_{Q_\alpha^2} w_{2\alpha}\, H_{ijk}^{\alpha\beta}\, H_{\ell mn}^{\alpha\beta}\, \mathrm{d}Q_\alpha^2 \\
\left\langle H_{ijk}^{\alpha\beta}, H_{\ell mn}^{\alpha\beta} \right\rangle_{Q_\beta^2} &= \int_{Q_\beta^2} w_{2\beta}\, H_{ijk}^{\alpha\beta}\, H_{\ell mn}^{\alpha\beta}\, \mathrm{d}Q_\beta^2 \\
\left\langle H_{ijk}^{\alpha\beta}, H_{\ell mn}^{\alpha\beta} \right\rangle_{Q_{\alpha\beta}^1} &= \int_{Q_{\alpha\beta}^1} w_1\, H_{ijk}^{\alpha\beta}\, H_{\ell mn}^{\alpha\beta}\, \mathrm{d}Q_{\alpha\beta}^1
\end{aligned} \tag{5.141}$$

with $(\alpha, \beta) = (a, b)$ or (c, d), and where the weight functions are those given in (5.131), (5.132), (5.137), and (5.138).

The normalization used for the hierarchical polynomials (5.133) and (5.139) of the previous subsubsections is summarized as follows:

- Edge-based polynomials:

$$\left\langle E_p^{\alpha\beta}, E_p^{\alpha\beta} \right\rangle_{Q^1_{\alpha\beta}} = 1 \tag{5.142}$$

- Face-based polynomials:

$$\left\langle F_{0jk}^{ab}, F_{0jk}^{ab} \right\rangle_{Q^2_a} = \left\langle F_{i0k}^{ab}, F_{i0k}^{ab} \right\rangle_{Q^2_b} = \left\langle F_{i0k}^{cd}, F_{i0k}^{cd} \right\rangle_{Q^2_c} = 1/3 \tag{5.143}$$

$$\left\langle F_{ij0}^{cd}, F_{ij0}^{cd} \right\rangle_{Q^2_d} = 1 \tag{5.144}$$

- Volume-based polynomials:

$$\left\langle V_{ijk}^{\alpha\beta}, V_{ijk}^{\alpha\beta} \right\rangle_{Q^3} = 1 \tag{5.145}$$

In numerical applications, the hierarchical polynomials can be normalized differently whenever convenient. However, if they are re-scaled, one should also re-scale the polynomials used for the tetrahedral or hexahedral element by the same factors, to guarantee conformity.

The following inner product results are also easily obtained:

- for inner products over one face of the prism:

$$\left\langle E_p^{ab}, E_p^{ab} \right\rangle_{Q^2_a} = \left\langle E_p^{ab}, E_p^{ab} \right\rangle_{Q^2_b} = \left\langle E_p^{cd}, E_p^{cd} \right\rangle_{Q^2_c} = 1/3 \tag{5.146}$$

$$\left\langle E_p^{cd}, E_p^{cd} \right\rangle_{Q^2_d} = \frac{1}{2(p+1)} \tag{5.147}$$

- for inner products over the prism volume:

$$\left\langle E_p^{ab}, E_p^{ab} \right\rangle_{Q^3} = 1/9 \tag{5.148}$$

$$\left\langle E_p^{cd}, E_p^{cd} \right\rangle_{Q^3} = \frac{1}{6(p+1)} \tag{5.149}$$

$$\left\langle F_{0jk}^{ab}, F_{0jk}^{ab} \right\rangle_{Q^3} = \left\langle F_{i0k}^{ab}, F_{i0k}^{ab} \right\rangle_{Q^3} = 1/9 \tag{5.150}$$

$$\left\langle F_{i0k}^{cd}, F_{i0k}^{cd} \right\rangle_{Q^3} = \frac{1}{6(i+1)} \tag{5.151}$$

$$\left\langle F_{ij0}^{cd}, F_{ij0}^{cd} \right\rangle_{Q^3} = 1/3 \tag{5.152}$$

Finally, we observe that, in general, a polynomial of one group is not orthogonal to those of a different group.

5.3.4 Condition Number Comparison

The curl-conforming hierarchical bases described in this section are intended for use in obtaining numerical solutions of the vector Helmholtz equation

$$\nabla \times \nabla \times \boldsymbol{H} = k^2 \boldsymbol{H} \tag{5.153}$$

for applications such as antennas, scattering, or resonant cavities. The details of the numerical solution procedure are presented in Chapter 6. Here, we briefly consider resonant cavities bounded by perfectly conducting walls (a homogeneous Neumann boundary) in order to study the linear independence of the basis functions. The magnetic field can be expressed in the form

$$\boldsymbol{H} = \sum_i \alpha_i \boldsymbol{B}_i \tag{5.154}$$

where $\boldsymbol{B}_i$ denote the vector basis functions. The process of forming a numerical solution involves computing the local element matrices **S** and **T**, which have entries of the form

$$\mathrm{S}_{mn} = \int_{\mathcal{D}} \nabla \times \boldsymbol{B}_m \cdot \nabla \times \boldsymbol{B}_n \, \mathrm{d}\mathcal{D} \tag{5.155}$$

$$\mathrm{T}_{mn} = \int_{\mathcal{D}} \boldsymbol{B}_m \cdot \boldsymbol{B}_n \, \mathrm{d}\mathcal{D} \tag{5.156}$$

These matrices may be local (confined to the interactions within a single cell) or global (involving all the interactions within the entire problem domain). Because of the nullspace of the curl operator, the local element matrix **S** and its global counterpart are singular. However, both the local and global **T** are non-singular and their condition numbers ($\mathrm{CN} = ||\mathbf{T}|| \, ||\mathbf{T}^{-1}||$) provide a measure of the degree of linear independence of the basis functions, which in turn gives an indication of the performance of the basis functions in numerical applications (see for example [27,47]).

The linear independence of the basis set is an important attribute of a good basis. Unfortunately, many hierarchical families of basis functions do not exhibit good linear independence as their order is increased. In the following, we consider a few examples of the condition number of the **T**-matrix obtained using the various vector bases proposed in this section. Numerical results obtained using these functions for several applications are presented in Chapter 6. We note that the hierarchical bases presented here and the interpolatory bases of Chapter 4 span the same reduced Nédélec spaces, and their numerical results for a fixed Nédélec order are identical to the precision maintained in the computations. Thus, the convergence behavior of the hierarchical bases as a function of cell size and order is the same as that of the interpolatory bases of Chapter 4 and it will not be studied in this subsection.

5.3.4.1 Triangular Bases

As an initial example, consider the triangular-cell bases. Reference 47 presented several examples comparing the (element and global) **T**-matrix condition numbers arising from several families of hierarchical vector basis functions to those of the hierarchical triangular basis functions presented here. That study, which treated the solution of (5.153) as an eigenvalue problem for the resonant frequencies of the cavity, concluded that the hierarchical basis functions of order 2.5 and 3.5 presented in this section produce lower matrix condition numbers over a range of meshes than those of most other families. Those results were obtained after attempts were made to improve the condition numbers for the other families by an appropriate choice of scale factors (the interested reader is referred to [47] for more details). From [47], we report here only the results summarized in Table 5.14. Table 5.14 shows the condition numbers of the global **T**-matrix obtained from five different triangular-cell meshes. (The mesh numbers indicate the number of cells.) Mesh #40b and mesh #34 were deliberately designed to have cells with a large aspect ratio, resulting in a poor matrix condition number. A singular value decomposition algorithm was used to find the largest and smallest singular values of the global **T**-matrix, and the ratio of those two parameters is reported.

Condition numbers for the Quadratic tangential/cubic normal (QT/CuN) bases in Table 5.14 suggest that the basis functions maintain better linear independence than many of the other proposed hierarchical vector bases for triangles. It was observed in [47] that two other hierarchical families also performed very well over a wide range of mesh geometries, the bases of Ingelström [17] and the orthogonalized bases of Webb [11]. (The Webb functions, however, do not exactly span the reduced-gradient Nédélec spaces.) In Table 5.14, we used the hierarchical bases of Section 5.3.1 in their original form, while scale factors were introduced to substantially improve the matrix condition numbers for the other hierarchical families. The use of optimizing scale factors should be considered for any hierarchical basis family, for any cell shape.

5.3.4.2 Tetrahedral Bases

For the tetrahedral bases, we compare the individual element **T**-matrix condition numbers obtained using the hierarchical basis functions of the first five lowest orders presented in this section.

The best possible tetrahedral cell has equilateral shape, edge-length ℓ, and height $h=\sqrt{6}\ell/3$. Lower quality cells may be obtained by scaling the height of this tetrahedron by h_n, thereby obtaining cells of the same (equilateral) base but of different height $h \times h_n$, h_n being the *normalized* height of the tetrahedron. The element condition numbers CNH and CNI obtained with the hierarchical and the interpolatory family, respectively, depend on the value of the scale factor used to modify the cell shape; for these bases, however, the condition numbers CNH and CNI are not modified by changing the value of ℓ while keeping h_n fixed.

Figure 5.12 shows, in log-log scale, the behavior of the condition numbers CNH versus the normalized height h_n in the range ($0.25 < h_n < 4$), for bases of half-integer orders 0.5, 1.5, 2.5, 3.5, and 4.5. It is clear from this figure that, for a given order, the lowest condition numbers are obtained with an equilateral tetrahedron. The condition numbers increase with the base order and the greatest increment occurs between order 0.5 and order 1.5; in fact, in the entire range $1/2 \leq h_n \leq 2$, the CNH of the 1.5 order base is roughly 30 times higher than that of the 0.5 order base; the CNH of the 2.5 order base is $\approx$12 times higher than that of the 1.5 order base; the condition number of the 3.5 order base is $\approx$8 times higher than that of the 2.5 order base; the condition number of the 4.5 order base is $\approx$4 times higher than that of the 3.5 order base. The condition number CNI for the interpolatory bases is lower than that obtained with hierarchical bases (CNH), with the exception of the hierarchical and interpolatory bases of order 0.5 that are the same and yield the same condition numbers. For example, in case of rectilinear tetrahedral cells of good quality, and for element order 2.5, the individual element condition number CNH is expected to be four to five times larger than the individual element condition number CNI

Table 5.14 Curl-conforming vector bases on triangular cells – condition number comparison for triangular bases of order 2.5.

The matrix condition number of the global **T**-matrix in (5.156) obtained by taking the ratio of the largest singular value to the smallest, for five triangular-cell meshes of different quality. For each basis family, the 15 curl-conforming basis functions making up the Nédélec reduced-gradient QT/CuN representation were used with optimal scale factors*.

Family	Mesh 12	Mesh 42	Mesh 40a	Mesh 40b	Mesh 34
Interpolatory, Chapter 4	528	352	597	6,075	3,319
Hierarchical, Chapter 5	720	459	855	1,923	1,366
Ingelström [17]	1,040	510	1,085	5,022	3,284
Webb orthogonal [11]	1,095	517	1,121	4,345	3,120
Lee, Lee, and Lee [15]	1,456	1,791	4,612	21,883	12,994
Preissig and Peterson [16]	1,137	1,510	4,359	27,974	26,765
Andersen and Volakis [9]	2,920	2,084	5,965	31,100	49,956
	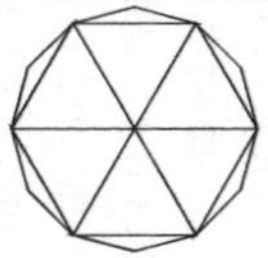	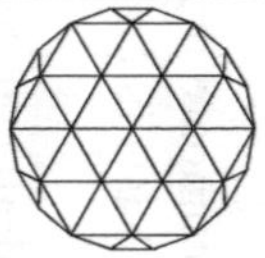	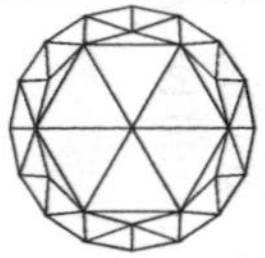		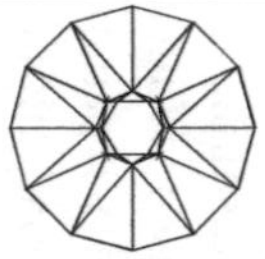
Number of cells:	12	42	40	40	34
Order of **T**:	114	468	444	444	384

*The "unscaled" bases of Section 5.3.1 (using the original scaling reported in Table 5.5) were used since those usually outperformed our "optimally" scaled hierarchical bases reported in [47]. For the other families, the optimally scaled bases outperformed the original, unscaled bases for these meshes, and those results are reported.

Adapted from A. F. Peterson and R. D. Graglia, "Scale factors and matrix conditioning associated with triangular-cell hierarchical vector basis functions" *IEEE Antennas Wireless Propag. Lett.*, vol. 9, pp. 40–43, 2010, doi:10.1109/LAWP. 2010.2042423

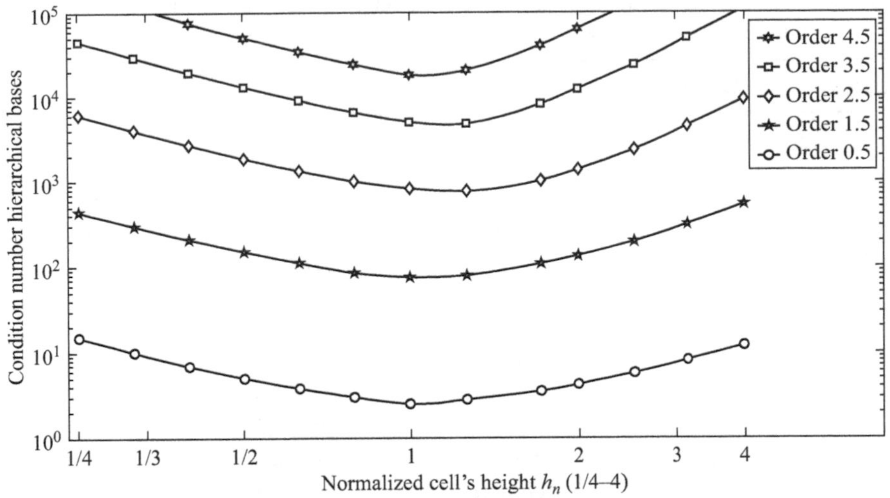

Figure 5.12 Individual element **T**-matrix condition numbers for the hierarchical (CNH) vector basis of order 0.5, 1.5, 2.5, 3.5, and 4.5 obtained by considering tetrahedral cells with the same (equilateral) base but of different *normalized* height h_n. The equilateral tetrahedron has height $h = \sqrt{6}\ell/3$ and normalized height $h_n = 1$.

obtained with interpolatory polynomials [25]. Conversely, for poor quality cells, the condition number obtained from the hierarchical vector bases could be much higher (say, by a factor of 7) than that obtained by using interpolatory polynomials [25]; however, this result is still within an order of magnitude and cells of such poor quality are usually avoided whenever possible. The numerical results presented here and in the above-quoted references suggest that the proposed hierarchical bases yield reasonably well-conditioned matrices.

5.3.4.3 Quadrilateral and Brick Bases

Table 5.15 presents element matrix condition numbers for vector bases of increasing order on square and cubic reference cells of unitary edge-length. The proposed hierarchical bases are compared to the interpolatory bases of Chapter 4. The CNH and C$\widetilde{\text{n}}$H data are relative to the hierarchical bases of Table 5.10 and to the hierarchical bases of Section 5.3.2.4, respectively. The CNI data obtained with the interpolatory basis functions of Chapter 4 are reported for reference. The hierarchical condition numbers clearly grow at a much slower rate as their order increases than those of the interpolatory set. It is interesting to observe that the bases of Table 5.10 and the alternative bases of Section 5.3.2.4 yield the same condition numbers for a 2D square cell, while for a cubic cell, the alternative bases reduce the condition numbers given by Table 5.10 bases by a factor of 2. This suggests that Table 5.10 bases are almost optimal. (In practice, it is not convenient to use the alternative bases of Section 5.3.2.4 because the global polynomial order of each of their terms is higher than that of the equivalent term of Table 5.10 bases.)

Additional numerical results for quadrilateral-cell bases are available in [22] and [27], where the global **T**-matrix condition numbers arising from basis families [13,19,34–36] are compared to the hierarchical ones of Section 5.3.2. These results showed that of the previous families,

Table 5.15 Vector bases on quadrilateral and brick cells – individual element **T**-matrix condition numbers for square and cubic cells of unitary edge-length.

Basis order	Order of T	CNH	CñH	CNI
0.5	4	3	3	3
1.5	12	22.956	22.956	20.639
2.5	24	22.956	22.956	109.720
3.5	40	60.483	60.483	486.676
4.5	60	60.483	60.483	2.5579×10^3
5.5	84	114.658	114.658	1.60285×10^4
6.5	112	114.658	114.658	1.16907×10^5
7.5	144	185.495	185.495	9.99106×10^5
8.5	180	185.495	185.495	9.71008×10^6
9.5	220	272.996	272.996	1.04921×10^8
10.5	264	272.996	272.996	1.23156×10^9
11.5	312	377.164	377.164	1.533906×10^{10}
12.5	364	377.164	377.164	1.978809×10^{11}
0.5	12	9	9	9
1.5	54	526.998	279.792	141.988
2.5	144	526.998	279.792	1.02709×10^3
3.5	300	3.65825×10^3	1.88093×10^3	6.88114×10^3
4.5	540	3.65825×10^3	1.88093×10^3	6.04903×10^4
5.5	882	1.31464×10^4	6.69883×10^3	7.08665×10^5
6.5	1344	1.31464×10^4	6.69883×10^3	1.08765×10^7
7.5	1944	3.44082×10^4	1.74670×10^4	2.19471×10^8

Individual element **T**-matrix condition numbers for hierarchical (CNH) and interpolatory (CNI) *vector bases* of different order, obtained by considering a square (top of the table) and a cubic (bottom of the table) cells of unitary edge-length. The CñH data were obtained with the alternative hierarchical bases of Section 5.3.2.4. For the square cell, the condition numbers CñH and CNH are equal.

Adapted from R. D. Graglia and A. F. Peterson, "Hierarchical curl-conforming Nédélec elements for quadrilateral and brick cells," *IEEE Trans. Antennas Propag.*, vol. 59, no. 8, pp. 2766–2773, Aug. 2011.

only the functions of [36] were comparable to the condition numbers of the hierarchical set of Section 5.3.2; the other families exhibited higher condition numbers.

5.3.4.4 Prism Bases

Table 5.16 presents element **T**-matrix condition numbers for complete subspaces of increasing order on a reference cell of unitary edge-length. The results obtained with the hierarchical bases proposed here are denoted by CNH, while those obtained with the interpolatory basis functions of Chapter 4 are denoted by CNI and are reported for comparison. Ideally, we would compare the condition numbers from the proposed hierarchical bases to other hierarchical functions, suitable for p-adaptive refinement procedures, but the only other available hierarchical basis set for prisms is that of [19], which do not span the same Nédélec space for each order, making a direct comparison questionable.

Table 5.16 Curl-conforming vector bases on prism cells – individual element **T**-matrix condition numbers for a cell of unitary edge-length.

Basis order	Order of T	CNH	CNI	$\frac{\text{CNH}}{\text{CNI}}$
0.5	9	6	6	1
1.5	36	734.117	25.7222	28.5
2.5	90	4048.44	420.924	9.62
3.5	180	2.49827×10^4	5.33134×10^3	4.69
4.5	315	6.55165×10^4	5.93356×10^4	1.10
5.5	504	2.60924×10^5	5.75117×10^5	0.45
6.5	756	4.97756×10^5	7.50997×10^6	0.07

Individual element **T**-matrix condition numbers for hierarchical (CNH) and interpolatory (CNI) *vector bases* of different order, obtained by considering a rectilinear prism cell with all its nine edges of equal unitary length. The bases considered in the table are those obtained by discarding from the volume and the triangular face-based function sets the high-order functions associated with the two parallel edge vectors $\mathbf{\Omega}_{j4}$ and $\mathbf{\Omega}_{j5}$, with $j = 1$, 2, or 3.

Adapted from R. D. Graglia and A. F. Peterson, "Hierarchical curl-conforming Nédélec elements for triangular-prism cells," *IEEE Trans. Antennas Propag.*, vol. 60, no. 7, pp. 3314–3324, Jul. 2012.

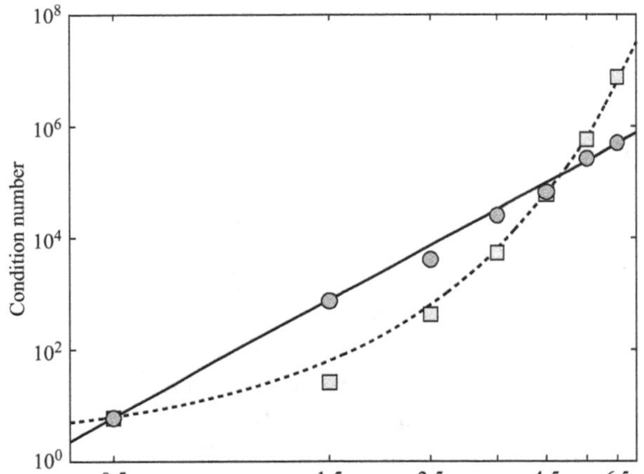

Figure 5.13 Comparison (using logarithmic scales) of the condition numbers of Table 5.16 with *reference* growth-rate lines. The hierarchical condition numbers are reported by circles, and the interpolatory condition numbers are reported by squares. The solid-line depicts a growth rate $g_1 = 130\times\,(\text{order})^{4.4}$; the dashed-line represents an exponential growth rate $g_2 = 2 \times 10^{\text{order}}$.

Table 5.16 shows that for orders higher than 4.5 the hierarchical condition numbers are smaller than the interpolatory ones. In fact, the hierarchical condition numbers of Table 5.16 have a polynomial growth rate whereas the interpolatory ones have exponential growth rate, as apparent from Figure 5.13 where we compare those condition numbers with the reference growth rate $g_1 = 130 \times (\text{order})^{4.4}$ and the exponential growth rate $g_2 = 2 \times 10^{\text{order}}$. The growth

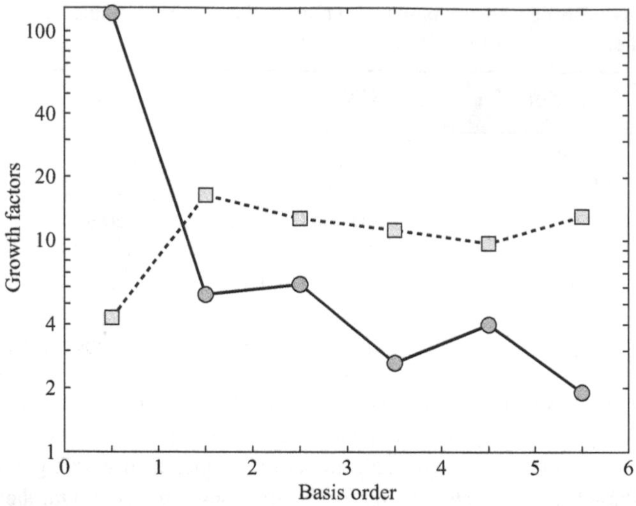

Figure 5.14 Condition-number incremental growth factors computed with the test-case results of Table 5.16. The hierarchical growth factors are reported by circles, and the interpolatory growth factors are reported by squares.

rate may be judged by the behavior of the condition-number *incremental growth factors*

$$\mathrm{GF(order)} = \frac{\mathrm{CN(order+1)}}{\mathrm{CN(order)}} \tag{5.157}$$

for the test-case results of Table 5.16 as reported in Figure 5.14 versus the basis order. Figure 5.14 shows that the hierarchical condition number has its biggest increment when the basis order changes from 0.5 to 1.5; apart from this, the hierarchical condition numbers grow at a much slower rate as their order increases than those of the interpolatory set.

Additional results for matrix condition numbers for the prism bases of this section may be found in [28].

5.4 Hierarchical Divergence-Conforming Vector Bases

The construction technique of the previous section is extended in this section to build hierarchical divergence-conforming bases for *volumetric* elements. In fact, there is no need to discuss the functions for surface elements any further since the hierarchical (as well as the interpolatory) divergence-conforming basis functions of the triangular and quadrilateral cells are easily obtained by a 90 degree rotation of the functions, or equivalently by cross-multiplying the corresponding curl-conforming functions with the unit vector $\hat{n}$ normal to the cell surface, as already discussed in Chapter 4.

Although papers proposing hierarchical vector basis functions began appearing in the electromagnetics literature in the early 1990s, there are relatively few scientific papers on hierarchical divergence-conforming functions for volumetric elements [8, 19, 48–50]. By contrast, the hierarchical divergence-conforming families proposed by Zaglmayr [19] for the three cell shapes considered in this section (tetrahedral, prism, and brick cells) span polynomial-complete spaces and not the mixed-order Nédélec spaces spanned by our functions, while Botha [49] proposed a hierarchical family for tetrahedral cells, but not for prism or brick shapes.

Normalized *interpolatory* higher order vector functions

$$\mathbf{\Lambda}^{i}_{\text{interp.}}(r) = \hat{\beta}^{i}_{\text{interp.}}(p, \boldsymbol{\xi})\, \mathbf{\Lambda}_i(r) \tag{5.158}$$

of arbitrary pth order are generated in Chapter 4 by multiplying the zeroth-order vector functions $\mathbf{\Lambda}_i$ with the normalized Silvester–Lagrange interpolatory polynomials $\hat{\beta}^{i}_{\text{interp.}}(p, \boldsymbol{\xi})$ (see Tables 4.5, 4.11, 4.13, 4.14, 4.17, 4.18, and 4.21 in Chapter 4). In the following, appropriate linear combinations of these polynomials are used to obtain normalized *hierarchical* polynomials $H^{i}_{\text{hierar.}}(p, \boldsymbol{\xi})$ that generate normalized *hierarchical* higher order vector functions of the form

$$\mathbf{\Lambda}^{i}_{\text{hierar.}}(r) = H^{i}_{\text{hierar.}}(p, \boldsymbol{\xi})\, \mathbf{\Lambda}_i(r) \tag{5.159}$$

The *generating* polynomials are subdivided from the outset into two different groups of face- (F) and volume-based (V) functions. In each group, all the hierarchical polynomials are properly scaled and constructed *a priori* to be mutually orthogonal independent of the definition domain of the relevant inner product, i.e. either the face or the volume of the cell. The hierarchical vector bases are then constructed by using the same technique given in Chapter 4 where we simply substitute the new scalar hierarchical polynomials for the interpolatory functions $\hat{\beta}^{i}_{\text{interp.}}$. The *face-based* divergence-conforming vector functions obtained in this manner are all independent; on the contrary, some of the vector functions obtained by multiplying the zeroth-order divergence-conforming vector functions with the volume-based generating polynomials are linearly dependent on the others. Using a procedure similar to that discussed for the interpolatory case, the dependent volume-based hierarchical polynomials from the generating scalar set are eliminated. After removing the redundant functions, the space spanned by the interpolatory and the hierarchical vector bases is exactly the same and, for any given polynomial order p, the total number of the new hierarchical vector basis functions is equal to the number of the interpolatory basis set and coincides with the number of DoFs necessary to span the Nédélec mixed-order spaces [5, 38]. These bases were first reported in [51–53].

As discussed in Chapter 4, we associate with each volumetric cell bounded by n_f faces a local system of normalized parent coordinates $\boldsymbol{\xi} = (\xi_1, \xi_2, \ldots, \xi_{n_f})$. The ith face of a given cell is the zero-coordinate surface for the normalized coordinate ξ_i, and each edge is the locus of points where two *local* parent coordinates of the 3D cell vanish. The coordinate ξ_i varies linearly across a cell, attaining a value of unity at the face, edge, or vertex opposite the zero-coordinate surface. Therefore, the coordinate gradient $\nabla\xi_i$ is perpendicular to the ith cell face and, on this face, it points toward the element's interior.

For the tetrahedral and the brick elements, we take ξ_1, ξ_2, and ξ_3 as independent coordinates indexed in a "right-handed" sense such that $(\nabla\xi_1 \times \nabla\xi_2) \cdot \nabla\xi_3$ is strictly positive. Conversely, the right-handed triad of independent coordinates of the prism element is $\{\xi_1, \xi_2, \xi_4\}$. With the coordinates ξ_i of the right-handed triad as the independent coordinates, the remaining coordinates become *dependent* coordinates, and each is related to the independent coordinates via a dependency relation (see Chapter 4). The so-called unitary basis vectors are then defined

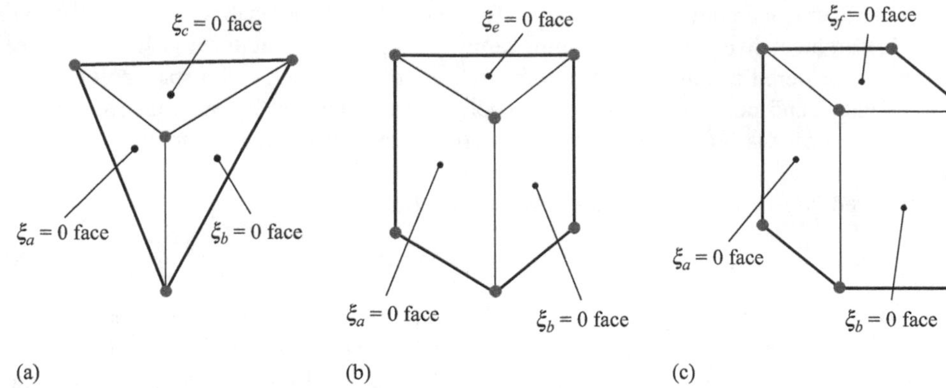

Figure 5.15 (a) The triangular faces of a tetrahedron are the ξ_a, ξ_b, ξ_c, and ξ_d zero-coordinate surfaces. (b) The quadrilateral faces of a prism are the ξ_a, ξ_b, and ξ_c zero-coordinate surfaces; its triangular faces are the ξ_d and ξ_e zero-coordinate surfaces. (c) The quadrilateral faces of a brick are the ξ_a, ξ_b, ξ_c, ξ_d, ξ_e, and ξ_f zero-coordinate surfaces.

© 2012 IEEE. Reprinted with permission from R. D. Graglia and A. F. Peterson, "Hierarchical divergence-conforming Nédélec elements for volumetric cells," *IEEE Trans. Antennas Propag.*, vol. 60, no. 11, pp. 5215–5227, Nov. 2012.

as $\boldsymbol{\ell}^i = \partial r/\partial \xi_i$, where ξ_i is one of the three independent coordinates. For rectilinear elements, each unitary basis vector coincides with (at least) one of the element's edge vectors.

In order to avoid reporting at length all the generating polynomials, we exploit the symmetry of the cells and denote the normalized coordinates of each cell by dummy parent variables, distinguished by a literal (i.e. not numerical) subscript. Thus, as shown in Figure 5.15a, the triangular faces of a tetrahedron are the ξ_a, ξ_b, ξ_c, and ξ_d zero-coordinate surfaces, and each tetrahedral cell is obtained by mapping the parent simplex T^3 defined by $\{0 \le \xi_a, \xi_b, \xi_c, \xi_d \le 1 : \xi_a + \xi_b + \xi_c + \xi_d = 1\}$ into the true 3D space, which we call the *child* space.

The quadrilateral faces of a prism are the ξ_a, ξ_b, and ξ_c zero-coordinate surfaces, while its triangular faces are the ξ_d and ξ_e zero-coordinate surfaces (Figure 5.15b). Each prism is obtained by mapping the parent cell V_{prism} defined by $\{0 \le \xi_a, \xi_b, \xi_c, \xi_d, \xi_e \le 1 : \xi_a + \xi_b + \xi_c = 1, \xi_d + \xi_e = 1\}$ into the child space.

The quadrilateral faces of a brick are the ξ_a, ξ_b, ξ_c, ξ_d, ξ_e, and ξ_f zero-coordinate surfaces (Figure 5.15c) and each brick is obtained by mapping the parent cell V_{brick} defined by $\{0 \le \xi_a, \xi_b, \xi_c, \xi_d, \xi_e, \xi_f \le 1 : \xi_a + \xi_d = 1, \xi_b + \xi_e = 1, \xi_c + \xi_f = 1\}$ into the child space.

Table 5.17 reports, for each volumetric cell, the parent variables, their dependency relations, and the number of DoFs together with the vector dependency relations used to eliminate any redundancy in higher order vector sets.

5.4.1 Reference Variables on the Face in Common to Adjacent Cells

As noted above, the hierarchical divergence-conforming vector bases for tetrahedral, prism, and brick cells are derived at the same time by denoting the normalized coordinates of each cell by dummy parent variables, distinguished by a literal (i.e. not numerical) subscript (see Figure 5.15).

Table 5.17 Parent variables, dependency relations, and number of DoFs for divergence-conforming Nédélec bases of order p on volumetric cells.

Cell	Coordinate surface of each quadrilateral face	Coordinate surface of each triangular face	Dependency relations for the parent variables	Dependency relations for the higher order vector functions	Number of dependent (gray) vector components $\xi_j \mathbf{\Lambda}_j$
		$\xi_a = 0$; $\xi_c = 0$; $\xi_b = 0$; $\xi_d = 0$.	$\xi_a + \xi_b + \xi_c + \xi_d = 1$.	$\xi_a \mathbf{\Lambda}_a + \xi_b \mathbf{\Lambda}_b + \xi_c \mathbf{\Lambda}_c + \xi_d \mathbf{\Lambda}_d = 0$.	1
	$\xi_a = 0$; $\xi_b = 0$; $\xi_c = 0$.	$\xi_d = 0$; $\xi_e = 0$.	$\xi_a + \xi_b + \xi_c = 1$; $\xi_d + \xi_e = 1$.	$\xi_a \mathbf{\Lambda}_a + \xi_b \mathbf{\Lambda}_b + \xi_c \mathbf{\Lambda}_c = 0$; $\xi_d \mathbf{\Lambda}_d + \xi_e \mathbf{\Lambda}_e = 0$.	2
	$\xi_a = 0$; $\xi_d = 0$; $\xi_b = 0$; $\xi_e = 0$; $\xi_c = 0$; $\xi_f = 0$.		$\xi_a + \xi_d = 1$; $\xi_b + \xi_e = 1$; $\xi_c + \xi_f = 1$.	$\xi_a \mathbf{\Lambda}_a + \xi_d \mathbf{\Lambda}_d = 0$; $\xi_b \mathbf{\Lambda}_b + \xi_e \mathbf{\Lambda}_e = 0$; $\xi_c \mathbf{\Lambda}_c + \xi_f \mathbf{\Lambda}_f = 0$.	3

For each cell, the number of faces is equal to the number of ξ-parent variables and to the number of independent zeroth-order divergence-conforming vector functions $\mathbf{\Lambda}$. At the zeroth order, for $p = 0$, the basis set is formed only by face-based vector functions. All the hierarchical (or interpolatory) face-based high-order vector functions are independent. The *face-based* vector functions associated with the $\xi_i = 0$ face contain only the zeroth-order vector component $\mathbf{\Lambda}_i$ as a factor; they all have a zero normal component on all the cell faces with the exception of the $\xi_i = 0$ face. Conversely, each higher order *volume-based* vector function (obtained for $p \geq 1$) vanishes completely only on a given coordinate surface $\xi_i = 0$ because of the presence of the vector factor $\xi_i \mathbf{\Lambda}_i$. A sub-set of *independent* higher order *volume-based* vector functions is obtained by eliminating from the full volume-based vector set all the functions that contain (for example) the factor $\xi_j \mathbf{\Lambda}_j$ reported in gray in the vector dependency relations. The number of independent (black) vector components $\xi_i \mathbf{\Lambda}_i$ is always equal to three for each volumetric cell.

Cell	DoF-Q DoF associated with each quadrilateral face	DoF-T DoF associated with each triangular face	DoF-VQ Number of volume-based functions that vanish on the same quadrilateral face	DoF-VT Number of volume-based functions that vanish on the same triangular face	Total number of independent DoFs
		$\frac{(p+1)(p+2)}{2}$		$\frac{p(p+1)(p+2)}{6}$	$\frac{(p+1)(p+2)(p+4)}{2}$
	$(p+1)^2$	$\frac{(p+1)(p+2)}{2}$	$\frac{p(p+1)^2}{2}$	$\frac{p(p+1)(p+2)}{2}$	$\frac{(p+1)(3p^2+12p+10)}{2}$
	$(p+1)^2$		$p(p+1)^2$		$3(p+1)^2(p+2)$

The total number of independent DoFs reported in the right-hand column is (4× DoF-T + 3× DoF-VT) for the tetrahedral cell, (3× DoF-Q + 2× DoF-T + 2× DoF-VQ + 1× DoF-VT) for the prism cell, and (6× DoF-Q + 3× DoF-VQ) for the brick cell.

© Adapted from R. D. Graglia and A. F. Peterson, "Hierarchical divergence-conforming Nédélec elements for volumetric cells," *IEEE Trans. Antennas Propag.*, vol. 60, no. 11, pp. 5215–5227, Nov. 2012.

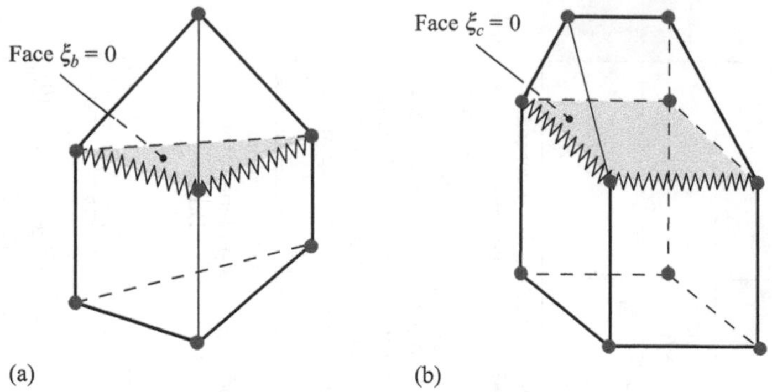

Figure 5.16 Pivoting edges for the face in common with two adjacent cells. Each node of a mesh has a different *global* node number while the edges are distinguished by different *global* edge numbers. Each face in common with two cells has two *pivoting* edges (denoted in the figure by snaked lines) which are those that depart from the face corner node with the lowest *global* node number. For convenience, the triangular face in common with two adjacent cells is always assumed to be the $\xi_d = 0$ face of both cells (a tetrahedron and a prism in Figure 5.16a), while the quadrilateral face in common with two adjacent elements is the $\xi_c = 0$ face of both cells (a prism and a brick in Figure 5.16b).

The use of dummy parent variables allows us to consider just the $\xi_d = 0$ triangular face in common with two adjacent tetrahedral or prism cells, or the $\xi_c = 0$ quadrilateral face in common with two adjacent prism or brick elements (see Figure 5.16). Furthermore, the common triangular face $\xi_d = 0$ is always described by the parent variables (ξ_a, ξ_b, ξ_c) and is obtained by mapping the parent simplex T^2 defined by $\{0 \le \xi_a, \xi_b, \xi_c \le 1\colon\ \xi_a + \xi_b + \xi_c = 1\}$ into the child 3D space. Similarly, the common quadrilateral face $\xi_c = 0$ is always described by four parent variables and is obtained by mapping a parent square cell Q into the child 3D space. On any face in common with two adjacent elements, two parent coordinates are considered as the reference variables used to write the generating orthogonal polynomials; the other face variables are obtained from the dependency relations of Table 5.17. The reference variables are easily distinguished by the *pivoting* edges of the face. As shown in Figure 5.16, the pivoting edges depart from the face corner node with the lowest *global* node number, and each reference variable vanishes only on one of the two pivoting edges, as shown in Figure 5.17.

The following subsections describe the procedure to obtain the hierarchical high-order divergence-conforming vector bases for tetrahedral, prism, and brick cells.

5.4.2 Tetrahedral Bases

To ensure the continuity of the normal component of the vector basis functions on the triangular face $\xi_d = 0$ of a tetrahedral cell shared by an adjacent cell (that has either tetrahedral or triangular prismatic shape), it is important to recognize the reference variables ξ_a and ξ_b of this face, as discussed in the previous subsection. The generating scalar polynomials are then obtained from orthogonal polynomial sets written in terms of the three variables $(\xi_{ab}, \chi_{ab}, \xi_d)$, with

$$\xi_{ab} = \xi_a - \xi_b, \quad \chi_{ab} = \xi_a + \xi_b \tag{5.160}$$

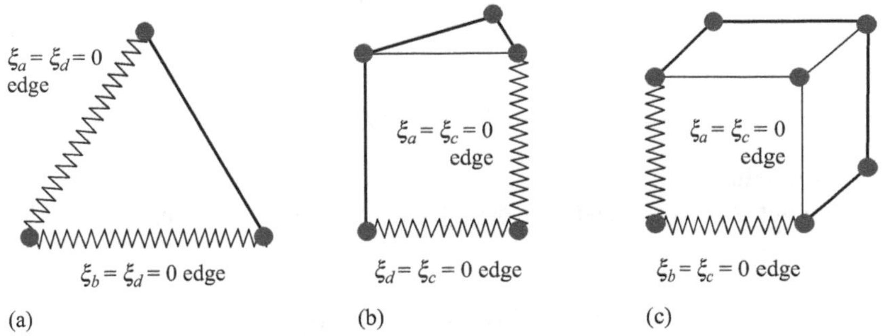

Figure 5.17 The two reference variables on the face in common with two adjacent cells. Two different *local* parent coordinates vanish on each pivoting edge: one is the parent coordinate that individuates the face, the other is assumed to be a reference parent variable associated with that face. The triangular face in common with two adjacent cells is always assumed to be the $\xi_d = 0$ face of a tetrahedron or a prism, while the quadrilateral face in common with two adjacent elements is the $\xi_c = 0$ face of a prism or a brick. (a) The reference variables for the $\xi_d = 0$ triangular face are ξ_a and ξ_b. (b) The reference variables for the $\xi_c = 0$ quadrilateral face of a prism are ξ_a and ξ_d, with $\xi_a = 0$ individuating another quadrilateral face of the prism. (c) The reference variables for the $\xi_c = 0$ quadrilateral face of a brick are ξ_a and ξ_b.

The first polynomial set is obtained from

$$\begin{cases} \mathcal{R}_0(\xi_d) = \sqrt{6} \\ \mathcal{R}_1(\xi_d) = 2\sqrt{15}\,\xi_d \end{cases} \tag{5.161}$$

by using the following recurrence relations with respect to the degree ℓ

$$a_{1\ell}\,\mathcal{R}_{\ell+1}(\xi_d) = a_{2\ell}\,(2\,\xi_d - 1)\,\mathcal{R}_\ell(\xi_d) - a_{3\ell}\,\mathcal{R}_{\ell-1}(\xi_d) \tag{5.162}$$

with

$$\begin{cases} a_{1\ell} = \sqrt{\ell(\ell+4)(2\ell+1)} \\ a_{2\ell} = \sqrt{(2\ell+1)(2\ell+3)(2\ell+5)} \\ a_{3\ell} = \sqrt{(\ell-1)(\ell+3)(2\ell+5)} \end{cases} \tag{5.163}$$

With the exception of $\mathcal{R}_0$, the remaining polynomials $\mathcal{R}_\ell$ (for all $\ell \geq 1$) are mutually orthogonal on the tetrahedral simplex T^3 and contain a common ξ_d factor which make them vanish for $\xi_d = 0$; these polynomials are normalized over the tetrahedral simplex T^3 by setting

$$\iiint_{T^3} \mathcal{R}_i(\xi_d)\,\mathcal{R}_j(\xi_d)\,\mathrm{d}T^3 = \delta_{ij}, \quad \text{for } i \text{ and } j \neq 0 \tag{5.164}$$

where

$$\delta_{ij} = \begin{cases} 1 & \text{for } i = j \\ 0 & \text{for } i \neq j \end{cases} \tag{5.165}$$

is the Kronecker delta.

The second set $\mathcal{F}_{mn}$ is obtained by orthogonalizing over the triangular simplex T^2 the polynomials

$$\mathcal{Q}_{mn}(\xi_a, \xi_b) = P_m(\chi_{ab})\, P_n(\xi_{ab}) \tag{5.166}$$

of order $m+n$ by running a double-nested loop for

- $t = 0, 1, \ldots, p$ (outer loop on the total order $t = m+n$);
- $n = 0, 1, \ldots, t$ (inner loop), with $m = (t-n)$ fixed in the inner loop;

where $P_q(z)$ indicates the qth-order Legendre polynomial. The $\mathcal{Q}_{mn}$ and $\mathcal{F}_{mn}$ polynomials are symmetric in (ξ_a, ξ_b) for even values of n (antisymmetric otherwise), with $\mathcal{F}_{mn}(\xi_a, \xi_b) = (-1)^n \mathcal{F}_{mn}(\xi_b, \xi_a)$.

The third set is formed by the volume-based generating polynomials $V_{\ell mn}$ which vanish for $\xi_d = 0$; these polynomials are obtained by orthogonalizing over the tetrahedral simplex T^3 the following polynomials of order $\ell + m + n$, for $\ell \geq 1$

$$\mathcal{V}_{\ell mn}(\boldsymbol{\xi}) = \mathcal{R}_\ell(\xi_d)\, \mathcal{F}_{mn}(\xi_{ab}, \chi_{ab}) \tag{5.167}$$

by running a triple-nested loop for

- $g = 1, 2, \ldots, p$ (outer loop on the global order $g = \ell + m + n$);
- $t = 0, 1, \ldots, g-1$ (second loop on $t = m+n$), with $\ell = (g-t)$ fixed in the second loop;
- $n = 0, 1, \ldots, t$ (inner loop), with $m = (t-n)$ fixed in the inner loop.

The $\mathcal{V}_{\ell mn}$ and $V_{\ell mn}$ polynomials are symmetric in (ξ_a, ξ_b) for even values of n (antisymmetric otherwise), with $V_{\ell mn}(\xi_a, \xi_b) = (-1)^n V_{\ell mn}(\xi_b, \xi_a)$. The $V_{\ell mn}$ polynomials are mutually orthogonal over the tetrahedral simplex and normalized so that

$$\iiint_{T^3} V_g^2(\boldsymbol{\xi})\, \mathrm{d}T^3 = 1 \tag{5.168}$$

where $g = \ell + m + n$ indicates the global polynomial order, equal to the sum of its subscripts.

In order to obtain a hierarchical family of face-based polynomials $F_{mn}(\boldsymbol{\xi})$ mutually orthogonal on both the T^2 and the T^3 simplices, it is sufficient to add to $\mathcal{F}_{mn}(\boldsymbol{\xi})$ an appropriate linear combination of the volume-based polynomials $V_q(\boldsymbol{\xi})$ of global order $q \leq (m+n)$, and which share the same symmetry properties of $\mathcal{F}_{mn}$ with respect to the ξ_a and ξ_b variables. The polynomials F_{mn} are symmetric in (ξ_a, ξ_b) for even values of n (antisymmetric otherwise). To produce a more compact expression, we substitute $(1 - \chi_{ab} - \xi_c)$ for ξ_d, and normalize as

$$\iint_{T^2} F_q^2(\boldsymbol{\xi})\, \mathrm{d}T^2 = 1 \tag{5.169}$$

$$\iiint_{T^3} F_q^2(\boldsymbol{\xi})\, \mathrm{d}T^3 = \frac{1}{2q+3} \tag{5.170}$$

where $q = m + n$ indicates the global order of the polynomial, equal to the sum of its subscripts.

The vector hierarchical divergence-conforming bases are obtained by multiplying the generating polynomials defined above with the *normalized* zeroth-order vector functions $\widetilde{\boldsymbol{\Lambda}}_1$, $\widetilde{\boldsymbol{\Lambda}}_2$, $\widetilde{\boldsymbol{\Lambda}}_3$, and $\widetilde{\boldsymbol{\Lambda}}_4$ of the tetrahedral cell (see Tables 4.5 and 4.13 in Chapter 4). In forming these products, recall that the zeroth-order function $\widetilde{\boldsymbol{\Lambda}}_i$ is associated with the $\xi_i = 0$ face; this implies that the ξ_d dummy variable must be set equal to ξ_i in the products involving $\widetilde{\boldsymbol{\Lambda}}_i$, while the three dummy variables

$$\boldsymbol{\xi}_{t_i} = \{\xi_{a_i}, \xi_{b_i}, \xi_{c_i}\} \tag{5.171}$$

that describe the ith triangular face are chosen accordingly to the procedure described in the previous subsection.

Throughout this section, a tilde is used to denote a normalized zeroth-order vector basis while a second subscript is added to denote the dummy parent variables, since the reference variables ξ_a and ξ_b of the considered triangular face are face dependent.

The hierarchical pth-order complete vector set consists of the following functions

$$\begin{cases} \boldsymbol{\Lambda}^1_{\ell mn}(r) = \xi_1\, \mathcal{U}_{\ell-1,m,n}(\boldsymbol{\xi}_{t_1}, \xi_1)\, \widetilde{\boldsymbol{\Lambda}}_1(r) \\ \boldsymbol{\Lambda}^2_{\ell mn}(r) = \xi_2\, \mathcal{U}_{\ell-1,m,n}(\boldsymbol{\xi}_{t_2}, \xi_2)\, \widetilde{\boldsymbol{\Lambda}}_2(r) \\ \boldsymbol{\Lambda}^3_{\ell mn}(r) = \xi_3\, \mathcal{U}_{\ell-1,m,n}(\boldsymbol{\xi}_{t_3}, \xi_3)\, \widetilde{\boldsymbol{\Lambda}}_3(r) \end{cases} \tag{5.172}$$

$$\begin{cases} \boldsymbol{\Lambda}^1_{0mn}(r) = F_{mn}(\boldsymbol{\xi}_{t_1})\, \widetilde{\boldsymbol{\Lambda}}_1(r) \\ \boldsymbol{\Lambda}^2_{0mn}(r) = F_{mn}(\boldsymbol{\xi}_{t_2})\, \widetilde{\boldsymbol{\Lambda}}_2(r) \\ \boldsymbol{\Lambda}^3_{0mn}(r) = F_{mn}(\boldsymbol{\xi}_{t_3})\, \widetilde{\boldsymbol{\Lambda}}_3(r) \\ \boldsymbol{\Lambda}^4_{0mn}(r) = F_{mn}(\boldsymbol{\xi}_{t_4})\, \widetilde{\boldsymbol{\Lambda}}_4(r) \end{cases} \tag{5.173}$$

obtained for $\ell = 1, 2, \ldots, p$; $m, n = 0, 1, \ldots, p$, with $(\ell + m + n) \le p$. The vector sets (5.172) and (5.173) are hierarchical in the sense that the $(p+1)$th-order set contains all the functions of the pth-order set. The $p(p+1)(p+2)/2$ functions (5.172) are volume based. The $2(p+1)(p+2)$ functions (5.173) are face based, and the sign of these functions must be adjusted for conformity to the adjacent cell.

The 28 face-based hierarchical polynomials used to construct the tetrahedral-face-based divergence-conforming functions up to the sixth order are reported in Table 5.18, while the 56 volume-based hierarchical polynomials used to construct the tetrahedral-volume-based functions up to the sixth order are given in Table 5.19.

Table 5.18 Divergence-conforming vector bases on tetrahedral and prism cells – triangular-face-based hierarchical polynomials up to the sixth order.

Face-based polynomials $F_{mn}(\boldsymbol{\xi}_t)$ of global order $p=(m+n)$, mutually orthogonal on the T^2 and T^3 simplices, and on the volume $V=\{0\leq\xi_a,\xi_b,\xi_c\leq 1:\xi_a+\xi_b+\xi_c=1;\ 0\leq\xi_d,\xi_e\leq 1:\xi_d+\xi_e=1\}$ of the parent prism

$F_{00}=\sqrt{2}$	$F_{50}=2\sqrt{3}\left(\chi_{ab}^5-30\,\chi_{ab}^4\,\xi_c+150\,\chi_{ab}^3\,\xi_c^2-200\,\chi_{ab}^2\,\xi_c^3\right.$ $\left.+75\,\chi_{ab}\,\xi_c^4-6\,\xi_c^5\right)$
$F_{10}=2\left(\chi_{ab}-2\xi_c\right)$ $F_{01}=2\sqrt{3}\,\xi_{ab}$	$F_{41}=6\,\xi_{ab}\left(\chi_{ab}^4-28\,\chi_{ab}^3\,\xi_c+126\,\chi_{ab}^2\,\xi_c^2-140\,\chi_{ab}\,\xi_c^3+35\,\xi_c^4\right)$ $F_{32}=\sqrt{15}\left(3\,\xi_{ab}^2-\chi_{ab}^2\right)\left(\chi_{ab}^3-24\,\chi_{ab}^2\,\xi_c+84\,\chi_{ab}\,\xi_c^2-56\,\xi_c^3\right)$
$F_{20}=\sqrt{6}\left(\chi_{ab}^2-6\,\chi_{ab}\,\xi_c+3\,\xi_c^2\right)$ $F_{11}=3\sqrt{2}\,\xi_{ab}\left(\chi_{ab}-4\xi_c\right)$ $F_{02}=\sqrt{\tfrac{15}{2}}\left(3\,\xi_{ab}^2-\chi_{ab}^2\right)$	$F_{23}=\sqrt{21}\,\xi_{ab}\left(5\,\xi_{ab}^2-3\,\chi_{ab}^2\right)\left(\chi_{ab}^2-18\,\chi_{ab}\,\xi_c+36\,\xi_c^2\right)$ $F_{14}=\tfrac{3}{4}\sqrt{3}\left(35\,\xi_{ab}^4-30\,\xi_{ab}^2\,\chi_{ab}^2+3\,\chi_{ab}^4\right)\left(\chi_{ab}-10\,\xi_c\right)$ $F_{05}=\tfrac{\sqrt{33}}{4}\,\xi_{ab}\left(63\,\xi_{ab}^4-70\,\xi_{ab}^2\,\chi_{ab}^2+15\,\chi_{ab}^4\right)$
$F_{30}=2\sqrt{2}\left(\chi_{ab}^3-12\,\chi_{ab}^2\,\xi_c+18\,\chi_{ab}\,\xi_c^2-4\,\xi_c^3\right)$ $F_{21}=2\sqrt{6}\,\xi_{ab}\left(\chi_{ab}^2-10\,\chi_{ab}\,\xi_c+10\,\xi_c^2\right)$ $F_{12}=\sqrt{10}\left(3\,\xi_{ab}^2-\chi_{ab}^2\right)\left(\chi_{ab}-6\,\xi_c\right)$ $F_{03}=\sqrt{14}\,\xi_{ab}\left(5\,\xi_{ab}^2-3\,\chi_{ab}^2\right)$	$F_{60}=\sqrt{14}\left(\chi_{ab}^6-42\,\chi_{ab}^5\,\xi_c+315\,\chi_{ab}^4\,\xi_c^2-700\,\chi_{ab}^3\,\xi_c^3\right.$ $\left.+525\,\chi_{ab}^2\,\xi_c^4-126\,\chi_{ab}\,\xi_c^5+7\,\xi_c^6\right)$
$F_{40}=\sqrt{10}\left(\chi_{ab}^4-20\,\chi_{ab}^3\,\xi_c+60\,\chi_{ab}^2\,\xi_c^2\right.$ $\left.-40\,\chi_{ab}\,\xi_c^3+5\,\xi_c^4\right)$ $F_{31}=\sqrt{30}\,\xi_{ab}\left(\chi_{ab}^3-18\,\chi_{ab}^2\,\xi_c+45\,\chi_{ab}\,\xi_c^2-20\,\xi_c^3\right)$ $F_{22}=\tfrac{5}{\sqrt{2}}\left(3\,\xi_{ab}^2-\chi_{ab}^2\right)\left(\chi_{ab}^2-14\,\chi_{ab}\,\xi_c+21\,\xi_c^2\right)$ $F_{13}=\sqrt{\tfrac{35}{2}}\,\xi_{ab}\left(5\,\xi_{ab}^2-3\,\chi_{ab}^2\right)\left(\chi_{ab}-8\,\xi_c\right)$ $F_{04}=\tfrac{3}{4}\sqrt{\tfrac{5}{2}}\left(35\,\xi_{ab}^4-30\,\xi_{ab}^2\,\chi_{ab}^2+3\,\chi_{ab}^4\right)$	$F_{51}=\sqrt{42}\,\xi_{ab}\left(\chi_{ab}^5-40\,\chi_{ab}^4\,\xi_c+280\,\chi_{ab}^3\,\xi_c^2-560\,\chi_{ab}^2\,\xi_c^3\right.$ $\left.+350\,\chi_{ab}\,\xi_c^4-56\,\xi_c^5\right)$ $F_{42}=\sqrt{\tfrac{35}{2}}\left(3\,\xi_{ab}^2-\chi_{ab}^2\right)\left(\chi_{ab}^4-36\,\chi_{ab}^3\,\xi_c+216\,\chi_{ab}^2\,\xi_c^2\right.$ $\left.-336\,\chi_{ab}\,\xi_c^3+126\,\xi_c^4\right)$ $F_{33}=\tfrac{7}{\sqrt{2}}\,\xi_{ab}\left(5\,\xi_{ab}^2-3\,\chi_{ab}^2\right)\left(\chi_{ab}^3-30\,\chi_{ab}^2\,\xi_c+135\,\chi_{ab}\,\xi_c^2-120\,\xi_c^3\right)$ $F_{24}=\tfrac{3}{4}\sqrt{\tfrac{7}{2}}\left(35\,\xi_{ab}^4-30\,\xi_{ab}^2\,\chi_{ab}^2+3\,\chi_{ab}^4\right)\left(\chi_{ab}^2-22\,\chi_{ab}\,\xi_c+55\,\xi_c^2\right)$ $F_{15}=\tfrac{1}{4}\sqrt{\tfrac{77}{2}}\,\xi_{ab}\left(63\,\xi_{ab}^4-70\,\xi_{ab}^2\,\chi_{ab}^2+15\,\chi_{ab}^4\right)\left(\chi_{ab}-12\,\xi_c\right)$ $F_{06}=\tfrac{1}{8}\sqrt{\tfrac{91}{2}}\left(231\,\xi_{ab}^6-315\,\xi_{ab}^4\,\chi_{ab}^2+105\,\xi_{ab}^2\,\chi_{ab}^4-5\,\chi_{ab}^6\right)$

With $\xi_{ab}=\xi_a-\xi_b$, $\chi_{ab}=\xi_a+\xi_b$, and $\boldsymbol{\xi}_t=(\xi_a,\xi_b,\xi_c)$. The reported expressions hold for both the tetrahedral and the prism cells since they involve only the triangular parent variables $\boldsymbol{\xi}_t$ which, on the considered triangular face, are related by the simplified dependency relation $\xi_a+\xi_b+\xi_c=1$. (Recall that the tetrahedron and the prism cells have different dependency relations which also involve other parent variables.) The reported polynomials are normalized as follows

$$\iint_{T^2}F_p^2(\boldsymbol{\xi}_t)\,dT^2=1;\quad \iiint_{T^3}F_p^2(\boldsymbol{\xi}_t)\,dT^3=\frac{1}{2p+3};\quad \iiiint_V \xi_d^2\,F_p^2(\boldsymbol{\xi}_t)\,dV=\iiiint_V \xi_e^2\,F_p^2(\boldsymbol{\xi}_t)\,dV=\frac{1}{3}$$

where T^2 and T^3 are the triangular and the tetrahedral simplices, respectively; V is the parent prism, while p indicates the global order of the polynomial, equal to the sum of its subscripts.

© 2012 IEEE. Reprinted with permission from R. D. Graglia and A. F. Peterson, "Hierarchical divergence-conforming Nédélec elements for volumetric cells," *IEEE Trans. Antennas Propag.*, vol. 60, no. 11, pp. 5215–5227, Nov. 2012.

5.4.3 Prism Bases

The $(p+1)(p+2)/2$ orthogonal polynomials based on the triangular face $\xi_d=0$ (or $\xi_e=0$) of the prism coincide with the $F_{mn}(\boldsymbol{\xi}_t)$ face-based polynomials described in the previous subsection for the tetrahedral cell, since those polynomials are orthogonal on the triangular simplex T^2 and their expression involves only the triangular parent variables ξ_a, ξ_b, and ξ_c.

The higher order face-based vector functions associated with the $\xi_d=0$ (or the $\xi_e=0$) triangular face of the prism are obtained by multiplying the *reference* zeroth-order divergence-conforming vector function $\boldsymbol{\Lambda}=\pm\widetilde{\boldsymbol{\Lambda}}_d$ (or $\boldsymbol{\Lambda}=\pm\widetilde{\boldsymbol{\Lambda}}_e$) associated with the $\xi_d=0$ (or

Table 5.19 Divergence-conforming vector bases on tetrahedral cells – volume-based hierarchical polynomials $V_{\ell mn} = \xi_d\, \mathcal{U}_{\ell-1,m,n}$ for the tetrahedral cell up to the sixth order.

$\mathcal{U}_{000} = 2\sqrt{15}$	$\mathcal{U}_{100} = 2\sqrt{105}\,(2\,\eta+1)$	$\mathcal{U}_{010} = 2\sqrt{105}\,(2\,\eta+3\,\chi_{ab})$	$\mathcal{U}_{001} = 6\sqrt{35}\,\xi_{ab}$
$\mathcal{U}_{200} = 6\sqrt{5}\,[3+14\,\eta\,(\eta+1)]$	$\mathcal{U}_{101} = 6\sqrt{21}\,\xi_{ab}\,(8\,\eta+5)$	$\mathcal{U}_{011} = 18\sqrt{14}\,\xi_{ab}\,(4\,\eta+5\,\chi_{ab})$	
$\mathcal{U}_{110} = 6\sqrt{7}\,(8\,\eta+5)\,(2\,\eta+3\,\chi_{ab})$	$\mathcal{U}_{020} = 6\sqrt{42}\,(3\,\eta^2+12\,\chi_{ab}\,\eta+10\,\chi_{ab}^2)$	$\mathcal{U}_{002} = 3\sqrt{210}\,(3\,\xi_{ab}^2-\chi_{ab}^2)$	
$\mathcal{U}_{300} = 2\sqrt{1{,}155}\,(2\,\eta+1)\,[1+6\,\eta\,(\eta+1)]$	$\mathcal{U}_{102} = 3\sqrt{110}\,(10\,\eta+7)\left(3\,\xi_{ab}^2-\chi_{ab}^2\right)$		
$\mathcal{U}_{210} = 2\sqrt{462}\left(5+18\,\eta+15\,\eta^2\right)(2\,\eta+3\,\chi_{ab})$	$\mathcal{U}_{030} = 6\sqrt{110}\left(4\,\eta^3+30\,\chi_{ab}\,\eta^2+60\,\chi_{ab}^2\,\eta+35\,\chi_{ab}^3\right)$		
$\mathcal{U}_{201} = 6\sqrt{154}\,\xi_{ab}\left(5+18\,\eta+15\,\eta^2\right)$	$\mathcal{U}_{021} = 6\sqrt{330}\,\xi_{ab}\left(10\,\eta^2+30\,\chi_{ab}\,\eta+21\,\chi_{ab}^2\right)$		
$\mathcal{U}_{120} = 6\sqrt{22}\,(10\,\eta+7)\left(3\,\eta^2+12\,\chi_{ab}\,\eta+10\,\chi_{ab}^2\right)$	$\mathcal{U}_{012} = 15\sqrt{22}\,(6\,\eta+7\,\chi_{ab})\left(3\,\xi_{ab}^2-\chi_{ab}^2\right)$		
$\mathcal{U}_{111} = 6\sqrt{66}\,\xi_{ab}\,(10\,\eta+7)\,(4\,\eta+5\,\chi_{ab})$	$\mathcal{U}_{003} = 3\sqrt{770}\,\xi_{ab}\left(5\,\xi_{ab}^2-3\,\chi_{ab}^2\right)$		
$\mathcal{U}_{400} = 2\sqrt{2{,}730}\left[1+3\,\eta\,(\eta+1)\left(4+11\,\eta+11\,\eta^2\right)\right]$	$\mathcal{U}_{112} = 5\sqrt{858}\,(4\,\eta+3)\,(6\,\eta+7\,\chi_{ab})\left(3\,\xi_{ab}^2-\chi_{ab}^2\right)$		
$\mathcal{U}_{310} = 6\sqrt{130}\left(7+42\,\eta+77\,\eta^2+44\,\eta^3\right)(2\,\eta+3\,\chi_{ab})$	$\mathcal{U}_{103} = \sqrt{30{,}030}\,\xi_{ab}\,(4\,\eta+3)\left(5\,\xi_{ab}^2-3\,\chi_{ab}^2\right)$		
$\mathcal{U}_{301} = 6\sqrt{390}\,\xi_{ab}\left(7+42\,\eta+77\,\eta^2+44\,\eta^3\right)$	$\mathcal{U}_{040} = 2\sqrt{2{,}145}\,\big(5\,\eta^4+60\,\chi_{ab}\,\eta^3$ $+210\,\chi_{ab}^2\,\eta^2+280\,\chi_{ab}^3\,\eta+126\,\chi_{ab}^4\big)$		
$\mathcal{U}_{220} = 6\sqrt{65}\left(14+44\,\eta+33\,\eta^2\right)\left(3\,\eta^2+12\,\chi_{ab}\,\eta+10\,\chi_{ab}^2\right)$			
$\mathcal{U}_{211} = 6\sqrt{195}\,\xi_{ab}\left(14+44\,\eta+33\,\eta^2\right)(4\,\eta+5\,\chi_{ab})$	$\mathcal{U}_{031} = 6\sqrt{715}\,\xi_{ab}\left(20\,\eta^3+105\,\chi_{ab}\,\eta^2+168\,\chi_{ab}^2\,\eta+84\,\chi_{ab}^3\right)$		
$\mathcal{U}_{202} = 15\sqrt{13}\left(14+44\,\eta+33\,\eta^2\right)\left(3\,\xi_{ab}^2-\chi_{ab}^2\right)$	$\mathcal{U}_{022} = 5\sqrt{429}\left(3\,\xi_{ab}^2-\chi_{ab}^2\right)\left(21\,\eta^2+56\,\chi_{ab}\,\eta+36\,\chi_{ab}^2\right)$		
$\mathcal{U}_{130} = 2\sqrt{4{,}290}\,(4\,\eta+3)\left(4\,\eta^3+30\,\chi_{ab}\,\eta^2+60\,\chi_{ab}^2\,\eta+35\,\chi_{ab}^3\right)$	$\mathcal{U}_{013} = \sqrt{15{,}015}\,\xi_{ab}\,(8\,\eta+9\,\chi_{ab})\left(5\,\xi_{ab}^2-3\,\chi_{ab}^2\right)$		
$\mathcal{U}_{121} = 6\sqrt{1{,}430}\,\xi_{ab}\,(4\,\eta+3)\left(10\,\eta^2+30\,\chi_{ab}\,\eta+21\,\chi_{ab}^2\right)$	$\mathcal{U}_{004} = 3\sqrt{2{,}145}\left(35\,\xi_{ab}^4-30\,\xi_{ab}^2\,\chi_{ab}^2+3\,\chi_{ab}^4\right)/\,4$		
$\mathcal{U}_{500} = 6\sqrt{70}\,(2\,\eta+1)\left[3+11\,\eta\,(\eta+1)\left(4+13\,\eta+13\,\eta^2\right)\right]$	$\mathcal{U}_{131} = 30\sqrt{13}\,\xi_{ab}\,(14\,\eta+11)\,\big(20\,\eta^3+105\,\chi_{ab}\,\eta^2$ $+168\,\chi_{ab}^2\,\eta+84\,\chi_{ab}^3\big)$		
$\mathcal{U}_{410} = 6\sqrt{5}\left(70+616\,\eta+1{,}848\,\eta^2+2{,}288\,\eta^3+1{,}001\,\eta^4\right)$ $\times(2\,\eta+3\,\chi_{ab})$	$\mathcal{U}_{122} = 5\sqrt{195}\,(14\,\eta+11)\left(3\,\xi_{ab}^2-\chi_{ab}^2\right)$ $\times\left(21\,\eta^2+56\,\chi_{ab}\,\eta+36\,\chi_{ab}^2\right)$		
$\mathcal{U}_{401} = 6\sqrt{15}\,\xi_{ab}\left(70+616\,\eta+1{,}848\,\eta^2+2{,}288\,\eta^3+1{,}001\,\eta^4\right)$			
$\mathcal{U}_{320} = 6\sqrt{55}\left(42+216\,\eta+351\,\eta^2+182\,\eta^3\right)$ $\times\left(3\,\eta^2+12\,\chi_{ab}\,\eta+10\,\chi_{ab}^2\right)$	$\mathcal{U}_{113} = 5\sqrt{273}\,\xi_{ab}\,(14\,\eta+11)\,(8\,\eta+9\,\chi_{ab})\left(5\,\xi_{ab}^2-3\,\chi_{ab}^2\right)$		
	$\mathcal{U}_{104} = 15\sqrt{39}\,(14\,\eta+11)\left(35\,\xi_{ab}^4-30\,\xi_{ab}^2\,\chi_{ab}^2+3\,\chi_{ab}^4\right)/\,4$		
$\mathcal{U}_{311} = 6\sqrt{165}\,\xi_{ab}\left(42+216\,\eta+351\,\eta^2+182\,\eta^3\right)$ $\times(4\,\eta+5\,\chi_{ab})$	$\mathcal{U}_{050} = 6\sqrt{455}\,\big(6\,\eta^5+105\,\chi_{ab}\,\eta^4+560\,\chi_{ab}^2\,\eta^3$ $+1{,}260\,\chi_{ab}^3\,\eta^2+1{,}260\,\chi_{ab}^4\,\eta+462\,\chi_{ab}^5\big)$		
$\mathcal{U}_{302} = 15\sqrt{11}\left(42+216\,\eta+351\,\eta^2+182\,\eta^3\right)\left(3\,\xi_{ab}^2-\chi_{ab}^2\right)$	$\mathcal{U}_{041} = 6\sqrt{1{,}365}\,\xi_{ab}\,\big(35\,\eta^4+280\,\chi_{ab}\,\eta^3+756\,\chi_{ab}^2\,\eta^2$ $+840\,\chi_{ab}^3\,\eta+330\,\chi_{ab}^4\big)$		
$\mathcal{U}_{230} = 2\sqrt{330}\left(45+130\,\eta+91\,\eta^2\right)$ $\times\left(4\,\eta^3+30\,\chi_{ab}\,\eta^2+60\,\chi_{ab}^2\,\eta+35\,\chi_{ab}^3\right)$	$\mathcal{U}_{032} = 15\sqrt{91}\left(3\,\xi_{ab}^2-\chi_{ab}^2\right)\big(56\,\eta^3+252\,\chi_{ab}\,\eta^2$ $+360\,\chi_{ab}^2\,\eta+165\,\chi_{ab}^3\big)$		
$\mathcal{U}_{221} = 6\sqrt{110}\,\xi_{ab}\left(45+130\,\eta+91\,\eta^2\right)$ $\times\left(10\,\eta^2+30\,\chi_{ab}\,\eta+21\,\chi_{ab}^2\right)$	$\mathcal{U}_{023} = 21\sqrt{65}\,\xi_{ab}\left(5\,\xi_{ab}^2-3\,\chi_{ab}^2\right)$ $\times\left(36\,\eta^2+90\,\chi_{ab}\,\eta+55\,\chi_{ab}^2\right)$		
$\mathcal{U}_{212} = 5\sqrt{66}\left(45+130\,\eta+91\,\eta^2\right)(6\,\eta+7\,\chi_{ab})\left(3\,\xi_{ab}^2-\chi_{ab}^2\right)$	$\mathcal{U}_{014} = 9\sqrt{455}\,(10\,\eta+11\,\chi_{ab})$ $\times\left(35\,\xi_{ab}^4-30\,\xi_{ab}^2\,\chi_{ab}^2+3\,\chi_{ab}^4\right)/\,4$		
$\mathcal{U}_{203} = \sqrt{2{,}310}\,\xi_{ab}\left(45+130\,\eta+91\,\eta^2\right)\left(5\,\xi_{ab}^2-3\,\chi_{ab}^2\right)$			
$\mathcal{U}_{140} = 10\sqrt{39}\,(14\,\eta+11)\,\big(5\,\eta^4+60\,\chi_{ab}\,\eta^3$ $+210\,\chi_{ab}^2\,\eta^2+280\,\chi_{ab}^3\,\eta+126\,\chi_{ab}^4\big)$	$\mathcal{U}_{005} = 3\sqrt{5{,}005}\,\xi_{ab}\left(63\,\xi_{ab}^4-70\,\xi_{ab}^2\,\chi_{ab}^2+15\,\chi_{ab}^4\right)/\,4$		

The table reports the functions $\mathcal{U}_{ijk}$ up to the order $(i+j+k)=5$. These functions form the set of the tetrahedral-volume-based polynomials $V_{\ell mn} = \xi_d\, \mathcal{U}_{\ell-1,m,n}$ up to the sixth order. It is understood that $\xi_a+\xi_b+\xi_c+\xi_d=1$; $\xi_{ab}=\xi_a-\xi_b$; $\chi_{ab}=\xi_a+\xi_b$; and $\eta=\xi_d-1$. All the volume-based polynomials are mutually orthogonal on the T^3 simplex and are normalized so that $\iiint_{T^3} V_p^2(\boldsymbol{\xi})\,dT^3=1$, where p is the global order of the volume polynomial, equal to the sum of its subscripts.

$\xi_e = 0$) face of the prism with the F_{mn} polynomials. The sign of $\mathbf{\Lambda}$ is adjusted to correspond to an arbitrarily selected reference direction across the face at issue. The reference vector function $\mathbf{\Lambda}$ is proportional to $\xi_e \nabla\xi_a \times \nabla\xi_b$ and $\xi_d \nabla\xi_a \times \nabla\xi_b$ for the $\xi_d = 0$ and the $\xi_e = 0$ face of the prism, respectively (see Chapter 4). Since $\mathbf{\Lambda}$ contains either the factor $\sqrt{w} = \xi_e$ or $\sqrt{w} = \xi_d$, the normalization of these face-based polynomials over the volume V of the prism uses w as a weighting function, with

$$\iiint_V w^2 F_p^2 \, dV = \frac{1}{3} \tag{5.174}$$

The same weight function w yields, by inspection, the orthogonal volume-based polynomials

$$V_{\ell mn}(\boldsymbol{\xi}) = \xi_d f_{\ell-1}(2\xi_d - 1) F_{mn}(\boldsymbol{\xi}_t) \tag{5.175}$$

which vanish on the $\xi_d = 0$ triangular face of the prism, with

$$\iiint_V \xi_e^2 V_{\ell mn}^2 \, dV = \frac{1}{3} \tag{5.176}$$

and where

$$f_q(z) = (-1)^q f_q(-z) = \sqrt{\frac{(2q+5)(q+3)(q+4)}{3(q+1)(q+2)}} \, P_q^{(2,2)}(z) \tag{5.177}$$

is the re-scaled Jacobi polynomial of order q already given in (5.94). The $V_{\ell mn}$ polynomials are normalized with respect to the weight function $w = \xi_e^2$ to enhance the linear independence of the hierarchical higher order vector functions $V_{\ell mn} \mathbf{\Lambda}$.

For the $\xi_c = 0$ quadrilateral face of the prism, the sign of the zeroth-order vector function $\widetilde{\mathbf{\Lambda}}_c$ associated with this face is adjusted to correspond to an arbitrarily selected reference direction across it, to obtain the reference zeroth-order divergence-conforming vector function $\mathbf{\Lambda}$ proportional to $(\xi_a \nabla\xi_b - \xi_b \nabla\xi_a) \times \nabla\xi_d$.

In this case, the $p(p+1)^2/2$ orthogonal volume-based polynomials

$$V_{ijk}(\boldsymbol{\xi}) = \xi_c \, \mathcal{U}_{i-1,j}(\xi_{ab}, \chi_{ab}) E_k(\xi_{de}) \tag{5.178}$$

that vanish on the $\xi_c = 0$ quadrilateral face of the prism and the $(p+1)^2$ orthogonal polynomials

$$F_{jk}(\boldsymbol{\xi}) = \widetilde{E}_j(\xi_a, \xi_b, \xi_c) E_k(\xi_{de}) \tag{5.179}$$

based on the quadrilateral face $\xi_c = 0$ of the prism are obtained for $i = 1, 2, \ldots, p$, $j, k = 0, 1, \ldots, p$, with $i + j \leq p$ and

$$E_k(z) = \sqrt{(2k+1)} \, P_k(z) \tag{5.180}$$

$$\iiint_V V_{ijk}^2 \, dV = 1 \tag{5.181}$$

$$\iint_Q F_{jk}^2 \, dQ = 1; \quad \iiint_V F_{jk}^2 \, dV = \frac{1}{2(j+1)} \tag{5.182}$$

The polynomials $\xi_c\, \mathcal{U}_{i-1,j}$ in (5.178) are obtained by orthogonalizing in order the polynomials $\xi_c\, P_{i-1}(\chi_{ab})\, P_j(\xi_{ab})$ over the triangular simplex T^2 within a double-nested loop for

- $t=1,2,\ldots,p$ (outer loop on the total order $t=i+j$);
- $j=0,1,\ldots,t-1$ (inner loop) with $i=t-j$ fixed in the inner loop.

The orthogonal polynomials $\widetilde{E}_j$ appearing in (5.179) are obtained by adding to $E_j(\xi_{ab})$ an appropriate linear combination of the polynomials $\xi_c\, \mathcal{U}_{\ell m}$ which share the same symmetry properties of $E_j(\xi_{ab})$ with respect to the ξ_a and ξ_b variables, for $(\ell+m+1)\leq j$. Thus, the $F_{jk}(\boldsymbol{\xi})$ polynomials simplify into $E_j(\xi_{ab})E_k(\xi_{de})$ on the $\xi_c=0$ face of the prism and match the expression of the face-based polynomials of a brick cell possibly attached to this face (see Section 5.4.4).

The factors V_{ijk} are antisymmetric in (ξ_a, ξ_b) and in (ξ_d, ξ_e) for odd values of j and k, respectively; they are symmetric otherwise, with $V_{ijk}(\xi_a,\xi_b)=(-1)^j V_{ijk}(\xi_b,\xi_a)$; $V_{ijk}(\xi_d,\xi_e)=(-1)^k V_{ijk}(\xi_e,\xi_d)$.

The hierarchical divergence-conforming bases for the prism cell are obtained by multiplying the generating polynomials defined in this subsection with the *normalized* zeroth-order vector functions of the prism cell (see Tables 4.5 and 4.21 in Chapter 4). In forming these products, recall that the prism functions $\widetilde{\boldsymbol{\Lambda}}_1$, $\widetilde{\boldsymbol{\Lambda}}_2$, and $\widetilde{\boldsymbol{\Lambda}}_3$ are associated with the $\xi_1=0$, $\xi_2=0$, and $\xi_3=0$ quadrilateral faces of the prism, respectively, while the functions $\widetilde{\boldsymbol{\Lambda}}_4$ and $\widetilde{\boldsymbol{\Lambda}}_5$ are associated with the $\xi_4=0$ and the $\xi_5=0$ triangular faces of the prism, respectively. This implies that for the polynomial factors associated with $\widetilde{\boldsymbol{\Lambda}}_i$ (for $i=1,2,3$) the ξ_c dummy variable is set equal to ξ_i, with

$$\begin{cases} \boldsymbol{\xi}_{t_i} = \{\xi_{a_i}, \xi_{b_i}, \xi_{c_i}\} = \{\xi_{a_i},\ 1-\xi_i-\xi_{a_i},\ \xi_i\} \\ \xi_{de_i} = 2\xi_{d_i} - 1 \end{cases} \tag{5.183}$$

and where ξ_{a_i} and ξ_{d_i} indicate the reference variables of the ith quadrilateral face of the prism, chosen according to the procedure described in Section 5.4.1. Conversely, for $i=4$, 5, the polynomial factors associated with $\widetilde{\boldsymbol{\Lambda}}_i$ are obtained by setting $\xi_d=\xi_i$, with $\xi_{de_i}=2\xi_i-1$, while the three dummy variables $\boldsymbol{\xi}_{t_i}=\{\xi_{a_i},\xi_{b_i},\xi_{c_i}\}$ that describe the ith triangular face are chosen according to the procedure described in Section 5.4.1.

For the prism, the hierarchical pth-order complete vector set consists of the following functions

$$\begin{cases} \boldsymbol{\Lambda}^1_{0jk}(r) = \widetilde{E}_j(\boldsymbol{\xi}_{t_1})\, E_k(\xi_{de_1})\, \widetilde{\boldsymbol{\Lambda}}_1(r) \\ \boldsymbol{\Lambda}^2_{0jk}(r) = \widetilde{E}_j(\boldsymbol{\xi}_{t_2})\, E_k(\xi_{de_2})\, \widetilde{\boldsymbol{\Lambda}}_2(r) \\ \boldsymbol{\Lambda}^3_{0jk}(r) = \widetilde{E}_j(\boldsymbol{\xi}_{t_3})\, E_k(\xi_{de_3})\, \widetilde{\boldsymbol{\Lambda}}_3(r) \\ \boldsymbol{\Lambda}^4_{0mn}(r) = F_{mn}(\boldsymbol{\xi}_{t_4})\, \widetilde{\boldsymbol{\Lambda}}_4(r) \\ \boldsymbol{\Lambda}^5_{0mn}(r) = F_{mn}(\boldsymbol{\xi}_{t_5})\, \widetilde{\boldsymbol{\Lambda}}_5(r) \end{cases} \tag{5.184}$$

$$\begin{cases} \boldsymbol{\Lambda}^1_{ijk}(r) = \xi_1\, \mathcal{U}_{i-1,j}(\xi_{ab_1}, \chi_{ab_1})\, E_k(\xi_{de_1})\, \widetilde{\boldsymbol{\Lambda}}_1(r) \\ \boldsymbol{\Lambda}^2_{ijk}(r) = \xi_2\, \mathcal{U}_{i-1,j}(\xi_{ab_2}, \chi_{ab_2})\, E_k(\xi_{de_2})\, \widetilde{\boldsymbol{\Lambda}}_2(r) \\ \boldsymbol{\Lambda}^4_{\ell mn}(r) = \xi_4\, f_{\ell-1}(\xi_{45})\, F_{mn}(\boldsymbol{\xi}_{t4})\, \widetilde{\boldsymbol{\Lambda}}_4(r) \end{cases} \tag{5.185}$$

obtained for $i, \ell = 1, 2, \ldots, p$; $j, k, m, n = 0, 1, \ldots, p$, with $i+j \le p$, $m+n \le p$, and with

$$\begin{cases} \chi_{ab_1} = \xi_{a_1} + \xi_{b_1} = 1 - \xi_1 \\ \chi_{ab_2} = \xi_{a_2} + \xi_{b_2} = 1 - \xi_2 \\ \xi_{ab_1} = \xi_{a_1} - \xi_{b_1} = 2\xi_{a_1} + \xi_1 - 1 \\ \xi_{ab_2} = \xi_{a_2} - \xi_{b_2} = 2\xi_{a_2} + \xi_2 - 1 \end{cases} \tag{5.186}$$

The vector sets (5.184) and (5.185) are hierarchical in the sense that the $(p+1)$th-order set contains all the functions of the pth-order set. The $p(p+1)(3p+4)/2$ functions (5.185) are volume based. The $(p+1)(4p+5)$ functions (5.184) are face based and, consequently, their sign must be adjusted for conformity to basis functions in the adjacent cell. The volume- and face-based hierarchical polynomials of the prism cell are reported in Tables 5.20 and 5.21, respectively, up to the sixth order.

5.4.4 Brick Bases

For the $\xi_c = 0$ face of the brick, the reference zeroth-order vector function $\mathbf{\Lambda} = \pm\widetilde{\mathbf{\Lambda}}_c$ is obtained by adjusting the sign of the zeroth-order vector function $\widetilde{\mathbf{\Lambda}}_c$ associated with the $\xi_c = 0$ face of the brick and proportional to $\xi_f \nabla\xi_a \times \nabla\xi_b$. At the pth order, there are $(p+1)^2$ hierarchical-face-based vector functions $F_{mn}(\boldsymbol{\xi})\,\mathbf{\Lambda}(r)$ associated with the $\xi_c = 0$ quadrilateral face of the brick, and $p(p+1)^2$ hierarchical-volume-based vector functions $V_{\ell mn}(\boldsymbol{\xi})\,\mathbf{\Lambda}(r)$ that vanish on the $\xi_c = 0$ quadrilateral face of the brick.

The hierarchical divergence-conforming bases of the brick are obtained by multiplying the zeroth-order reference vector function $\mathbf{\Lambda}$ with the factors

$$F_{mn}(\boldsymbol{\xi}) = E_m(\xi_{ad})\,E_n(\xi_{be}) \tag{5.187}$$

$$V_{\ell mn}(\boldsymbol{\xi}) = \xi_c f_{\ell-1}(2\xi_c - 1)\,F_{mn}(\boldsymbol{\xi}) \tag{5.188}$$

$$\text{for } \ell = 1, 2, \ldots, p \quad \text{and} \quad m, n = 0, 1, \ldots, p$$

where $f_q(z)$ and $E_k(z)$ are given in (5.177) and (5.180), respectively. In forming these products, recall that $\widetilde{\mathbf{\Lambda}}_i$ is associated with the $\xi_i = 0$ face of the brick. For the polynomial factors associated with $\widetilde{\mathbf{\Lambda}}_i$, this implies that the ξ_c dummy variable is set equal to ξ_i, while the (dummy) reference variables ξ_{a_i}, ξ_{b_i} that describe the ith quadrilateral face are chosen accordingly to the procedure described in Section 5.4.1, with

$$\begin{cases} \xi_{ad_i} = \xi_{a_i} - \xi_{d_i} = 2\,\xi_{a_i} - 1 \\ \xi_{be_i} = \xi_{b_i} - \xi_{e_i} = 2\,\xi_{b_i} - 1 \end{cases} \tag{5.189}$$

Thus, after defining

$$F_{mn}(\boldsymbol{\xi}_i) = E_m(\xi_{ad_i})\,E_n(\xi_{be_i}) \tag{5.190}$$

Table 5.20 Divergence-conforming vector bases on prism cells – mutually orthogonal-volume-based hierarchical polynomials for the prism cell up to the sixth order.

VOLUME-BASED POLYNOMIALS THAT VANISH ON THE $\xi_d = 0$ TRIANGULAR FACE OF THE PRISM

The first and second reference edges of the $\xi_d = 0$ triangular face are described by the dummy variables ξ_a and ξ_b, respectively; for $\{0 \le \xi_a, \xi_b \le 1\}$ with $\xi_c = 1 - \xi_a - \xi_b$. The sign of the zeroth-order vector function $\widetilde{\mathbf{\Lambda}}_d$ associated with the $\xi_d = 0$ face of the prism is adjusted to correspond to an arbitrarily selected reference direction across the $\xi_d = 0$ face. The hierarchical-volume-based vector function subset $V_{\ell mn}(\boldsymbol{\xi})\,\mathbf{\Lambda}(r)$ is obtained by multiplying $\mathbf{\Lambda} = \pm\widetilde{\mathbf{\Lambda}}_d$ with the polynomial factors $V_{\ell mn}$ reported below. For the pth order, the $p^2(p+1)/2$ volume-based polynomials factors are

$$V_{\ell mn} = \xi_d f_{\ell-1}(2\xi_d - 1)\, F_{mn}(\xi_a, \xi_b, \xi_c); \quad \text{for } \ell = 1, 2, \ldots, p;\ m = 0, 1, \ldots, p;\ n = 0, 1, \ldots, p \text{ (with } m + n \le p)$$

where F_{mn} are the polynomials reported in Table 5.18 up to the global order $(m+n) = 6$, and

$$f_q(z) = (-1)^q f_q(-z) = \sqrt{\frac{(2q+5)(q+3)(q+4)}{3(q+1)(q+2)}}\, P_q^{(2,2)}(z)$$

is the qth-order re-scaled Jacobi polynomial (5.177). The factors $V_{\ell mn}$ are antisymmetric in (ξ_a, ξ_b) for odd values of n and symmetric otherwise, with $V_{\ell mn}(\xi_a, \xi_b) = (-1)^n V_{\ell mn}(\xi_b, \xi_a)$. The orthogonal polynomials defined here are normalized as follows

$$\int_0^1 \left[\xi_d f_q(2\xi_d - 1)\right]^2 \mathrm{d}\xi_d = \frac{1}{3}, \quad \iint_{T^2} F_{mn}^2\, \mathrm{d}T^2 = 1, \quad \iiint_V \xi_e^2\, V_{\ell mn}^2\, \mathrm{d}V = \frac{1}{3}$$

where $T^2 = T^2(\xi_a, \xi_b, \xi_c)$ is the triangular simplex and V is the volume of the parent prism cell. To increase the linear independence of the higher order vector functions, the $V_{\ell mn}$ polynomials are normalized with the weight $w = \xi_e^2$ since $\sqrt{w}$ is a factor of $\mathbf{\Lambda}$.

VOLUME-BASED POLYNOMIALS THAT VANISH ON THE $\xi_c = 0$ QUADRILATERAL FACE OF THE PRISM

The first and second reference edges of the $\xi_c = 0$ quadrilateral face are described by the dummy variables ξ_a and ξ_d, respectively; for $\{0 \le \xi_a, \xi_d \le 1\}$, with $\xi_b = 1 - \xi_a$ and $\xi_e = 1 - \xi_d$. The sign of the zeroth-order vector function $\widetilde{\mathbf{\Lambda}}_c$ associated with the $\xi_c = 0$ face of the prism is adjusted to correspond to an arbitrarily selected reference direction across the $\xi_c = 0$ face. The hierarchical-volume-based vector function subset $V_{ijk}(\boldsymbol{\xi})\,\mathbf{\Lambda}(r)$ is obtained by multiplying $\mathbf{\Lambda} = \pm\widetilde{\mathbf{\Lambda}}_c$ with the factors V_{ijk} reported below. At the pth order, the $p(p+1)^2/2$ volume-based polynomials factors are

$$V_{ijk} = \xi_c\, \mathcal{U}_{i-1,j}(\xi_{ab}, \chi_{ab})\, E_k(\xi_{de}); \quad \text{for } i = 1, 2, \ldots, p;\ j = 0, 1, \ldots, p-1 \text{ (with } i + j \le p);\ k = 0, 1, \ldots, p$$

with $E_k(\xi_{de}) = \sqrt{(2k+1)}\, P_k(\xi_{de})$ and $\mathcal{U}_{mn}(\xi_{ab}, \chi_{ab})$ reported below up to $m + n = 5$. The polynomials $\xi_c\, \mathcal{U}_{i-1,j}$ are obtained by orthogonalizing in order the polynomials $\xi_c\, P_{i-1}(\chi_{ab})\, P_j(\xi_{ab})$ over the triangular simplex $T^2(\xi_a, \xi_b, \xi_c)$ with a nested loop: for $t = 1, 2, \ldots, p$ (outer loop on the total order $t = i + j$); for $j = 0, 1, \ldots, t-1$ (inner loop) with $i = t - j$ fixed in the inner loop. The factors V_{ijk} are antisymmetric in (ξ_a, ξ_b) and in (ξ_d, ξ_e) for odd values of j and k, respectively; they are symmetric otherwise, with $V_{ijk}(\xi_a, \xi_b) = (-1)^j V_{ijk}(\xi_b, \xi_a)$; $V_{ijk}(\xi_d, \xi_e) = (-1)^k V_{ijk}(\xi_e, \xi_d)$. The orthogonal polynomials defined here are normalized as follows (where V indicates the volume of the parent prism cell)

$$\int_0^1 E_k^2(\xi_{de})\, \mathrm{d}\xi_d = \int_0^1 E_k^2(\xi_{de})\, \mathrm{d}\xi_e = 1, \quad \iint_{T^2} \left[\xi_c\, \mathcal{U}_{i-1,j}\right]^2 \mathrm{d}T^2 = 1, \quad \iiint_V V_{ijk}^2\, \mathrm{d}V = 1$$

Table 5.20 (*Continued*).

$\mathcal{U}_{00} = 2\sqrt{3}$

$\mathcal{U}_{10} = 2\sqrt{3}\left(2 - 5\chi_{ab}\right)$

$\mathcal{U}_{01} = 6\sqrt{5}\,\xi_{ab}$

$\mathcal{U}_{20} = 2\sqrt{30}\left(1 - 6\chi_{ab} + 7\chi_{ab}^2\right)$

$\mathcal{U}_{11} = 2\sqrt{30}\,\xi_{ab}\left(4 - 7\chi_{ab}\right)$

$\mathcal{U}_{02} = \sqrt{210}\left(3\xi_{ab}^2 - \chi_{ab}^2\right)$

$\mathcal{U}_{30} = 2\sqrt{15}\left(2 - 21\chi_{ab} + 56\chi_{ab}^2 - 42\chi_{ab}^3\right)$

$\mathcal{U}_{21} = 2\sqrt{105}\,\xi_{ab}\left(5 - 20\chi_{ab} + 18\chi_{ab}^2\right)$

$\mathcal{U}_{12} = 5\sqrt{42}\left(2 - 3\chi_{ab}\right)\left(3\xi_{ab}^2 - \chi_{ab}^2\right)$

$\mathcal{U}_{03} = 3\sqrt{70}\,\xi_{ab}\left(5\xi_{ab}^2 - 3\chi_{ab}^2\right)$

$\mathcal{U}_{40} = 2\sqrt{105}\left(1 - 16\chi_{ab} + 72\chi_{ab}^2 - 120\chi_{ab}^3 + 66\chi_{ab}^4\right)$

$\mathcal{U}_{31} = 6\sqrt{70}\,\xi_{ab}\left(4 - 27\chi_{ab} + 54\chi_{ab}^2 - 33\chi_{ab}^3\right)$

$\mathcal{U}_{22} = 3\sqrt{10}\left(3\xi_{ab}^2 - \chi_{ab}^2\right)\left(21 - 70\chi_{ab} + 55\chi_{ab}^2\right)$

$\mathcal{U}_{13} = 3\sqrt{35}\,\xi_{ab}\left(8 - 11\chi_{ab}\right)\left(5\xi_{ab}^2 - 3\chi_{ab}^2\right)$

$\mathcal{U}_{04} = 3\sqrt{165}\left(35\xi_{ab}^4 - 30\xi_{ab}^2\chi_{ab}^2 + 3\chi_{ab}^4\right)/4$

$\mathcal{U}_{50} = 2\sqrt{42}\left(2 - 45\chi_{ab} + 300\chi_{ab}^2 - 825\chi_{ab}^3 + 990\chi_{ab}^4 - 429\chi_{ab}^5\right)$

$\mathcal{U}_{41} = 6\sqrt{70}\,\xi_{ab}\left(7 - 70\chi_{ab} + 231\chi_{ab}^2 - 308\chi_{ab}^3 + 143\chi_{ab}^4\right)$

$\mathcal{U}_{32} = 3\sqrt{35}\left(3\xi_{ab}^2 - \chi_{ab}^2\right)\left(28 - 154\chi_{ab} + 264\chi_{ab}^2 - 143\chi_{ab}^3\right)$

$\mathcal{U}_{23} = 7\sqrt{165}\,\xi_{ab}\left(5\xi_{ab}^2 - 3\chi_{ab}^2\right)\left(6 - 18\chi_{ab} + 13\chi_{ab}^2\right)$

$\mathcal{U}_{14} = 3\sqrt{77}\left(10 - 13\chi_{ab}\right)\left(35\xi_{ab}^4 - 30\xi_{ab}^2\chi_{ab}^2 + 3\chi_{ab}^4\right)/4$

$\mathcal{U}_{05} = \sqrt{3,003}\,\xi_{ab}\left(63\xi_{ab}^4 - 70\xi_{ab}^2\chi_{ab}^2 + 15\chi_{ab}^4\right)/4$

The global order of each volume-based polynomial is equal to the sum of its subscripts. It is understood that $\xi_a + \xi_b + \xi_c = 1$; $\xi_d + \xi_e = 1$; $\xi_{ab} = \xi_a - \xi_b$; $\chi_{ab} = \xi_a + \xi_b$; and $\xi_{de} = \xi_d - \xi_e$.

© 2012 IEEE. Reprinted with permission from R. D. Graglia and A. F. Peterson, "Hierarchical divergence-conforming Nédélec elements for volumetric cells," *IEEE Trans. Antennas Propag.*, vol. 60, no. 11, pp. 5215–5227, Nov. 2012.

Table 5.21 Divergence-conforming vector bases on prism cells – face-based hierarchical polynomials for the prism cell up to the sixth order.

TRIANGULAR-FACE-BASED POLYNOMIALS

For the pth order, the $(p+1)(p+2)/2$ polynomials $F_{mn}(\boldsymbol{\xi})$ based on the triangular face $\xi_d = 0$ (or $\xi_e = 0$) of the prism are those already reported in Table 5.18 up to the global order $p = (m+n) = 6$. They are mutually orthogonal on the $\xi_d = 0$ (and the $\xi_e = 0$) triangular face T^2, and on the volume V of the parent prism cell. They are normalized as follows

$$\iint_{T^2} F^2_{mn}(\boldsymbol{\xi})\, \mathrm{d}T^2 = 1; \qquad \iiint_V F^2_{mn}(\boldsymbol{\xi})\, \mathrm{d}V = 1$$

QUADRILATERAL-FACE-BASED POLYNOMIALS

For the pth order, the $(p+1)^2$ polynomials based on the quadrilateral face $\xi_c = 0$ of the prism are

$$F_{jk}(\boldsymbol{\xi}) = \widetilde{E}_j(\xi_a, \xi_b, \xi_c)\, E_k(\xi_{de}); \quad \text{for } j = 0, 1, \ldots, p;\ k = 0, 1, \ldots, p$$

with $E_k(\xi_{de}) = \sqrt{(2k+1)}\, P_k(\xi_{de})$ and $\widetilde{E}_j$ reported below up to $j = 6$. The polynomials $\widetilde{E}_j$ are normalized and mutually orthogonal on the triangular face $T^2(\xi_a, \xi_b, \xi_c)$. They are obtained by adding to $E_j(\xi_{ab})$ an appropriate linear combination of Table 5.20 polynomials $\xi_c\, \mathcal{U}_{\ell m}$ which share the same symmetry properties of $E_j(\xi_{ab})$ with respect to the ξ_a and ξ_b variables, for $(\ell + m + 1) \le j$. Therefore, the $F_{jk}(\boldsymbol{\xi})$ polynomials simplify into $E_j(\xi_{ab}) E_k(\xi_{de})$ on the $\xi_c = 0$ face of the prism to match the expression of the face-based polynomials of a brick cell possibly attached to this face. The $F_{jk}(\boldsymbol{\xi})$ polynomials are mutually orthogonal on the $\xi_c = 0$ quadrilateral face Q, and on the volume V of the parent prism cell. They are normalized as follows

$$\iint_{Q} F^2_{jk}(\boldsymbol{\xi})\, \mathrm{d}Q = 1; \qquad \iiint_V F^2_{jk}(\boldsymbol{\xi})\, \mathrm{d}V = \frac{1}{2(j+1)}$$

$\widetilde{E}_0 = E_0(\xi_{ab}) = P_0(\xi_{ab}) = 1$	$\widetilde{E}_4 = \sqrt{9}\,\{P_4(\xi_{ab}) + 3\,\xi_c\,(2-\xi_c)\left[10\xi_{ab}^2 - (1+\chi_{ab}^2)\right]/8\}$
$\widetilde{E}_1 = E_1(\xi_{ab}) = \sqrt{3}\,P_1(\xi_{ab}) = \sqrt{3}\,\xi_{ab}$	$\widetilde{E}_5 = \sqrt{11}\,\{P_5(\xi_{ab}) + 5\,\xi_{ab}\,\xi_c\,(2-\xi_c)\left[14\xi_{ab}^2 - 3\,(1+\chi_{ab}^2)\right]/8\}$
$\widetilde{E}_2 = \sqrt{5}\left[P_2(\xi_{ab}) + \xi_c\,(2-\xi_c)/2\right]$	$\widetilde{E}_6 = \sqrt{13}\,\{P_6(\xi_{ab}) + 5\,\xi_c\,(2-\xi_c)$
$\widetilde{E}_3 = \sqrt{7}\left[P_3(\xi_{ab}) + 3\,\xi_{ab}\,\xi_c\,(2-\xi_c)/2\right]$	$\times \left[63\xi_{ab}^4 - (1+\chi_{ab}^2)\,(21\xi_{ab}^2 - 1) + \chi_{ab}^4\right]/16\}$

These expressions apply for a prism with triangular faces described by the parent coordinates (ξ_a, ξ_b, ξ_c), with $\xi_a + \xi_b + \xi_c = 1$, $\xi_d + \xi_e = 1$, $\xi_{ab} = \xi_a - \xi_b$, $\chi_{ab} = \xi_a + \xi_b$, and $\xi_{de} = \xi_d - \xi_e$; $P_n(z)$ is the nth-order Legendre polynomial.

the hierarchical pth-order complete vector set for the brick consists of the functions

$$\begin{cases} \boldsymbol{\Lambda}^1_{0mn}(r) = F_{mn}(\boldsymbol{\xi}_1)\, \widetilde{\boldsymbol{\Lambda}}_1(r) \\ \boldsymbol{\Lambda}^2_{0mn}(r) = F_{mn}(\boldsymbol{\xi}_2)\, \widetilde{\boldsymbol{\Lambda}}_2(r) \\ \boldsymbol{\Lambda}^3_{0mn}(r) = F_{mn}(\boldsymbol{\xi}_3)\, \widetilde{\boldsymbol{\Lambda}}_3(r) \\ \boldsymbol{\Lambda}^4_{0mn}(r) = F_{mn}(\boldsymbol{\xi}_4)\, \widetilde{\boldsymbol{\Lambda}}_4(r) \\ \boldsymbol{\Lambda}^5_{0mn}(r) = F_{mn}(\boldsymbol{\xi}_5)\, \widetilde{\boldsymbol{\Lambda}}_5(r) \\ \boldsymbol{\Lambda}^6_{0mn}(r) = F_{mn}(\boldsymbol{\xi}_6)\, \widetilde{\boldsymbol{\Lambda}}_6(r) \end{cases} \tag{5.191}$$

$$\begin{cases} \boldsymbol{\Lambda}^1_{\ell mn}(r) = \xi_1 f_{\ell-1}(2\xi_1 - 1)\, F_{mn}(\boldsymbol{\xi}_1)\, \tilde{\boldsymbol{\Lambda}}_1(r) \\ \boldsymbol{\Lambda}^2_{\ell mn}(r) = \xi_2 f_{\ell-1}(2\xi_2 - 1)\, F_{mn}(\boldsymbol{\xi}_2)\, \tilde{\boldsymbol{\Lambda}}_2(r) \\ \boldsymbol{\Lambda}^3_{\ell mn}(r) = \xi_3 f_{\ell-1}(2\xi_3 - 1)\, F_{mn}(\boldsymbol{\xi}_3)\, \tilde{\boldsymbol{\Lambda}}_3(r) \end{cases} \tag{5.192}$$

Table 5.22 Divergence-conforming vector bases on brick cells – face- and volume-based hierarchical polynomials for the brick cell.

It is understood that the first and second reference edges of the $\xi_c = 0$ quadrilateral face of the brick are described by the variables ξ_a and ξ_b, respectively; for $\{0 \le \xi_a, \xi_b \le 1\}$, with $\xi_a + \xi_d = 1$, $\xi_b + \xi_e = 1$, and $\xi_c + \xi_f = 1$. For the $\xi_c = 0$ face, the reference zeroth-order vector function $\mathbf{\Lambda} = \pm\widetilde{\mathbf{\Lambda}}_c$ is obtained by adjusting the sign of the zeroth-order vector function $\widetilde{\mathbf{\Lambda}}_c$ associated with the $\xi_c = 0$ face of the brick. For the pth order, there are $(p+1)^2$ hierarchical-face-based vector functions $F_{mn}(\boldsymbol{\xi})\,\mathbf{\Lambda}(r)$ associated with the $\xi_c = 0$ quadrilateral face of the brick, and $p(p+1)^2$ hierarchical-volume-based vector functions $V_{\ell mn}(\boldsymbol{\xi})\,\mathbf{\Lambda}(r)$ that vanish on the $\xi_c = 0$ quadrilateral face of the brick. These vector functions are obtained by multiplying the zeroth-order reference vector function $\mathbf{\Lambda}$ with the factors

$$F_{mn}(\boldsymbol{\xi}) = E_m(\xi_{ad})\,E_n(\xi_{be}),\quad V_{\ell mn}(\boldsymbol{\xi}) = \xi_c f_{\ell-1}(2\xi_c - 1)\,F_{mn}(\boldsymbol{\xi}),\quad \text{for } \ell = 1, 2, \ldots, p, \text{ and } m, n = 0, 1, \ldots, p$$

with

$$E_k(z) = \sqrt{(2k+1)}\,P_k(z);\; f_q(z) = (-1)^q f_q(-z) = \sqrt{\frac{(2q+5)(q+3)(q+4)}{3(q+1)(q+2)}}\,P_q^{(2,2)}(z)$$

and where $P_k(z)$ is the Legendre polynomial of order k while $f_q(z)$ is the qth-order re-scaled Jacobi polynomial (5.177). The factors $V_{\ell mn}$ are antisymmetric in (ξ_a, ξ_b) for odd value of n and symmetric otherwise, with $V_{\ell mn}(\xi_a, \xi_b) = (-1)^n\, V_{\ell mn}(\xi_b, \xi_a)$. These orthogonal polynomials are normalized as follows

$$\int_0^1 \left[\xi_c f_q(2\xi_c - 1)\right]^2 \mathrm{d}\xi_c = \frac{1}{3},\quad \iint_Q F_{mn}^2\,\mathrm{d}Q = 1,\quad \iiint_V \xi_f^2\, V_{\ell mn}^2\,\mathrm{d}V = \frac{1}{3}$$

where V is the volume and Q is the $\xi_c = 0$ face of the parent brick cell. To enhance the linear independence of the higher order volume-based vector functions, the $V_{\ell mn}$ polynomials are normalized with the weight $w = \xi_f^2$ since $\sqrt{w}$ is a factor of $\mathbf{\Lambda}$. Notice that $\xi_f = 1$ on the $\xi_c = 0$ face of the brick.

The global order of each polynomial factor is equal to the sum of its subscripts.

obtained for $\ell = 1, 2, \ldots, p$; m, $n = 0, 1, \ldots, p$. The vector sets (5.191) and (5.192) are hierarchical in the sense that the $(p+1)$th-order set contains all the functions of the pth-order set. The $3p(p+1)^2$ functions (5.192) are volume based (recall that the high-order vector functions $\xi_i\mathbf{\Lambda}_i$, $\xi_{i+3}\mathbf{\Lambda}_{i+3}$ are dependent; see the third row of Table 5.17). The $6(p+1)^2$ functions (5.191) are face based and their sign must be adjusted for conformity to the adjacent cell. The face- and volume-based hierarchical polynomials of the brick cell are summarized in Table 5.22.

5.4.5 Numerical Results and Comparisons with Other Bases

The hierarchical vector bases seen above span the same spaces as the interpolatory divergence-conforming bases of Chapter 4, and their numerical results and convergence for a fixed Nédélec order are identical to the precision maintained in the computations. The Nédélec mixed-order spaces are sometimes denoted by half-integer orders; in the following, we use "order 2.5" to denote the $p = 2$ functions, for instance. We also note that hierarchical functions are intended for use with p-refinement procedures that produce a mixture of polynomial orders and not a uniform Nédélec order. Since the principal concern of hierarchical bases tends to be the matrix

Table 5.23 Divergence-conforming vector bases – individual element **T**-matrix condition numbers for unitary edge-length cells.

TETRAHEDRAL CELL

Basis order	Order of **T**	CNH	CNI	$\frac{\text{CNH}}{\text{CNI}}$
0.5	4	1.667	1.667	1.00
1.5	15	32.42	21.19	1.53
2.5	36	146.4	140.4	1.04
3.5	70	425.3	854.4	4.98×10^{-1}
4.5	120	1,087	5,314	2.05×10^{-1}
5.5	189	2,469	2.524×10^{4}	9.78×10^{-2}
6.5	280	4,847	1.260×10^{5}	3.85×10^{-2}

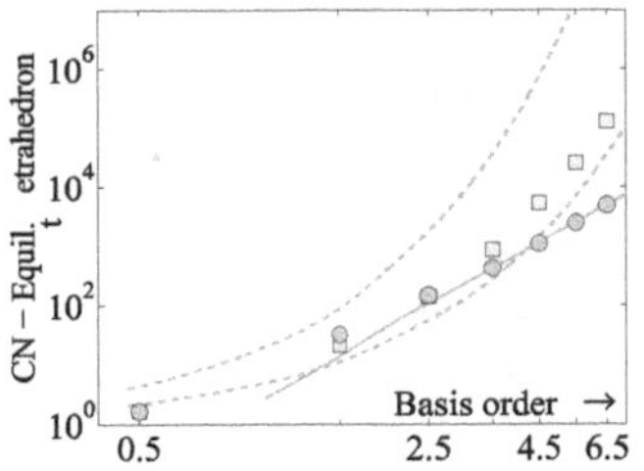

PRISM CELL

Basis order	Order of **T**	CNH	CNI	$\frac{\text{CNH}}{\text{CNI}}$
0.5	5	3	4	7.50×10^{-1}
1.5	25	40.57	74.19	5.47×10^{-1}
2.5	69	141.8	987.1	1.44×10^{-1}
3.5	146	423.8	1.226×10^{4}	3.46×10^{-2}
4.5	265	1,007	1.276×10^{5}	7.89×10^{-3}
5.5	435	2,238	1.585×10^{6}	1.41×10^{-3}
6.5	665	4,237	2.065×10^{7}	2.05×10^{-4}

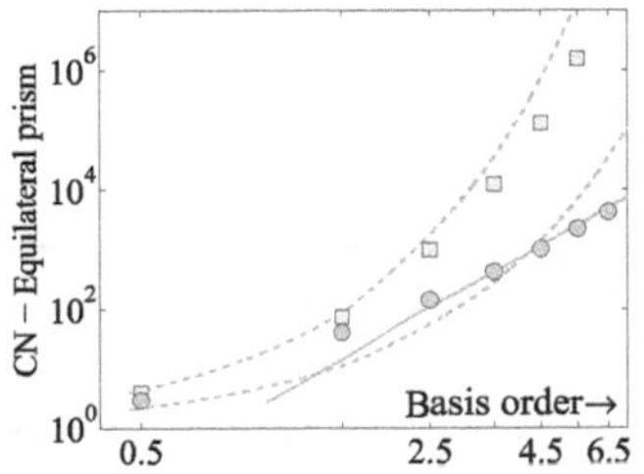

BRICK CELL

Basis order	Order of **T**	CNH	CNI	$\frac{\text{CNH}}{\text{CNI}}$
0.5	6	3	3	1.00
1.5	36	22.96	61.92	3.71×10^{-1}
2.5	108	22.96	1,286	1.79×10^{-2}
3.5	240	60.48	1.675×10^{4}	3.61×10^{-3}
4.5	450	60.48	2.767×10^{5}	2.19×10^{-4}
5.5	756	114.7	5.811×10^{6}	1.97×10^{-5}
6.5	1,176	114.7	1.469×10^{8}	7.81×10^{-7}

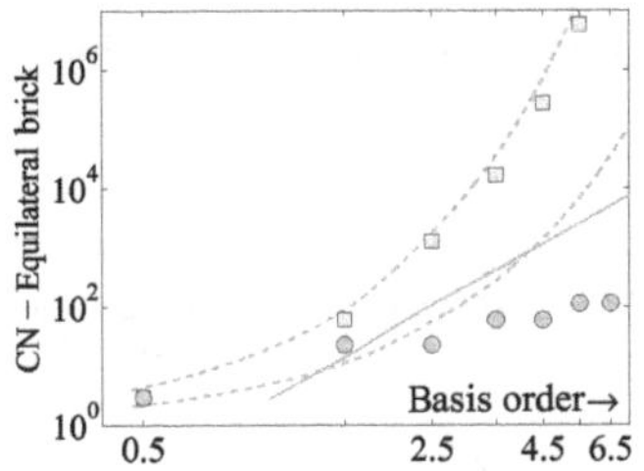

Individual element **T**-matrix condition numbers for our hierarchical (CNH) and interpolatory (CNI) *vector bases* up to the order 6.5, obtained by considering rectilinear cells with equal edges of unitary length. The interpolatory normalized bases are reported in Chapter 4. The figures on the right-hand side of the table compare (using logarithmic scales) the condition numbers with *reference* growth-rate lines. The hierarchical condition numbers are reported by circles, and the interpolatory condition numbers are reported by squares. The solid-line depicts a growth rate $g_1 = 2.8\times\ (\text{order})^4$; the dashed-line at top represents an exponential growth rate $g_2 = 20^{\text{order}}$; and the dashed-line at bottom represents an exponential growth rate $g_3 = 5^{\text{order}}$. The condition numbers growth rate is exponential for the interpolatory bases and polynomial for our hierarchical bases.

conditioning arising from their use, here we report some results for matrix condition numbers for the local Gram matrix

$$\mathrm{T}_{mn} = \iiint_V \boldsymbol{B}_m \cdot \boldsymbol{B}_n \, \mathrm{d}V \tag{5.193}$$

where $\boldsymbol{B}_n$ is a vector basis function, with the primary goal of establishing that the rate of growth in condition number is no worse than that of the interpolatory bases. For tetrahedral cells, we also compare the element matrix condition number to that of two other hierarchical basis families. Similar comparisons of **T**-matrix condition numbers were carried out for curl-conforming bases in the previous section.

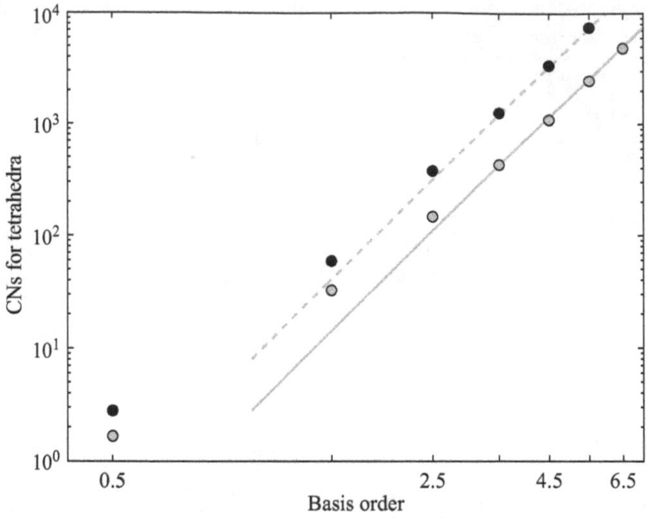

Figure 5.18 The individual element **T**-matrix condition numbers of the proposed hierarchical bases exhibit a polynomial growth rate. The condition numbers for an equilateral tetrahedron (gray circles) are compared with those obtained for a standard tetrahedral simplex (black circles) with vertices at (1,0,0), (0,1,0), (0,0,1), and (0,0,0). The solid-line represents the *reference* growth rate $g_1 = 2.8\times$ (order)^4; the dashed-line depicts a reference growth rate $g_{T^3} = 8\times$ (order)^4.

Table 5.23 presents element T-matrix condition numbers for unitary edge-length cells. The results obtained with the hierarchical bases proposed here are denoted by CNH, while those obtained with the interpolatory basis functions of Chapter 4 are denoted by CNI and are reported for comparison. It is of importance to observe that the condition number growth rate is exponential for the interpolatory bases and polynomial for our hierarchical bases; in particular, the condition number of our hierarchical bases is always lower than the condition number of the interpolatory bases for the prism and the brick cells, while for the tetrahedral cell it is lower than the condition number of the interpolatory bases for order $\geq$3.5. The polynomial growth rate of the hierarchical condition numbers is not strongly dependent on how the cell of the child space is distorted, as can be observed by the results of Figure 5.18 that show the growth rate of the condition numbers for two different tetrahedral cells.

Table 5.24 shows condition numbers arising from our functions, those of the Zaglmayr family [19], and those of the Botha family [49], for two tetrahedral cell shapes and basis orders up to 2.5. To attempt a fair comparison, these are computed after diagonal preconditioning is employed to scale the diagonal entries to unity and alleviate the effects of scale factors on the results [47]. In addition, for the comparison, we selected the subspaces of [19] that most closely span the mixed-order Nédélec spaces, since those functions do not exactly span those spaces for $p > 0.5$. These results show that the new functions have better matrix condition numbers than the Botha functions and are similar to those of the Zaglmayr family. We note that optimal

Table 5.24 Divergence-conforming vector bases on tetrahedral cells – comparison of individual element **T**-matrix condition numbers obtained with different hierarchical bases.

a) Equilateral tetrahedron with unitary-length edges					
Basis order	**Order of T**	**This book family**		**Zaglmayr family [19]**	**Botha family [49]**
0.5	4	**1.667**	1.667	1.667	1.667
1.5	15	**32.42**	36.71	47.20	143.8
2.5	36	**146.4**	156.6	188.1	303.1

b) Standard tetrahedral simplex with vertices at (1,0,0), (0,1,0), (0,0,1), and (0,0,0)					
Basis order	**Order of T**	**This book family**		**Zaglmayr family [19]**	**Botha family [49]**
0.5	4	**2.790**	2.816	2.816	2.816
1.5	15	**59.11**	71.48	50.82	252.2
2.5	36	**374.3**	410.8	309.6	480.4

The condition numbers (CNs) reported in dark black in the third column are obtained by using the original (scaled) expressions of our hierarchical bases as reported in the present book. To make a fair comparison with other bases, the other CNs (reported in black) are obtained after re-scaling all the vector functions to get **T**-matrices with unit entries along their main diagonal. This diagonal preconditioning slightly deteriorates the CNs of our basis functions while it substantially improves the CNs of the other families with respect to those (not reported in the table) obtained by using the original, unscaled expressions available in the quoted references.

Adapted from R. D. Graglia and A. F. Peterson, "Hierarchical divergence-conforming Nédélec elements for volumetric cells," *IEEE Trans. Antennas Propag.*, vol. 60, no. 11, pp. 5215–5227, Nov. 2012.

scale factors are not provided in either [19] or [49] but that the scale factors provided with the proposed bases actually produce better condition numbers than those obtained with diagonal preconditioning.

5.5 Conclusion

Hierarchical scalar bases and hierarchical vector bases that span the Nédélec mixed-order spaces have been proposed for the most common two- and three-dimensional cells. These bases maintain appropriate continuity in a mesh containing multiple cell shapes. In particular, the vector bases maintain the tangential or normal continuity in the curl- or divergence-conforming case, respectively. The use of orthogonal scalar polynomials to systematically construct the basis functions is believed to offer a simpler approach for enhancing their linear independence as their polynomial order increases than the partial orthogonalization of the final vector functions. The process by which the basis functions were obtained is described in detail. Numerical results for the matrix condition numbers suggest that these hierarchical bases exhibit reasonable linear independence for large orders.

References

[1] M. Salazar-Palma, T. K. Sarkar, L.-E. Garcia-Castillo, T. Roy, and A. Djordjevic, *Iterative and Self-Adaptive Finite-Elements in Electromagnetic Modeling*, Boston, MA: Artech House, 1998.

[2] L. Demkowicz, *Computing with hp-Adaptive Finite Elements*, vol. 1, Boca Raton, FL: Chapman & Hall/CRC Press, 2007.

[3] L. Demkowicz, *Computing with hp-Adaptive Finite Elements*, vol. 2, Boca Raton, FL: Chapman & Hall/CRC Press, 2008.

[4] Y. Zhu and A. Cangellaris, *Multigrid Finite Element Methods for Electromagnetic Field Modeling*. Piscataway, NJ, USA: Wiley-IEEE Press, 2006.

[5] J. C. Nédélec, "Mixed finite elements in R3," *Numer. Math.*, vol. 35, pp. 315–341, 1980.

[6] J. P. Webb and B. Forghani, "Hierarchal scalar and vector tetrahedra," *IEEE Trans. Magn.*, vol. 29, pp. 1495–1498, Mar. 1993.

[7] C. Carrié and J. P. Webb, "Hierarchal triangular edge elements using orthogonal polynomials," in *Digest of the 1997 IEEE International Antennas and Propagation Symposium*, Montreal, vol. 2, pp. 1301–1313, Jul. 1997.

[8] J. Wang and J. P. Webb, "Hierarchal vector boundary elements and p-adaption for 3-D electromagnetic scattering," *IEEE Trans. Antennas Propag.*, vol. 45, pp. 1869–1879, Dec. 1997.

[9] L. S. Andersen and J. L. Volakis, "Hierarchical tangential vector finite elements for tetrahedra," *IEEE Microwave Guided Wave Lett.*, vol. 8, pp. 127–129, Mar. 1998.

[10] L. S. Andersen and J. L. Volakis, "Development and application of a novel class of hierarchical tangential vector finite elements for electromagnetics," *IEEE Trans. Antennas Propag.*, vol. 47, pp. 112–120, Jan. 1999.

[11] J. P. Webb, "Hierarchal vector basis functions of arbitrary order for triangular and tetrahedral finite elements," *IEEE Trans. Antennas Propag.*, vol. 47, no. 8, pp. 1244–1253, Aug. 1999.

[12] D. K. Sun, J. F. Lee, and Z. Cendes, "Construction of nearly orthogonal Nedelec bases for rapid convergence with multilevel preconditioned solvers," *SIAM J. Sci. Comput.*, vol. 23, pp. 1053–1076, 2001.

[13] M. Ainsworth and J. Coyle, "Hierarchic hp-edge element families for Maxwell's equations on hybrid quadrilateral/triangular meshes," *Comput. Meth. Appl. Mech. Eng.*, vol. 190, pp. 6709–6733, 2001.

[14] M. Ainsworth and J. Coyle, "Hierarchic finite element bases on unstructured tetrahedral meshes," *Int. J. Numer. Meth. Eng.*, vol. 58, pp. 2103–2130, 2003.

[15] S. C. Lee, J. F. Lee, and R. Lee, "Hierarchical vector finite elements for analyzing waveguiding structures," *IEEE Trans. Antennas Propag.*, vol. 51, pp. 1897–1905, Aug. 2003.

[16] R. S. Preissig and A. F. Peterson, "A rationale for p-refinement with vector finite elements," *Appl. Comput. Electromagn. Soc. J.*, vol. 19, pp. 65–75, Jul. 2004.

[17] P. Ingelström, "A new set of H(curl)-conforming hierarchical basis functions for tetrahedral meshes," *IEEE Trans. Microwave Theory Tech.*, vol. 54, pp. 106–114, Jan. 2006.

[18] J. Schöberl and S. Zaglmayr, "High order Nédélec elements with local complete sequence properties," *Int. J. Comput. Math. Elect. Electron. Eng. (COMPEL)*, vol. 24, no. 2, pp. 374–384, 2005.

[19] S. Zaglmayr, *High order finite element methods for electromagnetic field computation,* Ph.D. Thesis, Johannes Kepler Universität, Linz, Austria, July 2006.

[20] R. D. Graglia and A. F. Peterson, "Fully conforming hierarchical vector bases for finite methods," *Abstracts of the 2009 URSI National Radio Science Meeting*, Charleston, SC, 1–5 June 2009.

[21] R. D. Graglia, A. F. Peterson, and F. P. Andriulli, "Hierarchical polynomials and vector elements for finite methods," *Proceedings of the International Conference on Electromagnetics in Advanced Applications (ICEAA 2009)*, Torino, Italy, vol. 1, pp. 1086–1089, Sept. 2009, doi:10.1109/ICEAA.2009.5297791

[22] A. F. Peterson and R. D. Graglia, "Evaluation and comparison of hierarchical vector basis functions for quadrilateral cells," Digest of the *14th Biennial IEEE Conference on Electromagnetic Field Computations,* Chicago, IL, May 2010.

[23] R. D. Graglia and A. F. Peterson, "Curl-conforming hierarchical vector elements for quadrilateral and brick meshes and their generating orthogonal polynomials," Abstracts of the *2010 CNC/USNC/URSI Radio Science Meeting*, Toronto, ON, July 2010.

[24] R. D. Graglia and A. F. Peterson, "Hierarchical vector polynomials for the triangular prism," *Proceedings of the International Conference on Electromagnetics in Advanced Applications (ICEAA 2010)*, Sydney, Australia, vol. 1, pp. 871–874, Sept. 2010.

[25] R. D. Graglia, A. F. Peterson, and F. P. Andriulli, "Curl-conforming hierarchical vector bases for triangles and tetrahedra," *IEEE Trans. Antennas Propag.*, vol. 59, no. 3, pp. 950–959, Mar. 2011.

[26] R. D. Graglia and A. F. Peterson, "Hierarchical curl-conforming Nédélec elements for quadrilateral and brick cells," *IEEE Trans. Antennas Propag.*, vol. 59, no. 8, pp. 2766–2773, Aug. 2011.

[27] A. F. Peterson, R. D. Graglia, "Evaluation of hierarchical vector basis functions for quadrilateral cells," *IEEE Trans. Magn.*, vol. 47, no. 5, pp. 1190–1193, May 2011.

[28] R. D. Graglia and A. F. Peterson, "Hierarchical curl-conforming Nédélec elements for triangular-prism cells," *IEEE Trans. Antennas Propag.*, vol. 60, no. 7, pp. 3314–3324, Jul. 2012.

[29] J. Xin and W. Cai, "A well-conditioned hierarchical basis for triangular H(curl)-conforming elements," *Commun. Comput. Phys.*, vol. 9, no. 3, pp. 780–806, Mar. 2011.

[30] J. Xin, N. Guo, and W. Cai, "On the construction of well-conditioned hierarchical bases for tetrahedral H(curl)-conforming Nédélec elements," *J. Comput. Math.*, vol. 29, no. 5, pp. 526–542, May 2011.

[31] R. Abdul-Rahman and M. Kasper, "Orthogonal hierarchical Nédélec elements," *IEEE Trans. Magn.*, vol. 44, no. 6, pp. 1210–1213, Jun. 2008.

[32] J. P. Webb, "Matching a given field using hierarchal vector basis functions," *Electromagnetics*, vol. 24, no. 1–2, pp. 113–122, Jan.–Mar. 2004.

[33] M. Abramowitz and I. Stegun, *Handbook of Mathematical Functions*. New York, NY: Dover, 1968.

[34] M. M. Ilic and B. M. Notaros, "Higher order hierarchical curved hexahedral vector finite elements for electromagnetic modeling," *IEEE Trans. Microwave Theory Tech.*, vol. 51, pp. 1026–1033, Mar. 2003.

[35] M. Djordjevic and B. M. Notaros, "Higher-order hierarchical basis functions with improved orthogonality properties for moment-method modeling of metallic and dielectric microwave structures," *Microwave Opt. Technol. Lett.*, vol. 37, no. 2, pp. 83–88, Apr. 2003.

[36] E. Jorgensen, J. L. Volakis, P. Meincke, and O. Breinbjerg, "Higher order hierarchical Legendre basis functions for electromagnetic modeling," *IEEE Trans. Antennas Propag.*, vol. 52, pp. 2985–2995, Nov. 2004.

[37] R. D. Graglia, D. R. Wilton, and A. F. Peterson, "Higher order interpolatory vector bases for computational electromagnetics," special issue on "Advanced Numerical Techniques in Electromagnetics," *IEEE Trans. Antennas Propag.*, vol. 45, no. 3, pp. 329–342, Mar. 1997.

[38] J. C. Nédélec, "A new family of mixed finite elements in R3," *Numer. Math.*, vol. 50, pp. 57–81, 1986.

[39] P. Dular, J.-Y. Hody, A. Nichlet, A. Genon, and W. Legros, "Mixed finite elements associated with a collection of tetrahedra, hexahedra, and prisms," *IEEE Trans. Magn.*, vol. 30, pp. 2980–2983, Sep. 1994.

[40] Z. S. Sacks and J. F. Lee, "A finite element time domain method using prism elements for microwave cavities" *IEEE Trans. Electromagn. Compat.*, vol. 37, pp. 519–527, Nov. 1995.

[41] T. Özdemir and J.-L. Volakis, "Triangular prisms for edge-based vector finite element analysis of conformal antennas," *IEEE Trans. Antennas Propag.*, vol. 45, pp. 788–797, May 1997.

[42] R. D. Graglia, D. R. Wilton, A. F. Peterson, and I.-L. Gheorma, "Higher order interpolatory vector bases on prism elements," *IEEE Trans. Antennas Propag.*, vol. 46, no. 3, pp. 442–450, Mar. 1998.

[43] K. Hirayama, Md. S. Alam, Y. Hayashi, and M. Koshiba, "Vector finite element method with mixed-interpolation-type triangular-prism element for waveguide discontinuities," *IEEE Trans. Microwave Theory Tech.*, vol. 42, pp. 2311–2316, Dec. 1994.

[44] J. Liu and J.-M. Jin, "A special higher order finite element method for scattering by deep cavities," *IEEE Trans. Antennas Propag.*, vol. 48, pp. 694–703, May 2000.

[45] D. I. Karatzidis and T. V. Yioultsis, "Efficient analysis of planar microwave circuits with mixed-order prism vector finite macroelements," *Int. J. Numer. Model.*, vol. 21, pp. 475–492, 2008.

[46] R. D. Graglia and I.-L. Gheorma, "Higher order interpolatory vector bases on pyramidal elements," *IEEE Trans. Antennas Propag.*, vol. 47, no. 5, pp. 775–782, May 1999.

[47] A. F. Peterson and R. D. Graglia, "Scale factors and matrix conditioning associated with triangular-cell hierarchical vector basis functions" *IEEE Antennas Wireless Propag. Lett.*, vol. 9, pp. 40–43, 2010, doi:10.1109/LAWP.2010.2042423

[48] M. M. Botha, "Solving the volume integral equations of electromagnetic scattering," *J. Comput. Phys.*, vol. 218, pp. 141–158, 2006.

[49] M. M. Botha, "Fully hierarchical divergence-conforming basis functions on tetrahedral cells, with applications," *Int. J. Numer. Meth. Eng.*, vol. 71, pp. 127–148, 2007.

[50] J. C. Eastwood and J. G. Morgan, "Higher-order basis functions for MoM calculations," *IET Sci. Meas. Technol.*, vol. 2, no. 6, pp. 379–386, 2008, doi:10.1049/iet-smt:20080056

[51] R. D. Graglia and A. F. Peterson, "Hierarchical vector basis functions for meshes with hexahedra, tetrahedra, and triangular prism cells," *Abstracts of ACES 2011*, Williamsburg, VA, 27–31 March 2011.

[52] R. D. Graglia and A. F. Peterson, "Well-conditioned hierarchical Nédélec elements for surface and volumetric cells," *Abstracts of the XXX URSI General Assembly and Scientific Symposium of International Union of Radio Science*, Istanbul, Turkey, 13–20 August 2011.

[53] R. D. Graglia and A. F. Peterson, "Hierarchical divergence-conforming Nédélec elements for volumetric cells," *IEEE Trans. Antennas Propag.*, vol. 60, no. 11, pp. 5215–5227, Nov. 2012.

CHAPTER 6

The Numerical Solution of Integral and Differential Equations

To demonstrate the use of vector bases developed in preceding chapters, here we consider their use in numerical solutions of the electric field integral equation (EFIE) for scattering from three-dimensional perfectly conducting objects, and the vector Helmholtz equation for modeling fields within three-dimensional cavities. The EFIE operator involves a divergence of the surface current density, and therefore the MoM or boundary element procedure will be illustrated in conjunction with divergence-conforming basis functions. Since the vector Helmholtz operator involves the curl operator, it will be treated using the finite element method in conjunction with curl-conforming bases. Select numerical results are presented for illustration, and the treatment of curved cells is described. Some of the following has been adapted from [1].

6.1 The Electric Field Integral Equation

Consider a perfect electric conducting (PEC) body in an infinite homogeneous environment of permittivity ϵ and permeability μ. The target is illuminated by a sinusoidal steady-state source of electromagnetic radiation, having radian frequency ω, and all the time-varying quantities are represented by phasors with suppressed time dependence $e^{j\omega t}$. In the absence of the target, the source produces an electric field $\boldsymbol{E}^{\text{inc}}$ throughout the surrounding space. (This is denoted the incident field.) In the presence of the conducting body, the fields are perturbed from these to the total field $\boldsymbol{E}^{\text{tot}}$. The perturbation can be accounted for by the presence of equivalent induced currents on the body.

The induced surface current density can be expressed as $\boldsymbol{J}(s,t)$, where s and t are parametric variables on the surface. The current density, if treated as a source function that exists in the absence of the conducting body, produces the scattered field $\boldsymbol{E}^{\text{s}}$. The fields are related by

$$\boldsymbol{E}^{\text{inc}} + \boldsymbol{E}^{\text{s}} = \boldsymbol{E}^{\text{tot}} \tag{6.1}$$

at points in space outside the conducting target. Since the scattered fields are produced in infinite homogeneous space, they are readily determined using any of the standard source-field relations, such as

$$\boldsymbol{E}^{\text{s}} = \frac{\nabla(\nabla \cdot \boldsymbol{A}) + k^2 \boldsymbol{A}}{j\omega\epsilon} \tag{6.2}$$

where the wavenumber of the medium is given by $k = \omega\sqrt{\mu\epsilon}$, and the magnetic vector potential function is

$$A(x, y, z) = \iint_{\text{surface}} J(s', t') \frac{e^{-jkR}}{4\pi R} \, ds' \, dt' \tag{6.3}$$

In (6.3), R is the distance from a point (s', t') on the surface to the point (x, y, z) where the field is evaluated. We employ primed coordinates to describe the "source" of the electromagnetic field (the current density), while unprimed coordinates denote the "observer" location where that field is evaluated.

The total field must satisfy the electromagnetic boundary condition on the surface of the perfect electric body:

$$\hat{n} \times E^{\text{tot}}\Big|_{\text{surface}} = 0 \tag{6.4}$$

where $\hat{n}$ is the outward normal unit vector. By combining (6.1)–(6.4), we obtain

$$\hat{n} \times E^{\text{inc}}\Big|_{\text{surface}} = -\hat{n} \times \frac{\nabla(\nabla \cdot A) + k^2 A}{j\omega\varepsilon}\Bigg|_{\text{surface}} \tag{6.5}$$

This equation is one form of the EFIE. (While it is in fact an integro-differential equation, the term "integral equation" is commonly used for simplicity.) Below, as is common practice, we actually use a rotated form of (6.5) where the cross product with the normal vector has been dropped in favor of a dot product with a tangent vector.

The EFIE can be solved in principle for the surface current density J appearing within the magnetic vector potential A. Once J is determined by a solution of (6.5), the fields and other observable quantities associated with the electromagnetic scattering problem can be found by direct calculation. (We note that there are closed structures for which the EFIE sometimes fails, see [2] for more information.)

We begin by constructing a weak form of the EFIE, with reduced differentiability requirements on the unknown current density. To this end, we multiply (by scalar product) the EFIE in (6.5) with a vector testing function $T(s, t)$, tangential to the surface, to obtain

$$\iint_{\text{surface}} T \cdot E^{\text{inc}} \, ds \, dt = -\iint_{\text{surface}} T \cdot E^{\text{s}} \, ds \, dt \tag{6.6}$$

where the integration is performed over the surface of the conductor. The incident electric field E^{inc} is the field in the absence of the structure (the excitation) and is assumed known. The function E^{s} is the "scattered" electric field produced by the surface current J, also in the absence of the structure, and may be obtained from the expression

$$E^{\text{s}} = -j\omega\mu \iint_{\text{surface}} J(s', t') G \, ds' \, dt' + \frac{1}{j\omega\epsilon} \nabla \iint_{\text{surface}} \nabla' \cdot J(s', t') G \, ds' \, dt' \tag{6.7}$$

where G is the free space Green's function

$$G(R) = \frac{e^{-jkR}}{4\pi R} \tag{6.8}$$

and

$$R=\sqrt{[x(s,t)-x(s',t')]^2+[y(s,t)-y(s',t')]^2+[z(s,t)-z(s',t')]^2} \tag{6.9}$$

To obtain (6.7) from (6.2), we employed the relation

$$\nabla\cdot\boldsymbol{A}=\iint_{\text{surface}}\nabla'\cdot\boldsymbol{J}(s',t')\,G(R)\,\mathrm{d}s'\,\mathrm{d}t' \tag{6.10}$$

The tested form of the scattered E-field can be expressed

$$\begin{aligned}-\iint_{\text{surface}}\boldsymbol{T}\cdot\boldsymbol{E}^{\text{s}}\,\mathrm{d}s\,\mathrm{d}t&=j\omega\mu\iint_{\text{surface}}\boldsymbol{T}\cdot\iint_{\text{surface}}\boldsymbol{J}(s',t')\,G\,\mathrm{d}s'\,\mathrm{d}t'\,\mathrm{d}s\,\mathrm{d}t\\&\quad-\frac{1}{j\omega\varepsilon}\iint_{\text{surface}}\boldsymbol{T}\cdot\nabla\iint_{\text{surface}}\nabla'\cdot\boldsymbol{J}(s',t')\,G\,\mathrm{d}s'\,\mathrm{d}t'\,\mathrm{d}s\,\mathrm{d}t\end{aligned} \tag{6.11}$$

Using the vector identity

$$\nabla\cdot(f\,\boldsymbol{T})=\boldsymbol{T}\cdot\nabla f+f\,\nabla\cdot\boldsymbol{T} \tag{6.12}$$

in conjunction with the divergence theorem, (6.11) can be manipulated into the form

$$\begin{aligned}-\iint_{\text{surface}}\boldsymbol{T}\cdot\boldsymbol{E}^{\text{s}}\,\mathrm{d}s\,\mathrm{d}t&=j\omega\mu\iint_{\text{surface}}\boldsymbol{T}\cdot\iint_{\text{surface}}\boldsymbol{J}(s',t')\,G\,\mathrm{d}s'\,\mathrm{d}t'\,\mathrm{d}s\,\mathrm{d}t\\&\quad+\frac{1}{j\omega\varepsilon}\iint_{\text{surface}}\nabla\cdot\boldsymbol{T}\iint_{\text{surface}}\nabla'\cdot\boldsymbol{J}(s',t')\,G\,\mathrm{d}s'\,\mathrm{d}t'\,\mathrm{d}s\,\mathrm{d}t\end{aligned} \tag{6.13}$$

The MoM procedure requires the choice of suitable basis and testing functions to reduce the continuous equation into a system of discrete equations [2, 3]. Constraints on the continuity of the primary unknown are dictated by the operator and usually play a large role in the choice of basis functions. Previous chapters have developed two classes of vector basis functions known, respectively, as divergence-conforming functions and curl-conforming functions. Divergence-conforming functions maintain normal-vector continuity from cell to cell on the surface, while curl-conforming functions maintain tangential-vector continuity between adjacent cells.

Since the EFIE involves a divergence operation, we approximate the unknown current density $\boldsymbol{J}$ by an expansion in linearly independent divergence-conforming vector basis functions

$$\boldsymbol{J}(s,t)\cong\sum_{n=1}^{N}I_n\boldsymbol{B}_n(s,t) \tag{6.14}$$

where I_n denote N complex-valued coefficients that henceforth are the unknowns to be determined. As discussed in Chapters 3–5, the divergence-conforming basis functions either straddle two adjacent cells or are entirely confined within a single cell. The connectivity arrays associated with the surface mesh can be used to provide a systematic way of organizing the basis functions within a cell and linking edge-based basis functions that straddle cells to appropriate basis functions in adjacent cells.

The discretized EFIE exhibits symmetry between the basis functions and testing functions (see (6.13)) and it is therefore convenient to define the testing functions to be the same as the divergence-conforming basis functions

$$\boldsymbol{T}_m(s,t) = \boldsymbol{B}_m(s,t) \tag{6.15}$$

The testing functions provide a means for obtaining N linearly independent equations from the EFIE. The equations can be organized into a matrix equation

$$\mathbf{E} = \mathbf{ZI} \tag{6.16}$$

where $\mathbf{E}$ and $\mathbf{I}$ are N by 1 column vectors and $\mathbf{Z}$ is an N by N matrix. The entries of $\mathbf{I}$ are the coefficients I_n, while those of $\mathbf{E}$ are given by

$$E_m = \iint_{\text{surface}} \boldsymbol{T}_m \cdot \boldsymbol{E}^{\text{inc}} \, ds \, dt \tag{6.17}$$

The entries of $\mathbf{Z}$ have the general form

$$\begin{aligned} Z_{mn} &= j\omega\mu \iint_{s,t} \iint_{s',t'} \boldsymbol{T}_m(s,t) \cdot \boldsymbol{B}_n(s',t') \, G \, ds' \, dt' \, ds \, dt \\ &\quad + \frac{1}{j\omega\varepsilon} \iint_{s,t} \nabla \cdot \boldsymbol{T}_m \iint_{s',t'} \nabla' \cdot \boldsymbol{B}_n \, G \, ds' \, dt' \, ds \, dt \end{aligned} \tag{6.18}$$

The matrix equation may be solved to produce the coefficients I_n, after which any other quantity of interest may be obtained by integrating over the current density in (6.14).

If the EFIE is used with a basis function that is not divergence conforming, the divergence operation would produce a Dirac delta function type of behavior at any locations where the function lacks normal-vector continuity. In special cases when it might be necessary to employ non-divergence-conforming basis functions (such as to incorporate lumped-element feeds or loads, junctions between surfaces and wires, etc.), the Dirac delta function must be included as part of the expression. The result is equivalent to including a line integral along with the surface integrals in (6.18).

6.2 Incorporation of Curved Cells

For generality, we consider the use of curved patch representations of the perfectly conducting targets under consideration. As described in Chapters 2 and 3, the procedure involves mapping the cell shape and the various basis functions residing on that cell from a standard reference (or parent) cell to the curved patch (or child cell). For our standard reference cell, we use the unit right triangle occupying the domain $0 \le \xi_1 \le 1$, $0 \le \xi_2 \le 1$, with $\xi_1 + \xi_2 \le 1$ (Figure 6.1). For this domain, the coordinates (ξ_1, ξ_2) happen to coincide with two of the three simplex coordinates (ξ_1, ξ_2, ξ_3) often used to represent quantities on triangles. Each simplex coordinate of a given point is the relative distance from that point to one side of the triangle, with value 0 at the side and value 1 at the opposite corner. The third coordinate can be obtained in this case as

$$\xi_3 = 1 - \xi_1 - \xi_2 \tag{6.19}$$

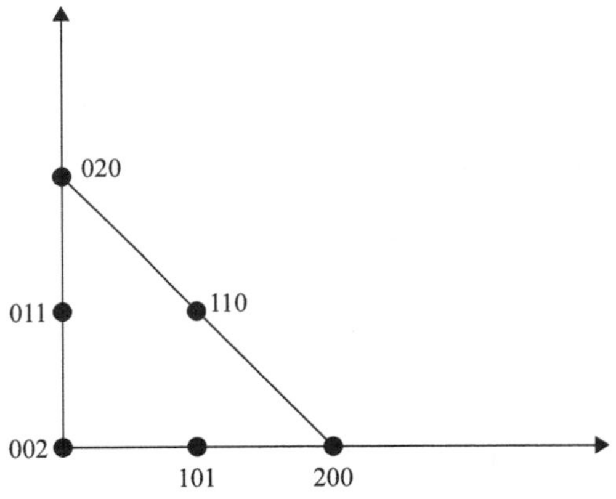

Figure 6.1 Standard reference (parent) triangle, showing the six-node locations associated with a quadratic Lagrangian mapping to a different cell shape.

and is clearly linearly dependent on the other two. As discussed in previous chapters, simplex coordinates offer a symmetrical way of describing a point within the triangle, and are convenient variables for use with triangular domains.

We illustrate the mapping procedure for a curvilinear patch defined in terms of six quadratic interpolation polynomials on a triangular domain, distributed around the reference cell as illustrated in Figure 6.1. The patch coordinates may be expressed as

$$x = \sum_{i=0}^{2} \sum_{j=0}^{2} x_{ijk} S_{ijk}(\xi_1, \xi_2, \xi_3) \tag{6.20}$$

$$y = \sum_{i=0}^{2} \sum_{j=0}^{2} y_{ijk} S_{ijk}(\xi_1, \xi_2, \xi_3) \tag{6.21}$$

$$z = \sum_{i=0}^{2} \sum_{j=0}^{2} z_{ijk} S_{ijk}(\xi_1, \xi_2, \xi_3) \tag{6.22}$$

where the index k is defined by $k = 2 - i - j$ and where the coefficients x_{ijk}, y_{ijk}, and z_{ijk} are the coordinates of points on the desired surface, in this case defining the boundary of the patch. The six interpolation polynomials are defined

$$S_{200} = (2\xi_1 - 1)\xi_1 \tag{6.23}$$

$$S_{020} = (2\xi_2 - 1)\xi_2 \tag{6.24}$$

$$S_{002} = (2\xi_3 - 1)\xi_3 \tag{6.25}$$

$$S_{110} = 4\xi_1\xi_2 \tag{6.26}$$

$$S_{101} = 4\xi_1\xi_3 \tag{6.27}$$

$$S_{011} = 4\xi_2\xi_3 \tag{6.28}$$

where (ξ_1, ξ_2, ξ_3) denote simplex coordinates. Since the dummy variable ξ_3 is dependent on the other two, (6.19) may be used to redefine these functions in terms of ξ_1 and ξ_2.

The explicit representation for S_{ijk} may be used to obtain

$$x = x_{002} + \xi_1(4x_{101} - 3x_{002} - x_{200}) + \xi_2(4x_{011} - 3x_{002} - x_{020}) + \xi_1^2(2x_{200} + 2x_{002} - 4x_{101}) + \xi_1\xi_2(4x_{002} + 4x_{110} - 4x_{101} - 4x_{011}) + \xi_2^2(2x_{020} + 2x_{002} - 4x_{011}) \quad (6.29)$$

where $y(\xi_1, \xi_2)$ and $z(\xi_1, \xi_2)$ have an identical form with x replaced by y or z throughout. Derivatives are obtained as

$$\frac{\partial x}{\partial \xi_1} = 4x_{101} - 3x_{002} - x_{200} + \xi_1(4x_{200} + 4x_{002} - 8x_{101}) + \xi_2(4x_{002} + 4x_{110} - 4x_{101} - 4x_{011}) \quad (6.30)$$

$$\frac{\partial x}{\partial \xi_2} = 4x_{011} - 3x_{002} - x_{020} + \xi_1(4x_{002} + 4x_{110} - 4x_{101} - 4x_{011}) + \xi_2(4x_{020} + 4x_{002} - 8x_{011}) \quad (6.31)$$

Derivatives of y and z are similar, with x replaced by y or z on the right-hand sides of (6.30) and (6.31).

Equations (6.30) and (6.31), and the corresponding equations obtained from $y(\xi_1, \xi_2)$ and $z(\xi_1, \xi_2)$, provide entries of the Jacobian matrix associated with the transformation. Since in general, we are mapping from a two-dimensional reference cell to a curved patch in three-dimensional space, the Jacobian relation has the form

$$\begin{bmatrix} \dfrac{\partial}{\partial \xi_1} \\ \dfrac{\partial}{\partial \xi_2} \end{bmatrix} = \begin{bmatrix} \dfrac{\partial x}{\partial \xi_1} & \dfrac{\partial y}{\partial \xi_1} & \dfrac{\partial z}{\partial \xi_1} \\ \dfrac{\partial x}{\partial \xi_2} & \dfrac{\partial y}{\partial \xi_2} & \dfrac{\partial z}{\partial \xi_2} \end{bmatrix} \begin{bmatrix} \dfrac{\partial}{\partial x} \\ \dfrac{\partial}{\partial y} \\ \dfrac{\partial}{\partial z} \end{bmatrix} \quad (6.32)$$

The parameter equivalent to the determinant of the 2 by 3 Jacobian matrix is given by

$$\mathcal{J}(\xi_1, \xi_2) = \sqrt{\left(\frac{\partial y}{\partial \xi_1}\frac{\partial z}{\partial \xi_2} - \frac{\partial z}{\partial \xi_1}\frac{\partial y}{\partial \xi_2}\right)^2 + \left(\frac{\partial z}{\partial \xi_1}\frac{\partial x}{\partial \xi_2} - \frac{\partial x}{\partial \xi_1}\frac{\partial z}{\partial \xi_2}\right)^2 + \left(\frac{\partial x}{\partial \xi_1}\frac{\partial y}{\partial \xi_2} - \frac{\partial y}{\partial \xi_1}\frac{\partial x}{\partial \xi_2}\right)^2} \quad (6.33)$$

and is a function of position for curved cells. These quantities will be used to scale the cell shape and to define vector basis functions on the curved cell.

Since each cell comprising the surface is defined by an independent mapping, the integral expression in (6.18) must be divided into sub-integrals that are evaluated for each cell. The observer cells where the testing functions reside are generally different from the source cells where the basis functions reside. A single-matrix entry may require integrals over as many as four cells, since each edge-based basis or testing function straddles one or two cells. In the following, an index (m or n) is used to denote specific testing and basis functions, and the discussion considers a single observer cell and a single source cell. The pointer arrays associated with the surface model are used to identify the appropriate cells where these functions reside.

Because the type of cell-to-cell continuity is different for divergence-conforming and curl-conforming basis and testing functions, the type of mapping is also different. For divergence-conforming functions, we define the vector basis functions in the x–y–z child space to have the Cartesian components

$$\begin{bmatrix} B_x \\ B_y \\ B_z \end{bmatrix} = \frac{1}{\mathcal{J}} \begin{bmatrix} \dfrac{\partial x}{\partial \xi_1} & \dfrac{\partial x}{\partial \xi_2} \\ \dfrac{\partial y}{\partial \xi_1} & \dfrac{\partial y}{\partial \xi_2} \\ \dfrac{\partial z}{\partial \xi_1} & \dfrac{\partial z}{\partial \xi_2} \end{bmatrix} \begin{bmatrix} R_{\xi_1}^{\text{div}} \\ R_{\xi_2}^{\text{div}} \end{bmatrix} = \frac{1}{\mathcal{J}} \mathbf{J}^{\text{T}} \begin{bmatrix} R_{\xi_1}^{\text{div}} \\ R_{\xi_2}^{\text{div}} \end{bmatrix} \tag{6.34}$$

where $\mathbf{J}^{\text{T}}$ is used to denote the transpose of the Jacobian matrix in (6.32), and $\mathcal{R}^{\text{div}}$ is the basis function in the reference cell. (The basis functions may be normalized so that desired components have unity value at certain locations on the curved cell, and basis functions that straddle two or more cells must be normalized and given the proper sign to ensure appropriate continuity across cell boundaries. For the present discussion, we omit the normalization constant.) The testing functions are also to be divergence-conforming functions, defined by a mapping of the same form. (Testing functions may also carry a normalization factor to ensure continuity.)

The evaluation of the integrals in (6.18) proceeds with the assistance of matrix relations such as

$$\begin{bmatrix} T_x^{\text{div}} \\ T_y^{\text{div}} \\ T_z^{\text{div}} \end{bmatrix}^{\text{T}} = \frac{1}{\mathcal{J}(s,t)|_{\text{observer}}} \begin{bmatrix} R_{\xi_1}^{\text{div}} & R_{\xi_2}^{\text{div}} \end{bmatrix} \mathbf{J}|_{\text{observer}} \tag{6.35}$$

For instance, the entries of the excitation column vector $\mathbf{E}$ in (6.17) may be obtained as

$$\iint_{\text{surface}} \boldsymbol{T} \cdot \boldsymbol{E}^{\text{inc}}\, \mathrm{d}s\, \mathrm{d}t = \iint_{\substack{\text{reference} \\ \text{cell} \\ \text{(observer)}}} \begin{bmatrix} R_{\xi_1}^{\text{div}} & R_{\xi_2}^{\text{div}} \end{bmatrix}_m \mathbf{J}|_{\text{observer}} \begin{bmatrix} E_x^{\text{inc}} \\ E_y^{\text{inc}} \\ E_z^{\text{inc}} \end{bmatrix} \mathrm{d}\xi_1\, \mathrm{d}\xi_2 \tag{6.36}$$

Note that the scale factor $\mathcal{J}(s,t)$ in (6.35) is canceled by the same factor within the differential surface area

$$\mathrm{d}s\, \mathrm{d}t = \mathcal{J}(s,t)|_{\text{observer}}\, \mathrm{d}\xi_1\, \mathrm{d}\xi_2 \tag{6.37}$$

Standard matrix multiplication is used to collect terms in the integrand of (6.36), reducing the chain of three matrices to a scalar quantity.

A similar approach can be used to express the dot product in the first integral of (6.18) as

$$\boldsymbol{T}_m \cdot \boldsymbol{B}_n = \begin{bmatrix} T_x^{\text{div}} & T_y^{\text{div}} & T_z^{\text{div}} \end{bmatrix}_m \begin{bmatrix} B_x^{\text{div}} \\ B_y^{\text{div}} \\ B_z^{\text{div}} \end{bmatrix}_n$$

$$= \frac{1}{\mathcal{J}(s,t)|_{\text{observer}}} \frac{1}{\mathcal{J}(s',t')|_{\text{source}}} \begin{bmatrix} R_{\xi_1}^{\text{div}} & R_{\xi_2}^{\text{div}} \end{bmatrix}_m \mathbf{J}|_{\text{observer}} \mathbf{J}^{\text{T}}|_{\text{source}} \begin{bmatrix} R_{\xi_1}^{\text{div}} \\ R_{\xi_2}^{\text{div}} \end{bmatrix}_n \quad (6.38)$$

In (6.38), it is explicit that the testing function is located in the observer cell, where the Jacobian matrix is $\mathbf{J}|_{\text{observer}}$, while the basis function resides in the source cell, where the matrix is $\mathbf{J}|_{\text{source}}$ (and the quantities are defined in terms of primed coordinates). The differential surface area for the source cell is

$$\mathrm{d}s'\,\mathrm{d}t' = \mathcal{J}(s',t')|_{\text{source}}\,\mathrm{d}\xi_1'\,\mathrm{d}\xi_2' \quad (6.39)$$

The index of the testing function, m, is also independent of the index of the basis function, n. (Depending on the context, these indices may refer to global numbering system throughout the entire surface or a local numbering system within the individual cell.)

Using (6.38), the first integral in (6.18) may be written in the reference cell coordinates as

$$\iint_{s,t} \iint_{s',t'} \boldsymbol{T}_m(s,t) \cdot \boldsymbol{B}_n(s',t')\, G\, \mathrm{d}s'\,\mathrm{d}t'\,\mathrm{d}s\,\mathrm{d}t$$

$$= \iint_{\substack{\text{reference}\\ \text{cell}\\ \text{(observer)}}} \iint_{\substack{\text{reference}\\ \text{cell}\\ \text{(source)}}} \begin{bmatrix} R_{\xi_1}^{\text{div}} & R_{\xi_2}^{\text{div}} \end{bmatrix}_m \mathbf{J}|_{\text{observer}} \mathbf{J}^{\text{T}}|_{\text{source}} \begin{bmatrix} R_{\xi_1}^{\text{div}} \\ R_{\xi_2}^{\text{div}} \end{bmatrix}_n G\,\mathrm{d}\xi_1'\,\mathrm{d}\xi_2'\,\mathrm{d}\xi_1\,\mathrm{d}\xi_2 \quad (6.40)$$

In the integrand of (6.40), the scale factors $\mathcal{J}|_{\text{observer}}$ and $\mathcal{J}|_{\text{source}}$ in (6.38) cancel with those in (6.37) and (6.39). This integral must be evaluated by numerical quadrature carried out in the reference coordinates. The basis and testing functions, and the mapping functions $x(\xi_1, \xi_2)$, $y(\xi_1, \xi_2)$, and $z(\xi_1, \xi_2)$, are computed at the quadrature points for both the source and observer reference cells. The entries of the two Jacobian matrices in (6.40) are also required at the quadrature points; these are easily obtained using the explicit expressions for the derivatives of x, y, and z in (6.30) and (6.31).

The second integral in (6.18) involves the divergence of the testing function and basis function. These can be obtained using (3.80), which is equivalent to

$$\nabla \cdot \boldsymbol{T}_m = \frac{1}{\mathcal{J}(s,t)|_{\text{observer}}} \left\{ \frac{\partial R_{\xi_1}^{\text{div}}}{\partial \xi_1} + \frac{\partial R_{\xi_2}^{\text{div}}}{\partial \xi_2} \right\}_m \quad (6.41)$$

$$\nabla' \cdot \boldsymbol{B}_n = \frac{1}{\mathcal{J}(s',t')|_{\text{source}}} \left\{ \frac{\partial R_{\xi_1}^{\text{div}}}{\partial \xi_1'} + \frac{\partial R_{\xi_2}^{\text{div}}}{\partial \xi_2'} \right\}_n \quad (6.42)$$

Therefore, the second integral in (6.18) can be written as

$$\iint_{s,t} \nabla \cdot \boldsymbol{T}_m \iint_{s',t'} \nabla' \cdot \boldsymbol{B}_n \, G \, \mathrm{d}s' \, \mathrm{d}t' \, \mathrm{d}s \, \mathrm{d}t$$

$$= \iint_{\substack{\text{reference}\\ \text{cell}\\ \text{(observer)}}} \left\{ \frac{\partial R_{\xi_1}^{\text{div}}}{\partial \xi_1} + \frac{\partial R_{\xi_2}^{\text{div}}}{\partial \xi_2} \right\}_m \iint_{\substack{\text{reference}\\ \text{cell}\\ \text{(source)}}} \left\{ \frac{\partial R_{\xi_1}^{\text{div}}}{\partial \xi_1'} + \frac{\partial R_{\xi_2}^{\text{div}}}{\partial \xi_2'} \right\}_n G \, \mathrm{d}\xi_1' \, \mathrm{d}\xi_2' \, \mathrm{d}\xi_1 \, \mathrm{d}\xi_2 \tag{6.43}$$

As in (6.40), the scale factors $\mathcal{J}|_{\text{observer}}$ and $\mathcal{J}|_{\text{source}}$ are canceled.

If it is desired to interpret the coefficients I_n in (6.14) as values of the surface current density (interpolatory basis functions), each divergence-conforming basis functions must be normalized so that the appropriate component has unity value at the desired location within each cell (in the curvilinear x–y–z space). This is accomplished by scaling each basis function by a constant equal to the magnitude of the appropriate base vector at that location, as explained in Section 3.11.2. For functions that straddle two cells, the normalization constant will also contain a sign that ensures that the basis function points in a consistent normal-vector direction from a global perspective. The normalization is incorporated into the matrix entries of the preceding section by multiplying the basis functions in the reference cell by the appropriate constants. These constants are cell specific, and can be determined at the start of the analysis and stored in an array for convenient reference as required during the matrix construction phase of the MoM procedure, and any post-processing that involves the basis functions.

6.3 Treatment of the Singularity of the Green's Function by Singularity Subtraction and Cancellation Techniques

When the source and observer cells in (6.40) and (6.43) coincide, there will generally be points where the function R within G vanishes. The resulting $1/R$ singularity complicates the evaluation of the integrals by quadrature. There are several possible approaches to evaluating these singular integrals; in this section, we illustrate two different approaches for the simple case of a flat patch, and return to a discussion of the curved patch in the following sections. The approaches are *singularity subtraction* and *singularity cancellation.*

Consider an integral

$$I(u, v) = \iint f(u', v') \frac{\mathrm{e}^{-jkR}}{R} \mathrm{d}u' \, \mathrm{d}v' \tag{6.44}$$

where the bounded function f incorporates the basis and testing functions or their derivatives, as well as any matrix products, etc. The function R is

$$R = \sqrt{(u - u')^2 + (v - v')^2} \tag{6.45}$$

When the observation point (u, v) falls within the domain of integration, the integrand is unbounded at $R = 0$. In addition, an expansion of the Green's function

$$\frac{\mathrm{e}^{-jkR}}{R} = \frac{1}{R} - jk - \frac{k^2 R}{2} + \frac{jk^3 R^2}{6} + \frac{k^4 R^3}{24} - \cdots \tag{6.46}$$

contains terms involving both even and odd powers of R. While the unbounded $1/R$ term is the most problematic, all terms involving odd powers of R are non-analytic at $R=0$ and potentially difficult to integrate by quadrature.

First-order singularity subtraction consists of subtracting a suitably-weighted $1/R$ term from the integrand in (6.44), and adding it back in as a separate integral, to obtain

$$I = \iint \left\{ f(u', v') \frac{e^{-jkR}}{R} - f(u, v) \frac{1}{R} \right\} du'\, dv' + f(u, v) \iint \frac{1}{R} du'\, dv' \qquad (6.47)$$

The term in brackets in (6.47) is bounded, and should be more amenable to evaluation by quadrature than (6.44) as long as the integrand is not sampled directly at the singular point. Observe that the $1/R$ term is weighted with a function $f(u, v)$ that is not a function of the integration variables, but coincides with the rest of the original integrand at the singular point. This permits the function f to be removed from the second integral, making it easier to evaluate that expression in closed form. The idea behind the approach is to evaluate the first integral in (6.47) by quadrature and evaluate the second by analytical means [4–11].

The first integrand in (6.47) is not an analytic function, however, due to the behavior of e^{-jkR} as $R \to 0$. This function has a derivative discontinuity at the singular point. This behavior often results in a substantial loss of accuracy in a typical quadrature evaluation, even if the point in question is located at an end of the domain of integration. To improve upon this approach, we consider an extension of the above where an additional term is subtracted, as suggested by Järvenpää et al. [12].

Second-order singularity subtraction consists of subtracting two terms from the integrand in (6.44), the first and third in (6.46), to obtain

$$\begin{aligned} I(u, v) = &\iint \left\{ f(u', v') \frac{e^{-jkR}}{R} - f(u, v) \left[\frac{1}{R} - \frac{k^2 R}{2} \right] \right\} du'\, dv' \\ &+ f(u, v) \iint \frac{1}{R} du'\, dv' - f(u, v) \frac{k^2}{2} \iint R\, du'\, dv' \end{aligned} \qquad (6.48)$$

The second-order subtraction removes the unbounded term and the leading order derivative discontinuity term in (6.46) from the original integrand. The remaining integrand is not analytic, however, due to higher order terms. (Additional terms may be removed if necessary, of course.)

To illustrate these two singularity subtraction approaches, we will consider a flat triangular domain $0 < x < 0.1, 0 < y < 0.1, x + y < 0.1$, the function $f(u, v) = 1$, and an observer location at $(0, 0.1)$. (This domain has been scaled to put it on the typical size of a cell in an MoM implementation, so we employ local variables $u = 10x$, $v = 10y$. There is no great loss of generality in this layout, since any triangular or quadrilateral cell can be divided into a number of triangles with the observer at one corner of each.) The integral to be computed is therefore given by

$$I = 0.01 \int_{v'=0}^{1} \int_{u'=0}^{1-v'} \frac{e^{-jkR}}{R} du'\, dv' \qquad (6.49)$$

where $k=2\pi$, and

$$R=0.1\sqrt{(0-u')^2+(1-v')^2} \tag{6.50}$$

There is a $1/R$ singularity at the upper corner of the triangular domain.

First-order singularity subtraction proceeds using

$$I=0.01\int_{v'=0}^{1}\int_{u'=0}^{1-v'}\left\{\frac{e^{-jkR}}{R}-\frac{1}{R}\right\}du'\,dv' \;+0.01\int_{v'=0}^{1}\int_{u'=0}^{1-v'}\frac{1}{R}\,du'\,dv' \tag{6.51}$$

where the second integral is evaluated exactly to produce

$$\begin{aligned}\int_{v'=0}^{1}\int_{u'=0}^{1-v'}\frac{1}{\sqrt{(u')^2+(1-v')^2}}\,du'\,dv' &= \int_{v'=0}^{1}\ln\left[u'+\sqrt{(u')^2+(1-v')^2}\right]\Bigg|_{u'=0}^{1-v'}dv'\\ &=\ln\left(1+\sqrt{2}\right)\int_{v'=0}^{1}dv'\\ &=\ln\left(1+\sqrt{2}\right)\end{aligned} \tag{6.52}$$

Thus, we obtain

$$I=0.01\int_{v'=0}^{1}\int_{u'=0}^{1-v'}\left\{\frac{e^{-jkR}}{R}-\frac{1}{R}\right\}du'\,dv' \;+0.1\ln\left(1+\sqrt{2}\right) \tag{6.53}$$

For second-order singularity subtraction, we employ

$$\begin{aligned}I &= 0.01\int_{v'=0}^{1}\int_{u'=0}^{1-v'}\left\{\frac{e^{-jkR}}{R}-\left[\frac{1}{R}-\frac{k^2R}{2}\right]\right\}du'\,dv'\\ &\quad+0.01\int_{v'=0}^{1}\int_{u'=0}^{1-v'}\frac{1}{R}\,du'\,dv' \;-0.01\frac{k^2}{2}\int_{v'=0}^{1}\int_{u'=0}^{1-v'}R\,du'\,dv'\end{aligned} \tag{6.54}$$

Since

$$\int_{v'=0}^{1}\int_{u'=0}^{1-v'} \sqrt{(u')^2+(1-v')^2}\,du'\,dv'$$

$$=\int_{v'=0}^{1}\left\{\frac{u'}{2}\sqrt{(u')^2+(1-v')^2}+\frac{(1-v')^2}{2}\ln\left[u'+\sqrt{(u')^2+(1-v')^2}\right]\right\}\Bigg|_{u'=0}^{1-v'} dv'$$

$$=\int_{v'=0}^{1}\frac{(1-v')^2}{2}\left\{\sqrt{2}+\ln\left(1+\sqrt{2}\right)\right\}dv'$$

$$=\frac{\sqrt{2}+\ln\left(1+\sqrt{2}\right)}{6} \tag{6.55}$$

we obtain

$$I=0.01\int_{v'=0}^{1}\int_{u'=0}^{1-v'}\left\{\frac{e^{-jkR}}{R}-\left[\frac{1}{R}-\frac{k^2R}{2}\right]\right\}du'\,dv'$$
$$+0.1\ln\left(1+\sqrt{2}\right)-0.001\frac{k^2}{2}\left(\frac{\sqrt{2}+\ln\left(1+\sqrt{2}\right)}{6}\right) \tag{6.56}$$

Table 6.1 shows a comparison of the accuracy in the real and imaginary parts of (6.53) and (6.56) when the integrals are computed using triangular-domain Gaussian quadrature rules involving 3, 7, 16, 33, and 61 sample points, taken from [13]. These are open rules that do not sample at the corners of the triangular domain.

Table 6.1 shows that the imaginary (non-singular) parts of the expressions are computed more accurately than the real (singular) parts. It is important to note that for this example, the subtracted parts of the integrand are larger than the remaining integral, and as a consequence no digits are lost to subtraction error when these results are recombined. Second-order singularity subtraction offers a more accurate result for a fixed number of quadrature samples.

As a second example, consider a cell one-half the size of the preceding example, with a similar shape and observer location. Table 6.2 presents the results. These are similar to the results of Table 6.1.

The results in Tables 6.1 and 6.2 suggest that first-order singularity subtraction is only suitable when low accuracy is desired in the integrations. However, second-order singularity subtraction works reasonably well. Better performance can be obtained by subtracting additional terms [12]. The procedure can be generalized to the case of curved cells by subtracting terms defined over the tangent plane at the singularity, and incorporating the appropriate Jacobian to account for the curved-cell mapping.

Next, we consider singularity cancellation techniques, which involve a transformation from the original triangular cell to a cell of different shapes, in such a manner that the Jacobian of the transformation cancels the singularity. We will consider the *Duffy transformation* [14] and the *Khayat–Wilton arcsinh transformation* [15] (see also [16, 17]). To begin, note that for

Table 6.1 Singularity subtraction.

Number of digits of accuracy obtained in the final result for (6.53) and (6.56) when integrals are evaluated using triangular-domain Gaussian quadrature rules, for a triangular domain with corners at (0,0), (0,0.1), (0.1,0) and the observer at (0,0.1). We employ a reference value of $I \cong 0.0807860997690 - j0.0300629968675$.

Samples in quadrature rule	First-order singularity subtraction		Second-order singularity subtraction	
	Re{(6.53)}	Im{(6.53)}	Re{(6.56)}	Im{(6.56)}
1	2	2	2	2
3	3	5	5	5
7	4	8	7	8
16	5	13+	8	13+
33	6	13+	10	13+
61	7	13+	11	13+

Table 6.2 Singularity subtraction for a smaller triangle.

Number of digits of accuracy obtained in the final result for (6.53) and (6.56) when integrals are evaluated using triangular-domain Gaussian quadrature rules for a triangular domain with corners at (0,0), (0,0.05), (0.05,0) and the observer at (0,0.05). A reference value of $I \cong 0.0431309994302 - j0.0077682487542$ is obtained.

Samples in quadrature rule	First-order singularity subtraction		Second-order singularity subtraction	
	Re{(6.53)}	Im{(6.53)}	Re{(6.56)}	Im{(6.56)}
1	2	2	2	2
3	3	5	6	5
7	4	9	8	9
16	5	13+	9	13+
33	6	13+	10	13+
61	7	13+	11	13+

any observer location within the domain of the integral, (6.44) can be divided into integrals over three or more triangular subcells with the singularity at one corner of each subcell. (We say "or more" because in practice it may be more efficient to separate the triangular cell into six subcells, each with a node at the singular point, to better preserve the cell shapes and facilitate accurate quadrature evaluations. Similarly, a quadrilateral cell may be divided into four or eight triangles.) This approach reduces the required integration to one that can always be expressed in the following form

$$I(0,1) = \int_{v'=0}^{1} \int_{u'=0}^{1-v} f(u',v') \frac{e^{-jkR}}{R} du' \, dv' \tag{6.57}$$

where the function f incorporates the basis and testing functions, matrix products such as those in (6.40), etc. The function R (under the assumption of a flat cell) is

$$R = \sqrt{(u')^2 + (1-v')^2} \tag{6.58}$$

The Duffy transformation involves a change of variables from u' to w', according to

$$u' = (1-v')w' \tag{6.59}$$

$$du' = (1-v')dw' \tag{6.60}$$

which yields new limits of integration

$$u' = 0 \rightarrow w' = 0 \tag{6.61}$$

$$u' = 1 - v' \rightarrow w' = 1 \tag{6.62}$$

The result can be expressed as the new integral, now over a square domain

$$I = \int_{v'=0}^{1} \int_{w'=0}^{1} \frac{e^{-jkR}}{R} (1-v') \, dw' \, dv' \tag{6.63}$$

In the vicinity of the original singularity at $u = 0$, $v = 1$, the original factor of $1/R$ in the integrand has been replaced by

$$\begin{aligned} \frac{(1-v')}{R} &= \frac{(1-v')}{\sqrt{(u')^2 + (1-v')^2}} \\ &= \frac{(1-v')}{\sqrt{(1-v')^2(w')^2 + (1-v')^2}} \\ &= \frac{1}{\sqrt{(w')^2 + 1}} \end{aligned} \tag{6.64}$$

which is bounded at that location. (We note that the exponential remaining in the integrand exhibits a derivative discontinuity there). In this specific case of a flat cell, the integral can be written as

$$I = \int_{v'=0}^{1} \int_{w'=0}^{1} \frac{e^{-jkR}}{\sqrt{(w')^2 + 1}} dw' \, dv' \tag{6.65}$$

The arcsinh transformation involves a similar change of variable from u' to w', according to

$$u' = (1 - v')\sinh w' \tag{6.66}$$

$$du' = \sqrt{(u')^2 + (1 - v')^2}\, dw' \tag{6.67}$$

which yields new limits of integration

$$u' = 0 \rightarrow w' = 0 \tag{6.68}$$

$$u' = 1 - v' \rightarrow w' = \sinh^{-1}(1) \cong 0.8814 \tag{6.69}$$

For this flat-cell example, the Jacobian arising from the arcsinh transformation exactly cancels the factor of R in the denominator, leaving a rectangular-domain integral

$$I = \int_{v'=0}^{1} \int_{w'=0}^{\sinh^{-1}(1)} e^{-jkR}\, dw'\, dv' \tag{6.70}$$

In the vicinity of the original singularity at $u = 0$, $v = 1$, the new integrand is bounded. (The exponential still exhibits a derivative discontinuity in the corner where R vanishes.)

Tables 6.3 and 6.4 illustrate the performance of the Duffy and arcsinh transformations for the preceding example of a triangular cell scaled to realistic sizes. Since these transformations convert the original triangular domain into a square or rectangular domain, product Gauss–Kronrod–Patterson quadrature rules were used to evaluate the integrals [18].

For this flat-cell example, the arcsinh transformation appears superior to the Duffy approach because the limiting form of R in the denominator is exactly canceled by the Jacobian of the transformation to the new variables. If other functions (basis functions, Jacobians associated with curved cells, etc.) are included in the integrand, both the Duffy and the arcsinh change of variables may still be implemented in the same way. Thus, the treatment of curved cells by singularity cancellation techniques is not substantially different from the above.

It is interesting to compare the singularity subtraction procedures to the singularity cancellation techniques. By comparing the results in Tables 6.1 and 6.2 to those in Tables 6.3 and 6.4, it appears that the second-order singularity subtraction yields the best accuracy for the fewest quadrature samples. In this case, the fact that we use product rules for the rectangular domains arising from the Duffy and arcsinh transformations tends to push the number of quadrature points up rapidly, so the comparison may be misleading.

The singularity subtraction and cancellation procedures enable the evaluation of singular integrals when the observer and source regions overlap. The matrix entries may also be difficult to evaluate when the observation point is outside the source cell but close to it [19]. Alternative transformations have been proposed for that situation [20, 21]. Because the actual functional dependence of the integrand is somewhat variable, a recommended approach is to employ adaptive quadrature algorithms that can estimate the error in the integrations, seek a prescribed error level, and inform the user when they fail! As a matter of fact, despite its convenience, the *arcsinh transformation* technique does not permit one to anticipate the precision of the numerical results as a function of the number of quadrature points even in the simpler case of static potential integrals. An alternative technique for the machine precision evaluation of singular and nearly singular potential integrals with $1/R$ singularities is available in [19]. The numerical quadrature scheme is based on a rational expression for the integrands, once

Table 6.3 Singularity cancellation.

Number of digits of accuracy obtained in (6.63) and (6.70) when integrals are evaluated using product Gauss–Kronrod–Patterson quadrature rules, for an original triangular domain with corners at (0,0), (0,0.1), (0.1,0) and the observer at (0,0.1). We employ a reference value of $I \cong 0.0807860997690 - y0.0300629968675$.

Samples in quadrature rule	Duffy transformation		Khayat/Wilton arcsinh transformation	
	Re{(6.63)}	Im{(6.63)}	Re{(6.70)}	Im{(6.70)}
1	1	2	1	2
$3 \times 3 = 9$	4	7	6	6
$7 \times 7 = 49$	8	13+	13	13
$15 \times 15 = 225$	13+	13+	13+	13+

Table 6.4 Singularity cancellation for a smaller triangle.

Number of digits of accuracy obtained in (6.63) and (6.70) when integrals are evaluated using product Gauss–Kronrod–Patterson rules, for an original triangular domain with corners at (0,0), (0,0.05), (0.05,0) and the observer at (0,0.05). A reference value of $I \cong 0.0431309994302 - j0.0077682487542$ is obtained.

Samples in quadrature rule	Duffy transformation		Khayat/Wilton arcsinh transformation	
	Re{(6.63)}	Im{(6.63)}	Re{(6.70)}	Im{(6.70)}
1	1	3	2	2
$3 \times 3 = 9$	4	9	7	7
$7 \times 7 = 49$	9	13+	13+	13+
$15 \times 15 = 225$	13+	13+	13+	13+

again obtained by a cancellation procedure. In particular, by using library routines for Gauss quadrature of rational functions available in the literature [22], this rational expression permits the exact numerical integration of singular static potentials associated with polynomial source distributions. It is good to know that the technique in [19] provides rules (number of quadrature points) that permits one to achieve machine precision results for singular and nearly singular static and dynamic potential integrals (similar to what happens when integrating polynomials with the Gauss–Legendre technique) although, at times, the computational time required by this alternative technique does not compare favorably with the computational time required by the *arcsinh transformation*.

6.4 Examples: Scattering Cross Section Calculations

A computer program was constructed that implements the discretization of the EFIE using the divergence-conforming functions of Chapters 3–5 with the quadratic curved-cell shapes described above. Consider a perfectly conducting sphere of radius 0.5λ, where λ is the

Table 6.5 Scattering cross section for a sphere for $p = 0.5$.

Results obtained using $p = 0.5$ basis and testing functions are given in dBλ^2 for a sphere of radius 0.5λ, for flat triangular-cell models and curved quadratic cells mapped from triangular cells. The results are organized by the number of edges in model (the number of unknowns) and by the observer angle θ, for an observer position with $\phi = 0$. After [1].

	Flat cells		Curved cells				
Edges	192	300	48	108	192	300	Exact
$\theta = 0$	9.38	9.50	8.32	9.49	9.59	9.62	9.66
30	6.59	6.69	5.49	6.61	6.75	6.79	6.83
60	4.21	4.19	4.04	4.19	4.16	4.16	4.15
90	−6.45	−6.52	−5.37	−6.67	−6.60	−6.60	−6.58
120	1.65	1.64	1.13	1.88	1.67	1.64	1.63
150	−1.95	−1.77	−2.03	−1.28	−1.33	−1.37	−1.42
180	−2.86	−2.62	0.12	−2.60	−2.35	−2.30	−2.26

free-space wavelength. It is convenient to examine the bistatic scattering cross section (SCS), defined by the following expression

$$\sigma(\theta, \phi) = \lim_{r \to \infty} 4\pi r^2 \frac{|\boldsymbol{E}^s(r, \theta, \phi)|^2}{|\boldsymbol{E}^{inc}|^2_{target}} \tag{6.71}$$

since that parameter is a composite that depends on the currents produced on the entire sphere in response to a plane wave excitation. Initially, consider several flat triangular-patch models of the sphere, as well as similar models where the cells were curved to conform to the spherical surface. These models were constructed by dividing the sphere uniformly along θ, and subdividing in ϕ so that along the equator the triangle sides have the same dimension as they do in θ. In all cases, the surface area of the models is scaled to the same surface area as the desired sphere.

As an initial example to illustrate the benefits of curved-cell modeling, Table 6.5 presents the SCS as a function of θ, for $\phi = 0$, in response to a plane wave propagating in the $\theta = 0$, $\phi = 0$ direction with the electric field polarized in the $\hat{\theta}$ direction. For these results, the lowest order ($p = 0.5$) divergence-conforming basis and test functions were used[1]. The table suggests that the curved-cell results converge to the exact solution as the model is refined. Furthermore, these results indicate that the 300-edge flat-cell model yields approximately the same accuracy as the 108-edge curved-cell model. The 300-edge model exhibits an average density of 96 unknowns/λ^2, roughly in the range where reasonably good solutions are expected from this type of formulation (at least for geometries as simple as a sphere). The 108-edge model only employs 34 unknowns/λ^2, but because of the curved cells are able to produce reasonable accuracy similar with only 36% of the unknowns of the flat-cell model. Results at other angles in ϕ exhibit a comparable accuracy.

[1] As explained in Section 3.2.1, the CN/LT divergence-conforming vector bases are sometimes denoted $p = 0$ (as in Chapters 4 and 5) and sometimes denoted $p = 0.5$. In this chapter, we freely employ both the half-integer ($p = 0.5$, 1.5, 2.5) and full integer ($p = 0$, 1, 2) designations for the vector basis functions in use.

Table 6.6 Scattering cross section for a sphere for $p = 1.5$ and $p = 2.5$.

Results given in $dB\lambda^2$ for a sphere of radius 0.5λ, for curved quadratic cells mapped from triangular cells, obtained from the EFIE

	$p = 1.5$		$p = 2.5$		
Edges	**108**	**192**	**108**	**192**	
Unknowns	**360**	**640**	**756**	**1,344**	**Exact**
$\theta = 0$	9.647	9.658	9.660	9.661	9.6604
30	6.821	6.828	6.827	6.830	6.8296
60	4.149	4.151	4.152	4.152	4.1519
90	−6.568	−6.583	−6.557	−6.579	−6.5846
120	1.617	1.632	1.623	1.633	1.6329
150	−1.412	−1.418	−1.427	−1.419	−1.4198
180	−2.324	−2.276	−2.300	−2.270	−2.2616

Table 6.6 presents additional results for the sphere of radius 0.5λ, for $p = 1.5$ and $p = 2.5$. The results improve in accuracy as the models are refined and as the polynomial order increases. For approximately the same number of unknowns, the results in Table 6.6 are more accurate than those in Table 6.5.

Figure 6.2 shows the two-norm error in the bistatic SCS of the same sphere obtained with basis functions of orders $p = 0.5$, 1.5, and 2.5. They illustrate improved accuracy as the polynomial order increases. (However, since they were obtained with quadratic curved patches, they may not be as accurate as possible with better models of the spherical surface.)

Results for the SCS of a torus of major radius $\lambda/3$ and minor radius $\lambda/6$ are tabulated in Table 6.7. The torus is centered at the origin, parallel to x–y plane, and is illuminated by a uniform plane wave propagating in the z-direction with the electric field polarized along $\hat{\theta}$. The torus surface was divided in an 8 by 16 mesh of curved quadrilateral cells, which were then divided across their diagonals into triangles. The final model used the quadratic mapping procedure described in Section 6.2. These results also show that the model may limit the accuracy of the results, since the SCS produced using bases of order $p = 0.5$ is almost identical to that produced using $p = 1.5$ bases. This suggests that it makes little sense to use basis functions with a uniform high order along with a model of limited accuracy.

In a recent article [23], Zha et al. employ the hierarchical vector basis functions introduced in Chapter 5 on curved triangular patches to model surface currents on complicated PEC targets, including hypothetical missiles and warheads. They employed a fast iterative solver, using a multi-level fast multipole method, and used the combined-field equation instead of the EFIE (for additional details, see [23]).

They initially present results for the relative computational effort required to solve the problem of scattering from a sphere of radius 5λ as the order of the vector bases varies (Table 6.8). These results suggest that the average patch size can grow considerably as the basis order increases, reducing the total number of unknowns without an overall loss of accuracy. However, their specific implementation seems to produce a fastest run time for functions of order $p = 1.5$.

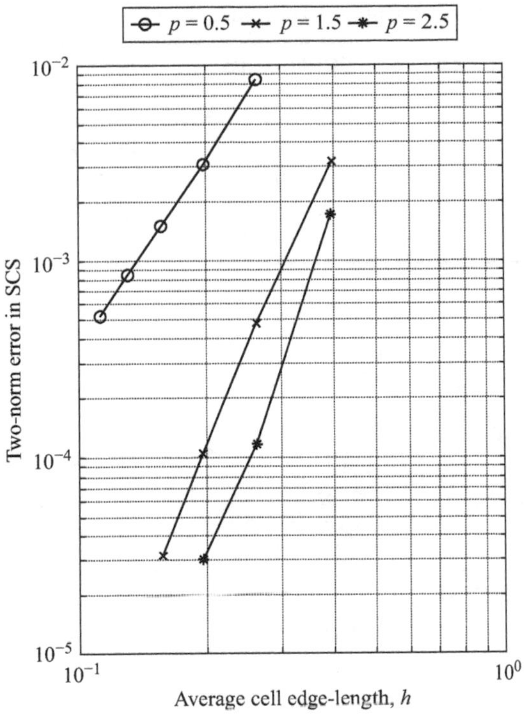

Figure 6.2 Error in the bistatic SCS for a sphere of radius 0.5λ, for basis and testing functions of various orders and quadratic curved cells.

Table 6.7 SCS of a torus.

Bistatic SCS results for a torus of major radius λ/3 and minor radius λ/6, for the excitation described in the text, along a $\phi = 0$ cut

	$p = 0.5$ 384 unknowns	$p = 1.5$ 1,280 unknowns	$p = 2.5$ 2,688 unknowns	Reference data
$\theta = 0$	10.757 dBλ^2	10.766	10.766	10.7641
30	8.087	8.099	8.099	8.0966
60	−0.152	−0.130	−0.129	−0.1310
90	−7.175	−7.214	−7.215	−7.2080
120	0.371	0.322	0.321	0.3185
150	6.004	5.979	5.978	5.9772
180	8.448	8.437	8.437	8.4366

Zha et al. present several examples, one of which is reproduced in Figures 6.3 and 6.4. Figure 6.3 shows a model of a 24λ Tomahawk missile, while Figure 6.4 shows the $\phi\phi$ polarization of the bistatic radar cross section obtained with functions of order $p = 0.5$ and $p = 2.5$. The authors do not report the number of unknowns required for the $p = 0.5$ solution, but used 15,372 for the $p = 2.5$ result. The two results appear to exhibit good agreement.

Table 6.8 Performance of basis functions versus order, for a PEC sphere of radius 5λ. After [23].

Order	Average patch size	Unknowns	RMS error	Fill/solve time
0.5	0.2 (λ)	25,962	0.29 dB	170/153 (s)
1.5	0.6	10,040	0.25	156/57
2.5	0.9	8,526	0.27	221/62
3.5	1.2	8,640	0.29	654/69

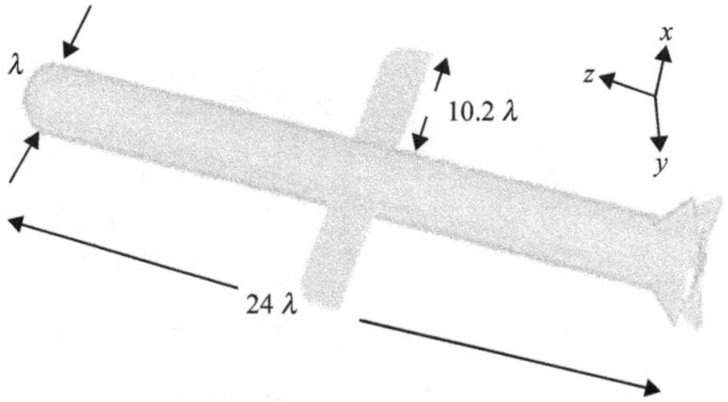

Figure 6.3 The model of a Tomahawk missile.

© 2012 IEEE. Reprinted with permission from L. P. Zha, Y. Q. Hu, and T. Su, "Efficient surface integral equation using hierarchical vector bases for complex EM scattering problems," *IEEE Trans. Antennas Propag.*, vol. 60, pp. 952–957, 2012.

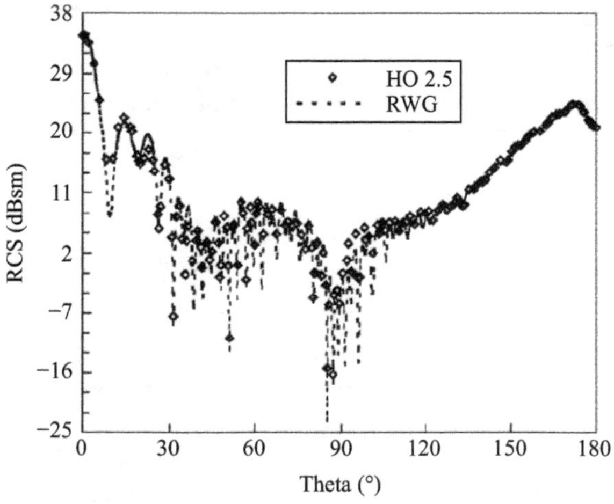

Figure 6.4 The $\phi\phi$ polarization bistatic radar cross section obtained with basis functions of order $p = 0.5$ (RWG) and $p = 2.5$, for the missile of Figure 3.

© 2012 IEEE. Reprinted with permission from L. P. Zha, Y. Q. Hu, and T. Su, "Efficient surface integral equation using hierarchical vector bases for complex EM scattering problems," *IEEE Trans. Antennas Propag.*, vol. 60, pp. 952–957, 2012.

In a recent article [24], Ludick et al. present results for these same basis functions, suggesting a considerable reduction in matrix solve times when using hierarchical bases of order up to $p = 3.5$ compared to a low-order solution. This is as expected since the number of unknowns should be greatly reduced with the higher order bases. Interestingly, the authors also report reduced matrix fill times when higher order functions are employed [24].

6.5 The Vector Helmholtz Equation

Consider a three-dimensional cavity formed by perfectly conducting walls, containing homogeneous or inhomogeneous dielectric and magnetic materials. The electromagnetic fields $\boldsymbol{E}$ or $\boldsymbol{H}$ that can exist within the cavity may be found from a solution of the vector Helmholtz equations

$$\nabla \times \left(\frac{1}{\mu_\mathrm{r}} \nabla \times \boldsymbol{E} \right) - k^2 \epsilon_\mathrm{r} \boldsymbol{E} = 0 \tag{6.72}$$

$$\nabla \times \left(\frac{1}{\epsilon_\mathrm{r}} \nabla \times \boldsymbol{H} \right) - k^2 \mu_\mathrm{r} \boldsymbol{H} = 0 \tag{6.73}$$

where $\epsilon_\mathrm{r}(x, y, z)$ and $\mu_\mathrm{r}(x, y, z)$ denote the relative permittivity and relative permeability functions, respectively. It is assumed that the time-varying fields are represented by phasors with suppressed time dependence $\mathrm{e}^{j\omega t}$. The wavenumber k is directly related to the resonant frequency by $k = \omega\sqrt{\mu_0 \epsilon_0}$, where ω is the radian frequency, and ϵ_0 and μ_0 are the permittivity and permeability of free space, respectively. The cavity interior will be denoted by Γ. Either equation, coupled with appropriate boundary conditions applied to the tangential fields on the boundary $\partial\Gamma$, is a sufficient eigenvalue equation for the resonant wavenumbers and cavity modes.

The "strong form" of the equations in (6.72) and (6.73) may be converted into "weak form" equations as follows, using (6.72) for illustration. By multiplying (6.72) with a transverse testing function $\boldsymbol{T}$ and employing the vector identities

$$\boldsymbol{T} \cdot \nabla \times \boldsymbol{E} = \nabla \times \boldsymbol{T} \cdot \boldsymbol{E} - \nabla \cdot (\boldsymbol{T} \times \boldsymbol{E}) \tag{6.74}$$

$$(\boldsymbol{T} \times \boldsymbol{E}) \cdot \hat{n} = -\boldsymbol{T} \cdot \left(\hat{n} \times \boldsymbol{E} \right) \tag{6.75}$$

and the divergence theorem

$$\iiint_\Gamma \nabla \cdot (\boldsymbol{T} \times \boldsymbol{E})\, \mathrm{d}v = \iint_{\partial\Gamma} (\boldsymbol{T} \times \boldsymbol{E}) \cdot \hat{n}\, \mathrm{d}S \tag{6.76}$$

the vector Helmholtz equation for $\boldsymbol{E}$ can be recast as the weak equation

$$\iiint_\Gamma \frac{1}{\mu_\mathrm{r}} \nabla \times \boldsymbol{T} \cdot \nabla \times \boldsymbol{E}\, \mathrm{d}v = k^2 \iiint_\Gamma \varepsilon_\mathrm{r} \boldsymbol{T} \cdot \boldsymbol{E}\, \mathrm{d}v - \iint_{\partial\Gamma} \frac{1}{\mu_\mathrm{r}} \boldsymbol{T} \cdot \hat{n} \times (\nabla \times \boldsymbol{E})\, \mathrm{d}S \tag{6.77}$$

where $\partial\Gamma$ denotes the cavity boundary. This is known as a "weak" equation since the differentiability requirements of the basis functions have been reduced by one degree.

Along perfect electric walls, the electric field must satisfy the essential boundary condition

$$\hat{n} \times \boldsymbol{E}\big|_{\partial\Gamma} = 0 \tag{6.78}$$

which eliminates tangential values of $\boldsymbol{E}$ on that boundary as unknowns. Since the tangential $\boldsymbol{E}$ field on the boundary is not an unknown, there is no need for testing functions with a non-zero tangential component along that part of $\partial\Gamma$ either. As a consequence, the boundary integral on the right-hand side of (6.77) does not contribute to the system of equations for a perfect electric boundary.

Along perfect magnetic walls, the natural boundary condition

$$\hat{n} \times (\nabla \times \boldsymbol{E})\big|_{\partial\Gamma} = 0 \tag{6.79}$$

may be imposed. Equation (6.79) does not eliminate the tangential electric fields as unknowns, but upon substitution into the boundary integral in (6.77) causes that integral to vanish for the magnetic part of the boundary. Thus, for perfect electric or perfect magnetic boundaries, the integral over $\partial\Gamma$ in (6.77) is omitted.

After the incorporation of boundary conditions on $\partial\Gamma$, the weak equation constitutes an eigenvalue equation for the resonant wavenumbers k (equivalently, the resonant frequencies ω) and the eigenfunctions $\boldsymbol{E}$ associated with cavity modes.

Similarly, the vector Helmholtz equation for $\boldsymbol{H}$ can be recast as the weak equation

$$\iiint_{\Gamma} \frac{1}{\varepsilon_r} \nabla \times \boldsymbol{T} \cdot \nabla \times \boldsymbol{H}\, dv = k^2 \iiint_{\Gamma} \mu_r \boldsymbol{T} \cdot \boldsymbol{H}\, dv - \iint_{\partial\Gamma} \frac{1}{\varepsilon_r} \boldsymbol{T} \cdot \hat{n} \times (\nabla \times \boldsymbol{H})\, dS \tag{6.80}$$

where $\partial\Gamma$ denotes the cavity boundary. Along perfect electric walls, the magnetic field must satisfy the natural boundary condition

$$\hat{n} \times (\nabla \times \boldsymbol{H})\big|_{\partial\Gamma} = 0 \tag{6.81}$$

while, on perfect magnetic walls, the field must satisfy the essential condition

$$\hat{n} \times \boldsymbol{H}\big|_{\partial\Gamma} = 0 \tag{6.82}$$

These conditions also eliminate the boundary integral on the right-hand side of (6.80) for perfect electric and perfect magnetic parts of the boundary. After boundary conditions are incorporated, (6.80) constitutes an eigenvalue equation for the resonant wavenumbers k and the eigenfunctions $\boldsymbol{H}$.

The preceding equations apply in the three-dimensional situation, but are easily modified for the two-dimensional case by reducing the dimension of the integrals by one.

6.6 Numerical Solution of the Vector Helmholtz Equation for Cavities

A general structure may be modeled by cells that include a combination of tetrahedra, hexahedra, or prisms. Consider a cavity bounded by perfect electric walls whose interior region is modeled with tetrahedral cells, each with constant ϵ_r and μ_r. The electric field within the cavity may be approximated by the following expansion

$$\boldsymbol{E}(\boldsymbol{r}) \cong \sum_{n=1}^{N} e_n \boldsymbol{B}_n(\boldsymbol{r}) \tag{6.83}$$

where $\boldsymbol{B}_n$ denotes a curl-conforming vector expansion function such as described in Chapters 3–5, and the global index n runs over all expansion functions, regardless of whether they reside on edges, faces, or cells of the mesh. A curl-conforming representation is used since the vector Helmholtz operator involves the curl of the field being expanded. The curl-conforming basis functions impose tangential-vector continuity and produce a bounded curl. The representation in (6.83) is substituted into the weak vector equation in (6.77). Since the weak equation also involves the curl of the testing function, it is convenient to use testing functions equal to the basis functions to form the system of equations, thus

$$\boldsymbol{T}_m(\boldsymbol{r}) = \boldsymbol{B}_m(\boldsymbol{r}) \tag{6.84}$$

The boundary condition $\hat{n} \times \boldsymbol{E} = 0$ on the PEC wall is enforced by omitting any basis function with a non-zero tangential value on the boundary from the system of equations. The procedure for constructing the system of equations may utilize a list of cell edges and faces that reside on the boundary of the mesh to implement boundary conditions. There may also be a master permutation list that assigns global unknown numbers to the various basis functions residing on edges, faces, and cells of the mesh, which may omit or otherwise flag edges and faces on perfect electric walls. In any event, since there are no basis functions remaining with non-zero tangential components along the boundary $\partial\Gamma$, there will be no need for testing functions with tangential components along $\partial\Gamma$ and the boundary integral in (6.77) will not contribute to the system of equations.

For a closed cavity, the discretization process yields a generalized matrix eigenvalue equation $\mathbf{Ae} = k^2\mathbf{Be}$ with entries

$$A_{mn} = \iiint_\Gamma \frac{1}{\mu_r} \nabla \times \boldsymbol{B}_m \cdot \nabla \times \boldsymbol{B}_n \, dv \tag{6.85}$$

and

$$B_{mn} = \iiint_\Gamma \varepsilon_r \boldsymbol{B}_m \cdot \boldsymbol{B}_n \, dv \tag{6.86}$$

In common with standard finite element implementations, the system of equations is usually constructed in a cell-by-cell manner where the integrals in (6.85) and (6.86) are evaluated throughout a single cell for all combinations of basis and testing functions, stored in a temporary "element matrix," and systematically transferred to the global system of equations.

Since the vector basis functions may reside on edges, faces, or cells, the process of building the system of equations requires "connectivity" pointer arrays that link cells to corresponding nodes, edges, and faces. The connectivity arrays are used in conjunction with the master permutation list to identify the appropriate locations of the element matrix entries within the global system. Some means must be imposed to ensure that basis functions that overlap two or more cells (edge based or face based) have a common orientation on either side of a given face. For example, vector basis functions that are primarily tangential to an edge may be defined to always point from the smaller node index to the larger node index along that edge, according to the global node numbering. A similar approach must be applied to face-based functions, which straddle two cells. The details of the element matrix calculations differ depending on the specific basis and testing functions in use. For rectilinear cells, it is often possible to evaluate the integrals in closed form or to construct universal tables that provide an efficient means of evaluation [25], with examples given in the literature (see [2, 26–28], etc.) If curvilinear cells are employed, the integrals may be evaluated by quadrature.

Table 6.9 shows numerical results for the resonant wavenumbers of an air-filled cavity of dimension 1.0 by 0.5 by 0.75 m, for a range of tetrahedral-cell models and LT/QN ($p = 1.5$) functions (after [26]). As the tetrahedral cell model is refined, the results converge toward the exact solutions, although not always in a monotonic fashion. Figure 6.5 shows the average

Table 6.9 Numerical results for resonant wavenumbers obtained with LT/QN basis functions.

Unknowns Edge length, h	204 0.44493	518 0.31161	668 0.29486	1058 0.25917	1430 0.23571	1882 0.21848	Exact wavenumber
TE101	5.26421	5.23886	5.23524	5.23498	5.23593	5.23671	5.23599
TE110	7.06509	7.03298	7.02931	7.03118	7.02600	7.02800	7.02481
TE011	7.56545	7.56938	7.55086	7.54626	7.55329	7.55216	7.55145
TE201	7.69411	7.57587	7.55620	7.55683	7.55896	7.55491	7.55145
TM111	8.22497	8.20124	8.18831	8.19219	8.18845	8.17913	8.17887
TE111	8.30736	8.21067	8.19727	8.19700	8.19189	8.18344	8.17887
TM210	8.81126	8.92389	8.89925	8.89649	8.89899	8.89752	8.88577
TE102	8.90346	8.97387	8.95721	8.93565	8.95150	8.95565	8.94726

© 1996 IEEE. Reprinted with permission from J. S. Savage and A. F. Peterson, "Higher-order vector finite elements for tetrahedral cells," *IEEE Trans. Microwave Theory Tech.*, vol. 44, pp. 874–879, Jun. 1996.

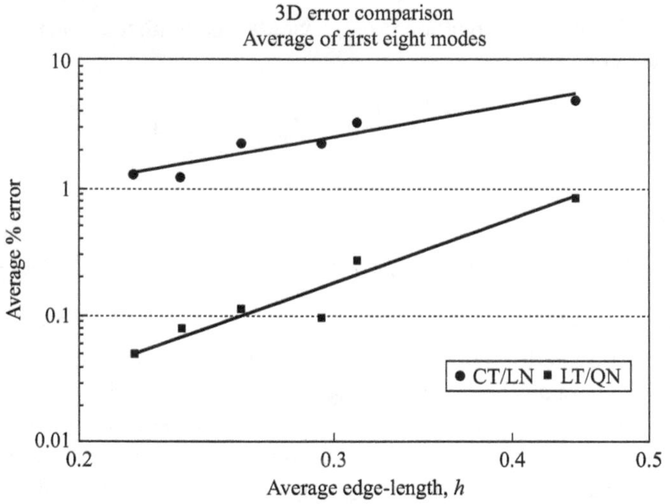

Figure 6.5 Average error in the first eight resonant wavenumbers of a rectangular cavity obtained using CT/LN and LT/QN basis functions.

© 1996 IEEE. Reprinted with permission from J. S. Savage and A. F. Peterson, "Higher-order vector finite elements for tetrahedral cells," *IEEE Trans. Microwave Theory Tech.*, vol. 44, pp. 874–879, Jun. 1996.

convergence rate of the lowest eight resonant frequencies of this same rectangular cavity for CT/LN and LT/QN basis and testing functions. This figure plots the convergence versus the average edge-length h of the meshes. The expected theoretical rates, as h tends to zero, are $O(h^2)$ and $O(h^4)$. The solid-lines in Figure 6.5 show a straight-line fit through each set of numerical data, which closely approximate the expected slopes. One readily observes that the accuracy is far better with the LT/QN functions. Although the specific form of the basis functions in [26] differs from those developed in Chapters 4 and 5, they span the same spaces as those functions and should produce identical numerical results, regardless of whether they are of interpolatory or hierarchical form.

Table 6.10 Lowest 11 non-zero resonant wavenumbers for an equilateral prism of side dimension 1.0 and height 1.0, obtained with bases of order 0.5 (top); 1.5 (middle), and 2.5 (bottom).

1 by 1 by 1	2 by 2 by 2	3 by 3 by 3	4 by 4 by 4	5 by 5 by 5	6 by 6 by 6	
9 unknowns	39 unknowns	102 unknowns	210 unknowns	375 unknowns	609 unknowns	Exact
		5.53198 (2)	5.40132 (2)	5.34137 (2)	5.30899 (2)	$\frac{5}{3}\pi \approx 5.23599$ (2)
		7.14416	7.22605	7.24414	7.25007	$\frac{4}{\sqrt{3}}\pi \approx 7.25520$
		8.59086 (2)	8.17237 (2)	7.95110 (2)	7.82869 (2)	$\frac{2\sqrt{13}}{3}\pi \approx 7.55145$ (2)
		7.86378	7.91217	7.91681	7.91585	$\sqrt{\frac{19}{3}}\pi \approx 7.90617$ (TM mode)
		9.09945	8.62477	8.36954	8.22782	$\sqrt{\frac{19}{3}}\pi \approx 7.90617$ (TE mode)
		9.62959 (2)	10.09035 (2)	9.67658 (2)	9.45031 (2)	$\frac{\sqrt{73}}{3}\pi \approx 8.94726$ (2)
		10.24886	10.01078	9.67658	9.78588	$\frac{2\sqrt{21}}{3}\pi \approx 9.59772$ (TM mode)
		11.22497	10.58300	10.23421	10.03992	$\frac{2\sqrt{21}}{3}\pi \approx 9.59772$ (TE mode)
6.0	*11.9*	*14.0*	*15.7*	*16.7*	*17.4*	$\Longleftarrow$ *Condition Number*

1 by 1 by 1	2 by 2 by 2	3 by 3 by 3	4 by 4 by 4	5 by 5 by 5	
36 unknowns	190 unknowns	546 unknowns	1,188 unknowns	2,200 unknowns	Exact
	5.26219 (2)	5.24158 (2)	5.23781 (2)	5.23674 (2)	5.23599 (2)
	7.65114	7.27849	7.26343	7.25873	7.25520
	8.10844 (2)	7.61131 (2)	7.57201 (2)	7.56021 (2)	7.55145 (2)
	8.27549	7.92853	7.91405	7.90954	7.90617 (TM mode)
	8.55640	7.97767	7.93104	7.91684	7.90617 (TE mode)
	9.48387 (2)	9.07678 (2)	8.99232 (2)	8.96668 (2)	8.94726 (2)
	10.32182	9.66021	9.61940	9.60698	9.59772 (TM mode)
	10.54837	9.70058	9.63339	9.61299	9.59772 (TE mode)
734.1	*935.8*	*974.8*	*988.7*	*995.2*	$\Longleftarrow$ *Cond. #*

1 by 1 by 1	2 by 2 by 2	3 by 3 by 3	
90 unknowns	525 unknowns	1,575 unknowns	Exact
5.25493 (2)	5.23654 (2)	5.23604 (2)	5.23599 (2)
7.57992	7.24715	7.25617	7.25520
8.81699 (2)	7.55320 (2)	7.55335 (2)	7.55145 (2)
8.20550	7.89887	7.90707	7.90617 (TM mode)
8.85909	7.90811	7.90843	7.90617 (TE mode)
10.35465 (2)	8.99043 (2)	8.95243 (2)	8.94726 (2)
10.83767	9.59279	9.59993	9.59772 (TM mode)
11.34056	9.60039	9.60105	9.59772 (TE mode)
4,048	*4,891*	*5,201*	$\Longleftarrow$ *Cond. #*

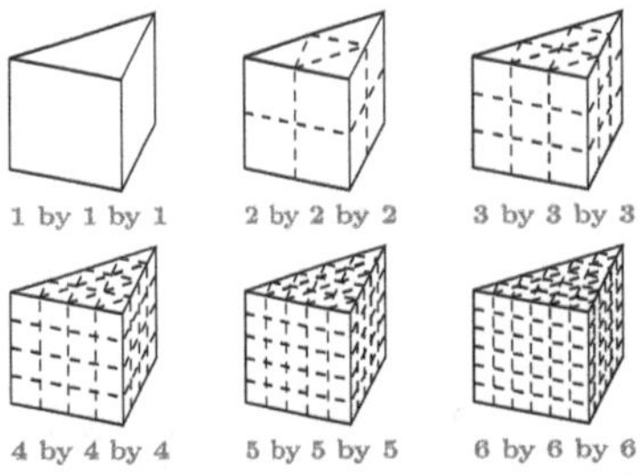

Lowest wavenumbers for an equilateral prism with a PEC bounding surface numerically obtained by solving the magnetic-field (H) vector Helmholtz equation with *natural* (Neumann) boundary conditions. The table also reports the condition numbers of the global **T**-matrices. Results for the 1 by 1 by 1 and the 2 by 2 by 2 equilateral prisms for $p = 0.5$, and results for the 1 by 1 by 1 equilateral prism for $p = 1.5$, contain an insufficient number of DoFs to easily distinguish the corresponding modes and are not reported. *Uniform* meshes are used, as illustrated by the figures in the lower right-hand corner of the table.

To illustrate a different family of vector basis functions, consider a cavity in the shape of a triangular prism and the use of hierarchical vector basis functions defined on prism cells. These basis functions have been discussed in Section 5.3.3 of Chapter 5. (The identical numerical results for resonant wavenumbers are obtained using the interpolatory bases of the same order from Section 4.7.3 of Chapter 4.) The construction of the matrix eigenvalue equation is entirely analogous to the approach of the preceding section.

Table 6.10 shows a range of numerical results obtained for the lowest resonant wavenumbers of a prism-shaped cavity with an equilateral triangle base of side dimension equal to the cavity

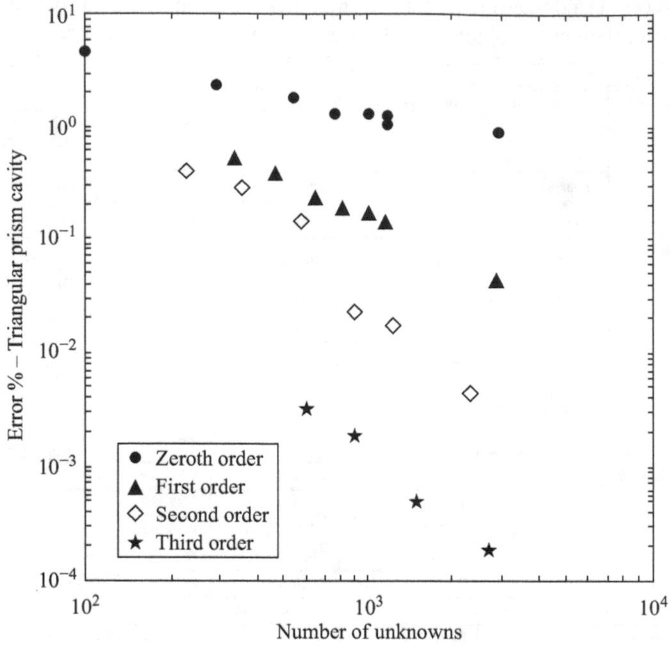

Figure 6.6 Average error in the lowest nine resonant frequencies for a cavity with an equilateral base and a height equal to the base dimension, for basis functions of orders $p = 0.5, 1.5, 2.5,$ and 3.5.

© 1998 IEEE. Reprinted with permission from R. D. Graglia, D. R. Wilton, A. F. Peterson, and I. L. Gheorma, "Higher-order interpolatory vector bases on prism elements," *IEEE Trans. Antennas Propag.*, vol. 46, pp. 442–450, Mar. 1998.

height [29]. These results are compared with the exact analytical solutions for the resonant wavenumbers [30]. It should be observed that the results obtained for the higher order bases ($p = 2.5$) are more accurate with fewer unknowns than those of the lower order bases. Since one consideration arising with hierarchical basis functions is the matrix condition number associated with their use, Table 6.10 also shows the matrix condition numbers for the B matrix in (6.86) for each result. It is observed that the condition number grows slowly as the meshes are refined.

In practice, it is expected that the prism bases will often be used in conjunction with compatible representations on tetrahedral and hexahedral cells and therefore they may occupy only a small portion of the computational domain. Furthermore, the hierarchical functions facilitate the use of p-refinement procedures that produce a mixture of polynomial orders and not a uniform Nédélec order throughout the mesh.

Figure 6.6 shows the average error in the first nine resonant frequencies for this same cavity, after [31]. These error curves are plotted versus the total number of unknowns, for vector expansion functions of orders $p = 0, 1, 2,$ and 3. The higher order functions produce lower error levels and steeper convergence slopes. Results in Figure 6.6 were obtained using the interpolatory form of the basis functions, but the identical results should be obtained with hierarchical bases of the same order on prism cells.

6.7 Avoiding Spurious Modes with Adaptive p-Refinement and Hierarchical Bases

The principal motivation for hierarchical bases is their use in adaptive p-refinement algorithms, because they facilitate mixing the polynomial order of the representation from place to place within the computational domain to enhance the efficiency of solution. The presence of sources, geometrical features, material density changes, etc., may cause more rapid field behavior in certain regions. Those regions may benefit from a higher polynomial order of representation than other regions where the field behavior is moderate. However, in practice, there are constraints that must be followed when mixing polynomial orders.

The vector Helmholtz equations are known to produce *spurious modes* when an inappropriate basis set is employed. For the cavity problem discussed in Section 6.6, spurious modes appear in the form of non-zero eigenvalues that do not correspond to physical solutions. These are actually nullspace eigenfunction/eigenvalue pairs that are distorted due to inappropriate DoFs or inappropriate continuity conditions in the expansion functions [2,28]. When the vector Helmholtz equation is solved using the curl-conforming bases described in preceding chapters, for a uniform polynomial order, spurious modes do not occur. When mixing polynomial orders from one region to another, however, spurious modes can be created if unbalanced DoFs are introduced into a cell.

Recall that the vector bases of previous chapters can be classified as either cell-based functions, face-based functions, or edge-based functions. Cell-based functions are local to a single cell, while face-based and edge-based functions straddle multiple cells. All these three types introduce various DoFs into the cells that they span. *To avoid spurious modes, the vector expansion within each cell must be complete to a given Nédélec order* (see Section 3.12 for a discussion of the Nédélec spaces). This implies that the cell-based functions in a given cell must be complete to the highest order of any of the edge-based or face-based functions that overlap that cell [32].

As an example, suppose we have a tetrahedral-cell mesh, and a basis of uniform order $p = 1$. Suppose further (as a hypothetical example) that we want to increase the representation order in one interior cell of that mesh to $p = 2$. To accomplish this without exciting spurious modes, we must increase the order of the cell-based functions assigned to that cell to $p = 2$, without changing the order of any of the face-based and edge-based functions surrounding that cell. This means that we augment the cell-based functions of order $p = 1$ with all the additional cell-based functions associated with order $p = 2$. We do not introduce any face-based or edge-based functions of order $p = 2$. In this manner, we preserve the order of the surrounding cells at $p = 1$.

If, on the other hand, we had also raised the order of the face-based basis functions surrounding the interior cell to $p = 2$, those functions "hang over" into some of the surrounding cells, and change (actually, corrupt) the representation in those cells. The resulting unbalanced representations in those cells are likely to cause spurious solutions. An example of this in two dimensions is illustrated in [32].

In practice, it is likely that there will be an entire region of the mesh where the order is raised to $p = 2$, as opposed to a single cell. In that case, we also raise the order of all the face-based and edge-based bases that reside entirely in that region to order $p = 2$. Analogous ideas apply to higher order regions.

As a general observation, it would be permissible to allow face-based or edge-based functions to overlap other cells if they were expressed entirely as gradients, since in that case they have an identically zero curl and do not cause spurious modes. The basis functions of Chapters 5 are not written in that form, but some of the alternate vector basis functions in the literature (such as Webb's hierarchical functions [33] and Zaglmayr's hierarchical functions [34]) include bases written explicitly as gradients.

A detailed discussion of adaptive refinement implementations is beyond the scope of the present text. To summarize, schemes that incorporate the hierarchical functions of Chapter 5 to provide a representation of varying order can avoid spurious modes by assigning a specific order to each cell of the mesh and transitioning the edge-based and face-based functions according to the above prescription, where the order of those basis functions is constrained to be equal to the smallest order of any of the cells they straddle.

6.8 Use of Curved Cells with Curl-Conforming Bases

Section 6.2 discussed the use of curved triangular patch surface representations; here we generalize that discussion to curved tetrahedral cells in three dimensions, and extend the discussion to the case of curl-conforming vector bases defined on those cells. The procedure involves mapping the cell shape and the various basis functions residing on that cell from a standard reference (or parent) cell to the curved patch (or child cell). For our standard reference or parent cell, we use the unit right tetrahedron occupying the domain $0 \le \xi_1 \le 1, 0 \le \xi_2 \le 1, 0 \le \xi_3 \le 1$, with $\xi_1 + \xi_2 + \xi_3 \le 1$. For this domain, the coordinates (ξ_1, ξ_2, ξ_3) are three of the four simplex coordinates on tetrahedrons. Each simplex coordinate of a given point is the relative distance from that point to one face of the tetrahedron, with value 0 at the face and value 1 at the opposite corner.

The coordinate mapping may be carried out in a manner almost identical to that described in Section 6.2, but in terms of three-dimensional scalar interpolation polynomials defined on tetrahedrons instead of on the triangular domain used in Section 6.2. The resulting Jacobian matrix has the form

$$\begin{bmatrix} \dfrac{\partial}{\partial \xi_1} \\ \dfrac{\partial}{\partial \xi_2} \\ \dfrac{\partial}{\partial \xi_3} \end{bmatrix} = \begin{bmatrix} \dfrac{\partial x}{\partial \xi_1} & \dfrac{\partial y}{\partial \xi_1} & \dfrac{\partial z}{\partial \xi_1} \\ \dfrac{\partial x}{\partial \xi_2} & \dfrac{\partial y}{\partial \xi_2} & \dfrac{\partial z}{\partial \xi_2} \\ \dfrac{\partial x}{\partial \xi_3} & \dfrac{\partial y}{\partial \xi_3} & \dfrac{\partial z}{\partial \xi_3} \end{bmatrix} \begin{bmatrix} \dfrac{\partial}{\partial x} \\ \dfrac{\partial}{\partial y} \\ \dfrac{\partial}{\partial z} \end{bmatrix} = \mathbf{J} \begin{bmatrix} \dfrac{\partial}{\partial x} \\ \dfrac{\partial}{\partial y} \\ \dfrac{\partial}{\partial z} \end{bmatrix} \tag{6.87}$$

where the various entries are computed from the explicit form of x, y, and z provided by the scalar mapping. This is a different mapping for each cell in the model.

Curl-conforming vector bases may be defined on the curved cells so that their Cartesian components satisfy

$$\begin{bmatrix} B_x \\ B_y \\ B_z \end{bmatrix} = \begin{bmatrix} \dfrac{\partial x}{\partial \xi_1} & \dfrac{\partial y}{\partial \xi_1} & \dfrac{\partial z}{\partial \xi_1} \\ \dfrac{\partial x}{\partial \xi_2} & \dfrac{\partial y}{\partial \xi_2} & \dfrac{\partial z}{\partial \xi_2} \\ \dfrac{\partial x}{\partial \xi_3} & \dfrac{\partial y}{\partial \xi_3} & \dfrac{\partial z}{\partial \xi_3} \end{bmatrix}^{-1} \begin{bmatrix} R_{\xi_1}^{\text{curl}} \\ R_{\xi_2}^{\text{curl}} \\ R_{\xi_3}^{\text{curl}} \end{bmatrix} = \mathbf{J}^{-1} \begin{bmatrix} R_{\xi_1}^{\text{curl}} \\ R_{\xi_2}^{\text{curl}} \\ R_{\xi_3}^{\text{curl}} \end{bmatrix} \tag{6.88}$$

This mapping maintains the tangential continuity of a basis function in the x–y–z (child) space, provided that the reference function exhibits tangential continuity along the corresponding boundary in the ξ_1–ξ_2–ξ_3 (parent) space. This mapping also preserves the tangential-vector interpolation properties of a basis function on corresponding cell boundaries. The inverse Jacobian matrix is normally computed by an explicit matrix inversion of the matrix in (6.87).

The integrand in (6.86) contains the dot product $\boldsymbol{B}_m \cdot \boldsymbol{B}_n$, which can be recast as the matrix operation

$$\boldsymbol{B}_m \cdot \boldsymbol{B}_n = \begin{bmatrix} B_x & B_y & B_z \end{bmatrix}_m \begin{bmatrix} B_x \\ B_y \\ B_z \end{bmatrix}_n$$

$$= \begin{bmatrix} R_{\xi_1}^{\text{curl}} & R_{\xi_2}^{\text{curl}} & R_{\xi_3}^{\text{curl}} \end{bmatrix}_m \mathbf{J}^{-\mathbf{T}} \mathbf{J}^{-1} \begin{bmatrix} R_{\xi_1}^{\text{curl}} \\ R_{\xi_2}^{\text{curl}} \\ R_{\xi_3}^{\text{curl}} \end{bmatrix}_n \tag{6.89}$$

The two Jacobian matrices in (6.89) are the same since the testing and basis functions are being evaluated in the same cell, at the same points. The resulting integral is performed over the reference (parent) cell in the form

$$B_{mn} = \iiint_{\substack{\text{parent}\\ \text{cell}}} \varepsilon_r \begin{bmatrix} R_u^{\text{curl}} & R_v^{\text{curl}} & R_w^{\text{curl}} \end{bmatrix}_m \mathbf{J}^{-\mathbf{T}} \mathbf{J}^{-1} \begin{bmatrix} R_{\xi_1}^{\text{curl}} \\ R_{\xi_2}^{\text{curl}} \\ R_{\xi_3}^{\text{curl}} \end{bmatrix}_n \mathcal{J} \, d\xi_1 \, d\xi_2 \, d\xi_3 \tag{6.90}$$

where $\mathcal{J}$ denotes the determinant of the Jacobian matrix.

The integrand in (6.85) requires $\nabla \times \boldsymbol{B}_n$, which can be evaluated in the reference coordinates using

$$\nabla \times \mathcal{R}_n^{\text{curl}} = \left\{ \hat{\xi}_1 \left(\frac{\partial R_{\xi_2}^{\text{curl}}}{\partial \xi_3} - \frac{\partial R_{\xi_3}^{\text{curl}}}{\partial \xi_2} \right) + \hat{\xi}_2 \left(\frac{\partial R_{\xi_3}^{\text{curl}}}{\partial \xi_1} - \frac{\partial R_{\xi_1}^{\text{curl}}}{\partial \xi_3} \right) + \hat{\xi}_3 \left(\frac{\partial R_{\xi_1}^{\text{curl}}}{\partial \xi_2} - \frac{\partial R_{\xi_2}^{\text{curl}}}{\partial \xi_1} \right) \right\}_n \tag{6.91}$$

The Cartesian components of $\nabla \times \boldsymbol{B}_n$ in the curved child cell are

$$\begin{bmatrix} \hat{\boldsymbol{x}} \cdot \nabla \times \boldsymbol{B}_n \\ \hat{\boldsymbol{y}} \cdot \nabla \times \boldsymbol{B}_n \\ \hat{\boldsymbol{z}} \cdot \nabla \times \boldsymbol{B}_n \end{bmatrix} = \frac{1}{\mathcal{J}} \mathbf{J}^{\mathbf{T}} \begin{bmatrix} \hat{\xi}_1 \cdot \nabla \times \mathcal{R}_n^{\text{curl}} \\ \hat{\xi}_2 \cdot \nabla \times \mathcal{R}_n^{\text{curl}} \\ \hat{\xi}_3 \cdot \nabla \times \mathcal{R}_n^{\text{curl}} \end{bmatrix} \tag{6.92}$$

The integral (6.85) can be expressed entirely in terms of the parent coordinates as

$$A_{mn} = \iiint_{\substack{\text{parent}\\ \text{cell}}} \frac{1}{\mu_r} \frac{1}{\mathcal{J}} \begin{bmatrix} \hat{\xi}_1 \cdot \nabla \times \mathcal{R}_m^{\text{curl}} \\ \hat{\xi}_2 \cdot \nabla \times \mathcal{R}_m^{\text{curl}} \\ \hat{\xi}_3 \cdot \nabla \times \mathcal{R}_m^{\text{curl}} \end{bmatrix}^{\mathbf{T}} \mathbf{J}\mathbf{J}^{\mathbf{T}} \begin{bmatrix} \hat{\xi}_1 \cdot \nabla \times \mathcal{R}_n^{\text{curl}} \\ \hat{\xi}_2 \cdot \nabla \times \mathcal{R}_n^{\text{curl}} \\ \hat{\xi}_3 \cdot \nabla \times \mathcal{R}_n^{\text{curl}} \end{bmatrix} d\xi_1 \, d\xi_2 \, d\xi_3 \tag{6.93}$$

Equations (6.93) and (6.90) generalize (6.85) and (6.86) for the element matrix entries of curved cells.

6.9 Application: Scattering from Deep Cavities

An interesting application of the finite element method is the problem of scattering from deep cavities. This problem has been investigated by Jin and his colleagues [35–37]. The formulation involved truncating the computational domain across the cavity exterior with an integral equation radiation boundary condition. The integral equation for the aperture can be expressed as

$$\hat{n} \times \boldsymbol{H}^{\text{tot}} = 2\hat{n} \times \boldsymbol{H}^{\text{inc}} + 2\hat{n} \times \frac{\nabla\nabla \cdot \boldsymbol{F} + k^2 \boldsymbol{F}}{j\omega\mu} \tag{6.94}$$

where $\hat{n}$ is the outward normal vector, the electric vector potential is defined by the two-dimensional convolution

$$\boldsymbol{F} = \boldsymbol{M} * G \tag{6.95}$$

and the equivalent magnetic current density on the cavity aperture is

$$\boldsymbol{M} = \boldsymbol{E} \times \hat{n} \tag{6.96}$$

Factors of 2 in (6.94) arise due to the nature of the aperture problem and the use of the method of images to reduce the problem to equivalent sources radiating in free space. In (6.94), the incident magnetic field is that of the original source in free space.

The weak vector Helmholtz equation in (6.77) can be modified by the substitution

$$\hat{n} \times (\nabla \times \boldsymbol{E}) = -j2\omega\mu\, \hat{n} \times \boldsymbol{H}^{\text{inc}} - 2\hat{n} \times \left(\nabla\nabla \cdot \boldsymbol{F} + k^2 \boldsymbol{F}\right) \tag{6.97}$$

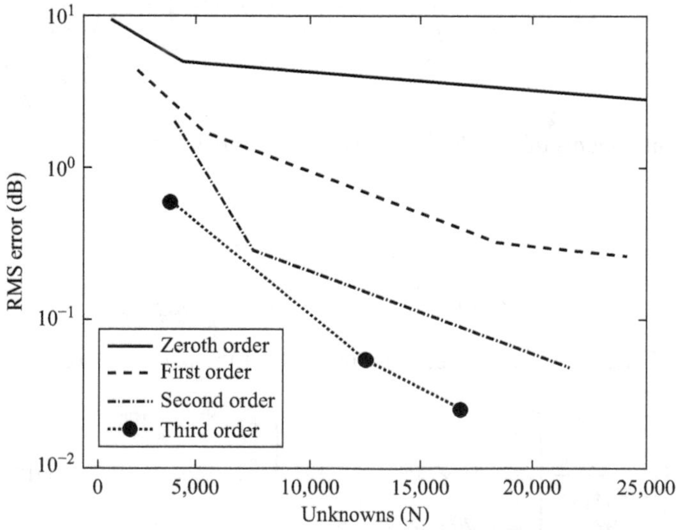

Figure 6.7 RMS error in the RCS of a rectangular cavity of dimensions 1λ by 1λ by 4λ, modeled with tetrahedral cells and the interpolatory vector bases of Chapters 3 and 4.

© 2000 IEEE. Reprinted with permission from J. Liu and J. M. Jin, "A special higher order finite element method for scattering by deep cavities," *IEEE Trans. Antennas Propag.*, vol. 48, pp. 694–703, 2000.

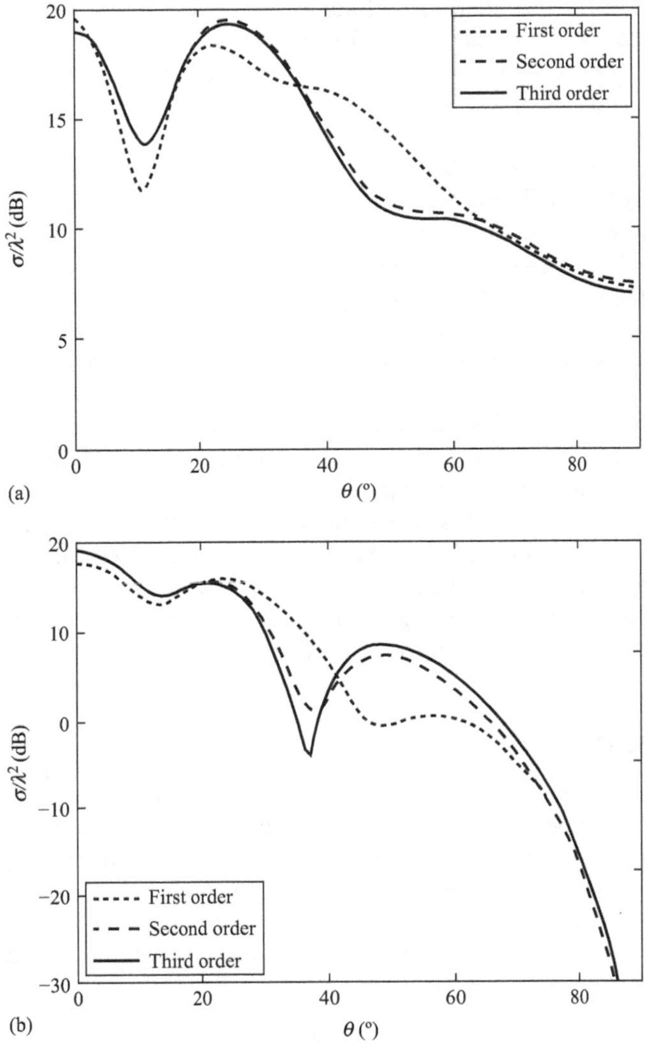

Figure 6.8 The monostatic RCS of a cavity of depth 10λ and circular cross section (diameter 2λ), obtained with tetrahedral cells and interpolatory vector basis of orders $p = 1$, $p = 2$, and $p = 3$. (a) $\theta\theta$ polarization. (b) $\phi\phi$ polarization.

© 2000 IEEE. Reprinted with permission from J. Liu and J. M. Jin, "A special higher order finite element method for scattering by deep cavities," *IEEE Trans. Antennas Propag.*, vol. 48, pp. 694–703, 2000.

into the boundary integral, producing the new equation

$$\begin{aligned}
&\iiint_{\Gamma} \left\{ \frac{1}{\mu_r} \nabla \times \boldsymbol{T} \cdot \nabla \times \boldsymbol{E} - k^2 \varepsilon_r \boldsymbol{T} \cdot \boldsymbol{E} \right\} \mathrm{d}v \\
&\quad -2k^2 \iint_{\partial\Gamma} \frac{1}{\mu_r} \boldsymbol{T} \times \hat{n} \cdot (\boldsymbol{M} * G)\, \mathrm{d}S + 2 \iint_{\partial\Gamma} \frac{1}{\mu_r} \nabla \cdot (\boldsymbol{T} \times \hat{n})\, (\nabla \cdot \boldsymbol{M} * G)\, \mathrm{d}S \\
&= j2\omega\mu \int_{\partial\Gamma} \frac{1}{\mu_r} \boldsymbol{T} \cdot \hat{n} \times \boldsymbol{H}^{\mathrm{inc}}\, \mathrm{d}S
\end{aligned} \tag{6.98}$$

This equation can be discretized using curl-conforming vector bases for the electric field. The same functions can be used for the testing function $\boldsymbol{T}$.

Initially, Jin and his colleagues studied this formulation with $p = 0$ vector basis functions of the type described in Chapter 3, for prism and tetrahedral cell shapes. (Prisms offer the possibility of representing uniform cavities with a simple, multilayered mesh constructed by extruding a single layer of cells, and thus are advantageous.) They reported that those functions worked adequately for small three-dimensional cavities, but not for deeper cavities. Low-order functions produce excessive dispersion error when used to represent fields over electrically large domains. Subsequently, Jin and his colleagues used basis functions of order $p = 3$ on tetrahedral cells (as well as a customized vector basis on prisms). Below, we reproduce some of the results obtained with tetrahedral cells.

As an example, Figure 6.7 shows the root-mean-square (RMS) error in the monostatic radar cross section (RCS) of a rectangular cavity, compared to a reference solution obtained from a dense mesh and $p = 3$ bases. The higher order basis functions clearly produce better accuracy for a given number of unknowns than the lower order functions. In addition, the results obtained using the higher order basis functions required less overall CPU time for a given accuracy [36].

As another example, Figure 6.8 shows the monostatic RCS of a cavity of circular cross section (diameter 2λ) and depth of 10λ. The $p = 1$ solution involved 14,218 unknowns, while the $p = 2$ result required 42,555 unknowns, and the $p = 3$ result involved 94,844 unknowns. The numerical results appear to converge as the basis order increases, showing that the higher order bases overcame the problem of excessive dispersion error.

6.10 Summary

This chapter has briefly considered some applications of the higher order vector bases to electromagnetic field problems cast in terms of integral and differential equations. We considered integral formulations for scattering problems and differential formulations for cavity resonator problems, although both types of formulation are applicable to either class of problem. Particular emphasis was placed on the details associated with the use of curved-cell models with either approach. We also briefly describe an approach for combining different polynomial orders within a mesh without exciting spurious modes.

References

[1] A. F. Peterson, *Mapped Vector Basis Functions for Electromagnetic Integral Equations*. San Rafael, CA: Morgan/Claypool, 2006.

[2] A. F. Peterson, S. L. Ray, and R. Mittra, *Computational Methods for Electromagnetics*. New York, NY: IEEE Press, 1998.

[3] R. F. Harrington, *Field Computation by Moment Methods*. New York, NY: IEEE Press, 1993.

[4] D. R. Wilton, S. M. Rao, A. W. Glisson, D. H. Schaubert, O. M. Al-Bundak, and C. M. Butler, "Potential integrals for uniform and linear source distributions on polygonal and polyhedral domains," *IEEE Trans. Antennas Propag.*, vol. 32, pp. 276–281, Mar. 1984.

[5] R. D. Graglia, "Static and dynamic potential integrals for linearly varying source distributions in two- and three-dimensional problems," *IEEE Trans. Antennas Propag.*, vol. AP-35, pp. 662–669, June 1987.

[6] R. D. Graglia, "The use of parametric elements in the moment method solution of static and dynamic volume integral equations," *IEEE Trans. Antennas Propag.*, vol. AP-36, pp. 636–646, May 1988.

[7] R. D. Graglia, P.L.E. Uslenghi, and R.S. Zich: "Moment method with isoparametric elements for three-dimensional anisotropic scatterers" (invited paper), "Proceedings of the IEEE," special issue on "*Radar Cross Sections of Complex Objects*," vol. 77, no. 5, pp. 750–760, May 1989. Also available in the *IEEE* book, "*Radar Cross Sections of Complex Objects*," pp. 206–216, 1990.

[8] L. Knockaert, "A general Gauss theorem for evaluating singular integrals over polyhedral domains," *Electromagnetics,* vol. 11, pp. 269–280, 1991.

[9] R. D. Graglia, "On the numerical integration of the linear shape functions times the 3-D Green's function or its gradient on a plane triangle," *IEEE Trans. Antennas Propag.*, vol. 41, pp. 1448–1456, Oct. 1993.

[10] S. Caorsi, D. Moreno, and F. Sidoti, "Theoretical and numerical treatment of surface integrals involving the free-space Green's function," *IEEE Trans. Antennas Propag.*, vol. 41, no. 9, pp. 1296–1301, Sep. 1993.

[11] L. Rossi and P. J. Cullen, "On the fully numerical evaluation of the linear-shape function times the 3-D Green's function on a plane triangle," *IEEE Trans. Microwave Theory Tech.*, vol. 47, pp. 398–402, Apr. 1999.

[12] S. Järvenpää, M. Taskinen, and P. Ylä-Oijala, "Singularity extraction technique for integral equation methods with higher order basis functions on plane triangles and tetrahedra," *Int. J. Numer. Methods Eng.*, vol. 58, pp. 1149–1165, 2003.

[13] D. A. Dunavant, "High degree efficient symmetrical Gaussian quadrature rules for the triangle," *Int. J. Numer. Methods Eng.*, vol. 21, pp. 1129–1148, 1985.

[14] M. G. Duffy, "Quadrature over a pyramid or cube of integrands with a singularity at a vertex," *SIAM J. Numer. Anal.*, vol. 19, pp. 1260–1262, 1982.

[15] M. A. Khayat and D. R. Wilton, "Numerical evaluation of singular and near-singular potential integrals," *IEEE Trans. Antennas Propag.*, vol. 53, pp. 3180–3190, Oct. 2005.

[16] P. R. Johnston, and D. Elliott, "A sinh transformation for evaluating nearly singular boundary element integrals," *Int. J. Numer. Methods Eng.*, vol. 62, pp. 564–578, 2005.

[17] B. M. Johnston, P. R. Johnston, and D. Elliott, "A sinh transformation for evaluating two-dimensional nearly singular boundary element integrals," *Int. J. Numer. Methods Eng.*, vol. 69, pp. 1460–1479, 2007.

[18] T. N. L. Patterson, "Generation of interpolatory quadrature rules of the highest degree of precision with preassigned nodes for general weight functions," *ACM Trans. Math. Softw.*, vol. 15, pp. 137–143, Jun. 1989.

[19] R. D. Graglia and G. Lombardi, "Machine precision evaluation of singular and nearly singular potential integrals by use of Gauss quadrature formulas for rational functions," *IEEE Trans. Antennas Propag.*, vol. AP-56, pp. 981–998, Apr. 2008.

[20] M. M. Botha, "A family of augmented Duffy transformations for near-singularity cancellation quadrature," *IEEE Trans. Antennas Propag.*, vol. 61, pp. 3123–3134, Jun. 2013.

[21] D. R. Wilton, F. Vipiana, and W. A. Johnson, "Evaluating singular, near-singular, and non-singular integrals on curvilinear elements," *Electromagnetics*, vol. 34, pp. 307–327, 2014.

[22] W. Gautschi, "Algorithm 793: GQRAT-Gauss Quadrature for Rational Functions," *ACM Trans. Math. Softw.,* vol. 25, no. 2, pp. 213–239, 1999.

[23] L. P. Zha, Y. Q. Hu, and T. Su, "Efficient surface integral equation using hierarchical vector bases for complex EM scattering problems," *IEEE Trans. Antennas Propag.*, vol. 60, pp. 952–957, 2012.

[24] D. J. Ludick, J. Van Tonder, and U. Jakobus, "Combining domain decomposition solution techniques with higher order hierarchical basis functions," *Digest of the International Conference on Electromagnetics in Advanced Applications (ICEAA)*, Torino, IT, pp. 70–73, Sep. 2013.

[25] P. P. Silvester and R. L. Ferrari, *Finite Elements for Electrical Engineers*. Cambridge: Cambridge University Press, 1996.

[26] J. S. Savage and A. F. Peterson, "Higher-order vector finite elements for tetrahedral cells," *IEEE Trans. Microwave Theory Tech.*, vol. 44, pp. 874–879, Jun. 1996.

[27] J. L. Volakis, A. Chatterjee, and L. C. Kempel, *Finite Element Method for Electromagnetics*. New York, NY: IEEE Press, 1998.

[28] J.M. Jin, *The Finite Element Method in Electromagnetics*. New York, NY: Wiley, 2014.

[29] R. D. Graglia and A. F. Peterson, "Hierarchical curl-conforming Nedelec elements for triangular-prism cells," *IEEE Trans. Antennas Propag.*, vol. 60, pp. 3314–3324, Jul. 2012.

[30] F. E. Borgnis and C. H. Papas, "Electromagnetic waveguides and resonators, Section C: Cylindrical waveguides," in *Encyclopedia of Physics: Electric Fields and Waves*, S. Flügge, ed. Berlin, Germany: Springer-Verlag, 1958, vol. XVI, pp. 336–345 [subsection 16 (equilateral triangular waveguide) and 17 (other cylindrical waveguides of simple cross-section)].

[31] R. D. Graglia, D. R. Wilton, A. F. Peterson, and I. L. Gheorma, "Higher-order interpolatory vector bases on prism elements," *IEEE Trans. Antennas Propag.*, vol. 46, pp. 442–450, Mar. 1998.

[32] A. F. Peterson, and R. D. Graglia, "Evaluation of hierarchical vector basis functions for quadrilateral cells," *IEEE Trans. Magn.*, vol. 47, no. 5, pp. 1190–1193, May 2011.

[33] J. P. Webb, "Hierarchal vector basis functions of arbitrary order for triangular and tetrahedral finite elements," *IEEE Trans. Antennas Propag.*, vol. 47, no. 8, pp. 1244–1253, Aug. 1999.

[34] S. Zaglmayr, *High order finite element methods for electromagnetic field computation*, Ph.D. Thesis, Johannes Kepler Universität, Linz, Austria, July 2006.

[35] J.-M. Jin, "Electromagnetic scattering from large, deep, and arbitrarily shaped open cavities," *Electromagnetics*, vol. 18, pp. 3–34, 1998.

[36] J. Liu and J.M. Jin, "A special higher order finite element method for scattering by deep cavities," *IEEE Trans. Antennas Propag.*, vol. 48, pp. 694–703, 2000.

[37] J.-M. Jin, J. Liu, Z. Lou, and C. S. T. Liang, "A fully high-order finite element simulation of scattering by deep cavities," *IEEE Trans. Antennas Propag.*, vol. 51, pp. 2420–2429, 2003.

CHAPTER 7

An Introduction to High-Order Bases for Singular Fields

For problems with smooth surfaces or other regular features, the high-order basis functions described in the previous chapters successfully improve accuracy and efficiency. However, for geometries with edges or corners where certain field components and the surface charge and current densities can be singular and sometimes infinite, high-degree polynomial expansion functions usually do not improve the solution accuracy. (In this context, the term "singular" is used to denote non-analytic behavior, even if the resulting function is bounded.) Although the singular behavior is localized at the edge or corner, it can affect the solution accuracy throughout the entire computational domain. Mitigating this error drives up the cost of computation, since one common remedy is to employ a fine mesh in the neighborhood of the singular region. Another approach is to utilize an adaptive h-refinement scheme, where cells near the singularity are reduced in size in a systematic attempt to obtain better accuracy. An alternate approach is to develop special singular basis functions that properly represent the non-analytic behavior. These functions may be extended to high order and should be suitable for adaptive p-refinement algorithms, or for use in a combination of h- and p-refinements [1–3].

There is a large body of literature on the numerical treatment of field singularities. Meixner and others characterized the specific behavior of electromagnetic fields near edges [4–7]. In 1947, Motz incorporated edge singularities into finite difference algorithms for Laplace's equation [8]. There were significant efforts in the 1970s to develop singular FEM models for various engineering applications [9–11]. Beginning at about the same time, electromagnetic field singularities were modeled within various MoM and FEM techniques [12–21], with many more approaches in recent years. The specific scalar and vector hierarchical basis functions proposed below are based on the recent publications [22–24]. The specific scalar and vector hierarchical basis functions proposed below are based on the recent publications [22–24], and much of the material in this chapter has been drawn from these references with permission from the IEEE and T&F.

In this chapter, we focus primarily on the corner singularity arising in two-dimensional domains such as waveguides. We introduce the type of corner singularity under consideration, review some of the singular basis functions that have been proposed, and describe hierarchical scalar and vector bases capable of modeling the non-analytic fields. The waveguide application is often treated using numerical solutions of the scalar or vector Helmholtz equation, and thus the vector bases under consideration are of the curl-conforming variety. We close this chapter with a brief look at the performance of divergence-conforming bases for treating knife-edge singularities arising in the analysis of thin conducting plates with the electric field integral equation. There are many other types of singularities that arise in electromagnetics, and the state-of-the-art is far from mature at the present time, so our survey is far from comprehensive.

7.1 Field Singularities at Edges

Consider a conducting or penetrable wedge of interior angle α with its vertex located at $\rho = 0$ in a cylindrical system (ρ, ϕ, z), and the z-axis parallel to the edge. For a conducting wedge, the field components associated with the TM-to-z polarization exhibit a leading-order behavior of the form [7]

$$E_z \sim A\rho^{\nu} f(\phi) \tag{7.1}$$

$$\boldsymbol{H}_t \sim \frac{A\rho^{\nu-1}}{-j\omega\mu} \left\{ \hat{\rho} f'(\phi) - \hat{\phi}\nu f(\phi) \right\} \tag{7.2}$$

where $\hat{\rho}$ and $\hat{\phi}$ denote unit vectors in the cylindrical system, a prime denotes the derivative with respect to the argument, and ν is

$$\nu = \pi/(2\pi - \alpha) \tag{7.3}$$

For the TE-to-z polarization, field components behave as

$$H_z \sim B\rho^{\nu} g(\phi) + C \tag{7.4}$$

$$\bar{E}_t \sim \frac{B\rho^{\nu-1}}{j\omega\epsilon} \left\{ \hat{\rho} g'(\phi) - \hat{\phi}\nu g(\phi) \right\} \tag{7.5}$$

where the most singular term also has an exponent given by (7.3). For a conducting wedge, the entire series of ν exponents is

$$\nu_{nm} = n\pi/(2\pi - \alpha) + 2m, \quad m = 0, 1, 2, \ldots \tag{7.6}$$

where $n = 1, 2, 3, \ldots$ for both the TM-to-z and TE-to-z polarizations.

The surface current density $\boldsymbol{J}_s$ on the wedge surface in the vicinity of the edge exhibits a leading-order behavior [7]

$$J_z \sim A\rho^{\nu-1} f'(\phi_0) \tag{7.7}$$

for the TM-to-z case and a behavior of

$$J_\rho \sim B\rho^{\nu} g(\phi_0) \tag{7.8}$$

for the TE-to-z polarization, where A and B are appropriate coefficients.

Figure 7.1 shows the behavior of the lower singularity coefficients ν_{nm} of a perfect electric conducting (PEC) wedge versus the wedge aperture angle α.

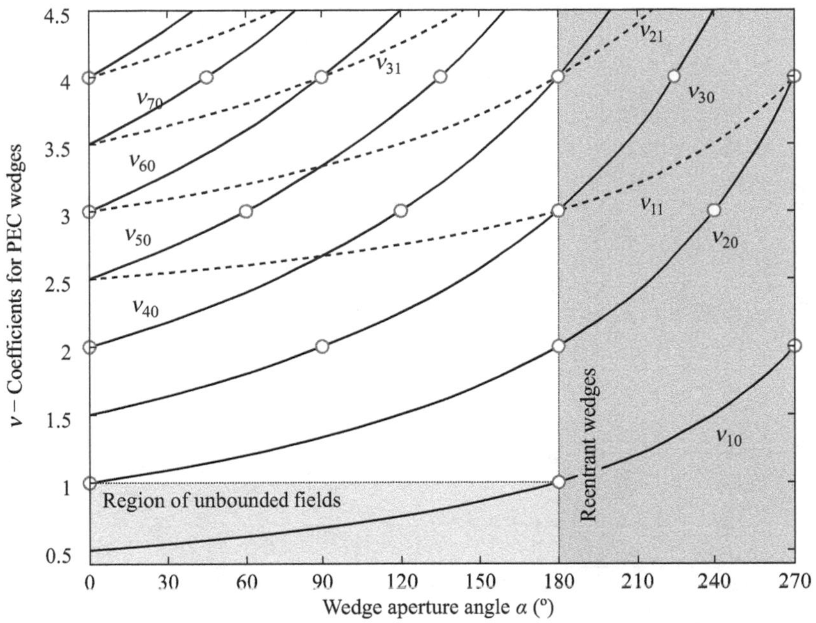

Figure 7.1 Singularity coefficients ν for a PEC wedge of aperture angle α, in the range $\{0° \leq \alpha \leq 270°, \nu \leq 4.5\}$. The solid-lines show the coefficients ν_{n0}; the dashed-lines show the coefficients ν_{nm} (with n and m positive integers). Integer values of the singularity coefficients are marked by circles; all the coefficients become integer at $\alpha = 180°$ (flat plate) and at $\alpha = 270°$. For even values of n, the coefficients ν_{nm} are integers at $\alpha = 0°$.

For a non-magnetic penetrable wedge, there are no singularities for the TM-to-z case. However, for a magnetic wedge, the fields behave as in (7.1) and (7.2), but with ν determined by a solution of [7][1]

$$\mu_r \tan\left(\frac{\nu\alpha}{2}\right) = -\tan\left(\nu\frac{2\pi - \alpha}{2}\right) \tag{7.9}$$

when $\alpha < \pi$, or from a solution

$$\frac{1}{\mu_r}\tan\left(\frac{\nu\alpha}{2}\right) = -\tan\left(\nu\frac{2\pi - \alpha}{2}\right) \tag{7.10}$$

when $\alpha > \pi$.

[1]In (7.9) and (7.10), μ_r is the relative permeability of the wedge measured with respect to the permeability of the medium in which the wedge is immersed.

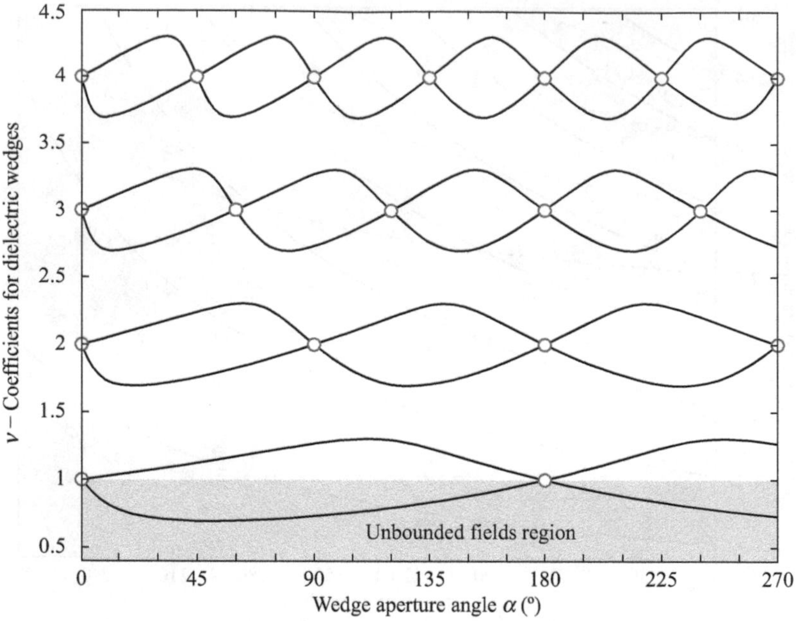

Figure 7.2 Singularity coefficients ν for a dielectric (or magnetic) wedge of aperture angle α and relative permittivity (or permeability μ_r) $\epsilon_r = 10$, in the range $\{0° \leq \alpha \leq 270°, \nu \leq 4.5\}$. Integer values of the singularity coefficients are marked by circles; all the coefficients become integer at $\alpha = 0°$, $180°$, and $360°$. In the limit for $\epsilon_r, \mu_r = 1$ the *eyes* are closed and the coefficients integer for all α values. Notice that the circles in Figures 7.1 and 7.2 are in the same locations.

For a dielectric wedge, the TE-to-z fields behave as in (7.4) and (7.5), but with an exponent determined from[2]

$$\epsilon_r \tan\left(\frac{\nu\alpha}{2}\right) = -\tan\left(\nu\frac{2\pi - \alpha}{2}\right) \tag{7.11}$$

when $\alpha < \pi$, or from a solution of

$$\frac{1}{\epsilon_r}\tan\left(\frac{\nu\alpha}{2}\right) = -\tan\left(\nu\frac{2\pi - \alpha}{2}\right) \tag{7.12}$$

when $\alpha > \pi$. For a magnetic wedge, there are no singularities in the TE-to-z case.

The singularity coefficients strongly depend on the wedge material properties. Figure 7.2 shows the behavior of the lower singularity coefficients for a dielectric wedge of relative dielectric permittivity $\epsilon_r = 10$.

[2]In (7.11) and (7.12), ϵ_r is the relative permittivity of the wedge measured with respect to the permittivity of the medium in which the wedge is immersed.

Notice also that singular terms of the form described above may or may not be excited by a particular source, and therefore may or may not be present in a particular solution even if a wedge angle is present [25].

7.2 Triangular-Polar Coordinate Transformation

In this chapter, we consider triangular cells, arranged in proximity to a sharp vertex as depicted in Figure 7.3. When working with triangular cells, it is usually convenient to express quantities of interest in terms of simplex coordinates (ξ_1, ξ_2, ξ_3), as used in the previous chapters. To aid in the development of singular basis functions on triangular cells, where the singularity is located at the $(\xi_i = 1, \xi_{i\pm1} = 0)$ corner (see Figure 7.3), we also adopt the use of triangular-polar coordinates (χ, σ), defined according to Figure 7.4 [10]. Each triangle of a meshed domain is obtained by mapping the same *parent* triangle $T^2 \equiv \{0 \leq \xi_1, \xi_2, \xi_3 \leq 1;\ \xi_1 + \xi_2 + \xi_3 = 1\}$ into the *x–y–z child space*. For singular triangles, it is more convenient to use pseudo-polar coordinates (χ, σ), requiring us to map the parent triangle $T^2(\boldsymbol{\xi})$ into a new *parent* triangle $T^2(\chi, \sigma)$, and then map the latter onto each *child* singular triangle of the meshed domain. The singular corner is located at $\chi = 0$ $(\xi_i = 1)$. The simplex and triangular-polar coordinates are related by

$$\xi_i = 1 - \chi \tag{7.13}$$

$$\xi_{i+1} = \chi \left(\frac{1-\sigma}{2} \right) \tag{7.14}$$

$$\xi_{i-1} = \chi \left(\frac{1+\sigma}{2} \right) \tag{7.15}$$

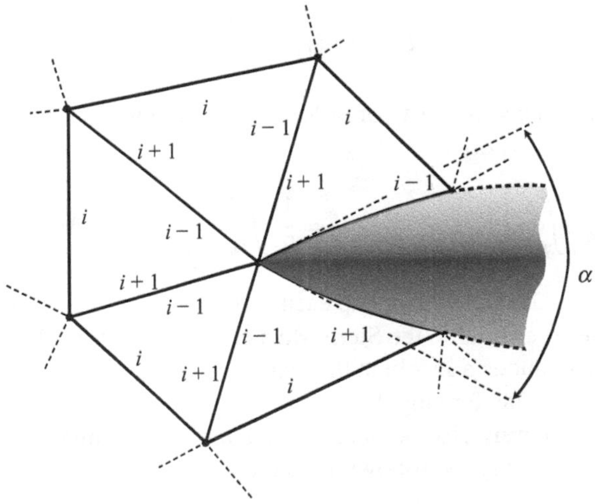

Figure 7.3 Cross-sectional view of the region around a sharp, but curved edge of aperture angle α meshed with curved triangular cells. The figure shows five sharp-edge elements attached to the sharp-edge vertex. In the figure, the triangular edges are labeled by local dummy indexes (i, $i-1$, or $i+1$), with ξ_i vanishing on the ith edge.

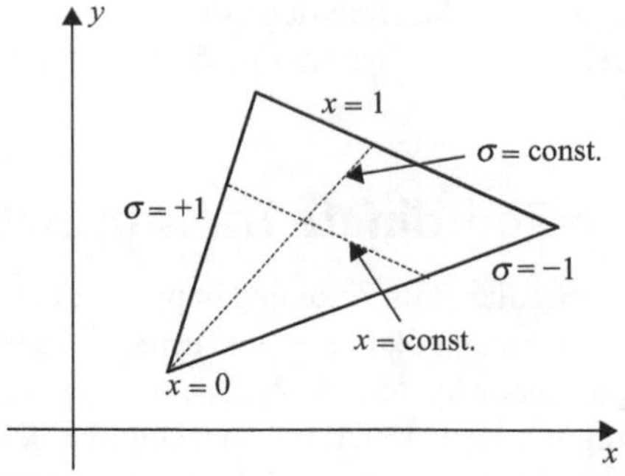

Figure 7.4 A local, pseudo-polar coordinate system (χ, σ) centered on the sharp-edge vertex $\xi_i = 1$ is defined for each singular triangle in the child space. The triangle edges attached to the sharp-edge vertex $\chi = 0$ are along the $\sigma = \pm 1$ coordinate lines; the edge opposite to the sharp-edge vertex is along the $\chi = 1$ coordinate line.

or, equivalently,

$$\chi = 1 - \xi_i \tag{7.16}$$

$$\sigma = \frac{\xi_{i-1} - \xi_{i+1}}{1 - \xi_i} \tag{7.17}$$

The differential area is related by

$$d\boldsymbol{\xi} = d\xi_i d\xi_{i+1} = \frac{\chi}{2} d\chi d\sigma \tag{7.18}$$

Integrable functions with a singularity of order δ at $\xi_i = 1$ can be numerically integrated over R^2 by using (7.13)–(7.18) to obtain

$$\iint_{T^2} (1 - \xi_i)^{-\delta} f(\boldsymbol{\xi})\, d\boldsymbol{\xi} = \frac{1}{2} \iint_R \chi^{1-\delta} f(\chi, \sigma)\, d\chi\, d\sigma \tag{7.19}$$

The transformations (7.13)–(7.15) that map each *singular* triangular cell (attached to the sharp edge) into a rectangle are similar to the Stern–Becker transformation [10]; as shown by (7.19), this transformation introduces a Jacobian that cancels the singular term and allows the use of standard quadrature formulas for singular vector functions.

The gradient function may also be obtained in simplex coordinates, after first noting that $\xi_{i-1} = 1 - \xi_i - \xi_{i+1}$, and using the following relation

$$\begin{bmatrix} \dfrac{\partial}{\partial \xi_i} \\ \dfrac{\partial}{\partial \xi_{i+1}} \end{bmatrix} = \begin{bmatrix} -1 & \dfrac{\sigma - 1}{\chi} \\ 0 & -\dfrac{2}{\chi} \end{bmatrix} \begin{bmatrix} \dfrac{\partial}{\partial \chi} \\ \dfrac{\partial}{\partial \sigma} \end{bmatrix} \tag{7.20}$$

We obtain the gradients as

$$\begin{cases} \nabla\chi = -\nabla\xi_i \\ \chi\,\nabla\sigma = (1-\sigma)\,\nabla\xi_{i-1} - (1+\sigma)\,\nabla\xi_{i+1} \end{cases} \tag{7.21}$$

with $\sigma = \pm 1$ at $\xi_{i\pm 1} = 0$, and the singular point located on the sharp-edge vertex ($\xi_i = 1$, $\chi = 0$).

The vector $\chi\nabla\sigma$ in the second row of (7.21) is normal to the two triangle edges ($\xi_{i-1} = 0$, $\sigma = -1$), ($\xi_{i+1} = 0, \sigma = +1$); equivalently, $\chi\nabla\sigma$ has a zero tangent component on the two edges departing from the singular point. However, $\nabla\chi$ is not orthogonal to $\chi\nabla\sigma$, although

$$[\nabla\chi] \times [\chi\,\nabla\sigma] = 2\,\frac{\hat{\boldsymbol{n}}}{\mathcal{J}} \tag{7.22}$$

where $\hat{\boldsymbol{n}}$ is the unit vector normal to the element, and $\mathcal{J}$ is the Jacobian of the transformation from the child space to the $\boldsymbol{\xi}$-parent coordinates. Apart from a normalization coefficient that depends on the size and shape of the element, the properties of $\chi\nabla\sigma$ are similar to those of the azimuthal unit vector of a polar coordinate frame centered at the sharp-edge vertex. Similarly, the properties of the vector $\nabla\chi$ resemble those of a radial vector. The pseudo-radial variable $\chi = 1 - \xi_i$ vanishes at the edge of the wedge ($\xi_i = 1$, $\chi = 0$), while σ is a dimensionless *azimuthal* variable which is undetermined at the singular vertex where the numerator and the denominator in its expression (7.17) both vanish.

7.3 Singular Scalar Basis Functions for Triangles

7.3.1 Lowest Order Bases of the Substitutive Type

The existing singular basis functions can usually be classified into two types: *substitutive* and *additive* functions. Substitutive basis functions are those for which one or more polynomial basis functions from the original set are removed and replaced by a basis function with an appropriate singular behavior. Additive functions, on the other hand, retain the entire original set and augment it with additional singular basis functions. The majority of proposed singular basis families are of the substitutive type.

As an example of substitutive functions, conventional scalar basis functions of linear order on triangles are defined by the simplex coordinates themselves, yielding three basis functions

$$B_1 = \xi_1, \quad B_2 = \xi_2, \quad B_3 = \xi_3 \tag{7.23}$$

In order to represent a behavior in the form of χ^{α}, where α is expected to be in the range $0 < \alpha < 1$, and the origin of the triangular-polar system is at node 1 of the triangle, the polynomial functions may be replaced by the set [14]

$$S_1 = 1 - \chi^{\alpha} \tag{7.24}$$

$$S_2 = \chi^{\alpha}\left(\frac{1-\sigma}{2}\right) \tag{7.25}$$

$$S_3 = \chi^{\alpha}\left(\frac{1+\sigma}{2}\right) \tag{7.26}$$

These functions interpolate to unit values at the three nodes, like the conventional non-singular basis functions in (7.23). Node 2 is located at $(\chi = 1, \sigma = -1)$, while node 3 is at $(\chi = 1, \sigma = 1)$. The singular basis functions are compatible with conventional basis functions at nodes 2 and 3 in the surrounding cells, and with singular functions of the same χ^α radial variation located at node 1 in the adjacent cells. When $\alpha = 1$, these functions revert back to the usual linear Lagrange interpolation functions in (7.23).

Gradients of these basis functions are needed for the FEM analysis of the Laplace and scalar Helmholtz equations, and may be obtained using (7.20) and (7.21). The gradient fields exhibit an $O(\chi^{\alpha-1})$ behavior. The vector function ∇S_2 contributes no tangential component along the edge between nodes 1 and 3 (edge 13), a constant tangential component along edge 23, and a tangential component that varies as $O(\chi^{\alpha-1})$ along edge 12. Similarly, the vector function ∇S_3 contributes no tangential component along edge 12, a constant tangential component along edge 23, and a tangential component that varies as $O(\chi^{\alpha-1})$ along edge 13. Explicit functions in the (χ, σ) system are

$$\nabla S_1 = \chi^{\alpha-1} \alpha \nabla \xi_1 \tag{7.27}$$

$$\nabla S_2 = \chi^{\alpha-1} \left\{ \left(\frac{1-\sigma}{2} \right) (1-\alpha) \nabla \xi_1 + \nabla \xi_2 \right\} \tag{7.28}$$

$$\nabla S_3 = \chi^{\alpha-1} \left\{ \left(\frac{1+\sigma}{2} \right) (1-\alpha) \nabla \xi_1 + \nabla \xi_3 \right\} \tag{7.29}$$

The gradients also satisfy the relation

$$\nabla S_1 + \nabla S_2 + \nabla S_3 = 0 \tag{7.30}$$

7.3.2 Higher Order Bases of the Substitutive Type

To illustrate the methodology behind the development of higher order functions, we also review a set of substitutive scalar basis functions of quadratic order that were proposed in [16, 20] to replace interpolatory scalar Lagrangian functions. The singular functions have the general form

$$B_{mn}^M(\chi, \sigma) = R_m^M(\chi) L_n^m(\sigma) \tag{7.31}$$

where M denotes the order of the function in the radial direction. L_n^m denotes Lagrange polynomials

$$L_n^m(\sigma) = \prod_{\substack{j=0 \\ j \neq n}}^{m} \frac{m\sigma + m - 2j}{2(n-j)} \tag{7.32}$$

where $L_0^0 = 1$, $L_0^1 = (1-\sigma)/2$, $L_1^1 = (1+\sigma)/2$, etc. R_m^M denotes a radial shape function of the form

$$R_m^M(\chi) = a_0 + \sum_{i=1}^{M} a_i \chi^{\alpha+i-1} \tag{7.33}$$

Note that R_m^M contains a term of the form χ^α, and possibly terms $\chi^{\alpha+1}$, $\chi^{\alpha+2}$, etc. The radial shape functions are also constrained by

$$\begin{cases} R_m^M(i/M) = 0, \quad i = 1, 2, \ldots, M \quad (i \neq m) \\ R_m^M(m/M) = 1 \end{cases} \tag{7.34}$$

to create the desired interpolatory properties.

As an example, the quadratic ($M = 2$) set of singular scalar functions is given in terms of the radial functions

$$\begin{cases} R_0^2 = 1 - (2^{\alpha+1} - 1)\chi^\alpha + (2^{\alpha+1} - 2)\chi^{\alpha+1} \\ R_1^2 = 2^{\alpha+1}\chi^\alpha(1 - \chi) \\ R_2^2 = -\chi^\alpha(1 - 2\chi) \end{cases} \tag{7.35}$$

The six basis functions are

$$\begin{aligned} B_{00}^2 &= R_0^2 L_0^0 = 1 - (2^{\alpha+1} - 1)\chi^\alpha + (2^{\alpha+1} - 2)\chi^{\alpha+1} \\ B_{10}^2 &= R_1^2 L_0^1 = 2^{\alpha+1}\chi^\alpha(1 - \chi)(1 - \sigma)/2 \\ B_{11}^2 &= R_1^2 L_1^1 = 2^{\alpha+1}\chi^\alpha(1 - \chi)(1 + \sigma)/2 \\ B_{20}^2 &= R_2^2 L_0^2 = -\chi^\alpha(1 - 2\chi)\sigma(1 - \sigma)/2 \\ B_{21}^2 &= R_2^2 L_1^2 = -\chi^\alpha(1 - 2\chi)(1 + \sigma)(1 - \sigma) \\ B_{22}^2 &= R_2^2 L_2^2 = -\chi^\alpha(1 - 2\chi)\sigma(1 + \sigma)/2 \end{aligned} \tag{7.36}$$

where index 00 denotes the singular node, 01 and 10 denote the two nodes halfway from the singular node to the other side of the cell, etc. In this case, the singular basis functions exhibit the same interpolatory properties as the usual Lagrangian shape functions, but with the functional dependence χ^α. In the situation where $\alpha \to 1$, these functions revert to the polynomial behavior of the Lagrangian shape functions on the cell boundaries. The gradient of any of the shape functions in (7.36) can be determined using (7.20) and (7.21); the gradients of these functions exhibit an $O(\chi^{\alpha-1})$ dependence.

In summary, substitutive basis sets such as (7.24)–(7.26) and (7.36) replace the conventional polynomial basis functions with singular basis functions having similar properties (interpolation and continuity) except that the functional dependence is chosen to incorporate the desired behavior, in this case χ^α. These bases incorporate a radial function with fractional powers of χ combined with polynomials in σ. A similar approach will be used below to define hierarchical singular functions of the additive type.

A number of different interpolatory or ad hoc sets of basis functions for treating corner singularities have been proposed in conjunction with low-order representations (constant, linear, and quadratic functional dependence in the absence of singularities). These are reviewed in papers such as [20, 23, 26, 27]. For polynomial orders higher than these, it is usually more efficient to employ hierarchical basis functions, which permit the order of representation to be different in different regions of the problem domain, and facilitate adaptive p-refinement techniques. Since we feel that the full benefit of (hierarchical) singular basis functions will not be realized through substitutive functions, we turn our attention to additive basis functions.

7.3.3 Additive Singular Basis Functions

When using a low-order representation for a problem containing unbounded fields, a substantial improvement in accuracy can often be obtained using a substitutive approach, where polynomial bases are replaced on a 1:1 basis with singular functions. The potential drawback to substitutive functions is that they introduce the singularity even if that behavior is not excited by the specific source [25], while removing polynomial DoFs that might be needed (especially for electrically large cells). We define the additive approach as one where basis functions with singular DoFs are added to the original polynomial set, *without removing any of the polynomial DoFs*. Additive functions are therefore more flexible, and can model appropriate field behavior under a wider range of conditions. The additive approach appears to be essential to obtain true high-order convergence behavior [27, 28]. References 19, 29, and 30 previously proposed additive basis families for triangular cells.

In the following, we build additive singular bases upon the family of scalar polynomial bases introduced in Section 5.2. Thus, at any order p the basis will consist of a polynomial subspace (complete to that order in the usual sense) and a singular subspace. The singular DoFs will be called the *Meixner* subspace. In addition, to facilitate adaptive p-refinement, we construct the bases in a hierarchical manner, so that a basis of order p contains all the functions in the basis of order $p-1$.

One difficulty with this process is the possible lack of linear independence between the singular functions and the members of the existing set. To reduce the growth of the matrix condition numbers (CNs) associated with the augmented set of functions, some orthogonalization is necessary. Thus, we will develop singular functions that have the sufficient flexibility to be made orthogonal to the polynomial bases in the original set.

For generality, the singular basis functions will incorporate an auxiliary radial function of the form

$$R_n(k,\nu,\chi) = a_{n1}\chi^{\nu_n} - \sum_{j=1}^{n+1} b_{nj}\chi^j \tag{7.37}$$

where χ is the triangular-polar coordinate defined previously. This form facilitates the use of arbitrary fractional exponents, so that a single set of functions can be used with a singularity of any order. In (7.37), the first term incorporates the desired fractional exponent, and the other (polynomial) terms are included to permit orthogonalization with the regular basis functions, as well as with R_j for $j < n$. The coefficients in (7.37) will be determined to provide orthogonality using one of the schemes discussed below.

The singular DoFs to be included in basis functions of some order may be organized into an ordered list of the exponents

$$\bar{\boldsymbol{\nu}} = \{\nu_1, \nu_2, \nu_3, \ldots, \nu_{n-1}, \nu_n, \nu_{n+1}, \ldots\} \tag{7.38}$$

Typically, the exponents in (7.38) are ordered by their increasing values at a given wedge aperture angle α, and not necessarily by the order of their indices in expressions such as (7.6). For the PEC wedge, where $\nu = \pi/(2\pi - \alpha)$, we order the singularity coefficients (7.6) at the onset in the list

$$\bar{\boldsymbol{\nu}} = \{\nu_{10}, \nu_{20}, \nu_{30}, \nu_{40}, \nu_{11}, \nu_{50}, \nu_{21}, \nu_{60}, \nu_{31}, \ldots\} \tag{7.39}$$

according to their increasing values at $\alpha \approx 0°$ (see Figure 7.1). In the derivations that follow, we consider the basis functions of the Meixner subset as functions of the lowest singularity coefficient ν rather than functions of the wedge angle α.

The Meixner basis functions are then defined in the pseudo-polar reference frame (χ, σ) centered on the sharp-edge vertex $\xi_i = 1$ by using appropriate functions of σ and χ. In particular, we use the qth-order polynomials

$$f_q(\sigma) = \sqrt{\frac{(2q+5)(q+3)(q+4)}{3\,(q+1)(q+2)}}\, P_q^{(2,2)}(\sigma) \tag{7.40}$$

mutually orthogonal on the σ-interval $[-1, 1]$ with respect to the weight function $w(\sigma) = (1-\sigma^2)^2$, with

$$\frac{1}{2}\int_{-1}^{1} \left[f_q(\sigma)\left(1-\sigma^2\right)/8\right]^2 \, d\sigma = \frac{1}{12} \tag{7.41}$$

where $\left(1-\sigma^2\right) f_q(\sigma)$ vanishes at $\sigma = \pm 1$. The polynomials (7.40) are defined in terms of the Jacobi polynomials $P_q^{(2,2)}(\sigma)$ which are constructed from

$$P_0^{(2,2)}(\sigma) = 1 \tag{7.42}$$

$$P_1^{(2,2)}(\sigma) = 3\,\sigma \tag{7.43}$$

and the recurrence relation

$$(q+1)(q+5)P_{q+1}^{(2,2)}(\sigma) = (q+3)\left[(2q+5)\,\sigma\, P_q^{(2,2)}(\sigma) - (q+2)P_{q-1}^{(2,2)}(\sigma)\right] \tag{7.44}$$

Therefore

$$f_0(\sigma) = \sqrt{10} \tag{7.45}$$

$$f_1(\sigma) = \sqrt{70}\sigma \tag{7.46}$$

$$f_2(\sigma) = \sqrt{\frac{15}{2}}\,(7\sigma^2 - 1) \tag{7.47}$$

We also define the mutually orthogonal, auxiliary radial functions

$$R_n(k, \nu, \chi) = \frac{N_n(k, \nu, \chi)}{D_n(k, \nu)} \tag{7.48}$$

with

$$N_n(k, \nu, \chi) = a_n \, \chi^{\nu_n} - \sum_{j=1}^{k+n} b_{nj} \, \chi^j \tag{7.49}$$

$$D_n(k, \nu) = \prod_{j=1}^{k+n} (\nu_n - j) \tag{7.50}$$

$$a_n = \sum_{j=1}^{k+n} b_{nj} \tag{7.51}$$

and where integers n and k are greater than or equal to unity, while the variable ν_n is the nth entry ($\nu_{st} = s\nu + 2\,t$) of the singularity coefficient list (7.39).

The functions (7.48) and (7.49) vanish at $\chi = 0$, $\chi = 1$ (at the singular vertex and along the triangle edge opposite the singular vertex, respectively). The $k + n$ coefficients b_{nj} that define the polynomial part of (7.49) are uniquely determined by imposing the orthogonality conditions[3]

$$\int_0^1 \chi \, R_n(k, \nu, \chi) \, R_j(k, \nu, \chi) \, \mathrm{d}\chi = 0 \tag{7.52}$$

$$\int_0^1 \chi \, R_n(k, \nu, \chi) \, E_{m2}(\chi) \, \mathrm{d}\chi = 0 \tag{7.53}$$

obtained for $j = 1, 2, \ldots, n-1$, and $m = 0, 1, \ldots, k-1$, and by setting

$$\int_0^1 \chi \, R_n^2(k, \nu, \chi) \, \mathrm{d}\chi = 1 \tag{7.54}$$

The integrals (7.52)–(7.54) converge and exist for all $n, k \geq 1$ since the lowest singularity coefficient ν_1 ($\geq 1/2$) of the ordered list (7.38) is not negative. The function $E_{m2}(\chi)$ in (7.53) is just the edge-based polynomial (5.34) introduced in Section 5.2.1.3, evaluated at $\boldsymbol{\xi} = \{\xi_a, \xi_b, \xi_c\} = \{1 - \chi, \chi, 0\}$; this polynomial is antisymmetric with respect to the point $\chi = 1/2$ for odd values of m and symmetric otherwise (see Figure 7.5). The parameter k appearing in (7.48) is the number of polynomials $E_{m2}(\chi)$ orthogonal to each $R_n(k, \nu, \chi)$, while the subscript n denotes the fact that the function $R_n(k, \nu, \chi)$ contains the irrational algebraic term χ^{ν_n}. Table 7.1 reports the first three radial functions (7.48) for two different sets, obtained with $k = 1$ and $k = 2$.

It is of importance to notice that for integer values of ν_n (for example, those marked by *circles* in Figure 7.1) the function (7.49) becomes a polynomial that can be expressed in terms of the *polynomial* bases discussed in the previous subsection, while from (7.48) and (7.50), it is clear

[3]In this section, by using (7.53), we follow the construction procedure first presented in [22]. In Section 7.6, we will generalize the procedure by using Legendre polynomials instead of the functions $E_{m2}(\chi)$.

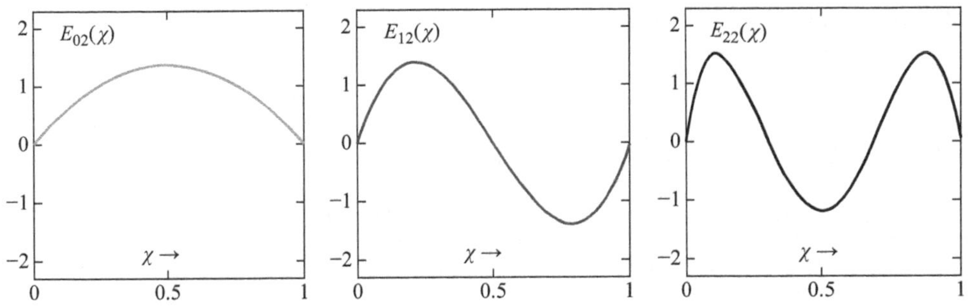

Figure 7.5 Behavior of the edge-based polynomial functions $E_{m2}(1-\chi, \chi, 0) = E_{m2}(\chi)$ on the interval $0 \le \chi \le 1$, for $m = 0, 1, 2$. As per (7.53), the auxiliary radial functions $R_n(k, \nu, \chi)$ are orthogonal to the functions $E_{m2}(\chi)$ for all $m = 0, 1, \ldots, k-1$. For example, with reference to Figure 7.6, the auxiliary radial functions $R_n(1, \nu, \chi)$ are orthogonal to $E_{02}(\chi)$; the functions $R_n(2, \nu, \chi)$ are orthogonal to $E_{02}(\chi)$ and $E_{12}(\chi)$; the functions $R_n(3, \nu, \chi)$ are orthogonal to $E_{02}(\chi)$, $E_{12}(\chi)$, and $E_{22}(\chi)$.

© 2013 IEEE. Reprinted with permission from R. D. Graglia, A. F. Peterson, and L. Matekovits, "Singular, hierarchical scalar basis functions for triangular cells," *IEEE Trans. Antennas Propag.*, vol. 61, no. 7, pp. 3674–3692, Jul. 2013.

that those integer ν_n values are *poles* of the irrational algebraic function $R_n(k, \nu, \chi)$ constructed by the orthogonalization process based on (7.52)–(7.54) (see also Table 7.1). Since ν is known in advance, if $R_j(k, \nu, \chi)$ has a pole at ν then $R_j(k, \nu, \chi)$ is a *degenerate* (singular) polynomial function that cannot be used at the given ν to form the irrational algebraic part of the basis set obtained with the procedure described in the rest of this section. Any functions constructed by multiplying the auxiliary functions $R_n(k, \nu, \chi)$ by a function of the σ variable contains a polynomial part in χ of order $n+k$; these irrational algebraic functions can therefore be used to form a basis set able to approximate a scalar field only in conjunction with the hierarchical polynomial set of order equal or higher than $n+k$.

With reference to Figure 7.6, observe that the radial function $R_j(k, \nu, \chi)$ does not change dramatically with ν even when ν approaches a pole, except around the $\chi = 0$ point (for example, consider the solid-line results of Figure 7.6 which are for ν almost equal to 1/2, a pole for $R_2(k, \nu, \chi)$).

To simplify the notation, in the following, the parameter k and the ν-value that define the auxiliary radial functions are understood, and the function $R_j(k, \nu, \chi)$ will be denoted by $R_j(\chi)$.

7.3.4 The Irrational Algebraic Scalar Basis Functions

Each auxiliary radial function $R_j(\chi)$ defined in the previous subsection (provided it does not have a pole at the given ν value) generates the basis functions

$$\begin{aligned} \phi_{j1}^{i+1}(\chi,\sigma) &= R_j(\chi)\,(1+\sigma)/4 \\ \phi_{j1}^{i-1}(\chi,\sigma) &= R_j(\chi)\,(1-\sigma)/4 \\ \phi_{j\ell}(\chi,\sigma) &= R_j(\chi) f_{\ell-2}(\sigma)\,(1-\sigma^2)/8 \end{aligned} \qquad (7.55)$$

Table 7.1 Auxiliary radial functions.

$$R_1(1,\nu,\chi)=\sqrt{c_{11}}\,\frac{a_{11}\,\chi^{\nu}-(b_{11}\,\chi+b_{12}\,\chi^2)}{(\nu-1)\,(\nu-2)}$$

$a_{11}=(\nu+3)\,(\nu+4)$ $\quad b_{11}=-2\,(\nu-2)\,(\nu+9)$

$b_{12}=(\nu-1)\,(\nu+8)$

$c_{11}=10\,(\nu+1)/(\nu+6)$

$d_{11}=2\,(\nu+1)\,(\nu+2)/(72+52\,\nu+13\,\nu^2+\nu^3)$

$$R_1(2,\nu,\chi)=\sqrt{c_{12}}\,\frac{a_{12}\,\chi^{\nu}-(b_{11}\,\chi+b_{12}\,\chi^2+b_{13}\,\chi^3)}{(\nu-1)\,(\nu-2)\,(\nu-3)}$$

$a_{12}=-3\,(\nu+3)\,(\nu+4)\,(\nu+5)$ $\quad b_{11}=-10\,(\nu-2)\,(\nu-3)\,(\nu+17)$

$b_{12}=35\,(\nu-1)\,(\nu-3)\,(\nu+16)$ $\quad b_{13}=-28\,(\nu-1)\,(\nu-2)\,(\nu+15)$

$c_{12}=2\,(\nu+1)/(\nu+10)$

$d_{12}=6\,(\nu+1)\,(\nu+2)/\left(4{,}840+24{,}436\,\nu+503\,\nu^2+27\,\nu^3\right)$

$$R_2(1,\nu,\chi)=\sqrt{c_{21}}\,\frac{a_{21}\,\chi^{2\nu}-(b_{21}\,\chi+b_{22}\,\chi^2+b_{23}\,\chi^3)}{(2\,\nu-1)\,(\nu-1)\,(2\,\nu-3)}$$

$a_{21}=-(\nu+2)\,(\nu+15)\,(2\,\nu+3)\,(3\,\nu+2)$

$b_{21}=-10\,(\nu-1)\,(2\,\nu-3)\,(\nu+3)\,(3\,\nu+14)$

$b_{22}=15\,(2\,\nu-1)\,(2\,\nu-3)\,(\nu+4)\,(3\,\nu+13)$

$b_{23}=-63\,(2\,\nu-1)\,(\nu-1)\,(\nu+4)\,(\nu+5)$

$c_{21}=8\,(2\,\nu+1)\,(2\,\nu+5)\,/\,(1{,}200+2{,}840\,\nu$
$\quad +1{,}169\,\nu^2+162\,\nu^3+9\,\nu^4)$

$d_{21}=2\,(2\,\nu+1)\,(\nu+1)\,/\,(5{,}520+14{,}764\,\nu$
$\quad +11{,}171\,\nu^2+3{,}067\,\nu^3+309\,\nu^4+9\,\nu^5)$

$$R_2(2,\nu,\chi)=\sqrt{c_{22}}\,\frac{a_{22}\,\chi^{2\,\nu}-(b_{21}\,\chi+b_{22}\,\chi^2+b_{23}\,\chi^3+b_{24}\,\chi^4)}{(2\,\nu-1)\,(\nu-1)\,(2\,\nu-3)\,(\nu-2)}$$

$a_{22}=(\nu+2)\,(\nu+24)\,(2\,\nu+3)\,(2\,\nu+5)\,(3\,\nu+2)$

$b_{21}=-40\,(\nu-1)\,(2\,\nu-3)\,(\nu-2)\,(\nu+3)\,(3\,\nu+23)$

$b_{22}=105\,(2\,\nu-1)\,(2\,\nu-3)\,(\nu-2)\,(\nu+4)\,(3\,\nu+22)$

$b_{23}=-1{,}008\,(2\,\nu-1)\,(\nu-1)\,(\nu-2)\,(\nu+5)\,(\nu+7)$

$b_{24}=84\,(2\,\nu-1)\,(\nu-1)\,(2\,\nu-3)\,(\nu+6)\,(3\,\nu+20)$

$c_{22}=5\,(2\,\nu+1)(\nu+3)/(3{,}600+8{,}040\,\nu+2{,}344\,\nu^2+228\,\nu^3+9\,\nu^4)$

$d_{22}=(2\,\nu+1)\,(\nu+1)\,/(45{,}600+113{,}792\,\nu+72{,}244\,\nu^2+14{,}180\,\nu^3$
$\quad +969\,\nu^4+18\,\nu^5)$

$$R_3(1,\nu,\chi)=\sqrt{c_{31}}\,\frac{a_{31}\,\chi^{3\,\nu}-(b_{31}\,\chi+b_{32}\,\chi^2+b_{33}\,\chi^3+b_{34}\,\chi^4)}{(3\,\nu-1)\,(3\,\nu-2)\,(\nu-1)\,(3\,\nu-4)}$$

$a_{31}=(\nu+1)\,(2\,\nu+1)\,(3\,\nu+4)\,(3\,\nu+5)\,(5\,\nu+2)\,(7{,}200+8{,}700\,\nu+2{,}603\,\nu^2+419\,\nu^3+37\,\nu^4+\nu^5)$

$b_{31}=-10\,(3\,\nu-2)\,(\nu-1)\,(3\,\nu-4)\,(\nu+3)\,(13{,}800+51{,}830\,\nu+54{,}759\,\nu^2+24{,}959\,\nu^3+5{,}261\,\nu^4+491\,\nu^5+20\,\nu^6)$

$b_{32}=15\,(3\,\nu-1)\,(\nu-1)\,(3\,\nu-4)\,(\nu+4)\,(46{,}200+171{,}170\,\nu+177{,}921\,\nu^2+80{,}143\,\nu^3+17{,}009\,\nu^4+1{,}647\,\nu^5+70\,\nu^6)$

$b_{33}=-56\,(3\,\nu-1)\,(3\,\nu-2)\,(3\,\nu-4)\,(\nu+3)\,(\nu+5)\,(2\,\nu+5)\,(420+1234\,\nu+625\,\nu^2+86\,\nu^3+5\,\nu^4)$

$b_{34}=280\,(3\,\nu-1)\,(3\,\nu-2)\,(\nu-1)\,(\nu+3)\,(\nu+6)\,(600+1{,}990\,\nu+1{,}575\,\nu^2+471\,\nu^3+57\,\nu^4+3\,\nu^5)$

$c_{31}=18\,(3\,\nu+1)\,(\nu+2)/\,(194{,}400{,}000+1{,}950{,}480{,}000\,\nu+7{,}849{,}170{,}000\,\nu^2+16{,}473{,}999{,}600\,\nu^3+20{,}004{,}681{,}360\,\nu^4+15{,}153{,}798{,}836\,\nu^5$
$\quad +7{,}613{,}999{,}687\,\nu^6+2{,}662{,}375{,}394\,\nu^7+668{,}617{,}133\,\nu^8+121{,}948{,}064\,\nu^9+16{,}012{,}253\,\nu^{10}+1{,}472{,}826\,\nu^{11}+90{,}387\,\nu^{12}$
$\quad +3{,}360\,\nu^{13}+60\,\nu^{14})$

$d_{31}=2\,(3\,\nu+1)\,(3\,\nu+2)/\,(4{,}104{,}000{,}000+42{,}855{,}840{,}000\,\nu+185{,}280{,}564{,}000\,\nu^2+434{,}441{,}142{,}800\,\nu^3+613{,}486{,}451{,}660\,\nu^4$
$\quad +554{,}105{,}545{,}028\,\nu^5+334{,}464{,}051{,}789\,\nu^6+139{,}794{,}085{,}605\,\nu^7+41{,}671{,}201{,}601\,\nu^8+9{,}053{,}141{,}909\,\nu^9+1{,}446{,}168{,}991\,\nu^{10}$
$\quad +168{,}500{,}595\,\nu^{11}+13{,}899{,}919\,\nu^{12}+758{,}163\,\nu^{13}+23{,}640\,\nu^{14}+300\,\nu^{15})$

$$R_3(2,\nu,\chi)=\sqrt{c_{32}}\,\frac{a_{32}\,\chi^{3\,\nu}-(b_{31}\,\chi+b_{32}\,\chi^2+b_{33}\,\chi^3+b_{34}\,\chi^4+b_{35}\,\chi^5)}{(3\,\nu-1)\,(3\,\nu-2)\,(\nu-1)\,(3\,\nu-4)\,(3\,\nu-5)}$$

$a_{32}=-3\,(\nu+1)\,(\nu+2)\,(2\,\nu+1)\,(3\,\nu+4)\,(3\,\nu+5)\,(5\,\nu+2)\,(176{,}400+173{,}280\,\nu+37{,}576\,\nu^2+4{,}643\,\nu^3+313\,\nu^4+6\,\nu^5)$

$b_{31}=-10\,(3\,\nu-2)\,(\nu-1)\,(3\,\nu-4)\,(3\,\nu-5)\,(\nu+3)\,(599{,}760+2{,}102{,}904\,\nu+1{,}835{,}516\,\nu^2+660{,}364\,\nu^3+104{,}026\,\nu^4+6{,}959\,\nu^5+210\,\nu^6)$

$b_{32}=35\,(3\,\nu-1)\,(\nu-1)\,(3\,\nu-4)\,(3\,\nu-5)\,(\nu+4)\,(1{,}330{,}560+4{,}605{,}984\,\nu+3{,}944{,}724\,\nu^2+1{,}398{,}248\,\nu^3+221{,}735\,\nu^4+15{,}362\,\nu^5+480\,\nu^6)$

$b_{33}=-84\,(3\,\nu-1)\,(3\,\nu-2)\,(3\,\nu-4)\,(3\,\nu-5)\,(\nu+5)\,(483{,}840+1{,}658{,}976\,\nu+1{,}400{,}620\,\nu^2+490{,}784\,\nu^3+78{,}297\,\nu^4+5{,}592\,\nu^5+180\,\nu^6)$

$b_{34}=3360\,(3\,\nu-1)\,(3\,\nu-2)\,(\nu-1)\,(3\,\nu-5)\,(\nu+3)\,(\nu+6)\,(13{,}020+39{,}982\,\nu+23{,}693\,\nu^2+4{,}961\,\nu^3+409\,\nu^4+15\,\nu^5)$

$b_{35}=-660\,(3\,\nu-1)\,(3\,\nu-2)\,(\nu-1)\,(3\,\nu-4)\,(\nu+7)\,(2\,\nu+7)\,(10{,}800+33{,}480\,\nu+20{,}736\,\nu^2+4{,}532\,\nu^3+391\,\nu^4+15\,\nu^5)$

$c_{32}=6\,(3\,\nu+1)(3\,\nu+7)/\left(112{,}021{,}056{,}000+1{,}072{,}201{,}536{,}000\,\nu+4{,}012{,}542{,}334{,}080\,\nu^2+7{,}532{,}520{,}459{,}264\,\nu^3+7{,}774{,}155{,}169{,}056\,\nu^4\right.$
$\quad +4{,}802{,}295{,}720{,}624\,\nu^5+1{,}914{,}595{,}342{,}400\,\nu^6+523{,}481{,}307{,}608\,\nu^7+102{,}068{,}338{,}264\,\nu^8+14{,}370{,}985{,}841\,\nu^9+1{,}445{,}736{,}919\,\nu^{10}$
$\quad \left.+101{,}034{,}800\,\nu^{11}+4{,}687{,}358\,\nu^{12}+131{,}640\,\nu^{13}+1{,}800\,\nu^{14}\right)$

$d_{32}=6\,(3\,\nu+1)\,(3\,\nu+2)/\left(10{,}818{,}033{,}408{,}000+107{,}419{,}810{,}176{,}000\,\nu+432{,}656{,}653{,}063{,}680\,\nu^2+918{,}537{,}036{,}044{,}544\,\nu^3\right.$
$\quad +1{,}132{,}715{,}110{,}075{,}776\,\nu^4+860{,}222{,}739{,}380{,}160\,\nu^5+422{,}605{,}664{,}736{,}560\,\nu^6+140{,}281{,}443{,}949{,}616\,\nu^7+32{,}685{,}327{,}710{,}488\,\nu^8$
$\quad \left.+5{,}498{,}017{,}902{,}056\,\nu^9+675{,}787{,}709{,}995\,\nu^{10}+60{,}267{,}483{,}845\,\nu^{11}+3{,}783{,}226{,}916\,\nu^{12}+155{,}363{,}064\,\nu^{13}+3{,}575{,}520\,\nu^{14}+32{,}400\,\nu^{15}\right)$

The functions are normalized so $\int_0^1 \chi\, R_n^2(k,\nu,\chi)\,\mathrm{d}\chi=1$ and are used to construct the potential functions. The coefficients d_{nk} define re-scaled functions $r_n(k,\nu,\chi)=\sqrt{d_{nk}/c_{nk}}\,R_n(k,\nu,\chi)$ normalized with respect to their first derivative $q_n(k,\nu,\chi)=\frac{\partial}{\partial\chi}r_n(k,\nu,\chi)$ to obtain $\int_0^1 \chi\, q_n^2(k,\nu,\chi)\,\mathrm{d}\chi=1$. The re-scaled functions $r_n(k,\nu,\chi)$ are those used to construct vector basis functions with a zero curl. The c_{nk} and d_{nk} coefficients are shown in Figure 7.7.

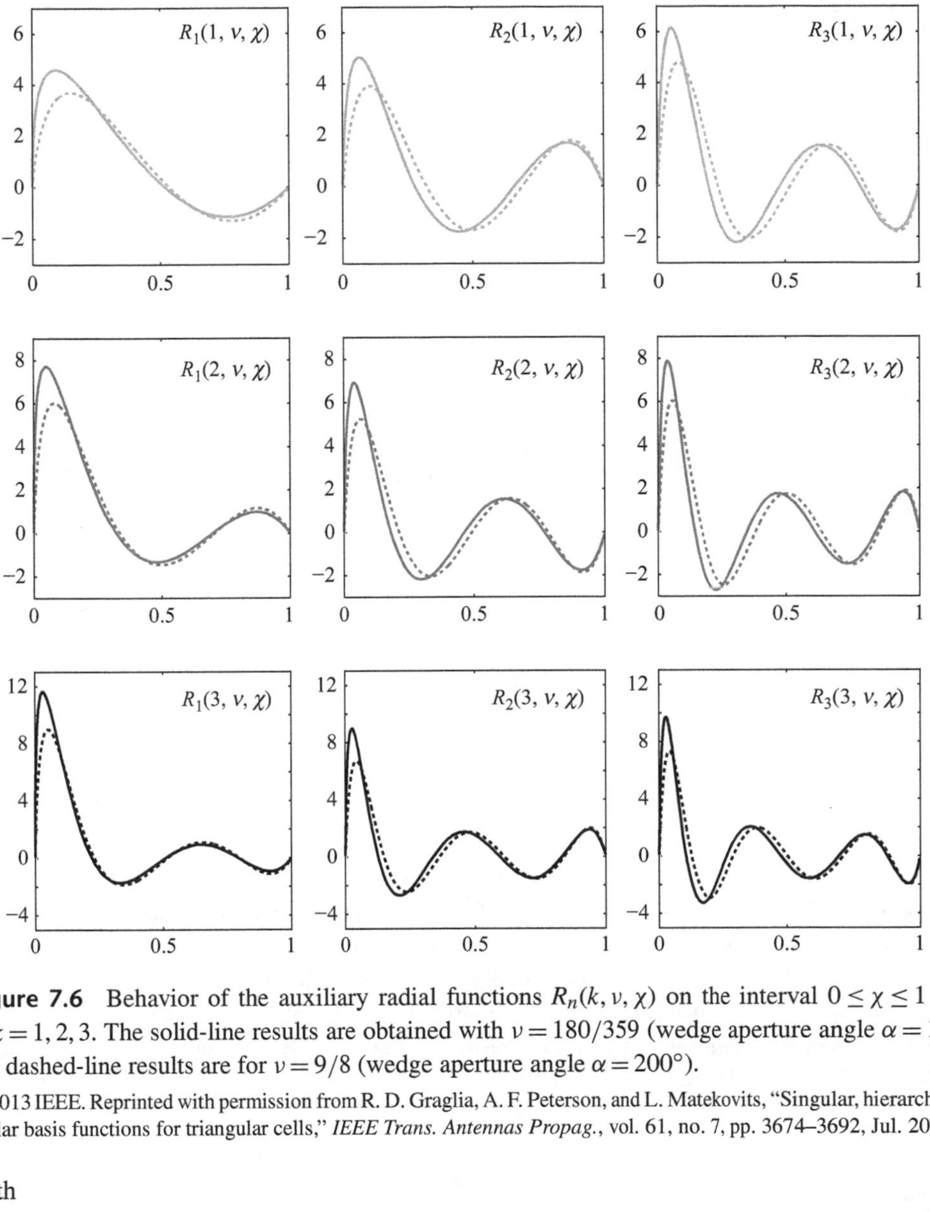

Figure 7.6 Behavior of the auxiliary radial functions $R_n(k, \nu, \chi)$ on the interval $0 \le \chi \le 1$ for $n, k = 1, 2, 3$. The solid-line results are obtained with $\nu = 180/359$ (wedge aperture angle $\alpha = 1°$); the dashed-line results are for $\nu = 9/8$ (wedge aperture angle $\alpha = 200°$).

with

$$\iint_{T^2} \left[\phi_{j1}^{i+1}\right]^2 \mathrm{d}T^2 = \iint_{T^2} \left[\phi_{j1}^{i-1}\right]^2 \mathrm{d}T^2$$
$$= \iint_{T^2} \phi_{j\ell}^2 \,\mathrm{d}T^2 = 1/12 \tag{7.56}$$

The functions ϕ_{j1}^{i+1}, ϕ_{j1}^{i-1} are based on the edge indicated in their superscript ($\xi_{i\pm1} = 0 \Rightarrow \sigma = \pm 1$) and vanish on the other two remaining edges. Conversely, $\phi_{j\ell}$ (for $\ell \ge 2$) are *bubble* functions that vanish along the three edges of the triangular cell.

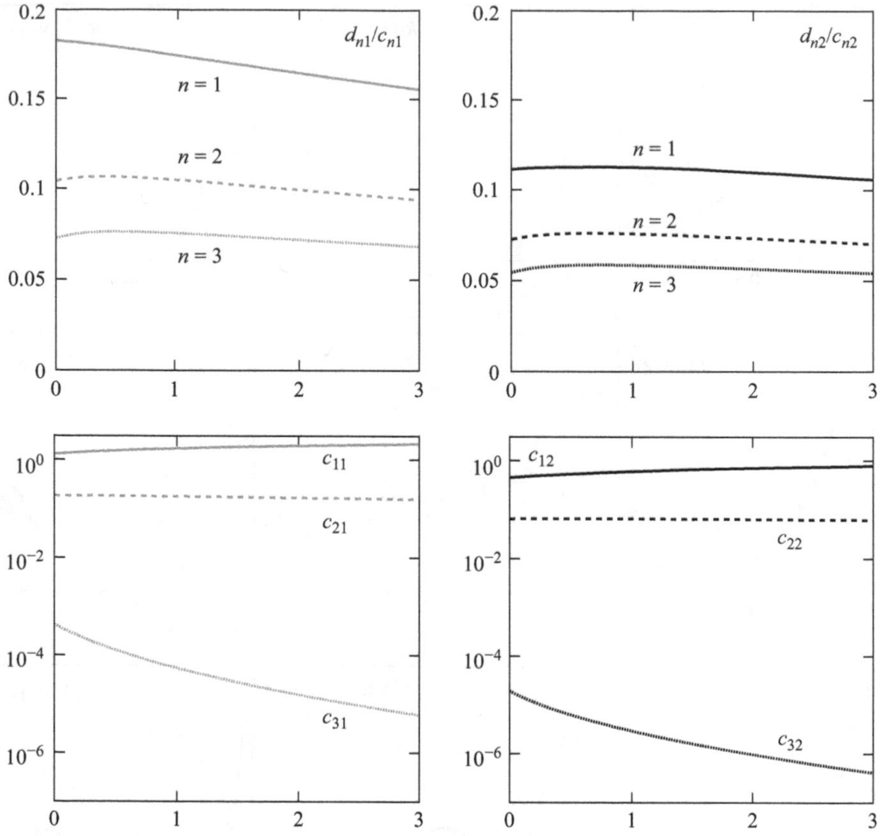

Figure 7.7 Behavior of the scaling coefficients of the auxiliary radial functions $R_n(k, \nu, \chi)$ and $r_n(k, \nu, \chi)$ of Table 7.1 on the interval $0 \leq \nu \leq 3$ (for $n = 1, 2, 3$ and $k = 1, 2$). The figures at top report the ratio of the radial functions $r_n(k, \nu, \chi)/R_n(k, \nu, \chi) = d_{nk}/c_{nk}$ which is practically constant in the range $\{0.5 \leq \nu \leq 1\}$. The figures at bottom show the behavior of the coefficients c_{nk}. In spite of their complex expression, the coefficients c_{1k} and c_{2k} are practically constant in the range $\{0.5 \leq \nu \leq 1\}$.
© 2013 IEEE. Reprinted with permission from R. D. Graglia, A. F. Peterson, and L. Matekovits, "Singular, hierarchical scalar basis functions for triangular cells," *IEEE Trans. Antennas Propag.*, vol. 61, no. 7, pp. 3674–3692, Jul. 2013.

The functions (7.55) are obtained by multiplying the radial function $R_j(\chi)$ with a qth-order complete set of $(q+1)$ pseudo-azimuthal polynomial functions of σ. The "lowest" order set in σ corresponds to $q = 1$; in this case, the set (7.55) is formed by the two edge-based functions $\phi_{j1}^{i+1}, \phi_{j1}^{i-1}$. Higher order sets in σ are obtained for $q \geq 2$; in this case, all the bubble functions having $\ell = 2, 3, \ldots, q$ are included in (7.55). The set (7.55) is clearly hierarchical, since the set obtained with $q = s + 1$ contains all the functions of the set obtained with $q = s$.

Different radial functions $R_j(\chi)$, for $j = 1, 2, \ldots, r$, generate different hierarchical sets of the form of (7.55). Equation (7.52) shows that each function of one set is orthogonal to all the functions of a different set. Furthermore, the *bubble* functions $\phi_{j\ell}(\chi, \sigma)$, $\phi_{mn}(\chi, \sigma)$ generated by the radial functions $R_j(\chi)$ and $R_m(\chi)$, respectively, are mutually orthogonal over the triangular simplex T^2, with

$$\iint_{T^2} \phi_{j\ell}(\chi, \sigma)\, \phi_{mn}(\chi, \sigma)\, \mathrm{d}T^2 = \frac{\delta_{jm}\, \delta_{\ell n}}{12} \tag{7.57}$$

where

$$\delta_{rq} = \begin{cases} 0 & \text{for } r \neq q \\ 1 & \text{for } r = q \end{cases} \tag{7.58}$$

is the Kronecker delta. The bubble- and edge-based potentials defined in this subsection form the Meixner subsets of the scalar basis functions used, for example, to expand in the numerical solution of a 2D electromagnetic problem the longitudinal component (E_z or H_z) of the electromagnetic field.

The Meixner subspace is naturally hierarchical and can be realized through many different basis sets. To maintain a reasonable conditioning of the system of equations obtained by combining the Meixner and regular subspaces, we imposed the orthogonality conditions (7.53) which involve k different hierarchical polynomials $E_{m2}(\chi)$. This orthogonalization process produces a unique set of Meixner basis functions for each specific k value being used. The Meixner basis functions obtained for a given k can be used with hierarchical polynomial bases of any order higher than k without modification.

Similarly, the Meixner subsets to be used in conjunction with *interpolatory* polynomial subsets could be formed by imposing orthogonality conditions between the Meixner and the *interpolatory* polynomial functions. However, all the functions of an interpolatory subset are polynomials of the same order and they all are different from the interpolatory polynomial functions of a different order. In principle, this implies that the Meixner set to be added to an interpolatory subset must be modified whenever the order of the interpolatory subset changes. In Section 7.6, we will generalize the procedure by using Legendre polynomials instead of the functions $E_{m2}(\chi)$.

7.3.5 Example: Quadratic Basis with One Singular Degree

For specificity, let us walk through the construction of the basis set for a quadratic expansion.

Following the above prescription, the lowest order radial function that can be utilized in the additive representation has the form

$$R_1(\nu, \chi) = a_{11}\chi^{\nu} - \left(b_{11}\chi + b_{12}\chi^2\right) \tag{7.59}$$

where the singular DoF has been combined with a quadratic polynomial behavior. (The specific coefficients for these and the functions to follow depend on the value of ν and may be obtained from Table 7.1; these are not necessary to explain the construction and use of the functions and we omit them here.) This function must be used with a regular expansion that is at least of quadratic order.

The singular basis function is a combination of the radial function of χ and an angular function of the variable σ; we define the singular basis functions associated with the two cell edges emanating from the singular node according to

$$\phi_{n1}^{i-1}(\nu,\chi,\sigma) = R_n(\nu,\chi)(1-\sigma)/4 \tag{7.60}$$

$$\phi_{n1}^{i+1}(\nu,\chi,\sigma) = R_n(\nu,\chi)(1+\sigma)/4 \tag{7.61}$$

For higher order singular functions, we also introduce bubble functions

$$\varphi_{n\ell}(\nu,\chi,\sigma) = R_n(\nu,\chi) f_{\ell-2}(\sigma)(1-\sigma^2)/4 \tag{7.62}$$

where $f_q(\sigma)$ is a Jacobi polynomial of order q.

Therefore, the smallest set of additive basis functions that can include singular functions begins with the set of six linear and quadratic polynomial basis functions

$$\xi_1, \xi_2, \xi_3, \sqrt{30}\xi_1\xi_2, \sqrt{30}\xi_2\xi_3, \sqrt{30}\xi_3\xi_1 \tag{7.63}$$

Those basis functions are augmented in the cells containing the singular node with two additional functions per cell (one per each edge associated with the singular node)

$$\phi_{11}^{i-1}(\chi,\sigma) = \left\{a_{11}\chi^{\nu} - \left(b_{11}\chi + b_{12}\chi^2\right)\right\}\left(\frac{1-\sigma}{4}\right) \tag{7.64}$$

$$\phi_{11}^{i+1}(\chi,\sigma) = \left\{a_{11}\chi^{\nu} - \left(b_{11}\chi + b_{12}\chi^2\right)\right\}\left(\frac{1+\sigma}{4}\right) \tag{7.65}$$

The functions in (7.64) and (7.65) are each zero on two edges of the cell and provide a functional dependence incorporating the χ^{ν} term on the third edge.

Since the background order of the expansion is quadratic, it is also possible to add a third function in each singular cell, the bubble function

$$\phi_{12}(\chi,\sigma) = \left\{a_{11}\chi^{\nu} - \left(b_{11}\chi + b_{12}\chi^2\right)\right\}\left(\frac{1-\sigma^2}{4}\right) \tag{7.66}$$

The addition of this singular function brings the azimuthal variation up to the same level (quadratic) as the polynomial background discretization, which is shown below to enhance the solution accuracy.

For this quadratic representation, each cell adjacent to a corner supports six polynomial and three singular basis functions. However, the singular functions in (7.64) and (7.65) are continuous functions shared by the adjacent cells, like the edge-based quadratic functions in (7.63), while the node-based linear functions $\{\xi_1, \xi_2, \xi_3\}$ in (7.63) are shared by all adjacent cells with the same nodes. Dirichlet boundary conditions imposed along the cell edges may reduce the number of unknowns.

7.3.6 Example: Cubic Basis with Two Singular Degrees

To improve the overall representation beyond the quadratic order, we must move to the cubic regular expansion obtained by including the cubic polynomial functions

$$\left\{\begin{array}{l}\sqrt{210}\xi_1\xi_2(\xi_1-\xi_2), \sqrt{210}\xi_2\xi_3(\xi_2-\xi_3), \\ \sqrt{210}\xi_3\xi_1(\xi_3-\xi_1), 3\sqrt{70}\xi_1\xi_2\xi_3\end{array}\right\} \tag{7.67}$$

in the basis set (with the functions of (7.63)). Since the polynomial background is now cubic, it should improve the accuracy to augment the singular functions in (7.64)–(7.66) with the additional bubble function

$$\varphi_{13}(\chi,\sigma)=\left\{a_{11}\chi^{\nu}-\left(b_{21}\chi+b_{22}\chi^{2}\right)\right\}\sqrt{70}\left(\frac{\sigma(1-\sigma^{2})}{4}\right) \tag{7.68}$$

having cubic variation in σ and the same singular exponent in (7.64)–(7.66). The combination of the basis functions in (7.63)–(7.68) provides a cubic expansion with one singular DoF.

However, with a cubic representation, the singular behavior can be improved by including a second fractional exponent. Suppose that we decide to include the exponent 2ν in the expansion (this exponent is usually the second singular term required to represent a field near a conducting wedge; in the event that a different exponent is needed one can easily be incorporated). A second radial function

$$R_2(\nu,\chi)=a_{21}\chi^{2\nu}-\left(b_{21}\chi+b_{22}\chi^{2}+b_{23}\chi^{3}\right) \tag{7.69}$$

may be employed, whose coefficients are chosen to impose orthogonality with $R_1(\nu,\chi)$ over the unit interval with respect to the weight function χ. The singular basis functions incorporating (7.69) have the form

$$\varphi_{21}^{i-1}(\chi,\sigma)=\left\{a_{21}\chi^{2\nu}-\left(b_{21}\chi+b_{22}\chi^{2}+b_{23}\chi^{3}\right)\right\}\left(\frac{1-\sigma}{4}\right) \tag{7.70}$$

$$\varphi_{21}^{i+1}(\chi,\sigma)=\left\{a_{21}\chi^{2\nu}-\left(b_{21}\chi+b_{22}\chi^{2}+b_{23}\chi^{3}\right)\right\}\left(\frac{1+\sigma}{4}\right) \tag{7.71}$$

$$\varphi_{22}(\chi,\sigma)=\left\{a_{21}\chi^{2\nu}-\left(b_{21}\chi+b_{22}\chi^{2}+b_{23}\chi^{3}\right)\right\}\left(\frac{1-\sigma^{2}}{4}\right) \tag{7.72}$$

$$\varphi_{23}(\chi,\sigma)=\left\{a_{21}\chi^{2\nu}-\left(b_{21}\chi+b_{22}\chi^{2}+b_{23}\chi^{3}\right)\right\}\sqrt{70}\left(\frac{\sigma(1-\sigma^{2})}{4}\right) \tag{7.73}$$

where the specific coefficients in (7.70)–(7.73) may be determined from Table 7.1 and ensure the orthogonality of (7.59) and (7.69) as well as the partial orthogonality of the singular and non-singular basis functions in the set. In this representation, the cubic polynomial expansion is augmented in the cells containing the singular node by two singular basis functions per radial edge, and four singular basis functions per cell.

If it is desired to increase the order of representation beyond two fractional exponents, we must expand the background expansion to one of degree 4, and augment the set of singular functions with a new collection that includes radial function R_3, which now contains polynomial terms up to χ^4 and is orthogonal to R_1 and R_2. In that case, the overall expansion involves three singular functions per radial edge, and nine singular functions per cell, in the cells containing the singular node (a total of 30 basis functions overlapping each cell). In cells not containing the singular node, a regular degree 4 expansion (15 overlapping basis functions) may be employed.

The preceding approach involves some flexibility. One may first select the number of singular exponents, and then choose any polynomial order as long as the polynomial degree is at least one greater than the number of different exponents. Thus, a polynomial order of quadratic, cubic, or higher may be used with a single exponent. For two exponents, the minimum order is cubic. User experimentation may be required to optimize the expansion for a specific problem. Alternatively, adaptive refinement procedures may offer a more systematic way to build appropriate expansions. Since the bases are hierarchical, the regular and singular parts may be adapted to the fields using p-refinement procedures that employ a local error estimator to guide the decision as to what DoFs to include [1, 2, 31]. That concept is beyond the scope of the present text.

7.3.7 Evaluation of Integrals of Singular Bases

In finite element applications, the coefficients of the mass and stiffness matrix are either evaluated in closed form or numerically by appropriate quadrature routines. The integrals pose no problems whenever they involve polynomial basis functions, but one may lose numerical accuracy if the integrals involve the Meixner functions (7.55) and are evaluated using conventional quadrature rules. In the latter situation, the integrals on a singular triangle may be performed in the pseudo-polar reference frame (χ, σ) discussed in Section 7.2.

In the following, we illustrate a procedure that can be used whenever the singular triangles are rectilinear to simplify the numerical evaluation of the most problematic integrals. The procedure is based on the following three facts:

1. The Meixner functions $\phi_{j\ell}(\chi, \sigma) = R_j(\chi)\, \Theta_\ell(\sigma)$ are the product of a radial function R_j of the χ variable times a polynomial function Θ_ℓ of the σ variable (see (7.55)).
2. On rectilinear triangles, the gradient vector $\nabla\chi$ is constant and the vector $\chi\nabla\sigma$ is only a function of the azimuthal variable σ (see (7.21)).
3. The radial integrals (7.74)–(7.78) reported below can be evaluated analytically (Table 7.2).

$$A_{jm}(k, \nu) = \frac{1}{2}\int_0^1 \chi \frac{\mathrm{d}R_j(k, \nu, \chi)}{\mathrm{d}\chi} \frac{\mathrm{d}R_m(k, \nu, \chi)}{\mathrm{d}\chi}\, \mathrm{d}\chi \tag{7.74}$$

$$B_{jm}(k, \nu) = \frac{1}{2}\int_0^1 \frac{R_j(k, \nu, \chi)\, R_m(k, \nu, \chi)}{\chi}\, \mathrm{d}\chi \tag{7.75}$$

$$C_{jm}(k, \nu) = \frac{1}{2}\int_0^1 R_j(k, \nu, \chi) \frac{\mathrm{d}R_m(k, \nu, \chi)}{\mathrm{d}\chi}\, \mathrm{d}\chi \tag{7.76}$$

$$D_j(k, \nu) = \frac{1}{2}\int_0^1 R_j(k, \nu, \chi)\, \mathrm{d}\chi = -\frac{1}{2}\int_0^1 \chi \frac{\mathrm{d}R_j(k, \nu, \chi)}{\mathrm{d}\chi}\, \mathrm{d}\chi \tag{7.77}$$

$$E_j(k, \nu) = \frac{1}{2}\int_0^1 \chi(1 - \chi) R_j(k, \nu, \chi)\, \mathrm{d}\chi \tag{7.78}$$

For singular rectilinear triangles, the accurate evaluation of the most problematic *stiffness*-matrix coefficients is accelerated by using the previous analytical pre-integrated results (7.74)–(7.77) together with polynomial quadrature routines. (The integral (7.78) is used to evaluate

Table 7.2 Remarkable integrals of the auxiliary radial functions (for $k = 1$).

$$A_{jm} = \frac{1}{2}\int_0^1 \chi\,\frac{\mathrm{d}R_j(1,\nu,\chi)}{\mathrm{d}\chi}\,\frac{\mathrm{d}R_m(1,\nu,\chi)}{\mathrm{d}\chi}\,\mathrm{d}\chi = \frac{a_{jm}}{g_j\,g_m}$$

$$a_{11} = 5\left(72 + 52\,\nu + 13\,\nu^2 + \nu^3\right)/2$$

$$a_{12} = 2\,(2,520 + 1,494\,\nu - 7,509\,\nu^2 - 5,006\,\nu^3 \\ -979\,\nu^4 - 60\,\nu^5)/h_{12}$$

$$a_{13} = 3\,(2,880,000 + 21,748,800\,\nu + 61,281,600\,\nu^2 + 81,211,352\,\nu^3 \\ +55,432,558\,\nu^4 + 23,698,751\,\nu^5 + 8,986,235\,\nu^6 + 3,424,428\,\nu^7 \\ +957,952\,\nu^8 + 155,899\,\nu^9 + 13,335\,\nu^{10} + 450\,\nu^{11})/h_{13}$$

$$a_{22} = 2\,(5,520 + 14,764\,\nu + 11,171\,\nu^2 + 3,067\,\nu^3 + 309\,\nu^4 + 9\,\nu^5)$$

$$a_{23} = 6\,(6,120,000 + 40,464,000\,\nu + 56,954,500\,\nu^2 - 177,610,260\,\nu^3 \\ -703,623,635\,\nu^4 - 981,188,656\,\nu^5 - 703,177,491\,\nu^6 \\ -290,004,528\,\nu^7 - 73,139,785\,\nu^8 - 11,552,704\,\nu^9 \\ -1,104,669\,\nu^{10} - 55,692\,\nu^{11} - 1,080\,\nu^{12})/h_{23}$$

$$a_{33} = 9\,(4,104,000,000 + 42,855,840,000\,\nu + 185,280,564,000\,\nu^2 \\ +434,441,142,800\,\nu^3 + 613,486,451,660\,\nu^4 + 554,105,545,028\,\nu^5 \\ +334,464,051,789\,\nu^6 + 139,794,085,605\,\nu^7 + 41,671,201,601\,\nu^8 \\ +9,053,141,909\,\nu^9 + 1,446,168,991\,\nu^{10} + 168,500,595\,\nu^{11} \\ +13,899,919\,\nu^{12} + 758,163\,\nu^{13} + 23,640\,\nu^{14} + 300\,\nu^{15})/2$$

$$B_{jm} = \frac{1}{2}\int_0^1 \frac{R_j(1,\nu,\chi)\,R_m(1,\nu,\chi)}{\chi}\,\mathrm{d}\chi = \frac{b_{jm}}{\nu\,g_j\,g_m}$$

$$b_{11} = 5\left(144 + 96\,\nu + 17\,\nu^2 + \nu^3\right)/4$$

$$b_{12} = (14,400 + 40,800\,\nu + 30,448\,\nu^2 + 9,887\,\nu^3 \\ +1,413\,\nu^4 + 72\,\nu^5)/h_{12}$$

$$b_{13} = (12,960,000 + 111,348,000\,\nu + 370,659,600\,\nu^2 + 620,989,890\,\nu^3 \\ +586,864,701\,\nu^4 + 342,277,674\,\nu^5 + 129,697,023\,\nu^6 + 32,309,940\,\nu^7 \\ +5,149,539\,\nu^8 + 495,306\,\nu^9 + 26,577\,\nu^{10} + 630\,\nu^{11})/h_{13}$$

$$b_{22} = (3,600 + 9,120\,\nu + 7,304\,\nu^2 + 2,603\,\nu^3 + 426\,\nu^4 + 27\,\nu^5)$$

$$b_{23} = 3\,(17,280,000 + 174,960,000\,\nu + 727,598,400\,\nu^2 + 1,629,202,800\,\nu^3 \\ +2,181,554,318\,\nu^4 + 1,853,860,933\,\nu^5 + 1,043,248,259\,\nu^6 \\ +397,058,040\,\nu^7 + 101,768,412\,\nu^8 + 17,042,097\,\nu^9 \\ +1,761,831\,\nu^{10} + 102,210\,\nu^{11} + 2,700\,\nu^{12})/h_{23}$$

$$b_{33} = 3\,(5,184,000,000 + 53,352,000,000\,\nu + 230,725,800,000\,\nu^2 \\ +553,996,752,000\,\nu^3 + 828,596,965,300\,\nu^4 + 825,912,414,440\,\nu^5 \\ +573,069,887,337\,\nu^6 + 284,514,805,753\,\nu^7 + 102,548,538,293\,\nu^8 \\ +26,912,874,657\,\nu^9 + 5,105,018,123\,\nu^{10} + 688,687,807\,\nu^{11} \\ +64,442,787\,\nu^{12} + 4,011,443\,\nu^{13} + 151,360\,\nu^{14} + 2,700\,\nu^{15})/4$$

$$C_{jm} = \frac{1}{2}\int_0^1 R_j(1,\nu,\chi)\,\frac{\mathrm{d}R_m(1,\nu,\chi)}{\mathrm{d}\chi}\,\mathrm{d}\chi = \frac{c_{jm}}{g_j\,g_m}$$

$$c_{11} = c_{22} = c_{33} = c_{jj} = 0$$

$$c_{21} = -c_{12} = 2\,(3,240 + 10,038\,\nu + 5,882\,\nu^2 \\ +1,273\,\nu^3 + 87\,\nu^4)/h_{12}$$

$$c_{31} = -c_{13} = 3\,(2,016,000 + 18,492,000\,\nu + 61,727,600\,\nu^2 \\ +95,787,354\,\nu^3 + 75,343,591\,\nu^4 + 32,173,767\,\nu^5 + 7,526,260\,\nu^6 \\ +876,756\,\nu^7 + 24,579\,\nu^8 - 4,117\,\nu^9 - 270\,\nu^{10})/h_{13}$$

$$c_{32} = -c_{23} = 6\,(5,400,000 + 58,788,000\,\nu + 249,471,900\,\nu^2 \\ +538,803,750\,\nu^3 + 649,280,778\,\nu^4 + 454,868,863\,\nu^5 \\ +192,046,949\,\nu^6 + 50,246,960\,\nu^7 + 8,306,552\,\nu^8 \\ +861,067\,\nu^9 + 51,741\,\nu^{10} + 1,440\,\nu^{11})/h_{23}$$

$$D_j = \frac{1}{2}\int_0^1 R_j(1,\nu,\chi)\,\mathrm{d}\chi = -\frac{1}{2}\int_0^1 \chi\,\frac{\mathrm{d}R_j(1,\nu,\chi)}{\mathrm{d}\chi}\,\mathrm{d}\chi = \frac{d_j}{g_j}$$

$$d_1 = \sqrt{5\,(2+\nu)}/\sqrt{2(1+\nu)}$$

$$d_2 = \left(20 + 27\,\nu + 3\,\nu^2\right)\,\sqrt{1+\nu}/\sqrt{8\,(1+2\,\nu)}$$

$$d_3 = 3\,(1,200 + 5,630\,\nu + 9,255\,\nu^2 + 7,775\,\nu^3 + 3,271\,\nu^4 \\ +609\,\nu^5 + 50\,\nu^6 + 2\,\nu^7)\,\sqrt{(2+3\,\nu)}/\sqrt{2\,(1+3\,\nu)}$$

$$E_j = \frac{1}{2}\int_0^1 \chi\,(1-\chi)\,R_j(1,\nu,\chi)\,\mathrm{d}\chi = \frac{e_j}{g_j}$$

$$e_1 = (12+\nu)\,\sqrt{1+\nu}/\sqrt{1,440\,(2+\nu)}$$

$$e_2 = \nu\,(17+3\,\nu)\,\sqrt{1+2\,\nu}/\sqrt{1,800\,(1+\nu)}$$

$$e_2 = \nu\,(7,800 + 32,530\,\nu + 38,609\,\nu^2 + 19,003\,\nu^3 + 3,901\,\nu^4 \\ +307\,\nu^5 + 10\,\nu^6)\,\sqrt{1+3\,\nu}/\sqrt{800\,(2+3\,\nu)}$$

Table 7.2 (*Continued*).

where:

$$h_{12} = 6\sqrt{5}\sqrt{(2+\nu)(1+2\nu)}$$
$$h_{13} = 4\sqrt{5}\sqrt{(1+\nu)(2+\nu)(1+3\nu)(2+3\nu)}$$
$$h_{23} = 10\sqrt{(1+\nu)(1+2\nu)(1+3\nu)(2+3\nu)}$$

$$g_1^2 = (2+\nu)(6+\nu)$$
$$g_2^2 = (1+\nu)(1,200 + 2,840\,\nu + 1,169\,\nu^2 + 162\,\nu^3 + 9\,\nu^4)/(5+2\,\nu)$$
$$g_3^2 = (2+3\,\nu)(194,400,000 + 1,950,480,000\,\nu + 7,849,170,000\,\nu^2 + 16,473,999,600\,\nu^3 + 20,004,681,360\,\nu^4 + 15,153,798,836\,\nu^5 + 7,613,999,687\,\nu^6 + 2,662,375,394\,\nu^7 + 668,617,133\,\nu^8 + 121,948,064\,\nu^9 + 16,012,253\,\nu^{10} + 1,472,826\,\nu^{11} + 90,387\,\nu^{12} + 3,360\,\nu^{13} + 60\,\nu^{14})/(2+\nu)$$

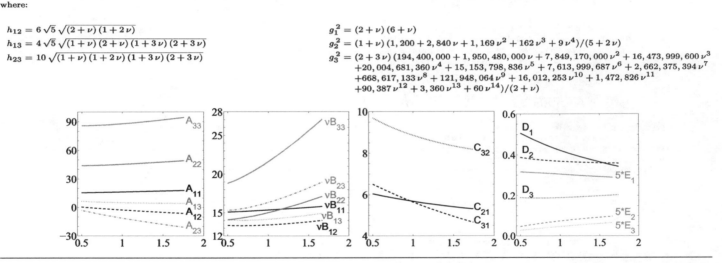

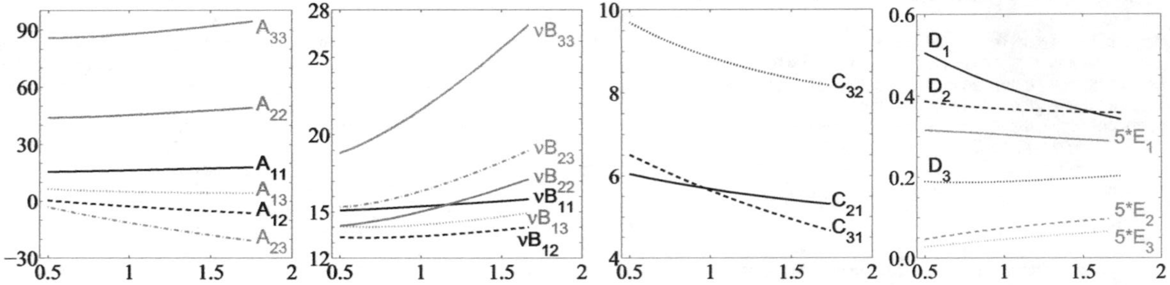

© 2013 IEEE. Reprinted with permission from R. D. Graglia, A. F. Peterson, and L. Matekovits, "Singular, hierarchical scalar basis functions for triangular cells," *IEEE Trans. Antennas Propag.*, vol. 61, no. 7, pp. 3674–3692, Jul. 2013.

the *mass*-matrix coefficients, as shown below.) In fact, for the Meixner functions $\phi_{j\ell}$, ϕ_{mn} one immediately obtains

$$\iint_S \nabla\phi_{j\ell}\cdot\nabla\phi_{mn}\,\mathrm{d}S = \mathcal{J}\,A_{jm}\nabla\chi\cdot\nabla\chi\int_{-1}^{1}\Theta_\ell\,\Theta_n\,\mathrm{d}\sigma$$
$$+\,\mathcal{J}\,B_{jm}\int_{-1}^{1}\frac{\mathrm{d}\Theta_\ell}{\mathrm{d}\sigma}\frac{\mathrm{d}\Theta_n}{\mathrm{d}\sigma}\,(\chi\,\nabla\sigma\cdot\chi\,\nabla\sigma)\,\mathrm{d}\sigma$$
$$+\,\mathcal{J}\int_{-1}^{1}\left[C_{jm}\frac{\mathrm{d}\Theta_\ell}{\mathrm{d}\sigma}\,\Theta_n + C_{mj}\,\Theta_\ell\,\frac{\mathrm{d}\Theta_n}{\mathrm{d}\sigma}\right](\chi\nabla\sigma\cdot\nabla\chi)\,\mathrm{d}\sigma \quad (7.79)$$

where S is the *rectilinear* triangular cell and $\mathcal{J}$ is the *constant* Jacobian of the transformation from the child-space triangle to the $\boldsymbol{\xi}$-parent coordinates. The last integral on the right-hand side of (7.79) vanishes for $\ell = n$, since $C_{mj} = -C_{jm}$. In this connection, recall that for $\ell, n \geq 2$ the azimuthal functions Θ_ℓ and Θ_n are given in terms of orthogonal Jacobi polynomials (see (7.40) and the last of (7.55)). If $\ell = n$, (7.41) holds with $q = \ell = n$, while, for Jacobi polynomials of different orders (i.e. for $\ell \neq n$), the first integral on the right-hand side of (7.79) vanishes.

To evaluate the integrals of the dot product of the gradient of the polynomial basis functions times the gradient of the Meixner functions on singular rectilinear triangles of surface S, we first observe that, on a *rectilinear* triangle, the gradient of the vertex function $V_1^i = \xi_i$, $V_1^{i\pm1} = \xi_{i\pm1}$ is a constant. This immediately yields

$$\iint_S \nabla\begin{bmatrix} V_1^i \\ V_1^{i+1} \\ V_1^{i-1}\end{bmatrix}\cdot\nabla\phi_{j1}^{i\pm1}\,\mathrm{d}S = -\mathcal{J}D_j\,\nabla\begin{bmatrix}\xi_i\\ \xi_{i+1}\\ \xi_{i-1}\end{bmatrix}\cdot\nabla\xi_{i\pm1} \quad (7.80)$$

where $\phi_{j1}^{i\pm1}$ are the two Meixner functions $\phi_{j\ell}$ obtained by setting $\ell = 1$. All the other Meixner functions obtained for $\ell \geq 2$ are bubble functions; therefore, for $\ell \geq 2$, one gets

$$\iint_S \nabla V_1^{i,i\pm1}\cdot\nabla\phi_{j\ell}\,\mathrm{d}S = 0 \quad (7.81)$$

Now, by observing that the gradient of a polynomial *scalar* basis function, $\mathcal{P}(\boldsymbol{\xi})$ is the sum of a constant ($\boldsymbol{P}$) plus a residue vector

$$\nabla\mathcal{P} = \boldsymbol{P} + (\nabla\mathcal{P} - \boldsymbol{P}) \quad (7.82)$$

with

$$\boldsymbol{P} = \nabla\mathcal{P}|_{\chi=0} \quad (7.83)$$

$$\nabla\mathcal{P} - \boldsymbol{P} = 0 \quad \text{at } \chi = 0 \quad (7.84)$$

and $\mathcal{P}$ being equal to one of the face- (F_g) or edge-based (E_g) polynomial functions given in (5.32) and (5.34) (Section 5.2), (7.81) yields on a rectilinear triangle

$$\iint_S \nabla\mathcal{P}\cdot\nabla\phi_{j\ell}\,\mathrm{d}S = \iint_S (\nabla\mathcal{P} - \boldsymbol{P})\cdot\nabla\phi_{j\ell}\,\mathrm{d}S \tag{7.85}$$

for all the bubble Meixner functions $\phi_{j\ell}$ obtained for $\ell \geq 2$. Conversely, for the Meixner functions $\phi_{j1}^{i\pm1}$ (i.e. at $\ell = 1$) one has

$$\iint_S \nabla\mathcal{P}\cdot\nabla\phi_{j1}^{i\pm1}\,\mathrm{d}S = -\mathcal{J}D_j\boldsymbol{P}\cdot\nabla\xi_{i\pm1} + \iint_S (\nabla\mathcal{P} - \boldsymbol{P})\cdot\nabla\phi_{j1}^{i\pm1}\,\mathrm{d}S \tag{7.86}$$

The numerical evaluation of the integrals appearing on the right-hand side of (7.85) and (7.86) is simplified by the fact that all the integrands vanish at the singular vertex $\chi = 0$, $\xi_i = 1$, although it is still convenient to perform those integrals in the pseudo-polar reference frame (χ, σ).

The integrals required to evaluate the *mass*-matrix coefficients are less problematic than those required to evaluate the stiffness-matrix coefficients discussed above. In this case, pre-integration along the radial direction can also be used to deal with curved (distorted) triangles. For example, for the vertex function $V_1^i = \xi_i$ attaining a unit value at the singular vertex $\chi = 0$, one obtains

$$\iint_S V_1^i\,\phi_{j\ell}\,\mathrm{d}S = \mathcal{J}E_j\int_{-1}^{1}\Theta_\ell\,\mathrm{d}\sigma \tag{7.87}$$

that vanishes for odd values of $\ell \geq 3$. In this connection, recall also that the orthogonality of the auxiliary radial functions implies

$$\iint_S \phi_{j\ell}\,\phi_{mn}\,\mathrm{d}S = \mathcal{J}\frac{\delta_{jm}}{2}\int_{-1}^{1}\Theta_\ell\,\Theta_n\,\mathrm{d}\sigma \tag{7.88}$$

which vanishes for $\ell, n \geq 2$ and $\ell \neq n$, and where δ_{jm} is the Kronecker delta (7.58).

In spite of the ν poles appearing in the expression of the auxiliary radial functions $R_j(k, \nu, \chi)$ and $R_m(k, \nu, \chi)$, the integral coefficients A_{jm}, B_{jm}, C_{jm}, D_j, and E_j given in (7.74)–(7.78) are never singular for $\nu \geq 1/2$. Table 7.2 reports, for the $k = 1$ case, the expressions of these coefficients (for $1 \leq j, m \leq 3$) as functions of the ν variable along with their graphical behavior on the interval $\{0.5 \leq \nu < 1.7\}$.

7.4 Numerical Results for Scalar Bases

This section compares numerical results obtained with hierarchical scalar bases of additive kind to those obtained with purely polynomial hierarchical bases. As noted before, a polynomial base is specified by its order $(p + 0.5)$, $p + 1$ being the maximum order of the polynomials forming that base. Conversely, an additive base also contains the Meixner subset (7.55), and it is therefore defined by an ordered list of numbers. The first entry in the list is the order $(p + 0.5)$ of the polynomial subset. The second entry is a sublist that reports the j indexes that indicate which auxiliary radial functions $R_j(k, \nu, \chi)$ are used to form the Meixner subset. If the Meixner subset is obtained by using the auxiliary radial functions $R_1(k, \nu, \chi)$, $R_2(k, \nu, \chi)$, and $R_3(k, \nu, \chi)$, then the sublist is $\{1, 2, 3\}_k$; similarly, if the second entry is the sublist $\{1, 3\}_k$ this means that we are using only the auxiliary radial functions $R_1(k, \nu, \chi)$ and $R_3(k, \nu, \chi)$ to form the Meixner subset. (The ν value depends on the wedge considered and is always understood.) The last entry in the list giving the order of the additive base is an integer A_σ indicating the maximum order of the azimuthal polynomial of the σ variable used to form the irrational algebraic basis functions (7.55). That is, $A_\sigma = 1$ means that the Meixner set contains only the edge functions $\phi_{j1}^{i\pm 1} = R_j(\chi)(1 \pm \sigma)/4$ having a linear variation in σ; similarly, $A_\sigma \geq 2$ means that we are using all the functions $\phi_{j\ell}$ given in (7.55) up to $\ell = A_\sigma$. The j values indicating the radial functions in use are already specified in the second entry of the list giving the order of the additive base. (One could obviously use different azimuthal order ℓ for the Meixner functions $\phi_{i\ell}$ and $\phi_{j\ell}$; in this case, the list of numbers specifying the basis order would be longer.)

To further clarify this notation, the example basis $[4.5, \{1, 3\}_1, 5]$ is formed by the polynomial subset of order 4.5 augmented by the Meixner subset $\{\phi_{1\ell}, \phi_{3\ell}\}_1$ formed by multiplying the radial function $R_1(1, \nu, \chi)$ and $R_3(1, \nu, \chi)$ with the azimuthal polynomials $(1 \pm \sigma)/4$ and $f_{\ell-2}(\sigma)(1 - \sigma^2)/8$, for $\ell = 2, 3, \ldots, 5$ (see (7.40 and 7.55)); this example uses an azimuthal polynomial order of the Meixner subset equal to 5.

7.4.1 Eigenvalues of Waveguiding Structures with Edges

The higher convergence rate of the proposed hierarchical bases for increasing order is validated by the results of Table 7.3 for the circular vaned waveguide (a homogeneous waveguide of normalized (unit) radius a with a metal, zero-thickness radial vane extending to its center) [20, 29]. In this case, the wedge aperture angle is $\alpha = 0°$ and the lowest singularity coefficient (7.6) one has to consider is $\nu = 1/2$. The other non-integer singularity coefficients are equal to $\nu + q$, with integer $q \geq 1$ (see Figure 7.1).

The zeros of the Bessel functions $J_{m/2}$ of half-integer order, and of the derivatives of these Bessel functions (reported in the left-hand column of Table 7.3) are the TM and TE eigenvalues, respectively (see [20, 29] and references therein). (The multiplicity of these eigenvalues is one, in contrast to most of the modes of the *standard* circular waveguide whose multiplicity is two.) The first subscript labeling these modes is $m/2$; the second subscript n indicates the order of the zero, as usual. Even values of m correspond to modes supported also by the circular waveguide, although the vane suppresses all the TM_{0n} circular waveguide modes. The modal fields exhibiting a $\nu = 1/2$ singularity coefficient at the edge of the vane are those of the $TE_{1/2,n}$ and the $TM_{1/2,n}$ modes, and the singular $TE_{1/2,1}$ mode is dominant. A mode is defined to be singular when its *vector* eigenfield exhibits unbounded behavior.

The eigenvalues (k_c^2) are obtained numerically by the classical FEM discretization (Galerkin approach) of the *scalar* differential equation

$$\nabla^2 \Phi + k_c^2 \Phi = 0 \tag{7.89}$$

with boundary condition $\Phi = 0$ or $\partial\Phi/\partial n = 0$ for the TM and the TE modes, respectively, where $\hat{\boldsymbol{n}}$ is the outward unit normal on the domain boundary. For the TE modes, the boundary condition is *natural*; conversely, for the TM modes, basis functions are not assigned to nodes or edges on the conductor boundary.

Notice that at $k_c^2 = 0$, the scalar problem (7.89) simplifies into $\nabla^2 \Phi = 0$ which, for the TE boundary condition, has the mathematical (albeit non-physical) solution $\Phi = \text{constant}$. Typically, the non-physical TE eigenvalue $k_{c0}^2 = 0$ obtained numerically is of the order of 10^{-7}

Table 7.3 Eigenvalues of the circular vaned waveguide computed with singular hierarchical scalar functions.

Magnitude of the percentage error of the computed k_c^2 eigenvalue of (7.89) for each of the first 21 modes of the circular vaned waveguide of normalized (unit) radius a. Error bars for modes with percentage errors below the 10^{-7} threshold value do not appear in the figures.

Order ⇓	# Mode	k_c Reference Values
1	$TE_{\frac{1}{2}1}$	1.165561185207
2	TE_{11}	1.841183781341
3	$TE_{\frac{3}{2}1}$	2.460535572190
4	TE_{21}	3.054236928227
5	$TM_{\frac{1}{2}1}$	3.141592653590
6	$TE_{\frac{5}{2}1}$	3.632797319832
7	TE_{01}	3.831705970208
8	TM_{11}	3.831705970208
9	TE_{31}	4.201188941211
10	$TM_{\frac{3}{2}1}$	4.493409457909
11	$TE_{\frac{1}{2}2}$	4.604216777201
12	$TE_{\frac{7}{2}1}$	4.762196386967
13	TM_{21}	5.135622301841
14	TE_{41}	5.317553126084
15	TE_{12}	5.331442773525
16	$TM_{\frac{5}{2}1}$	5.763459196895
17	$TE_{\frac{9}{2}1}$	5.868419863031
18	$TE_{\frac{3}{2}2}$	6.029292381615
19	$TM_{\frac{1}{2}2}$	6.283185307180
20	TM_{31}	6.380161895924
21	TE_{51}	6.415616375700

24 Triangle mesh

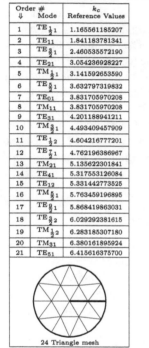

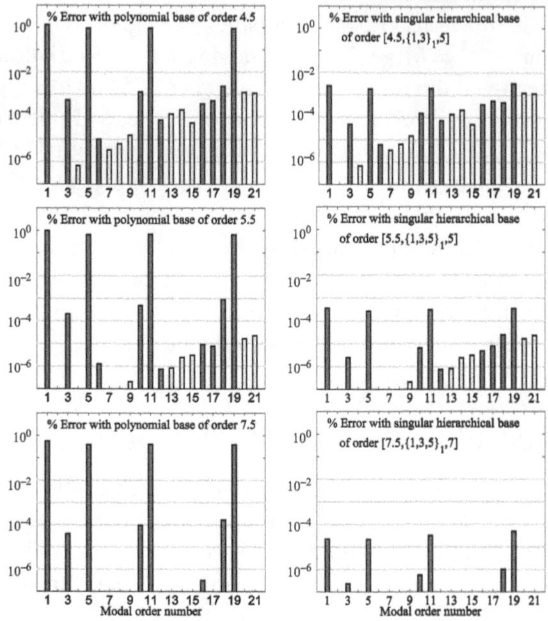

$$\% \text{ Error} = 100 \, \frac{k_c^2 - k^2_{c,\text{reference}}}{k^2_{c,\text{reference}}}$$

	Base order	Number of unknowns	CN mass matrix ΦΦ	k_c value for mode $TE_{1/2,1}$ (num. obtained)
TE problem	[4.5, –, –]	341	2.17×10^3	**1.173**
	$[4.5, \{1,3\}_1, 5]$	403	4.47×10^5	**1.165575**
	[5.5, –, –]	481	3.99×10^3	**1.171**
	$[5.5, \{1,3,5\}_1, 5]$	574	6.55×10^6	**1.1655632**
	[7.5, –, –]	833	1.07×10^4	**1.1688**
	$[7.5, \{1,3,5\}_1, 7]$	962	1.16×10^7	**1.16556131**

	Base order	Number of unknowns	CN mass matrix ΦΦ	k_c value for mode $TM_{1/2,1}$ (num. obtained)
TM problem	[4.5, –, –]	261	8.15×10^2	**3.156**
	$[4.5, \{1,3\}_1, 5]$	319	3.56×10^4	**3.14163**
	[5.5, –, –]	385	1.42×10^3	**3.152**
	$[5.5, \{1,3,5\}_1, 5]$	472	1.18×10^6	**3.1415967**
	[7.5, –, –]	705	3.88×10^3	**3.1476**
	$[7.5, \{1,3,5\}_1, 7]$	828	5.49×10^6	**3.14159298**

or lower, and this is indeed always associated with a numerically constant eigensolution Φ_0; in Table 7.3, the "zero" eigenvalue is not reported.

The numerical results were obtained by using the coarse mesh shown in Table 7.3, consisting of 12 curved and 12 rectilinear triangles, for a total of 24 cells, with only six singular triangles attached to the sharp-edge vertex. The curved edges of the triangular cells lying on the circular waveguide border are defined by parametric curves of eighth degree; that is, the triangles attached to these edges are distorted to the eighth degree. The circular border is therefore described by 12 curved segments of eighth degree, defined by 96 interpolation points in total.

When using cells of large size, the error due to a low-order description of a curved boundary should never be underestimated since it can easily be quite high, as it is for the present circular vaned waveguide test-case if the degree of distortion of the 12 curved triangles is less than the sixth. In this connection, observe that the principal advantage of high-order basis functions is only realized when they are used with high-order representations of curved geometries to allow using cells of large size.

The figures in Table 7.3 show the magnitude of the percentage error in the computed square value of k_c for each of the first 21 modes of the circular vaned waveguide. The results reported by light bars correspond to modes supported also by the standard circular waveguide. Dark bars show the modes supported only by the vaned waveguide. Error bars for modes with percentage errors below the 10^{-7} threshold value do not appear in the figures.

The three figures in the central column of Table 7.3 were obtained using purely polynomial (hierarchical) basis functions. The results in the right-hand column were obtained by augmenting the polynomial bases by the Meixner subset $\{\phi_{1\ell}, \phi_{3\ell}\}_1$ or $\{\phi_{1\ell}, \phi_{3\ell}, \phi_{5\ell}\}_1$. (Recall that R_2 and R_4 have a pole at $\nu = 1/2$ and cannot be used in the present case.) Table 7.3 clearly shows that the error for the modes supported also by the "standard" circular waveguide diminishes by increasing the order of the polynomial subset and it is rather independent of the order of the Meixner subsets. Conversely, the error for the 1st, 5th, 11th, and 19th mode of the vaned waveguide obtained with purely polynomial basis always "flies up into the sky." This is clearly due to the fact that the *vector*-field $\nabla\Phi$ for these singular modes (that correspond to $m/2 = 1/2$) is singular at the edge of the vane, and cannot be modeled by purely polynomial basis functions. Furthermore, in contrast to what one may think, the error for the singular modes obtained by using purely polynomial bases is reduced only slightly by diminishing the size of the mesh (results not reported). On the contrary, with additive bases, one can easily reduce the percentage errors for k_c^2 of the singular modes by a factor of 10^2 or 10^4 with respect to the error obtained with the purely polynomial bases.

The number of unknowns and the mass-matrix CN

$$g_{mn} = \int_{\mathcal{D}} B_m B_n \, d\mathcal{D} \tag{7.90}$$

obtained with singular hierarchical bases of different order are reported below the figures of Table 7.3. The CN remains roughly constant as the azimuthal order A_σ of the Meixner basis subset is increased; that is, by increasing the order of the Jacobi polynomials appearing in (7.55). This is clearly due to the fact that the azimuthal factors of our Meixner bases involve orthogonal Jacobi polynomials.

The percentage error in the computed value of k_c^2 for increasing azimuthal order A_σ of the Meixner subset is shown in Figure 7.8 for the singular modes $\text{TE}_{1/2,1}$, $\text{TM}_{1/2,1}$, and for the fifth ($\text{TM}_{1/2,1}$) and tenth ($\text{TM}_{3/2,1}$) mode of the circular vaned waveguide. (Recall that the error versus the azimuthal order is constant for all the modes that are also supported by the circular waveguide.) The results of Figure 7.8, obtained for $p = 4$, are "typical" and are reported to show that, with the mesh shown in Table 7.3, good convergence to the exact results is already obtained by using Meixner subsets of azimuthal order $A_\sigma = p$, where $p + 0.5$ is the order of the polynomial subset. In other words, for properly meshed domains, there is really no advantage to Meixner subsets with azimuthal order higher than the order of the polynomial subset.

Finally, Figure 7.9 shows the improvement in the results obtained by increasing just the order $(p + 0.5)$ of the polynomial subset while using the Meixner subset $\{\phi_{1\ell}\}_1$ (dashed-lines) or $\{\phi_{1\ell}, \phi_{3\ell}\}_1$ (solid-lines) of fifth azimuthal order (i.e. for all ℓ up to five). Although Figure 7.9

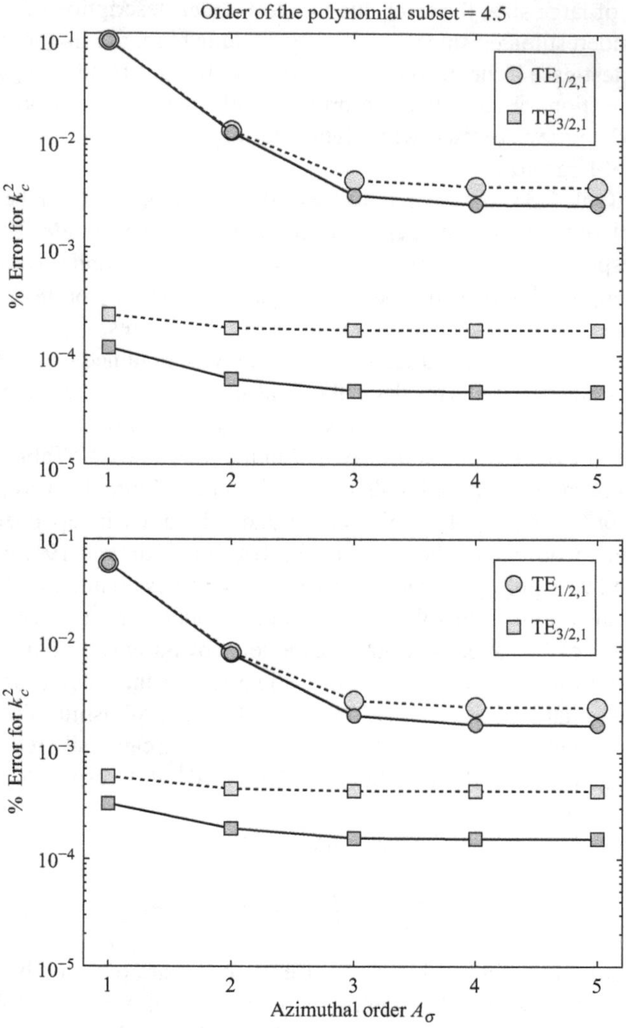

Figure 7.8 Convergence of the results for increasing azimuthal order A_σ of the Meixner subset obtained with the 24 triangle mesh of Table 7.3. The figure at top reports, in a semi-logarithmic scale, the percentage error in k_c^2 for the first $(\mathrm{TE}_{1/2,1})$ and third $(\mathrm{TE}_{3/2,1})$ modes of the circular vaned waveguide. The figure at bottom shows the error for the fifth $(\mathrm{TM}_{1/2,1})$ and tenth $(\mathrm{TM}_{3/2,1})$ modes. The dashed-line results were obtained with the hierarchical set $[4.5, \{1\}_1, A_\sigma]$; the solid-line results were obtained using the hierarchical set $[4.5, \{1, 3\}_1, A_\sigma]$.

© 2013 IEEE. Reprinted with permission from R. D. Graglia, A. F. Peterson, and L. Matekovits, "Singular, hierarchical scalar basis functions for triangular cells," *IEEE Trans. Antennas Propag.*, vol. 61, no. 7, pp. 3674–3692, Jul. 2013.

shows the percentage error of the first four modes only, the behavior of the error for the other modes is similar. That is, as the order of the polynomial subset is increased, the error for the *regular* modes (those supported by the circular waveguide) is independent of the Meixner subset and is greatly reduced, while the error for the other modes decreases at a much slower

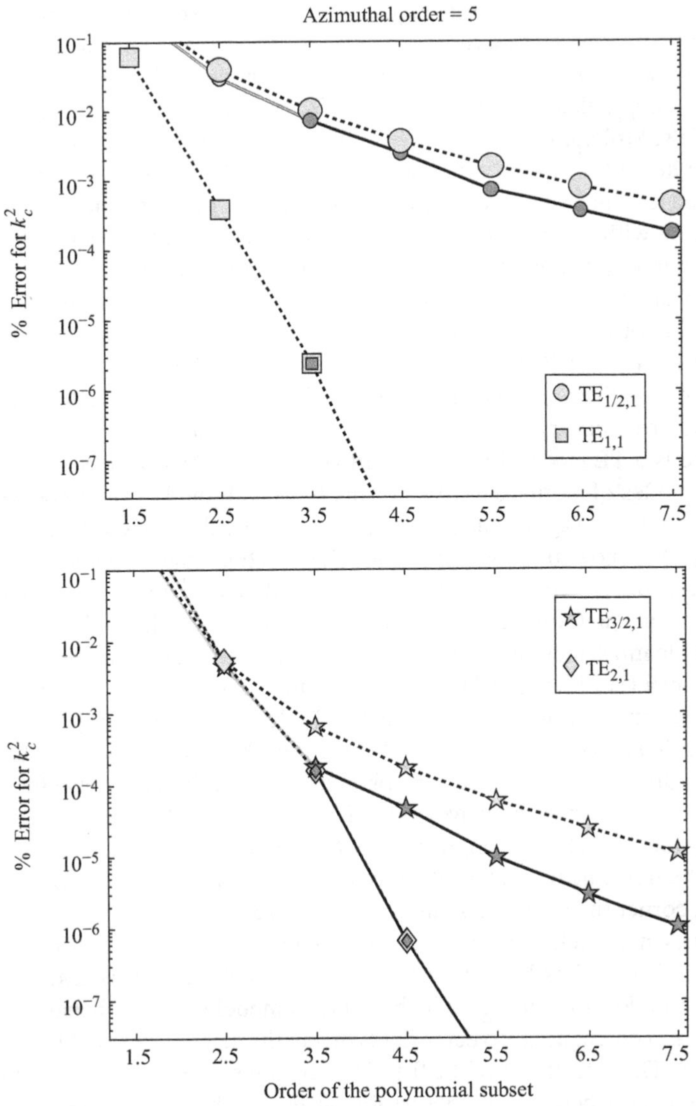

Figure 7.9 Percentage error in k_c^2 for increasing polynomial subset order obtained using the 24 triangle mesh of Table 7.3 and Meixner subsets of fifth azimuthal order. The figure at top shows the error for the dominant ($TE_{1/2,1}$) and second ($TE_{1,1}$) modes of the circular vaned waveguide; the figure at bottom reports the error for the third ($TE_{3/2,1}$) and fourth ($TE_{2,1}$) modes. The dashed-line results are for the Meixner subset $\{\phi_{1\ell}\}_1$. The solid-line results are obtained by using the Meixner subset $\{\phi_{1\ell}, \phi_{3\ell}\}_1$. Recall that to use the Meixner subset $\{\phi_{1\ell}, \phi_{3\ell}\}_1$ one needs to use a polynomial subset of order at least equal to 3.5.

© 2013 IEEE. Reprinted with permission from R. D. Graglia, A. F. Peterson, and L. Matekovits, "Singular, hierarchical scalar basis functions for triangular cells," *IEEE Trans. Antennas Propag.*, vol. 61, no. 7, pp. 3674–3692, Jul. 2013.

rate and is similar to those shown in Figure 7.9. (Recall also that the error for the singular modes obtained with a purely polynomial basis is quite large, as shown in Table 7.3.)

A waveguide whose cross section is the L-shaped union of three unit squares is a second, much-studied example that further proves the effectiveness of the singular hierarchical bases. (In the late 1970s, Moler used this example to illustrate the power of his new computer system MATLAB; see also [32–34] and references therein.) Table 7.4 compares the smallest values of the computed k_c eigenvalues reported by other authors (and obtained with non-traditional superelements or with other more sophisticated numerical techniques) with those we have obtained with purely polynomial and with singular hierarchical bases for the three different meshes (La, Lb, and Lc) shown in the table. As reported in [34] from symmetry considerations, the exact k_c value of the third TM mode is equal to $\sqrt{2}\,\pi$; a value we reproduce to the first 13 digits using the $[7.5, \{1, 2, 4\}_1, 7]$ singular basis with mesh Lb. For the third and fourth TE modes of Table 7.4, the sequences of the k_c numerical values appear to converge to π. By appealing to other symmetry considerations (omitted here for brevity), one can quite easily prove that there is a TE mode of the L-shaped waveguide whose $k_c = \pi$ value coincides with one of the TE modes of a square cavity. For the fourth TE mode, we reproduce $k_c = \pi$ to the first 13 digits using, once again, the $[7.5, \{1, 2, 4\}_1, 7]$ singular basis with mesh Lb.

Table 7.4 further shows that pyramidal (nodal) basis functions (order $p = 0$) on a very dense mesh (mesh Lc with 2,748 triangles and more than a thousand unknowns) can provide results which, at best, are accurate only to two or three digits, while singular high-order basis functions can easily provide much more accurate results (from four to six digits) even with an extremely coarse mesh (mesh La with four triangles and less than 250 unknowns). The mass-matrix CN obtained in each computation is also reported in Table 7.4, while the figure at the bottom-left corner of the table shows the CNs obtained using the Lb mesh with various basis orders. For the other two figures in the table at bottom, obtained with mesh La, we have first computed and normalized (to unit power in the waveguide) the electric field $\boldsymbol{E}_{\text{sing.}}$ and $\boldsymbol{E}_{\text{pol.}}$ of the first TM mode by using the $[7.5, \{1\}_1, 7]$ singular basis and the pure polynomial basis of order $p = 7$, respectively. The figure in the middle at bottom shows the $\boldsymbol{E}_{\text{sing.}}$ field topography. The figure at the lower right corner shows the magnitude of the vector difference $|\boldsymbol{E}_{\text{sing.}} - \boldsymbol{E}_{\text{pol.}}|$ (maximum value of 0.06 shown in black, zero value in white; obviously the field is not sampled on the edge tip where it is infinite). This latter figure shows that the field error is drastically reduced by adding just the first Meixner subset, which is able to model irrational behavior of the form χ^{ν} (with $\nu = 2/3$), and consequently, vector fields that in the neighborhood of the wedge approach infinity as $\chi^{-1/3}$. The potential Φ of the first TM singular mode is not reported in the table (this is shown in [34]) since there is little observable difference between the potential distributions $\Phi_{\text{sing.}}$ and $\Phi_{\text{pol.}}$, while there is an obvious difference between the fields $\boldsymbol{E}_{\text{sing.}}$ and $\boldsymbol{E}_{\text{pol.}}$. (For TM modes, $\boldsymbol{E}$ is proportional to $\nabla\Phi$ which, for this mode, is infinite at the corner.)

7.4.2 Effect of Varying the Number of Radial and Azimuthal Terms

In order to explore the behavior of the solution accuracy, matrix CN, and computational cost of using the singular scalar bases proposed in the preceding sections, we consider a specific example and systematically explore the behavior of these parameters as the number of radial and azimuthal basis functions is varied. Consider a circular cavity with a 30° wedge (mesh B in Figure 7.10). The dominant resonant frequencies of these cavities can be approximated by numerical solutions of the scalar Helmholtz equation using finite element procedures incorporating the additive singular basis functions. In this case, the list of exponents is $\nu = \{6/11, 12/11, 18/11, \ldots\}$.

Table 7.4 k_c Values of the first five TE and TM modes of the L-shaped waveguide.

Computed by using polynomial (regular) bases of order p

	Mesh La $p = 5$	Mesh La $p = 7$	Mesh Lb $p = 7$	Mesh Lc $p = 0$	
Ref. [35] 192 DoF	CN = 5.0×10^3 91 DoF	CN = 1.3×10^4 153 DoF	CN = 1.4×10^4 889 DoF	CN = 83.4 1,459 DoF	TE modes
1.2149	1.2171	1.2159	1.2150	1.2158	
1.8800	1.879946	1.879912	1.8799026	1.8805	
3.1423	3.1415988	3.141592660	3.141592653590	3.1445	
3.1423	3.141600	3.141592669	3.141592653591	3.1446	
3.3757	3.374887	3.3748392	3.37483076	3.3785	
	CN = 86.2 55 DoF	CN = 149.6 105 DoF	CN = 5.2×10^3 777 DoF	CN = 39.7 1,291 DoF	TM modes
3.1054	3.112	3.1084	3.1055	3.1098	
3.8989	3.89876	3.89844	3.8983685	3.9037	
4.4438	4.44294	4.4428830	4.442882938158	4.451	
5.4375	5.445	5.43351	5.4333683	5.449	
5.6551	5.676	5.654	5.650	5.670	

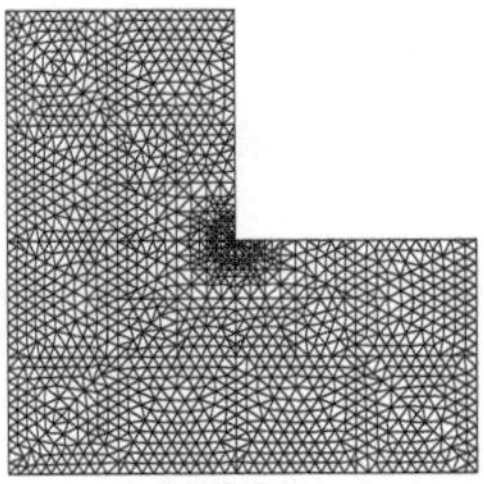

Mesh Lc
2,748 Triangles; 1,459 nodes

Computed by using the present singular hierarchical bases

	Mesh La [7.5, $\{1, 2, 4\}_1$, 7]	Mesh Lb [4.5, $\{1\}_1$, 5]	Mesh Lb [5.5, $\{1, 2, 4\}_1$, 5]	Mesh Lb [7.5, $\{1, 2, 4\}_1$, 7]	
	CN = 6.3×10^7 240 DoF	CN = 3.1×10^4 387 DoF	CN = 1.1×10^7 589 DoF	CN = 4.0×10^7 997 DoF	TE modes
	1.2147523	1.2147553	1.214751771	1.214751769205	
	1.8799021	1.8799029	1.879901957835	1.879901957825	
	3.141592660	3.141592678	3.141592653564	3.141592653589	
	3.141592661	3.141592681	3.141592653589	3.141592653589	
	3.37483047	3.3748311	3.374830277913	3.374830277903	
Ref. [36]	CN = 2.5×10^7 186 DoF	CN = 3.6×10^3 315 DoF	CN = 4.5×10^6 499 DoF	CN = 2.1×10^7 879 DoF	TM modes
$3.1047904\,^{61}_{73}$	3.1047929	3.10480	3.10479076	3.10479056	
$3.8983652\,^{10}_{78}$	3.8983671	3.898370	3.89836534	3.898365299	
$4.4428829\,^{04}_{72}$	4.442882986	4.4428849	4.442882945	4.442882938158	
$5.433367\,^{32}_{45}$	5.43348	5.433373	5.43336784	5.43336739	
$5.64912\,^{68}_{72}$	5.64930	5.649149	5.6491282	5.64912743	

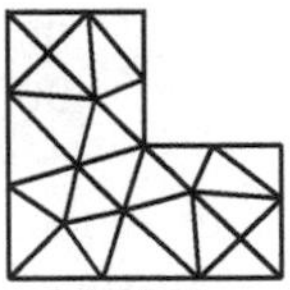

Mesh Lb
26 Triangles; 21 nodes

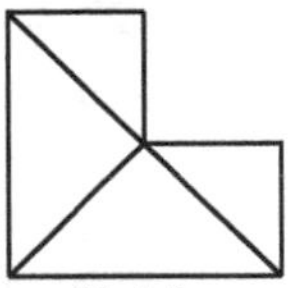

Mesh La
4 Triangles; 6 nodes

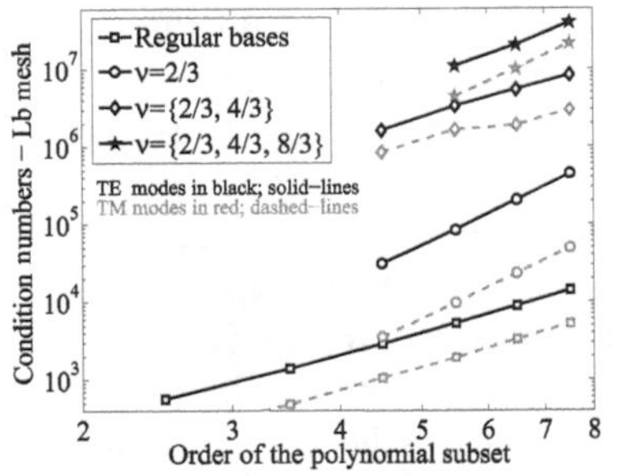

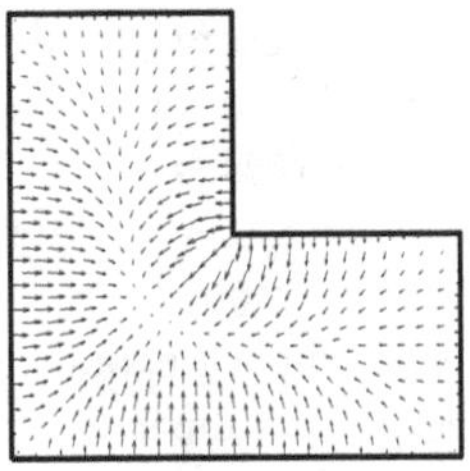

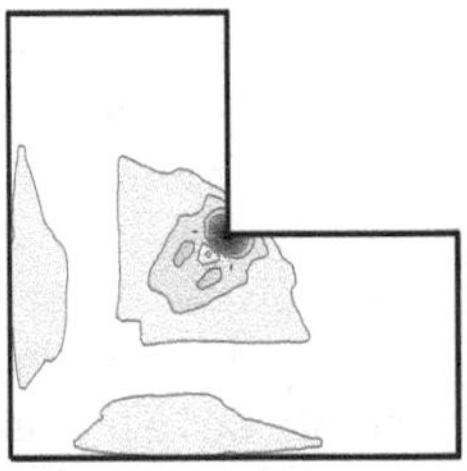

The smallest values of k_c for an L-shaped waveguide (consisting of a concatenation of three unit squares) reported by other authors are compared with those obtained (using the three meshes shown above) with purely polynomial bases (top sub-table) and singular hierarchical bases (bottom sub-table). The sub-table at top shows that pyramidal (nodal) basis functions (order $p = 0$) on a very dense mesh (Lc) can provide results which, at best, are accurate only to two or three digits; better results (with the exception of the first TE and TM singular modes) are obtained with high-order *purely* polynomial bases and very coarse meshes. Use of singular bases significantly improves the quality of the numerical results, as shown in the sub-table at bottom. The three figures below the tables are discussed in the text.

Table 7.5 shows the error in the resonant wavenumber obtained for the lowest order (dominant) TE mode, as a function of various combinations of singular and non-singular basis functions used in the representation of the H_z field. Table 7.6 shows the matrix CNs for the Gram matrix for each of the basis combinations, while Table 7.7 shows the number of unknowns required for each result. The first column in these tables describes the combination of basis

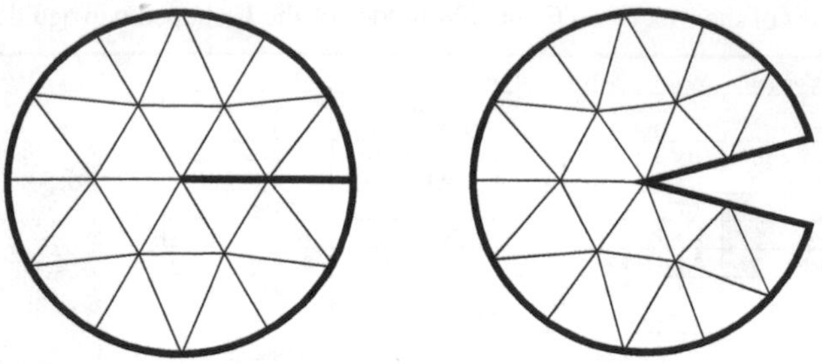

Mesh A
24 triangles

Mesh B
24 triangles

Figure 7.10 Two test problems, each with a wedge penetrating to the center of a cylindrical cavity of unit radius. Meshes A and B are used for problems with wedge angles of 0° and 30°, respectively.

functions in the singular cells; "poly" denotes the background polynomial order, while the number of singular functions per radial edge and cell (bubble functions) are given. The mode in question contains a fractional exponent, and the numerical results quickly improve in accuracy as singular DoFs are added to the representation. Despite the orthogonalization, the matrix CN grows as singular DoFs are added.

A close inspection of the data in Tables 7.5–7.7 suggests that (1) adding the first singular DoFs (the first radial function) reduces the error by at least one order of magnitude, (2) adding additional basis functions with that same singular behavior but additional azimuthal variation further reduces the error, and (3) while each additional radial function causes an increase of at least one order of magnitude in the matrix CN, additional azimuthal functions improve the accuracy but usually do not provide a significant increase in CN. It is noteworthy that adding additional radial functions without the corresponding azimuthal (bubble) singular functions does not reduce the error. As a specific example, for the degree 4 background (non-singular) expansion, the number of unknowns increases from 225 to 300 as the full range of singular functions is employed, while the error in the resonant frequency is reduced from 0.61% to 0.0005%.

For a given polynomial order, the matrix CN is primarily a function of the number of (singular) radial functions employed. Since the total number of unknowns increases only slowly with additional azimuthal functions, it is clear that the best accuracy for a given number of unknowns is obtained when the azimuthal variation is the same as that of the regular polynomial basis functions. In other words, for a given polynomial order, we should employ as many singular functions as possible up to the same azimuthal variation as the underlying polynomial order.

Specifically, for quadratic polynomial order, the best accuracy is obtained with one singular radial function per edge and one additional singular *bubble* function per cell in cells containing a singular node; for cubic order, the best accuracy requires two singular radial functions per edge and four singular bubble functions per cell (two per each radial function); etc. Since the additional unknowns are confined to cells around the singular node, the growth of unknowns is small. However, when following this approach, the matrix CN grows by roughly two orders of magnitude for each increment in order. This suggests that despite the orthogonality, the linear dependence of the singular functions remains an issue.

Results obtained using the suggested approach were used to compute the error in the lowest six TE and TM modes of a circular cavity containing 0° and 30° wedges in Figures 7.11– 7.14.

Table 7.5 Percent error in the resonant wavenumber obtained for the dominant TE mode of the circular cavity with a 30° wedge extending to the center (mesh B in Figure 7.10).

Order	Quadratic	Cubic	Degree 4	Degree 5
Poly only (no singular)	2.22	1.06	0.61	0.39
Poly + 1 radial/edge	0.17	0.061	0.034	0.022
Paly + 1 radial/edge + 1 bubble function	0.11	0.014	0.0052	0.0028
Poly + 1 radial/edge + 2 bubble functions		0.011	0.0028	0.0010
Poly + 1 radial/edge + 3 bubble functions			0.0027	0.00094
Poly + 1 radial/edge + 4 bubble functions				0.00093
Poly + 2 radial/edge		0.061	0.034	0.022
Poly + 2 radial/edge + 1 bubble function per exponent		0.010	0.0050	0.0025
Poly + 2 radial/edge + 2 bubble functions per exponent		0.0077	0.0020	0.00061
Poly + 2 radial/edge + 3 bubble functions per exponent			0.0019	0.00046
Poly + 2 radial/edge + 4 bubble functions per exponent				0.00046
Poty + 3 radial/edge			0.034	0.022
Poly + 3 radial/edge + 1 bubble functions per exponent			0.0037	0.0023
Poly + 3 radial/edge + 2 bubble functions per exponent			0.00062	0.00025
Poly + 3 radial/edge + 3 bubble functions per exponent			0.00050	0.000095
Poly + 3 radial/edge + 4 bubble functions per exponent				0.000089
Poly + 4 radial/edge				0.022
Poly + 4 radial/edge + 1 bubble function per exponent				0.0023
Poly + 4 radial/edge + 2 bubble functions per exponent				0.00019
Poly + 4 radial/edge + 3 bubble functions per exponent				0.000038
Poly + 4 radial/edge + 4 bubble functions per exponent				0.000033

Table 7.6 Matrix CN of the mass-matrix for the TE analysis of the circular cavity with a 30° wedge (mesh B in Figure 7.10).

Order	Quadratic	Cubic	Degree 4	Degree 5
Poly only (no singular)	121	428	1,060	2,190
Poly + 1 radial/edge	897	3,270	9,610	26,200
Poly + 1 radial/edge + 1 bubble function	905	3,290	9,680	26,400
Poly + 1 radial/edge + 2 bubble functions		3,300	9,680	26,400
Poly + 1 radial/edge + 3 bubble functions			9,710	26,500
Poly + 1 radial/edge + 4 bubble functions				26,500
Poly + 2 radial/edge		100,000	393,000	1,200,000
Poly + 2 radial/edge + 1 bubble function per exponent		100,000	397,000	1,330,000
Poly + 2 radial/edge + 2 bubble functions per exponent		100,000	397,000	1,330,000
Poly + 2 radial/edge + 3 bubble functions per exponent			397,000	1,330,000
Poly + 2 radial/edge + 4 bubble functions per exponent				1,330,000
Poly + 3 radial/edge			5,960,000	10,700,000
Poly + 3 radial/edge + 1 bubble function per exponent			6,080,000	10,700,000
Poly + 3 radial/edge + 2 bubble functions per exponent			6,080,000	10,700,000
Poly + 3 radial/edge + 3 bubble functions per exponent			6,080,000	10,700,000
Poly + 3 radial/edge + 4 bubble functions per exponent				10,700,000
Poly + 4 radial/edge				20,700,000
Poly + 4 radial/edge + 1 bubble functions per exponent				103,000,000
Poly + 4 radial/edge + 2 bubble functions per exponent				103,000,000
Poly + 4 radial/edge + 3 bubble functions per exponent				103,000,000
Poly + 4 radial/edge + 4 bubble functions per exponent				103,000,000

Table 7.7 Number of unknowns for the TE analysis of the circular cavity with a 30° wedge extending to the center (mesh B in Figure 7.10).

Order	Quadratic	Cubic	Degree 4	Degree 5
Poly only (no singular)	65	133	225	341
Poly + 1 radial/edge	72	140	232	348
Poly + 1 radial/edge + 1 bubble function	78	146	238	354
Poly + 1 radial/edge + 2 bubble functions		152	244	360
Poly + 1 radial/edge + 3 bubble functions			250	366
Poly + 1 radial/edge + 4 bubble functions				372
Poly + 2 radial/edge		147	239	355
Poly + 2 radial/edge + 1 bubble function per exponent		159	251	367
Poly + 2 radial/edge + 2 bubble functions per exponent		171	263	379
Poly + 2 radial/edge + 3 bubble functions per exponent			275	391
Poly + 2 radial/edge + 4 bubble functions per exponent				403
Poly + 3 radial/edge			246	362
Poly + 3 radial/edge + 1 bubble function per exponent			264	380
Poly + 3 radial/edge + 2 bubble functions per exponent			282	398
Poly + 3 radial/edge + 3 bubble functions per exponent			300	416
Poly + 3 radial/edge + 4 bubble functions per exponent				434
Poly + 4 radial/edge				369
Poly + 4 radial/edge + 1 bubble functions per exponent				393
Poly + 4 radial/edge + 2 bubble functions per exponent				417
Poly + 4 radial/edge + 3 bubble functions per exponent				441
Poly + 4 radial/edge + 4 bubble functions per exponent				465

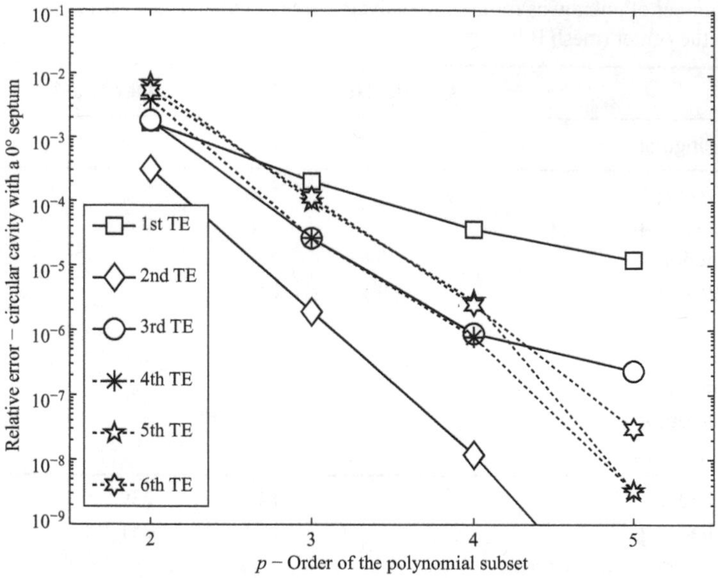

Figure 7.11 Error in the resonant frequency for the first six TE modes of the circular cavity with a 0° septum, obtained using a scalar expansion.

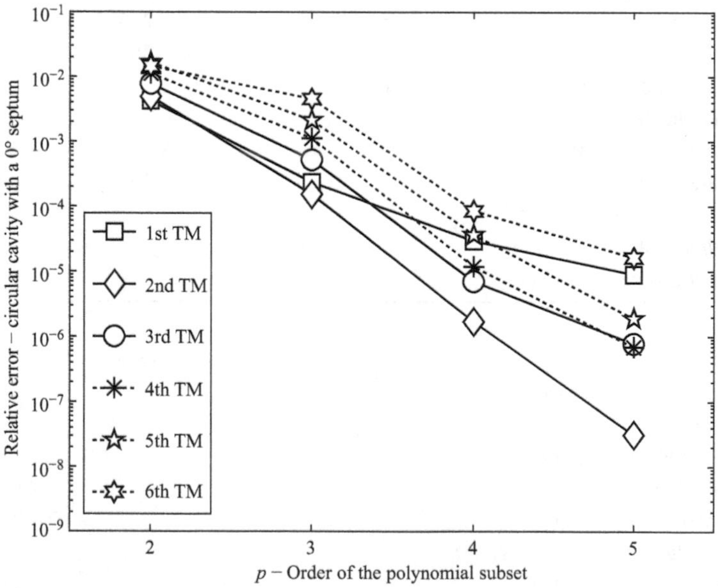

Figure 7.12 Error in the resonant frequency for the first six TM modes of the circular cavity with a 0° septum, obtained using a scalar expansion.

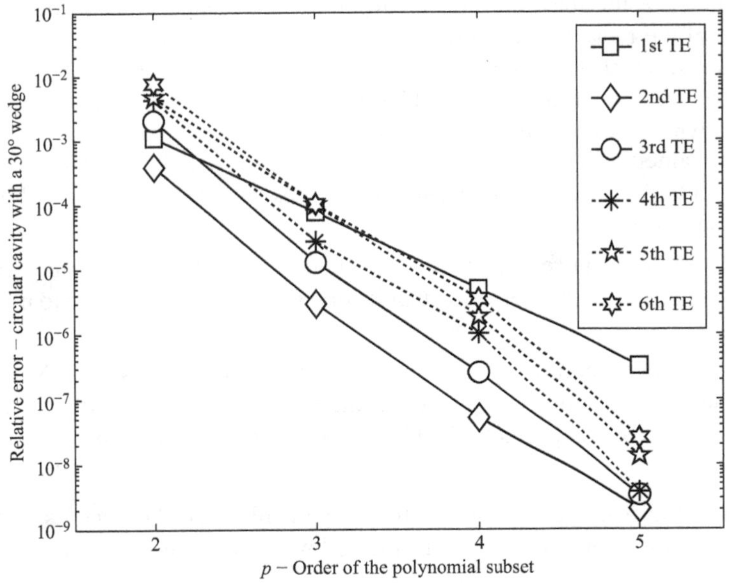

Figure 7.13 Error in the resonant frequency for the first six TE modes of the circular cavity with a 30° wedge, obtained using a scalar expansion.

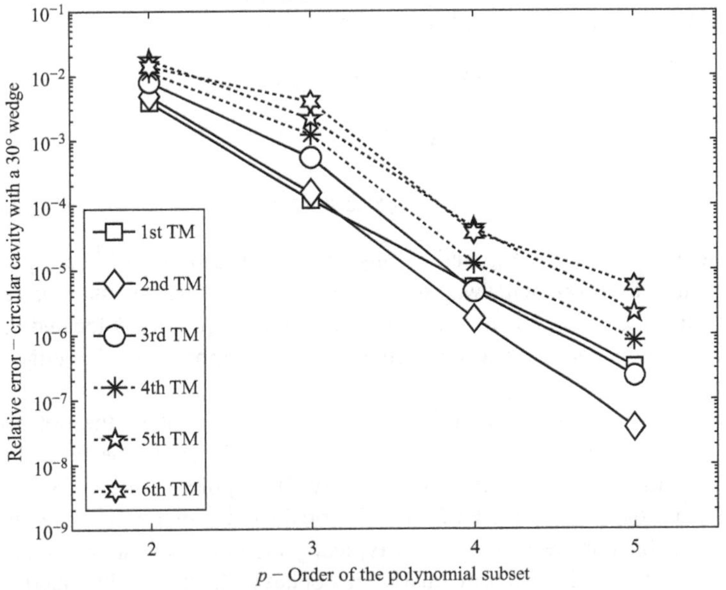

Figure 7.14 Error in the resonant frequency for the first six TM modes of the circular cavity with a 30° wedge, obtained using a scalar expansion.

In some of the 0° results, the full number of radial functions was not employed due to the fact that some of the radial functions developed using the procedure described above involve integer exponents, which duplicate the DoFs of the background representation.

In summary, the data of the tables and the results reported so far show that when dealing with problems where field singularities are excited one observes that

1. Results obtained with purely polynomial bases are restricted in practice to far less accuracy than additive singular bases, even if one uses very dense meshes and/or high-order polynomial bases.
2. To reduce the error in regions where unbounded fields are present one needs to use a fairly complete field representation that involves Meixner subsets able to model multiple irrational terms.
3. The cost of incrementing purely high-order polynomial bases with Meixner subsets of sufficiently high order is typically an increase by a factor of 10^2–10^3 in the CN of the system matrix, compared to the CN of the system matrix for the purely polynomial base.

In summary, the numerical results of this section illustrate the benefits of the singular scalar bases introduced in Section 7.3. In the following sections, we extend these ideas to vector expansions.

7.5 Singular Vector Basis Functions for Triangles

7.5.1 Substitutive Curl-Conforming Vector Bases

For vector representations, the lowest order (non-singular) curl-conforming bases are the three "edge-based" functions given in simplex coordinates by

$$\mathbf{\Omega}_1 = \xi_2 \nabla \xi_3 - \xi_3 \nabla \xi_2 \tag{7.91}$$

$$\mathbf{\Omega}_2 = \xi_3 \nabla \xi_1 - \xi_1 \nabla \xi_3 \tag{7.92}$$

$$\mathbf{\Omega}_3 = \xi_1 \nabla \xi_2 - \xi_2 \nabla \xi_1 \tag{7.93}$$

These functions have order $p = 0.5$ and belong to the mixed-order Nédélec spaces [35]. The substitutive singular functions will swap some or all of these with basis functions exhibiting the desired vertex singularity. As above, we assume that the singularity is located at node 1 with $\chi = 0$, and that the edge indices and basis functions are numbered so that edge 1 is opposite node 1 in the cell.

The FEM treatment of the vector Helmholtz equation involves the dot product of the basis and testing functions, as well as the dot product of curl of the basis functions with the curl of the testing functions. (In most of these 2D problems, the curls represent scalar longitudinal fields, the z-components E_z or H_z considered previously.) The singular vector basis functions for the transverse fields will typically exhibit a dependence like the gradients of the scalar functions considered previously. This behavior is often unbounded at the singular point. Vector basis functions can be designed so that their curl is (1) unbounded, (2) singular but bounded at the singular point, or (3) non-singular. Here, we consider examples where the curl exhibits a non-analytic behavior of $O(\chi^\alpha)$ when the transverse vector contains an unbounded term of $O(\chi^{\alpha-1})$. That dependence represents the physical behavior of fields near the tip of a perfectly conducting or dielectric wedge, for an appropriate excitation, as discussed in Section 7.1.

As an example of substitutive functions, Graglia and Lombardi [36] proposed that (7.91)–(7.93) be replaced by functions, expressed by a mixture of simplex and triangular-polar coordinates, with the form

$$\boldsymbol{B}_1 = \left[\chi^\alpha + \alpha(1-\chi)\right]\boldsymbol{\Omega}_1 \tag{7.94}$$

$$\boldsymbol{B}_2 = \boldsymbol{\Omega}_1 - \chi^{\alpha-1}\left\{\nabla\xi_3 + (1-\alpha)\left(\frac{1+\sigma}{2}\right)\nabla\xi_1\right\} \tag{7.95}$$

$$\boldsymbol{B}_3 = \boldsymbol{\Omega}_1 - \chi^{\alpha-1}\left\{\nabla\xi_2 + (1-\alpha)\left(\frac{1-\sigma}{2}\right)\nabla\xi_1\right\} \tag{7.96}$$

These functions are constructed so that they reduce to the conventional base vectors when $\alpha = 1$, and can be substituted for them when a transverse field behavior with an $O(\chi^{\alpha-1})$ dependence is desired at node 1. The function $\boldsymbol{B}_1$ has a constant tangential component along the edge opposite node 1, and can be scaled in order to maintain tangential continuity with a regular basis function from the set (7.91)–(7.93) in the adjacent cell. Basis function $\boldsymbol{B}_2$ has a non-zero tangential component only along the edge opposite node 2, where it varies as $\chi^{\alpha-1}$. Function $\boldsymbol{B}_3$ has a variation $\chi^{\alpha-1}$ along the edge opposite node 3, and contributes no tangential field to the other edges. These two functions can be adjusted to maintain tangential-vector continuity to similar singular functions in the adjacent cells.

The curls of (7.94)–(7.96) are given by

$$\hat{z}\cdot\nabla\times\boldsymbol{B}_1 = \left[(\alpha+2)(1-\xi_1)^\alpha + \alpha(3\xi_1 - 1)\right]/\mathcal{J} \tag{7.97}$$

$$\hat{z}\cdot\nabla\times\boldsymbol{B}_2 = \hat{z}\cdot\nabla\times\boldsymbol{B}_3 = 2/\mathcal{J} \tag{7.98}$$

where $\mathcal{J}$ is the Jacobian of the mapping to the child space. Thus, the first function provides a curl with $O(\chi^\alpha)$ behavior. The other two basis functions have a constant curl, like the original base vectors in (7.91)–(7.93).

A variety of other substitutive basis functions have been proposed to replace the $p = 0.5$ functions in (7.91)–(7.93) or their divergence-conforming counterparts [19–21, 26, 36, 37]. The literature shows differing philosophies about the nature of the singularity of the curl of these functions: some have curls that exhibit as strong a singularity as the transverse fields, while others have curls that are completely non-singular.

As in the scalar case, substitutive vector basis sets such as (7.94)–(7.96) replace the conventional basis functions with singular basis functions having similar interpolation and continuity properties. However, substitutive functions have several limitations as described previously. In the subsequent sections, we consider additive functions.

7.5.2 Additive Curl-Conforming Vector Bases

In this section, we review a recently proposed singular, high-order hierarchical vector basis family of the additive kind for triangular cells [23, 24]. These bases are applicable to two-dimensional problems such as microstrip structures or waveguides whose cross section contains edges. They constitute one possible extension of the scalar bases described above to the vector case.

The singular functions developed below build upon the hierarchical vector bases proposed in Chapter 5, originally proposed and investigated in [38–39]. In 2D, the functions of order $p = 0.5$ are the three base vectors in (7.91)–(7.93). To expand the space to order $p = 1.5$, the $p = 0.5$ functions are combined with three edge-based linear expansion functions

$$\begin{gathered}\sqrt{3}(\xi_2 - \xi_3)\mathbf{\Omega}_1 \\ \sqrt{3}(\xi_3 - \xi_1)\mathbf{\Omega}_2 \\ \sqrt{3}(\xi_1 - \xi_2)\mathbf{\Omega}_3\end{gathered} \tag{7.99}$$

and two cell-based quadratic basis functions

$$\begin{gathered}2\sqrt{3}\xi_1\mathbf{\Omega}_1 \\ 2\sqrt{3}\xi_2\mathbf{\Omega}_2\end{gathered} \tag{7.100}$$

The edge-based functions are shared by adjacent cells; the cell-based functions are local to the cell and are not shared by neighboring cells. The next order, $p = 2.5$, requires the addition of three edge-based functions

$$\begin{gathered}\sqrt{5}\left\{\frac{3(\xi_2 - \xi_3)^2 - 1}{2} - \frac{\xi_1}{2}(\xi_1 - 2)\right\}\mathbf{\Omega}_1 \\ \sqrt{5}\left\{\frac{3(\xi_3 - \xi_1)^2 - 1}{2} - \frac{\xi_2}{2}(\xi_2 - 2)\right\}\mathbf{\Omega}_2 \\ \sqrt{5}\left\{\frac{3(\xi_1 - \xi_2)^2 - 1}{2} - \frac{\xi_3}{2}(\xi_3 - 2)\right\}\mathbf{\Omega}_3\end{gathered} \tag{7.101}$$

and four cell-based functions

$$\begin{gathered}6\sqrt{5}(\xi_2 - \xi_3)\xi_1\mathbf{\Omega}_1 \\ 6\sqrt{5}(\xi_3 - \xi_1)\xi_2\mathbf{\Omega}_2 \\ 2\sqrt{3}\xi_1(5\xi_1 - 3)\mathbf{\Omega}_1 \\ 2\sqrt{3}\xi_2(5\xi_2 - 3)\mathbf{\Omega}_2\end{gathered} \tag{7.102}$$

These are hierarchical functions, so the functions in (7.101) and (7.102) are added to the $p = 1.5$ basis functions to bring the total number of basis functions in the $p = 2.5$ set to 15. Functions of order greater than $p = 2.5$ are described in Chapter 5.

The following section considers the construction of the hierarchical Meixner basis functions, using the pseudo-polar coordinates (χ, σ). Orthogonal polynomials of the σ variable and orthogonal functions of the χ variable are employed to improve the linear independence of the resulting additive basis.

7.6 Singular Hierarchical Meixner Basis Sets

7.6.1 The Singularity Coefficients

As in the scalar case, the Meixner basis functions are organized in a hierarchy by ordering at the onset, typically for increasing values, all the non-integer exponents one considers in the finite list (7.38). In common with the scalar development, the first Meixner subset in the hierarchy is made with functions that depend only on the first singularity coefficient ν_1 of the list, while the functions of the second subset depend on the first and the second singularity coefficients ν_1 and ν_2 to achieve orthogonality between the two Meixner subsets. The functions of the generic nth Meixner subset depend on the first n singularity coefficients (from ν_1 to ν_n).

In the following, for brevity, we sometimes omit the singularity coefficients from the list of the function's variables, or just use the symbol ν to indicate the first n singularity coefficients that define the nth subset of the Meixner functions.

7.6.2 Auxiliary Functions

Above, for the singular triangle of Figure 7.4, we derived the scalar hierarchical Meixner basis functions in (7.55). The gradients of these functions are

$$\nabla \phi_{n1}^{i\pm 1} = \frac{1}{4}\left[R_n'(\chi)\,(1\pm\sigma)\,\nabla\chi \pm \frac{R_n(\chi)}{\chi}\,(\chi\nabla\sigma)\right] \tag{7.103}$$

$$\nabla \phi_{n\ell} = \frac{1}{4}\left[R_n'(\chi) f_{\ell-2}(\sigma)\,\frac{(1-\sigma^2)}{2}\nabla\chi + \frac{R_n(\chi)}{\chi}\,g_{\ell-2}(\sigma)\,(\chi\nabla\sigma)\right] \tag{7.104}$$

with $\nabla\chi$ and $\chi\nabla\sigma$ given in (7.21). For ease of reference, the expression of the radial functions $R_n(\chi)$, with $R_n'(\chi)=\mathrm{d}R_n(\chi)/\mathrm{d}\chi$, is reported in the second row of Table 7.8.

The azimuthal polynomials $f_q(\sigma)$ in (7.55) and (7.104) are a re-scaled version of the Jacobi polynomials $P_q^{(2,2)}(\sigma)$ (see (7.40)) and are *mutually orthogonal* on the σ-interval $[-1, 1]$. The function

$$g_{\ell-2}(\sigma) = \left[\frac{(1-\sigma^2)}{2}\,\frac{\mathrm{d}f_{\ell-2}(\sigma)}{\mathrm{d}\sigma} - \sigma f_{\ell-2}(\sigma)\right] \tag{7.105}$$

is even for odd values of ℓ and odd otherwise, with zero integral mean value on the σ-interval $[-1, +1]$; its expression in terms of Jacobi polynomials is readily obtained, for instance

$$g_0(\sigma) = -\sqrt{10}\sigma \tag{7.106}$$

$$g_1(\sigma) = \frac{\sqrt{70}}{2}(1-3\sigma^2) \tag{7.107}$$

$$g_2(\sigma) = \sqrt{30}(4\sigma - 7\sigma^3) \tag{7.108}$$

Table 7.8 Method to Numerically Evaluate the Coefficients of the Auxiliary Radial Functions.

The coefficients that define the radial functions $R_n(k,\nu,\chi)$ and $\chi\, S_n(k,\nu,\chi)$ (for $n = 1,\ldots,N$) can be obtained numerically by a recursive algorithm that requires, at the nth step, the solution of a linear system plus the evaluation of the square root of a quadratic form. For example, to be clear, for the function $\chi\, S_n(k,\nu,\chi)$ the recursive algorithm requires to first find the coefficients of the function $\chi\, S_1(k,\nu,\chi)$, then those of the function $\chi\, S_2(k,\nu,\chi)$, and so on. The recursive algorithm is readily obtained (details are left to the reader) by imposing the orthogonality and the normalization conditions and by recognizing that each radial function can be written as the product of a row-vector $\boldsymbol{X}_n$ times a column-vector $\boldsymbol{U}_n^t$ divided by an appropriate scaling coefficient c_n. The entries of $\boldsymbol{X}_n$ are well-defined functions of the radial variable χ while the other coefficients (apart c_n) defining the radial function are ordered in the column-vector $\boldsymbol{U}_n^t$ (the superscript t indicates that $\boldsymbol{U}_n^t$ is the transpose of the row vector $\boldsymbol{U}_n$). The length of the vectors $\boldsymbol{X}_n$ and $\boldsymbol{U}_n$ is $(n+k)$, and $\boldsymbol{U}_n^t$ is the solution of the linear system $\boldsymbol{M}_n\,\boldsymbol{U}_n^t = \boldsymbol{V}_n^t$, with $\boldsymbol{V}_n = [0,0,\ldots,0,1]$ and the last row of $\boldsymbol{M}_n = [0,0,\ldots,0,1]$. The coefficient c_n is the square root of the quadratic form $c_n^2 = \boldsymbol{U}_n\,\boldsymbol{Q}_n\,\boldsymbol{U}_n^t$. The square matrices $\boldsymbol{M}_n$ and $\boldsymbol{Q}_n$ are of order $(n+k)$ and $\boldsymbol{Q}_n$ is symmetric.

The zeros of the scaling coefficient c_n are poles of the radial functions. $S_n(k,\nu,\chi)$ has $(n+k-1)$ poles at $\nu_n = 0,1,\ldots,n+k-2$ while $R_n(k,\nu,\chi)$ has $(n+k)$ poles at $\nu_n = 1,2,\ldots,n+k$.

$$R_n(k,\nu,\chi) = \frac{1}{c_n}\left(a_n\,\chi^{\nu_n} - \sum_{j=1}^{n+k} b_{nj}\,\chi^j\right) = \frac{1}{c_n}\,\boldsymbol{X}_n\,\boldsymbol{U}_n^t$$

$$\boldsymbol{X}_n = \left[(\chi^{\nu_n}-\chi),\,(\chi^{\nu_n}-\chi^2),\ldots,(\chi^{\nu_n}-\chi^{n+k})\right]$$

$$\boldsymbol{U}_n = [b_{n1}, b_{n2},\ldots,b_{nn}, b_{n,\,n+k}]$$

Recall that to get $R_n|_{\chi=1} = 0$ we set $a_n = \sum_{j=1}^{n+k} b_{nj}$

The entries of $\boldsymbol{M}_n$ reported below are obtained by setting $b_{n,\,n+k} = 1$ and by imposing the orthogonality conditions (7.109) plus the orthogonality conditions **A** (7.110), or **B** (7.111), or **C** (7.112).

$$\chi\, S_n(k,\nu,\chi) = \frac{1}{c_n}\left(a_n\,\chi^{1+\nu_n} - \sum_{j=1}^{n+k} b_{nj}\,\chi^j\right) = \frac{1}{c_n}\,\boldsymbol{X}_n\,\boldsymbol{U}_n^t$$

$$\boldsymbol{X}_n = \left[(\chi^{1+\nu_n}-\chi),\,(\chi^{1+\nu_n}-\chi^2),\ldots,(\chi^{1+\nu_n}-\chi^{n+k})\right]$$

$$\boldsymbol{U}_n = [b_{n1}, b_{n2},\ldots,b_{nn}, b_{n,\,n+k}]$$

Recall that to get $S_n|_{\chi=1} = 0$ we set $a_n = \sum_{j=1}^{n+k} b_{nj}$

The entries of $\boldsymbol{M}_n$ reported below are obtained by setting $b_{n,\,n+k} = 1$ and by imposing the orthogonality conditions (7.126) plus the orthogonality conditions **D** (7.127).

$\boldsymbol{M}_n$ Entries

For $j = 1,2,\ldots,n+k$ set:

$$\begin{cases} M_n[1+m,j] = d_{nmj}\,, \\ \text{for } m = 0,1,\ldots,k-1. \end{cases}$$

$$\text{If } n>1 \text{ then } \begin{cases} M_n[k+i,j] = f_{nij}\sum_{\ell=1}^{i+k} b_{i\ell}\,g_{nij\ell}\,, \\ \text{for } i = 1,2,\ldots,n-1. \end{cases}$$

$$M_n[n+k,p] = 0,\ \text{for } p = 1,2,\ldots,n+k-1.$$

$$M_n[n+k,n+k] = 1\,.$$

With

	f_{nij}	$g_{nij\ell}$
On cond. **A**	$\dfrac{\nu_n - j}{\nu_i + j + 2}$	$\dfrac{\nu_i - \ell}{\nu_n + \ell + 2}\left(\dfrac{1}{j+\ell+2} + \dfrac{1}{\nu_i + \nu_n + 2}\right)$
On cond. **B**	$\dfrac{\nu_n - j}{\nu_i + j}$	$\dfrac{\nu_i - \ell}{\nu_n + \ell}\left(\dfrac{j\ell}{j+\ell} + \dfrac{\nu_i\,\nu_n}{\nu_i + \nu_n}\right)$
On cond. **C**	$\dfrac{\nu_n - j}{\nu_i + j}$	$\dfrac{\nu_i - \ell}{\nu_n + \ell}\left(\dfrac{1}{j+\ell} + \dfrac{1}{\nu_i + \nu_n}\right)$
On cond. **D**	$\dfrac{\nu_n - j + 1}{\nu_i + j + 3}$	$\dfrac{\nu_i - \ell + 1}{\nu_n + \ell + 3}\left(\dfrac{1}{j+\ell+2} + \dfrac{1}{\nu_i + \nu_n + 4}\right)$

$$\text{and } d_{nmj} = \begin{cases} \mathrm{IntP}_m(\nu_n) - \mathrm{IntP}_m(j) & \text{for the } R(\chi) \text{ functions,} \\ \mathrm{IntP}_m(2+\nu_n) - \mathrm{IntP}_m(1+j) & \text{for the } S(\chi) \text{ functions.} \end{cases}$$

$\boldsymbol{Q}_n$ Entries

The scaling coefficients c_n are the square root of the quadratic form $c_n^2 = \boldsymbol{U}_n\,\boldsymbol{Q}_n\,\boldsymbol{U}_n^t$. The entries of $\boldsymbol{Q}_n$ for the various normalizations are given at right

	Normalization	$Q_n[i,j]$ (for $i,j = 1,2,\ldots,n+k$)
nA (7.113)	$\int_0^1 R_n^2(k,\nu,\chi)\,\chi\,d\chi = 1$	$\dfrac{(\nu_n - i)\,(\nu_n - j)}{(\nu_n + i + 2)\,(\nu_n + j + 2)}\left(\dfrac{1}{i+j+2} + \dfrac{1}{2\nu_n + 2}\right)$
nB (7.114)	$\int_0^1 \left[R_n'(k,\nu,\chi)\right]^2 \chi\,d\chi = 1$	$\dfrac{(\nu_n - i)\,(\nu_n - j)}{(\nu_n + i)\,(\nu_n + j)}\left(\dfrac{ij}{i+j} + \dfrac{\nu_n}{2}\right)$
nC (7.115)	$\int_0^1 \left[\dfrac{R_n(k,\nu,\chi)}{\chi}\right]^2 \chi\,d\chi = 1$	$\dfrac{(\nu_n - i)\,(\nu_n - j)}{(\nu_n + i)\,(\nu_n + j)}\left(\dfrac{1}{i+j} + \dfrac{1}{2\nu_n}\right)$
nD (7.128)	$\int_0^1 [\chi\, S_n(k,\nu,\chi)]^2\,\chi\,d\chi = 1$	$\dfrac{(\nu_n - i + 1)\,(\nu_n - j + 1)}{(\nu_n + i + 3)\,(\nu_n + j + 3)}\left(\dfrac{1}{i+j+2} + \dfrac{1}{2\nu_n + 4}\right)$

To improve the linear independence of the Meixner subsets, the number $(n+k)$ of the coefficients b_{nj} (with $j = 1,2,\ldots,n+k$) that define $R_n(k,\nu,\chi)$ in Table 7.8 increases with n. These coefficients are in fact obtained by imposing $(n+k-1)$ orthogonality conditions plus one normalization condition. The first k conditions impose the orthogonality of the radial function

$R_n(k, \nu, \chi)$ to a number k (≥ 1) of regular polynomial functions. Here, we generalize[4] what was done in Section 7.3.3, and impose the orthogonality of the radial functions to the first k-shifted Legendre polynomials $P_m^*(\chi) = P_m(2\chi - 1)$ obtained for $m = 0, 1, \ldots, k-1$, by imposing

$$\int_0^1 P_m(2\chi - 1)\, R_n(k, \nu, \chi) \mathrm{d}\chi = 0, \quad \forall n \tag{7.109}$$

The other $(n-1)$ orthogonality conditions require the orthogonality of the nth radial function to the previously defined ith radial function, for all $i = 1, 2, \ldots, n-1$. By noticing that the scalar functions (7.55) and the components of the gradient vectors (7.103) and (7.104) contain as a factor the radial function $R_n(\chi)$, $R_n'(\chi)$, or $R_n(\chi)/\chi$, one may choose one among the following three *orthogonality conditions*

$$\textbf{A:} \quad \int_0^1 R_i(k, \nu, \chi)\, R_n(k, \nu, \chi)\, \chi\, \mathrm{d}\chi = 0 \tag{7.110}$$

$$\textbf{B:} \quad \int_0^1 R_i'(k, \nu, \chi)\, R_n'(k, \nu, \chi)\, \chi\, \mathrm{d}\chi = 0 \tag{7.111}$$

$$\textbf{C:} \quad \int_0^1 \frac{R_i(k, \nu, \chi)}{\chi} \frac{R_n(k, \nu, \chi)}{\chi}\, \chi\, \mathrm{d}\chi = 0 \tag{7.112}$$

(Recall that χ is the factor appearing in the Jacobian ($\mathcal{J}_R = \chi/2$) of the transformation from ξ-parent to pseudo-polar coordinates – see (7.18).) The radial functions are finally normalized by using one of the following three *normalization conditions*

$$\textbf{nA:} \quad \int_0^1 R_n^2(k, \nu, \chi)\, \chi\, \mathrm{d}\chi = 1 \tag{7.113}$$

$$\textbf{nB:} \quad \int_0^1 \left[R_n'(k, \nu, \chi)\right]^2 \chi\, \mathrm{d}\chi = 1 \tag{7.114}$$

$$\textbf{nC:} \quad \int_0^1 \left[\frac{R_n(k, \nu, \chi)}{\chi}\right]^2 \chi\, \mathrm{d}\chi = 1 \tag{7.115}$$

$R_n(\chi)$ can be normalized with the condition **nA**, **nB**, or **nC** independently of the orthogonality condition one uses (**A**, **B**, or **C**). Notice that the Jacobian in (7.109) is unity while that in (7.110)–(7.115) is χ. Notice also that Section 7.3.3 considers only the orthogonalization (7.110) and the normalization (7.113).

With reference to Table 7.8, the polynomial component

$$R_{\text{polypart}}(n, k, \chi) = -\frac{1}{c_n} \sum_{j=1}^{n+k} b_{nj}\, \chi^j \tag{7.116}$$

of the radial function $R_n(k, \nu, \chi)$ depends on the orthogonality condition (**A**, **B**, or **C**). To form the Meixner base, R_n is multiplied by an azimuthal polynomial of order ℓ. Therefore, for a

[4]This generalization, first discussed in [24], permits one to numerically derive the radial functions $R_n(\chi)$ with the procedure explained in Section 7.6.7.

given k and for fixed (maximum) values of n and ℓ, the scalar (7.55) and vector (7.103) and (7.104) Meixner sets depend on the orthogonality conditions used to define the radial functions. The space spanned by the resulting basis is not a strong function of the chosen orthogonality condition (**A**, **B**, or **C**) provided that the regular polynomial subset added to form the entire base is of sufficiently high order. With reference to (7.55) and (7.116), this certainly occurs when using polynomial subsets of order higher than $(n+k+\ell)$ (values of $(n+k+\ell)>6$ will seldom be used in practice because they are quite high), although very good convergence to the exact results is normally obtained with complete polynomial subsets of order n or $n+1$, independently from the orthogonality condition used to form the Meixner subsets (**A**, **B**, or **C**; see the results of Section 7.7). The relative small influence of the orthogonality conditions on the precision of the numerical results is also due to the constraints set for the polynomial component (7.116) of the radial function that vanishes at $\chi=0$, while $R_n(k,\nu,\chi)$ vanishes at $\chi=1$ (see [22, 24]).

7.6.3 Representation of Singular Fields

In the following, we introduce three types of basis functions, each with the goal of representing singular behavior in a different component of the vector field. The first category, the *singular scalar bases*, is used to provide the correct behavior for the $\hat{z}$-component of the primary field (the field being expanded in basis functions). The second basis type, the *singular static vector bases*, provides the desired singular behavior of the primary transverse vector field, *without introducing singular behavior into the curl of that field*. The third type, the *singular non-static vector bases*, provides singular behavior in the curl of the primary transverse vector field (the $\hat{z}$-component of the secondary field). Depending on the problem, and what DoFs are desired, one might not use all three types of singular basis function in a given analysis.

7.6.4 Singular Scalar Bases

The singular part of the longitudinal field component is expanded in terms of the hierarchical scalar basis functions (7.55), which in turn are obtained from the radial functions $R_n(k,\nu,\chi)$ derived by imposing the orthogonality condition **A** in (7.109) and (7.110), plus the normalization condition **nA** in (7.113).

The first subscript n used for $\phi_{n\ell}$ in (7.55) indicates that these functions model a field component that vanishes at the singular point as χ^{ν_n}; the second subscript is the azimuthal order of the function, with $\ell\geq 2$ in the last of (7.55). The functions ϕ_{n1}^{i+1} and ϕ_{n1}^{i-1} are based on the edge indicated in their superscript ($\xi_{i\pm1}=0\Rightarrow\sigma=\pm1$) and vanish on the other two remaining edges. Conversely, $\phi_{n\ell}$ (for $\ell\geq 2$) are *bubble* functions that vanish along the three edges of the triangular cell.

The rest of this section discusses the Meixner *vector* set which is organized into two subsets formed by static curl-free functions (Section 7.6.5) and by non-static (Section 7.6.6) vector functions whose curl vanishes at the edge of the wedge as χ^{ν}.

7.6.5 Singular Static Vector Bases

The basis functions to expand the singular *curl-free* component of the transverse field are the gradient functions (7.103) and (7.104). These functions are defined by using the radial functions $R_n(k,\nu,\chi)$ as derived by imposing the orthogonality conditions (7.109) together with any of the orthogonality conditions **A**, **B**, or **C** (7.110)–(7.112). That is, the $R_n(k, \nu, \chi)$ that define the singular *vector* bases can be different (as far as their orthogonalization properties are concerned) from those defining the singular scalar bases.

Our numerical experiments indicate that the orthogonality conditions **A**, **B**, or **C** produce little difference in the CN of the system matrices (see Section 7.7) while, typically, the most convenient normalization condition to be used to define the static vector bases is condition **nB**.

For $\ell \geq 2$, the gradients of $\phi_{n\ell}$ (7.104) are *bubble* vector functions, with a vanishing tangent along the three edges of the singular triangular cell. On the other hand, the gradients of $\phi_{n1}^{i\pm1}$ (7.103) are edge-based functions with a non-vanishing tangent component only along the edge indicated in their superscript, with

$$\nabla\phi_{n1}^{i-1}\cdot\boldsymbol{\ell}_{i-1}\Big|_{\sigma=-1} = -\nabla\phi_{n1}^{i+1}\cdot\boldsymbol{\ell}_{i+1}\Big|_{\sigma=+1} = \frac{R_n'(\chi)}{2} \tag{7.117}$$

and where $\boldsymbol{\ell}_{i\pm1}$ and $\boldsymbol{\ell}_i$ are the edge vectors defined in Section 4.5.1 of Chapter 4 (see also [3]), with $\boldsymbol{\ell}_i+\boldsymbol{\ell}_{i+1}+\boldsymbol{\ell}_{i-1}=0$. The gradients (7.103) and (7.104) contain a singular term proportional to χ^{ν_n-1} (because both $R_n'(\chi)$ and $R_n(\chi)/\chi$ exhibit that behavior); this singularity is generally required by the physics of the wedge problem, although sometimes it may not be excited by the source.

The first of the conditions (7.109) (obtained for $m=0$ and $P_0^*(\chi)=1$) together with the fact that R_n vanishes at $\chi=1$ implies

$$\int_0^1 R_n(\chi)\mathrm{d}\chi = \int_0^1 \left[\frac{R_n(\chi)}{\chi}\right]\chi\,\mathrm{d}\chi = \int_0^1 R_n'(\chi)\,\chi\,\mathrm{d}\chi = 0 \tag{7.118}$$

which, in turn, makes the mean value of each vector component of the gradient functions (7.103) and (7.104) equal to zero on singular rectilinear triangles (i.e. when $\nabla\xi_{i\pm1}$ and $\nabla\xi_i$ do not vary on the triangular cell). Similarly, for rectilinear triangles, it is easily verified that the first two conditions (7.109) (obtained with $m=0, 1$) impose the orthogonality of the functions (7.103) and (7.104) to the zeroth-order regular vector basis functions $\boldsymbol{\Omega}_i$, $\boldsymbol{\Omega}_{i\pm1}$. This follows from the fact that χ and $(1-\chi)$ are the two radial factors appearing in the pseudo-polar expression of the zeroth-order regular vector basis functions

$$\boldsymbol{\Omega}_{i\pm1} = \pm\left[\chi\frac{(1\pm\sigma)}{2}\nabla\xi_i - (1-\chi)\nabla\xi_{i\mp1}\right] \tag{7.119}$$

$$\boldsymbol{\Omega}_i = \chi\left[\frac{(1-\sigma)}{2}\nabla\xi_{i-1} - \frac{(1+\sigma)}{2}\nabla\xi_{i+1}\right] \tag{7.120}$$

(On a rectilinear triangle of area A, the integral mean value of the zeroth-order function $\boldsymbol{\Omega}_j$ is $A[\nabla\xi_{j-1}-\nabla\xi_{j+1}]/3$, for $j=1, 2, 3$.)

As discussed in Section 7.3.3, $R_n(k, \nu, \chi)$ has $(n+k)$ poles at $\nu_n=1, 2, \ldots, n+k$ that originate from the normalization condition being used (**nA**, **nB**, or **nC**). These poles do not present a problem since the Meixner set is defined by excluding *a priori* all the integer values of the singularity coefficients ν.

7.6.6 Singular Non-Static Vector Bases

The non-static singular vector functions have a non-zero curl and are bounded. They are designed to introduce non-analytic DoFs into the curl of the primary transverse field, if that behavior is desired. They take the form

$$\begin{aligned}\boldsymbol{\mho}_{n\ell}(k,\chi,\sigma) &= S_n(k,\nu,\chi)\,\sqrt{2\,\ell+1}\,P_\ell(\sigma)\,\boldsymbol{\Omega}_i \\ &= \chi\,S_n(k,\nu,\chi)\,\sqrt{2\,\ell+1}\,P_\ell(\sigma)\,\frac{\boldsymbol{\Omega}_i}{\chi}\end{aligned} \tag{7.121}$$

$$\nabla\times\boldsymbol{\mho}_{n\ell}(k,\chi,\sigma) = \frac{\hat{\boldsymbol{n}}}{\mathcal{J}}\,T_n(k,\nu,\chi)\,\sqrt{2\,\ell+1}\,P_\ell(\sigma) \tag{7.122}$$

with $\boldsymbol{\Omega}_i$ given in (7.120)

$$S_n(k,\nu,\chi) = \frac{1}{c_n}\sum_{j=1}^{n+k} b_{nj}\left[\chi^{\nu_n} - \chi^{(j-1)}\right] \tag{7.123}$$

$$T_n(k,\nu,\chi) = \left[2\,S_n + \chi\,\frac{\mathrm{d}S_n}{\mathrm{d}\chi}\right] = \frac{1}{c_n}\sum_{j=1}^{n+k} b_{nj}\left[(2+\nu_n)\,\chi^{\nu_n} - (1+j)\,\chi^{(j-1)}\right] \tag{7.124}$$

where $P_\ell(\sigma)$ indicates the Legendre polynomial of order ℓ, $\hat{\boldsymbol{n}}$ is a unit vector normal to the element, and $\mathcal{J}$ is the Jacobian of the transformation from the child space to the $\boldsymbol{\xi}$-parent coordinates. The procedure to derive the coefficients c_n and b_{nj} of (7.123) is discussed in the next subsection and in Table 7.8. According to (7.123), the radial functions $\chi\,S_n(k,\nu,\chi)$ that define the singular non-static components contain a polynomial part that vanishes at $\chi=0$, which is included to improve the linear independence of the whole additive basis.

The function $S_n(k,\nu,\chi)$, and therefore $\boldsymbol{\mho}_{n\ell}(k,\chi,\sigma)$, vanishes at $\chi=1$. $\boldsymbol{\mho}_{n\ell}(k,\chi,\sigma)$ also has a vanishing tangential component along the other edges of the triangular cell, and is a bubble function.

The first subscript n used for $\boldsymbol{\mho}_{n\ell}$ indicates that these functions model a field whose curl vanishes at the singular point as χ^{ν_n}. This curl behavior is required because, in a source free region, the curl of the electric field $\nabla\times\boldsymbol{E}_t$ (or of the magnetic field $\nabla\times\boldsymbol{H}_t$) is proportional to the longitudinal component of the magnetic field H_z (or to E_z). The second subscript ℓ used to denote $\boldsymbol{\mho}_{n\ell}$ is the azimuthal order of the function.

The non-static singular basis functions must only be used in conjunction with basis functions that properly model the relevant transverse vector space, otherwise spurious solutions may arise in solutions of the vector Helmholtz equation. While these basis functions provide the singular DoFs desired in the $\hat{z}$-component of the secondary field (the curl of the primary vector field), they also introduce additional DoFs in the primary transverse space that must be made complete to properly represent the fields in that space. To complete the representation in the transverse space, and avoid the spurious solutions that the singular non-static vector functions (7.121) may produce, one has to include in the Meixner set the gradient functions (7.103) and (7.104) obtained using the radial function R containing the singularity coefficient (ν_n+2).

These gradient functions $\nabla\Theta_{j\ell}$, with[5]

$$\begin{aligned}\Theta_{j1}^{i\pm1}(\chi,\sigma) &= R_j(\nu+2,\chi)\,(1\pm\sigma)/4\\ \Theta_{j\ell}(\chi,\sigma) &= R_j(\nu+2,\chi)f_{\ell-2}(\sigma)\,(1-\sigma^2)/8\end{aligned} \tag{7.125}$$

contain a vector component proportional to χ^{ν_n+1} and a (vanishing) curl with a component proportional to χ^{ν_n}, as for the curl (7.122).

By recalling that χ is proportional to the Jacobian ($\mathcal{J}_R=\chi/2$) of the transformation from ξ-parent to pseudo-polar coordinates and that the vector function $\mathbf{\Omega}_i$ contains a χ factor (see (7.120)), the linear independence of the whole additive set is improved by setting the following orthogonality (7.126) and (7.127) and normalization (7.128) conditions

$$\int_0^1 P_m(2\chi-1)\,[\chi\,S_n(k,\nu,\chi)]\,\chi\,\mathrm{d}\chi=0 \tag{7.126}$$

$$\mathbf{D}:\quad \int_0^1 [\chi\,S_i(k,\nu,\chi)]\,[\chi\,S_n(k,\nu,\chi)]\,\chi\,\mathrm{d}\chi=0 \tag{7.127}$$

$$\mathbf{nD}:\quad \int_0^1 [\chi\,S_n(k,\nu,\chi)]^2\,\chi\,\mathrm{d}\chi=1 \tag{7.128}$$

obtained for $m=0,1,\ldots,k-1$ and $i=1,2,\ldots,n-1$. Notice that the Jacobian in (7.126) is χ, while in (7.109) it is unity.

The first of the conditions (7.126) (obtained with $m=0$) renders the mean value of the vector functions $\mathbf{\mho}_{n\ell}=0$ on *rectilinear* triangles, or whenever the gradient vectors $\nabla\xi_i$ and $\nabla\xi_{i\pm1}$ are constant. Similarly, for rectilinear triangles, the first two conditions (7.126) (obtained with $m=0,1$) impose the orthogonality of the function $\mathbf{\mho}_{n\ell}$ to the zeroth-order regular functions $\mathbf{\Omega}_i$ and $\mathbf{\Omega}_{i\pm1}$ given in (7.119) and (7.120).

For given k, the orthogonalization process discussed above and defined by (7.126)–(7.128) produces a unique Meixner subset (7.121). The radial function $S_n(k,\nu,\chi)$ has $(n+k)$ poles at $\nu_n=0,1,\ldots,n+k-1$ that originate from the normalization condition (7.128). (This is not a problem since the Meixner set is defined by excluding *a priori* all the integer values of the singularity coefficients ν.) Notice that the radial functions S are not made orthogonal to the static radial functions R to avoid further increasing the order of the polynomial basis subsets. The order p of the polynomial vector subset to be used with the Meixner subset incorporating $S_n(k,\nu,\chi)$ and $R_n(k,\nu,\chi)$ must be at least $(n+k-1)$ (see also [22]).

7.6.7 Numerical Evaluation of the Radial Functions R_n and S_n

Because of the orthogonality conditions, the auxiliary radial functions R_n and S_n used to build the Meixner subsets are functions of all the (non-integer) singularity coefficients from ν_1 to ν_n (although this is not explicitly specified in the expression reported above). Explicit expressions for the radial functions can be derived for the simpler case of metal wedges (see Table 7.1 and [22]) while for dielectric, magnetic, or composite wedges, the singularity coefficients are

[5] The functions $\Theta_{j1}^{i\pm1}$ and $\Theta_{j\ell}$ are different from the functions Θ_ℓ discussed in Section 7.3.7.

completely arbitrary to the point that it is not practical to derive explicit expressions for the radial functions beyond $n > 3$. A faster and much more efficient solution is to numerically evaluate the auxiliary radial functions R_n and S_n. As summarized in Table 7.8, the coefficients that define each radial function can be obtained by a recursive algorithm that involves the solution of linear systems and the evaluation of quadratic forms. In particular, the coefficients of the first k rows of the square matrix $\boldsymbol{M}_n$ in Table 7.8 that defines the linear system are obtained using the following fundamental integral that holds for $\beta > 0$

$$\mathrm{IntP}_m(\beta) = \int_0^1 P_m(2\chi - 1)\,\chi^\beta\,\mathrm{d}\chi = \begin{cases} \dfrac{1}{\beta+1} & \text{for } m = 0 \\ \dfrac{1}{\beta+1+m}\displaystyle\prod_{q=1}^{m}\dfrac{\beta+1-q}{\beta+q} & \text{for } m > 0 \\ 0 & \text{iff } \beta \in \mathbb{N} \text{ and } m > \beta \end{cases} \tag{7.129}$$

For example, with reference to Table 7.8, (7.129) yields

$$\int_0^1 P_m(2\chi - 1)\,(\chi^{\nu_n} - \chi^j)\,\mathrm{d}\chi = \mathrm{IntP}_m(\nu_n) - \mathrm{IntP}_m(j) \tag{7.130}$$

$$\int_0^1 P_m(2\chi - 1)\,(\chi^{1+\nu_n} - \chi^j)\,\chi\,\mathrm{d}\chi = \mathrm{IntP}_m(2+\nu_n) - \mathrm{IntP}_m(1+j) \tag{7.131}$$

The matrices $\boldsymbol{M}_n$ and $\boldsymbol{Q}_n$ of Table 7.8 tend to be ill conditioned for large values of k and n (say, for $n \geq 6$); this implies that one should use algorithms of sufficiently high numerical precision to find the radial function coefficients.

Table 7.8 considers the construction of the functions $R_n(k, \nu, \chi)$ (at left) and of the functions $S_n(k, \nu, \chi)$ (at right). In this connection, it is of importance to observe that, in applications, the maximum order of the polynomial part in χ of the vector functions $\nabla\phi_{n1}^{i\pm1}$, $\nabla\phi_{n\ell}$, and $\boldsymbol{\mho}_{n1}^{i\pm1}$, $\boldsymbol{\mho}_{n\ell}$ must be equal. This order is equal to $n + k - 1$ (see (7.103), (7.104), and (7.121)).

7.6.8 Example: $p = 1.5$ Basis with One Singular Exponent

As an example, let us develop the set of vector bases for treating a single singular exponent ν. A minimum background order of $p = 1.5$ (linear tangential/quadratic normal polynomial functions) is required. In cells containing a singular node, the regular basis functions are augmented with one radial function per edge ($\nabla\phi_{11}^{i-1}$ and $\nabla\phi_{11}^{i+1}$), and may also include one azimuthal function ($\nabla\phi_{12}$) per cell. If it is desired to represent the singularity in the curl, one or both of the functions

$$\boldsymbol{\mho}_{10}(\nu, \chi) = S_1(\nu, \chi)\boldsymbol{\Omega}_1 \tag{7.132}$$

$$\boldsymbol{\mho}_{11}(\nu, \chi) = S_1(\nu, \chi)\sqrt{3}\sigma\boldsymbol{\Omega}_1 \tag{7.133}$$

may be included. If $\boldsymbol{\mho}_{10}$ is included, the functions $\nabla\Theta_{11}^{i-1}$ and $\nabla\Theta_{11}^{i+1}$ must also be included (see (7.125)). If $\boldsymbol{\mho}_{11}$ is included, $\nabla\Theta_{12}$ must also be used.

7.6.9 Example: $p = 2.5$ Basis with Two Singular Exponents

For improved accuracy, an expansion of background order $p = 2.5$ may be considered. For that order, it is possible to include all the singular functions described in the previous subsection, incorporating the dominant fractional exponent. In addition, one additional azimuthal order may be included by including the functions $\nabla\phi_{13}$, $\boldsymbol{\mho}_{12}$, and $\nabla\Theta_{13}$. Then, a second fractional exponent may be introduced through radial function R_2 and the functions $\nabla\phi_{21}^{i-1}$, $\nabla\phi_{21}^{i+1}$, $\nabla\phi_{22}$, and $\nabla\phi_{23}$. A second exponent in the curl may be introduced through functions $\boldsymbol{\mho}_{20}$, $\boldsymbol{\mho}_{21}$, and $\boldsymbol{\mho}_{22}$. If $\boldsymbol{\mho}_{20}$ is included in the basis set, $\nabla\Theta_{21}^{i-1}$ and $\nabla\Theta_{21}^{i+1}$ must be included as well. Similarly, $\nabla\Theta_{22}$ must accompany $\boldsymbol{\mho}_{21}$ and $\nabla\Theta_{23}$ must be used with $\boldsymbol{\mho}_{22}$.

The previous paragraphs describe the full possible extent of the singular representation. However, it is also possible to use just the basis functions $\nabla\phi_{n1}^{i-1}$, $\nabla\phi_{n1}^{i+1}$, and $\nabla\phi_{n\ell}$ in conjunction with the regular (non-singular) vector basis functions to represent the transverse vector fields in a problem, if it is not desired to introduce fractional exponents into the curl space (the longitudinal fields). To introduce one or more fractional exponents into the curl, the $\boldsymbol{\mho}$ and $\nabla\Theta$ functions must be employed. However, these do not necessarily have to be introduced as early or as systematically as described above.

7.7 Numerical Results

To illustrate the performance of the vector representations, we begin by considering the same example considered previously (the cavity resonator with a 30° septum, for which exact resonant wavenumbers are provided in Table 7.9). Tables 7.10–7.12 show numerical results obtained for the dominant resonant wavenumber of the cavity, based on the triangular-cell mesh in Figure 7.15, and the singular vector representations. Various combinations of radial and azimuthal functions are included in order to assess the relative accuracy they produce. The specific set of exponents used for the results in these tables is $\nu = 6/11, 12/11, 2 + 6/11$. The coefficients of the various basis functions are obtained numerically, as are the entries of the element matrices for the vector Helmholtz equation. The specific combination of basis functions used in the cells containing the singular node is given in the first column of the tables, where "poly" denotes the background representation from Chapter 5.

Table 7.9 Smallest six exact resonant wavenumbers for the circular cavity of unit radius with a 30° wedge. Parameter ν is defined in (7.6), and p denotes the zero of the corresponding Bessel function.

ν	p	TE	TM
6/11	1	1.23133932138672	3.20587245491204
12/11	1	1.95655873019426	3.95379237868849
18/11	1	2.62439279978557	4.67022948640560
24/11	1	3.26604454018465	5.36539039582788
0	1	3.83170597020751	
30/11	1	3.89221742547959	6.04500834141272
6/11	2		6.35088699385497

Table 7.10 Percent error in the resonant wavenumber obtained for the dominant TE mode of the circular cavity with a 30° wedge (Figure 7.15), vector expansion.

Order	$p=1.5$	$p=2.5$	$p=3.5$
Poly only (no singular)	2.20	1.05	0.61
$\text{Poly} + \nabla\phi_{11}$	0.15	0.060	0.034
$\text{Poly} + \nabla\phi_{11} + \nabla\phi_{12}$	0.11	0.015	0.0055
$\text{Poly} + \nabla\phi_{11} + \nabla\phi_{12}, \nabla\phi_{13}$		0.014	0.0036
$\text{Poly} + \nabla\phi_{11} + \nabla\phi_{12}, \nabla\phi_{13}, \nabla\phi_{14}$			0.0035
$\text{Poly} + \nabla\phi_{11}, \nabla\phi_{21}$		0.060	0.034
$\text{Poly} + \nabla\phi_{11}, \nabla\phi_{21} + \nabla\phi_{12}, \nabla\phi_{22}$		0.0095	0.0055
$\text{Poly} + \nabla\phi_{11}, \nabla\phi_{21} + \nabla\phi_{12}, \nabla\phi_{22}, \nabla\phi_{13}, \nabla\phi_{23}$		0.0081	0.0025
$\text{Poly} + \nabla\phi_{11}, \nabla\phi_{21} + \nabla\phi_{12}, \nabla\phi_{22}, \nabla\phi_{13}, \nabla\phi_{23}, \nabla\phi_{14}, \nabla\phi_{24}$			0.0024
$\text{Poly} + \nabla\phi_{11}, \nabla\phi_{21}, \nabla\phi_{31}$			0.034
$\text{Poly} + \nabla\phi_{11}, \nabla\phi_{21}, \nabla\phi_{31} + \nabla\phi_{12}, \nabla\phi_{22}, \nabla\phi_{32} + \mho_{10}$			0.0044
$\text{Poly} + \nabla\phi_{11}, \nabla\phi_{21}, \nabla\phi_{31} + \nabla\phi_{12}, \nabla\phi_{22}, \nabla\phi_{32}, \nabla\phi_{13}, \nabla\phi_{23}, \nabla\phi_{33} + \mho_{10}, \mho_{11}$			0.0013
$\text{Poly} + \nabla\phi_{11}, \nabla\phi_{21}, \nabla\phi_{31} + \nabla\phi_{12}, \nabla\phi_{22}, \nabla\phi_{32}, \nabla\phi_{13}, \nabla\phi_{23}, \nabla\phi_{33}, \nabla\phi_{14}, \nabla\phi_{24}, \nabla\phi_{34} + \mho_{10}, \mho_{11}, \mho_{12}$			0.0012

Table 7.11 Matrix CN obtained for the circular cavity with a 30° wedge (Figure 7.15), vector expansion, TE.

Order	$p=1.5$	$p=2.5$	$p=3.5$
Poly only (no singular)	42	150	637
$\text{Poly} + \nabla\phi_{11}$	53	151	640
$\text{Poly} + \nabla\phi_{11} + \nabla\phi_{12}$	53	155	650
$\text{Poly} + \nabla\phi_{11} + \nabla\phi_{12}, \nabla\phi_{13}$		348	1,120
$\text{Poly} + \nabla\phi_{11} + \nabla\phi_{12}, \nabla\phi_{13}, \nabla\phi_{14}$			2,700
$\text{Poly} + \nabla\phi_{11}, \nabla\phi_{21}$		760	1,650
$\text{Poly} + \nabla\phi_{11}, \nabla\phi_{21} + \nabla\phi_{12}, \nabla\phi_{22}$		820	1,860
$\text{Poly} + \nabla\phi_{11}, \nabla\phi_{21} + \nabla\phi_{12}, \nabla\phi_{22}, \nabla\phi_{13}, \nabla\phi_{23}$		2,200	4,200
$\text{Poly} + \nabla\phi_{11}, \nabla\phi_{21} + \nabla\phi_{12}, \nabla\phi_{22}, \nabla\phi_{13}, \nabla\phi_{23}, \nabla\phi_{14}, \nabla\phi_{24}$			10,000
$\text{Poly} + \nabla\phi_{11}, \nabla\phi_{21}, \nabla\phi_{31}$			7,100
$\text{Poly} + \nabla\phi_{11}, \nabla\phi_{21}, \nabla\phi_{31} + \nabla\phi_{12}, \nabla\phi_{22}, \nabla\phi_{32} + \mho_{10}$			125,000
$\text{Poly} + \nabla\phi_{11}, \nabla\phi_{21}, \nabla\phi_{31} + \nabla\phi_{12}, \nabla\phi_{22}, \nabla\phi_{32}, \nabla\phi_{13}, \nabla\phi_{23}, \nabla\phi_{33} + \mho_{10}, \mho_{11}$			133,000
$\text{Poly} + \nabla\phi_{11}, \nabla\phi_{21}, \nabla\phi_{31} + \nabla\phi_{12}, \nabla\phi_{22}, \nabla\phi_{32}, \nabla\phi_{13}, \nabla\phi_{23}, \nabla\phi_{33}, \nabla\phi_{14}, \nabla\phi_{24}, \nabla\phi_{34} + \mho_{10}, \mho_{11}, \mho_{12}$			270,000

Table 7.12 Number of unknowns for the mesh of Figure 7.15 (circular cavity with a 30° wedge), vector expansion, TE.

Order	$p = 1.5$	$p = 2.5$	$p = 3.5$
Poly only (no singular)	104	228	400
Poly + $\nabla\phi_{11}$	109	233	405
Poly + $\nabla\phi_{11}$ + $\nabla\phi_{12}$	115	239	411
Poly + $\nabla\phi_{11}$ + $\nabla\phi_{12}$, $\nabla\phi_{13}$		245	417
Poly + $\nabla\phi_{11}$ + $\nabla\phi_{12}$, $\nabla\phi_{13}$, $\nabla\phi_{14}$			423
Poly + $\nabla\phi_{11}$, $\nabla\phi_{21}$		238	410
Poly + $\nabla\phi_{11}$, $\nabla\phi_{21}$ + $\nabla\phi_{12}$, $\nabla\phi_{22}$		250	422
Poly + $\nabla\phi_{11}$, $\nabla\phi_{21}$ + $\nabla\phi_{12}$, $\nabla\phi_{22}$, $\nabla\phi_{13}$, $\nabla\phi_{23}$		262	434
Poly + $\nabla\phi_{11}$, $\nabla\phi_{21}$ + $\nabla\phi_{12}$, $\nabla\phi_{22}$, $\nabla\phi_{13}$, $\nabla\phi_{23}$, $\nabla\phi_{14}$, $\nabla\phi_{24}$			446
Poly + $\nabla\phi_{11}$, $\nabla\phi_{21}$, $\nabla\phi_{31}$			415
Poly + $\nabla\phi_{11}$, $\nabla\phi_{21}$, $\nabla\phi_{31}$ + $\nabla\phi_{12}$, $\nabla\phi_{22}$, $\nabla\phi_{32}$ + $\mho_{10}$			439
Poly + $\nabla\phi_{11}$, $\nabla\phi_{21}$, $\nabla\phi_{31}$ + $\nabla\phi_{12}$, $\nabla\phi_{22}$, $\nabla\phi_{32}$, $\nabla\phi_{13}$, $\nabla\phi_{23}$, $\nabla\phi_{33}$ + $\mho_{10}$, $\mho_{11}$			463
Poly + $\nabla\phi_{11}$, $\nabla\phi_{21}$, $\nabla\phi_{31}$ + $\nabla\phi_{12}$, $\nabla\phi_{22}$, $\nabla\phi_{32}$, $\nabla\phi_{13}$, $\nabla\phi_{23}$, $\nabla\phi_{33}$, $\nabla\phi_{14}$, $\nabla\phi_{24}$, $\nabla\phi_{34}$ + $\mho_{10}$, $\mho_{11}$, $\mho_{12}$			487

© 2014 T&F. Reprinted with permission from R. D. Graglia, A. F. Peterson, L. Matekovits, and P. Petrini, "Hierarchical additive basis functions for the finite-element treatment of corner singularities," special issue on "Finite Elements for Microwave Engineering," *Electromagnetics*, vol. 34, pp. 171–198, Mar. 2014.

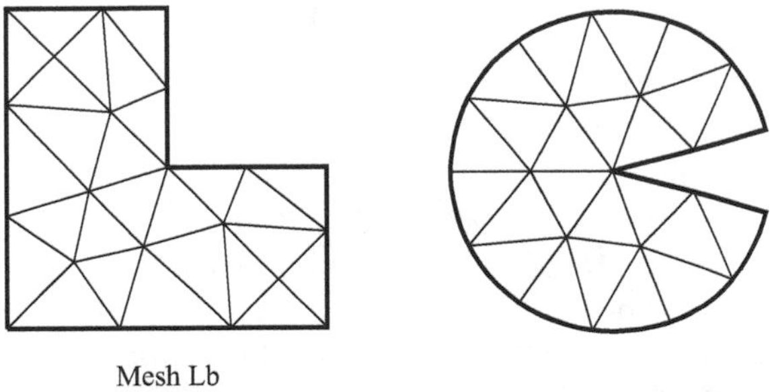

Mesh Lb
26 Triangles;
21 nodes

Figure 7.15 Two-dimensional cavities used for the test-case studies. The L-shaped cavity at left (consisting of a concatenation of three unit squares) is meshed with 26 triangles and 46 edges, with 14 edges on the border. The wedge aperture angle of the circular cavity at right is 30° and the cavity is meshed with 24 curvilinear triangles and 44 edges. To enhance the accuracy, the 12 edges of the cells lying on the border of the circular cavity are defined by curved segments of eighth degree, using (96 + 1) interpolation points in total.

In these results, we employ the singular dynamic vector bases $\boldsymbol{\mho}$ to a lesser extent than possible for a given background order; specifically, we begin to use those functions with background order of $p = 3.5$. For those expansions, the associated $\nabla\Theta$ functions coincide with $\nabla\phi$ functions also used in the representation, and are therefore omitted.

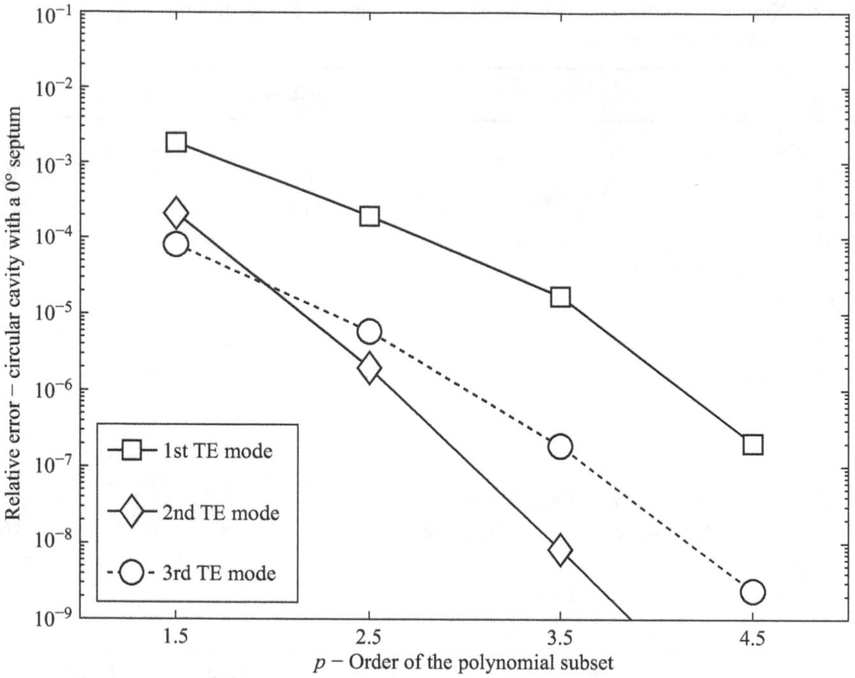

Figure 7.16 Error in the resonant frequency for the first three TE modes of the circular cavity with a 0° septum, obtained using a vector expansion.

© 2014 T&F. Reprinted with permission from R. D. Graglia, A. F. Peterson, L. Matekovits, and P. Petrini, "Hierarchical additive basis functions for the finite-element treatment of corner singularities," special issue on "Finite Elements for Microwave Engineering," *Electromagnetics*, vol. 34, pp. 171–198, Mar. 2014.

As with the singular scalar basis functions described previously, it is clear that the best accuracy (for a given background order) is obtained by including as many radial functions and as much azimuthal variation as provided by the background representation. This approach requires a relatively small increase in the number of unknowns for additional azimuthal variation, but a substantial enhancement in accuracy. As in the scalar case, this improvement comes at the cost of an increase in the matrix CN. With this approach, we observe an order of magnitude improvement in accuracy for each increment in p.

Figures 7.16–7.19 show results for the error in the first three resonant wavenumbers for the TE and TM modes of the cavities with a 0° septum and 30° wedge, using the maximum number of fractional exponents and (static) azimuthal functions for a given background order. In these figures, an "order" of 1.5 denotes a background order of $p=1.5$ with the complete collection of the static singular functions to match that azimuthal variation. The exponents considered are $\nu = 6/11, 12/11, 2+6/11, 2+12/11$. The dynamic singular $\mho$ functions are added beginning with order $p=3.5$. The results suggest that, with this approach, modes that exhibit strong singularities in the transverse field are represented to about the same accuracy as modes that do not excite the wedge singularity.

Next, we compare numerical results obtained with the singular hierarchical vector bases of the additive kind to those obtained with the purely polynomial hierarchical vector bases

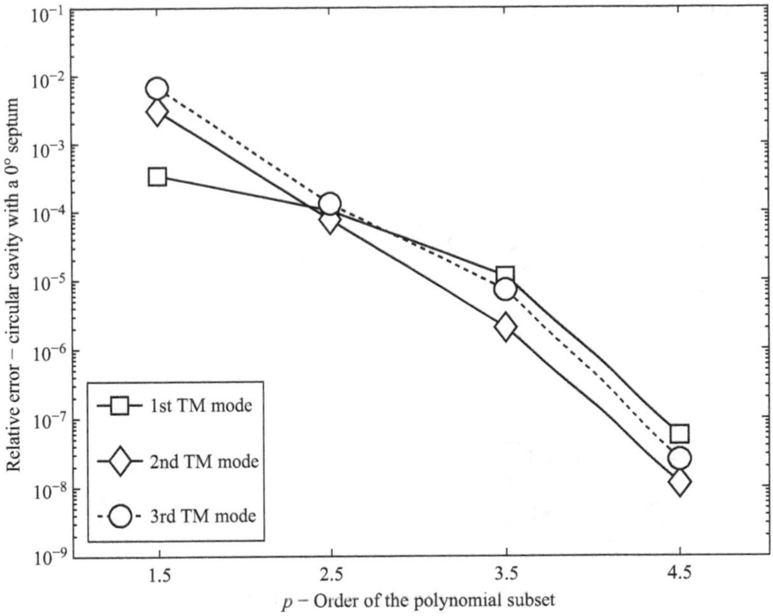

Figure 7.17 Error in the resonant frequency for the first three TM modes of the circular cavity with a 0° septum, obtained using a vector expansion.

© 2014 T&F. Reprinted with permission from R. D. Graglia, A. F. Peterson, L. Matekovits, and P. Petrini, "Hierarchical additive basis functions for the finite-element treatment of corner singularities," special issue on "Finite Elements for Microwave Engineering," *Electromagnetics*, vol. 34, pp. 171–198, Mar. 2014.

of Chapter 5. To limit the order of the polynomial subset of the additive bases, all the Meixner sets used here are defined for $k = 1$. The polynomial base is specified by an integer p, where $(p + 0.5)$ is the order of the vector base. The additive singular base also contains the Meixner subset of the gradient (curl-free) functions (7.103) and (7.104) and, at times, the non-static singular vector functions (7.121).

To evaluate the relative performance of the singular hierarchical bases described previously, the basis families were tested in a code that computed the resonant wavenumbers (k_c) of a two-dimensional cavity bounded by perfectly conducting walls, by solving the two-dimensional vector Helmholtz equation. Part of this procedure involves the computation of triangular-cell element matrices

$$S_{mn} = \iint_S \nabla \times \boldsymbol{B}_m \cdot \nabla \times \boldsymbol{B}_n \, \mathrm{d}S \tag{7.134}$$

$$T_{mn} = \iint_S \boldsymbol{B}_m \cdot \boldsymbol{B}_n \, \mathrm{d}S \tag{7.135}$$

where $\boldsymbol{B}_i$ is the ith basis function. In the course of the computation, element matrices are aggregated to form a global eigensystem that is solved for the resonant wavenumbers. Since additive expansions have the possibility of being poorly conditioned, we also investigate the relative linear independence of the basis functions. Because of the nullspace of the curl operator,

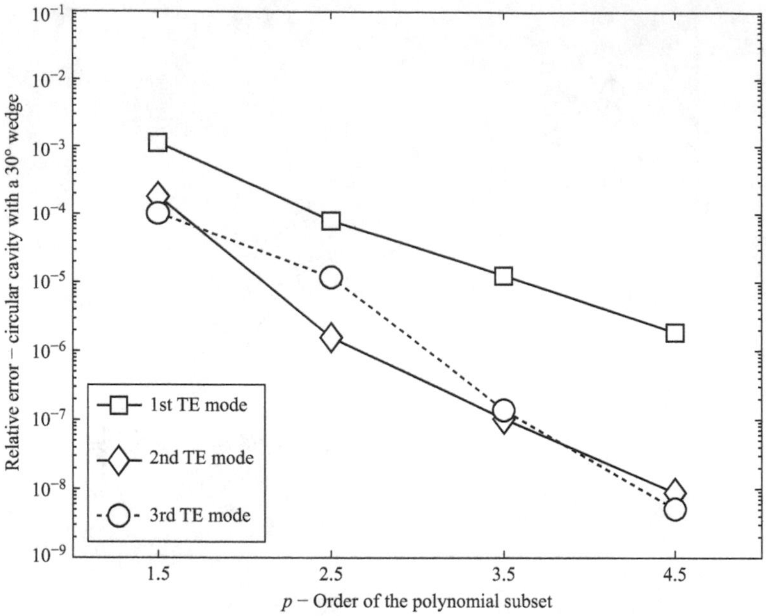

Figure 7.18 Error in the resonant frequency for the first three TE modes of the circular cavity with a 30° wedge, obtained using a vector expansion.

the global matrix corresponding to **S** is singular. However, the global **T** matrix is non-singular, and its CN provides an indication of the relative linear independence of the specific set of basis functions being used.

In order to show the robustness of our present technique, all the results discussed below were obtained by numerically defining the Meixner sets for $k = 1$ as explained in Section 7.6.7 and in Table 7.8 using a 64 bit-precision arithmetic (this implies that the auxiliary radial functions are quasi-orthogonal, that is the orthogonality integral of two quasi-orthogonal functions is not exactly equal to zero, but is of the order of 10^{-12}). The integrals to compute the elements of the **S** and **T** matrices were also performed by standard quadrature routines; in case of integrals on singular triangular cells, the integrals were performed using the pseudo-radial transformation discussed at length in Section 7.2. Because of this, the maximum expected accuracy of the computed wavenumbers obtained with singular bases is of order 10^{-7}.

Consider the L-shaped metal cavity meshed with 26 triangles as shown in Figure 7.15 at left. This structure contains a 90° wedge and the first six non-integer singularity coefficients are

$$\overline{\nu} = \left\{ \frac{2}{3}, \frac{4}{3}, \frac{8}{3}, \frac{10}{3}, \frac{14}{3}, \frac{16}{3} \right\} \tag{7.136}$$

The Meixner subsets used for this test-case are obtained from auxiliary radial functions $R_j(\chi)$ made orthogonal according to condition **B** and normalization (7.114). Figure 7.20 reports the

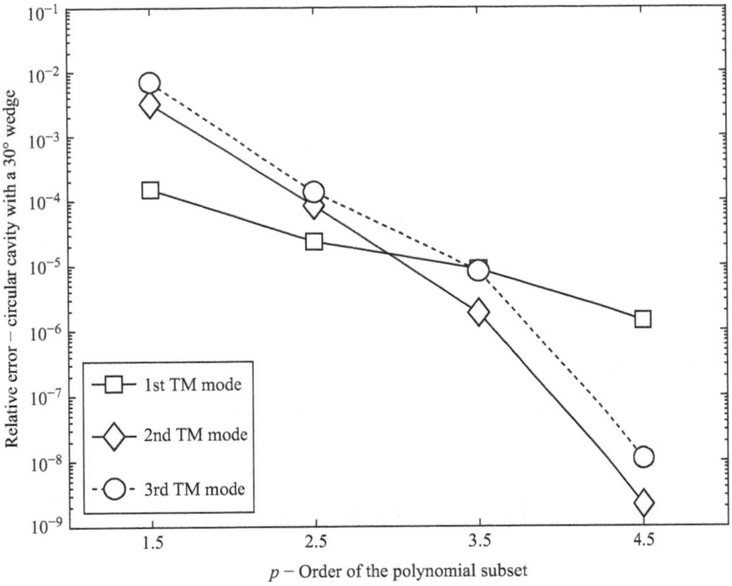

Figure 7.19 Error in the resonant frequency for the first three TM modes of the circular cavity with a 30° wedge, obtained using a vector expansion.

© 2014 T&F. Reprinted with permission from R. D. Graglia, A. F. Peterson, L. Matekovits, and P. Petrini, "Hierarchical additive basis functions for the finite-element treatment of corner singularities," special issue on "Finite Elements for Microwave Engineering," *Electromagnetics*, vol. 34, pp. 171–198, Mar. 2014.

relative error in k_c for the first and second TE mode (both singular) versus the order p of the polynomial subsets. The results marked by squares were obtained by using purely polynomial vector bases of order p. The other results were obtained using a polynomial basis of order p to which we added different Meixner subsets formed by gradient basis functions only. The additive bases are denoted in the figure by three digits where the first indicates the polynomial order, the second indicates the number of the non-integer singularity coefficients included, and the third indicates the azimuthal order of the Meixner subsets. Specifically, the basis labeled by $[p, n, a]$ is formed by a vector polynomial subset of order p to which we added Meixner vector basis functions that model the first n singularity coefficients from the $\overline{\nu}$ list (7.136) up to the azimuthal order a.

The results suggest that for a given base order p, the "best" accuracy is obtained with combination $[p, p, p]$. We also observed that results obtained with bases of order $[p, n, 0]$ for $n > 1$ (not shown in the figure) are not more accurate than those obtained with the $[p, 1, 0]$ base; in other words, the degree of the azimuthal expansion is important. In Figure 7.20, for the second TE mode, we report only the relative error obtained by using purely polynomial bases and the additive bases of order $[p, p, p]$.

Since the exact k_c values of these modes of the L-shaped cavity are not available, we have used as a reference the best results reported in Table 7.4 (1.2147517692; 1.8799019578) which were obtained numerically using the singular hierarchical *scalar* basis functions described in Section 7.3. The number of DoFs needed for the "reference" scalar problem is 997 (see Table 7.4) whereas the DoF arising from the vector TE problem is 1,520 for the $[6, 6, 6]$ base.

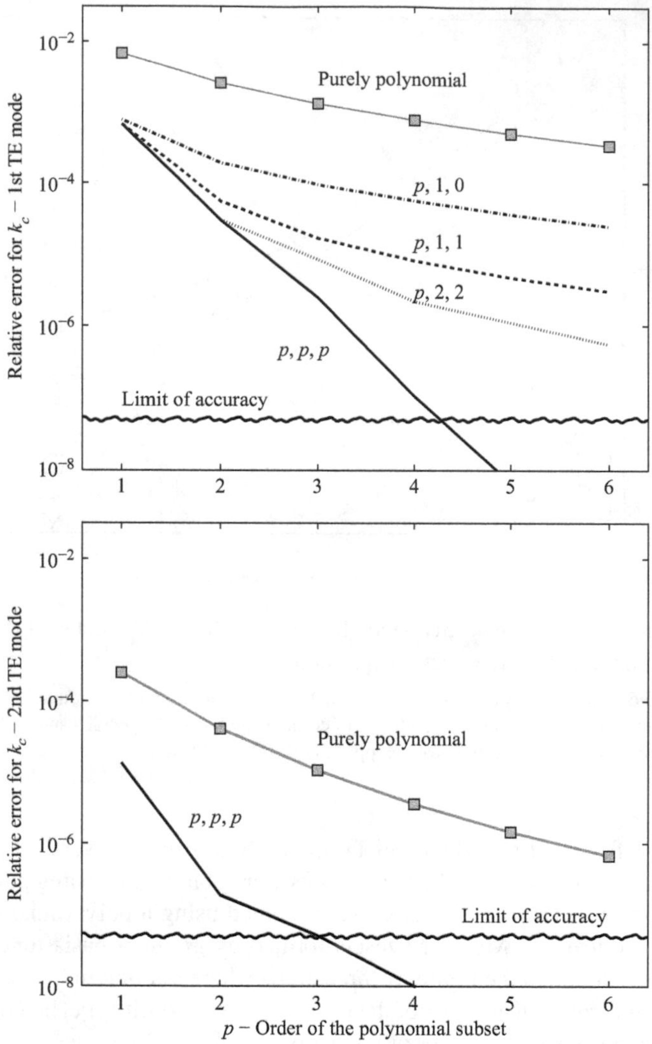

Figure 7.20 L-shaped cavity: relative error for the k_c of the first (at top) and second (at bottom) TE modes. These modes are both singular.

© 2014 IEEE. Reprinted with permission from R. D. Graglia, A. F. Peterson, L. Matekovits, and P. Petrini, "Singular hierarchical curl-conforming vector bases for triangular cells," *IEEE Trans. Antennas Propag.*, vol. 62, no. 7, pp. 3632–3644, Jul. 2014.

Figure 7.20 shows that the relative error depends on the mode considered. In fact, the eigenfield of the second TE mode of the L-shaped cavity (Figure 7.20 at bottom) has a singular behavior connected with the second singularity coefficient $\frac{4}{3}$. This can be observed by noticing that the error drastically decreases as soon as the Meixner set models the singular behavior coupled with the second singularity coefficient of the list (7.136).

Figure 7.21 compares the results obtained with the 26-cell mesh shown in Figure 7.15 and with the 4-cell mesh shown in Table 7.4. The results at top are for the first (singular)

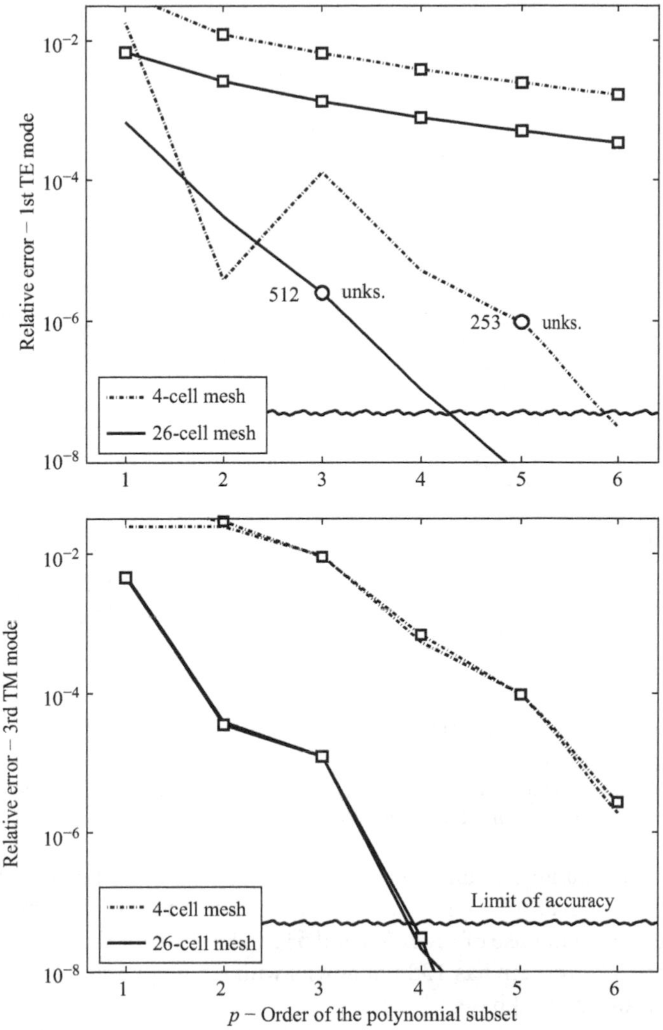

Figure 7.21 L-shaped cavity: relative error for k_c of the first TE mode (at top) and of the third TM mode (at bottom) obtained with the 26-cell mesh shown in Figure 7.15 and with the 4-cell mesh shown in Table 7.4 [22, Table VII]. Results obtained with purely polynomial bases are marked by squares.

TE mode; at bottom, the figure shows the results for the (regular) third TM mode whose $k_c = \sqrt{2}\pi$ (see [22]). The singular expansion results are obtained with bases of order $[p, p, p]$ (purely polynomial results are denoted by squares). Note that the error for the regular mode (Figure 7.21 at bottom) does not improve as Meixner subsets are added to the expansion, as already noted previously. It is clear from these results that higher order bases on coarse meshes

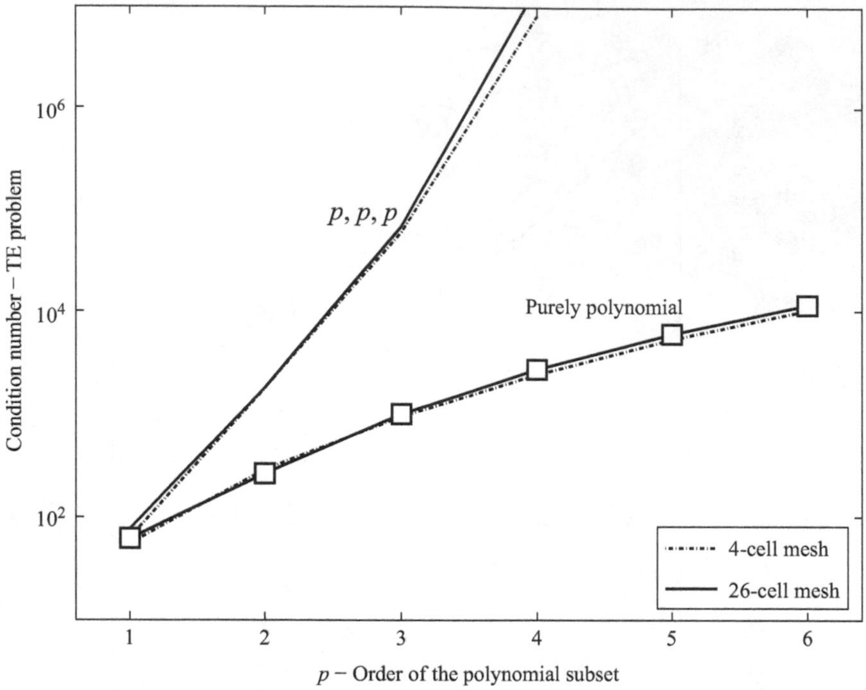

Figure 7.22 L-shaped cavity: CNs of the TE-problem mass matrices.

© 2014 IEEE. Reprinted with permission from R. D. Graglia, A. F. Peterson, L. Matekovits, and P. Petrini, "Singular hierarchical curl-conforming vector bases for triangular cells," *IEEE Trans. Antennas Propag.*, vol. 62, no. 7, pp. 3632–3644, Jul. 2014.

require, for the same accuracy, a smaller number of unknowns than required by low-order bases on denser meshes. For example, with the 26-cell mesh, the TE problem has 440 unknowns with the purely polynomial base of order 3.5 and 512 unknowns for the [3, 3, 3] base. With the four-cell mesh, the TE problem has 138 unknowns with the polynomial base of order 5.5 and 253 unknowns for the [5, 5, 5] base.

The error also depends on the mesh; this is evident by observing that the error reported in Figure 7.21 (at top) obtained with the [2, 2, 2] base on the coarse mesh is smaller than the one obtained with the denser mesh for the same order. Apart from this anomaly, the errors obtained with both meshes decrease with the same slope.

Figure 7.22 shows the behavior of the CN of the global mass matrix **T** for the TE problem. From this figure, it is clear that the CN is mainly affected by the number of singular cells attached to the tip wedge while the mesh density has little effect on the CN.

The second test-case considers the circular cavity with a 30° wedge shown in Figure 7.15 (at right). This geometry is of interest because:

- its exact k_c values for unit radius are available and given by the zeros of the Bessel functions $J_{m\tau}$ in the TM case, and by the zeros of derivatives of these Bessel functions in the TE case [42] (for $J_{m\tau}$ one has to set

$$\tau = 180/(360 - 30) = 6/11 \tag{7.137}$$

where $m \geq 0$ is an integer, with $m \neq 0$ for the TM modes);

- this cavity supports a huge number of modes with a non-analytic behavior in the neighborhood of the wedge tip (out of the first 22 modes only the sixth mode, that is the fifth TE mode, has analytic behavior at the tip, while the first TM mode with analytic behavior at the tip is the 38th mode);
- four modes out of the first 19 supported by this cavity have unbounded fields at the tip;
- the cavity does not have a straight border.

This test-case is then well suited to study the effects of the geometrical approximation of the structure due to the use of a finite mesh, as well as the three different orthogonalizations (7.110)–(7.112) one may use to define the auxiliary radial function R. As far as the geometrical approximation is concerned, once again we observe that when using cells of large size, the error due to a low-order description of a curved boundary should never be underestimated since it can easily be quite high, as it is for the present test-case if the degree of distortion of the 12 curved triangles is less than the sixth. The principal advantage of high-order basis functions is only realized when they are used with high-order representations of curved geometries to allow cells of large size.

An important aspect of the second test-case is that our technique for constructing the singular bases permits one to easily try different sets of the singularity coefficients to solve the same problem, because the auxiliary radial functions are obtained numerically with the very fast iterative procedure described in Section 7.6.7 and in Table 7.8. In fact, for this second test-case, the first seven non-integer singularity coefficients are (see (7.6))

$$\overline{\nu} = \left\{ \boxed{\frac{6}{11}}, \boxed{\frac{12}{11}}, \frac{18}{11}, \frac{24}{11}, \boxed{\frac{28}{11} = 2 + \frac{6}{11}}, \frac{30}{11}, \boxed{\frac{34}{11} = 2 + \frac{12}{11}} \right\} \tag{7.138}$$

but we skip at the onset the third, fourth, and sixth and consider the following four singularity coefficients

$$\overline{\nu} = \left\{ \frac{6}{11}, \frac{12}{11}, \frac{28}{11}, \frac{34}{11} \right\} \tag{7.139}$$

to obtain the results of Figures 7.23–7.28.

Figure 7.23 reports the relative error in k_c for the first and second TE (at top) and TM (at bottom) modes versus the order p of the polynomial subsets. The exact wavenumbers are (1.23133932138672, 1.95655873019426) and (3.20587245491204, 3.95379237868849) for the TE and TM modes, respectively. (The TE modes considered are the first two waveguide modes; the first and second TM modes are the fourth and eighth waveguide modes, respectively.) The results labeled A, B, and C in Figure 7.23 are obtained using the singular additive bases of order $[p, p, p]$ with orthogonalizations **A**, **B**, and **C**, respectively (for sake of comparison, the radial functions were all normalized using (7.114) for the three different orthogonalizations under consideration). For the first TE and TM modes reported in Figure 7.23, the relative error for the [4,4,4] base is of order 10^{-6} using orthogonalizations **A** and **B** although, because of the mesh and the eigenfield topography, the orthogonalization used to define the Meixner subsets has a stronger effect on the error of the first TM mode than it has for the other modes shown in Figure 7.23. The results of Figure 7.23 were obtained using Meixner subsets formed by gradient functions only.

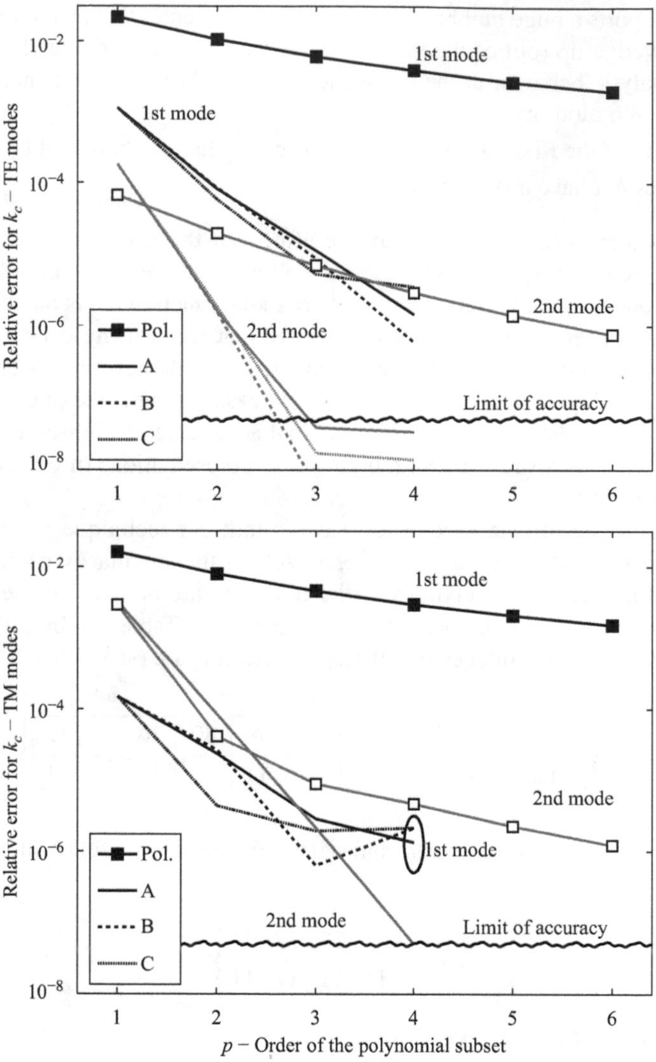

Figure 7.23 Results from the circular cavity with a 30° wedge. Relative error for k_c for the first and second TE and TM modes versus the expansion order. These four modes are singular. The TE results are reported at top; TM results at bottom. The results marked with squares are obtained using purely polynomial (regular) bases. The A, B, and C results are obtained with the singular additive bases of order $[p, p, p]$ and orthogonalizations **A**, **B**, and **C**, respectively.

© 2014 IEEE. Reprinted with permission from R. D. Graglia, A. F. Peterson, L. Matekovits, and P. Petrini, "Singular hierarchical curl-conforming vector bases for triangular cells," *IEEE Trans. Antennas Propag.*, vol. 62, no. 7, pp. 3632–3644, Jul. 2014.

The error behavior for two *singular* modes of Figure 7.23 obtained with Meixner gradient functions is shown by the eigenfield topographies in Figure 7.24, obtained by normalizing each eigenfield to unit power (the field is not sampled on the edge tip where it is infinite). Because of the boundary conditions, the magnetic field (obtained by solving the TM problem) circulates

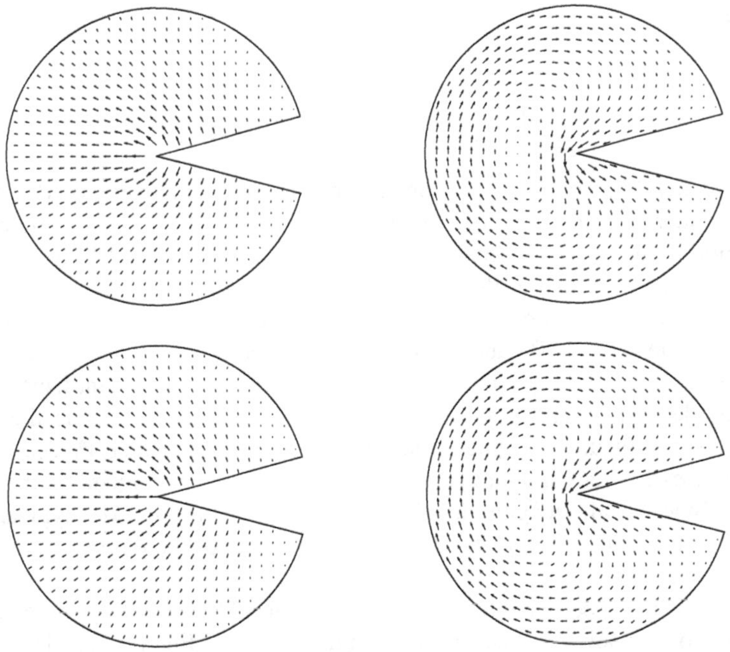

Figure 7.24 Circular cavity with a 30° wedge. The plots on the left-hand side of the figure are the transverse E-field for the first singular (dominant) TE mode. The plots at right are the transverse H-field of the first singular TM mode (the fourth mode supported by this cavity). The figure shows the exact (at top) and the numerically obtained (at bottom) field topographies computed using the [4, 4, 4] basis (Meixner gradient functions only).

around the wedge and therefore has a strong curl component in the immediate vicinity of the tip; conversely, the electric field (obtained by solving the TE problem) is normal to the wedge and hence, at least for low-order modes, usually has a relatively small curl component in this same region. In this connection, recall that the primary unknowns in the TE and TM problems are the transverse-electric and the transverse-magnetic fields, respectively. The exact analytical solution for the curl fields [42]

$$\nabla \times \begin{bmatrix} \boldsymbol{H}_{\text{TM}m} \\ \boldsymbol{E}_{\text{TE}m} \end{bmatrix} = \hat{z} \begin{bmatrix} \sin\{m\tau\,(\phi - 15°)\} \\ \cos\{m\tau\,(\phi - 15°)\} \end{bmatrix} \times \left[k_c^2 J''_{m\tau}(k_c\rho) + \frac{k_c}{\rho} J'_{m\tau}(k_c\rho) - \left(\frac{m\tau}{\rho}\right)^2 J_{m\tau}(k_c\rho) \right] \tag{7.140}$$

depends on the wavenumber k_c of the mode. The radial factors of the curl of the first two singular TM

$$\hat{z} \cdot \nabla \times \begin{bmatrix} \boldsymbol{H}_{\text{TM}1} \\ \boldsymbol{H}_{\text{TM}2} \end{bmatrix} \propto \begin{bmatrix} \rho^{6/11}\,(1 - 1.6626\rho^2 + \mathrm{O}[\rho]^4) \\ \rho^{12/11}\,(1.1031 - 1.3556\rho^2 + \mathrm{O}[\rho]^4) \end{bmatrix} \tag{7.141}$$

and TE modes

$$\hat{z} \cdot \nabla \times \begin{bmatrix} \boldsymbol{E}_{\mathrm{TE1}} \\ \boldsymbol{E}_{\mathrm{TE2}} \end{bmatrix} \propto \begin{bmatrix} \rho^{6/11}\left(0.0875 - 0.0215\rho^2 + \mathrm{O}[\rho]^4\right) \\ \rho^{12/11}\left(0.2398 - 0.1097\rho^2 + \mathrm{O}[\rho]^4\right) \end{bmatrix} \tag{7.142}$$

show that in the neighborhood of the wedge tip the curl of the magnetic field of the first two singular TM modes is about 11 times greater than the curl of the electric field of the first singular TE mode. The equally normalized expressions (7.141) and (7.142) also show, for this test-case, that the curls vanish at the tip wedge ($\rho = 0$) and do not contain any polynomial function of ρ in the neighborhood of the tip. Notice also that the radial factors (7.141) and (7.142) valid for the second TM and TE modes are proportional to $\rho^{12/11}$; these factors are more easily represented by polynomial approximations than the dominant factors for the first TM and TE modes, which are proportional to $\rho^{6/11}$ (in fact, by increasing the order of the polynomial subset in Figure 7.23, the errors for the second TM and TE modes exponentially diminish). Series expansions similar to (7.141) and (7.142) can be obtained for the other higher order modes; obviously, the magnitude of each coefficient of these series typically increases with the mode wavenumber k_c together with the complexity of the field topographies around the tip of the wedge.

Thus, it may be possible to reduce the error in the first TM mode by adding singular non-static functions to the Meixner bases used to get the results of Figure 7.23. This can be done for bases that contain a polynomial vector subset of order greater than or equal to the third, as shown in Figure 7.25. Specifically, the results of Figure 7.25 were obtained by augmenting the $[3, 3, 3]$ basis with the non-static functions containing the radial DoF $\chi^{6/11}$ in the curl while, for the $[4, 4, 4]$ basis, we added the non-static functions incorporating curl behavior of the form $\chi^{6/11}$ and $\chi^{12/11}$. (The gradient and non-static function sets we used are of the same azimuthal order.) In Figure 7.25, the results are compared with those of Figure 7.23 (at bottom) obtained with the orthogonalization **A**. It is clear that a true benefit is obtained only in case of the $[4, 4, 4]$ base. Notice that the error in the first TM mode obtained using only Meixner gradient-function sets with the coefficients listed in (7.139) and the error obtained with the singularity coefficients

$$\overline{\nu} = \left\{ \frac{6}{11}, \frac{12}{11}, \frac{18}{11}, \frac{24}{11} \right\} \tag{7.143}$$

are almost identical. (Therefore, the latter are not reported in Figure 7.25.) This justifies use of the coefficient list (7.139) to define the non-static functions being added to the Meixner gradient-function set. Very good results can be obtained modeling singular fields using only the Meixner gradient functions, as in Figure 7.23, with errors that are two to three orders of magnitude lower than those obtained for the first singular mode with purely polynomial bases. The non-static Meixner functions merely model the non-analytical *vanishing* behavior of the curl at the tip of the wedge and hence they have little influence on the quality of the field topographies of the test-case problems considered in this paper, although they could be useful when solving other driven problems. (The source of a driven problem typically does not excite all the expected singularities but only a few *selected* ones [25] that can be more easily modeled by using only the relevant gradient and non-gradient Meixner basis functions.)

In fact, as far as the field topographies are concerned, very good accuracy is already obtained using Meixner gradient functions only. For example, Figure 7.26 shows the behavior of the

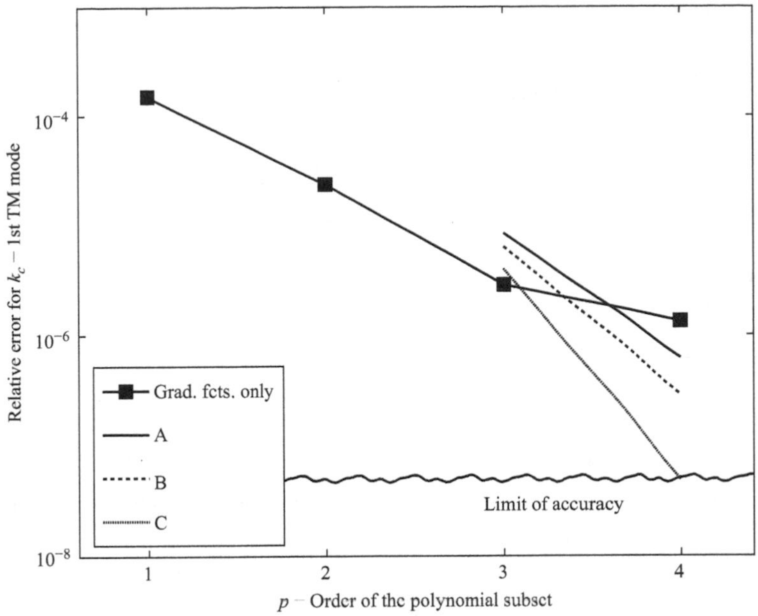

Figure 7.25 Circular cavity with a 30° wedge. Relative error in k_c for the first TM mode versus the order of the polynomial basis subset. The results marked with squares were obtained using the $[p, p, p]$ basis incorporating only singular gradient functions (orthogonalization **A**). The other results are obtained by adding singular non-static functions to the $[p, p, p]$ basis and using orthogonalizations **A**, **B**, and **C**, respectively.

© 2014 IEEE. Reprinted with permission from R. D. Graglia, A. F. Peterson, L. Matekovits, and P. Petrini, "Singular hierarchical curl-conforming vector bases for triangular cells," *IEEE Trans. Antennas Propag.*, vol. 62, no. 7, pp. 3632–3644, Jul. 2014.

x and y components of the electric field of the first TE mode and of the magnetic field of the first TM mode along the negative (horizontal) x-axis of Figure 7.24. This figure compares the exact (solid-line) solution, normalized to a unit power eigenfield, to the solutions obtained with the $[4, 4, 4]$ base and the purely polynomial base of order 4. These latter numerical solutions are obtained directly from the eigenvalues computed by the Lapack library routine DGGEVX (no further normalization needed). For these modes, the y-directed component of the electric TE field and the x-directed component of the magnetic TM-field are zero (see bottom part of Figure 7.26), although the cross-pol fields obtained with the $[4, 4, 4]$ base are unbounded in the neighborhood of the tip (say, for $\rho < 1/100$) because the numerically computed eigenvalues associated with the singular Meixner basis functions are not exactly equal to zero. Conversely, our singular bases model the unbounded dominant (co-pol) field components very well (top part of Figure 7.26) while the pure polynomial results are clearly incorrect in the neighborhood of the tip wedge (the polynomial base of order 4 yields $E_x \approx -3.79$ and $H_y \approx -1.73$ at $\rho = 0$). At any rate, at $\rho = 1/100$ the $[4, 4, 4]$ singular base yields co-pol/cross-pol ratios equal to $E_y/E_x \approx 0.11$, $H_x/H_y \approx 0.21$. The numerically obtained y-directed field components were computed by considering only the contributions of the basis functions associated with the two triangular cells located above the x-axis of Figure 7.24. These components

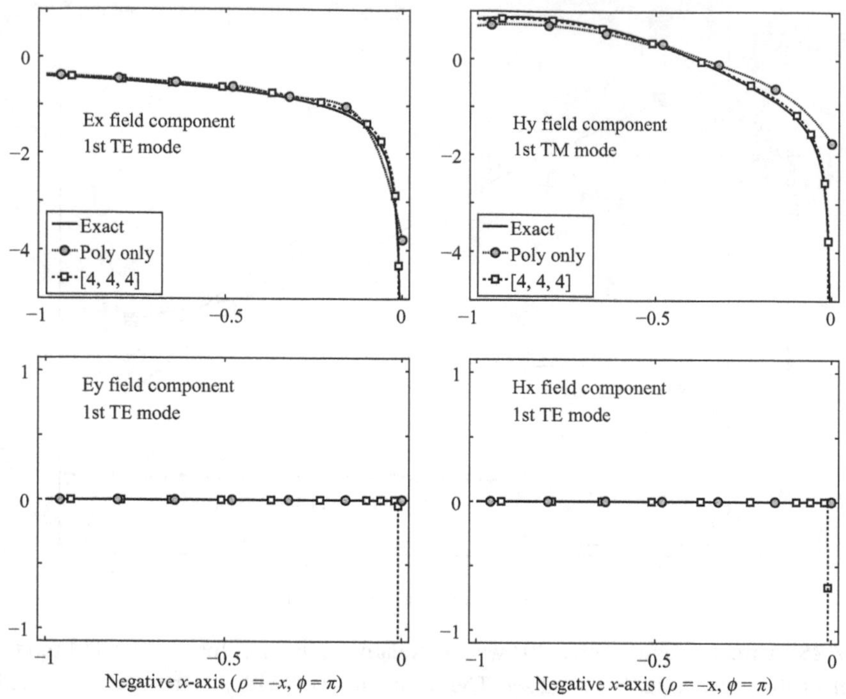

Figure 7.26 Circular cavity with a 30° wedge. Dominant (co-pol) and cross-pol field components along the negative x-axis of Figure 7.24 for the first TE (at left) and TM (at right) modes.

© 2014 IEEE. Reprinted with permission from R. D. Graglia, A. F. Peterson, L. Matekovits, and P. Petrini, "Singular hierarchical curl-conforming vector bases for triangular cells," *IEEE Trans. Antennas Propag.*, vol. 62, no. 7, pp. 3632–3644, Jul. 2014.

are normal to the triangular edges located along the negative x-axis and discontinuous if one crosses these edges. Obviously, if one considers the average value of the normal components across the edges the results for the y-directed components are even better than those shown in Figure 7.26.

Figure 7.27 reports the behavior of the CN of the global **T**-matrix (7.135) for the TE case using singular bases containing only Meixner gradient functions (as in Figure 7.23). The CNs for the TM problem are similar, although they tend to be a little bit higher than the TE CNs because the TM problem requires more unknowns, see Figure 7.28. It is interesting to observe that purely polynomial hierarchical vector bases [38–39] yield CNs that increase polynomially with the base order, while those of the singular bases containing the Meixner subsets increase exponentially with order (see Figures 7.22 and 7.27). In summary, the orthogonalization used to define the auxiliary radial functions does not seem to substantially influence the CNs of the system matrix; it appeared to have some influence on the precision of the numerical results, depending on the mesh in use as well as on the field topography of the specific mode.

In conclusion, the results reported in this section suggest that the best accuracy (for a given background order) is obtained by including as many radial functions and as much azimuthal variation as provided by the background polynomial representation. This approach requires a relatively small increase in the number of unknowns for additional azimuthal variation, but

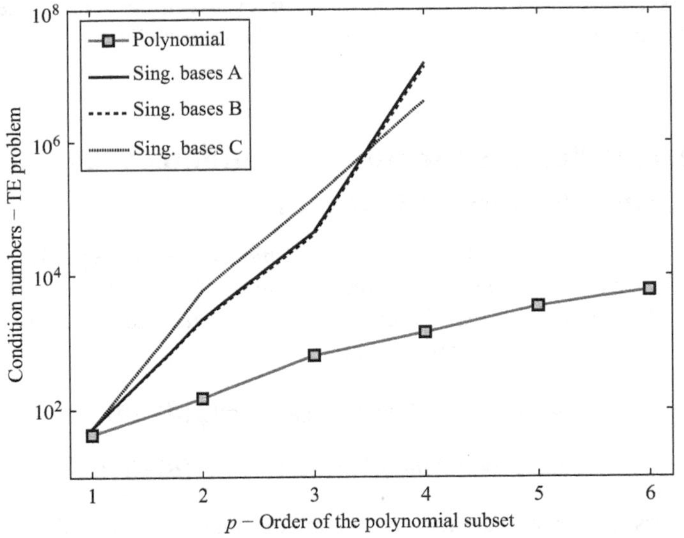

Figure 7.27 Circular cavity with a 30° wedge. CNs of the TE mass matrix for different orthogonalization schemes.

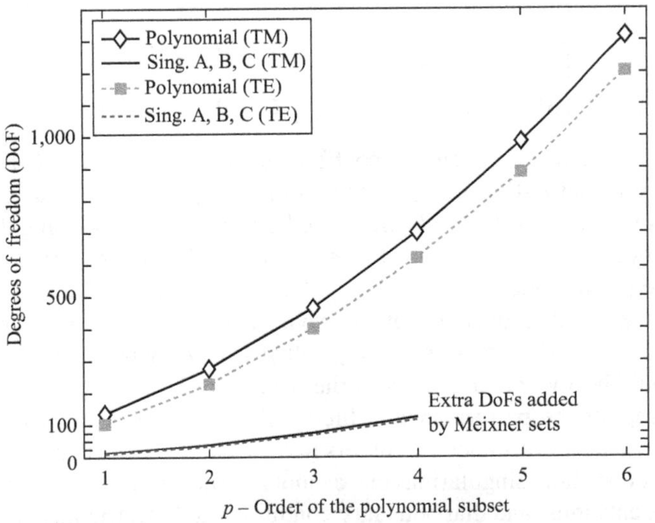

Figure 7.28 Circular cavity with a 30° wedge. DoFs used to solve the TE and TM problems. The figure shows the DoFs used by the purely polynomial vector bases of order p and the additional DoFs associated with the Meixner gradient functions needed to form $[p, p, p]$ bases.

usually produces a substantial improvement in accuracy. As in the scalar case (Section 7.4), this improvement comes at the cost of an increase in the matrix CN.

7.8 Numerical Results for Inhomogeneous Waveguiding Structures Containing Corners

The previous sections considered as test-cases *homogeneous* waveguides or 2D cavities with corners, thereby using the scalar formulation [22] or a transverse-field (T-field) formulation [23, 24]. In these cases, the eigenvalues (k_c^2) are obtained numerically by solving the following generalized eigenvalue problems

$$\text{Scalar formulation: } [A]_S\,[\boldsymbol{\phi}] = -k_c^2\,[B]_S\,[\boldsymbol{\phi}] \tag{7.144}$$

$$\text{T-field formulation: } [A]_T\,[\boldsymbol{e}_t] = -k_c^2\,[B]_T\,[\boldsymbol{e}_t] \tag{7.145}$$

whose eigenvectors ($[\boldsymbol{\phi}]$ or $[\boldsymbol{e}_t]$) are the expansion coefficients of the scalar potential $\boldsymbol{\phi}$, or of the transverse electric field $\boldsymbol{e}_t$.

This section considers the analysis of inhomogeneous waveguide structures containing one or more conducting or dielectric corners where field singularities are introduced. Inhomogeneous waveguides require the use of the transverse-field longitudinal-field (TL-field) formulation, where the problem unknowns are the transverse and the longitudinal E-field (or H-field) components. The inhomogeneous waveguide problem can be approached using the techniques described in [40] or [41]. In this section, we use the second one because it allows us to directly compute the waveguide propagation constants (k_z) for any given operating frequency by solving the eigenvalue problem [41]

$$\begin{bmatrix} A & 0 \\ 0 & 0 \end{bmatrix}_{TL} \begin{bmatrix} \boldsymbol{e}_t \\ e_z \end{bmatrix} = -k_z^2 \begin{bmatrix} B & C^t \\ C & D \end{bmatrix}_{TL} \begin{bmatrix} \boldsymbol{e}_t \\ e_z \end{bmatrix} \tag{7.146}$$

whose eigenvector stores the expansion coefficients of the transverse E-field ($\boldsymbol{e}_t$) and of the re-scaled longitudinal E-field component (e_z) [41]. In general, k_z^2 is a complex number; for homogeneous waveguides, k_c and k_z are related by $k_z^2 = k^2 - k_c^2$, k being the wavenumber of the homogeneous medium. In the equations above, the A-matrices are *singular* while the B-matrices are frequency-independent non-singular mass-matrices. (The D-matrix in (7.146) is singular at the resonant frequencies of the cavity being modeled. The CN of the matrix that multiplies k_z^2 on the right-hand side of (7.146) strongly depends on k^2 and, at certain frequencies, it could be lower than the CN of the B-matrix.)

This section reports the performance of hierarchical singular vector bases applied to an inhomogeneous structure with dielectric corners [42]. Results suggest that corners in dielectric structures can introduce field singularities that cannot be efficiently modeled with conventional polynomial representations, and that boundary conditions at dielectric interfaces near corners are not well satisfied by polynomial expansions. Instead, a combination of the scalar and vector singular basis functions provides a more efficient means of treating this type of field singularity. The results reported in this section are for purely polynomial bases of order p and for singular bases of order $[p, p, p]$. Furthermore, the singular vector basis function sub-sets that we used are those formed entirely from gradient vector basis functions.

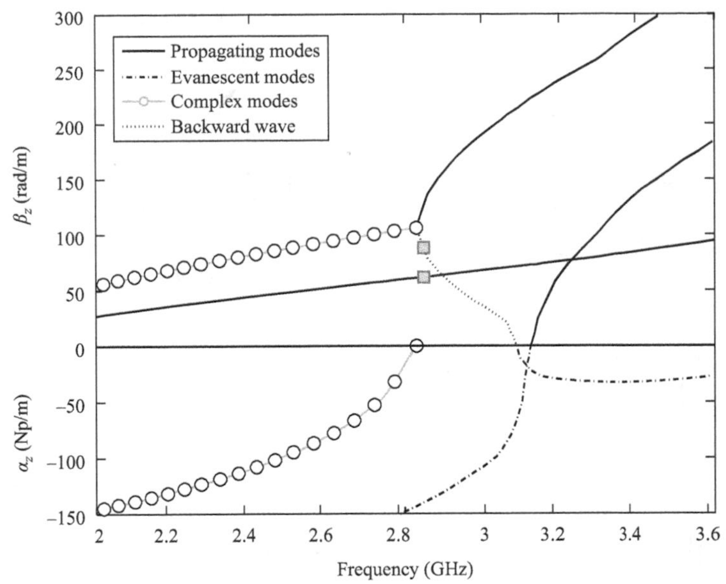

Figure 7.29 Mode diagram for the first few modes of a shielded dielectric-rod waveguide to show some of the associated complex modes, evanescent modes, and backward waves. The side length of the square shield is 24 mm, the side length of the square dielectric rod ($\varepsilon = 37.13$) is 12 mm [43, Figure 7]. The square markers indicate the modes and frequency considered to produce the results of Figures 7.30–7.32.

© 2014 IEEE. Reprinted with permission from R. D. Graglia, P. Petrini, A. F. Peterson, and L. Matekovits, "Full-wave analysis of inhomogeneous waveguiding structures containing corners with singular hierarchical curl-conforming vector bases," *IEEE Antenn. Wireless Propag. Lett.*, vol. 13, pp. 1701–1704, 2014.

As a test-case, consider a square metallic waveguide (side length = 24 mm) with a square dielectric rod inside (side length = 12 mm) of relative permittivity $\varepsilon_r = 37.13$; results for this waveguide are available in [43]. Figure 7.29 shows the dispersion diagram of the first few modes (with $k_z = \beta_z + j\alpha_z$) obtained with singular basis functions, although the differences between the dispersion diagrams obtained with purely polynomial and with singular bases cannot be detected in this figure. The results of Figure 7.29 are in excellent agreement with those reported in [43], where a detailed discussion of these modes is provided. We notice in passing that two complex conjugate eigenvalues k_{z1}^2 and k_{z2}^2 are associated with a *complex mode* (for example, the one marked by circles in Figure 7.29), with $k_{z_1} = \beta_z + j\alpha_z$ and $k_{z_2} = -\beta_z + j\alpha_z$; that is, *complex modes* exist always in pairs and are coupled in such a way that the total power transmitted by the two modes is purely reactive [44].

Figures 7.30 and 7.31 show the behavior of the transverse displacement vector $\boldsymbol{D}$ for the dominant (first propagating) mode and for the backward mode at 2.85 GHz. The results of these figures were obtained using the two meshes shown at top of Figure 7.30 and with bases of polynomial subset order $p = 4$. The transverse field topographies for the modal $\boldsymbol{D}$ field are reported at top of Figure 7.31; the field topographies do not change with the two meshes and can also be obtained from purely polynomial bases of sufficiently high order. The results in the bottom part of these figures show the behavior of the normal component of the displacement

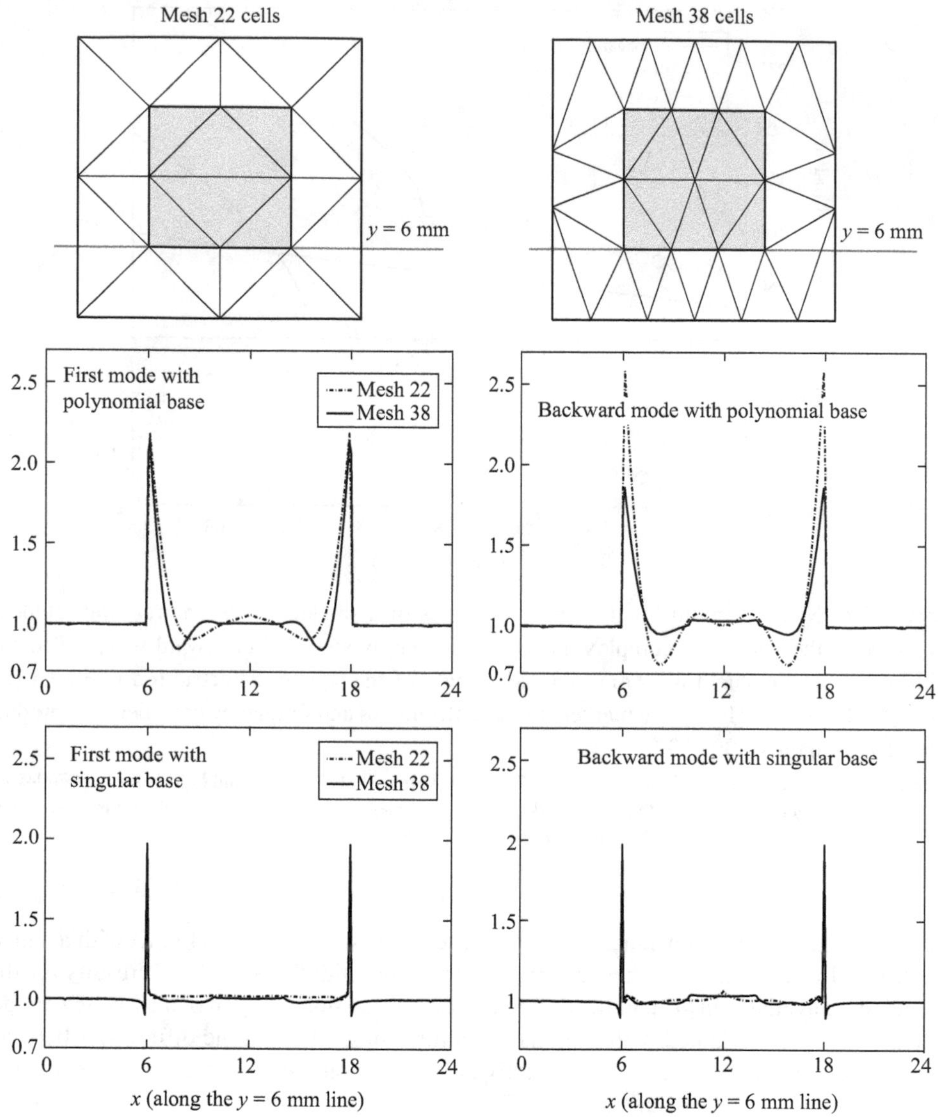

Figure 7.30 Square shielded dielectric-rod waveguide of Figure 7.29. The meshes used are shown at top. The ratio of the two D_y values (at 2.85 GHz) evaluated above and below the boundary line $y = 6$ mm are shown in the second and bottom row of the figure. The result obtained with the purely polynomial base of order $p = 4$ is reported in the second row; the result at bottom is obtained with the singular [4, 4, 4] base (using only singular curl-free vector basis functions) with the singularity coefficients [0.68578277, 1.31421723, 2.68578277, 3.31421723].

© 2014 IEEE. Reprinted with permission from R. D. Graglia, P. Petrini, A. F. Peterson, and L. Matekovits, "Full-wave analysis of inhomogeneous waveguiding structures containing corners with singular hierarchical curl-conforming vector bases," *IEEE Antenn. Wireless Propag. Lett.*, vol. 13, pp. 1701–1704, 2014.

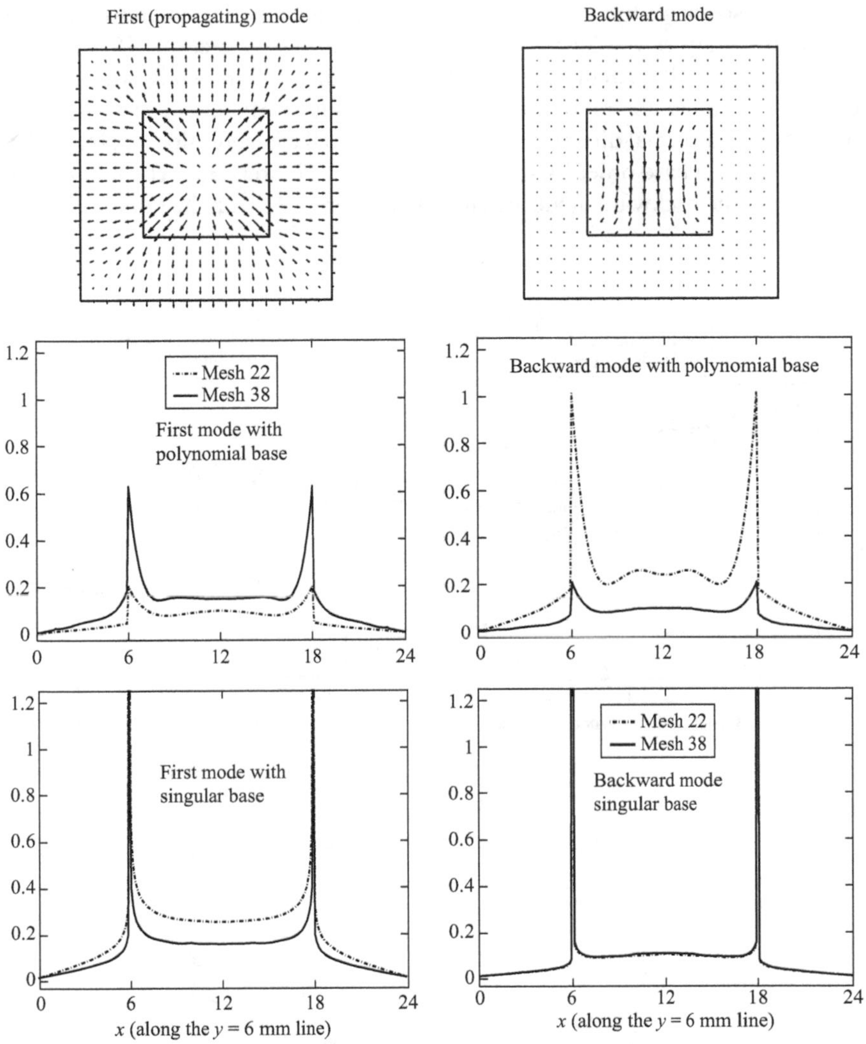

Figure 7.31 Square shielded dielectric-rod waveguide of Figure 7.29. The figures at top show the transverse $\boldsymbol{D}$-field topographies of the dominant (at left) and of the backward (at right) mode at 2.85 GHz. The central and bottom row of the figure show the $\boldsymbol{D}_y$ value evaluated right above the boundary line $y = 6$ mm. The result obtained with the purely polynomial base of order $p = 4$ is reported in the central row; the result at bottom is obtained with the singular [4, 4, 4] base.

© 2014 IEEE. Reprinted with permission from R. D. Graglia, P. Petrini, A. F. Peterson, and L. Matekovits, "Full–wave analysis of inhomogeneous waveguiding structures containing corners with singular hierarchical curl–conforming vector bases," *IEEE Antenn. Wireless Propag. Lett.*, vol. 13, pp. 1701–1704, 2014.

vector ($\boldsymbol{D}_y$) along the line $y = 6$ mm (bottom of the dielectric rod). The unbounded behavior of the $\boldsymbol{D}$ vector at the rod corners is properly modeled only by using the singular bases, as evident from the results of Figure 7.31. Very good convergence is obtained by using the base [4, 4, 4] (see the figures at bottom of Figures 7.30 and 7.31).

Figure 7.30 also shows the ratio of the normal $\boldsymbol{D}_y$ values evaluated above and below the air–dielectric boundary at $y = 6$ mm. The ratio oscillates near the corner if one uses purely polynomial bases, regardless of the mesh and base order in use. Conversely, with singular bases of sufficiently high order, the ratio of the two $\boldsymbol{D}_y$ values is nearly unity everywhere, except at the corners where the normal direction is not defined, and the best results are obtained with the 38-cell mesh and with base order [4, 4, 4] (see bottom of Figure 7.30). The relative errors for k_z for the first and the backward mode are shown in Figure 7.32 with the relevant CN and DoF.

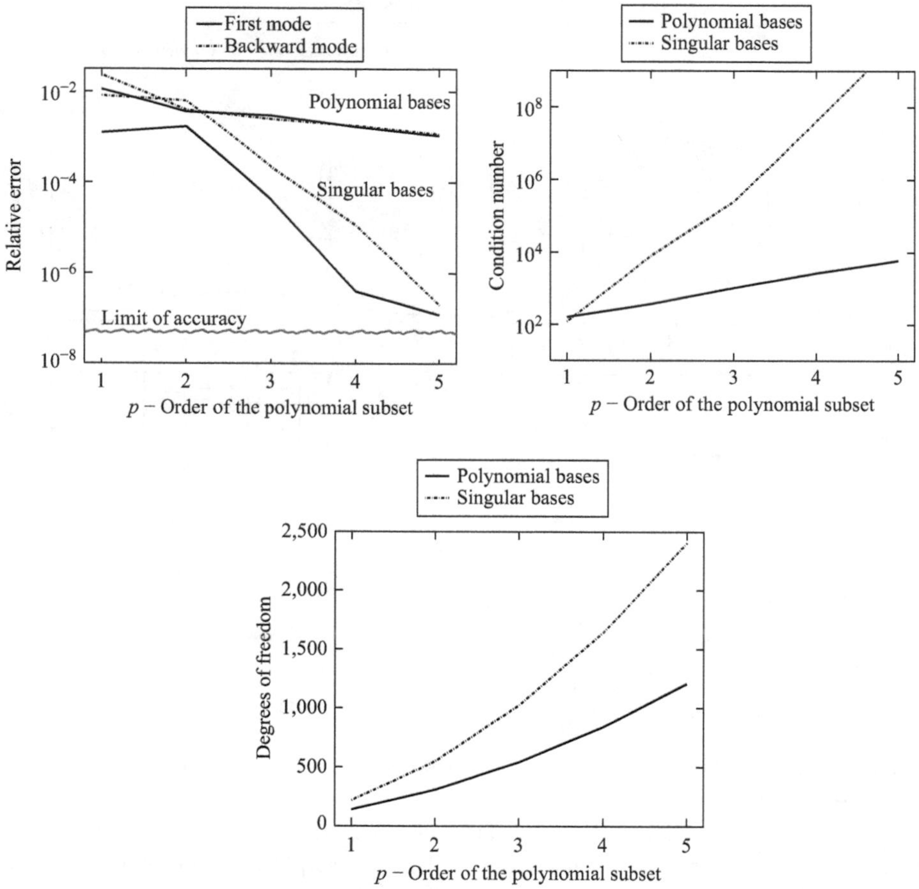

Figure 7.32 Square shielded dielectric-rod waveguide described in Figures 7.30 and 7.31. The relative errors for the eigenvalue k_z of the first (dominant) and of the backward mode shown at left are obtained with the coarse 22-cell mesh. (The reference solution to compute the errors used the [5, 5, 5] base with the denser 38-triangle mesh.) The mass-matrix CNs obtained with purely polynomial bases of order p and singular bases of order $[p, p, p]$ are shown at right. The number of unknowns (TL formulation) using the singular and the purely polynomial bases is reported at bottom.

7.9 Numerical Results for Thin Metallic Plates with Knife-Edge Singularities

We close this chapter with a brief look at the representation of edge singularities in 3D problems involving perfectly conducting plates. Thin plates have knife edges where current and field singularities are often excited by external sources. The results reported here are a selection of those shown in [30,45] as obtained from numerical solutions of the electric field integral equation and the method of moments. The specific basis functions used are the high-order singular divergence-conforming vector bases of the additive kind originally developed in [29, 30]. Although the polynomial basis subsets in [30] are formed using interpolatory polynomials, the results reported in this section should be identical to those obtained with hierarchical bases since the singular Meixner subset added to the polynomial subset is always that of the lowest possible order.

Figure 7.33 considers a zero-thickness circular PEC-plate, with diameter equal to one wavelength λ, normally illuminated by a linearly polarized plane wave with incident electric field in the x-direction, and incident magnetic field with unity amplitude H^i in the y-direction. The mesh used for this problem is shown as an inset in the figure and contains 64 (quadratic curvilinear) triangular elements. The current density is represented using the polynomial vector base of order $p = 2$ (yielding 648 unknowns) and also by incrementing the $p = 2$ base with singular vector basis functions in the triangles along the rim of the plate (increasing the number of unknowns to 696). In both cases, the density of unknowns is relatively high (825 unknowns/λ^2), which should guarantee extremely accurate numerical results in the near-field region.

Figure 7.33 shows the magnitude of the x-component of the current density (J_x) along the vertical ($x = 0$) axis (at left), and along the horizontal ($y = 0$) axis (at right). Along the vertical

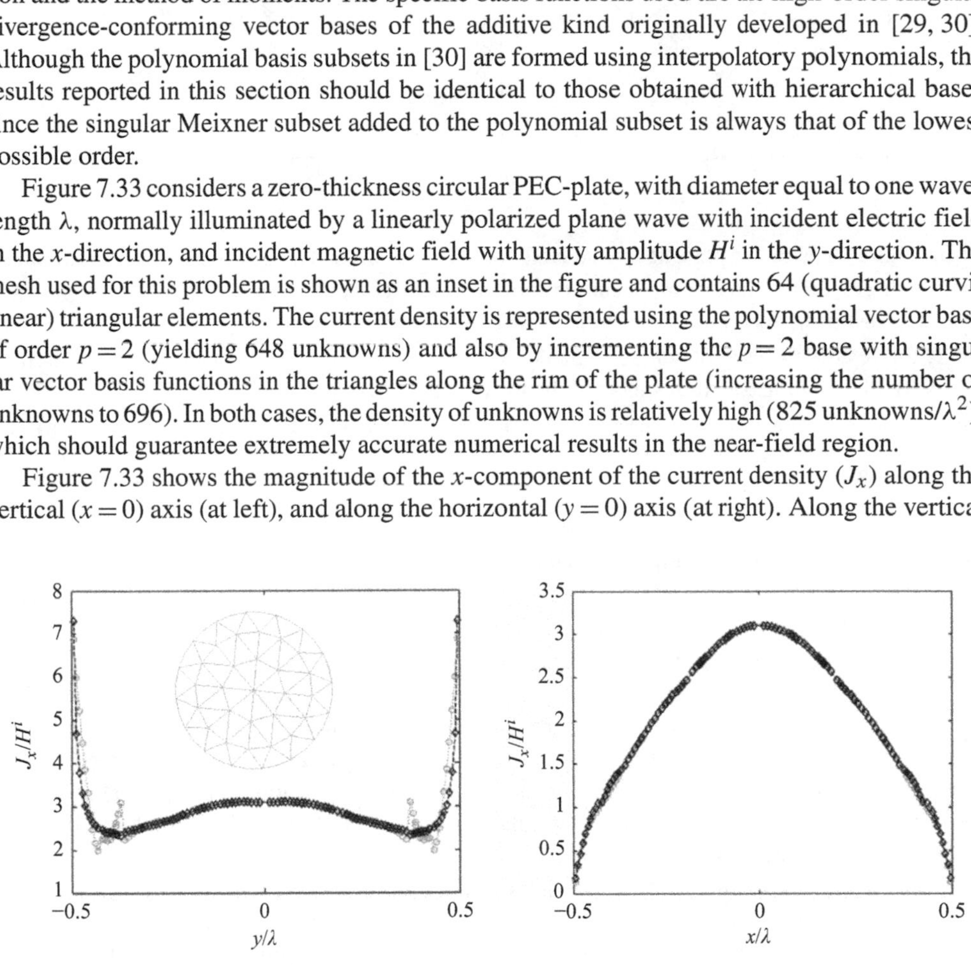

Figure 7.33 Current density magnitude induced on a circular PEC-plate 1λ in diameter by a normally incident plane wave. The results in gray denoted by circles were computed using the purely polynomial vector base of order $p = 2$; the results in black denoted by diamonds were computed by incrementing the polynomial base with the singular Meixner subset of the lowest possible order. The normalized magnitude (J_x/H^i) of the current component J_x along the vertical ($x = 0$) axis and along the horizontal ($y = 0$) axis is shown at left and right, respectively.

Adapted from R. D. Graglia and G. Lombardi, "Singular higher order divergence-conforming bases of additive kind and moments method applications to 3D sharp-wedge structures," *IEEE Trans. Antennas Propag.*, vol. 56, no. 12, pp. 3768–3788, Dec. 2008.

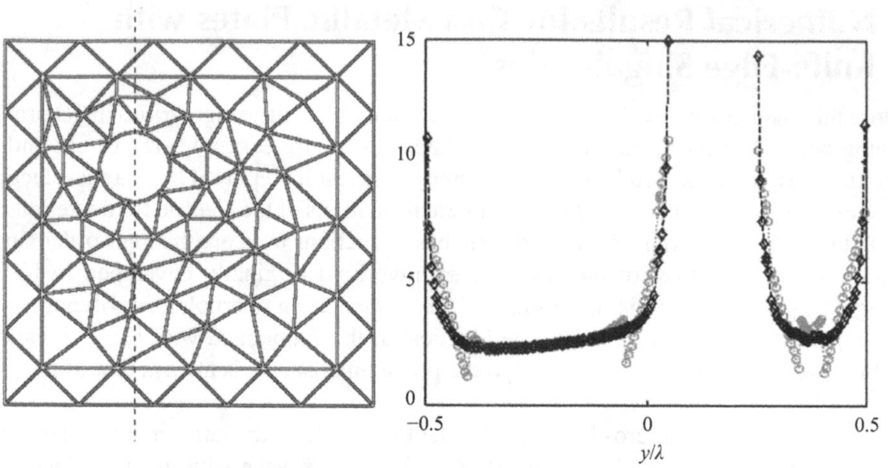

Figure 7.34 Perforated plate at normal incidence. The mesh used is shown at left. The normalized magnitude (J_x/H^i) of the x-component of the current density along the $x = -0.15\lambda$ axis is shown at right. The results in gray denoted by circles were computed using the purely polynomial base of order $p = 1$; the results in black denoted by diamonds were computed by incrementing the polynomial base with the singular Meixner subset of the lowest possible order.

Adapted from R. D. Graglia and G. Lombardi, "Singular higher order divergence-conforming bases of additive kind and moments method applications to 3D sharp-wedge structures," *IEEE Trans. Antennas Propag.*, vol. 56, no. 12, pp. 3768–3788, Dec. 2008.

axis, J_x is the azimuthal current component whereas, along the horizontal axis, it corresponds to the radial current component. The results clearly show that regular bases produce numerical solutions with non-physical oscillating behavior in the vicinity of the edge profile.

Figure 7.34 considers a ($1\lambda \times 1\lambda$) square PEC-plate of zero thickness with a hole of radius $r = \lambda/10$ centered at ($x = -0.15\lambda$, $y = +0.15\lambda$), relative to an origin at the plate center. The plate is normally illuminated by a linearly polarized plane wave with incident electric field in the x-direction, and incident magnetic field of unity amplitude H^i in the y-direction. The triangular mesh shown at left in Figure 7.34 contains 128 (quadratic curvilinear) triangular elements. The current density was represented using the polynomial vector basis of order $p = 1$ (with 614 unknowns) and also by incrementing the $p = 1$ base with singular vector basis functions in the triangles along the hole rim and on the external rim of the plate (producing 696 unknowns). The (numerically obtained) normalized magnitude of the x-component of the current density along the $x = -0.15\lambda$ axis shown at right in Figure 7.34 shows that regular bases generate solutions with non-physical oscillating behavior in the vicinity of the edge profiles.

A procedure to handle junctions in the presence of sharp edges in surface integral equation methods together with several other results for thin metallic plates connected together are reported in [45], and show the superior modeling capability of singular basis functions.

For example, the T-structure shown in the inset of Figure 7.35 (meshed with 168 right-angled triangular cells) is the union of three zero-thickness metal plates of equal size (0.1 m × 0.3 m). By introducing a Cartesian (and a spherical) reference frame with z-axis along the *line of junction* and origin in the middle of the line of junction, the three plates share one of their shorter edges along the z-axis with the T-structure *leg* lying on the half plane $\{x=0, y \leq 0\}$, and its two *arms* lying on the $y=0$-plane. The T-structure is illuminated by a 300 MHz, unit amplitude, linearly polarized plane wave

$$\boldsymbol{E}^i = \hat{\boldsymbol{\theta}} \exp\left(jk_o\hat{k}^i \cdot r\right)\exp(-j\omega t) \tag{7.147}$$

$$\hat{k}^i = -\left(\hat{\boldsymbol{x}} \cos\phi^i \sin\theta^i + \hat{\boldsymbol{y}} \sin\phi^i \sin\theta^i + \hat{\boldsymbol{z}} \cos\theta^i\right) \tag{7.148}$$

$$\theta^i = 45°, \quad \phi^i = -75° \tag{7.149}$$

that produces a non-zero-induced current density along the line of junction, on each of the three $0.1\lambda \times 0.3\lambda$ plates.

Figure 7.35 reports the induced current density normal to the z-axis computed at a distance of $\lambda/1,000$ from the bottom border of the T-structure. This current component is unbounded along the bottom edge of the structure while, at the same time, being normal to the junction, it is strongly affected by the bifurcation of the structure. Figure 7.35 clearly shows that purely polynomial vector basis functions are unable to correctly model this current, since the results contain unphysical oscillations and an incorrect magnitude, even relatively far from the junction. Conversely, the basis set that incorporates the first, lowest order Meixner functions appears to produce reasonable current density behavior [45].

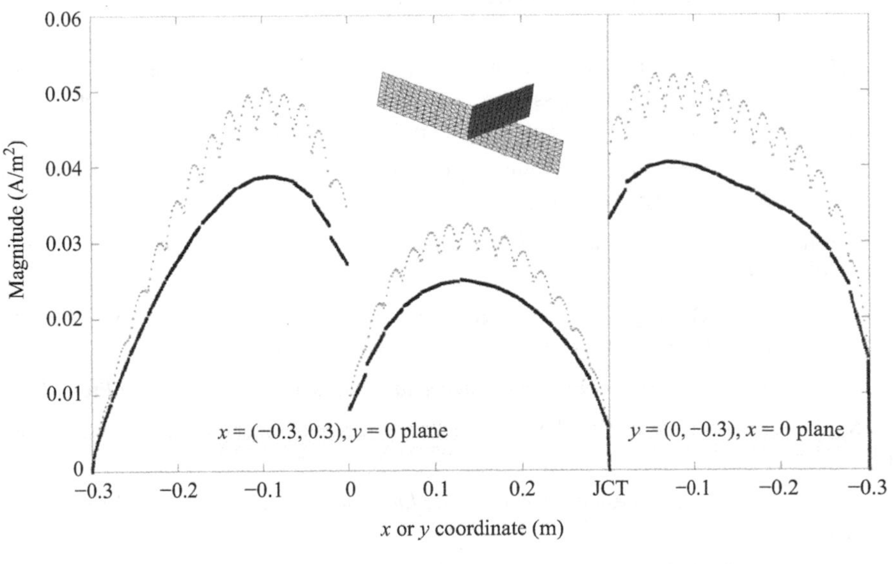

Figure 7.35 T-shape structure at 300 MHz. Magnitude of the current component normal to the z-axis computed at a distance of $\lambda/1,000$ from the bottom border of the structure. The results in gray obtained with the $p=2$ purely polynomial basis functions are reported by circles; the results in black obtained by adding to the $p=2$ polynomial base the lowest order Meixner basis functions are reported by dark crosses.

Adapted from G. Lombardi and R. D. Graglia, "Modeling junctions in sharp edge conducting structures with higher order method of moments," *IEEE Trans. Antennas Propag.*, vol. 62, no. 11, pp. 5723–5731, Nov. 2014.

7.10 Conclusion

Singular scalar and vector basis functions have been described and used to analyze two-dimensional cavity resonators and waveguide structures. The singular functions are designed to provide non-analytic DoFs to efficiently model fields near sharp corners and edges. These functions are hierarchical, in the sense that a representation of order p contains the members of order $p-1$. These functions are also additive, which provide maximum flexibility at the cost of somewhat higher matrix CNs. The philosophy behind these functions is discussed and appropriate combinations of singular and non-singular functions are proposed. The possibility of numerically constructing these singular basis functions should facilitate their use in problems containing field singularities arising from very general situations such as wedge tips made of penetrable material.

Several types of singular vector bases have been proposed in the recent literature for application to 3D problems. While this topic is far from mature at the time of this writing, we briefly report the performance of the functions proposed in [30, 45] for illustration.

References

[1] L. Demkowicz, *Computing with hp-Adaptive Finite Elements*, vol. 1. Boca Raton, FL: Chapman & Hall/CRC Press, 2007.

[2] L. Demkowicz, *Computing with hp-Adaptive Finite Elements*, vol. 2. Boca Raton, FL: Chapman & Hall/CRC Press, 2008.

[3] R. D. Graglia, D. R. Wilton, and A. F. Peterson, "Higher order interpolatory vector bases for computational electromagnetics," special issue on "Advanced Numerical Techniques in Electromagnetics," *IEEE Trans. Antennas Propag.*, vol. 45, no. 3, pp. 329–342, Mar. 1997.

[4] J. Meixner, "The behavior of electromagnetic fields at edges," *IEEE Trans. Antennas Propag.*, vol. AP-20, no. 4, pp. 442–446, Jul. 1972.

[5] R. Mittra and S. W. Lee, *Analytical Techniques in the Theory of Guided Waves*. New York: Macmillan, 1971.

[6] J. Van Bladel, "Field singularities at metal-dielectric wedges," *IEEE Trans. Antennas Propag.*, vol. AP-33, pp. 450–455, Apr. 1985.

[7] J. Van Bladel, *Singular Electromagnetic Fields and Sources*. Oxford: Clarendon Press, 1991.

[8] H. Motz, "The treatment of singularities of partial differential equations by relaxation methods," *Q. Appl. Math.*, vol. 4, pp. 371–377, 1947.

[9] G. Strang and G. E. Fix, *An Analysis of the Finite Element Method*. Englewood Cliffs, NJ: Prentice-Hall, 1973.

[10] M. Stern and E. B. Becker "A conforming crack tip element with quadratic variation in the singular fields," *Int. J. Numer. Methods Eng.*, vol. 12, pp. 279–288, 1978.

[11] M. Stern, "Families of consistent conforming elements with singular derivative fields," *Int. J. Numer. Methods Eng.*, vol. 14, pp. 409–421, 1979.

[12] D. R. Wilton and S. Govind, "Incorporation of edge conditions in moment method solutions," *IEEE Trans. Antennas Propag.*, vol. 25, no. 6, pp. 845–850, Nov. 1977.

[13] J. Richmond, "On the edge mode in the theory of TM scattering by a strip or strip grating," *IEEE Trans. Antennas Propag.*, vol. 28, no. 6, pp. 883–887, Nov. 1980.

[14] Z. Pantic and R. Mittra, "Quasi-TEM analysis of microwave transmission lines by the finite-element method," *IEEE Trans. Microwave Theory Tech.*, vol. 34, no. 11, pp. 1096–1103, Nov. 1986.

[15] J. P. Webb, "Finite element analysis of dispersion in waveguides with sharp metal edges," *IEEE Trans. Microwave Theory Tech.*, vol. 36, no. 12, pp. 1819–1824, Dec. 1988.

[16] Z. Pantic-Tanner, C. H. Chan, and R. Mittra, "The treatment of edge singularities in the full wave finite-element solution of waveguiding problems," Abstracts of the 1988 URSI Radio Science Meeting, Syracuse, NY, p. 336, 1988.

[17] T. Andersson, "Moment-method calculations on apertures using basis singular functions," *IEEE Trans. Antennas Propag.*, vol. 41, no. 12, pp. 1709–1716, Dec. 1993.

[18] J. M. Gil and J. Zapata, "Efficient singular element for finite element analysis of quasi-TEM transmission lines and waveguides with sharp metal edges," *IEEE Trans. Microwave Theory Tech.*, vol. 42, no. 1, pp. 92–98, Jan. 1994.

[19] J. M. Gil and J. P. Webb, "A new edge element for the modeling of field singularities in transmission lines and waveguides," *IEEE Trans. Microwave Theory and Tech.*, vol. 45, no. 12, Part 1, pp. 2125–2130, Dec. 1997.

[20] Z. Pantic-Tanner, J. S. Savage, D. R. Tanner, and A. F. Peterson, "Two dimensional singular vector elements for finite element analysis," *IEEE Trans. Microwave Theory Tech.*, vol. 46, pp. 178–184, Feb. 1998.

[21] W. J. Brown and D. R. Wilton, "Singular basis functions and curvilinear triangles in the solution of the electric field integral equation," *IEEE Trans. Antennas Propag.*, vol. 47, no. 2, pp. 347–353, Feb. 1999.

[22] R. D. Graglia, A. F. Peterson, and L. Matekovits, "Singular, hierarchical scalar basis functions for triangular cells," *IEEE Trans. Antennas Propag.*, vol. 61, no. 7, pp. 3674–3692, Jul. 2013.

[23] R. D. Graglia, A. F. Peterson, L. Matekovits, and P. Petrini, "Hierarchical additive basis functions for the finite-element treatment of corner singularities," special issue on "Finite Elements for Microwave Engineering," *Electromagnetics*, vol. 34, pp. 171–198, Mar. 2014.

[24] R. D. Graglia, A. F. Peterson, L. Matekovits, and P. Petrini, "Singular hierarchical curl-conforming vector bases for triangular cells," *IEEE Trans. Antennas Propag.*, vol. 62, no. 7, pp. 3632–3644, Jul. 2014.

[25] P. Ya. Ufimtsev, B. Khayatian, and Y. Rahmat-Samii, "Singular edge behavior: to impose or not impose – that is the question," in *Microwave Opt. Tech. Lett.*, vol. 24, pp. 218–223, Feb. 2000.

[26] D.-K. Sun, L. Vardapetyan, and Z. Cendes, "Two-dimensional curl-conforming singular elements for FEM solutions of dielectric waveguiding structures," *IEEE Trans. Microwave Theory Tech.*, vol. 53, pp. 984–992, 2005.

[27] M. M. Bibby, A. F. Peterson, and C. M. Coldwell "High-order representations for singular currents at corners," *IEEE Trans. Antennas Propag.*, vol. 56, no. 8, pp. 2277–2287, Aug. 2008.

[28] M. M. Bibby, A. F. Peterson, and C. M. Coldwell "Optimum cell size for high order singular basis functions at geometric corners," *ACES J.*, vol. 24, pp. 368–374, Aug. 2009.

[29] R. D. Graglia and G. Lombardi, "Singular higher order complete vector bases for finite methods," *IEEE Trans. Antennas Propag.*, vol. 52, no. 7, pp. 1672–1685, Jul. 2004.

[30] R. D. Graglia and G. Lombardi, "Singular higher order divergence-conforming bases of additive kind and moments method applications to 3D sharp-wedge structures," *IEEE Trans. Antennas Propag.*, vol. 56, no. 12, pp. 3768–3788, Dec. 2008.

[31] M. Salazar-Palma, T. K. Sarkar, L.-E. Garcia-Castillo, T. Roy, and A. Djordjevic, *Iterative and Self-Adaptive Finite-Elements in Electromagnetic Modeling*. Boston, MA: Artech House, 1998.

[32] B. Schiff and Z. Yosibash, "Eigenvalues for waveguides containing re-entrant corners by a finite-element method with superelements," *IEEE Trans. Microwave Theory Tech.*, vol. 48, no. 2, pp. 214–220, Feb. 2000.

[33] L. Fox, P. Henrici, and C. Moler, "Approximations and bounds for eigenvalues of elliptic operators," *SIAM J. Numer. Anal.*, vol. 4, no. 1, pp. 89–102, 1967.

[34] L. N. Trefethen and T. Betcke "Computed eigenmodes of planar regions," Recent advances in differential equations and mathematical physics, *Contemp. Math.*, vol. 412, American Mathematical Society, Providence, RI, 2006, pp. 297–314. MR 2259116 (2008a:35042), doi:10.1090/conm/412/07783

[35] J. C. Nédélec, "Mixed finite elements in R3," *Numer. Math.*, vol. 35, pp. 315–341, 1980.

[36] R. D. Graglia and G. Lombardi, "Vector functions for singular fields on curved triangular elements, truly defined in the parent space," *IEEE Antennas and Propagation International Symposium Digest*, San Antonio, TX, pp. 62–65, 2002.

[37] J. Masoni, G. Pelosi, and S. Selleri, "Substitutive divergent bases for FEM modeling of field singularities near a wedge," *Microwave Opt. Technol. Lett.*, vol. 44, pp. 327–328, 2005

[38] R. D. Graglia, A. F. Peterson, and F. P. Andriulli, "Curl-conforming hierarchical vector bases for triangles and tetrahedra," *IEEE Trans. Antennas Propag.*, vol. 59, no. 3, pp. 950–959, Mar. 2011.

[39] A. F. Peterson, R. D. Graglia, "Scale factors and matrix conditioning associated with triangular-cell hierarchical vector basis functions" *IEEE Antennas Wireless Propag. Lett.*, vol. 9, pp. 40–43, 2010, doi:10.1109/LAWP.2010.2042423

[40] P. Savi, I.-L. Gheorma, and R. D. Graglia, "Full-wave high-order FEM model for lossy anisotropic waveguides," *IEEE Trans. Microwave Theory Tech.*, vol. 50, no. 2, pp. 495–500, Feb. 2002.

[41] J.-F. Lee, D.-K. Sun, and Z. J. Cendes, "Full-wave analysis of dielectric waveguides using tangential vector finite elements," *IEEE Trans. Microwave Theory Tech.*, vol. 39, no. 8, pp. 1262–1271, Aug. 1991.

[42] R. D. Graglia, P. Petrini, A. F. Peterson, and L. Matekovits, "Full–wave analysis of inhomogeneous waveguiding structures containing corners with singular hierarchical curl–conforming vector bases," *IEEE Antennas and Wireless Propagation Letters*, vol. 13, pp. 1701–1704, 2014.

[43] C. G. Wells and J. A. R. Ball, "Mode-matching analysis of a shielded rectangular dielectric-rod waveguide," *IEEE Trans. Microwave Theory Tech.*, vol. 53, no. 10, pp. 3169–3177, Oct. 2005.

[44] A. S. Omar and K. F. Schünemann, "The effect of complex modes at finline discontinuities," *IEEE Trans. Microwave Theory Tech.*, vol. 34, no. 12, pp. 1508–1514, Dec. 1986.

[45] G. Lombardi and R. D. Graglia, "Modeling junctions in sharp edge conducting structures with higher order method of moments," *IEEE Trans. Antennas Propag.*, vol. 62, no. 11, pp. 5723–5731, Nov. 2014.

About the Authors

Roberto D. Graglia received his Laurea degree (*summa cum laude*) in electronic engineering from the Polytechnic of Turin in 1979 and his Ph.D. degree in electrical engineering and computer science from the University of Illinois at Chicago in 1983. From 1980 to 1981, he was a research engineer at CSELT, Italy. From 1985 to 1992, he was with the Italian National Research Council (CNR), where he supervised international research projects. In 1991 and 1993, he was an associate visiting professor at the University of Illinois at Chicago. Since 1992, he has been a faculty member with the Department of Electronics and Telecommunications, Polytechnic of Turin, where he is now Professor of Electrical Engineering. His areas of interest comprise numerical methods for high- and low-frequency electromagnetics, theoretical and computational aspects of scattering and interactions with complex media, waveguides, antennas, electromagnetic compatibility, and low-frequency phenomena. He has organized and offered several short courses in these areas.

Since 1997, he has been a member of the Editorial Board of *Electromagnetics*. He has served as an IEEE AP-S distinguished lecturer (2009–2012), as an associate editor of *IEEE Transactions on Antennas and Propagation*, *IEEE Transactions on Electromagnetic Compatibility*, and *IEEE Antennas and Wireless Propagation Letters*, and as a member of IEEE AP-S AdCom. He was the guest editor of a special issue on Advanced Numerical Techniques in Electromagnetics for *IEEE Transactions on Antennas and Propagation* in March 1997. He has been Invited Convener at URSI General Assemblies for special sessions on Field and Waves in 1996, Electromagnetic Metrology in 1999, and Computational Electromagnetics in 1999. He served the International Union of Radio Science (URSI) for the triennial International Symposia on Electromagnetic Theory as organizer of the Special Session on Electromagnetic Compatibility in 1998 and was the co-organizer of the special session on Numerical Methods in 2004. Since 1999, he has been General Chairperson of the International Conference on Electromagnetics in Advanced Applications (ICEAA), and, since 2011, he has been General Chairperson of the IEEE-APS Topical Conference on Antennas and Propagation in Wireless Communications (IEEE-APWC). He is a fellow of the IEEE, and he served as President of the IEEE Antennas and Propagation Society during 2015.

Andrew F. Peterson received his B.S., M.S., and Ph.D. degrees in electrical engineering from the University of Illinois at Urbana-Champaign in 1982, 1983, and 1986, respectively. Since 1989, he has been a member of the faculty of the School of Electrical and Computer Engineering at the Georgia Institute of Technology, Atlanta, where he is now Professor and Associate Chair for Faculty Development. He teaches electromagnetic field theory and computational electromagnetics and conducts research in the development of computational techniques for microwave frequency electromagnetic applications. He is the principal author of *Computational Methods for Electromagnetics* (IEEE Press, 1998) and several volumes in the Morgan/Claypool Synthesis Lectures.

Dr. Peterson is a past recipient of the ONR Graduate Fellowship and the NSF Young Investigator Award. He has served as an associate editor of *IEEE Transactions on Antennas and Propagation*, and *IEEE Antennas and Wireless Propagation Letters*, as General Chair of the 1998 IEEE AP-S International Symposium and URSI/USNC Radio Science Meeting, and as a member of IEEE AP-S AdCom. He also served for six years as a director of the Applied Computational Electromagnetics Society (ACES) and two years as Chair of the IEEE Atlanta Section. He was President of the IEEE AP-S during 2006 and of ACES from 2011 to 2013. He is a fellow of the IEEE, and the ACES, and a member of the International Union of Radio Scientists (URSI) Commission B, the American Society for Engineering Education, and the American Association of University Professors. He is also a recipient of the IEEE Third Millennium Medal.

Index